# ON THE SHOULDERS OF
# TITANS

*The first space rendezvous*

NASA SP-4203

# ON THE SHOULDERS OF TITANS

# A History of Project Gemini

*by*
BARTON C. HACKER
*and*
JAMES M. GRIMWOOD

*Scientific and Technical Information Office*        1977
NATIONAL AERONAUTICS AND SPACE ADMINISTRATION
*Washington, D.C.*

# Foreword

GEMINI was the intermediate manned space flight program between America's first steps into space with Mercury and the amazing and unprecedented accomplishments achieved during the manned lunar expeditions of Apollo. Because of its position between these two other efforts, Gemini is probably less remembered. Still, it more than had its place in man's progress into this new frontier.

Gemini accomplishments were manyfold. They included many firsts: first astronaut-controlled maneuvering in space; first rendezvous in space of one spacecraft with another; first docking of one spacecraft with a propulsive stage and use of that stage to transfer man to high altitude; first traverse of man into the Earth's radiation belts; first extended manned flights of a week or more in duration; first extended stays of man outside his spacecraft; first controlled reentry and precision landing; and many more.

These achievements were significant in ways one cannot truly evaluate even today, but two things stand out: (1) it was the time when America caught up and surpassed the Soviet Union in manned space flight, and (2) these demonstrations of capability were an absolute prerequisite to the phenomenal Apollo accomplishments then yet to come.

America's first manned space flight program, Mercury, involved a careful buildup of flight duration to slightly beyond one day with accompanying concerns about man's physiological response to weightlessness and other aspects of his safety and well being. In the meantime, the Russian effort had achieved durations of five days, flight of a multiple crew shortly after the Mercury Program had terminated, and the first extravehicular operation by a cosmonaut shortly before the first manned Gemini flight. The question at that time was who would perform the first rendezvous, seen as a very complex operation but absolutely needed for future space endeavors.

About the time Gemini started, the Russian effort slowed down as they attempted to develop and flight qualify their second-generation manned spacecraft, Soyuz. In the meantime, Gemini, America's second-generation spacecraft, reeled off ten manned flights in less than twenty months—a flight rate yet to be surpassed in any space program. The last five manned launches were accompanied by nearly simultaneous and precisely timed launches of rendezvous target vehicles. During this period, rendezvous demonstrations and many other activities took place which were not to be matched by corresponding Soviet accomplishments for years to come, and more than five years passed before the two-week long mission of *Gemini VI* was exceeded by the Russians with their Salyut spacecraft.

However, these Gemini mission spectaculars were not aimed at "beating the Russians"; rather, their purpose was to support and demonstrate needed mission capability for the upcoming Apollo flights to the Moon. Apollo needed a reliable rendezvous and docking operation if the astronauts were to get back from the Moon. Could this be done? Gemini demonstrated such a capability with great success six straight times and with many different techniques. The Apollo missions required a duration of a week or two. Could this be done? Gemini demonstrated mission durations of one and two weeks with no major untoward effects on the astronauts. The Apollo astronauts would spend hours outside their spacecraft exploring the lunar surface. Could this be done? Of the five EVA missions conducted in Gemini, four of them lasted from two to four hours. Tired astronauts returning from the Moon would want to land as close as possible to the recovery aircraft carrier. Could this be done? Indeed, it was accomplished seven straight times during the last two-thirds of the Gemini Program. Apollo needed to develop advanced reliable systems. Could this be done? Their names probably still sound strange to many, but fuel cells, cryogenic storage of hydrogen and oxygen, ablative thrusters using hypergolic propellants, an onboard digital computer, an inertial guidance system, and a rendezvous radar were developed and demonstrated in Gemini. One must admit to considerable difficulty in these developments, but, in the end, they provided a high degree of confidence that systems embodying high reliability could be obtained.

Equally important to Apollo was the training provided by the Gemini missions to the flight and ground crews. The mission control center techniques and the flight control team procedures were largely implemented during Gemini. Of the astronaut complement assigned to the first four flights to the Moon, ten of the twelve had prior Gemini flight experience and the other two had been members of Gemini backup crews. In all, over half of the Apollo crew members had direct Gemini flight experience.

Gemini also carried forward a major experiment program in space science and applications. Over 50 such experiments were carried out involving astronomy, biology, atmospheric sciences, medicine, radiation effects, micrometeoroid investigations, space environmental effects, and others. Technical and operational experimentation involved such things as low light level TV observations, special photography, special communications tests, tethering of two vehicles, and gravity gradient stabilization. The hundreds of synoptic weather and Earth terrain color photographs taken contributed greatly to the development of the meteorological and Earth resources programs which are now bringing important benefits from rapid global observations of the Earth to people here on the surface.

Lest one think that the Gemini flights were carried forward with great smoothness, be assured that most of them encountered real cliff-hanging incidents. On *Gemini IV*, the astronauts had great difficulty in closing the hatch after their EVA which was accomplished only after great physical exertion and almost complete exhaustion. Needless to say, corrections were made before the next flight. *Gemini V*, which was planned to fly for eight days, was almost called back after a few hours because of loss of pressure in the cryogenic tanks supplying fuel for the new electrical power devices called fuel cells. But the crew and flight controllers nursed the spacecraft along for the full mission duration by powering down the spacecraft and using just a few watts of electrical power. Their problems were compounded when some of the attitude stabilization rockets failed late in the mission.

After loss of the first rendezvous target vehicle, caused by an explosion during launch, *Gemini VI* and *VII* were reconfigured so that *Gemini VII* served as the rendezvous target for *Gemini VI* in the "Spirit of 76" mission just before Christmas in 1965, after which *Gemini VII* continued to struggle along with balky fuel cells for a record duration of 14 days in space. *Gemini VIII* spun out of control just after accomplishing the first docking in space. The crew was able to correct this condition in spite of rotating nearly one revolution every second. But the spacecraft had to be returned to Earth rapidly and landed in the western Pacific Ocean. Astronauts became exhausted from EVA exertions on *Gemini IX* and *XI*. Only the last mission, *Gemini XII*, (and perhaps *Gemini X*) could be called really smooth, carried out pretty much as planned.

In spite of all these exciting mission events, problems, and accomplishments, the thing that stands out in my own mind is the way in which the effort and dedication of many individuals and groups coalesced into an extremely effective team. This cliché is often voiced whenever an activity is successful, but, in Gemini, the observed capacity for accomplishment proved to be well beyond a program manager's most optimistic hopes. Although not so visible from a program manager's level, this cooperation and support had to extend to the level of the NASA Administrator and his interfaces with the President, congressional leaders, heads of other agencies, industry, and the public in general.

Most certainly, this same situation occurred during the Apollo Program and, no doubt, has occurred in connection with most major achievements of man. However, Gemini—though a complex undertaking—was small enough for this to stand out very clearly. Such an experience leads one to believe that man can accomplish almost anything if sufficient dedication and cooperation exists between and within the groups involved.

In Gemini, this esprit de corps was actually enhanced by the mistakes made or the problems encountered because of the positive approach to dealing with them. A prime example of this occurred when a critical hydraulic system failed on the launch vehicle just at engine start prior to liftoff on the first full systems test of Gemini. The response of the people involved was truly outstanding. Even though recovery from this problem involved trying work over the Christmas holidays, everyone involved put forward a maximum effort, including the small job shop that built a new casting, the hydraulic valve contractor, the prime contractor, the Air Force, its support contractors, and NASA. As a result of such effort, the cause of the failure was isolated, a completely new component designed, built, tested, qualified, installed, and checked out so that a second attempt could be made only six weeks after this major difficulty occurred. There may have been evidences of parochialism and vested interests early in the program, but after an event and accomplishment such as that, the whole team concentrated on the program in the spirit of an elite group.

I believe that this is a lesson and a legacy that our space programs have left to future generations just as other eras of great accomplishment have done. Admittedly, Mercury, Gemini, and Apollo had very clear objectives. But even in more complex and confusing situations an integrated and dedicated striving to solutions of problems would seem to be an approach well worth taking. In today's world, there seems to be an undue degree of second-guessing and lack of cooperation in many endeavors. Gemini was far from perfect, but, although its people recognized and encountered imperfection, they strove as a group for perfection.

*Charles W. Mathews*
*Associate Administrator for Applications*
*July 1975*

# Contents

# List of Illustrations

# Preface

PROJECT Gemini is now little remembered, having vanished into that special limbo reserved for the successful intermediate steps in a fast-moving technological advance. Conceived and approved in 1961, the second major project in the American manned space flight program carried men into orbit in 1965 and 1966. Gemini thus kept Americans in space between the path-breaking but limited Earth-orbital missions of Project Mercury and the far more ambitious Project Apollo, which climaxed in 1969 when two men first set foot on the Moon. Although keeping the nation in space was one of the motives that induced the National Aeronautics and Space Administration (NASA) to go ahead with Project Gemini, it was not the overriding one. It furnished the setting in which a new project could be approved, but the precise character of that project grew out of two distinct lines of development that converged during 1961.

President John F. Kennedy's decision in May 1961 to commit the United States to landing on the Moon before the end of the decade gave Gemini its central objective. NASA planners had been thinking about the Moon, an obvious goal for manned space flight, almost from the moment the agency itself was created in 1958. The Moon, however, was seen as a target for the 1970s, pending development of a huge rocket, called Nova. It would launch a spacecraft that would fly directly to the Moon, land there, and then return. This direct approach was widely accepted on the grounds that it was almost certain to work.

Some NASA engineers had advocated an alternative method, in which two or more spacecraft might rendezvous in orbit rather than proceed directly to the Moon. This approach promised enormous savings in fuel and weight; the lunar mission based on rendezvous might be launched with much smaller rockets, and therefore much sooner, than the direct mission. The greatest drawback of this approach was its novelty. No one knew how hard a rendezvous in space might be. So long as time was ample, the direct method offered by far the safer prospect. When the President imposed a deadline, however, support for rendezvous waxed. It promised a quicker and cheaper road to the Moon, if it could be achieved. The "if" was a big one in 1961, big enough to justify the expense of a full-fledged manned space flight project to resolve it. Gemini was first and foremost a project to develop and prove equipment and techniques for rendezvous.

That the project turned out to be Gemini, however, rather than something else, resulted from a second distinct chain of causes. Government and industry engineers who worked in Project Mercury saw innumerable ways to improve their product. Constrained by the limited power of the Atlas rocket that launched Mercury, they had been forced to design a spacecraft with integrated systems; the inside of the capsule was crammed with layered components, filling every cranny and making it hard to build, hard to test, and hard to prepare for flight. As a first step it might do, but it could never be much more. Throughout 1959 and 1960, while the main effort centered on making Mercury work, thinking turned more and more to the kind of spacecraft that should come next; it should be based on the lessons learned in working on the essentially handcrafted, experimental machine that was Mercury, but modified to permit something more closely resembling routine building, testing, and operation than Mercury allowed. By mid-1961, these ideas coalesced into a concrete proposal for a new spacecraft, just when NASA was casting about for a means of working out the problems of rendezvous. Gemini's second taproot was an engineering concern to improve spacecraft technology beyond the first step that was Mercury.

Project Gemini owed its origins both to its predecessor—it built on the technology and experience of Project Mercury—and to its successor—it derived its chief justification from Project Apollo's concerns. The new project acquired other objectives as well: testing the concept of controlled landing, determining the effects of lengthy stays in space, and training ground and flight crews. The process through which a broad range of ideas and concerns came together in a clearly defined space flight program is the main theme of this book's first three chapters.

By December 1961, when the new project received its formal stamp of approval from NASA Headquarters in Washington, much of the design work had been done, many of the major decisions had already been made. A Gemini Project Office at the Manned Spacecraft Center (renamed Lyndon B. Johnson Space Center in February 1973) in Houston took charge of overseeing the effort. Just a week after project approval, the first major contract went to McDonnell Aircraft

Corporation for the Gemini spacecraft. A separate contract with North American Aviation had already initiated work on the paraglider landing system that was intended to allow Gemini to alight on land rather than water. Other key contracts were soon awarded through the Air Force Space Systems Division for the project's several rocket boosters: to Martin Company for the Titan II to launch the spacecraft, to Lockheed Missiles & Space Company for the Agena to serve as rendezvous target, and to General Dynamics Corporation for the Atlas to boost Agena into space. A matter of months sufficed to erect the whole structure of contracts and subcontracts that united the efforts of government and industry in Project Gemini.

Gemini thus moved quickly into its development phase, the central effort of 1962 and 1963. It was an unsettled period, as such times always are for high-technology projects. Although Gemini, perhaps more than most such undertakings, rested on already tested technologies, it still strained the limits of the known at some points. Inevitably this produced problems not always easy to resolve, the more so since Gemini was bound by severe time constraints. It could not, whatever happened, be allowed to overlap or interfere with Project Apollo. In one major instance, the paraglider, answers could not be found in time, and that goal had to be dropped.

Gemini's difficulties in its first two years were not solely technical, nor were technical problems perhaps even the most pressing. Gemini labored under a sharply restricted budget. The project faced a severe financial crisis during its first year and lesser such crises throughout its life. Within NASA and without, Apollo and the trip to the Moon always held center stage. Gemini got more than crumbs—its final cost exceeded a billion dollars—but the margin remained narrow. More than once, lack of funds threatened the loss of one or another of its major goals, and money problems played a key role in managerial changes in 1963. That year, however, also saw Gemini's development completed, the worst of its technical problems (except the paraglider) resolved or clearly on the way to solution. Project Gemini's development troubles and their outcome provide the central thread for Chapters IV through VII.

By the end of 1963, Gemini was moving into its qualifying trials, which extended into 1965. The road was far from easy, but the worst was past, as reflected in the slow decline in the number of workers directly assigned to Gemini. Early 1964 saw the first of Gemini's 12 missions, an unmanned test of spacecraft and booster that was flawless. The long delay that followed was a reflection not on Gemini but on the Florida climate, as the launch site was buffeted by hurricanes. The second unmanned mission, in January 1965, proved that Gemini was ready to carry men aloft. Some two months later, Virgil I. Grissom and

John W. Young flew *Gemini 3* through three circuits of Earth, and the project office set out after its planned goals. Gemini's qualification is the subject of Chapters VIII through X.

In striking contrast to the endless difficulties that had frustrated attempts to keep Project Mercury on schedule, Project Gemini came close to achieving a routine launch every other month throughout 1965 and 1966. *Gemini XII* closed out the program in November 1966. Gemini's operational phase was hardly so free of trouble as such a schedule might suggest, but the design that had been geared to easier testing and checkout proved its worth when coupled with the experience derived from earlier efforts. One by one, Gemini achieved its objectives, proving that astronauts could leave the shelter of their vehicle and function in space, that they could closely control spacecraft flight and landing, that they could survive up to two weeks in orbit without ill effects, and that they could rendezvous with a target in orbit. This is the story told in Chapters XI through XV.

The teams who serviced, flew, directed, and supported Gemini missions opened the near-space environment of Earth as a potential workshop and stilled some nagging fears about what might happen to men on the way to the Moon. They did not do it alone. Just as surely as the Gemini spacecraft rested on the shoulders of its Titan II launch vehicle, those who combined to make Project Gemini succeed stood on the shoulders of the giants who preceded them. Isaac Newton, who first formulated the laws of motion that Gemini applied in orbit three centuries later, wrote, "If I have seen farther, it is by standing on the shoulders of giants." So, too, did Project Gemini, not the least on those of Newton himself.

And so, too, did the authors of this history of Project Gemini: Barton C. Hacker, who wrote the first ten chapters on design, development, and qualification; and James M. Grimwood, who described operations in the last five chapters. Although this book will not be the last word on Gemini, we enjoyed an access to its documentary remains and to its participants not likely to be duplicated.

Aid in threading a path through this embarrassment of riches came from many sources at the Manned Spacecraft Center, elsewhere in NASA, other government agencies (especially the Air Force), the Gemini contractors, and others. Their numbers preclude individual thanks, but the authors gratefully acknowledge their help. Combing the records and interviewing the actors proved an arduous and challenging task. The contemporary historian must beware the sensitivities of the many people he writes about who are still very much alive. This may be especially true of a project so successful as Gemini proved to be, since the afterglow of accomplishment tends to dim memories of things that went wrong. Yet the advantage of having the counsel of the participants in weighing the mass of evidence more than compensates for any concomitant handicaps. They cheerfully endured lengthy interviews, cleared up technical points, ransacked their files, and commented on drafts.

This help was all the more important because Project Gemini never attracted as much attention as either Mercury or Apollo. Having neither the novelty of the first nor the enormously exciting goal of the second, Gemini prompted relatively little outside description or analysis. Journalistic interest was largely confined to Gemini's manned missions in 1965 and 1966, and even that coverage was slight after the first two. Never as high in public consciousness as Mercury or Apollo, Gemini now lives mainly in the memories of those who worked on it. This in part reflects Gemini's ambiguous status even within NASA—important to be sure but somehow outside the mainstream that flowed from Mercury to Apollo. Gemini seemed less touched by outside events than its brother programs. In writing its history, we have adopted what in the history of science is often called an internalist approach. The course of Gemini's history was clearly dictated by internal technical demands, and the focal point of the story is the work of the Gemini Program Office at the Manned Spacecraft Center. Picking out the particular individual whose contribution was unique is seldom possible, not because such contributions were lacking but because Gemini was so much a team effort. Many of those team members, both from government and from industry, have remarked on the sense of unity and *elan* they enjoyed in those days and have suggested that Gemini might have achieved a good deal more than it was called upon to do. However true that may be, Project Gemini, in terms of its actual costs, schedules, and performance, must rank among the most successful research and development projects ever conducted by the United States.

We would like to extend special thanks to those whose efforts in behalf of this book significantly lightened our burdens. Sally D. Gates, Historical Office Archivist-Editor, served indispensably in a multitude of roles: research assistant, editor, coordinator of the comment draft, compiler of appendixes, typist, proofreader, and friendly critic. Billie D. Rowell and Corinne L. Morris, both of the Historical Office, at various times organized and managed the office's archives and performed a variety of other services. Jewell Norsworthy, Center records management officer, helped retrieve documents that had been retired to holding areas. Ivan D. Ertel, former Center assistant historian, and Peter J. Vorzimmer, former contract historian, conducted a number of interviews on behalf of the Gemini history. This book was written under the auspices of the NASA historical program through a contract with the History Department of the University of Houston; it has benefited from the advice and assistance of NASA historians Monte D. Wright, Frank W. Anderson, Jr., Eugene M. Emme, and William D. Putnam, as well as University of Houston professors James A. Tinsley and Loyd

S. Swenson, Jr. Although it is officially sponsored, its authors alone must bear full responsibility for whatever defects it contains.

B.C.H.
J.M.G.

*Houston*
*September 1974*

NOTE: NASA has placed itself in the forefront of the effort to convert the United States to the metric system. In 1973, use of all English weights and measures was prohibited in all NASA publications, including historical. This did present certain problems, since NASA engineers during the 1960s normally expressed themselves in feet, miles, pounds, etc. In general, where round figures are clearly intended, we have substituted round metric figures. Precise figures are converted precisely. For the reader's convenience, one metric unit requires a word of explanation. In the English system, "pound" is a unit of both mass and force, but the metric system is more rational and uses two distinct units: the familiar gram for mass, the less familiar newton for force—thus, for example, pounds of weight become grams, but pounds of thrust become newtons.

I

# Between Mercury and Apollo

IN Houston, Texas, December temperatures in the low sixties seem cool.[1] And so it must have seemed to Robert R. Gilruth when he landed in the city on 7 December 1961, especially in contrast to the muggy end-of-summer heat that had greeted him on his first visit two and a half months before. Gilruth's September visit had followed close on the heels of the announcement that the Space Task Group (STG) he headed was moving to Houston. With several of his colleagues, he had come to look over the new site for his fast-growing branch of the National Aeronautics and Space Administration (NASA). Now he was back in Houston to tell the city's business community something about his group and its work—putting American astronauts into space and eventually landing them on the Moon. The occasion was what the Houston Chamber of Commerce billed, with a bit of Texas hyperbole, as its 121st annual meeting. True, a chamber of commerce had been formed in 1840, but it soon vanished without a trace. Seventy years later, the 15-year-old Houston Business League voted to rename itself the "Chamber of Commerce."[2] Whether the 1961 session was the 121st, the 66th, or the 51st, it was still a big event. Houston "was a businessman's town."[3]

And it was a booming town, sprawling over more than 480 square kilometers (300 square miles) of Texas Gulf Coast "like a bucket of spilled water."[4] In the same month that Gilruth first visited Houston, the city's population had passed the million mark. And that, according to the president of the Chamber of Commerce, was one of the "most significant milestones of Houston's progress in 1961."[5] Houston and its people blended, not always smoothly, the South and the West. Chicanos joined blacks as part of the "problem" that sometimes troubled the ruling Anglos, who were "conservative, cautious, and business-oriented . . . because they reflect community attitudes."[6] September 1961 was also the month when the first black pupils, twelve of them, entered Houston's white school system.[7]

But Houston's leaders, in a pattern that has marked American development at least since the 19th century, coupled social conservatism with economic opportunism. Founded as a lucky real-estate venture, the city had grown by exploiting the resources of a vast hinterland. Freewheeling promotion was, and remained, the order of the day, and nowhere more so than in the multibillion-dollar oil industry that Houston headquartered.[8] The hotel to which Gilruth repaired was a perfect symbol of the city and a fitting site for the "121st" annual meeting of the Chamber of Commerce. Brainchild of Glenn Mc-Carthy—oil millionaire, land speculator, and all-round promoter—the Shamrock Hotel had taken five years to build and cost $21 million. It opened grandly on St. Patrick's Day 1949, with 50,000 people gathered to eat $42-a-plate dinners. Six shades of green garnished its outer walls, a prospect otherwise so dull that Frank Lloyd Wright refused to comment on it, though glimpsing the interior did move him to muse, "I always wondered what the inside of a juke box looked like." Mc-Carthy lost the hotel when his oil empire collapsed five years later, and it ended up in the hands of another Texas entrepreneur, Conrad Hilton. So it was the Shamrock Hilton, with Hilton's portrait gracing the lobby instead of McCarthy's, when Gilruth arrived.[9]

Gilruth himself symbolized another of the "milestones of Houston's progress in 1961." On 19 September, just a day after the city officially topped a million, NASA had announced its choice for the site of a new multimillion-dollar manned space flight laboratory.[10] It was to be near Clear Lake, some 32 kilometers southeast of the city on a tract of land donated by the Humble Oil and Refining Company. This, too, fit the pattern of Houston's growth, at least since World War I, as federal funds had begun to flow into the city like the oil that much of that money financed. The president of the Chamber of Commerce welcomed NASA's new move as "one of the Houston's most meaningful developments since the opening of the Ship Channel for deep sea shipping in 1915." Gilruth directed the new facility, the Manned Spacecraft Center (MSC), which came officially into being on 1 November 1961.[11]

The Center was, in fact, merely the renamed Space Task Group (STG), created in 1958 to put Americans in space via Project Mercury. So far, STG had managed to loop two astronauts over the fringes of the atmosphere on Redstone boosters and to orbit with an Atlas rocket a chimpanzee named Enos. But the much-delayed attempt to orbit a man still receded. On the same day that Gilruth spoke to the Houston Chamber of Commerce, he announced that the scheduled 19 December launch of Mercury-Atlas 6, with John H. Glenn, Jr., aboard, was now postponed until 1962. The United States was not going to match, at least in the same year, the Soviet Union's feat of sending a man into orbit. Nonetheless, optimism prevailed. The causes of the delay were minor, and success seemed just around the corner.[12]

STG, like Houston, had boomed in 1961. Two largely successful manned suborbital flights, followed by Mercury-Atlas 4 with its "mechanical man" and the ape-bearing Mercury-Atlas 5, had eased the worries caused by Mercury's technical problems during 1960. In the meantime, STG had added the manned lunar landing program, Project Apollo, to its responsibilities. It had outgrown its makeshift facilities at Langley Research Center in Virginia and its old name as well. After a painstaking search, NASA settled on Houston for STG's new location and soon furnished the group with a new name to match its larger role.[13]

For Houston, it was love at first sight, but the 750 NASA workers faced with moving 2400 kilometers from Tidewater Virginia to Gulf Coast Texas in the midst of Project Mercury were less enthusiastic. Gilruth himself had qualms after his first view of the new site in September, shortly after it had been swept by Hurricane Carla.[14] The decision had been made, however, and the space fever that promptly seized Houston helped smooth the changeover. A crowd of some 900 greeted Gilruth with a standing ovation when he stepped to the dais at the Shamrock Hilton to begin his remarks.[15]

What Gilruth had to say turned out to be headline news and earned him another standing ovation when he finished. NASA, he revealed, planned to launch a third manned space flight program to fill the gap between Mercury and Apollo. He outlined a half-billion-dollar project to orbit a two-man Mercury capsule via the Air Force's new Titan II booster. The key goal was to develop orbital rendezvous, a novel technique NASA planned to use in the Apollo mission to the Moon. Once in orbit, the crewmen would steer their rocket-powered craft to a meeting with an unmanned Agena spacecraft, boosted into orbit separately by an Atlas.[16] Gilruth had learned only that day of NASA Headquarters' approval of the new project.[17]

Still something of a puzzle was what to call it. In making it public, Gilruth labeled it a "two-man Mercury." Inside NASA, at one time or another, it had gone by the name of Advanced Mercury, Mercury Mark II (the one-man capsule being Mark I), or simply Mark II. Within three months, however, an ad hoc "program-naming" committee in NASA Headquarters decided on "Gemini" for the new project. Recognition for having picked that name, along with a bottle of scotch as prize, went to Alex P. Nagy in NASA Headquarters. Gemini, "The

*Emblem adopted for Gemini program.*

Twins," was one of the 12 constellations of the zodiac. Nagy thought that " 'the Twins' seems to carry out the thought nicely, of a two-man crew, a rendezvous mission, and its relation to Mercury. Even the astronomical symbol (II) fits the former Mark II designation."[18]

By an unlikely coincidence, since Nagy disclaims any knowledge of astrology, Gemini as a sign of the zodiac is controlled by Mercury. Its spheres of influence include adaptability and mobility—two features

4

the spacecraft designers had explicitly pursued—and, through its link with the third house of the zodiac, all means of communication and transportation as well. Astrologically, at least, Gemini was a remarkably apt name, the more so since the United States is said to be very much under its influence.[19] To those with no more than a passing knowledge of astrology, however, Gemini must have seemed a most obscure choice. To this day, its proper pronunciation has not been settled in NASA. Although an informal survey of astronomical opinion came down on the side of a terminal "ee" sound, many still opt for "eye."[20] The new program publicly became Project Gemini on 3 January 1962.[21]

## THE BACKGROUND OF RENDEZVOUS

The project that Gilruth announced on 7 December 1961 had not just then sprung into being. A year of planning, work, and advocacy had gone before, and more than three years of intense effort lay ahead before Gemini carried men into space. Even so, Gemini was something of an afterthought in the American manned space flight program. Gemini did fly after Mercury had achieved its major goal of putting an American into orbit and bringing him back safely and before Apollo first bore men aloft on the path that led eventually to the surface of the Moon. But that is misleading. One of the reasons for Gemini, in fact, was to keep Americans in space during the time when Mercury had run its course but Apollo had yet to be launched.

Gemini took shape after Apollo had begun, in part to answer a crucial question for Apollo: Was rendezvous and docking in orbit a feasible basis for a manned lunar landing mission? When NASA officials appeared before Congress early in 1962 to justify the new program, the heart of the case they argued was the need to develop and prove the techniques of orbital rendezvous.[22] Project Gemini was intended to show that a piloted spacecraft could meet an unmanned target in space—the orbit of the spacecraft matching that of the target so that there was no significant difference in speed and no significant distance between the two, in much the same way that two aircraft might fly in formation.

Many aspects of modern space flight were first suggested in the sometimes fanciful but often profound space-travel writings of the early 20th century. One was the value of rendezvous in orbit. It first emerged as part of the space-station concept, which can be traced through the works of the Russian pioneers of astronautics—K. E. Tsiolkovskii, Yu. V. Kondratyuk, and F. A. Tsander—and in the writings of their Central European counterparts—Hermann Oberth, Walter Hohmann, Guido von Pirquet, and "Hermann Noordung." Their goal was flight to the Moon and planets, but their calculations suggested that chemically propelled rockets might lack the power to launch such journeys directly from Earth's surface. If a journey were carried out in stages, however, the problem might be surmounted.

They proposed using a space station, a stopover point in orbit. Once such a station was built, any number of rockets might be launched to meet it, each bearing its cargo of fuel or supplies to be transferred to the station. When enough had been gathered, fuel and supplies might then be loaded aboard an interplanetary vessel, perhaps itself constructed in orbit, and the real journey to the planets could begin. In effect, the trip would be launched from orbit, the greater part of the velocity needed to escape Earth's gravitational field having been already attained. This concept had been widely accepted in space-travel circles by 1929.[23]

While rendezvous was clearly a key technique in this scheme, it failed to receive any special emphasis. That changed after 1949, when two members of the British Interplanetary Society pointed out that orbital staging need not depend on first building a space station. The new concept was called "orbital technique" or "orbital operations." The pieces of an interplanetary vessel might simply be assembled in Earth orbit without troubling to construct a space station, or several rockets might meet in orbit and transfer their fuel to one of their number, which would then embark on the final mission.[24] As Wernher von Braun, later one of NASA's leading advocates of orbital operations, remarked, the space station really amounted to no more than "a space rigger's hotel."[25]

The rapid spread of this idea brought rendezvous into sharp focus. Unlike the space-station concept, to which rendezvous was a sometimes neglected adjunct, orbital operations moved rendezvous to center stage. The first paper specifically addressed to the problem of "Establishing Contact Between Orbiting Vehicles" appeared in 1951.[26] One result was a renewed attention to orbital mechanics, a topic that had languished since the path-breaking work of Walter Hohmann in 1925. By the end of the 1950s, a theoretical framework for rendezvous techniques had been largely erected.[27]

When NASA planners began to grapple with the problem of picking long-range goals for the American space program, however, they tended to overlook the part rendezvous might play except as it related to space stations. This may have reflected, as much as anything else, the imprint on NASA of the National Advisory Committee for Aeronautics (NACA). When NASA began its career on 1 October 1958, its core was the 43-year-old NACA, to which had been added several military and quasi-military space projects. NASA was designed to be, and in time became, something larger, wealthier, and more adventurous than NACA had been. But for a time much remained unchanged or

6

changed only slowly. The habits of mind, the viewpoints, the styles, the biases fostered by the old setting did not vanish overnight with the old name. The same NACA engineers, scientists, managers, and technicians who left work on 30 September 1958 were back on the job for NASA the next morning. Time would bring new faces and fresh viewpoints, thin the ranks of the old NACA hands, and weaken the grip of old habits; but NACA left an enduring mark on NASA and its programs.[28]

NACA had existed to serve—to solve problems for military and industrial aircraft programs. Its field, in which it was very good, was applied research—solving general engineering and technical problems in aeronautics. NACA laboratories had produced many of the technological innovations that transformed the post-World War I airplane, a slow and inefficient machine of small military and no commercial importance, into the major weapon and economic giant of mid-century. Langley Memorial Laboratory was the first and, until the eve of World War II, the only NACA laboratory; Langley research pioneered many prewar innovations in aeronautical design. Lewis Flight Propulsion Laboratory and Ames Aeronautical Laboratory went into operation early in the Second World War, the Pilotless Aircraft Station in 1945, and the High Speed Flight Station in 1947. In 1940, NACA had 650 employees and a budget of $4.37 million; five years later it employed 6800 and spent $40.5 million. But NACA still focused its research in those areas where lack of knowledge hindered aviation progress, spending little effort on basic research—expanding scientific knowledge—and steering clear of development, which meant seeing a specific project through design, building, and testing.[29]

During the 1950s, some of the most pressing problems in aeronautics arose from the little studied and poorly understood effects of high temperatures on very fast-moving aircraft and rockets. This made the focus of NACA research in that decade transonic and hypersonic flight, with special stress on aerodynamic heating phenomena.[30] When *Sputnik I* on 4 October 1957 transformed space from a region of scientific curiosity to an arena for national rivalry and spurred planning for manned space flight, this background stood NACA in good stead.

A small group of engineers at Langley began working informally on a manned orbital satellite. At the start of October 1958, in one of his opening moves as NASA's first Administrator, T. Keith Glennan approved the project. He formed the Space Task Group to run it and announced its name as Mercury two months later. STG started with 45 people led by Robert Gilruth and they had only one job: the most direct and speedy achievement of manned orbital flights.[31] It was a complex but straightforward engineering task. Project Mercury "did not require and does not require any major technological breakthroughs."[32] What it did need was just what a NACA background

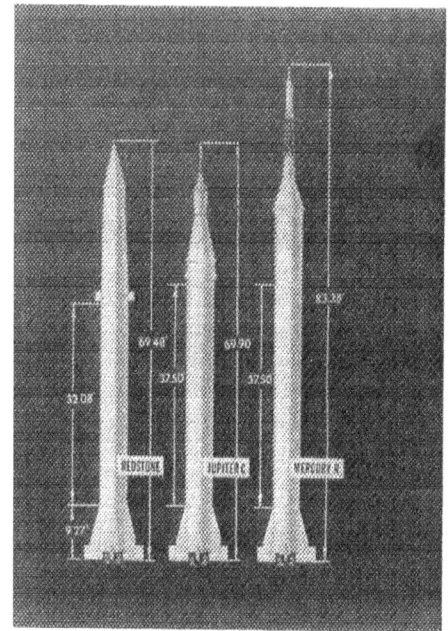

Modifications to ballistic missiles by 1961 made two vehicles safe for adaptation to manned space flight. Top left, Redstone and its modified versions for unmanned and manned space missions. Top right, Mercury-Redstone 3 being prepared for launch of Astronaut Alan Shepard on his suborbital space flight. Right, Mercury-Atlas 4 on pad at Cape Canaveral in 1961. This unmanned mission was to qualify the tracking network and the spacecraft for the upcoming manned orbital mission of John Glenn.

8

provided, the skills of applied research and aeronautical engineering and particularly experience in the aerodynamics of hypersonic flight.

Manned space flight beyond Mercury, however, was another matter. The crucial role of boosters in setting the limits of what could be done in space prompted NASA to its first long-range planning venture, "A National Space Vehicle Program," issued in January 1959.[33] This report surveyed existing boosters and proposed developing a series of new ones. It did no more than suggest a range of missions suited to each of them. What could be done, however, was one thing; what should or would be done was something else. Choosing among the possible goals now became NASA's central planning concern.

This concern produced "The Ten Year Plan of the National Aeronautics and Space Administration" in December 1959. Ultimately spacecraft would carry explorers to the Moon and planets, but for the 1960s, NASA chose the more modest goal of circumlunar flight—a trip to the Moon, a few passes in orbit, and a return to Earth. "Manned exploration of the moon and the nearer planets must remain as major goals for the ensuing decade."[34]

NASA planners assumed that a trip to the Moon would be launched directly from Earth's surface. That required the giant Nova booster, the largest of the four new vehicles proposed in January 1959. Nova was a concept built on an engine (the F-1) designed to produce 6.7 meganewtons (1.5 million pounds of thrust). Air Force contracts with Rocketdyne had begun F-1 development in mid-1958. This was one of the military projects turned over to NASA when it was formed. Four of these engines were planned for Nova's first stage to provide 27 meganewtons (6 million pounds of thrust) at a time when the most powerful existing American booster required three engines to generate 1.6 meganewtons (360 000 pounds of thrust).[35] The belief expressed in the January report that, "with Nova, a manned lunar landing first becomes possible,"[36] pervaded NASA planning throughout 1959 and 1960. Even when refueling or assembly in orbit were discussed as alternatives worthy of study, they were discarded as a basis for planning, since "it is assumed that the Nova approach will be followed."[37]

The choice was by no means final, but NASA was leaning strongly toward direct ascent, perhaps more by default than by decision. To the extent that they had been compared at all, the merits of direct ascent and orbital operations had been merely asserted rather than studied. The question had been cited as a major one, and some of the problems involved in "the all-the-way approach versus the assembly-in-orbit approach" had been aired at meetings of the Research Steering Committee on Manned Space Flight, more commonly known as the Goett Committee after its chairman, Harry J. Goett of Ames, during

1959.*[38] But, as NASA's 10-year plan showed, the question had yet to exert much effect on NASA policy.

Notably absent from NASA's budget request for fiscal year 1961 was money to study rendezvous, nor did NASA spokesmen mention rendezvous when they defended the budget before Congress early in 1960.[39] There was also little talk of space stations. That had not been true the year before, when NASA asked for funds to study both a small orbiting space laboratory and rendezvous techniques. These were closely related. NASA's 1959 choice of lunar landing over a space station as its long-range goal caused rendezvous to fade into the background, since the agency had yet to conceive rendezvous for any purpose other than supporting a space station.[40]

### CHALLENGE FROM THE FIELD

Although rendezvous ceased to seem very important to NASA Headquarters, 1960 saw that viewpoint challenged in the field. Several NASA field centers had begun to look more closely at the possibilities, and two, in particular, began to urge strongly an open-minded reassessment of the merits of rendezvous. One was the George C. Marshall Space Flight Center, in Huntsville, Alabama; the other was Langley.

Marshall was unique in NASA for its background and outlook. It was the former Development Operations Division of the Army Ballistic Missile Agency, which joined NASA and received its new name in March 1960.[41]

Marshall's Director, Wernher von Braun, and his chief lieutenants had been responsible for the German Army's rocket development programs before and during World War II, coming to the United States after the Nazi regime collapsed in 1945.[42] They had known the heady atmosphere of Weimar Germany's dreams of space travel, and they

---

*This phrase became the standard shorthand for the controversy between direct ascent and rendezvous for the lunar mission in the minutes of the Goett Committee, which was formed in April 1959. The members were Milton B. Ames, Jr. (NASA Office of Aeronautical and Space Research), De E. Beeler (High Speed Flight Station), Alfred J. Eggers, Jr. (Ames), Maxime A. Faget (STG), Laurence K. Loftin (Langley), George M. Low (NASA Office of Space Flight Development), Bruce T. Lundin (Lewis), Harris M. Schurmeier (Jet Propulsion Laboratory), and Ralph W. May, Jr. (NASA Office of Advanced Research Programs), secretary. The committee intended both to "take a reasonably long term look at man-in-space problems leading eventually to recommendations as to what future mission steps should be" and to recommend appropriate research programs to support these steps. This function recalled that of the technical advisory committees that had been NACA's instrument for promoting the exchange of information and recommending needed research, although unlike them its membership was drawn entirely from within the organization. NASA research was to be aligned with NASA development, just as NACA research had been aligned with military and industrial development in the past. The Goett Committee was chiefly responsible for choosing lunar landing as NASA's appropriate long-term goal.

had a long head start on their American colleagues in the hard, practical work of making these dreams real. They had studied space stations long before they joined NASA. Von Braun had moved on to the notion of orbital operations. As early as December 1958, he was urging NASA to base its lunar mission planning on rendezvous techniques. In a presentation to top-level NASA officials, von Braun dismissed direct flight as very difficult, then described four alternative rendezvous schemes, two requiring only Earth orbital operations and two calling for rendezvous in lunar orbit as well.[43]

Von Braun and his colleagues had been working since 1957 on the concept of using a cluster of relatively small rocket engines to build a booster of 6.7 meganewtons (1.5 million pounds of thrust) as the basis for a space flight program leading to manned lunar landing.[44] The booster project was approved by the Advanced Research Projects Agency of the Department of Defense in August 1958.[45] Then known as Juno V, the vehicle became Saturn in February 1959 and studies began on suitable upper stages in a complete system for a military lunar mission.[46] Whether there was any military need for Saturn was the question of 1959, and the answer was no. The decision to shift Saturn to NASA was behind the transfer of von Braun's group.*[47]

Spokesmen for von Braun's group led the defense of the "assembly-in-orbit approach" at Goett Committee meetings during 1959, with strong backing from George M. Low, who urged study of "vehicle staging so that Saturn could be used for manned lunar landing without complete reliance on Nova." The committee supported von Braun's request for a NASA contract to study orbital operations (his group then still belonged to the Army), and Low, who was highly placed in the NASA Headquarters Office of Space Flight Development, helped push it through.[48] Von Braun's group studied Saturn's role in lunar landing missions, both manned and unmanned, under NASA auspices during the last half of 1959. The new findings confirmed what an earlier report had concluded, "that a manned circumlunar satellite could be launched from the earth's surface, but some other technique will have to be used for a manned lunar landing with the present state of the art." Most of the chapter on "Manned Circumlunar Flights and Lunar Landings" in the 1959 study report was devoted to the role of orbital operations in these missions.[49]

Joining NASA did nothing to alter this Center's viewpoint. Until well into 1960, however, Marshall's leanings toward orbital operations produced little work specifically on rendezvous.[50] Concerned mainly

---

*The clustered-small-engine booster eventually became Saturn I, then Saturn IC. Saturn V, which lifted Apollo to the Moon, clustered five of the much larger F-1 engines in its first stage, making it a kind of small Nova.

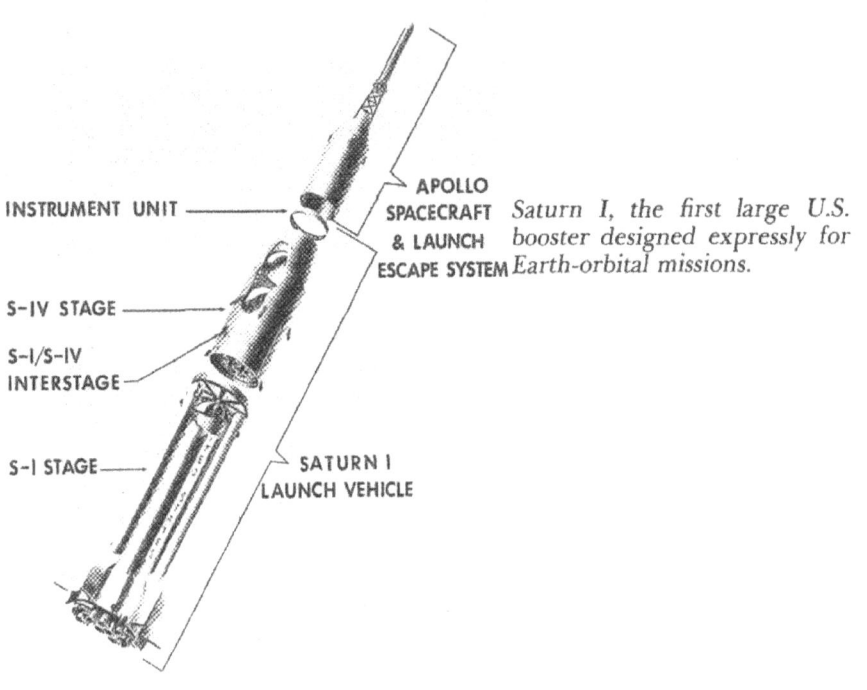

INSTRUMENT UNIT

APOLLO
SPACECRAFT
& LAUNCH
ESCAPE SYSTEM

S-IV STAGE

S-I/S-IV
INTERSTAGE

S-I STAGE

SATURN I
LAUNCH VEHICLE

*Saturn I, the first large U.S. booster designed expressly for Earth-orbital missions.*

## SATURN I LAUNCH SUMMARY

RESEARCH AND DEVELOPMENT FLIGHTS

SA-1   1. LAUNCHED—OCT. 27, 1961
        2. S-1 STAGE PROPULSION SYSTEM SATISFACTORY

SA-2   1. LAUNCHED—APR. 25, 1962
        2. PROJECT HIGHWATER RELEASED 22,900 GAL. $H_2O$ INTO IONOSPHERE

SA-3   1. LAUNCHED—NOV. 16, 1962
        2. 2ND PHASE PROJ HIGHWATER
        3. FULL PROPELLANT LOADING

SA-4   1. LAUNCHED—MAR. 28, 1963
        2. ENGINE OUT CAPABILITY DEMONSTRATED

SA-5   1. LAUNCHED—JAN. 29, 1964
        2. FIRST LIVE S-IV STAGE AND INSTRUMENT UNIT

SA-6   1. LAUNCHED—MAY 28, 1964
        2. FIRST ACTIVE GUIDANCE FLIGHT
        3. FIRST FLIGHT APOLLO BOILERPLATE AND LES
        4. ENGINE OUT (UNPLANNED)

OPERATIONAL FLIGHTS

SA-7   1. LAUNCHED—SEPT. 18, 1964
        2. COMPLETELY ACTIVE ST-124 GUIDANCE

SA-9   1. LAUNCHED—FEB. 16, 1965
        2. FIRST PEGASUS (METEOROID TECHNOLOGY SATELLITE) ORBITED
        3. FIRST UNPRESSURIZED INSTRUMENT UNIT

SA-8   1. LAUNCHED—MAY 25, 1965
        2. SECOND PEGASUS SATELLITE ORBITED

SA-10  1. LAUNCHED—JULY 30, 1965
        2. THIRD PEGASUS SATELLITE ORBITED
        3. SATURN I PROGRAM COMPLETED

with development programs, especially Saturn, Marshall had few resources to devote to the kind of research needed to locate and solve
basic problems of technique. Such studies, in any case, more properly
fell to one of NASA's research centers, which could focus on rendezvous itself rather than on the missions that the technique might open
up. This was where Langley entered the picture, for whatever these
missions might be, in true space flight "there will undoubtedly be space
rendezvous requirements."[51]

Rendezvous research centered on guidance and propulsion at
Langley, where two groups were working more or less independently
during 1959. In the Aerospace Mechanics Division, John M. Eggleston
and his colleagues were looking at the mechanics of orbital rendezvous. And in the Theoretical Mechanics Division, a group headed by
John D. Bird was studying launch windows and trajectories for rendezvous.[52] The spokesman for Langley in the Goett Committee
agreed that lunar landing ought to be "the 'ultimate' manned mission
for present consideration." But he also voiced Langley's belief that
some form of manned space laboratory was "a necessary intermediate
step" as a focus for research. That meant a space ferry, and a space
ferry meant rendezvous.[53]

Late in 1959 this concern generated a space station commmittee at
Langley, with a subcommittee on rendezvous headed by John C. Houbolt, then assistant chief of the Dynamic Loads Division. Houbolt was
fresh from a successful attack on the problems that had caused several
Lockheed Electras to crash. Despite, or perhaps because of, his inexpe-
rience in spacecraft technology, Houbolt zealously espoused rendezvous. Although his subcommittee had been formed to look at rendezvous in the context of space stations, Houbolt insisted from the start
that it study rendezvous in the broadest terms, since that technique
would play a large role in almost any advanced space mission. Loosely
organized and largely unscheduled, the subcommittee became a meeting ground for everyone at Langley concerned with any aspect of rendezvous.*[54]

When Langley hosted the Goett Committee in December 1959,
Houbolt was among the space-station committee members invited to
describe their work. He concluded by urging a rendezvous-satellite
experiment "to define and solve the problems more clearly,"[55] the first
of many such pleas Houbolt was to make with as little response. Space-
station thinking still guided rendezvous work at Langley over the next
six months.

In May 1960, Langley was once more host to a meeting, this time

---

*This included, among others, John M. Eggleston, John D. Bird, Arthur W. Vogeley, Max C.
Kurbjun, John A. Dodgen, William C. Mace, W. Hewitt Phillips, and Clinton E. Brown.

of lesser scope but greater impact. Bernard Maggin, from the Office of Aeronautical and Space Research in NASA Headquarters, had called the meeting to discuss space rendezvous and served as its chairman; he was the only member from Headquarters. Maggin had intended to invite to the meeting only the NASA research centers—Langley, Ames, and Lewis—which his office directed. He soon learned, however, that rendezvous had excited wider interest, so he invited the development centers—Marshall and Goddard—as well. The meeting was designed to give the centers a chance to acquaint each other with current research and to exchange thoughts on future prospects.[56]

Most of the first day was given over to a series of technical papers on propulsion, guidance, and trajectories, which mainly reviewed work in progress.[57] They revealed two salient facts about NASA rendezvous research in mid-1960; work centered on rendezvous between space station and ferry, and Langley was doing most of it.

All NASA rendezvous research was in-house; NASA had yet to provide contract funds for industrial or academic studies. This was one of the chief topics at the round-table talks on future rendezvous requirements that took up the second day of the meeting. Lack of funding was ascribed to strong resistance within NASA to any program aimed solely at the modest goal of proving a new technique or advancing the state of the art. To win funds, a research program on rendezvous needed larger ends. Everyone at the meeting believed that NASA ought to begin to develop and prove rendezvous techniques, because all were convinced that the need for rendezvous was going to become urgent within the next few years. What had to be done, then, was to find a context for rendezvous, and the best choice for the task was Marshall, since "resistance to . . . rendezvous [was] currently strong" in both Goddard Space Flight Center and the Space Task Group, NASA's other two development organizations.[58]

This may have been the most important by-product of the conference—the conclusion that Marshall had both the capacity and the desire to carry through an orbital operations and rendezvous program. In September 1960, Marshall's Future Projects Office was able to tell a gathering of industrial representatives that it had $3.1 million in study contracts to award during fiscal year 1961, a number of them related to rendezvous and orbital operations.[59] By the end of the fiscal year, the office had issued $817 422 in contracts to ten corporations and four universities for studies ranging from the broad problems of satellite rendezvous to the design of orbital refueling systems for Saturn.[60]

Marshall's commitment to the principle of orbital operations began to produce in late 1960 specific studies of rendezvous and orbital mechanics, much as the first proposal of the idea in 1949 had done. As befitted a development center, Marshall's research was mission oriented. Its role in the study of rendezvous hinged on how the technique might best be used in manned space missions, in particular a manned landing on the Moon.

The focus of work at Langley also shifted, as Houbolt and his co-workers succumbed to the fascination of a novel application of rendezvous technique, rendezvous in lunar orbit. The essence of the idea was to leave that part of the equipment and fuel needed for the return to Earth in lunar orbit while only a small landing craft descended to the lunar surface, later to rejoin the orbiting mother ship before starting the trip home. In one form or another, this idea had appeared in the work of Oberth, Kondratyuk, and the British Interplanetary Society, to say nothing of later writers. But it reached Langley's rendezvous subcommittee via a brief paper by William H. Michael, Jr., little more than a week after the rendezvous conference at Langley had adjourned.

Michael was part of a small group in the Theoretical Mechanics Division that had been working on trajectories for lunar and planetary missions. The group outlined some of its findings in a pamphlet that made the local rounds near the end of May 1960. Michael's contribution was a brief calculation of the amount of weight that might be saved in a lunar landing mission by parking the return propulsion and part of the spacecraft in lunar orbit.[61] The idea hit Houbolt like revealed truth:

> I can still remember the "back of the envelope" type of calculations I made to check that the scheme resulted in a very substantial savings in earth boost requirements. Almost spontaneously, it became clear that lunar orbit rendezvous offered a chain reaction simplification on all back effects: development, testing, manufacturing, erection, count-down, flight operations, etc. . . . All would be simplified. The thought struck my mind, "This is fantastic. If there is any idea we have to push, it is this one!" I vowed to dedicate myself to the task.[62]

And dedicate himself he did. Houbolt and a band of disciples embarked on a crusade to convert the rest of NASA to the truth that lunar orbit rendezvous was the quickest and cheapest road to the Moon.

Rendezvous found an important ally in NASA Headquarters late in 1960, when Robert C. Seamans, Jr., arrived in Washington to fill the post of Associate Administrator. Seamans, whose formal appointment dated from 1 September, came to NASA from the Radio Corporation of America, where he had been chief engineer of the Missile Electronics and Controls Division in Burlington, Massachusetts.[63] Seamans' division had been one of two Air Force contractors to study requirements for an unmanned satellite interceptor (Saint) during 1959. In 1960, when Saint moved from study to development, RCA got the Air Force contract to develop its final stage and inspection payload and to demonstrate its rendezvous and inspection capability.[64]

Saint was part of a quiet but far-reaching Air Force program, much of it concerned with rendezvous and orbital operations, intended to carve out a larger military role in space. Reading the minutes of a November 1960 meeting of the Air Force Scientific Advisory Board, at which both the Air Force and Marshall reviewed rendezvous work and plans, convinced a Space Task Group observer that Air Force planning and progress toward orbital operations "is much further ahead (2 to 3 years) than the NASA Program at MSFC."[65]

Seamans thus came to NASA with a solid background in rendezvous work. He spent most of his first month as Associate Administrator touring NASA's field centers. At Langley, he talked to Houbolt. Seamans was deeply impressed by Houbolt's account of the weight savings to be achieved even if only the spacecraft heatshield remained in a lunar parking orbit.[66] Seamans invited Houbolt to Washington for a more formal hearing before the Headquarters staff. Houbolt and some of his Langley colleagues presented the case for putting rendezvous into the national space program in a mid-December briefing at NASA Headquarters.*[67]

So by the end of 1960 NASA Headquarters had been exposed to the idea of orbital operations, to the potential value of rendezvous techniques in manned space missions other than those related to space stations. It had also been introduced to the case for lunar orbit rendezvous as a basis for manned flight to the Moon. These ideas had worked their way up from the field, chiefly from the von Braun group at Marshall and Houbolt and his colleagues at Langley. The once unchallenged assumption that a lunar mission, if it were to be undertaken, would be launched directly from Earth's surface had now been called into question; and the questions multiplied in the following months.

## MERCURY AS PROLOGUE

Throughout 1959 and 1960, Mercury was the first and only approved American manned space flight program. From the very start, however, few people expected it to be last. The Mercury capsule was essentially experimental, an attempt to master the problems of manned space flight. Someday spacecraft would do more than go up, circle

---

*Houbolt stressed the general utility of rendezvous in future space missions; John Bird, the advantages of orbital operations; Max Kurbjun, the problems of visual rendezvous; and Clinton Brown, the lunar-orbit-rendezvous concept. In addition to those who spoke formally, the Langley delegation included Eggleston and Phillips. Besides Robert C. Seamans, Jr., Headquarters was represented by Ira H. A. Abbott, Milton Ames, Hermann H. Kurzweg, and Bernard Maggin of the Office of Advanced Research Programs; Eldon W. Hall, Launch Vehicle Programs; George Low, Space Flight Programs; Berg Paraghamian, Program Planning and Evaluation; Alfred M. Mayo, Life Sciences Programs; and Donald H. Heaton, Seamans' assistant.

Earth a few times, and then come down. They would have to be maneuverable, both in space and after they returned to the air. They should be able to fly to a landing, and preferably on land rather than in the water. They should be easy to test and repair, if space flight were ever to be put on something like a routine basis. NASA was ready to suggest research along these lines in its first hastily prepared budget for fiscal year 1960, submitted to Congress early in 1959.

Mercury was an engineering project. Its major goal was "to achieve at the earliest practicable date orbital flight and successful recovery of a manned satellite."[68] This dictated utmost reliance on the best-known techniques: a ballistic reentry capsule—blunt, cone-shaped, with almost no aerodynamic lift, recovered by parachute after it returned to the atmosphere.[69] But it also excluded some promising alternatives, two of which took tentative shape in NASA's 1960 budget. One was the so-called environmental satellite, a kind of small temporary space station able to sustain one or more men in orbit for several weeks or even months. The other was a maneuverable spacecraft, one equipped with rocket motors to change its path in orbit and endowed with enough aerodynamic lift to alter its flight-path in the atmosphere.

NASA asked for $300 000 to study design changes that might turn Mercury into an orbiting laboratory and for $1 million to study a Mercury refined to make it maneuverable and flyable. Looking toward a real space station, NASA also asked for $3 million to study space rendezvous techniques.[70] These modest sums signalled no great commitment. When NASA ran into budget problems, this effort was simply shelved and the money diverted to more pressing needs.[71]

The view from Space Task Group, the Mercury team, was different. Even during the first hectic months, while Mercury was still moving from the drawing boards into the laboratories, some people in STG were turning their thoughts to what might come next. Although a ballistic capsule might get the job done quickly, it also had patent shortcomings, not the least of which was "that it will be very difficult to control the landing point within a distance of perhaps the order of a hundred miles each way."[72] The ballistic capsule had been only one of three basic types under study in 1958 for a manned satellite program. The others were a winged glider and a lifting body, so shaped that even without wings it still had enough lift to allow the pilot some control.[73] For later missions, either offered a clear edge over Mercury. The winged glider, which could be flown much like an airplane once it was back in the atmosphere, had been preempted by the Air Force in its Dyna-Soar program.

Dyna-Soar was a development project of the Air Research and Development Command (ARDC). The project received its name in October 1957 and Air Force Headquarters approval in November, some four years after study had begun on vehicles boosted into orbit

*17*

by rocket and gliding back to Earth under pilot control. Much of the work had been done under contract by Bell Aircraft Company. NACA joined the project in May 1958 to provide technical advice and help to the Air Force-directed and -funded program, an arrangement re-affirmed by NASA in November 1958. ARDC's consolidated Dyna-Soar development plan in October 1958 aimed the project specifically at developing a winged glider for return from orbit. Later X-20 replaced Dyna-Soar as the project's name.[74] Leaving gliders to the Air Force was no hardship since many in NASA, especially in the research centers, preferred the lifting-body approach.[75] As early as June 1959, STG could report promising results from studies of building some lift into a Mercury capsule.[76]

STG was also looking into a more radical approach to controlled spacecraft landing. Between 1945 and 1958, a Langley engineer named Francis M. Rogallo had been working at home on a flexible kite, its lifting surface draped from an inflated fabric frame. In con-

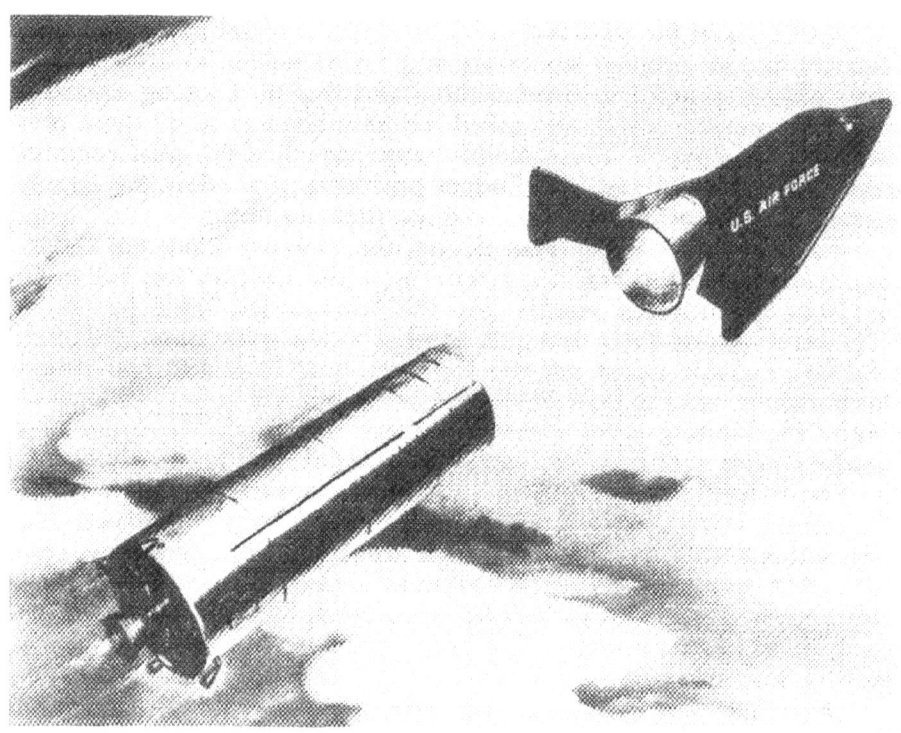

*Dyna-Soar spacecraft shown in artist's drawing separating from second stage of its Titan I booster.*

trast to other flexible aerial devices like parachutes, a load-bearing Rogallo wing produced more lift than drag, though not as much as a conventional wing. But rigid wings could not be folded neatly away when not in use, and they were inherently far heavier. Rogallo first realized what this might mean in 1952, when he chanced across an article on space travel

> with beautiful illustrations depicting rigid-winged gliders mounted on top of huge rockets. I thought that the rigid-winged gliders might better be replaced by vehicles with flexible wings that could be folded into small packages during the launching.[77]

Rogallo's efforts to promote his insight met scant success until late 1958, when the new American commitment to explore space furnished him a willing audience. In December, the Langley Committee on General Aerodynamics heard him describe his flexible wing and how it might be used in "space ship landing."[78] The group responded warmly, and work on the concept moved from Rogallo's home to laboratories at Langley.

A few months later, STG asked Rogallo for an informal meeting to discuss his research. Some of STG's top people, Manager Gilruth among them, showed up on 30 March 1959 to hear what Rogallo had to say.[79] Gilruth was impressed enough to suggest at a staff meeting two months later that some study go into a follow-on Mercury using maneuverable capsules for land landing.[80]

In the meantime, STG was spreading the news about its "preliminary thinking about Project Mercury follow-ups." H. Kurt Strass of STG's Flight Systems Division reported to the Goett Committee on some ideas for a larger, longer-lived Mercury capsule. STG's thinking ranged from an enlarged capsule to carry two men in orbit for three days, through adding a three-meter cylinder behind the capsule to support a two-week mission, to cabling the combined capsule and cylinder to a booster's final stage and rotating them to provide artificial gravity. This was modest compared to the more sophisticated "environmental satellite" favored by Langley, "a true orbiting space laboratory with crew and equipment exchangeable" via ferry.[81]

The Goett Committee divided on just how large the next step ought to be but agreed that some such step belonged between Mercury and a lunar mission.[82] So did the NASA planners, who, during 1959, were drawing up a long-range plan for manned space flight. Although NASA's future program was "directed heavily toward manned lunar exploration," there was still a place in it for developing maneuverability and a long-life capsule, both based on modifying Mercury.[83]

In seeking to explore the possibilities of improving Mercury to fit it for more advanced missions, STG was moving beyond the limits of its charter. It had been formed for only one purpose: to manage Pro-

*Francis Rogallo of the Langley Research Center adjusts a model of his paraglider, often called a "Rogallo wing," in the 480-kilometer-per-hour wind tunnel. In this 1959 test, the paraglider was being considered as a device to recover stages of the Saturn booster following launch.*

ject Mercury. By mid-1959, the initial group of 45 had grown eightfold, and Gilruth's title had changed from Manager to Director of Project Mercury. Despite this rapid expansion, STG felt understaffed. An STG study in June 1959 concluded that 223 people should be added to the 388 authorized, just "to maintain the schedule set for PROJECT MERCURY." But simply keeping pace was not enough.

> In addition, . . . some attention should be given to advanced or follow-on systems to MERCURY. It is estimated that a staff of approximately 20 additional professional personnel should be built up during the next year in order that a year or more gap will not occur in NASA manned space flight operations at the conclusion of the presently planned MERCURY Program.*[84]

*In 1959, STG comprised three divisions: Flight Systems under Max Faget; Operations, Charles W. Mathews; and Engineering and Contract Administration, Charles H. Zimmerman (replaced in August by James A. Chamberlin).

Gilruth foresaw a total strength of some 900 by 1 July 1960, less than half of them working directly on Project Mercury. The rest would be divided among three other projects—a maneuverable manned satellite, a manned orbiting laboratory, and a manned lunar expedition—and a supporting program in biotechnology and human factors. The maneuverable manned satellite project accounted for 302 of the 485 new positions, showing which goal STG though should be pursued immediately after Mercury.[85]

During the same month, June 1959, Kurt Strass argued that the time had come to stop just thinking about these projects and to start actually designing one. He proposed forming a group to work out the preliminary design of "a relatively sophisticated space laboratory providing living accommodations for two men for two weeks," ready to fly by late 1962.[86] Strass found a sympathetic ear in the chief of the Flight Systems Division (FSD), Maxime A. Faget, who appointed him to head a New Projects Panel within the Division.* It met for the first time on 12 August 1959, and Strass told his fellow panelists they were there to plan a manned lunar landing through a series of graded steps, the first of which was to define "an intermediate practical goal to focus attention on problems to be solved, and thus serve to guide new technological developments."[87]

The panel floundered a bit, not quite certain of the direction it should take, but soon zeroed in on the design of an advanced spacecraft suited to the lunar mission, the first step on the road that led to the Apollo spacecraft. That still left a sizable gap in the manned space flight program, which a new engineering report by McDonnell Aircraft Corporation, prime contractor for the Mercury capsule, suggested some ways to fill. The panel decided to take a close look.[88]

The McDonnell report of September 1959, "Follow On Experiments, Project Mercury Capsules," was the result of a summer's work by a small advanced project group.†[89] It proposed six experiments that might be conducted with practical modifications of the Mercury capsule, to explore some problems of space flight beyond those to be attacked in Project Mercury.[90] The New Projects Panel found none of the McDonnell ideas wholly satisfactory but agreed that parts of the

---

*Besides H. Kurt Strass, the panel included Alan B. Kehlet, Head, Aerodynamics Section, Performance Branch; Jack Funk, Head, Space Mechanics Section, Dynamics Branch; Harry H. Ricker, Jr., Head, On Board Systems Branch; Robert G. Chilton, Head, Dynamics Branch; Stanley C. White, Head, Life Systems Branch; William S. Augerson, Life Systems Branch; and Caldwell C. Johnson, Head, Engineering Branch, Engineering and Contract Administration Division (the only non-FSD member of the panel). The meetings of the panel were attended by nonmembers, as well, again largely from FSD.

†The group, headed by E. M. Flesh, McDonnell engineering manager for Mercury, included Fred J. Sanders, William J. Blatz, Darrell B. Parke, and Walter D. Pittman.

first three "could be combined into a new proposal which could offer increased performance and an opportunity to evaluate some advanced mission concepts at the earliest opportunity."[91]

All three experiments dealt with spacecraft maneuverability and guidance. The first sought to achieve some control of landing by adding an external trim-flap device to the capsule, coupled with a simple radar guidance technique or, alternatively, with a more sophisticated inertial guidance system to reduce the capsule's dependence on ground facilities. The second aimed at maneuvering in orbit by adding to the capsule a special adapter to carry a propulsion system, with guidance provided by either a Mercury system or an inertial guidance system. The third experiment was designed to test the inertial guidance system that might be used with either of the first two experiments. The system—inertial platform, computer, and star tracker— would allow the capsule to guide itself toward an orbital rendezvous, to control its touchdown point more precisely, and to navigate on lunar and interplanetary missions. All three experiments used a modified one-man Mercury launched by an Atlas, with minimum changes.[92]

The panel saw the prospect of a useful test vehicle in joining an adapter-borne propulsion system to an inertial guidance system. Maneuverable in both space and atmosphere, a capsule so equipped might then be used to develop advanced system components, such as environmental systems for long-term missions, auxiliary power systems, and photographic reconnaissance. These were parts of McDonnell's suggested fourth and fifth experiments. The fourth was a 14-day mission, using an adapter to carry both a propulsion system and the extra supplies and equipment to support the extended time in orbit, with fuel cells substituted for batteries to supply electrical power. The fifth mainly involved adding a camera to the Mercury periscope system to allow the pilot to photograph Earth's surface from orbit.* The panel asked for "authority to initiate this program" to "continue with the least possible delay" after the Mercury program.[93]

The time, however, was not yet ripe. The attractive possibilities of experimenting with a modified Mercury capsule paled in comparison with the far more exciting prospect of designing an advanced spacecraft for a trip to the Moon. When STG's top management met a month later, on 2 November 1959, it was the advanced spacecraft rather than the modified Mercury that they decided to pursue.†[94]

---

*The panel ignored the sixth McDonnell experiment, which differed radically from the other five. It projected the use of a heavily instrumented unmanned Mercury capsule to study the problems of stability and heating during reentry from lunar orbit, simulated by launching the capsule into a highly elliptical orbit with the Atlas-Centaur.

†At the meeting were Robert R. Gilruth, his special assistant Paul E. Purser, Kurt Strass, Robert O. Piland, John D. Hodge, Caldwell Johnson, Charles J. Donlan, Max Faget, Charles W. Mathews, and James A. Chamberlin.

That was the story of STG planning for better than a year. Although engineers were still thinking about an improved Mercury, that thought took second place to work on a new lunar spacecraft.[95] Lifting reentry was still seen as an important objective, a point stressed by NASA witnesses in budget hearings early in 1960, but not necessarily as part of the Mercury program.[96] By April 1960, the central aim of advanced vehicle development had become "lunar reconnaissance." The possibility of a lifting Mercury received only passing mention, as advanced planning focused on a spacecraft able to orbit the Moon, "a logical intermediate step toward future goals of landing men on the moon and other planets."[97] This was the program that officially became "Apollo" in July 1960. As then conceived, it did not go beyond circumlunar flight, although lunar landing was the ultimate goal.[98]

What was becoming clear was that any advanced Mercury program, such as lifting reentry, was likely to become a major undertaking in its own right.[99] In March 1960, STG's summary of projected funding needs for manned space flight programs put the cost of a lifting Mercury project at over $34 million during fiscal years 1960 through 1962.[100] STG did go on with its lifting Mercury plans into April 1960, getting as far as a preliminary specification for the reentry control system and plans to solicit contractor proposals for the system.[101]

Lifting reentry, in principle, had NASA Headquarters approval. Still lacking was a firm commitment based on a specific proposal with clearly defined costs.[102] That commitment failed to materialize. In May 1960, Administrator Glennan's budget analysis team turned down STG's request for funds to pursue advanced technical development of Mercury-type capsules. Glennan conceded the probability of Mercury flights beyond the three-orbit mission then authorized, to avoid a break in manned space flights, if nothing else. But thinking about somewhat longer missions was one thing; approving a lifting capsule was something else.[103]

That decision put a temporary halt to STG efforts to improve Mercury. Mounting problems in the project itself, especially during the last quarter of 1960, kept STG busy, and such advanced work as time allowed was limited to Apollo.

## NASA AFTER TWO YEARS

As 1960 drew to a close, NASA's manned space flight program was still limited to Project Mercury, but plans and hopes for a larger enterprise were rife. At the center of NASA's aspirations was a lunar landing program, endorsed by the Goett Committee in mid-1959 and

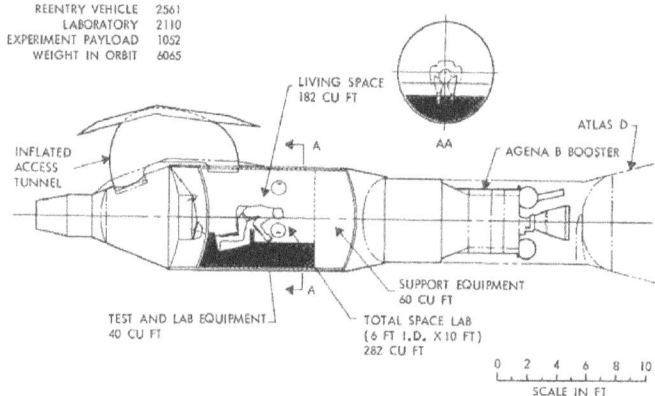

WEIGHT DATA (LB)
(14-DAY MISSION)

REENTRY VEHICLE 2561
LABORATORY 2110
EXPERIMENT PAYLOAD 1052
WEIGHT IN ORBIT 6065

LIVING SPACE
182 CU FT

ATLAS D

INFLATED
ACCESS
TUNNEL

A

AA

AGENA B BOOSTER

SUPPORT EQUIPMENT
60 CU FT

TEST AND LAB EQUIPMENT
40 CU FT

A

TOTAL SPACE LAB
(6 FT I.D. X 10 FT)
282 CU FT

0 2 4 6 8 10
SCALE IN FT

NASA's planners in 1960 and early 1961 aimed higher than just an improved Mercury spacecraft. In St. Louis, McDonnell proposed a 14-day space laboratory. In Houston, Robert Gilruth (second from left), Director of the Space Task Group, and his chief assistants, Charles Donlan (left), Maxime Faget, and Robert Piland discuss selection of contractors to study the feasibility of a manned circumlunar mission (August 1960). In Washington, NASA's second administrator, James Webb (center), and George Low (right) of NASA Headquarters, receive a model of the vehicle proposed by General Electric (April 1961).

written into the agency's ten-year plan at the end of the year. This goal was framed on technical grounds, as a legitimate end in its own right and as the best means to focus further work on manned space flight after Mercury. Questions of politics, economics, and the other external forces that would decide whether the United States should actually undertake such a program played no part in the choice of the goal.[104] NASA engineers were convinced that they could reach the Moon and that reaching the Moon made sense in technical terms. But the technical facts also forced NASA to settle for planning a lesser program for the 1960s. A landing on the Moon remained the long-range goal, but plans were scaled down for a partway effort, a trip around the Moon and back in Project Apollo.

## NASA TEN YEAR PLAN

| Calendar Year | Event |
|---|---|
| 1960 | First launching of a Meterological Satellite |
| | First launching of a Passive Reflector Communications Satellite |
| | First launching of a Scout vehicle |
| | First launching of a Thor-Delta vehicle |
| | First launching of an Atlas-Agena-B vehicle (by the Department of Defense) |
| | First suborbital flight of an astronaut |
| 1961 | First launching of a lunar impact vehicle |
| | First launching of an Atlas-Centaur vehicle |
| | Attainment of manned space flight, Project Mercury |
| 1962 | First launching in the vicinity of Venus and/or Mars |
| 1963 | First launching of the two-stage Saturn vehicle |
| 1963-1964 | First launching of unmanned vehicle for controlled landing on the Moon |
| | First launching of Orbiting Astronomical and Radio Astronomy Observatory |
| 1964 | First launching of unmanned lunar circumnavigation and return to Earth vehicle |
| | First reconnaissance of Mars and/or Venus by an unmanned vehicle |
| 1965-1967 | First launching in a program leading to manned circumlunar flight and to permanent near-Earth space station |
| Beyond 1970 | Manned flight to the Moon |

The main factor in this less ambitious program was the limited weight-lifting capability of existing boosters, as well as those expected to be ready for the 1960s. The real force of this restriction rested on the widely held assumption that a flight to the Moon would be launched directly from Earth's surface on a very large booster. Outside NASA, workers in the new field of astronautics, picking up a lead from early space-travel writers, had proposed rendezvous as an alternative to direct ascent. Within NASA, this idea was slow to take hold, although a few isolated voices supported it and grew louder. The pressure for change came mainly from the field.

NASA's field centers, though under tighter rein than NACA's had been, nevertheless were far from being mere agents of Headquarters. The precise ordering of relationships between Washington and the field has, in fact, been a continuing source of tension and a factor in the frequent reorganizations that NASA has undergone. Policy and long-range planning have tended to center in NASA Headquarters, design and development at lower levels. But what goes on at one level has not always seemed to mesh with what goes on at another. Headquarters policy has sometimes appeared to be nothing more than a belated ratification of work already under way in the field. This is the way rendezvous entered the space program.

Some form of rendezvous in Earth or lunar orbit appeared to offer the prospect of making do with lesser boosters than the giant Nova. While simple in theory, however, orbital rendezvous might well present problems in practice. A program designed to test the technique was beginning to look like a prudent move. This pointed to another aspect of NASA activity during 1959 and 1960, and to a still smaller step between Project Mercury and a lunar landing. Suitably altered, the Mercury capsule might become the basis for a new program. Given a certain eager optimism, such changes might be seen as nothing more than an effort to improve the experimental machine and convert it to an operational model. By 1960, proving rendezvous techniques was beginning to emerge as a logical task for the improved Mercury.

Prospects for a larger program at the end of 1960, whether lunar landing, circumlunar flight, or even rendezvous development, were not, in fact, good. During the last quarter of the year, Project Mercury suffered setbacks that strained STG morale and raised questions about the American manned space flight program.[105] The political climate was bleak. President Eisenhower rejected NASA's request for Apollo funds in the coming year's budget and leaned toward the view that Project Mercury was the only manned space flight program the United States needed. NASA's prospects under newly elected President John F. Kennedy seemed not much better.[106] Policy, however, was one thing, technology another. NASA could, and did, pursue its technical planning. When the climate changed, NASA was ready.

# II

# The Transmutation of Mercury

DURING January 1961, NASA's manned space flight program altered course. At the policy-making level in Headquarters, thinking shifted from lunar reconnaissance to lunar landing. This change was crucial, not only for the lunar program itself but also for what was to become Project Gemini; before 1961 was over that shift would provide justification for a rendezvous development program. In the field, the newly independent Space Task Group stopped talking about an improved Mercury capsule and began working on it. Plans for a lunar landing mission and work on an advanced Mercury proceeded through the summer of 1961 at different levels and varying rates. These separate paths converged in the autumn to give birth to a new program.

Whether these efforts would have borne fruit without a sharp change in the political climate is anyone's guess. The past two years had seen their share of false starts, dashed hopes, and aborted plans. But the climate did change. Within months after taking office, President Kennedy and his advisors found compelling reasons to support an American manned space flight program far larger than Project Mercury. One factor was certainly the renewed clamor about a space race between the United States and the Soviet Union. Informed opinion might discount Soviet accomplishments or stress American sophistication against Russian brute force; that smacked of quibbling to the American public, especially after 12 April 1961, when Cosmonaut Yuri A. Gagarin aboard *Vostok I* became the first human being to orbit in space. Two days later, the chairman of the House Committee on Science and Astronautics was not merely speaking for himself when he asserted, "My objective . . . is to beat the Russians." The President announced his decision on 25 May 1961, in a speech to Congress on "Urgent National Needs." He committed the United States to landing an American on the Moon before the end of the decade.[1]

## NEW DIRECTIONS

NASA had long since begun to lay plans for lunar flights, although throughout 1960 it had tended to focus on flying around, rather than landing on, the Moon. A new direction in NASA thinking surfaced at the quarterly meeting of the Space Exploration Program Council (SEPC) on 5-6 January 1961. The council was a NASA device for smoothing out technical and managerial problems at the highest level. Its members were the heads of the field development centers and Headquarters program offices,*[2] with the Associate Administrator serving as chairman.[3] The January meeting was the first presided over by Robert Seamans in his new assignment, and it marked a decisive turning point in the manned space flight program. The first day was devoted to manned lunar landing.

The meeting began with a series of presentations arranged by George Low, Chief of Manned Space Flight in the Office of Space Flight Programs, to provide "a 'first cut' at a NASA Manned Lunar Landing Program."[4] Low, an early advocate of orbital staging techniques as an alternative to the Nova direct approach, made sure that the council heard about Earth orbit and lunar orbit rendezvous as well as direct ascent.†[5] The next step was setting up a study team to devise

---

*NASA Headquarters had been reorganized in December 1959, largely in anticipation of the transfer of Wernher von Braun's Development Operations Division from the Army. The major change was the establishment of a new program office, the Office of Launch Vehicle Programs, which assumed jurisdiction over the Huntsville facility (later the George C. Marshall Space Flight Center) as well as substantial launch facilities at Cape Canaveral. This launch facility, the Missile Firing Laboratory, was combined with NASA's Atlantic Missile Range Operations Office (a liaison group between NASA and the Air Force) in June 1960 to form the Launch Operations Directorate, a semi-autonomous unit of Marshall. Director of the new Headquarters office was Don R. Ostrander, an Air Force major general who had been acting head of the Advanced Research Projects Agency, the Department of Defense unit responsible for Saturn. Ostrander's staff consisted of some 25 people from the Office of Space Flight Development, which now became the Office of Space Flight Programs, still directed by Abe Silverstein. Ira Abbott's Office of Aeronautical and Space Research now became the Office of Advanced Research Programs. In March 1960 NASA established a fourth technical program office under Clark T. Randt, the Office of Life Sciences Programs. Albert F. Siepert's Office of Business Administration changed neither its name nor its function during this period.

†In October 1960, Low had formed a small working group to lay out a preliminary program for manned lunar landing. This group comprised Eldon Hall (Office of Launch Vehicle Programs), Oran W. Nicks, and John H. Disher (both of the Office of Space Flight Programs). At the SEPC meeting in January 1961, Maxime Faget (Space Task Group) spoke on Apollo, Melvyn

a more complete plan. This the council did, naming Low its chairman. Unable to agree on the best approach, the council simply asked for "an answer to the question 'What is NASA's Manned Lunar Landing Program?' "[6]

The Low Committee began its work a week later.* Low himself drafted its report, revised it on the basis of comments from other members, and submitted it to Seamans early in February.[7] The report set out the two themes that came to dominate NASA lunar-mission planning throughout 1961. First, Low argued that both orbital operations and large boosters were going to be needed in the long run. NASA must include Nova-class boosters in the national space program, but "orbital operation techniques must be developed as part of the space program, whether or not the manned lunar landing mission is considered." Second, he insisted that, barring unforeseen problems, rendezvous "could allow us to develop a capability for the manned lunar mission in less time than by any other means."[8]

In Space Task Group, the question of rendezvous took a different form. It was seen as one of several classes of missions around which a follow-on Mercury program might be built. This was one of the subjects at a meeting on 20 January 1961 between Director Robert Gilruth and his chief lieutenants.† Max Faget, aided by his Flight Systems Division staff, led the discussion and outlined hardware and booster requirements for several possible types of missions.[9] Two broad classes came in for particular attention: one was labeled extended time in orbit, the other was rendezvous.

Extended time in orbit covered two possible missions. The first was an 18-orbit manned Mercury mission based on augmented capsule power supply and environmental control systems. The standard Atlas booster already slated for Mercury seemed adequate for this mission, but Gilruth suggested that the group think about using an Atlas-Agena. Atlas-Agena was a two-stage vehicle. The Atlas, which served as first stage, was a product of the Astronautics Division of General Dynamics Corporation in San Diego, California, and the Agena was built by the Lockheed Missiles & Space Company, Sunnyvale, California. Agena development began in 1957 under the Air Force Ballistic

---

Savage (Office of Launch Vehicle Programs) on direct ascent, Wernher von Braun (Marshall Space Flight Center) on Earth orbit rendezvous, and John Houbolt (Langley Research Center) on lunar orbit rendezvous.

*Other members of the Low Committee were Eldon Hall, Max Faget, John Houbolt, Oran Nicks, Alfred Mayo (Office of Life Sciences Programs), Earnest O. Pearson, Jr., and Heinz H. Koelle (Marshall).

†Associate Directors Charles Donlan and Walter C. Williams; Flight Systems and Flight Operations Division chiefs Max Faget and Charles Mathews, respectively; assistant Engineering Division chief William M. Bland, Jr.; and special assistant Paul Purser.

Missile Division. An improved model, Agena B, with a restartable engine and larger propellant tanks, entered development in June 1959 and flew on 12 November 1960.[10] Atlas might or might not have enough power to carry aloft the capsule modified for the mission; but if a primate were to pave the way for a manned mission of 7 to 14 days, then Atlas was clearly lacking. It could not lift the required weight.

Atlas was even more doubtful for rendezvous missions. Faget and his colleagues discussed two types, which differed chiefly in their targets. Both used Mercury capsules modified to make them maneuverable, but the target in the first instance was Saint; in the second, an as-yet-undeveloped space laboratory. Discussion centered on the need for a much "refined capsule with better operational and maintenance capabilities, better door, better wiring, possibly a bi-propellant control system, etc." All this meant weight, more than an Atlas could lift. But the basic objection to the rendezvous mission was that it "might be considered too hazardous for a one-man operation."[11]

Whatever their merits, all these possibilities were too vague. Before proposing a Mercury follow-on program to NASA Headquarters, STG had to be "more specific with regard to particular flights needed, funding, management, etc." This was the task assigned to Faget,* who had only a week to complete it before a scheduled visit to STG on 26-27 January by Abe Silverstein, head of Space Flight Programs in NASA Headquarters. The meetings with Silverstein resulted in a shift in focus to "the question of capsule redesign to speed up check-out and maintenance."[12]

With a good deal more work clearly needed, Gilruth turned to James A. Chamberlin. Canadian-born and trained at the University of Toronto and the Imperial College of Science and Technology in London, Chamberlin had been working in aeronautical engineering and design since 1939 for several Canadian firms. By March 1959 he had become chief of design for AVRO Aircraft, Inc., of Toronto, where he worked on the CF-105 Arrow, an advanced interceptor aircraft.[13] When that project was canceled, NASA was able to recruit Chamberlin and several of his colleagues.[14]

Chamberlin joined STG in April 1959; by August he had become acting chief of the Engineering and Contract Administration Division.[15] For the next year and half, he directed STG's technical monitoring of Mercury development and production. When, on 1 February 1961, Gilruth assigned him to work on an improved Mercury, Chamberlin remained titular chief of what had since become the Engineer-

---

*Faget was assisted by Mathews, Bland, and Kenneth S. Kleinknecht (Gilruth's technical assistant).

ing Division but turned over most of his organization's administrative, technical, and operational matters to his assistants, André J. Meyer, Jr., and William M. Bland, Jr.[16] Chamberlin himself went to St. Louis in mid-February; during the next months he actually worked from an office in the McDonnell Aircraft Corporation plant two or three days a week.[17]

STG's change in status at the beginning of 1961 may have sparked its renewed pursuit of a post-Mercury program. Although located at Langley Research Center in Virginia, STG belonged administratively to Goddard Space Flight Center in Maryland. This clumsy arrangement served no very useful purpose, since the Space Task Group was largely self-directed in any case. So NASA Administrator Keith Glennan announced on 3 January 1961 that STG was henceforth an independent field element, charged not only with managing Mercury but also with planning and carrying out programs "in the general area of manned space flight."[18] This was more hope than fact, however; Mercury was still the only approved program, and independence was largely formal. STG stayed at Langley, on which it still depended for much of its support, both technical and administrative.

The union with Langley was the next to go, for a number of compelling reasons: the threatened impact on Langley research of a full-fledged development effort, the strain of fitting a much expanded STG into already cramped Langley quarters, the chance to spread NASA more widely across the country, and the need to move before new programs had progressed to the point where moving would disrupt them.[19] These reasons anticipated, rightly as it proved, the President's lunar landing decision. Where to move was settled during the summer of 1961, after a special committee visited 19 possible sites.*[20] Houston won the prize, and the booming space agency joined forces with the booming city.

That massive expansion, which saw the tripling of both the manned space flight program and the center in charge of it, had been well prepared. NASA's first two years had seen most of the relevant issues raised, many of the answers suggested. Nothing had been decided beyond recall, but the channels were carved into which later events flowed. In the first half of 1961, some channels broadened, others dwindled and vanished. Before the summer was over, a far larger, far more complex, and far more costly manned space flight program emerged. An enormous lunar project had joined Mercury and a third project stood in the wings, justified by the needs of Apollo but growing out of the technology of Mercury.

---

*Locations surveyed were: in Louisiana, New Orleans, Baton Rouge, Shreveport, and Bogalusa; in Texas, Houston, Beaumont, Corpus Christi, Victoria, Liberty, and Harlingen; in Florida, Tampa and Jacksonville; in California, Los Angeles, San Diego, Richmond, Moffett Field, Berkeley, and San Francisco; and, in Missouri, St. Louis.

## STG PLUNGES AHEAD

The report of the Low Committee early in February 1961 produced no immediate action. As outgoing Administrator Glennan had warned his colleagues in the January meeting of the Space Exploration Program Council, lunar landing was not something NASA could undertake on its own hook; so large and costly a program needed backing at the highest levels.[21] In the uncertain political climate of early 1961, planning for a lunar landing remained temporarily in abeyance, though work on the Apollo spacecraft went ahead in STG. But renewed interest in rendezvous and orbital operations in NASA Headquarters, as shown in the Low report, led to a second inter-center meeting on rendezvous at the end of February. This time the site was Washington, instead of one of the field centers. The agenda reflected the changing nature of rendezvous research within NASA. Though Langley still dominated the discussions on rendezvous studies, Marshall took a full session to describe aspects of the rendezvous and orbital operations program it had under contract. This meeting saw the lunar orbit rendezvous idea introduced to NASA as a whole.[22] Until then, it had been limited to Langley circles and NASA Headquarters.

Rendezvous and orbital operations also figured prominently in congressional hearings on NASA's proposed budget for fiscal year 1962 during the first months of 1961.[23] The House Committee on Science and Astronautics, in particular, displayed a marked interest in the prospect of orbital rendezvous and scheduled a special hearing on the subject for May.[24] NASA's budget included some $2 million for further rendezvous studies. This was much less than NASA had wanted, but the Bureau of the Budget had sliced $6 million from the agency's initial request. The House committee recommended the full $8 million and NASA did eventually get the money.[25] In sharp contrast to the marked concern for space station logistics in 1959 hearings, the testimony in 1961 consistently stressed the role of rendezvous in mounting lunar and planetary expeditions and the broad value of rendezvous applications.[26]

While NASA spokesmen were telling Congress how important rendezvous was going to be, a working group in NASA Headquarters was drawing up guidelines for a full-fledged orbital operations development program. The resulting staff paper, ready in May, presented the case for the immediate "establishment of an integrated research, development and applied orbital operations program." Stressing the need for orbital operations in future space programs, the report urged NASA to set up "an aggressive program," coordinated with other NASA programs and with the Department of Defense, but separate from either. Such a program, the report concluded, would buy for the United States at a cost of roughly $1 billion three important skills: the

32

ability to intercept and inspect orbiting satellites, to support a space station, and to launch from orbit.

Bernard Maggin, who had arranged the first NASA rendezvous meeting a year earlier, headed the working group.* He sent copies of the report to the program office directors in NASA Headquarters and to the director of Program Planning and Evaluation. His request for comments, however, went unanswered.[27] By early May, NASA knew that President Kennedy was ready to approve a lunar landing program. The decision for a speeded up and expanded program transformed the context of NASA planning and made the kind of program Maggin suggested seem far too modest.

In the meantime, James Chamberlin followed his own course. He had arrived in St. Louis in February convinced that his job was to redesign the Mercury capsule from the bottom up. This was a belief not widely shared. The common view had it that Mercury only needed to be improved. Chamberlin felt, and as engineering director of Project Mercury he was surpassingly well qualified to judge, that the Mercury design precluded simple upgrading.[28] The Mercury capsule was merely a first try at a manned spacecraft. It clearly took too long to build, test, check out, and launch. The heart of the trouble was Mercury's integrated design, which packed the most equipment into the least space with the smallest weight. This could hardly have been avoided, given the limited weight-lifting capacity of the boosters available for the Mercury program. But integration also meant that reaching parts to test, repair, or replace was harder than it should be.

Chamberlin first met with McDonnell engineers to discuss the improved Mercury on 13 February. Little more than a month later, he had the chance to present some of his ideas to the head of Space Flight Programs, Abe Silverstein. On 17 March, Gilruth and his top-ranking staff journeyed to Wallops Island, Virginia, for a weekend retreat, where they were joined by Silverstein.[29] Mercury problems took up some time, but the meeting's main purpose was to discuss advanced programs. This chiefly meant Apollo. Chamberlin did, however, have a chance to describe his approach to redesigning the Mercury capsule.

He had attended the meeting mainly to discuss Mercury's progress. But after Silverstein outlined a series of desirable future Mercury missions, ranging from the one- and three-orbit manned missions already planned to rendezvous development, Chamberlin launched into a largely impromptu blackboard lecture on the program's future, which he saw as very limited. The trouble with trying anything more

---

*Its members were Joseph E. McGolrick and Eldon Hall (Office of Launch Vehicle Programs), John Disher and John L. Sloop (Office of Space Flight Programs), and Alfred M. Nelson and Berg Paraghamian (Office of Program Planning and Evaluation).

ambitious with Mercury than had been planned was that even these relatively modest goals could only be achieved at the expense of the most painstaking and arduous care in testing and checkout. This was not a manned spacecraft problem so much as it was a Mercury design problem. Drawing on his experience with fire control and weapons delivery systems for fighter aircraft, Chamberlin sketched a new capsule structure with its equipment located outside the cockpit in self-contained modules easy to install and check out. Although Chamberlin focused his remarks on capsule modification, he had obviously given some thought to a suitable mission for the new design. He had, in fact, prepared a brochure dealing with an audacious circumlunar flight for the improved Mercury, which Silverstein looked at and dismissed without comment.[30]

Both Silverstein and Gilruth, however, saw the need for changes along the lines Chamberlin had suggested. Gilruth asked Chamberlin to pursue the ideas in more detail with McDonnell, as the basis for specific proposals. Silverstein authorized STG to prepare a work statement to cover a McDonnell study of modifying the Mercury capsule for enhanced equipment accessibility. STG was also to place an order with McDonnell for parts to be used in several capsules beyond the 20 already contracted for. Looking back, Chamberlin was sure that was where it started: "As far as I was concerned, the meeting at Wallops was the initiation of Gemini."[31]

On 14 April STG and McDonnell signed an amendment to the original contract for the Mercury capsule. This amendment authorized McDonnell to procure so-called long-lead-time items—those parts that took longest to get—for six extra Mercury capsules. The parts and material so obtained would be used in what was now termed the Mercury Mark II spacecraft, once the design had been agreed upon by NASA and McDonnell. Specifically excluded from this procurement effort were capsule structure, ablation heatshield, and escape-tower systems, but all other capsule systems were covered up to a cost of $2.5 million.[32]

The design of the Mark II spacecraft was the subject of a second contract. After talks with STG, McDonnell submitted a study proposal on 12 April.[33] McDonnell proposed to spend $126 385 for 9000 hours of engineering study, with two objectives: first, to reduce the time needed to build and check out a Mark II capsule by improving the location of equipment and the way it was installed; second, by means of these changes to make the new capsule easy to modify to meet new program objectives. Capsule shape and heat protection were not to be altered, nor were capsule systems to be replaced or greatly modified. The focus of change was to be rearrangement; moving equipment from inside to outside the cabin and putting it in modular subassemblies, with special concern for escape, retrograde, and recovery systems.[34] McDonnell was authorized on 14 April to proceed with the engineering study, and a contract for $98 621 was signed on 24 April.[35]

By then, the study was already well under way. Chamberlin began calling on others in STG to help him. The first was James T. Rose, a recent transfer to Engineering from Flight Systems Division.[36] McDonnell created a small project group for the study, headed by William J. Blatz, with Winston D. Nold as chief assistant project engineer. Although they brought with them several engineers from McDonnell's advanced design section, the new group drew most heavily on Project Mercury, particularly a team led by Fred J. Sanders, for its staff.[37] Chamberlin regarded Mercury experience as indispensable. "That was the point," he recalled, "to use and build on experience, to gain and not to start over again . . . without the benefit of the detailed hardware experience."[38]

The guiding idea shared by Chamberlin and his McDonnell colleagues was "to make a better mechanical design"; capsule parts would be more accessible, leading to "a more reliable, more workable, more practical capsule."[39] The experimental Mercury capsule was to be transformed into an operational spacecraft. At this point, neither Chamberlin nor the McDonnell group were much concerned with the purpose such a redesigned capsule might serve. The subject arose, of course, as Chamberlin's lunar scheme shows, but it took a back seat. For the moment, the urgent question was strictly one of improving the engineering design. Working out the objectives for a program based on the improved capsule could wait.

## DIRECT ASCENT VERSUS RENDEZVOUS

While Chamberlin, Blatz, and their co-workers were eyeing the Mercury capsule and seeing, as engineers always can, any number of ways to make it better, events in the upper reaches of NASA were moving during the spring of 1961 toward the conclusion that would eventually give the engineers their chance to put ideas into practice. Enough of a case had been made for rendezvous in the lunar program during the past year to make it seem worth a closer look. But President Kennedy's decision to call for a lunar landing before the end of the decade transformed the context of lunar mission planning.

When NASA planning had first focused on flight around the Moon rather than landing on it, rendezvous lacked any urgency. Orbital operations seemed a matter of expedience, a way of making do with smaller boosters than direct ascent demanded. Circumlunar flight, too, could be launched with smaller boosters, but without any need for rendezvous, and a lunar landing appeared to be a long way off. Nobody denied that larger launch vehicles would be an asset to the

American space program, and nothing suggested that building such vehicles would pose any special problem other than time and money. Rendezvous, on the other hand, was an unknown. How hard it might be, how dangerous, could not be predicted. Nobody denied that rendezvous could be a useful and important technique, but planning the lunar mission around it appeared unnecessarily risky. Under the circumstances, direct ascent could be defended as more prudent.

Kennedy's decision changed all that. Gone were the long stretches of time that had allowed the choice between rendezvous and direct ascent to seem less than urgent. NASA now had to select the method that offered the best prospect for meeting the deadline. Even before it was announced, but knowing that a decision was imminent, NASA began seeking the answer.

On 2 May, Associate Administrator Seamans formed a task group to explore "for NASA in detail a feasible and complete approach to the accomplishment of an early manned lunar mission."[40] Most members of the ad hoc group came from NASA Headquarters, as did its chairman, William A. Fleming, then acting as Assistant Administrator for Programs.*[41] Fleming had been working closely with Seamans for several months and had, in fact, drafted the Seamans memorandum that created the task group.

The Fleming Committee had four weeks to size up the scope of the task that NASA faced. This was a tall order for so short a time, and the committee felt compelled to limit itself to one approach.[42] It elected direct ascent as "the simplest possible approach—the approach of least assumptions and least unknowns."[43] Rendezvous, much the biggest unknown, had no place in the lunar landing program, although it was "an essential program in its own right."[44] Having dismissed rendezvous, the Fleming group devoted most of its effort to choosing between solid and liquid propellants for the first stages of Nova-class boosters.[45] While this did permit the group to pinpoint some crucial decisions that needed to be made quickly—especially the importance of an early choice of sites for the large ground facilities the lunar mission required[46]—it merely avoided the question of rendezvous versus direct ascent. Convinced, as Fleming later remarked, "that it was always possible to 'build something bigger and make it work,' "[47] his committee saw no reason to base its study on a risky and untried alternative.

---

*Of the 23 members of the Fleming Committee, 18 were from NASA Headquarters: Fleming, Addison M. Rothrock, Albert J. Kelley, Berg Paraghamian, Walter W. Haase, John Disher, Merle G. Waugh, Eldon Hall, Melvyn Savage, William L. Lovejoy, Norman Rafel, Alfred Nelson, Samuel Snyder, Robert D. Briskman, Secrest L. Berry, James P. Nolan, Jr., Ernest Pearson, and Robert Fellows. Remaining members were Koelle, Marshall; Kleinknecht and Alan Kehlet, STG; A. H. Schwichtenberg, Lovelace Foundation; and William S. Shipley, Jet Propulsion Laboratory.

Others in NASA were not so sure. On 19 May, while the Fleming Committee was still meeting, John Houbolt wrote Seamans from Langley deploring the state of the launch vehicle program and urging more serious attention to rendezvous. He denied any wish to argue for rendezvous against direct ascent but insisted that, "because of the lag in launch vehicle development, it would appear that the only way that will be available to us in the next few years is the rendezvous way. For this very reason I feel it mandatory that rendezvous be as much in future plans as any item, and that it be attacked vigorously."[48]

This was a viewpoint that Seamans, long a student of orbital rendezvous and openly receptive to such ideas since joining NASA, must have shared. On 25 May, he called on Don R. Ostrander, Director of Launch Vehicle Programs, and Ira H. A. Abbott, Director of Advanced Research Programs, to name "a group of qualified people . . . to assess a wide variety of possible ways for executing a manned lunar landing." Seamans wanted their report quickly, "at about the same time as the one under way by the Ad Hoc Task Group on Manned Lunar Landing." NASA Headquarters furnished none of the six members of this committee, led by Bruce T. Lundin of Lewis Research Center.*[49] Lundin regarded his committee as speaking for the field centers, in contrast to the Headquarters viewpoint expressed by the Fleming group.[50] The Lundin report was ready by 10 June, a week before the Fleming report.

Although Lundin's committee discussed other matters, its main concern was to compare the several rendezvous schemes with each other. It pointedly excluded any specific comparison of rendezvous with direct ascent but noted two inherent advantages in rendezvous that promised an earlier manned lunar landing. One was the relative capacity of a rendezvous-based program to absorb increases in payload weight, which meant that early decisions on booster design and development might not so critically affect the program. The other was the smaller size of launch vehicles required by a rendezvous mission, a size which would not call for the development of large new engines.[51]

Time limited the Lundin Committee to a brief qualitative survey, which could not compare in scope or detail to the elaborate quantitative assessment provided by the Fleming Committee.†[52] Clearly, however, the choice between solid or liquid propellants in the first stage or

---

*Lundin's Committee consisted of Walter J. Downhower (Jet Propulsion Laboratory), Alfred Eggers (Ames Research Center), Laurence Loftin (Langley), Harry O. Ruppe (Marshall), and Lt. Col. George W. S. Johnson (Air Force).

†The Lundin Committee met during the week of 5 June 1961. Most of its sessions were devoted to presentations by Ames, Langley, Lewis, and Marshall on Earth orbit rendezvous, by Langley and Marshall on lunar orbit rendezvous, and to a general discussion of rendezvous proposals.

two of a Nova booster was too restricted; the proper alternative to direct ascent was some form of rendezvous. This proposition won unanimous agreement at a meeting between Seamans and the program directors* on 18 June, though only after considerable discussion. They decided to pursue two courses. Ostrander would form a team from NASA Headquarters and Marshall to define an overall plan for using orbital operations to achieve manned lunar landing. At the same time, the Fleming Committee study of direct ascent would be paralleled by an equally intensive investigation of the rendezvous and orbital operations approach.[53]

The first line of action under Ostrander produced a preliminary project development plan for orbital operations by mid-September.[54] For the second, Seamans formed still another ad hoc group that was "to establish program plans and supporting resources necessary to accomplish the manned lunar landing mission by the use of rendezvous techniques" with as much rigor as the Fleming report. He named Donald H. Heaton, his former assistant who had become Assistant Director for Vehicles in Ostrander's office, as chairman of the new group.[55]

Heaton's group was about the same size as Fleming's, but its members were more evenly divided between Headquarters and the field centers.† Its findings, issued late in August, concluded that "*rendezvous offers the earliest possibility for a successful manned lunar landing.*"[56] Despite this parade of studies, as future events were to show, the issue had only been joined, not settled. But the view that rendezvous techniques were important enough to pursue "*whether or not* rendezvous is selected as an operating mode" for the lunar mission[57] was clearly gaining strength. And this viewpoint was crucial to the fate of Mercury Mark II, which had in the meantime taken on a much more sharply defined form.

## THE ADVANCED CAPSULE DESIGN

Chamberlin and Blatz were ready to report progress toward an advanced capsule design early in June 1961. Chamberlin had conceived his task in terms that diverged widely from what was generally expected. Adept at keeping his ideas to himself until they matured, he

---

*The meeting was attended by Seamans, Silverstein, Abbott, Ostrander, Siepert, DeMarquis D. Wyatt, and Charles H. Roadman (who had replaced Clark Randt as Director of Life Sciences Programs).

†The members were Heaton, Richard B. Canright, L.I. Baird, Rafel, McGolrick, Louis H. Glassman, John L. Hammersmith, Briskman, Nolan, Warren J. North, and William H. Woodward, from NASA Headquarters; Wilson B. Schramm, R. Voss, Koelle, Peter J. deFries, and Harry Ruppe, of Marshall; John Houbolt and Hewitt Phillips, from Langley; Hubert M. Drake, from Flight Research Center; and J. Yolles, Air Force System Command.

was not much of a talker. As far as Space Task Group knew, at least officially, McDonnell was studying an advanced version of the Mercury capsule for just two reasons: to extend the capsule's lifetime in orbit to one day (or 18 orbits) and to make the capsule easier to check out and test before flight.[58] The extent of the changes that Chamberlin and Blatz revealed to STG leaders on Friday afternoon, 9 June, took some of them aback.* Chamberlin explained that the primary aim "of the design was to increase component and system accessibility to reduce manufacturing and checkout time." That was no surprise. But to do it, he had packaged and relocated almost every capsule system. Those closest to Project Mercury tended to share Chamberlin's view that the Mercury capsule was inherently limited because of its design—making it better meant making it over. This was, after all, the heart of the case Chamberlin had presented at the Wallops Island meeting in March, and he had followed through along the lines he had then suggested. But others in STG, more distant from the daily problems of working with Mercury, were likely to assume that the capsule needed only relatively minor changes to improve it, not the nearly complete new design that Chamberlin offered.[59]

Chamberlin later justified this approach in an enlightening lecture on the design philosophy of the Gemini spacecraft (which Mercury Mark II was to become).[60] The main trouble with the Mercury capsule was that

> most system components were in the pilot's cabin; and often, to pack them in this very confined space, they had to be stacked like a layer cake and components of one system had to be scattered about the craft to use all available space. This arrangement generated a maze of interconnecting wires, tubing, and mechanical linkages. To replace one malfunctioning system, other systems had to be disturbed; and then, after the trouble had been corrected, the systems that had been disturbed as well as the malfunctioning system had to be checked out again.[61]

Mercury designers had been preoccupied with solving such basic problems of manned space flight as reentry heating and human tolerance of both high acceleration and zero gravity, for "the sole purpose of placing a man in orbit in a minimum time." Thus they paid no great attention to making a convenient, serviceable spacecraft. That, however, was precisely what the new design offered. In it,

> systems are modularized and all pieces of each system are in compact packages. The packages are so arranged that any system can be

---

*Those who attended the Capsule Review Board meetings of 9 and 12 June were Gilruth, Walter Williams, Paul Purser, Max Faget, Charles Mathews, Robert Piland, Wesley L. Hjornevik, George Low, and John Disher.

removed without tampering with any other system, and most of the packages ride on the outside walls of the pressurized cabin for easy access. This arrangement allows many technicians to work on different systems simultaneously.[62]

The Mercury capsule, in contrast, could only be worked on from the inside, which meant, as a rule, only one person working at a time.

The new design attacked a number of other Mercury trouble spots. Perhaps the most troublesome was the sequencing system. Chamberlin argued that one of his chief motives for keeping systems in the new design separated was to avoid the endless complications Mercury experienced because so many sequentially controlled operations were built into it. Most of Mercury's flight operations could be controlled by the pilot, but safety demanded that they also be automatic, each complex series of events triggered by an appropriate signal and ordered through a predetermined sequence by a tangle of electrical circuitry.[63] So complex was Mercury sequencing that Chamberlin recalled it as "the root of all evil and anybody that really worked on Mercury—that's all they talked about."[64] The new design relied on pilot control, instead of merely allowing it and backing it up with automatic sequencing. The result was a much simpler machine; the 220 relays in Mercury, for example, were reduced to 60 in Mark II.[65]

What may have been the most complex sequencing of all was demanded by the automatic abort modes in Mercury, which depended on a rocket-propelled escape tower to pull the capsule away from the booster in an emergency during or just after liftoff.[66] In Chamberlin's mind, "the sequencing of the escape system was one of the major problem areas in Mercury in all its aspects—its mechanical aspects in the first part of the program, and the electronic aspects later."[67] What made this peculiarly frustrating was that the escape tower added hundreds of kilograms to the capsule's weight, even though it was essentially irrelevant to the function of the capsule itself; in a successful flight it was jettisoned shortly after launch. Yet its many relays and complex wiring, besides making it inherently untrustworthy, were major factors in prolonging checkout time. To make matters worse, the Mercury abort modes—NASA shorthand for the methods that allowed the pilot to escape when a booster malfunction threatened his life—were automatic. Some circumstances not actually calling for an aborted mission—including a malfunction of the abort system itself—could trigger one, as happened more than once in the Mercury development program.[68]

The new design put the pilot in an ejection seat and eliminated the escape tower.[69] This change, if installed, excluded Atlas as a booster for the new capsule. Atlas propelled itself with liquid oxygen and a mixture of hydrocarbons called RP-1, a highly explosive combination if the booster broke up. No ejection seat had the power to kick a pilot away from an exploding Atlas quickly enough, particularly if escape were not automatically triggered. Safety was thus a key reason for the escape tower and for its automatic features in Mercury. But Chamberlin had just become aware of a new booster that might relax these constraints.

Its name was Titan II, and the Martin Company was developing it as an intercontinental ballistic missile for the Air Force and as a manned booster in the Air Force Dyna-Soar program.[70] Albert C. Hall, general manager of Martin's Baltimore Division, had proposed it to Associate Administrator Seamans, an old MIT classmate, for a role in NASA's lunar mission. Although Seamans was skeptical, he arranged for Martin spokesmen to present their case at NASA Headquarters on 8 May 1961. The visit was strictly unofficial, since Titan II was an Air Force project. Any formal contact between NASA and Martin required Air Force sanction. Among those who heard about Titan II that day was Abe Silverstein, who saw enough in the new missile to ask Gilruth to look into the possibility of using it somewhere in the manned space flight program.[71] Silverstein dismissed any thought of a role for Titan II in the lunar program.

To Chamberlin, however, Titan II looked very good for the improved Mercury. Weight was the most serious constraint in spacecraft design. An improved Mercury meant a heavier Mercury, since the price for packaged components was extra kilograms. This, in turn, meant that the new design called for a launcher more powerful than Atlas. Titan II had power to spare, its total thrust being almost two and a half times that of Atlas. Not only could it easily lift the heavier spacecraft, but it could also carry the redundant systems that would make it a safer booster for manned space flight. This, in a way, merely augmented what may have been Titan II's outstanding features—simplicity and reliability.[72]

Titan II ran on storable hypergolic propellants: a blend of hydrazine and unsymmetrical dimethyl hydrazine (UDMH) as fuel with nitrogen tetroxide as oxidizer. Because this combination is hypergolic—fuel and oxidizer burn spontaneously on contact—Titan II needed no ignition system. Since both fuel and oxidizer can be stored and used at normal temperatures—instead of the supercold required by the liquid oxygen of Atlas or Titan I—Titan II required no cold storage and handling facilities. The design and the lessons learned from Titan I combined to reduce the 172 relays, umbilicals, valves, and regulators in the first version of the missile to 27 in Titan II.[73] This simplification struck a responsive chord in Chamberlin, who saw in it something to match what he had been trying to achieve in redesigning the Mercury capsule. Booster and spacecraft seemed almost to have been made for each other.[74]

Titan II's self-igniting propellants had still another advantage.

They reacted much less violently with each other than did the cryogenic propellants of Atlas or Titan I. In June 1961, there was still some question about whether a Titan II explosion would be sufficiently less violent, compared to Atlas, to permit the use of an ejection seat. Chamberlin was not yet ready to spell out his plans for using Titan II, but that was the way he was thinking. And his active distaste for escape towers made him eager to include ejection seats in his design.

Ejection seats not only promised to relieve a major source of trouble by getting rid of the escape tower, but they also furthered the concept of modularization, keeping each spacecraft system, so far as possible, independent. "The paramount objective in the program," according to Chamberlin, "was to dissociate systems." Ejection seats, in what he called "a very happy coincidence that was fully realized at the time," also fitted in nicely with another design change, substituting paraglider for parachute recovery.[75]

STG had not displayed much active interest in Francis Rogallo's flexible wing concept after the initial flurry in early 1959.[76] Rogallo and his co-workers at Langley had pushed ahead with their studies in the meantime. By mid-1960, they had convinced themselves that a controllable, flexible wing could carry a returning spacecraft safely to land, thus providing "a lightweight controllable paraglider for manned space vehicles."[77] STG rediscovered the paraglider at the start of 1961 as a by-product of work on Apollo. A technical liaison group on Apollo configuration and aerodynamics met at Langley on 12 January.* In the course of describing his center's work for Apollo, the Langley representative mentioned the paraglider landing system: "The feeling at Langley is that if the paraglider shows the same type of reliability in large-scale tests . . . that it has achieved in small-scale tests, the potential advantages of this system outweigh other systems." Engineering design of large paragliders appeared to be no problem and would be demonstrated in manned and unmanned drop tests.[78]

Space Task Group engineers met informally with Rogallo and his colleagues in February, March, and April to explore the use of a paraglider in the Apollo program.† The STG team was less than enthusiastic. They believed much work was yet to be done before the device

---

*The group comprised Alan Kehlet as chairman, and William W. Petynia as secretary (both of STG), Hubert Drake (Flight Research Center), Edward L. Linsley (Marshall), Eugene S. Love (Langley), Edwin Pounder (JPL), and Clarence A. Syvertson (Ames). During January and April meetings of the group, visitors were John Disher (Headquarters), Alvin Seiff (Ames), and John B. Lee and Bruce G. Jackson (STG). The large-scale program got under way in April, using a fully deployed 19-foot paraglider. Tests with partially deployed and packaged paragliders were to follow.

†The STG engineers were John W. Kiker, Richard C. Kennedy, Fred J. Pearce, Jr., and Gerard J. Pesman. Rogallo's team consisted of Delwin R. Croom, Robert T. Taylor, Donald E. Hewes, Lloyd J. Fisher, Jr., and Lou S. Young.

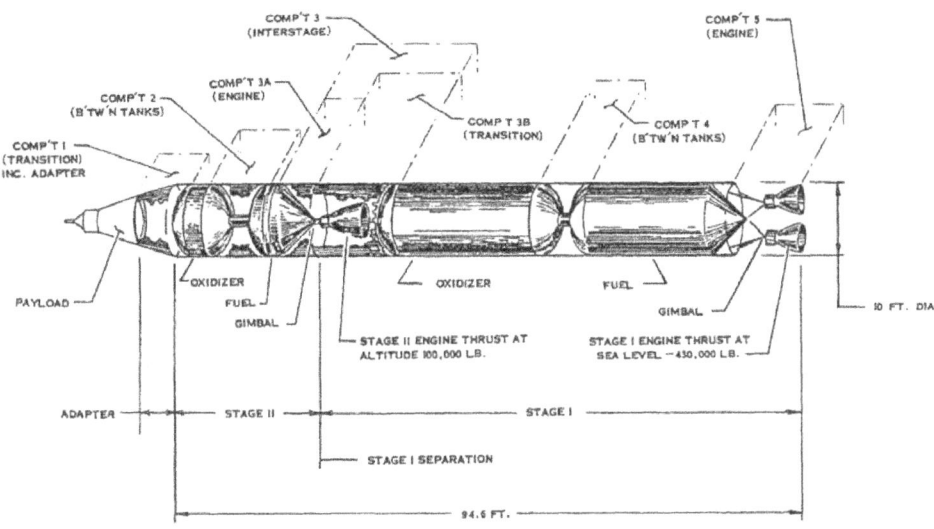

Above, a drawing of the Titan II, built by the Martin-Marietta Corporation, as proposed to be adapted for manned space flight. Below, artist's sketch of ejection seats propelling the astronauts to escape distance from a launch failure. They would be used in emergencies before launch (pad-abort) and in flight to about 18 000 meters altitude.

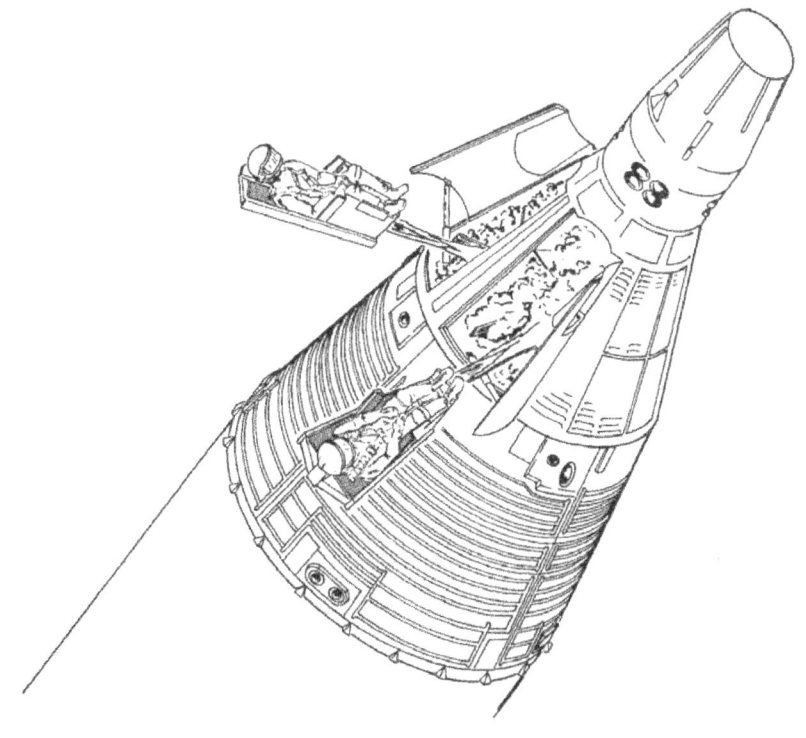

could be seriously considered as a landing system for Apollo. The biggest unknown was the deployment characteristics of an inflatable wing; no inflatable structure had ever been successfully deployed in flight. Other questions—how the paraglider was to be packaged, whether the pilot's view from the capsule would be good enough for flying and landing with it—were nearly as important and also largely unanswered. The STG team advised gathering at least six months of data before awarding any paraglider development contract.[79] At the same time, however, McDonnell engineers were looking at a paraglider for the modified Mercury, and Marshall Space Flight Center had already let two contracts to study paraglider as a booster recovery system. The idea clearly had promise, and in May 1961 Gilruth decided to contract for further study.

Three contractors each got $100 000 for two and a half months to design a paraglider landing system and define potential problem areas.* The best design was expected later to become the basis for a development contract to "provide the modified [Mercury] spacecraft with the capability of achieving a controlled energy landing through the use of aerodynamic lift."[80] In fact, the design studies soon received a new name—Phase I of the Paraglider Development Program.[81] Observed by a small technical monitoring group from STG, the paraglider design studies were under way before May ended.†[82] McDonnell engineers also maintained close liaison with paraglider work, independent though it was of the Mercury Mark II study contract.[83] The redesigned Mercury, as presented by Chamberlin and Blatz to the Capsule Review Board in June, could be adapted to a paraglider landing system, once it was developed.[84]

One other significant innovation marked the new design, an enlarged overhead mechanical hatch, which would allow the pilots to get in the spacecraft more easily and to get out more quickly in an emergency. It was another way of making the new spacecraft a truly operational machine, one that could be entered and left like an airplane. Such a hatch was also needed if ejection seats were to be used. But it also had a special virtue that its designers were well aware of, though they did not talk about it. A large mechanical hatch would enable the pilot to leave and return to the spacecraft while it was in orbit and

---

*They were Goodyear Aircraft Corporation, Akron, Ohio; Ryan Aeronautical Company, San Diego, California; and North American Aviation Space & Information Systems Division, Downey, California. Goodyear was an experienced builder of inflatable aerial devices, and Ryan and North American were already working on the Marshall contracts.

†The technical monitors were Rodney G. Rose, Harry C. Shoaf, Kenneth W. Christopher, and Lester A. Stewart; in mid-June, they visited each of the contractors' plants to review progress on the study. The group continued to meet with the contractors at regular intervals until the studies were completed.

thus permit what later became known as extravehicular activity, or EVA.[85]

The many changes proposed by Chamberlin and Blatz did not make the redesigned spacecraft a totally new machine. Though somewhat enlarged, it retained the fully tested and proved shape and heat protection of the Mercury capsule. It was still to be a one-man craft, and its designers expected to use mostly Mercury parts, packaged and rearranged but not otherwise substantially altered. The new design would not be much longer-lived in orbit than Mercury, 18 orbits (or one day) being the most the designers were aiming at.[86] Nevertheless, members of the Capsule Review Board seemed staggered by the scope of the changes presented to them. They refused to accept the complete Chamberlin-Blatz package but agreed to reconvene after the weekend to decide if any of the new features might be worth pursuing.[87]

Chamberlin came back again Monday morning, since he was a regular member of the board, but Blatz had returned to St. Louis.* The board talked over the design of the ejection seat and hatch, simpler sequencing, better accessibility, and an 18-orbit capability. Each of these ideas had its own appeal, but most of them carried a price tag far too high to fit within the scope of the follow-on Mercury program STG was then thinking about, a program budgeted for less than $10 million in the coming fiscal year.[88]

Although reaching no clear-cut decision, the board still hesitated to endorse Chamberlin's plans in full. Instead, he was allowed to continue working on alternative approaches to an improved Mercury, while McDonnell studied "the minimum modifications that could be made to the present capsule to provide 18-orbit capability" and looked into "a larger retro and posigrade pack."[89] This amounted to little more than reviving an early Mercury objective, once the ultimate goal of the program. Growing capsule weight and power requirements, as well as the limitations of the manned space flight tracking network, had forced STG to scrap the 18-orbit mission by October 1959.[90] The idea lived on, however, in the form of a proposal to fit the capsule with its own rocket motors to provide the final increment of velocity needed to attain an orbit high enough to resist Earth's gravity for 18 revolutions.[91] This was the idea the Capsule Review Board again endorsed at its meeting on 12 June.

### MODIFICATION OR TRANSMUTATION

The matter of a post-Mercury manned space flight program was far from settled in the Capsule Review Board meetings of 9 and 12

---

*Hjornevik, Low, and Disher, all of NASA Headquarters, had also gone home.

June. Chamberlin was not giving up, and McDonnell, despite the board's injunction to limit its work to minor modifications, was still pressing for a more radical effort. At the beginning of July 1961, top STG officials were looking at an ephemeral "Hermes Plan," calling for a new Mark II design much along the lines proposed by Chamberlin and Blatz a few weeks before. This Mark II was contrasted with a minimally redesigned capsule for an 18-orbit mission, now termed Mark I. The question of Mark II design, as Gilruth's special assistant Paul E. Purser noted, was still very much "up in the air."[92] Still unclear was the scope of a follow-on Mercury program. A choice in favor of the extensively redesigned Mark II would impose a far greater effort than the slightly altered Mark I.[93]

A McDonnell group led by Mercury manager Walter F. Burke attended a senior staff meeting at STG on 7 July to outline the company's studies of an advanced Mercury capsule that took three distinct forms. One version, the "minimum change capsule," involved not much more than cutting some hatches in the side of the capsule for better access. Although it could be ready to launch relatively quickly and cheaply (11 months, $79.3 million), it had some obvious drawbacks. Better access only accented the capsule's cramped interior, and the hatches themselves weakened the capsule's structure and heat protection. As Chamberlin later remarked, "It was clear that this mod. was too little to inspire any additional confidence in the design, and hence make it worth doing. Thus, the merits of the greater modifications became apparent."[94] The second McDonnell advanced design, called a "reconfigured Mercury capsule," adhered closely to the Chamberlin-Blatz proposal of June. It would take longer to build and cost more than the minimum change capsule (20 months and $91.303 million), but it might very well be worth the expense. And for another two months and $12.248 million, NASA might do even better with McDonnell's third version, a "two-man Mercury capsule."[95]

The notion of putting more than one man in a modified Mercury capsule was not new, having been suggested at least as early as January 1959.[96] That idea had gone nowhere, but Faget revived the possibility at the review board meeting on 9 June 1961. Blatz recalled that, after he and Chamberlin had made their pitch, Faget's comment was, "If we're going to go to all of this trouble to redesign Mercury, why not make it a multiplace spacecraft in the process?"[97] Faget's interest in a two-man spacecraft was prompted, in part, by the prospect of extravehicular operations. As early as March 1961, he had asked John F. Yardley, McDonnell's manager for Mercury operations at Cape Canaveral, to look into the possibility "of expanding Mercury into a two-man version" for this purpose.[98] Others saw reason for a two-man spacecraft in the rigors of long missions. If the Mark II were to be in space for more than a few orbits, then having two men to share the strain and support each other's activities made good sense.[99] There was also a certain compelling logic in building a two-man spacecraft for a program falling between the one-man Mercury and three-man Apollo.[100]

NASA Headquarters seemed uncertain about the size of the changes STG was thinking about during July 1961. George Low told Associate Administrator Seamans and the Washington program directors on 6 July that McDonnell and STG were working on a minimally modified 18-orbit capsule. He reported that

> McDonnell originally looked upon the 18-orbit capsule as a development of a new flight article with substantial increase in size and weight, and incorporating rendezvous capabilities. McDonnell has been advised, however, to proceed on the basis of minimal changes to the existing hardware and to approach design modifications on this basis.[101]

But a master plan for orbital operations, dated 19 July, included, besides four 18-orbit Mercury flights during 1963, eight one-man Mercury Mark II flights to be launched at two-month intervals—from October 1963 through December 1964—and to perform rendezvous and docking tests in orbit.[102]

Whatever confusion may have existed, however, was resolved before the end of the month. On 27 July, Abe Silverstein joined Gilruth and other STG leaders, as well as several astronauts, at the McDonnell plant in St. Louis. McDonnell engineers displayed quarter-scale models of four basic spacecraft configurations: an Eighteen Orbit MK I, a Minimum Change MK II, a Reconfigured MK II, and a Two Man MK II. Also on display was a full-size wood and plastic mockup of the cockpit for a two-man spacecraft—Astronaut Walter M. Schirra, Jr., sat in it and exclaimed, "You finally found a place for a left-handed astronaut!"* [103]

Although the ideas for an advanced Mercury presented by the McDonnell study team were much the same as they had been 20 days earlier,[104] the audience on 27 July now represented NASA Headquarters as well as STG. Silverstein had long been convinced of the importance of Mercury missions more ambitious than merely circling Earth three times. What he saw in St. Louis was apparently enough to tip the scales toward a decision that many in NASA were ready to welcome. On 28 July, during the second day of the St. Louis meeting, Silverstein directed McDonnell to focus all further effort to improve Mercury solely on the two-man approach.†[105] The choice had been made for a larger, rather than a smaller, follow-on Mercury program.

---

*Ironically, Schirra flew in Gemini as spacecraft commander, occupying the left seat and using his right hand for most operations.

†McDonnell was also told to go ahead with work on the 18-orbit Mark I; this directive became official on 25 October 1961. The 18-orbit Mercury was no longer deemed an improved version. As *Faith 7*, it eventually carried L. Gordon Cooper, Jr., through the 22-orbit Mercury-Atlas 9 mission in May 1963.

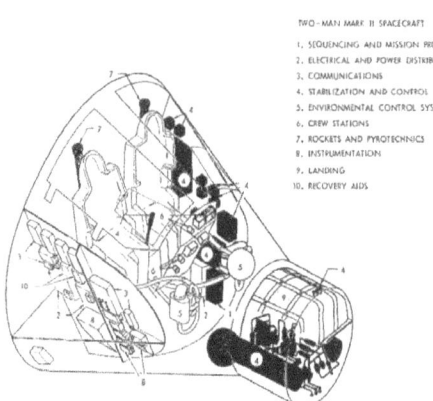

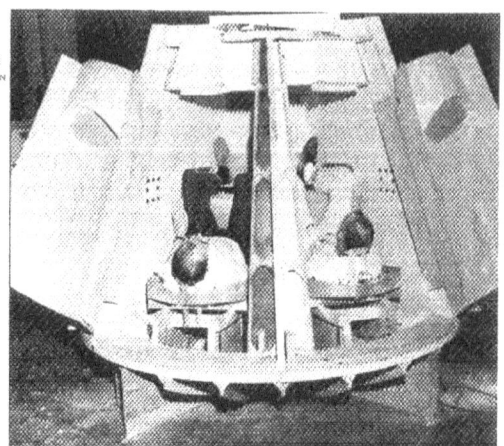

The NASA-McDonnell meeting in St. Louis on 27-28 July 1961 featured the unveiling of the mockup of the Mark II Spacecraft; above left, sketch of the modularized systems in the two-man spacecraft; above right, subjects in wooden mockup of two-man spacecraft in position it would rest on launch pad; below left, two-man spacecraft in normal orientation for orbital flight; and below right, subjects check open hatch characteristics to evaluate feasibility of extravehicular activity while in orbital flight. The harness attached to the subject at left simulated the weightlessness of orbital flight.

In what was to become a familiar pattern, that program had already grown far beyond its original bounds. The McDonnell study contract, the basis for the company's design work on advanced Mercury, had outlined a relatively modest effort. By the time that contract was signed, on 24 April, the work was well along. In just over three weeks, McDonnell requested and received a contract increase from $98 621 to $187 189.[106] McDonnell efforts soon far surpassed that limit. By 6 August, the company had assigned 45 engineers to the study, and the original 9000 engineering manhours called for in the contract had climbed to almost 23 000; added to that figure were 6000 shop manhours for building and testing models not even mentioned in the contract. The estimated cost now topped $535 000.[107]

Since STG had agreed that advanced Mercury needed more study, McDonnell had not felt obliged to wait until its contract had been amended to provide the extra funds. The company spent its own money. This was the kind of initiative that earned the firm a good deal of respect in NASA circles. Where others refused to move without money in hand, McDonnell focused on the task and relied on the good faith of its customer to make up the cost. It was seldom disappointed. In this instance, the company proposed a new contract to cover the extra engineering study and shop work done since 19 June, when contract funds had been exhausted, and to pay its projected expenses through the end of September.[108] The original contract and the new request together totaled over $670 000, nearly seven times the figure first approved in April. STG did not issue a new contract but, instead, amended the procurement contract to authorize the additional funds.[109]

## A TECHNOLOGICAL IMPERATIVE

Before the end of July 1961, the joint efforts of Chamberlin, Blatz, and their co-workers in STG and McDonnell had produced the design of an advanced Mercury capsule, Mercury Mark II. Space Flight Programs Director Silverstein had endorsed it. Although the final verdict was not yet in, the larger program seemed to be in the works, something that could scarcely have been predicted when the year opened. The situation was transformed on 25 May, when the President asked the country to assume the burden and the glory of reaching for the Moon.

The metamorphosis of Space Task Group into Manned Spacecraft Center, followed by its move from Virginia to Texas, flowed directly from this decision. STG had been created solely to manage Project Mercury; as a single-purpose task force, it was outmoded. Project Mercury now became only the first step on the path that was to lead Americans to the Moon before 1970.

As always, the lunar mission, in whatever form, held center stage. This was just as true in Headquarters as it was in the field. Although

Washington's chief planning concern was the voyage to the Moon, research and development in the field focused on specific problems raised by a lunar mission and the hardware needed to surmount them. STG, of course, had Project Mercury to worry about; but when it had time to look ahead, what it looked at was the Moon. Even before the President's decision for a lunar landing, STG engineers were hard at work on the spacecraft that would ultimately carry men there.

Once a deadline had been set, the question of rendezvous as part of the lunar mission took on a new guise. By holding out the prospect of using smaller and thus more quickly developed boosters, rendezvous offered a chance to reach the Moon sooner than did a direct approach. During the spring and summer of 1961, discussion of this promise became widespread, and support for some form of rendezvous mission gathered strength. Even those who objected to chancing a lunar mission on an unproved technique were quite willing to admit that the technique needed to be developed, if only for its intrinsic value in future manned space flight. The growing conviction of the need for rendezvous, still further bolstered by studies during the fall of 1961, provided the framework for what became Project Gemini.

By the time NASA decided that it needed a rendezvous development program, a freshly designed spacecraft was on the drawing boards. Mercury Mark II was not so much the product of planning as it was of a kind of technological imperative, the ceaseless and unquenchable desire of working engineers to perfect their machines. Some features of Mark II did, of course, spring from thinking about the objectives of a program to follow Mercury. But most of the changes in the new design suggested improvement in the abstract, rather than means to defined goals.

When Chamberlin talked about the design, it was in terms of accessibility and convenience, serviceability and simplification, "a better mechanical design" that was "more reliable, more workable, more practical." These are qualities that can never be absolutely realized, though they may be endlessly pursued. During the first half of 1961, Chamberlin, Blatz, and the others pursued them far beyond the intent of those who had set them the task. By July they had reached a point where they were willing to pause, although, as the later career of Gemini was to show, it was not a point at which they could long rest content.

When Silverstein endorsed the two-man Mark II, its designers faced a new task. The gap between a spacecraft design, whatever its merits, and a manned space flight project was a wide one. Early in 1961, NASA Headquarters had set up a formal procedure for planning and carrying out new projects.[110] The first step for such large and complex projects as Mercury Mark II now promised to be was a preliminary project development plan. This was the task to which Chamberlin and his colleagues now turned.

# III

# From Spacecraft to Project

WHEN August 1961 began, James Chamberlin, backed by the Space Task Group and McDonnell Aircraft Corporation, had produced the makings of a post-Mercury manned space flight program. The major task, rethinking the design of the Mercury capsule, was finished, although many details had yet to be worked out.[1] A Mercury Mark II project had attained a kind of shadow being and had the support of Abe Silverstein, Director of Space Flight Programs in NASA Headquarters. Only NASA's highest echelon remained to be convinced.

So far, the working engineers in STG and McDonnell had been more concerned with an improved spacecraft than with larger goals. To give their ideas substance, they now faced the task of fitting the spacecraft within the framework of a NASA project. This meant finding those larger goals to justify the cost in time and money that turning concept into practice required. It also meant putting together more pieces; a project was more than a spacecraft.

## MORE THAN A SPACECRAFT

Neither Chamberlin and his staff nor the McDonnell designers had specified a booster for their improved versions of the Mercury capsule, although they had mentioned several prospects at one time or another and Chamberlin himself was more than a little taken with the Titan II. During June and July, STG Director Robert Gilruth and his staff had met often, but always informally, with Martin spokesmen,

*51*

chiefly James L. Decker, to talk about Titan II as the booster for the scaled-up Mercury.[2]

The first formal meeting came on 3 August 1961, when Decker briefed Gilruth and his colleagues on "A Program Plan for a Titan Boosted Mercury Vehicle."[3] The Martin plan was decidedly optimistic. For just under $48 million, NASA could buy nine boosters, developed, tested, and launched, the first launch to be within 18 months.[4] What made this proposal so startling was that Titan II was still mostly promise. Martin's contract with the Air Force to develop the missile was scarcely a year old (June 1960), and Titan II's maiden flight was almost a year in the future. But the company had reason to believe that rapid progress was likely.

For one thing, much of the work and expense of Titan II development would be provided by the Air Force missile program. For another, some of the design and testing of changes needed to convert the missile to a booster for manned space flight had already been done, and more could be expected, as part of the Air Force Dyna-Soar program. The same simplicity and reliability that so appealed to Chamberlin in the Titan II, augmented by the redundant systems its greater power permitted it to carry, likewise promised a quick and successful development program.[5]

By the end of July 1961, when Silverstein approved the two-man Mark II, STG was all but ready to put that spacecraft on Titan II. Many of the rough spots had already been smoothed away; Martin had been talking not only to STG but to NASA Headquarters and the Air Force. The formal meeting of 3 August simply confirmed a nearly accomplished fact. At a senior staff meeting four days later, Gilruth commented on the vehicle's promise, particularly the greater power that made it "a desirable booster for a two-man spacecraft."[6]

The choice of a Titan to carry Mercury aloft may have done some violence to classical mythology. The giants of Greek myth were far removed in time and space from the Roman god. Those who first named Atlas and Titan in the mid-1950s were thinking of the symbolism of power, strength, and invincibility, qualities no less appropriate when their missiles were turned to more peaceful uses.[7] Yet, in scouring classical mythology to name their missiles, and setting a precedent that NASA followed, they tapped a vein of symbolism far richer than they knew. Just as Atlas, though he bore heaven and Earth on his shoulders, was but a puny shadow of the Titans themselves, so was the Atlas booster far less powerful than the Titan II that succeeded it. Titan II could carry men to new heights, allowing them to say with Isaac Newton, "If I have seen farther, it is by standing on the shoulders of giants."[8] Titan might also help to underscore the living relevance of Newtonian science in an age dominated by Einsteinian relativity and quantum mechanics. For if "the 'sputniks' constitute[d] the first *experimental* proof of Newtonianism on a cosmic scale,"[9] then the spacecraft carried aloft by Titan, shifting its orbital path in response to the commands of its pilots, offered an applied demonstration of Newtonian orbital mechanics. Eventually Titan II would carry the renamed Mercury on its shoulders in flights that soared far beyond the limits previously attained by mankind and would allow them to see farther than they had ever seen before.

At about the same time that Gilruth was endorsing Titan II, Chamberlin was looking at Agena for use as a rendezvous target. On 8 August 1961, he made his first contact with the Lockheed Missiles & Space Company of Sunnyvale, California.[10] The Agena was a highly successful second-stage vehicle that Lockheed had developed for the Air Force. In its then-current version, Agena B, it had flown for the first time in 1960. It was powered by a pump-fed rocket engine made by Bell Aerosystems Company of Buffalo, New York. Like Titan II, Agena used storable hypergolic propellants—in this case, unsymmetrical dimethyl hydrazine as fuel, inhibited red fuming nitric acid as oxidizer. The engine had a dual-burn capability; that is, it could be fired, shut off, then fired again.[11] This feature, plus its impressive string of successes, gave Agena the look of a winner. It not only seemed reliable, but its extra power offered a chance to practice really large-scale maneuvers once spacecraft and target had docked.[12]

Chamberlin's talks with Lockheed about Agena as a rendezvous target reflected the new orientation of Mark II work, toward a project rather than a spacecraft. Rendezvous was now a matter of intense concern within NASA. Despite its great promise, as stressed by the several committees that had discussed the subject during the spring and summer of 1961, it was still an unknown. Whether rendezvous would be as simple and useful in practice as it appeared to be in theory was a question that Mercury Mark II might well be able to answer.

Of other questions looking for answers, one of the most pressing involved the effects of extended stays in space on the human body. Mercury might lay some fears to rest, but its short missions could not allay doubts about long-term space dangers. Those doubts would become crucial in the Apollo program. A trip to the Moon and back demanded at least a week, compared to the four and a half hours of the longest Mercury mission then scheduled. Here was another area that Mark II might explore. The large increase in payload weight permitted by Titan II and the greater size of Mark II would allow the spacecraft to carry the extra supplies and batteries or fuel cells to provide electrical power for a mission of one or two weeks.

The end of the first phase of the paraglider development program in mid-August, which proved the feasibility of the concept for recovery of manned spacecraft,[13] pointed to still another part Mark II might play. Mercury came back to Earth's surface via parachute. Uncon-

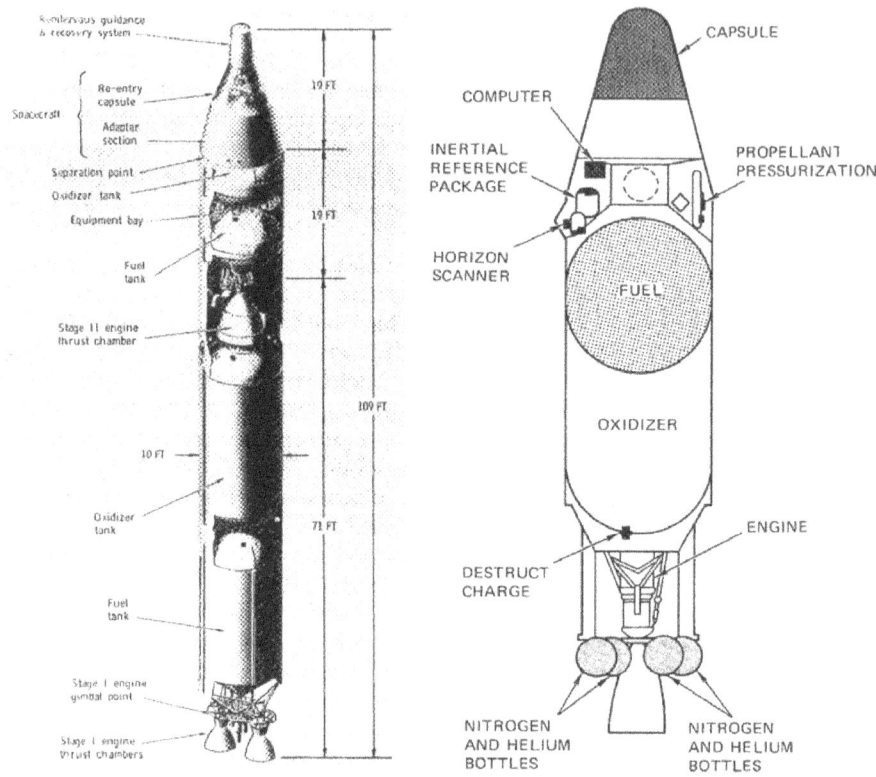

Left figure labels (Titan II):
- Rendezvous guidance & recovery system
- Spacecraft
  - Re-entry capsule
  - Adapter section
- Separation point
- Oxidizer tank
- Equipment bay
- Fuel tank
- Stage II engine thrust chamber
- Oxidizer tank
- Fuel tank
- Stage I engine gimbal point
- Stage I engine thrust chambers
- 19 FT
- 19 FT
- 109 FT
- 10 FT
- 71 FT

Right figure labels (Agena B):
- CAPSULE
- COMPUTER
- INERTIAL REFERENCE PACKAGE
- PROPELLANT PRESSURIZATION
- HORIZON SCANNER
- FUEL
- OXIDIZER
- ENGINE
- DESTRUCT CHARGE
- NITROGEN AND HELIUM BOTTLES
- NITROGEN AND HELIUM BOTTLES

*Gemini launch vehicles were (above left) the Titan II and (above right) the Agena B. Below is a 1962 drawing of the rendezvous and docking role envisioned for Agena.*

# PROJECT GEMINI
## FLIGHT MISSION

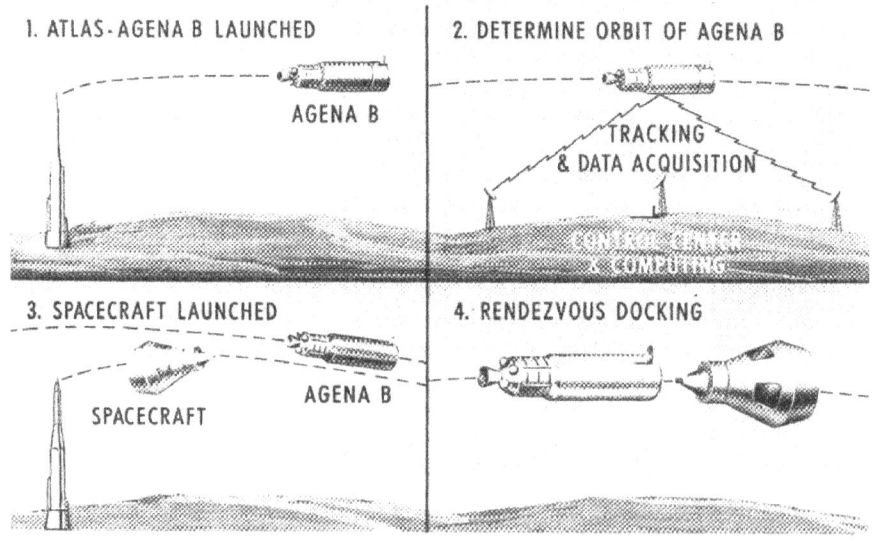

1. ATLAS-AGENA B LAUNCHED — AGENA B

2. DETERMINE ORBIT OF AGENA B — TRACKING & DATA ACQUISITION — CONTROL CENTER & COMPUTING

3. SPACECRAFT LAUNCHED — SPACECRAFT — AGENA B

4. RENDEZVOUS DOCKING

trolled return made the ocean the best landing field. But this meant that each landing was a major undertaking in its own right, with fleets of ships and aircraft deployed to ensure the safe recovery of pilot and spacecraft. This clearly would not do if space flight were ever to become a routine enterprise. Fitted out with a paraglider system, Mercury Mark II might show the way to controlled recovery on land.

These were all, however, only ideas that needed to be hammered into specific proposals with goals, costs, and timetables. This was the purpose of the preliminary project development plan that Chamberlin and his co-workers began to prepare early in August 1961. The focus of their effort now shifted from the engineering design of an improved Mercury to framing the program such a capsule might serve. McDonnell reoriented work under its NASA study contract toward "basic and alternate missions for the MK-II Spacecraft" and increased the number of engineers assigned from 45 to 74.[14] At the same time, three McDonnell engineers, led by Fred Sanders, journeyed to Langley, where Chamberlin, aided by James Rose and several contracting and scheduling specialists,* was getting started on the preliminary plan for a new project, using the Mercury Mark II two-man spacecraft.[15] The first result was ready 14 August 1961.

REACHING FOR THE MOON

The "Preliminary Project Development Plan for an Advanced Manned Space Program Utilizing the Mark II Two Man Spacecraft"[16] framed six objectives. They were to be achieved in 10 flights, the first in March 1963 and the rest to follow once every two months until September 1964. The first objective was long-duration flights, with men in orbit for up to 7 days, animals for up to 14. The two extended manned flights, scheduled third and fourth in the program, came first. Two animals flights were then to provide "completely objective physiological data which could not be obtained otherwise." These were to be the sixth and eighth flights, because the planners were not sure that some of the spacecraft components, especially the retrofire system, could be relied upon over so long a time; the required reliability would be shown in the earlier manned flights, when manual backup was available. Otherwise, the only purpose of the manned flights was to

---

*Sander's team stayed at Langley for two weeks; the other two members were Ervin S. Kisselburg and Gilbert G. Munroe. Munroe, who came to Virginia to work on spacecraft weight analyses, soon returned to an earlier assignment on the aircraft side of McDonnell. Frank G. Morgan, Jr., the company's marketing engineer for Mercury, was a frequent visitor to STG at this time, helping with cost estimates. Chamberlin's contract and scheduling help came from George F. MacDougall, Jr., Joseph V. Piland, Walter D. Wolhart, Lester Stewart, Nicholas Jevas, William C. Muhly, Richard F. Baillie, Donald L. Jacobs, Allen L. Grandfield, Paul M. Sturtevant, and Paul H. Kloetzer.

test the ability of the crew to function in space for as long as a week. Russian Cosmonaut Gherman S. Titov completed his 17-circuit, 25-hour mission aboard *Vostok II* on 6-7 August 1961; although he complained of nausea, he proved that a man could last a day in space.[17]

A look at the Van Allen radiation belts was the second objective. The first flight was to be an unmanned test to make sure that spacecraft and booster would be compatible for manned missions, but it would also carry biological experiments. Titan II would boost the spacecraft into a highly elliptical orbit, 160 kilometers above Earth at its lowest point but 1400 kilometers out at its highest and through the Van Allen belts to acquire data on radiation.

Controlled landing was the third goal, to be pursued on all seven manned flights. This meant that the pilot had to have some means of flying the spacecraft toward a relatively limited landing area. The most direct method was to offset the spacecraft's center of gravity to yield some degree of aerodynamic lift, using the attitude control system to roll the spacecraft during its flight through the air and thus control the amount and direction of lift to correct any errors in the predicted landing point. Controlled land landing also demanded some way to cushion touchdown impact. This was a harder problem, but one to which the paraglider seemed to promise an answer.

Rendezvous and docking stood fourth in the list of objectives. The fifth, seventh, ninth, and tenth flights in the program each required two launches, so the Titan II-launched Mark II could meet and dock with the Atlas-launched Agena B in orbit. The planners foresaw the major problem in the first rendezvous missions to be the size of the "launch window," the length of time during which a spacecraft could be launched to rendezvous with its target. The larger the launch window, the greater the difference in speed between spacecraft and target that had to be made good. That was beyond the powers of the spacecraft alone, but the difference might be made up, in part, by the target. Later, with more experience, the engineers expected to reduce the size of the launch window. Then the extra power provided by the target might find other uses, perhaps in "deep space and lunar missions with the target vehicle being used as a booster following rendezvous." The fifth objective was astronaut training, mainly a useful byproduct of the program.[18]

The plan stressed extensive use of vehicles and equipment on hand, altered as little as possible. The Mark II spacecraft retained what the Mercury capsule had proved, its aerodynamic shape, thermal protection, and systems components. Some changes were demanded by new goals. In the longer flights, crew members needed improved pressure suits, fuel cells to replace batteries, and more stable propellants than hydrogen peroxide in the attitude control system. Although Mercury carried none of the gear required for rendezvous missions, planners expected to meet these needs with little or no modification of existing inertial platforms, radar, and computers; of the major requirements, only a rendezvous propulsion system was not on hand.

Other major changes were limited to ejection seats instead of Mercury's escape tower and the environmental control system, in essense two Mercury systems hooked together. Since everything else in Mark II differed little if any from the equipment flight-tested in Mercury, the engineers looked forward to only a modest testing effort in the new project. They guessed that it would cost only $177 million to develop, procure, and test eight Mark II spacecraft, two of which were to be reused.[19]

The Mark II planners were just as sanguine when it came to launch vehicles. Atlas-Agena B could be used almost as it came off the assembly line, at a cost to the program of only $38 million for the four required. Titan II demanded more in the way of changes, but the Air Force would bear most of the cost. The chief exception was lengthened second-stage propellant tanks to increase the payload by 300 kilograms. As a manned booster, Titan II promised to be so simple and reliable that only one extra feature was needed to leave all decisions to abort a mission in the hands of the pilots. That was a redundant guidance and control system. Titan II's most dangerous potential failing, and the only one that demanded an automatic abort system, was first-stage engine hardover. A malfunction in the guidance and control system could drive the gimbaled engines to their extreme positions—hardover—their thrust vector then being directed at the farthest possible angle from the proper flight path, accelerating the booster away from the correct course in the region where it would be subjected to the greatest dynamic pressure. The danger lay in the possibility of the booster's breaking up before the pilots could react. By adding a second first-stage guidance and control system, the hazards of this failing were all but erased. Since the booster demanded little in the way of new parts, testing could be quite limited. The best estimate of the price of the boosters was $86 million.

The cost of the entire program from drawing board through the last flight came to $347.8 million. It would be managed by a project office that would also take charge of the rest of the Mercury program, the three-orbit flights already planned and the proposed 18-orbit mission using the minimum-change capsule. Forming the core of the new project office would be the 76 members of STG's Engineering Division, at the time chiefly engaged in Project Mercury and largely outside the mainstream of Apollo. The planners were careful to stress that the new office could be fully staffed to a total of 175 and the new program could be carried on without threat to other programs. Mercury would not be hindered, Apollo would not be interrupted.[20]

Should the proposed project meet with complete success, the stage would then be set for the sixth objective, which might supplant much of Project Apollo. If the Mark II spacecraft showed itself able to support a crew for seven days or more and if rendezvous proved to be practical, then the advanced program based on the Mark II might "accomplish most of the Apollo mission at an earlier date than with the Apollo program as it is presently conceived." By taking full advantage of the new spacecraft and rendezvous technique, "it is a distinct possibility that lunar orbits may be accomplished by the interim spacecraft after rendezvous with an orbiting Centaur." This prospect was the subject of an appendix to the development plan.

Centaur was a second-stage vehicle then under development that would use high-energy liquid hydrogen as its fuel. If Centaur were inserted into orbit by Titan II, it would have enough power after docking to boost the spacecraft to escape velocity. The deep-space version of Mark II differed from the rendezvous type only in having backup navigation gear and extra heat and radiation protection, 270 kilograms more on a 2900-kilogram spacecraft. The appendix explored two possible mission sequences. One simply added four flights to the ten in the Mark II program. The first two extra flights were deep-space missions, with Centaur boosting the spacecraft into an elliptical orbit with an apogee of some 80 000 kilometers to study navigation and reentry problems. The last two flights, scheduled for March and May 1965, were circumlunar, and the whole package added only $60 million to the cost of the basic Mark II program.

The alternative was an accelerated program, nine flights in all. The first three flights were the same in both programs—an 18-orbit unmanned qualification and radiation test, an 18-orbit manned qualification test, and a manned long-duration test. In the speeded up program, the fourth and fifth flights developed the techniques of rendezvous and docking with Agena B as the target. Centaur launched by Titan II then replaced the Agena for the rest of the program—two deep-space missions and two flights around the Moon. This faster program put the first Mark II in lunar orbit in May 1964 for a cost not much greater than the basic 10-flight program: $356.3 million versus $347.8 million.[21]

During the week after its release, the Mark II plan had STG buzzing.[22] A second version of the plan came out just a week later, on 21 August. It differed from the first in only one notable respect. All mention of a lunar mission for Mark II had vanished, leaving behind only a circumspect suggestion that, "if a vehicle such as the Centaur were used as the rendezvous target, the spacecraft would then have a large velocity potential for more extensive investigations."[23] Even this hint dropped out of later versions of the plan.

The appeal of going to the Moon with Mark II, however, was not

Atlas-Centaur was rated operational after this launch on 26 October 1966. The tar̶get goal of launching a Centaur in 1961 was missed. The first try came on 8 May 1962 and failed, putting off a successful launch until 27 November 1963—a date that would have been out of phase with Gemini launch schedules of 1962.

so easily quashed. After cutting circumlunar flight from the Mark II plan, Chamberlin revived the even more daring idea of using the spacecraft in a lunar landing program.[24] The booster was Saturn C-3 and the key technique was lunar orbit rendezvous. The scheme involved a lunar landing vehicle that was little more than a 680-kilogram skeleton, to which a propulsion system and propellants were attached. Fully fueled, it weighed either 3284 kilograms or 4372 kilograms, depending on choice of propellants. The lighter version used liquid hydrogen, the heavier used hypergolic propellants. Total Earth-launched payload in this mission fell between 11 000 and 13 000 kilograms, one-sixth to one-fifth of the 68 000-kilogram payload then in prospect for the direct ascent lunar mission. The cost was low, $584.3 million plus the expense of two Saturn C-3 boosters, but the risk was high.

The flight plan for this lunar landing program derived from the speeded up circumlunar proposal appended to the Mark II plan of 14 August. The first two flights, in March and May of 1964, were to be unmanned and manned qualification tests of the spacecraft and Titan II. The next two flights put the spacecraft in orbit for extended periods of time. Three flights then developed and demonstrated rendezvous and docking techniques with Agena as target. The eighth and ninth missions had Centaur boosting the spacecraft into an 80 000-kilometer deep-space orbit. Next came three flights to test rendezvous between the manned spacecraft and the unmanned lunar landing craft in Earth orbit, culminating with the crew transferring from one to the other. Flights 13 and 14 had Centaur boosting the spacecraft to escape velocity for an early demonstration of circumlunar capability. Saturn was to launch the 15th flight, a Moon orbital mission. Men would land on the Moon in the final flight, slated for January 1966.[25]

When Chamberlin proposed this scheme to Gilruth's senior staff at the start of September 1961, he was the first in STG to offer a concrete plan for manned lunar landing that depended on the technique of rendezvous in lunar orbit.[26] STG so far had seen little merit in any form of rendezvous for lunar missions, but it reserved its greatest disdain for the lunar orbit version. The Langley partisans of lunar orbit rendezvous had first put their scheme before STG on 10 December 1960, when they rehearsed what they planned to say to Associate Administrator Robert Seamans and his staff a week later.[27] On 10 January, John Houbolt and some of his colleagues met with three STG engineers and tried to convince them that lunar orbit rendezvous belonged in the Apollo program. The response was reserved, the scheme dismissed as too optimistic.*[28]

---

*Houbolt, Clinton Brown, Manuel J. Queijo, and Ralph W. Stone, Jr., described the lunar orbit rendezvous idea to Kurt Strass, Owen E. Maynard, and Robert L. O'Neal. O'Neal's report to Associate Director Charles Donlan was distinctly skeptical of Langley's claims on weight saving.

Three months later, Houbolt was back for another briefing, this time supported by a printed circular on "Manned Lunar Landing via Rendezvous." It included one project called MORAD (for Manned Orbital Rendezvous And Docking), a modest two-flight effort to be completed by mid-1963, intended as a quick proof of the feasibility of rendezvous. A small unmanned payload would propel itself to a link-up with a Mercury capsule, its maneuvers under the control of the Mercury pilot. The key project, however, was MALLIR (Manned Lunar Landing Involving Rendezvous). Chamberlin, who attended this briefing, had known about Langley's rendezvous work, but had not before heard about the lunar orbit version. He asked Houbolt for a copy of the circular and for anything else he had on rendezvous.[29]

Others in STG had yet to be convinced. Gilruth saw rendezvous as a distant prospect, not something for the near future. Mercury was proving so troublesome that rendezvous, however simple in theory, seemed very far away. He strongly insisted on the need for large boosters:

> Rendezvous schemes are and have been of interest to the Space Task Group and are being studied. However, the rendezvous approach itself will, to some extent, degrade mission reliability and flight safety. I am concerned that rendezvous schemes may be used as a crutch to achieve early planned dates for launch vehicle availability, and to avoid the difficulty of developing a reliable NOVA class launch vehicle.[30]

This viewpoint was widespread in NASA, leading some to resist rendezvous, not because they believed it a poor idea but because it threatened to subvert another goal seen to be more important.

The efforts of Houbolt and his Langley colleagues to sell rendezvous in general, and lunar orbit rendezvous in particular, may have been frustrated less because their concept was faulty than because, as Chamberlin has suggested, "they were considered to be pure theorists with no practical experience." The major trouble with the lunar orbit rendezvous scheme may well have been that it simply looked too good to be true. Paper-and-pencil calculations did yield striking figures, but what looked good in theory might not stand up so well in practice. Chamberlin and his co-workers, although fully alive to the weight-saving features of rendezvous, stressed another aspect—it made a lunar spacecraft easier to design. Direct ascent posed a particularly thorny design problem because the spacecraft had both to land on the Moon and to reenter Earth's atmosphere. A rendezvous mission, however, allowed one design for a lunar lander, a second for a reentry capsule—a distinct spacecraft to meet the special demands of each of these

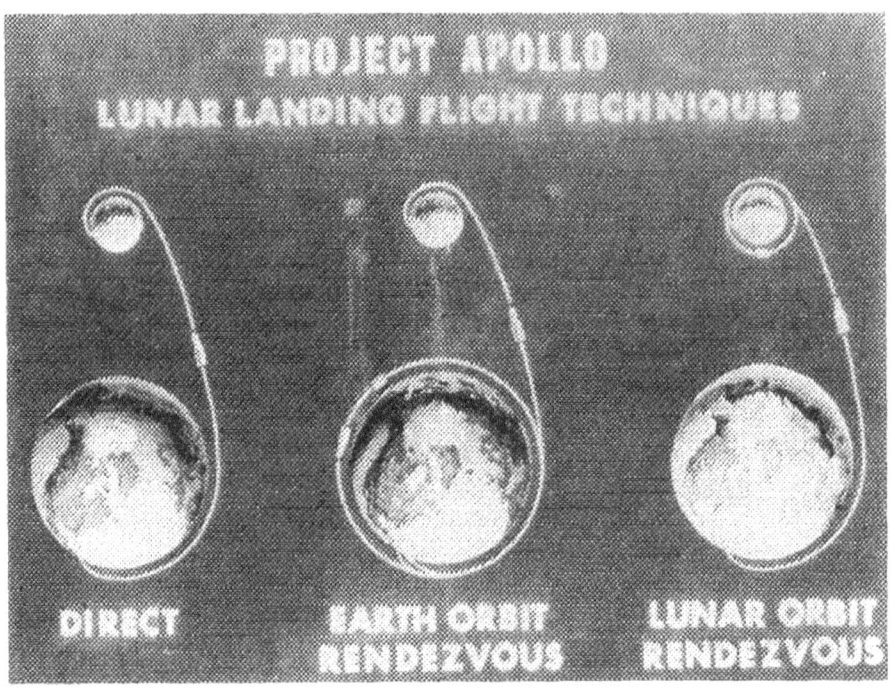

*Shown here are three competitive modes for landing a man on the Moon. The direct mode, which would fly directly from Earth to the Moon, was early favored by NASA Headquarters and the Space Task Group. Earth orbit rendezvous was first favored by the Marshall Space Flight Center, wherein two vehicles would rendezvous in Earth orbit and then fly to the Moon. Langley Research Center first championed the lunar orbit rendezvous mode in which the spacecraft would go into lunar orbit and a small ferry vehicle would take the crew to the lunar surface and then back to the mother ship in lunar orbit for the return to Earth.*

two most critical phases of the lunar flight. Chamberlin's group had, in fact, centered its work on a lunar-lander design, since reentry problems were already well in hand. Stressed as an answer to design constraints rather than a weight-saving expedient and sponsored by men with plenty of practical experience in Mercury, lunar orbit rendezvous in Chamberlin's plan for a Mark II lunar landing mission received its first serious hearing from STG.[31]

Toward the end of September 1961, Chamberlin's plan showed up as part of an "Integrated Apollo Program" STG presented to Silverstein and his staff at NASA Headquarters. What "integrated" meant was adding a Mark II orbital rendezvous project to the Apollo program. Much of the presentation was drawn from the Mark II preliminary plan, but part of it was based on Chamberlin's lunar landing scheme of 30 August. Some of the figures were new: the lunar landing system, complete with propulsion unit and fuel, weighed little more than 1800 kilograms, roughly half what the first version had. The cost now included the Saturn boosters for a total of $706.4 million, but the flight development plan had not changed.[32]

Silverstein proved to be no more excited by a Mark II lunar mission in September than he had been by an improved Mercury lunar mission in March. But he was willing to go along with the idea of a rendezvous development project. On 6 October, Silverstein asked for, and got, Associate Administrator Seamans' formal approval for the "preparation of a preliminary development plan for the proposed orbital flight development program." Seamans now granted STG sanction to begin talks with McDonnell on buying the Mark II spacecraft, with the Department of Defense on Titan II boosters and launch-stand alterations, and with the NASA Office of Launch Vehicle Programs on the Atlas-Agena.[33]

The Mark II project itself, however, had yet to be approved, even though Seamans remarked that "our present plans call for a Mercury Mark II for test of orbital operations during 1963 and 1964."[34] Still lacking was an approved project development plan. Such a plan, in fact, had yet to be submitted, although copies of Chamberlin's preliminary plan had been making the rounds of NASA Headquarters in search of comments. With his Mark II lunar landing scheme rejected, Chamberlin now set out to revise the Mark II plan and put it in shape for Seamans to sign.

## MERCURY MARK II BEGINS

Chamberlin finished the revised project development plan on 27 October 1961.[35] The bulk of it followed the August versions word for word, although some new material appeared, some old ideas vanished, and some accents changed. Most striking was the greatly increased stress on the development of rendezvous techniques. Long duration retained first place on the list of objectives, but rendezvous had moved into second, with controlled land landing third, and astronaut training (still incidental) fourth.

Gone were the radiation study and the animal flights; no trace remained of a lunar mission, nor even of a deep-space sortie. The focus became developing the technique, rather than applying it. More of the text dwelt on rendezvous, with several new paragraphs to describe in detail the special equipment needed for rendezvous navigation, maneuvering, and docking systems. A closing statement of expected "Project Results," new in the October plan, clearly showed that rendezvous now had priority. The August plan "... for an Advanced Manned Space Program Utilizing the Mark II Two Man Spacecraft" became, in October, a "Project ... for Rendezvous Development."

The new flight plan also reflected the shift in focus. Although the total number of flights in the program only expanded from 10 to 12, the rendezvous flights doubled—from four to eight. The first flight had become strictly a qualification test of the unmanned spacecraft and booster, the spacecraft to be launched into a 160-kilometer circular orbit. An 18-orbit manned qualification was still second, followed as before by two extended manned missions, though these might now last up to 14 days. All other flights were designed to develop and test rendezvous techniques.

Logically enough, the October plan proposed a starting date of 1 November 1961, instead of 1 September. Two months still separated each flight from the next, the first now scheduled for May 1963, the last for March 1965. Program costs, however, climbed higher than only two more flights might suggest. The new figure was $529.45 million, more than one and a half times the August estimate. Two factors accounted for the seeming discrepancy. One was the new provision for spare spacecraft and boosters: 12 spacecraft rather than 8 (the first plan had called for 2 spacecraft to be re-used in later missions; the revised version planned for 3 spacecraft to be refurbished, but only as spares); 15 Titan IIs instead of 10, the extra 3 to serve as backups; and 11 Atlas-Agenas instead of 4, 8 to fly and 3 spares. The combined effect of these changes added $140.45 million to the program's costs. Most of the remainder of the increase came from a new $29-million item, "Supporting Development," for paraglider.[36]

STG forwarded the revised project development plan to NASA Headquarters on 30 October 1961.[37] Its approval expected as a matter of course, Chamberlin got busy setting up the program. Since McDonnell was obviously going to get the prime contract for the Mark II spacecraft, the company ought to be told to organize itself for the effort, to assign key people to the new program, and to make sure that the staff would be available.[38] Chamberlin proposed to amend the letter contract between NASA and McDonnell that had authorized the contractor to procure long-lead-time items for Mark II.[39]

Chamberlin wanted the McDonnell effort tailored to making a general-purpose spacecraft. This meant that Mark II should not only be able to perform its assigned long-duration and rendezvous missions, but also that it ought to be easy to adapt for other missions. Two other design objectives were only slightly less important, both springing from the notion of Mark II as a truly operational spacecraft (in contrast to the chiefly experimental Mercury): it should be simple to test realistically on the ground, leaving actual flights free to focus on major goals that could only be achieved in space; and it should be easy to check out, so a faulty spacecraft was less likely to cause a mission failure. To achieve these goals, Chamberlin thought McDonnell had three central tasks to tackle at once: for systems inherited mainly unchanged from Mercury, utmost refinement; for new systems, engineering analysis; and for special problems, like the integration of a paraglider system, special study groups.[40]

Chamberlin himself formed a Mark II rendezvous group, whose five members were, by mid-October, already talking to people in Langley's Aerospace Mechanics Division about some theoretical aspects of rendezvous.*[41] They had also approached (and been approached by) prospective contractors about what equipment might be needed, which allowed them to rough out a set of guidelines for rendezvous development by 10 November 1961.[42] The group then began a series of technical coordination meetings with McDonnell spokesmen in St. Louis, 14-15 November.

McDonnell engineers themselves had been looking at rendezvous for several months, and the meetings showed that company and NASA thinking had diverged sharply. McDonnell had assumed that the target would not be maneuverable and that control of the spacecraft during maneuvers could be either automatic or manual (or some mixture), the choice hinging on how much fuel the spacecraft could carry. The company, in other words, thought the spacecraft it was going to build should be the active agent in rendezvous. In contrast, Chamberlin's group from Manned Spacecraft Center (MSC; Space Task Group had changed its name on 1 November 1961) had approached the rendezvous system as a whole, spacecraft and target, and assumed a highly maneuverable target, with pilot control of the spacecraft and ground control of the Agena.

McDonnell's approach, which favored a combination of automatic and semi-automatic control, required a spacecraft target-tracking radar, and a digital computer and inertial platform for guidance, as well as a high-capacity propulsion system. MSC's preference for semi-manual control for the spacecraft—automatically stabilized but steered by the pilot—combined with target control under ground command stressed changes in the Agena rather than spacecraft equipment: a restartable engine, a data communication system to link the Agena to ground controllers, an optical tracking aid of some kind, a radar transponder, and an attitude stabilization system.

McDonnell and MSC decided to combine their approaches, fitting the spacecraft with the equipment the company believed necessary and

---

*The Mark II rendezvous group comprised Jerome B. Hammack, Orton L. Duggan, James T. Rose, Jean L. Petersen, and Harry Shoaf. Among those the group talked to were Thomas J. Voglewede, Arthur Vogeley, Max Kurbjun, and Edgar C. Lineberry.[40]

altering the Agena to conform to what MSC wanted. This "would allow the most flexibility in the choice of rendezvous techniques without equipment change."[43]

By mid-November 1961, McDonnell had completed most of the documents that spelled out the company's view of what should be in the expected contract with NASA to build the two-man spacecraft. The most important was a detail specification of the Mark II spacecraft, issued 15 November.[44] The McDonnell design was deliberately conservative, notably in retaining both the launch escape tower and the impact bag used in Mercury.

McDonnell engineers who drew up the specification could not yet be sure that safety permitted striking the escape tower from the design. Still under study was what might happen if a Titan II exploded on the launch pad while the crew was aboard the spacecraft. Whether ejection seats could in fact propel the two men away from an exploding booster fast enough to outdistance the expanding fireball remained in doubt. Speed and range of the ejection seat were both critical. As a hedge, the Mark II design included the escape tower.

The presence of a Mercury-type impact bag in the specification was another cautious note. The Mercury capsule had an inflatable bag that served to cushion the impact of landing. Although the paraglider promised greatly reduced landing stresses, the designers felt that work on the concept was not far enough advanced to allow them to rely on it entirely. No one really believed that either the tower or landing bag was going to be necessary but, faced with drawing up a specification for Mark II, McDonnell engineers chose to put on paper something they *knew* would work.[45]

Planning for the second phase of the paraglider program, a two-part system research and development effort, had already begun. In Phase II, Part A, the contractor was to spend eight months in further study of the design concept, chiefly to settle on what configuration would yield the best performance. The second part of Phase II called for the as-yet-unnamed contractor to build a prototype paraglider landing system, to conduct a series of unmanned and manned flight tests, and to complete a final design. The third and final phase of the program would see a paraglider system in production and pilots being trained to fly it.[46] On 20 November, North American received official word that it had been awarded the contract and was authorized to begin work.[47]

The same team that had monitored the paraglider design study for STG* now joined spokesmen for North American, Langley, and Flight Research Center to discuss putting Phase-II A into motion. They soon agreed that the half-scale models and full-size vehicle for

---

*Rodney Rose, Harry Shoaf, and Lester Stewart.

this phase should be based on the Mark II design. "Power requirements, control actuation, landing gear, etc., should be compatible with the MK II spacecraft, where MK II is sufficiently firmed up for this to be practical without delaying the full-scale test program." Most wind tunnel testing would be done by Langley, while Flight Research Center, at Edwards Air Force Base, California, was to take charge of flight testing, all under the aegis of MSC. Even at this early date, the interface—that useful term for the region where two or more things share a common boundary—between paraglider and spacecraft was beginning to pose questions: how the glider and its gear were to be stowed; how it was to be deployed, sequenced, and jettisoned; what kind of cockpit controls and displays it would need; and how it would fit with the emergency escape system.[48]

When Gilruth and Chamberlin visited NASA Headquarters in late November 1961 to see Associate Administrator Seamans and report on the Mark II program, they had a good deal to talk about. Spacecraft design was just about settled, paraglider development was beginning, and some basic approaches to developing rendezvous techniques had been decided. Although Gilruth's and Chamberlin's meeting with Seamans did nothing to dampen their belief that project approval was only a matter of time, that time was not yet. Seamans was not quite ready to take the final step. November had been a busy month in NASA Headquarters, and the turmoil had touched the Mark II project.

### THE LAST OBSTACLES

One source of delay was the still unsettled question of the place of rendezvous in NASA planning. The key factor was the size of boosters. The persistent appeal of orbital rendezvous for many NASA and Defense Department planners was its promise (and, in 1961, only its promise) of making do with lesser boosters. Even they were a long way from ready; the most powerful in operation in the United States at that time, the Atlas-Agena, could only put about 1800 kilograms in Earth orbit. The smallest payload required for a lunar landing mission, even with rendezvous techniques, was thought to be ten times that figure. This was a matter of concern to both NASA and the Department of Defense, leading them to form a joint Large Launch Vehicle Planning Group in July 1961 under Nicholas E. Golovin for NASA and Lawrence L. Kavanau for the Defense Department.*[49]

---

*Serving under Golovin and Kavanau were Eldon Hall, Harvey Hall, Milton W. Rosen, Kurt R. Stehling, and William A. Wolman (NASA Headquarters); Warren H. Amster and Edward J. Barlow (Aerospace); Aleck C. Bond (STG); Seymour C. Himmel (Lewis); Wilson Schramm and Francis L. Williams (Marshall); Colonel Mathew R. Collins (Army); Rear Admiral Levering Smith and Captain Lewis J. Stecher, Jr. (Navy); and Colonel Otto J. Glaser, Lieutenant Colonel David L. Carter, and Heinrich J. Weigand (Air Force).

Golovin, technical assistant to Seamans, spelled out what he believed to be the group's central goal. "The primary basis for organizing information and preparing recommendations for a National Large Launch Vehicle Program will be the assumption that this program will provide vehicle systems for the attainment of a manned lunar landing and return during the fourth quarter of calendar year 1967 or before."[50] The group worked from July through October, its efforts yielding a massive preliminary report in November.[51]

The team, often referred to as the "Golovin Committee," essayed a detailed, quantitative comparison of direct ascent with several forms of rendezvous-based missions, and each of the rendezvous missions with the others. A subcommittee under Harvey Hall, Chief of Advanced Development in NASA's Office of Launch Vehicle Programs, took charge of this phase of the study and asked each of three field centers to prepare a brief for one form of rendezvous mission. Marshall was to work on Earth orbit, Langley on lunar orbit, and the Jet Propulsion Laboratory (JPL) on lunar surface rendezvous. The lunar surface rendezvous scheme grew out of JPL's experience in the unmanned lunar exploration program. It proposed to automatically assemble unmanned modules on the Moon; this assembly would then serve as the return vehicle for a crew carried to the Moon via direct ascent from Earth. Hall's own office furnished data for direct ascent.*[52]

By mid-September, preliminary analysis strongly supported some type of rendezvous over direct ascent as the best basis for a lunar mission, though no single rendezvous scheme had a clear edge over the others. The smaller boosters that could be used in such a mission would be ready sooner, which meant more flight tests and greater reliability for less money.[53] When Hall reported to the full committee on 10 October, after the field center studies were in,[54] lunar surface rendezvous was out of the running and direct ascent nearly so. The choice was narrowing to rendezvous in Earth or lunar orbit, with Hall's subgroup tending to favor some combination of the two.[55]

This view had the full, even vigorous, support of the committee as a whole.[56] In its report, the Large Launch Vehicle Planning Group, after a detailed analysis of the rival schemes, found that orbital rendezvous promised the best chance for an early lunar landing, the lunar orbit version perhaps the quickest.[57] Either form of rendezvous in orbit, or some hybrid of the two, would beat a direct ascent mission to the Moon, because the smaller boosters they needed could be ready sooner.[58]

Despite its elaborate quantitative analysis, the Golovin Committee

---

*John Houbolt was technical supervisor of Langley's effort; Peter deFries, of Marshall's; and John W. Small, Jr., of JPL's.

did not have the last word in the controversy over direct ascent versus rendezvous for an early manned lunar landing. Too many questions remained open, too many answers were equivocal, pleasing neither NASA nor Defense, and the committee had failed to produce the integrated national launch vehicle program it had been created for.[59] So boosters remained the first order of business.

Early in November, Milton W. Rosen, author of NASA's first launch vehicle program in 1959 and Director of Launch Vehicles and Propulsion in NASA's new Office of Manned Space Flight, set up a working group to decide on a large booster program geared to manned space flight.*[60] Drawing on the findings of the other committees that had been chewing on the problem since May, Rosen's 12-man group was able to submit its recommendations by 20 November.[61]

The intense two-week study centered on the technical and operational problems posed by rendezvous. The group decided that rendezvous looked good but preferred direct ascent for the lunar mission because rendezvous was still an unknown. That was something the group insisted had to be corrected. Rendezvous had too much promise, both generally for a broad range of future missions and specifically for an early lunar landing, to permit the techniques to go on being ignored. Prudence dictated planning based on direct ascent, but "vigorous high priority rendezvous development effort must be undertaken immediately."[62]

November 1961 also saw the structure of NASA revamped.[63] Almost eight months had gone into a reorganization of the agency to handle a program the size of Apollo. Shortly after he took over the reins as NASA's second administrator, James E. Webb, at a retreat in Luray, Virginia, on 8-10 March 1961, met with his key people from Headquarters and the field centers. Webb stated that the three top leaders of NASA would act as a team in running the agency. He and Dryden would serve as co-equals and Seamans would function as the "operating vice-president," presiding over the daily affairs of NASA. Essentially, Webb said, Dryden would be concerned with "what to do" and Seamans with "how to do it."

After the retreat, the problems of getting Apollo defined, approved, and pieces of its hardware under contract, and to acquire land suitable for the erection of development, test, and operational facilities, gave rise to a surfeit of committees to study and recommend action on one phase of the program or another. By September, however, Webb

---

*Rosen's group began with Richard Canright, Eldon Hall, Elliott Mitchell, Norman Rafel, Melvyn Savage, Adelbert O. Tischler, and John Disher, of NASA Headquarters; and William A. Mrazek, Hans H. Maus, and James B. Bramlet, of Marshall, who were soon joined by David M. Hammock of MSC.

knew that NASA could no longer afford to wait on committees to convene and make recommendations. He needed decision makers at the program office levels. Moreover, the field centers seemed to be competing among themselves too much. So Webb, Dryden, and Seamans searched the country for someone who could come into NASA Headquarters and take charge of Apollo and the new Office of Manned Space Flight, an offshoot of the Office of Space Programs now to be a program office in its own right. Radio Corporation of America, which had earlier sent Robert Seamans to become NASA's Associate Administrator, now furnished the Director of Manned Space Flight in the person of D. Brainerd Holmes.[64]

The old program offices vanished. The four new offices—Space Sciences, Advanced Research and Technology, Manned Space Flight, and Applications—were not the semi-autonomous bureaus their predecessors had been nor did they retain control of the field centers. They became less operating line offices, more advising staff offices. The field centers, including the new Manned Spacecraft Center, now reported to the Associate Administrator rather than to Headquarters program officers.

These changes furthered the cause of rendezvous but delayed the Mark II project. Seamans, a longtime supporter of rendezvous, won a stronger hand in NASA programming and a useful ally in Holmes. Silverstein, most powerful of the former program directors and foremost advocate of direct ascent, left Washington. His old office was gone, and, unwilling to accept the leadership of the new Office of Manned Space Flight, he instead assumed directorship of Lewis Research Center.

STG had reported to Silverstein's office. He himself favored the Mark II project, but he also knew that he was going to be leaving Washington after the reorganization. He was understandably reluctant to commit his successor to a large new program. Holmes, who arrived at NASA Headquarters in October, had little to do with the Mark II decision, anyway. The new order left that squarely in Seamans' hands.

Although the reorganization caused some delay, a larger obstacle loomed from another quarter. NASA still depended upon the Air Force for its boosters. In November 1961, smooth progress toward using a modified Titan II in the Mark II project hit an abrupt snag. John H. Rubel, Assistant Secretary of Defense and Deputy Director of Defense Research and Engineering, informed Seamans that the Air Force was now developing a

> TITAN III Standard Launch Vehicle System. This vehicle is intended to serve as the single standardized TITAN vehicle to be used in support of both NASA and DOD programs as appropriate. We expect the design to meet any or all need which NASA may have for space application of the TITAN ICBM.

*Artist's concept of a Titan III boosting a Dyna-Soar spacecraft into Earth-orbital flight.*

Rubel asked Seamans to see that all NASA studies of Titan be routed through the Air Force Systems Command, which had just begun a design analysis as the first phase of the Titan III program.[65]

Titan III differed from Titan II chiefly in adding two very large solid propellant rocket motors. These motors, 3 meters across, were to be strapped to a core, a much strengthened Titan II, to become in effect the booster's first stage. Their firing would carry the booster aloft, where they would be dropped and the liquid propellant engines of what had been the Titan II first stage would ignite. The much more powerful Titan III was to replace Titan II as the booster in the Air Force's Dyna-Soar program. Its use in NASA's Mark II project might further justify its development.[66]

That the Air Force planned to develop Titan III as a standardized vehicle to meet both its own and NASA's needs for launching payloads of up to 14 000 kilograms into low-Earth orbit came as no surprise to NASA. Seamans and Rubel had discussed the project, and the Golovin Committee had endorsed it and recommended launching Mark II with the Titan III core. NASA's response, at first favorable, had since cooled. By November 1961, NASA officials evinced little desire to adopt Titan III for any program, least of all Mark II.[67] This may have been the real source of friction. NASA had expected to use a modified Titan II in the Mark II project, but Rubel's letter implied that the Titan III core was what NASA would get, like it or not. Not only was the reinforced core likely to be too heavy, but the central logic of the Mark II project demanded that it be done quickly because any delay raised the prospect of conflict with Apollo. Titan III development meant a major new program, which could hardly be completed in time to meet the tight Mark II schedule.[68]

The Department of Defense countered by claiming that the modifications NASA wanted in Titan II—lengthened tanks and redundant systems—also implied a new development program. This version of Titan II was now unofficially labeled Titan II-½. Efforts to resolve this impasse led to a top-level meeting of NASA and Defense officials on 16 November. They decided to recall the Large Launch Vehicle Planning Group expressly to study the place of Titan III in the long-term national launch vehicle program and to decide whether the Mark II project really needed Titan II-½.[69] The order went out two days later, and the planning group reconvened on 20 November.[70]

When the Golovin Committee had finished its brief but intense study, Seamans and Rubel agreed that the Department of Defense should go ahead with Titan III. Titan II-½ they deemed unnecessary. The Mark II project could be adequately served by "TITAN II missiles, virtually unmodified"; the only changes to be permitted were those that mechanically adapted the booster to the spacecraft and others "specifically aimed at and limited to the marriage of payload and launch vehicle." Major changes in structure or tankage, or "the addition of new or the extensive modification of existing subsystems internal to the missile," were specifically excluded.[71]

Although NASA failed to get the lengthened tanks and redundant systems it wanted in Titan II, it did get Titan II. Until the day Rubel and Seamans made these recommendations, even that issue was in doubt. But, with the decision of 5 December, the last obstacle to the approval of Mark II vanished. And, as events were soon to show, NASA was not going to have to make do with "TITAN II missiles, virtually unmodified."

Seamans' approval of Mark II took the form of a note at the foot of a three-page memorandum from Holmes' Office of Manned Space Flight on 6 December, which offered a concise statement of Chamberlin's project development plan. The statement identified the development of rendezvous techniques as "the primary objective of the Mark II project," with long-duration flights, controlled land landing, and astronaut training as "important secondary objectives." It went beyond Chamberlin's plan to point out that rendezvous would permit manned lunar landing to be achieved more quickly and that rendezvous took on special importance when it became part of the lunar landing maneuver itself, an oblique reference to the lunar orbit rendezvous scheme.

Holmes asked for $75.8 million from current fiscal year 1962 funds to start the project at once and promised a formal project development plan in short order. Seamans wrote "Approved" and signed it on 7 December 1961.[72] The promised plan appeared the next day. Only the date on the cover and title page distinguished it from the plan of 27 October, copies of which now bore a large red "PRELIMINARY" stamp.[73] On 3 January 1962, NASA unveiled the first pictures of the new spacecraft and announced that it had been christened Gemini.[74]

## JUSTIFICATION FOR GEMINI

When Chamberlin and his co-workers in STG and McDonnell began to devise a program to fit the new spacecraft they had already designed, the choice of goals open to them was wide. How well and how long man could survive and function beyond the reach of the gravity in which the species had evolved and beyond the shield of air which had sheltered it from the harsh extremes of space had long been matters of concern. Project Mercury could not—and before May 1961 had not even started to—resolve these questions, and answers were essential before men ventured into deep space. The spacecraft's return to Earth was another concern. Landing that could be controlled and directed by the crew to an area more nearly on the order of an airport than the ocean-sized zones required by Mercury was clearly something to be worked for. Neither of these goals, however, was itself enough to justify a program for Mark II. Any post-Mercury program would support longer flights, and controlled landing was more convenience than absolute necessity. Rendezvous, however, presented quite a different picture.

The exciting potential of orbital rendezvous in future manned space flight had largely ceased to be a matter for dispute in NASA after the middle of 1961. Some planners still hesitated to endorse rendezvous techniques as the basis for a lunar landing mission in Apollo, but none denied its long-term importance. Theory and experiment alike suggested that guiding two spacecraft to a meeting in orbit ought to present no special problems, but until the technique could be demonstrated doubts remained. Should rendezvous prove to be as trouble free in practice as it seemed to be in theory, then it might be worked into planning for the trip to the Moon and allow that journey to be mounted sooner and more cheaply. In late summer 1961, this prospect inspired Chamberlin to propose a program for Mark II that would beat Apollo to the Moon.

Chamberlin first proposed using rendezvous in Earth orbit to allow Mark II to circumnavigate the Moon and followed that up with an even more daring scheme based on rendezvous in lunar orbit to land men on the Moon. This succession reflected the trend of thinking in NASA as a whole. The last half of 1961 saw the technique of lunar orbit rendezvous gain growing support as a means to achieve early manned lunar landing. But Chamberlin was moving far more quickly than his colleagues. Perhaps the greatest defect in his plans was that they assumed the rendezvous technique itself to need no special work, that a few flights would suffice to prove the technique before going on to apply it to larger ends. This was an assumption not widely shared, and both plans were rejected for Mark II, although Chamberlin may well have blazed the trail for rendezvous in Apollo.

This still left the development and demonstration of rendezvous maneuvers as a proper goal for the Mark II project, and that became the basis for Chamberlin's revised plan. This fitted NASA's clearly growing inclination to see a place for rendezvous in its lunar mission. Pressure for a rendezvous development program of some kind was becoming intense. Thinking about lunar orbit rendezvous for Project Apollo could only make the matter seem more urgent. There might be some room for error in Earth orbit, where a failure need not mean the loss of the crew. But that margin did not exist in lunar orbit; sound and fully proved techniques would be crucial.

By late 1961, a rendezvous development program may well have become inevitable, and Mark II was not the only candidate in the field. Phase A of Project Apollo itself and Marshall's orbital operations development program were likely rivals. The Mark II project, however, had a clear edge: a spacecraft already designed and very nearly ready to go into production and a set of sharply defined and suitably limited objectives. When NASA decided late in 1961 that it needed a rendezvous development program, Mark II was there.

# IV

# Organizing Project Gemini

WHEN Mark II was approved on 7 December 1961, much of the groundwork had already been laid. Aside from the paraglider, however, whose development was not directly tied to the Mark II project, none of the pieces was yet under contract. The Manned Spacecraft Center itself was not going to build spacecraft, booster, target, or paraglider. In line with the practice pioneered by the Air Force after World War II, NASA relied on private firms to develop and produce most of its hardware. The first priority, even before getting the project office fully in order, was putting the spacecraft under contract and making arrangements with the Air Force for booster and target vehicles.

## THE PRIME CONTRACTS

Because so much of the preliminary design work had been done, MSC had a letter contract for the spacecraft prepared by 15 December.[1] Since it called for a "Two-Man Spacecraft" to be developed from "the present Mercury Spacecraft, retaining the general aerodynamic shape and basic system concepts," there was no question of seeking competitive bids. The choice clearly fell to the McDonnell Aircraft Corporation, which had not only developed and was building Mercury but had also been an active partner in drawing up the new design. The company's president, James S. McDonnell, Jr., signed the contract on 22 December.[2]

The contract did spell out some major changes demanded by the broad goal of ending up with "a versatile general purpose spacecraft for the accomplishment of space missions of increasing complexity." There were, of course, more specific goals: 14 days in Earth orbit, controlled land landing, rendezvous and docking in orbit, and simplified countdown procedures. All this meant that the new spacecraft had to be larger to carry two men; include ejection seats; have an adapter section that stayed with the spacecraft in orbit to house stores and special equipment; carry systems that would allow it to be maneuvered and docked in orbit and to be controlled in flight and landing; and have its equipment packaged in modules, each independent of the others and located outside the cabin so they would be easy to reach while the spacecraft was being tested and readied for launch.[3]

Despite these changes, the two-man spacecraft was still viewed as an improved Mercury. The contract required McDonnell to outfit the new spacecraft chiefly with equipment that had already been developed so that in most instances expected changes were small. This permitted a much compressed schedule. McDonnell was to provide full-scale mockups of spacecraft and adapter within six months and of the target vehicle docking adapter (TDA) within ten. The TDA, though McDonnell-built, was to be mounted on the target; it carried the gear needed to connect spacecraft and target in orbit. McDonnell had 15 months to produce the first spacecraft, with others due every 60 days until 12 had been delivered. Because docking came later in the program, the contractor had 23 months for the first TDA.[4]

The new contract between NASA and McDonnell replaced the earlier contract that had authorized the company to procure long-lead-time items for extra Mercury capsules. Since it was a temporary device to cover expenses during the time it took to negotiate a final contract, the letter contract had a ceiling of $25 million. The final contract was expected by 20 April 1962.[5]

Although NASA could deal directly with McDonnell for spacecraft development, launch vehicles were another matter. Titan II and Atlas-Agena belonged to the Air Force, and the Air Force was clearly going to have to serve the new project in some role. Just what that role was to be, in fact, may have been the first question tackled after formal approval. On 7 December 1961, the same day that NASA Associate Administrator Robert Seamans approved the project, he and John Rubel, Assistant Secretary of Defense and Deputy Director of Defense Research and Engineering, issued a joint statement on "the division of effort between the NASA and the DOD in the development of space rendezvous and capabilities."

Seamans and Rubel agreed that the program belonged to NASA but that using the Air Force, in essence, as a NASA contractor could help the civilian agency achieve its goals and permit the Air Force (and other Defense elements) "to acquire useful design, development and operational experience." The Air Force, acting as contractor, would see that NASA got its Titan II launch vehicles and Atlas-Agena target vehicles. (As in the case of the spacecraft, the nature of the project precluded any choice of vehicles to be used.) The Department of Defense also intended to provide launch and recovery support for Mark II missions (the project had not yet been named Gemini) and to help NASA in choosing and training astronauts. Making "detailed arrangements ... directly between the NASA Office of Manned Space Flight and the Air Force and other DOD organizations" was the next step.[6]

This task was turned over to an ad hoc group that met for the first time on 13 December. Paul Purser, special assistant to MSC Director Robert Gilruth, headed the MSC contingent, and Colonel Keith G. Lindell led the Air Force team.* Group cooperation was so marked that a first draft of the plan was ready two days later.[7] It was passed around in both NASA and the Air Force, and two weeks were enough to put it in final form as the "NASA-DOD Operational and Management Plan" of 29 December 1961.[8]

The plan assigned launch vehicle development—Titan II and Atlas-Agena—to the Los Angeles-based Space Systems Division (SSD) of the Air Force Systems Command. The set-up was simple for Titan II: SSD would simply act as MSC contractor. Like NASA, SSD itself developed and built nothing. Its role was to manage the "associate industrial contractors" who actually provided the vehicles, with help from the non-profit Aerospace Corporation of El Segundo, California, in general systems engineering and technical direction.[9]

Arrangements for Atlas-Agena added another organizational layer, however, because NASA was already using the vehicle in its unmanned space flight programs and there was a working Agena Project Office at Marshall Space Flight Center. NASA's newly created Management Council for Manned Space Flight† simply decided to let the Marshall office take care of Atlas-Agena for the manned program as well.

---

*Representing NASA were Dave W. Lang, Sigurd A. Sjoberg, Charles F. Bingman, Warren North, and Colonel Daniel D. McKee; Air Force members were Lieutenant Colonel Robert R. Hull, Majors Edward H. Peterson, William E. Haynes, James E. Fasolas, and Earl W. Anderson, and civilians Herbert L. Repetti and John F. Bankert, Jr.

†D. Brainerd Holmes, Director of NASA's Office of Manned Space Flight, had established the council and called its first meeting on 21 December 1961. It met once a month to coordinate manned space flight activities and to help overcome the obstacles to communications inherent in the fact that neither Marshall nor MSC reported directly to Holmes' office. Holmes served as chairman. Its membership comprised the two top officials of Marshall (Wernher von Braun and Eberhard F. M. Rees) and MSC (Gilruth and Walter Williams) and Holmes' five principal subordinates: Charles Roadman (Director, Aerospace Medicine), Joseph F. Shea (Deputy Director, Systems Engineering), George Low (Director, Spacecraft and Flight Missions), Milton Rosen (Director, Launch Vehicles and Propulsion), and William E. Lilly (Director, Program Review and Resources Management).

MSC, in other words, had to order the vehicles from Marshall, which, in turn, procured them from SSD.[10]

MSC set the guidelines for launch vehicle development and had the last word in any technical dispute, but the day-to-day direction of the work belonged to SSD. MSC was to be allowed only limited contact with SSD's contractors, watching but not touching. If MSC saw something that needed to be done, it told SSD, which would pass the word on to the contractor.*

The "Operational and Management Plan" assigned two other major functions to the Department of Defense, with SSD acting as agent. One required SSD to oversee the modification of launch facilities at Cape Canaveral, Florida, to meet the needs of the new program. The other involved SSD in the support of program operations—launching, tracking, recovery—along the same lines already worked out for the Mercury program.[11]

On 26 January 1962, the plan was endorsed as a working arrangement between NASA's Office of Manned Space Flight and the Air Force Systems Command by the heads of the two agencies, Brainerd Holmes and General Bernard A. Schriever.[12] At the next step up the ladder, Seamans and Rubel were not so sure that everything had been taken care of. They had questions about the plan's provisions for Defense operational support and its failure to define in detail a pilot safety program, the astronaut selection and training process, and project scheduling and funding. These matters seemed less pressing, however, than getting on with the development of Titan II and Atlas-Agena. Seamans and Rubel decided to let the plan stand as an interim measure, until a better defined version could be worked out.[13] That took another six months and largely confirmed the arrangements already in force.[14]

Contracting for launch vehicles was in motion even while NASA and Air Force spokesmen were framing the Gemini Operational and Management Plan. NASA Headquarters juggled its fiscal year 1962 research and development funds to come up with $27 million, which it allotted to MSC for Titan II on 26 December 1961. As soon as notice came that funds were on hand, MSC wired SSD that work on the Titan II could start. SSD told the Martin Company's Baltimore Division to go ahead on 27 December.[15]

In the meantime, the MSC group that was to take charge of Gemini was writing a formal statement of work for Titan II. Ready on 3

---

*Scott H. Simpkinson, James A. Chamberlin's technical assistant, spent about a month as liaison at the Martin-Baltimore plant before turning these duties over to Harle L. Vogel, who served until the end of the Gemini program. A. B. Triche was the liaison with Lockheed at Sunnyvale throughout the program.

January 1962, it went to SSD with a formal request to buy 15 launch vehicles for Gemini. Although it could hardly have been a surprise, Titan II now appeared to require many more changes than had been allowed in the NASA-Air Force agreement only a month earlier. The terms of the memorandum that Seamans and Rubel had signed on 5 December 1961 explicitly limited changes to the fewest needed to adapt the missile to its spacecraft payload. But that was not going to be enough. To fit Titan II for Gemini would require new or modified systems to ensure the safety of the crew during countdown and launch. This included specifically a system to detect existing or impending malfunctions and signal them to the crew. MSC also expected changes in Titan II to enhance the probability of a successful mission, though what these were to be was not spelled out. The Air Force had Martin-Baltimore under letter contract by 19 January 1962.[16]

Putting Atlas-Agena under contract took longer, despite just as quick a start. The first steps had been taken before the Mark II project was approved. After its mid-November meeting with McDonnell,[17] the MSC rendezvous group had been able to define what would be required of Agena in greater detail and to check back with Lockheed Missiles & Space Company, its builder, about how these needs might be met. The MSC group outlined its views on Agena requirements in a note on 19 December 1961[18] and requested that Lockheed be asked to assess Agena's role in a rendezvous mission. Lockheed responded on 26 January 1962 with a report on Agena systems related to rendezvous—propulsion, communications and control, and guidance—and some informed guesses about further development that might be needed.[19]

By the end of January, MSC had evolved a fairly clear idea of the rendezvous techniques it planned for Gemini[20] and had prepared a statement of work for Atlas-Agena. This was forwarded to Marshall on 31 January, along with a request to buy 11 Atlas-Agenas. Atlas as launch vehicle for Agena was no problem, since it was already being used for just that purpose in other programs. But Agena needed a good many changes to adapt it to its rendezvous role—radar and other tracking aids, a restartable engine, better stabilization, more elaborate controls, and a docking unit were only the more important. Fortunately, time was not so pressing for Atlas-Agena as for the spacecraft and Titan II since it was not scheduled until later in the program. MSC wanted the first target vehicle delivered in 20 months, or about September 1963.[21] MSC did not pay its first installment to Marshall for buying Atlas-Agena until early March 1962, and another two weeks elapsed before SSD told Lockheed to go ahead with Gemini-Agena development.[22]

By March 1962, all major Gemini systems—spacecraft, booster, target, and paraglider—were under contract. This reflected the care and forethought that had gone into the project plan. It also mirrored the absence of any competition for major Gemini contracts. The project had been designed around an improved Mercury spacecraft, which made the company that built Mercury the only reasonable choice to receive the contract for Gemini. Of boosters powerful enough to lift the new spacecraft, only Titan could be ready in time for Gemini schedules. Atlas-Agena was the only likely target. And paraglider, the only major system to undergo the competition and elimination process and not really tied (on paper) into Gemini, had been under contract before the Mark II project was approved.

## RUNNING THE NEW PROJECT

Informal working arrangements and ad hoc groups had carried the Mark II project through its formative stages and handled the first steps in putting it under contract. But something more settled would be needed to oversee the future career of Gemini. By the end of December 1961, a Gemini Project Office was taking shape, though without official status as yet.[23] Its first report,* issued on 5 January 1962, was little more than an educated guess at potential problems in meeting Gemini launch schedules. Original launch dates were revised, with the first flight optimistically set for late July or early August 1963 (instead of May). One notable, but unremarked, change spaced the first, second, and third launches only six weeks apart—mid-September for the second, late October or early November for the third—while the remaining flights remained at two-month intervals. Since hard data for real analysis did not yet exist, the report did little more than point up the need for placing subcontracts promptly.[24]

Setting up the project office was only part of the complicated task of reorganizing the Manned Spacecraft Center and moving it from Virginia to Texas. On 15 January 1962, Director Gilruth announced the formation of separate Mercury, Apollo Spacecraft, and Gemini Project Offices.[25] The old Engineering Division was abolished, its staff divided between the new Gemini and Mercury offices. Chamberlin, former head of Engineering and prime mover of the Mark II project, took over as Manager of Project Gemini. Kenneth S. Kleinknecht, Gilruth's technical assistant, became head of the Mercury Project Office (then in the throes of trying to launch John Glenn into orbit aboard Mercury-Atlas 6, an event that took place on 20 February).[26] Chamberlin's deputies separated—William Bland remained with the ongoing Mercury program and André Meyer moved into Gemini with

---

*Compiled principally by Nicholas Jevas and William C. Muhly, scheduling specialists who had worked on the project development plan.

Chamberlin. Meyer recalled that he and Kleinknecht "split the Engineering Division in half. Just about as evenly as we could split it, put half the talent in one group and half the talent in the other group . . . just the two of us sitting across a desk and arguing—'No, I don't want this man.' 'We want this man.' "

Gemini came out of these sessions with a roster of 44, Mercury with one of 42.* The 18-person staff of MSC's liaison office at the McDonnell plant in St. Louis, headed by Wilbur H. Gray, was assigned to Gemini but served both projects. Meyer took over as chief of project administration in the new office with a staff of 10. The other members of the project office were temporarily grouped in spacecraft management, launch vehicles integration, and flight operations support.[27]

The first members of what was to become the Gemini Project Office (GPO) arrived in Houston during December 1961; the transfer was largely complete by February 1962. Gemini was among the first MSC elements to be resettled in Houston, once it was fully divorced from Mercury. Meyer's chief task during this period was to recruit, interview, and hire people to fill out the project office, specifically seeking experts with at least ten years' experience in each of the essential disciplines required to manage work on both spacecraft and launch vehicles. This was the central function of the project office: to plan, direct, and coordinate all aspects of the Gemini program and, more specifically, to see that Gemini contractors produced systems that allowed the program to meet its objectives. GPO enjoyed a degree of autonomy that permitted Chamberlin to deal directly with McDonnell and Air Force Space Systems Division. He reported only to MSC Director Gilruth, and that was chiefly a matter of keeping Gilruth informed on the status of the project.[28]

One of Chamberlin's first concerns was choosing his key staff members. He had Meyer, but for his other two chief lieutenants he turned to the Astronautics Division of General Dynamics Corporation, in San Diego. When interviews with Duncan R. Collins and Willis B. Mitchell, Jr., convinced Chamberlin that these were the men he needed, he got NASA Headquarters to approve his choice, a necessary step because both Mitchell and Collins were appointed at salaries above civil-service levels—so-called excepted positions. Collins became spacecraft systems manager and Mitchell launch vehicle systems manager. Mitchell also took over most of the personnel and functions of "flight operations support" when that branch of the project office quietly disappeared.[29]

---

*The division was actually 43 and 43; Walter J. Kapryan, in charge of engineering at Cape Canaveral, was transferred to Gemini on paper but was assigned full-time to Mercury until further notice.

When GPO officially settled in Houston in March 1962, the Manned Spacecraft Center was an organization without a home. Plans were under way for building a physical plant for the new center at the Clear Lake site south of Houston, but during most of its first two years MSC was housed in rented buildings (eventually a total of 13) scattered over much of the city and at Ellington Air Force Base, about halfway between Houston and Clear Lake. GPO, minus its manager, was installed in offices at the Houston Petroleum Center, a sprawling set of one-story buildings just off the Gulf Freeway. Chamberlin's desk was some distance away on the other side of the Freeway in the Farnsworth & Chambers building, which served as MSC's interim headquarters.[30] Such mundane matters as getting from one office to another, phoning a colleague, or even finding a desk complicated life but scarcely slowed the pace of the program.

Coordination meetings between GPO and its prime contractors were already beginning.[31] These meetings were Gemini's central management device. Chamberlin and Meyer set up six coordination panels, three for the spacecraft—mechanical systems, electrical systems, and flight operations—and one each for paraglider, Atlas-Agena, and Titan II. The panels provided a setting where design and engineering problems could be talked out and settled as they arose. They also helped to short-circuit such complex chains of command as might have slowed, for example, the target vehicle program, in which GPO had to deal with Marshall, the Air Force, and Lockheed—spokesmen for each sat on the panel and were able to resolve problems with far greater dispatch than might otherwise have been possible. Panel membership was not fixed, but shifted with items on the agenda for each meeting. But the essential experts were permanent, and outside help could be called in as needed.

Decisions reached at each panel meeting, usually once a week, were submitted to Chamberlin. They could be implemented only after he or Meyer had signed the minutes. This had the double advantage of letting those most familiar with the specific problems work out the technical details and, at the same time, keeping the project manager fully informed about what was going on. These coordination meetings remained the heart of the day-to-day decision-making process throughout Gemini's developmental phase. The number of panels grew as problems mounted and new areas needed closer attention. Later in the program, panels concerned mainly with development programs tended to give way to panels oriented more toward operations. At the same time, panels met less often, since there were fewer technical problems to reconcile as development faded into production and operation.[32]

GPO's function was to manage Project Gemini, not to build spacecraft or boosters. That was the task of the contractors who, early in 1962, were gearing up for their part.

## THE CONTRACTORS GET MOVING

Gemini management at McDonnell comprised six functional divi-sions corresponding, for the most part, to divisions within the company as a whole, each under a manager who reported to Walter Burke, company vice-president and general manager for spacecraft.*[33] The key position was that of the Gemini Engineering Manager. Robert N. Lindley, like Chamberlin, had found himself without a job when the Arrow project was canceled and had also moved from Canada to the United States. Unlike Chamberlin, however, Lindley found a place in industry.[34] As engineering manager for Gemini, his central responsibility was the design and development of the spacecraft. This included not only the work that McDonnell itself was to do but also the specification and technical management of the effort to be farmed out to subcontractors. Under Lindley were three project engineers: Raymond D. Hill, Jr., had charge of electrical and electronic design, Fred Sanders of mechanical design, and William Blatz of design integration and testing.[35]

The first engineering task was to define the spacecraft as a whole and each major subsystem to conform to the job required by the terms of the NASA contract. Since the basic form and function of the vehicle had already been decided by the time the contract was awarded, the definition phase centered chiefly on refining details and was largely complete by the end of March 1962. The products of this effort were SCDs for each major spacecraft system. The SCD, or Specification Control Drawing, was not the simple document its name implied. Often running to several hundred pages, it set out precisely what McDonnell expected the final system to look like and to do. After each SCD was discussed and cleared with NASA, McDonnell sent it out to potential subcontractors for bids. With minor exceptions, McDonnell developed and built only the spacecraft structural shell and electrical system. All other major spacecraft systems were developed under subcontract, with McDonnell acting as supervisor and integrator.

Like so much else in Gemini, subcontracting plans were well along before the project received formal sanction. McDonnell had convened

---

*Three of the six managers handled both Mercury and Gemini: William Dubusker for manufacturing, William D. Eckert for program administration, and John F. Yardley for launch operations. The other three worked only on Gemini: Robert F. Cortinovis for procurement, A.S. Torgerson for reliability and quality assurance, and Robert N. Lindley for engineering. A seventh manager reporting to Burke, Logan T. MacMillan, was assigned solely to Project Mercury.

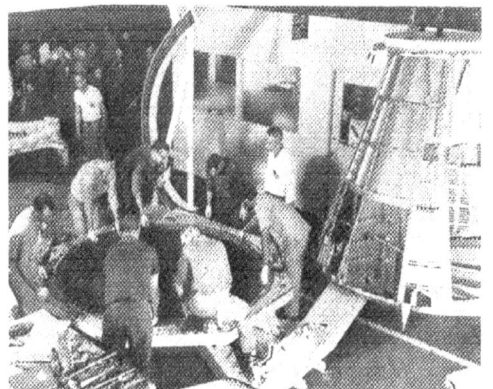

### Gemini Takes Shape in St. Louis

*Heatshield ready to join spacecraft shell.*

*Two men work inside the two-man spacecraft.*

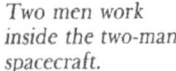

*Buildup of instrument panel.*

*Panel emplaced in the spacecraft.*

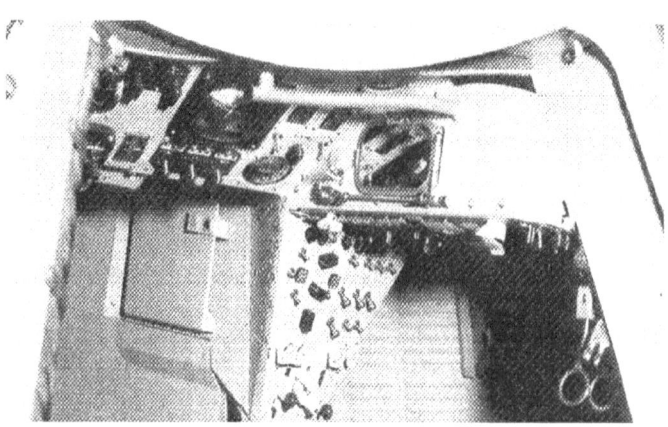

Spacecraft
assembly line.

Gemini's long
nose—rendezvous
and recovery unit.

Final assembly.

The rear end—
instrument units.

Final test.

a review board early in November 1961, at which procurement and engineering specialists began going over the spacecraft to decide which parts to buy.[36] Within a week after James McDonnell signed the contract with NASA, his company was able to present MSC with a list of the major items it planned to procure rather than make and to propose a set of bidders for each item.[37]

This was the prelude to a January 1962 meeting between Chamberlin, Burke, and Gray to reach an understanding on a standard procedure for securing NASA approval in the company's choice of subcontractors.[38] This could become a delicate matter, since a number of Gemini systems were to follow Mercury closely enough to suggest sole-source procurement—that is, asking only one company for a bid instead of seeking competitive proposals from several firms.

McDonnell awarded its first subcontract on a sole-source basis. It was for the development of the spacecraft environmental control system, which supplied the oxygen, regulated the temperature, and disposed of wastes for the crew. In broad terms, it was to be little more than two Mercury systems hooked together, so McDonnell simply selected the company that had developed the Mercury system, AiResearch Manufacturing Company of Los Angeles, California.[39] NASA agreed, and McDonnell told AiResearch to go ahead on 19 February 1962.[40]

McDonnell's second subcontract set the pattern for those systems that had no real Mercury counterpart. The Gemini spacecraft was going to have to maneuver in orbit to achieve rendezvous, and this meant that it had to carry a propulsion system (called OAMS for Orbit Attitude and Maneuvering System). Besides letting a pilot steer the spacecraft, the OAMS also held the ship steady in orbit and, at the start of the mission, provided the power to push the spacecraft away from the spent second stage of the launch vehicle and to insert the craft into orbit—or, in case of trouble, to abort the mission. The complete OAMS had 16 small engines, which burned hypergolic propellants fed under pressure from one fuel (monomethylhydrazine) and one oxidizer (nitrogen tetroxide) tank. All engines were mounted in fixed positions and were run at fixed levels of thrust. Eight of the OAMS engines were rated at 111 newtons (25 pounds of thrust) and fired in pairs, allowing the pilot to pitch, roll, and yaw the spacecraft and so control its attitude. The other eight engines were rated at 444 newtons (100 pounds of thrust); two were oriented to fire forward, two backward, and two to each side. This was the maneuvering part of the system. In July 1962, the rated thrust of the two forward-firing engines was reduced to 378 newtons (85 pounds).

A second spacecraft rocket system, the reentry control system, was functionally distinct from the OAMS but used the same kind of engines, so the same contractor would develop them. The reentry control system comprised two rings of eight 111-newton (25-pound) thrusters located forward of the crew cabin. Either of the rings alone could handle the job, but the function was crucial enough—holding the spacecraft attitude steady during its reentry into the atmosphere—to justify complete duplication.[41]

McDonnell decided that any of four companies might supply the OAMS and the reentry control system and asked each of them to submit a technical proposal. The prime contractor rated the bids and sent a survey team of engineering, quality control, and procurement personnel to grade each of the prospective subcontractors on resources and capabilities. North American Aviation's Rocketdyne Division in Canoga Park, California, won the highest combined rating. Although Rocketdyne's quoted cost was highest, it included an extensive test program unusually early in development, a feature that particularly impressed NASA, which made the choice. McDonnell told Rocketdyne to commence work on 26 February 1962.[42]

By the end of March, most of the major subcontractors had been instructed to proceed, and all had been selected by the end of May. The Air Force Space Systems Division, acting as NASA's contractor for Gemini launch vehicles, moved just as quickly. SSD set up a Gemini Launch Vehicle Directorate to manage booster development, naming Colonel Richard C. Dineen as director and Colonel Ralph C. Hoewing as deputy.*[43] General systems engineering and technical direction of development, with special stress on man-rating—making sure that Titan II was a safe and reliable booster for manned launches—was contracted to the Aerospace Corporation, which filled much the same role in Mercury for the Atlas booster. Aerospace set up its own Gemini launch vehicle program office under James A. Marsh.[44]

Gemini launch vehicle development was assigned to Martin's Baltimore plant, although the Titan II missile was developed and built in Denver. Baltimore got the nod chiefly to avoid any conflict between booster and missile work, although the decision did also help to sustain a facility that might otherwise have had to shut down.[45] Bastian Hello took over as Gemini Program Manager, reporting directly to Albert Hall, Martin vice-president and general manager of one of the three Martin divisions located in Baltimore.

Martin did not set up a Gemini project organization as such. Rather, each of the nine functional departments in Hall's division ap-

---

*The directorate had four branches: programs, under Major Roland D. Foley; engineering, under Lieutenant Colonel Alfred J. Gardner; safety and test, under Lieutenant Colonel Emmett J. Kelly; and procurement, under William Fried.

pointed a Gemini manager, who took charge of the program work in his area but remained in the normal departmental chain of command.* Hello also had the help of a program manager at Denver, where the booster's propellant tanks would be built since the tooling required was too costly to duplicate in Baltimore, and a Martin-Canaveral program manager responsible for launch facilities and operations.† Subcontracts played a much smaller part in the Martin than in the McDonnell scheme of things, largely because the booster differed much less from the missile than the Gemini spacecraft did from the Mercury capsule. For the most part, Martin could simply buy what it needed.[46]

Those systems that did need to be developed—engines, airborne guidance, ground computers—were not handled by Martin through subcontracts. Instead, they became the subjects of separate SSD direct contracts. The contract for propulsion systems went to Aerojet-General Corporation's Liquid Rocket Operations plant in Sacramento, California, in March. Two other major contracts followed later, one with General Electric in Syracuse, New York, to furnish the booster radio guidance system (the missile used inertial guidance), the other with the Burroughs Corporation of Paoli, Pennsylvania, to supply ground computers and implement launch vehicle guidance equations.[47]

The target vehicle for Gemini required even less in the way of special arrangements. Both Atlas and Agena were ongoing programs, already well established, and there seemed little need at the outset for anything more than fitting them to Gemini. The Agena Project Office at Marshall, headed by Friedrich Duerr, bought these vehicles for all NASA programs, and Gemini was simply another customer.†† For the target as for the booster, SSD acted as NASA's contractor. Atlas-Agena programs were managed by SSD's SLV-3 Directorate, commanded by Colonel F. E. Brandeberry. The Directorate's Program Integration Division, under Major John G. Albert, took care of NASA Agena programs.§[48] SSD authorized Lockheed to proceed with Gemini Agena development on 19 March 1962, and Lockheed assigned Herbert J. Ballard to manage the Gemini program.[49]

At the time NASA was arranging to buy Agena for Gemini, the model in use was Agena B. Agena B was essentially hand tailored for

---

*They were Edward D. Tarmey, Contracts; Lee J. Knight, Finance; George A. Biddle, Planning; Eddie Ball, Sales and Requirements; Jeremie U. LaFrance, Engineering; Martin Barrett, Materiel and Procurement; Francis O. Furman, Manufacturing; Haggai "Guy" Cohen, Quality; and Gordon T. Chambers, Logistic Support.

†Howard J. Jansen was the Denver manager; O. E. Tibbs had the Cape job.

††Duerr assigned George J. Detko as chief project engineer to monitor the target vehicle program in behalf of MSC.

§Captain Norbert J. Walecka became project engineer for Gemini Agena.

each of its missions, but the Air Force had decided to develop a more advanced Agena D, needing only to have the proper equipment modules installed to carry out any particular mission. On 10 May Brockway McMillan, Assistant Secretary of the Air Force for Research and Development, invited NASA to join in this program. This appealed to the engineers, but the managers hesitated for much the same reasons that had obtained in the case of Titan III. Agena D was a distinctly less ambitious effort than Titan III had been, however, and Duerr wired Albert on 11 June that Gemini would use Agena D.[50]

The Atlas for Gemini was also to be a standardized vehicle, the SLV-3. This improved version of the Atlas included many mechanical and electrical changes designed to make it more reliable, less troublesome. Its total engine thrust was upped by about 10 percent, mainly to offset the weight added by these changes.[51] On 23 July Seamans notified Rubel that NASA would support the SLV-3 program and planned to use the standard booster in all NASA actitivies that required an Atlas. For its projected role in Gemini, Atlas needed nothing that resembled development. The Air Force bought it from the Convair Division of General Dynamics Corporation right off the production line in its San Diego, California, plant.[52]

## THE PARAGLIDER CONTROVERSY

The one real exception to Gemini's smooth progress through its first half year was paraglider. Its development was a step ahead of the rest of Gemini, North American having been authorized to begin work on 20 November 1961, and the headstart may have accounted for the earlier signs of trouble.

Paraglider was controversial. Although GPO, and Chamberlin in particular, stoutly defended the concept, others in MSC had strong doubts. The Engineering and Development Directorate under Max Faget had been notably cool to the idea from the outset. The key question had been, and still was, "whether the deployment reliability of a single paraglider will equal that of a main and back-up chute system."[53] The long-time efforts of Langley's Francis Rogallo, inventor of the paraglider, to sell his concept had been repeatedly countered by the argument that parachutes had proved they could be relied upon to recover spacecraft. Instead of wasting time on an untried concept, Faget's group favored efforts to improve parachute technology to permit land landing. They advocated using a new form of parachute that could be steered, with landing rockets to cushion the final impact as the spacecraft touched down.[54]

Another source of opposition to paraglider was the Flight Operations Division under Christopher C. Kraft, Jr. Questions of reliability here took second place to concern for the operational problems posed

## Building Gemini Launch Vehicles

*Titan II first-stage engines, Aerojet-General, Sacramento.*

*Titan II systems in clean rooms, Martin-Baltimore.*

*Titan II assembly, Martin-Baltimore.*

Titan II
production line,
Martin-Denver.

Titan II
assembled.

Agena in alignment
checks,
Lockheed, Sunnyvale.

by paraglider in the Gemini program. For Kraft's division, using paraglider and using ejection seats were two sides of the same coin: one required the other, neither was reliable, and both promised immense practical obstacles to the safe return of the astronauts.[55] Kraft himself urged on Chamberlin, and later on MSC Director Gilruth, his objections to both systems.[56]

Paraglider critics found plenty of ammunition in North American's slow progress toward a working system. At first, paraglider development aimed at a landing system for manned spacecraft in general. Early in 1962, however, GPO decided that the program ought to be oriented explicity to Gemini. North American faced a large new effort and a major delay, and not just because the Gemini spacecraft was much larger than the generalized model first planned for. The half-scale free-flight test vehicle would have to be redesigned to carry a flight control system, just as the full-scale model did. North American had to join with McDonnell to design a compatible landing gear system and check it out in a test program. And, finally, North American now had to develop and qualify emergency parachute systems for both half-scale and full-size test vehicles.[57]

This last demand, in particular, delayed North American, and it was mid-March before a subcontract for the emergency parachute system could be placed.[58] Norbert F. Witte, North American's project manager for paraglider, planned to begin free-flight tests of the half-scale model toward the end of May. With its wing inflated and deployed before it left the ground, the test vehicle needed no emergency parachute. It would be towed into the air by a helicopter and released to fly under radio control. This series of tests would allow North American engineers to see how well the paraglider flew, how precise flight control could be, and whether the vehicle could flare—raise its nose to increase wing lift and drag and slow its rate of descent—just before landing.[59]

These were all questions that needed answers, but the most crucial was still whether or not the wing would deploy in flight. That had to wait for the emergency parachutes, since the test vehicles were too costly to risk without a backup system. Witte expected to have the half-scale emergency system tested by the start of June, when deployment tests could begin. The full-size emergency parachute would take longer but ought to be ready by mid-July. There still seemed to be a reasonable chance to complete this phase of the development program by September 1962.[60]

Timing was critical for paraglider development, since its place in the Gemini program depended upon its meeting the very tight launch schedule. Despite snags in the current phase of the program, Chamberlin decided that North American needed to get started on the next phase, a 14-month effort to design, build, and test an advanced two-man paraglider trainer, to start a flight simulation program, and to design and develop a fully man-rated prototype Gemini paraglider landing system.[61] That was in March 1962; by May the task was scaled down to require only the design of the prototype system, rather than its complete development. This was expected to reduce the time to five months from the date of the contract award.[62]

The project office still expected the paraglider to be ready on time, but warned in a 4 May schedule analysis that the program "will require close monitoring to prevent slippage." Paraglider was scheduled to be installed in the second Gemini spacecraft, which would be the first to carry a crew. The first spacecraft, since it was unmanned, was slated to come down by parachute. A prudent response to delays already incurred dictated that plans be laid for using a parachute system in the second spacecraft as well. By mid-June, GPO conceded that the paraglider would not be ready until the third flight.[63]

## A QUICK SMOOTH START

Despite some doubts about the paraglider, Project Gemini was moving smoothly in the spring of 1962. GPO noted a certain tightness in launch vehicle schedules that might constrict the time needed to resolve any unexpected problems but concluded that close monitoring would help to bring the modified Titan II out on time. Late delivery of some components from McDonnell subcontractors threatened schedules for building the first two spacecraft, but the threat seemed modest. The target vehicle and its booster, Atlas-Agena, appeared to present no problems, even after a slow start, since a target was not needed until the fifth mission.[64]

Overall, August 1963 still seemed like a reasonable prospect for the first launch. But the ambitious timing of the second launch (the first manned flight in Gemini, earlier scheduled just six weeks after the first),[65] was now adjusted to allow a more realistic three months and set for November 1963. The rest of the program held to an every-other-month schedule, the 12th and final flight to be in July 1965.[66] From the viewpoint of the project office as it surveyed Gemini progress and prospects in its first half-year, there were no serious problems.[67]

Project Gemini had won approval in late 1961 over several competing rendezvous development proposals because its design was further along than those of its competitors and because its scope seemed to be limited enough to fit the relatively compressed span of time between the last flights in Mercury and the first mission in Apollo. That these reasons were valid appeared amply borne out by the rapid placement of contracts during the first months of the project's official exist-ence. Within a matter of six months, most major contracts had been awarded and a firm organizational framework had been established.

Even Congress appeared unperturbed that NASA had embarked on a large new project with scarcely any advance warning to those expected to furnish the money for it. In doing so, NASA had not ex-ceeded its authority. Although obliged to lay out its spending plans during budgetary hearings, NASA at that time received a single appro-priation for research and development and was largely free to distrib-ute the money as it saw fit. The $75 million in fiscal year 1962 funds needed to get Gemini started were provided simply by shifting money from one account to another inside NASA.[68]

In hearings early in 1962 on the upcoming fiscal year 1963 budg-et, NASA spokesmen felt no need to apologize for the new project. Quite the contrary: from Administrator James E. Webb on down, they described it in glowing terms, stressing its role in the development of rendezvous techniques and in extending the length of man's stay in space—but all within the context of a merely enlarged (or advanced) Mercury. This was, of course, a fair picture of the thinking that lay behind Project Gemini, and none of the listening congressmen chal-lenged it.[69]

Chamberlin summed up the optimism that pervaded Gemini dur-ing its first half year in his monthly report on project office activities as of 28 May 1962. He saw no problems that might imply delays for the program, although "all elements of the schedule are extremely tight." There were no technical problems that contractors and project office could not handle. "As technical problems arise they are being assigned to capable organizations for solution with close project office monitor-ing to assure progress. No technical problems are particularly out-standing at this time."[70]

Despite its complexity, Project Gemini was meeting only success. The project office remained silent about any doubts it may have had that Gemini's objectives could be achieved on time.

# V

# Expansion and Crisis

AS summer gave way to fall in 1962, the smooth progress that Project Gemini had enjoyed during its first half year roughened. Concern mounted over the steady expansion and rising costs of the project as a whole. Hopes for using much of Mercury's technology in Gemini eroded. One system after another became the subject of full-scale development, rather than modification or simple transfer from Mercury. The scope of launch vehicle development likewise grew far beyond first expectations. Costs kept climbing until they collided with an unexpectedly restricted budget toward the year's end.

These concerns were virtually unknown outside NASA. But the striking dual mission launched by the Soviet Union in August led some to wonder if the United States had any hope of flying the first rendezvous mission. *Vostok III,* piloted by Major A. G. Nikolayev, lifted off on 11 August, followed a day later by Lieutenant Colonel P.R. Popovich in *Vostok IV.* The two spacecraft came close enough to each other to spur some talk of rendezvous. With no means of maneuvering their spacecraft, however, the two cosmonauts could not match orbits or speeds. The Soviet Union had shown only that it could launch two spacecraft in quick succession, so there was still hope for the maneuverable Gemini spacecraft to achieve the first rendezvous, if it survived its troubles.[1]

## CHANGING PLANS AND RISING COSTS

Preliminary cost estimates from Gemini contractors began reaching the Gemini Project Office in March 1962. These rough figures pointed toward a large but not yet clearly defined increase in the projected total cost of the program. Air Force Space Systems Division (SSD) discussed finances with the project office at the first launch vehicle coordination meeting on 1 March and furnished its first budget estimate for the program at a meeting in Houston later that month. Boosters now appeared likely to cost Project Gemini a good deal more than had been supposed. The development plan of December 1961 had assumed $113 million for modified Titan II launch vehicles. But the March 1962 figure was half again as much—something over $164 million.[2]

The statement of work for Titan II that SSD had received early in January called for more than the limited modifications first proposed. It required a malfunction detection system and other unspecified changes to improve the missile. Making sure that Titan II could safely launch manned spacecraft—referred to as man-rating—was crucial, and it was going to cost money. A revised statement of work in mid-May 1962 spoke of "an adaptation of the Titan II ICBM," rather than "a development of the present Titan II ICBM," and spelled out the changes required in greater detail. They included not only a fully redundant malfunction detection system but also a backup flight control system; an electrical system with backup circuits for guidance, engine shutdown, and staging; inertial instead of radio guidance; and a new launch tracking system.[3]

The target vehicle likewise soon seemed to demand more than had first been expected. Even though Agena work was moving at a slower pace, by May the $88 million programmed for Atlas-Agena development in the December 1961 plan had climbed above $106 million.[4]

The project development plan had the Gemini spacecraft costing $240.5 million. This figure, like those for launch and target vehicles, could not have been more than an educated guess, with a natural bias toward guessing on the low side to make the program more palatable. But McDonnell's first formal cost proposal for the Gemini spacecraft still came as something of a shock. The first step in negotiations between the project office and McDonnell to convert the letter contract of December 1961 into a definitive contract was a series of technical meetings in Houston between 19 April and 24 May 1962, to make sure that both sides agreed on plans and specifications.[5] McDonnell's "Gemini Spacecraft Cost and Delivery Proposal," prepared for these meetings, raised the spacecraft ante to $391.6 million.[6]

This new and higher estimate was based in part on McDonnell's more careful study of the cost of what the contract called for, in part on its enlarged view of what the program ought to include. The letter contract, for example, had mentioned the need for flight simulators and trainers as well as test spacecraft but included no specifics. A new feature of engineering development for Gemini was to be the use of a number of test articles—spacecraft built for early static and dynamic testing—for want of which Mercury had sometimes been delayed. GPO admitted that building them might slow spacecraft construction at first but believed that the data they provided would more than make up for the temporary setback.[7] McDonnell proposed four boilerplate spacecraft (metal models designed to be used chiefly in escape and recovery system testing) and four static articles (non-flying spacecraft to be used in structural tests). McDonnell also proposed to add to Gemini a test program that it had worked out in Mercury. Known as "Project Orbit," this entailed building an extra spacecraft and target docking adapter for an extended series of laboratory-simulated orbital missions "to investigate potential problems and to evaluate engineering changes generated during the life of the program."[8]

A major part of crew training for Gemini depended on simulating in great detail every aspect of a mission, to expose the astronauts before they left the ground to anything they might meet during a flight. The basic device was a flight trainer, a precise duplicate of the real spacecraft, in which crews could fly a complete simulated mission from launch through touchdown, seeing through its windows what they would see in flight, hearing the noises—even feeling the vibrations— they could expect. There were to be two flight trainers, one in Houston and the other at Cape Canaveral, each hooked up to mission control and remote displays to form a complete mission simulator.

Three aspects of a mission were outside the scope of the flight trainers. One involved the forces imposed upon the astronauts by high acceleration during launch and by rapid deceleration during reentry. These stresses could be matched on a man-carrying centrifuge. The project office planned to use the one at the Naval Air Development Center in Johnsville, Pennsylvania, its gondola fitted out with a mockup of the inside of the spacecraft. Maneuvering in orbit to rendezvous was the second aspect. This was to be simulated by a translation and docking trainer, in which the crews would practice techniques of rendezvous and docking.[9] The third, extended weightlessness, was then beyond human ingenuity to imitate.

Training equipment and test articles together, increased in detail and enlarged in scope, came to just under $39 million in McDonnell's cost proposal. McDonnell also needed money to cover its roles in mission planning and launch operations support and for spare parts and checkout gear, to name only some of the more costly items. And all this aside from the expense of developing and building the spacecraft ($242.7 million), which alone exceeded the December budget ($240.5 million).[10] Even at that, McDonnell's estimate was still little more than guesswork. Few of the company's subcontractors had yet provided any hard financial data. The chiefs of procurement and financial management at MSC jointly deplored both the size of the McDonnell estimate and the lack of data on which it was based, a viewpoint that echoed Paul Purser's marginal note on SSD's interim financial plan for boosters in April 1962: "This is still up in the air. Attempts are being made to bring down these costs."[11]

On 12 May 1962, in a review of Project Gemini for NASA Administrator James Webb, the Office of Manned Space Flight revealed for the first time the pattern of rising costs that was beginning to mark the program. Since the project development plan was issued, little more than five months earlier, Gemini's expected cost had climbed from $529.5 million to $744.3 million.[12] Given the shaky data on which the new total depended, it could not be the last word. The program kept growing and technical problems began to appear, not all of them in areas where they had been expected.

### SOME FORESEEABLE PROBLEMS AND A SURPRISE

As Project Gemini moved from design into testing during the spring and summer of 1962, problems multiplied, although not (with one exception) beyond what might be seen as the normal headaches of a large-scale research and development project. Those areas that demanded the longest step beyond current practice were those where trouble threatened. The paraglider program, with its early start, began running into marked delays in planning and design before the rest of Project Gemini. When actual testing began in May 1962, only two contract months remained to settle on the best design for a paraglider landing system.

The first task was qualifying an emergency parachute recovery system for the half-scale vehicle. North American began on 24 May with a successful drop test at the Naval Parachute Facility in El Centro, California, near the Mexican border. Two failures followed before a second success, on 20 June. What should have been the final drop to complete qualification failed on 26 June, when the vehicle's electrical system shortcircuited. North American shuttled the vehicle 260 kilometers back to its plant in Downey for a closer.look, which revealed a design flaw. The company reworked the test vehicle and returned it to El Centro for another try, on 10 July, with no better luck. This time the drogue designed to pull out the main parachute failed to do so. After another round trip to Downey for changes, everything worked on 4 September. GPO agreed with North American that the half-scale emergency landing system was now qualified. But two and a half months had been lost.

The full-scale emergency system proved even harder to qualify. First came design problems, then the parachutes were late in arriving.

North American could not ship the test capsule to El Centro until 20 July. The Air Force's 6511th Test Group, which ran the El Centro test range, demanded a special test to be certain the vehicle's pyrotechnic devices were safe—that delayed the first qualification flight until 2 August. It was a success, but more delays followed—first bad weather, then the lack of a launch aircraft. The second drop, on 21 August, was marred by one of the three main parachutes breaking loose. Damage was only minor, as it was in the next test, on 7 September, when two parachutes failed. Efforts to correct this problem took over two months. On 15 November, some four months after the full-scale emergency recovery system was supposed to have been qualified, the fourth drop was a disaster. When all three parachutes failed, the test vehicle was destroyed as it hit the ground. Clearly the system could not be relied upon. GPO directed McDonnell to furnish North American with a boilerplate spacecraft for further tests at some later date.[13]

These problems, however disheartening, should not have cast any shadow on the concept of a paraglider. The emergency parachute systems were intended only to back up testing; they were not part of the Gemini landing system. Yet the pattern of delays, errors, and malfunctions that marked North American's efforts to qualify the emergency system proved to be symptomatic of a lingering malaise. Paraglider advocates knew that the program would be made or broken, so far as Gemini was concerned, by the success or failure of flight testing, and time was limited. North American had been chosen over Ryan and Goodyear because of its first-rate job in testing the design during the summer of 1961.[14] But on 28 November, scarcely a week after North American received word to go ahead with paraglider development, NASA notified the company that it had been selected as prime contractor for the Apollo spacecraft. The impact on paraglider was catastrophic. North American froze the number of engineers assigned to paraglider, then allowed even that group to decline. The quality of work suffered as well, becoming, in the opinion of one NASA engineer assigned to the program, "abysmal."[15]

The pattern of trouble sketched in emergency system testing persisted when North American began testing the paraglider itself by flying half-scale models with wings inflated and deployed before they left the ground. Scheduled to begin in May 1962, these trials got under way in mid-August at Edwards Air Force Base, 100 kilometers north of Downey. North American's first try, on 14 August, got nowhere. Because a plug pulled loose inside the capsule, the wing, which was tied down for takeoff, failed to release after a helicopter had towed it to the proper height. The wing released too soon in the second try, three days later, although the capsule did go briefly into a stable glide. North American also achieved a stable glide in the third flight, on 23 August, but an erroneous radio command caused the vehicle to come down too fast and suffer some damage in landing. The fourth flight was postponed twice, each time because someone forgot to charge the battery. Towed aloft on 17 September, the vehicle failed to release on command, voiding the test. Twice in a row, short circuits forced the contractor to call off the fifth flight test, the second time on 21 September.[16]

That same day, James Chamberlin, MSC Gemini Project Manager, ordered North American to halt flight tests of the half-scale paraglider. He expressed "growing concern" over "the repeated unsuccessful attempts of S&ID [North American's Space and Information Systems Division] to conduct satisfactory predeployed half-scale paraglider tests." Flights were not to resume until the contractor had reorganized its paraglider project and could spell out just what it intended to do about the test vehicle's electronics and pyrotechnics and the company's own checkout and inspection procedures.[17]

North American had already made some moves along the lines Chamberlin demanded. The paraglider effort was raised to the status of a major program, and George W. Jeffs was named Paraglider Program Manager on 1 September 1962. Norbert Witte, the former project manager, stayed on as Jeffs' assistant.[18] Jeffs was something of a corporate troubleshooter, and he had the respect of the NASA engineers working on paraglider.[19] This augured well for the future, but, in the meantime, a fully successful flight test had yet to be performed.

North American reworked the half-scale vehicle in its plant, then shipped it back to Edwards Air Force Base on 15 October for another try. A bad ground transmitter stalled matters for a while but, on 23 October, the fifth test flight was a complete success.[20] Even with all its problems, the series of tests had met its main goal, showing that the paraglider was stable in free flight.[21] But predeployed flight testing ended more than two months late, and the crucial deployment flight tests—spreading the paraglider wing in flight—had not even begun.

In the meantime, other problems were beginning to compete for the attention of the overworked project office. Like the paraglider, ejection seats had been a controversial innovation in manned spacecraft, and their development problems also gave critics an early opening. The reasons were much the same. Both systems were a long step beyond current practice, both presented test problems not clearly related to their final roles, and both were subject to changing requirements that imposed makeshift adjustments, further complicating matters.

Although ejection seats were widely used in military aircraft, they were designed to give pilots a chance to survive, not to guarantee that survival. Manned spacecraft levied more stringent demands. Most critical was the "off-the-pad abort mode." Before liftoff, the spacecraft perched some 45 meters from the ground atop a shell filled with potentially explosive chemicals, the Titan II launch vehicle. However rigorous the precautions, there was always the danger of some mischance setting it off. For a length of time that might stretch into hours before they were airborne, the crew would be aboard with no recourse, should that mishap occur, save their ejection seats. The Gemini seat had to be able to propel itself from a starting point 45 meters in the air in a trajectory stable enough to get clear of an exploding booster and high enough to allow parachutes to open. No existing seat could do that, and developing one that could was the crux of the Gemini effort.[22]

McDonnell chose Rocket Power, Inc., of Mesa, Arizona, to supply the rocket catapult (or rocat) for the Gemini escape system.[23] For the seat itself, McDonnell turned to Weber Aircraft, of Burbank, California.[24] As luck would have it, the Naval Ordnance Test Station at China Lake in the middle of California's Mojave Desert had earlier constructed a 45-meter tower for Sidewinder missile tests. This tower was admirably suited for simulated off-the-pad ejection (or, acronymically, Sope) tests.[25] Kenneth F. Hecht, who left the ordnance test station in January 1962 to take charge of Gemini escape and recovery systems, set up a special working group to oversee seat development and qualification.* He was convinced, and in this he was seconded by those who knew most about ejection seats, that the key problem was finding ways to control the relationship between the rocat's line of thrust and the shifting center of gravity of the seat-man combination while the rocket was burning. Without this control, a trajectory of the proper height and stability could not be achieved. This was one of the reasons why Hecht insisted the tests be conducted with a dummy in the seat, rather than with a solid mass. He also knew that haste was vital, since the seat design could not be settled until the answers were in.[26]

The first Sope test came off on schedule 2 July 1962, followed by four more over the next month. All produced their share of problems and mechanical failures, each dealt with as quickly as possible to get on with the next test. None of these mechanical problems much bothered Hecht and his colleagues, because they had their eyes on the dynamic problem of rocket thrust and center of gravity. They were concerned with ejection at this point, not the complete escape sequence through recovery, and thought they were close to solving that key problem.[27] From this viewpoint, the first five tests were a success. But if the goal were seen as a complete system with all parts working as they should in the final version, the tests left much to be desired. The seat seemed to be turning into a maze of makeshift fixes, and the personnel recovery parachute system (the crewman's landing device) had failed twice.

*Hecht's group included Edward A. Armstrong, Louis A. Bernardi, Frederick T. Burns, Paul R. Penrod, Hilary A. Ray, and Stanley White.

For simulating aborts from the spacecraft on the launch pad, ejection seats were tested from the 45-meter tower (above) at the Naval Ordance Test Station, China Lake, California. Aborts in flight were simulated on the rocket-propelled test sled at China Lake. Below left, a dummy is hoisted into the Gemini spacecraft mockup mounted on the sled. Below right, three high-speed photos (reading up) show ejection, seat propulsion, and descent by parachute.

At an extended meeting in Houston on 6 and 7 August, the total system viewpoint prevailed. Sope testing was halted until a complete design of the whole system was ready and the personnel parachute had been fully tested.[28]

A month elapsed before McDonnell was able to report on 6 September that seat design and testing were complete, clearing the way for a new round of Sope trials. Tests on 12 and 26 September went well but highlighted a set of problems with the rocket motor. Some were functional and some structural, but all affected, however slightly, the direction of thrust and so made accurate control impossible. Testing stopped again, pending the availability of the rocat in its final form.[29] This delay was much prolonged, lasting well into 1963.

Other major Gemini systems seemed less troublesome. Through the summer and early fall of 1962, such problems as appeared could be, and were, regarded as nothing more than the routine hurdles in a large program. One possible exception was the fuel cell, which, like paraglider and ejection seats, was new to manned spacecraft and had aroused some debate, at least in its General Electric version.

The basic source of electrical power in the spacecraft was to be batteries. The weight of ordinary batteries, however, became prohibitive as missions increased in length. Something more was needed, and the choice was fuel cells. That choice was resolved in January 1962. After analyzing the merits and defects of competing approaches, Robert Cohen of MSC strongly recommended the General Electric fuel cell as lighter, simpler, and more generally suited to Gemini needs than other designs he had investigated.[30]

In a fuel cell, hydrogen and oxygen react to produce water and heat. The unique feature of the General Electric design was its use of a solid ion-exchanging membrane in which electrolyte and water were chemically bound; most other cells diffused gases into a liquid electrolyte. A separate stream of coolant condensed the water produced at the cell, then removed it through a series of wicks to keep the reaction going at a constant rate. This used little of the cell's own power, in contrast to the gas-diffusion cells that required a complex self-powered process of flushing with hydrogen, condensation, and centrifuging to remove the water produced. General Electric had devoted intense research to the design since 1959 and had already set up a fuel-cell facility, the Direct Energy Conversion Operation in West Lynn, Massachusetts.[31] McDonnell shared Cohen's view and formally recommended General Electric for a subcontract, to which NASA agreed.[32]

Nonetheless, in early 1962 the General Electric fuel cell was still no more than a laboratory device, however promising.[33] NASA Headquarters was looking into fuel cells for Apollo, which raised some questions about Gemini's choice of General Electric. The Office of Manned Space Flight's survey of General Electric alleged that the company was understaffed, slow in getting started, and unlikely to meet Gemini schedules—all this in addition to what seemed to be an untested and questionable design concept.[34] Cohen responded to these charges for GPO. He saw no reason to doubt that General Electric would meet its commitments: the company was adding to its staff and improving its effort, which had only begun with an order from McDonnell two and a half weeks earlier. More important, the much tested General Electric design was at least as far along as any other and was inherently simpler to boot.[35] That settled the issue.

As development got under way, General Electric began to run into problems that seemed to suggest that theory had outpaced practice. The most serious in mid-1962 was how to achieve a satisfactory bond between cell membrane and frame. Solving these problems appeared more likely to tighten the schedules than to threaten the program as a whole. In any case, the worst appeared to be over by the end of August.[36]

During the last half of 1962, the paraglider's troubles probably posed the greatest threat to an approved Gemini objective, that of land landing, although ejection seats and, to a lesser extent, fuel cells were also worrisome. The paraglider was a major new system that demanded a large-scale effort. Ejection seats and fuel cells, though not so novel, were still major innovations in manned space flight. In all three cases, the novelty of the application and the advance beyond current practice imposed a greater development effort than required for other Gemini systems. Given that fact, the problems should have come as no surprise. A quite unexpected source of trouble loomed in another quarter. The suitability of Titan II as a launch vehicle for manned space flight came into question.

Responsibility for developing the Titan II missile belonged to the Ballistic Systems Division (BSD), like SSD a part of Air Force Systems Command. Titan II research and development test flights began on 16 March 1962, with a launch from the Atlantic Missile Range in Florida. In its first flight, Titan II displayed a disquieting characteristic. A minute and a half after it lifted off, while the first-stage engine was still firing, the missile began to vibrate lengthwise like an accordian about 11 times a second for roughly 30 seconds. This was not likely to bother a missile too much, but it implied real trouble for a launch vehicle with a manned payload. The steady acceleration of a booster like Titan II pressed a crewman to his couch with about two and a half times the force of gravity at that point in a normal flight. Bouncing at an extra two and a half gravities ($\pm$ 2.5g) could badly hamper a pilot's efforts to respond to an emergency, a matter of special concern in Gemini since the crew played so large a role in flying the spacecraft.[37]

Titan II's longitudinal oscillations quickly acquired the nickname "pogo stick," soon simply Pogo. Its cause remained unclear, how to get rid of it a matter of guesswork. By July, Pogo was becoming a regular topic at MSC's weekly senior staff meetings, and BSD had formed a special Committee for Investigation of Missile Oscillations.* [38] The problem turned out to be surprisingly easy to solve for the missile: higher pressure in the first-stage fuel tank cut Pogo in half during the fourth test flight, on 25 July, although nobody was quite sure why.[39]

There were some ideas, however. Martin engineers thought the culprit might be oscillating pressure in propellant feedlines, analogous to the chugging of water in pipes, or "water hammer." This suggested the use of something like the surge tanks familiar as devices to stabilize pressure in the flow lines of hydroelectric plants and pumping stations. Martin proposed to install a surge-suppression standpipe in the oxidizer line of a later Titan II. MSC endorsed the plan, and BSD agreed. By the end of August, GPO was cautiously optimistic. The lowered Pogo level of $\pm1.25g$ achieved in the fourth Titan II test flight was still too high for manned space flight, but the water hammer analogy at least suggested an answer.[40]

GPO was also watching another problem. In two of its first four test flights, Titan II's second-stage engine failed to reach full thrust. The causes appeared to be different in each case and unrelated to one another. Just how serious this might be could not be foreseen. Much depended upon whether or not it recurred, and GPO adopted a wait-and-see stance.[41]

Project Gemini's technical problems in the summer and fall of 1962 might have aroused more concern if a far more serious threat had not intruded. The financial structure of the program began to totter. Two circumstances combined to produce a major crisis. On one hand, Gemini contractors were spending money at a much faster rate than the project office had expected. On the other, Congress was slow to approve NASA's appropriation for fiscal year 1963, which restricted the funds available to Gemini. However serious development problems might be, or become, they could always be resolved if there were enough money. But now the question was how to spread limited funds over an ever more costly program.

## THE BUDGET CRISIS

The pattern of program growth and cost increase revealed during the spring of 1962 presisted, and with the same shortage of dependable information. To NASA's repeated pleas for more funding data, McDonnell regularly denied that any existed. In mid-July 1962, three

*Chairman of the special committee was Abner Rasumoff of Space Technology Laboratories.

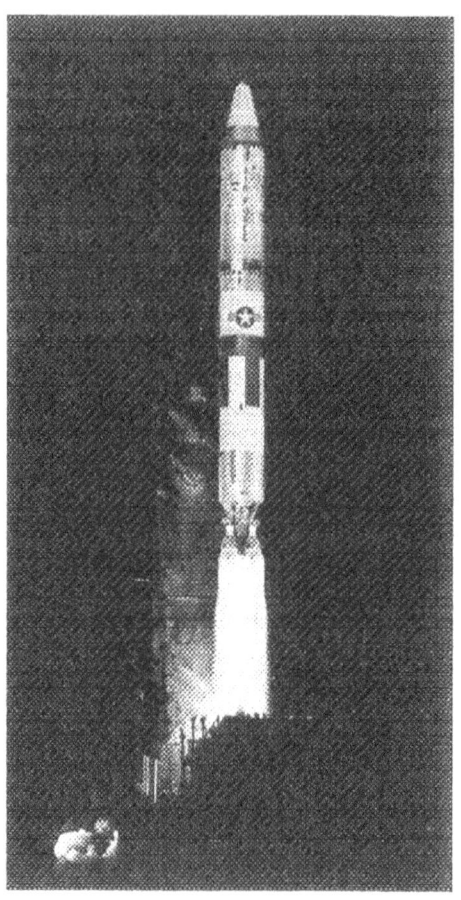

*Titan II missile N-15, launched from Cape Canaveral in January 1963, was the second Titan to show a substantial reduction in longitudinal oscillations after pressures were increased in the propellant tanks.*

months after its first budget proposal, the company could still not provide a detailed forecast of program costs because "cost projections from suppliers and subcontractors are currently unavailable as purchase order values continue to change and negotiated costs have not been established."[42] In August, when MSC and McDonnell began working out the final terms of the spacecraft contract, the contractor proposed a startling total of $498.8 million, double NASA's first estimate in December 1961 and more than $100 million higher than the company's own April 1962 proposal.[43] Hard negotiation brought the new figure down to $464.1 million,[44] but efforts to agree on a final price were suspended before the end of August because the whole Gemini program was in trouble.

Other costs were also on the upswing during the summer and early fall of 1962, though not as spectacularly as those for the spacecraft. SSD's March 1962 figure of $164 million for the launch vehicles topped $170 million by September.[45] Less than a month later, SSD submitted to NASA a formal revised budget of $172.61 million.[46] Word reached MSC in July that the Atlas-Agena for Gemini now had a price tag of $12.3 million over its earlier total,[47] and this despite the fact that NASA had deleted the three spares to cut the number of Atlas-Agenas on order from 11 to 8.[48] A special briefing for NASA Administrator Webb on 28 September revealed that Project Gemini might cost as much as $925 million before it was over, 25 percent higher than Webb had been told in May it was going to cost and 75 percent more than MSC's first estimate.[49]

Such fast-rising costs would have been bad enough at any time. Now they presaged disaster, since Congress had not yet acted on NASA's appropriation for fiscal year 1963 (which began on 1 July 1962). Without an approved money bill, NASA was compelled to carry on under a joint congressional resolution that provided enough money to support projects at roughly the same level they had enjoyed the year before but not enough to cover increases.[50] Gemini's status was all the more threatened because it had not even appeared in the 1962 budget. NASA had found enough money to get Gemini started, but that was a makeshift that could not support an ongoing program.

The bill that authorized NASA's funds was signed into law on 14 August, but the bill to appropriate that money was yet to come. Congressional action on NASA's 1963 appropriation was not completed until 25 September. The figure was $3 774 115 000, $113 161 000 less than NASA had asked for and $70 000 000 under the total authorized in August.[51]

This delay prevented the Office of Manned Space Flight in Washington from giving MSC the normal authority to spend money on the basis of the full year's budget. Instead, that authority was being granted on a month-to-month basis.[52] Monthly funding brought anguished complaints from contractors, as expenses constantly threatened to outstrip the resources available to pay for them. By October, MSC was being bombarded with telegrams, each with urgent demands for full and quicker funding.[53]

Lack of an appropriation also prevented NASA from adopting a final financial operating plan (FOP) for fiscal year 1963. Each center prepared an annual FOP to be approved by NASA Headquarters for alloting funds at the start of the fiscal year.[54] To meet the impending crisis, Associate Administrator Seamans imposed a ceiling of $1.51 billion on NASA research and development expenditures for the coming year. By this time, however, estimated funding needs for this purpose had already exceeded the figure first presented to Congress and now stood at $1.91 billion. Manned space flight chief Brainerd Holmes warned Seamans that current schedules could only be met by a supplemental appropriation from Congress.

In the meantime, Holmes directed MSC to prepare two separate fiscal-year 1963 FOPs: one staying within the Seamans-imposed ceiling, the other geared to actual needs. For Gemini, this meant a limit of $234.1 million against a needed $299 million. Holmes predicted a severe setback to program schedules if the smaller budget prevailed: a three-month delay in the first launch and in the first long-duration flight, an extra ten-month wait for the initial rendezvous mission, and no paraglider before the third flight.[55]

Hopes for meeting the higher budget were dashed when President Kennedy rejected NASA's case for extra funding. Holmes notified MSC on 9 October that its funds for fiscal year 1963 would be limited to $660.1 million. He directed the center to prepare new schedules to reflect this limit, voicing the somewhat forlorn hope that the unavoidable delay of several months might be made good if "later developments make it possible for the Administrator to obtain a FY 63 supplemental."[56]

The new ceiling was $27 million less than MSC had planned for under the earlier Seamans ceiling. The situation was now critical. Already tight at the level of $687 million, a budget of $660 million was nearly crippling. And Project Gemini bore the full brunt. Upon first hearing of the newly reduced budget, MSC planned to split the $27-million cut between Gemini and Apollo. Washington, however, ordered Gemini to take all the losses. Wesley L. Hjornevik, MSC's Assistant Director for Administration, evaluating the situation for the senior staff on 19 October, saw no way out of this dilemma except to curtail Gemini sharply. "It appears," he glumly remarked, "that the consequent reduction to Gemini can only come by dropping paraglider, Agena, and all rendezvous equipment."

Further complicating matters was the rate at which Gemini was piling up costs, a rate much higher than expected. Hjornevik pointed out that the program seemed to be costing $15 million a month, rather than the planned $11 million.[57] A budget memorandum that reviewed Gemini funding during the first quarter of fiscal year 1963 described as "an area of growing concern and one which can no longer be left unattended" the speed at which costs for spacecraft, paraglider, launch vehicle, and target vehicle were growing. The FOP could not "support acceleration of cost rates so projected by these contractors. Unless appropriate direction is given to the contractors to restrict this buildup or a Gemini reprogramming action is effected immediately then funding difficulties will commence during the second quarter."[58]

## REPROGRAMMING GEMINI

The project office had already moved to reprogram Gemini, to alter the course of the program and compel the contractors to conform to the newly limited budget. Reprogramming was much more drastic in some areas than in others. Paraglider escaped almost untouched. McDonnell's spacecraft effort took some trimming but remained much what it had been. The launch and target vehicle programs, the Air Force portion of Gemini, endured the most far-reaching changes. Plans for testing the Gemini launch vehicle were sharply cut back. Target vehicle testing was even more drastically curtailed; for some months, in fact, whether Agena still had a Gemini role was an open question.

Realignment of McDonnell's work began first. Spokesmen from McDonnell and its subcontractors met in Houston at MSC on 24-26 August and again on 6-8 September. They agreed to limit the scope of development for some spacecraft systems and gound equipment.[59] But MSC Director Gilruth told Walter Burke, McDonnell's spacecraft chief, not to do anything right away. When Gilruth talked to Burke on 8 September, the financial situation was still fluid enough to warn against too-hasty action. By the end of the month, however, prospects for any quick easing of the money crisis were fading. Burke flew to Houston to see Gilruth and Chamberlin on 28 September. Gilruth told Burke to carry out the earlier agreement on the revised scope of the program. Burke set his staff to work that same day on the necessary paperwork, wiring the subcontractors formal notice of their altered responsibilities and drawing up the required purchase order changes.[60]

Reprogramming at McDonnell in St. Louis was mainly a matter of making some adjustments. The company cut back its own and its subcontractors' quality assurance and reliability programs, reduced the number of published reports, decreased the number of spare parts to be maintained, trimmed the amount of engineering data and support required of subcontractors, and limited its support at Cape Canaveral. The net result of these changes was to slice $26 million from the $464 million that McDonnell thought its part of the project would cost, bringing the total down to $438.2 million.[61]

The largest savings in spacecraft development were to come through lessened testing by subcontractors. Teams from GPO spent much of October on two-day trips to major spacecraft subcontractors.* At each plant, they reviewed in detail the effect of various forms of

---

*The teams included Richard R. Carley, Robert Cohen, Duncan R. Collins, Paul L. Charvoz, William H. Douglas, John R. Hoffman, Clifford M. Jackson, Lemuel S. Menear, Jean Petersen, and William F. Smith. Companies visited were Minneapolis-Honeywell, St. Petersburg, Florida (inertial measuring unit); Minneapolis-Honeywell, Minneapolis (attitude control and maneuver electronics); Electro-Mechanical Research, Inc. (data transmission systems); IBM, Owego, New York (computer); Westinghouse, Baltimore, Maryland (rendezvous radar); Motorola, Scottsdale, Arizona (digital command system); Collins Radio Company, Cedar Rapids, Iowa (voice communications); Advanced Technology Laboratories, Mountain View, California (horizon sensor); and General Electric, West Lynn, Massachusetts (fuel cells).

systems failures, plans for qualification and reliability testing, and test facilities required. In general, they agreed that reliability testing could be sharply curtailed at the expense of slightly increased qualification testing. Qualification tests ensured that something worked; they usually preceded reliability tests, which made sure that something worked consistently. Assured reliability could thus be gained from augmented qualification tests.[62] Concerned by the way the program had grown, GPO also asked McDonnell for prompt notice of any future action that might affect contract costs or schedules.[63]

Spacecraft reprogramming was largely complete by mid-October, but the project office saw some further trimming possible in McDonnell's test program. After a review of its plans for structural tests of the spacecraft, the contractor concluded that one of the four programmed static articles might be dispensed with, and GPO agreed.[64] The project office also suggested that Project Orbit might be canceled, a view McDonnell opposed. The dispute was eventually resolved with Orbit restricted to testing the spacecraft's heat balance and renamed "spacecraft thermal qualification test."[65]

Another casualty of Gemini's financial straits was a lately revived lunar landing scheme. This time the impetus had come from NASA Headquarters in the person of Joseph F. Shea, newly appointed Deputy Director for Systems in the Office of Manned Space Flight. Shea wanted McDonnell to study using a Gemini spacecraft as a lunar logistics and rescue vehicle, a possibility also under study during that summer by the Space Technology Laboratories.[66] The eight-week McDonnell effort explored the concept of a two-man command module, evaluated using a Gemini spacecraft to land two men on the lunar surface, and looked at the design changes needed for such a mission.[67] Meanwhile, GPO computed the cost of buying extra spacecraft.[68] McDonnell submitted its findings to NASA Headquarters in November 1962.[69] Whatever chance the scheme may have had, however, vanished in the wake of Gemini's money problems, and the idea once again came to nothing.[70]

With the spacecraft taken care of by mid-October, the project office turned to launch vehicle programming. Limited funds compelled GPO to restrict 1963 costs to $59.28 million, some $10 million below its earlier plan and $18 million less than the $77.5 million SSD now claimed to need.[71] Chamberlin wired Richard Dineen, SSD's chief of Launch Vehicle Development, on 19 October to apprise him of the new funding limits. GPO believed that Gemini's major goals might still be met despite shortage of funds. The key was a sharp cutback in testing, especially where it involved repeated engine firing.[72] To Dineen, these changes seemed drastic, and he asked Chamberlin for a fuller explanation.[73] Chamberlin insisted that there was no hope of more than $59.28 million for 1963, which meant the planned test program had to be reduced and, in part, canceled. He asked Dineen for an early meeting to decide how to put these changes into effect.[74] SSD still objected.[75] Chamberlin persisted, wiring Dineen on 16 November that a meeting to review the launch vehicle test program was urgent and "should take precedence over other SSD/Aerospace/Martin/Aerojet Gemini commitments."[76] The meeting finally convened on 27 November.

The proposed changes were indeed drastic. The revised engine program called for only 34 test firings, less than a fifth of the number originally planned. This would yield all the data needed at a saving of several million dollars, if effort were focused on thorough development and qualification to make sure each part worked and would keep on working.[77] Sound engineering, in other words, made reliability a natural product of development and qualification. SSD and its contractors could scarcely quarrel with this view, but they tended to see reliability in more statistical terms—a part was reliable if it failed no more than some very small percentage of the times it was tested. The issue was not merely philosophical. Proving reliability statistically meant more tests, more equipment, and more money.

What was true for engines was also true for other parts of the launch vehicle. Martin's reliability program was budgeted for $2.7 million, but the GPO approach, by concentrating dollars on qualification rather than on reliability testing, could cut that figure in half.[78] Further study convinced Chamberlin that most of the planned prelaunch firings of the complete launch vehicle could also be safely discarded, and they were.[79]

NASA's budget crisis in the fall of 1962 never posed any real danger to Project Gemini itself. Work on spacecraft and launch vehicle was simply adjusted to meet an unexpected funding squeeze. Whether the Gemini that emerged from reprogramming would be the same project that had been planned, however, was another question. Tight money threatened to deprive Gemini of its chief objective, the development of orbital rendezvous techniques. For several months the role of Atlas-Agena in the program was in jeopardy, as NASA Headquarters debated dropping it, cutting it back, or keeping it with whatever slippage restricted funding entailed. The choice was not made any easier by the complex management structure of the target vehicle program. Two organizations, Marshall and SSD, stood between GPO and Lockheed, Agena's builder.

Word of tight budgets and a need to cut costs had reached Marshall's Agena Project Office by early October 1962 but was slower getting to SSD.[80] The first firm notice that the Atlas-Agena program was to endure something more than a routine economy drive came on 23 October, when Chamberlin wired Friedrich Duerr, Agena systems manager at Marshall, "to reshape and reschedule the Atlas-Agena to conform to budget limitations. MSFC is further directed to establish accounting procedures and funds expended monitoring procedures to assure that Agena development is prosecuted within the established fund limitations."

GPO had just completed a detailed study of changes that might be made in the Agena program to keep costs within budget limits. It concluded that $16.7 million could be sliced from the 1963 Atlas-Agena budget, dropping it from $27 million to $10.3 million. Chamberlin presented Duerr with the $10.3-million figure as a funding limit for fiscal year 1963, as part of an overall goal to reduce the cost of development by a third. For Agena, like Titan II, the savings were to be found mainly in less engine test firing and more built-in reliability. But Agena faced sterner sanctions—no more money and all work stopped until reprogramming was complete.[81]

Duerr passed the word to the Air Force,[82] although, as he informed Chamberlin, GPO's view of the savings that might be achieved was "optimistic" and the changes could only mean a long delay in the development program.[83] Reprogramming began with a meeting in Houston on 25 October to discuss plans and schedules. What reliability meant emerged as the central issue, just as it did for Titan II. A second meeting, to agree on a specific plan, was set for 2 November.[84]

Before that meeting convened, however, the real need for Agena in the Gemini program was called into question. In mid-1962, NASA had decided in favor of the lunar orbit rendezvous scheme for the Apollo lunar landing. That tentative decision was confirmed on 24 October by the findings of a manned lunar landing comparison study.[85] At a meeting of the Manned Space Flight Management Council six days later, Holmes raised the issue of Gemini objectives in light of this decision. Shea reviewed Gemini's aims and claimed "that all of these objectives appear to be possible of achievement without use of the Agena in the program." MSC Director Gilruth disagreed, and an inconclusive debate over the fate of Agena followed. Although he knew that time was running out, Holmes asked Gilruth to study the matter further.[86]

Meanwhile, the second reprogramming session convened at the Lockheed plant in Sunnyvale, California. The monthly spending rate under the Gemini-Agena contract had reached $2 million during October. The limit for November, however, was fixed at $650 000, and Lockheed was instructed to stay within it. Lockheed spokesmen protested, claiming that Bell Aerosystems, the engine subcontractor, could not produce engines for an October 1964 launch if funds were so restricted. Chamberlin told them they had no choice—they must find ways to stay within the fixed limits. Lockheed had a week to provide a rough cost estimate for the revised program to SSD, which would turn its findings over to Marshall's Agena Project Office, which, in turn, would pass its findings up the line to GPO. A final meeting to coordinate the changes was scheduled for 20 November.[87]

Duerr reminded Chamberlin that limited funding was bound to cost time, perhaps as much as 14 months, in Agena development. Extra money—$12.7 million instead of $10.3 million for the current fiscal year—could hold the loss to a less painful five and a half months.[88] But even at that, it would still be "a maximum risk program. That is to say that the target vehicle program has been minimized and no allowance is made for contingencies that may arise which would adversely affect costs and schedules."[89] Chamberlin knew as well as anyone that time was being traded for money, but his hands were tied. A financial operating plan for 1963 had yet to be approved. Whether Agena could even be kept in the Gemini program—and not the precise level of funding—was the crucial question.

At a meeting of MSC's senior staff on 9 November, Chamberlin strongly objected to Shea's claims at the Management Council meeting on 30 October. Shea, and others in NASA Headquarters, believed that rendezvous goals might be met by using a "piggyback" rendezvous package, carried aloft in the adapter section of the spacecraft and then ejected in orbit to serve as a stable but non-maneuverable target. Chamberlin dismissed the piggyback technique as inherently limited in contrast to the stabilized and maneuverable Agena. He also believed that the package would be far heavier than its proponents claimed. André Meyer, chief of GPO administration, figured its weight at 180 kilograms, twice the Headquarters estimate. If that were true, it could mean the end of paraglider. Meyer thought the package would cost as much as Agena, although without the problems and expenses of separate launches.[90]

MSC had been thinking along similar, but much more modest, lines. A study issued on 28 March 1962 had concluded that a piggyback rendezvous target could provide useful data. A month later, McDonnell had suggested testing the spacecraft rendezvous radar and maneuvering systems on an early Gemini flight with what it called a "Rendezvous Evaluation Pod (REP)." This was a small battery-powered module with a radar transponder, radar beacon, and flashing light, the whole package weighing about 30 kilograms and designed to give the pilots a chance to practice terminal rendezvous maneuvers with their spacecraft. In June, MSC had told McDonnell to go ahead with design and development. The REP would be carried on the second and third Gemini flights. Planning was largely complete by the end of 1962, with Westinghouse, the rendezvous radar subcontractor, responsible for components and McDonnell for the package and its ejection.[91] This, however, amounted to little more than an experiment, intended to prepare for, not supplant, the Agena rendezvous missions.

On 16 November, Wesley Hjornevik, chief of MSC administration, reported to the senior staff that a financial operating plan for fiscal year 1963 had finally been approved. Agena funding, however, had been withheld.[92] Target vehicle reprogramming went ahead, with the final meeting on 20 November in Houston. Lockheed's new program was accepted. The major changes made reliability demonstration part of development and qualification testing, cut engine development testing to the bone, and trimmed production lead times to keep down 1963 expenses. This last meant chiefly that Lockheed was to work at a reduced level through the rest of calendar year 1962, then return to full effort on 2 January 1963. The program would be four months late, but its total cost could be as low as $44.1 million, $32.7 million less than estimated before reprogramming began.[93]

Gilruth outlined the revised Atlas-Agena plans to the Management Council on 27 November, with a sharp reminder that "it is very critical that a decision as to the inclusion of the Atlas-Agena in the program is reached soon if the Agena target schedule is to be maintained." Holmes promised a ruling by 10 December.[94] Not only had the fate of Agena become a matter of public speculation, but lack of funds threatened to stop the target vehicle even before anything was decided.[95]

The decision came early but turned out to be only a stopgap: $900 000 for another month. This brought the total for fiscal year 1963 to $4.9 million; the balance of the planned $10.3 million for Atlas-Agena remained in abeyance.[96] Shea, who had proposed dropping Agena from Gemini, told a reporter that NASA was thinking about several alternatives to simplify the rendezvous concept, with a decision due shortly. He gave Agena only a 50-50 chance of staying in the program.[97] Agena's fate was in the hands of a NASA-wide committee, which Shea himself headed. A thorough investigation, bolstered by the well-informed and forceful case presented by James Rose, the GPO member, decided the committee in favor of Agena. A wire from Washington on 21 December authorized MSC to spend the full $10.3 million needed for the reprogrammed Agena in fiscal year 1963.[98]

MSC also took over management of the Gemini Agena program. NASA decided to transfer all its Agena programs from Marshall so that that Center could focus on the Saturn launch vehicle for Apollo. Lewis Research Center in Cleveland, Ohio, assumed control of all NASA Agena programs except Gemini, which went to MSC.[99] MSC, now dealing directly with SSD,[100] took formal charge of the Gemini Atlas-Agena program on 14 January 1963, with active advice from the Marshall office for the next month and a half.[101] Lockheed and SSD also adjusted their management relationships. The Gemini manager at Lockheed, Herbert Ballard, moved up a notch; he now reported directly to the head of Lockheed's Medium Space Vehicles Programs. SSD followed suit by upping the rank of its program manager from captain to major; and Major Charles A. Wurster took over the reins.[102]

Since the only function for Atlas in Project Gemini was launching the target, its fate waited on Agena's. But Atlas, too, suffered in NASA's fall budget crisis. On 25 July 1962, NASA Associate Administrator Seamans had agreed to support Air Force development of a standard Atlas launch vehicle, SLV-3.[103] By the time the Department of Defense had drafted a formal Memorandum of Agreement and forwarded it to NASA on 21 August, NASA's funding outlook had so deteriorated that it could no longer contribute to the program. Seamans restated NASA's interest in SLV-3 development but declined to commit the roughly $10 million that was to have been its share of the cost.[104]

Reprogramming raised the possibility of using surplus Atlas boosters from the Mercury program in Gemini. Chamberlin asked SSD for an opinion. A report to the Atlas-Agena reprogramming meeting of 20 November was favorable. Chamberlin then asked the Atlas contractor, General Dynamics/Astronautics, for a formal proposal.[105] The results made conversion look promising economically. Three converted Mercury boosters could be had for a net cost of $3.364 million, as opposed to $5.4 million for three new standard Atlases.[106] But by the time those figures were submitted on 13 February 1963, Gemini's budget crisis was over, and NASA was back in the standard Atlas development program. In December, Seamans had formally committed NASA to pay its $10-million share.[107]

## THE PROSPECT FOR 1963

With reprogramming completed, Gemini's prospects looked reasonably bright as 1962 gave way to 1963. The crisis through which the program passed in the last quarter of 1962 was monetary, not technical. That crisis weathered, the technical problems looked less menacing as well. In his report to the Management Council on 18 December, Gilruth noted that Gemini still had a number of technical problems, but all, he judged, "are being actively pursued and none appear to be unresolvable."[108]

Gemini had lost time, though. The new Gemini program was chiefly a response to budget limits imposed from outside, compounded by sharply rising costs. Its immediate goal was cutting back expenses during the current fiscal year, and this meant slowing down the program. But a longer program, despite the curtailed and streamlined development that emerged from Gemini's fall crisis, was likely to cost more in the long run. Whether the total cost of the program would really rise, and how much, could only be answered with the passage of time.

The effects of reprogramming on Gemini schedules were easier to define. Gemini was going to lose four months. The new date for the first launch was December instead of August 1963. It was now an unmanned suborbital qualification test. McDonnell had proposed in July 1962 an extra mission that it called Flight No. 0, a suborbital shot to precede the first planned mission. On 20 July, Burke and Chamberlin agreed to replace the planned unmanned orbital flight with the suborbital flight as the first mission (a slightly revised version of the Mission 0 plan). It was to be a ballistic test to investigate spacecraft heat protection, to integrate launch vehicle and spacecraft preflight and launch operations, and to obtain data on spacecraft structure and systems.[109] All other launch dates were set back four months. The second flight—manned orbital qualification—followed the first by three months, in March 1964, with the rest of the missions coming every two months until the 12th and last, now scheduled for November 1965.[110]

By December 1962, everything seemed to be under control again. But while the project office and MSC were wrestling with the hard tasks of fitting development work to the limited money available, NASA Headquarters found itself fending off quite a different threat—perhaps the least expected of all. The Department of Defense was making gestures toward taking over Project Gemini.

# VI

# Challenge and Change

GOING into its second full year, the Gemini Project Office had just finished moving into new quarters. The office had been split between two sites, with project manager James Chamberlin at the Farnsworth & Chambers building (interim headquarters for the Manned Spacecraft Center) and the rest of the Gemini office across the Gulf Freeway in the Houston Petroleum Center. By December 1962, the office had doubled its original staff of 44 and outgrown its former space. Chamberlin and all of his people moved into the old Veterans Administration building on the edge of downtown Houston by 10 December, and the Gemini Procurement Office of MSC's Procurement and Contracts Division followed in March 1963.[1]

Putting all of Gemini under one roof no doubt helped as the program became more taxing. The early months of 1963 soon showed that many technical problems were far from resolved and that the question of money was not fully settled by the reprogramming efforts. But Gemini's first big worry of the new year had little to do with technology or funding. The Air Force had long been interested in orbital rendezvous and manned space flight, as reflected in its unmanned satellite interceptor project (Saint) and the maneuverable manned Dyna-Soar program. That interest now expanded to include Project Gemini.

## BLUE GEMINI

"Blue Gemini" was the tag name for an Air Force manned space flight program to develop rendezvous, docking, and transfer for military purposes, using Gemini-type spacecraft. The germ of the idea first surfaced in February 1962, during congressional hearings on the defense budget, as part of a far-ranging Air Force Space Plan for the development of military space technology over the next 10 years. The concept became firmer in June, when the Air Force Space Systems Division (SSD) began working on plans to use Gemini hardware as the first step in a new Air Force man-in-space program called Mods (Manned Orbital Development System), a kind of military space station with Gemini spacecraft as ferry vehicles. The term Blue Gemini first showed up in August as part of a more specific proposal to fly six Gemini missions with Air Force pilots in a preliminary orientation and training phase of Mods.[2]

Blue Gemini was neither clearly defined nor officially sanctioned. Air Force opinion was divided on the best approach to the goal of military manned space flight. Some, like Air Force Chief of Staff Curtis E. LeMay, wanted nothing to do with Gemini, fearing that entanglement in the NASA program might jeopardize Dyna-Soar. Others, like Major General Osmond J. Ritland, deputy for manned space flight in Air Force Systems Command, urged a more active Air Force role in Gemini, since Dyna-Soar would not fly for at least two years. Civilian officials in the Pentagon remained skeptical of any military man-in-space proposals, for much the same reason that had tended to block such efforts all along: the absence of any clear-cut military need for manned operations in space.[3]

By the fall of 1962, the situation was in flux. The Saint program suffered a sharp cutback in December, following cost overruns and schedule slippages. This made Gemini look even more attractive to those Air Force planners still convinced of the military importance of orbital rendezvous but now lacking a program to test their ideas. Techniques for rendezvous between remote-controlled machines, as in Saint, would differ from those suited for manned rendezvous, but manned work in space looked more exciting anyway. Dyna-Soar, a winged glider boosted into space by a Titan III to orbit Earth and fly back to an airfield landing, had lost much of its promise as a result of changes and delays. The exciting potential of such a program, when it took shape in the late 1950s, looked much less impressive by the end of 1962, especially in contrast to Gemini. No decision had yet been made in the Department of Defense, but the entire military manned space role was under review and forecasts of Dyna-Soar's extinction were rife.[4]

Meanwhile, the Air Force role in Project Gemini was limited to the one set out in the "NASA/DOD Operational and Management Plan" of December 1961, SSD acting as contractor to NASA for launch and target vehicles.[5] The idea of Blue Gemini—a larger part for the Air Force in the program—had a good deal of support within NASA, especially from MSC Director Gilruth. Gemini had been designed as an operational spacecraft, and the Air Force was the most likely customer. The Air Force could also be expected to pay for what it wanted, and Gemini could use an infusion of Defense funds. At a meeting in November 1962, Chamberlin and some of his staff described salient aspects of Gemini to a group of SSD representatives.* This meeting was intended to lay the groundwork for coordinating Air Force planning with MSC and to set up channels for future collaboration.[6]

NASA Administrator Webb and Associate Administrator Seamans visited the Pentagon for a talk with Roswell L. Gilpatric, Deputy Secretary of Defense, in an effort to convince Pentagon planners that an augmented role for the Air Force in Project Gemini was a good idea. Chance brought Secretary of Defense Robert S. McNamara to the meeting. His response to their offer was more than the two NASA spokesmen had bargained for; it took the Air Force by surprise as well. McNamara not only welcomed the idea of cooperation—he proposed merging the NASA Gemini program with the Air Force project and moving the combined effort to the Department of Defense.[7]

That was too much for NASA. W. Fred Boone, a retired admiral who had become NASA Deputy Associate Administrator for Defense Affairs on 1 December, took charge of building the case against Gemini's transfer to the Air Force. In NASA's view, not surprisingly, "the Gemini program should continue under the direction of NASA." The keystone of NASA's case was that Gemini was integral to the step-by-step climb from the first moves into space in Mercury to the final landing on the Moon in Apollo. Any delay in Gemini might delay the lunar landing. Increased Air Force participation "to further DOD objectives in space" was all right, but it must not hamper NASA in promptly carrying out the Gemini program.[8]

To support his position, Boone asked each of the NASA staff offices for a statement on the effects of an Air Force takeover of Gemini. The replies stressed the clear threat that such a move might disrupt NASA's manned space flight effort in general and the manned lunar landing program in particular. Beyond this most pressing danger, they feared nasty responses from outside NASA: increased criticism from a Congress already perturbed by signs of military influence in NASA programs; rising concern from a public disturbed by questions about the viability of a civilian space program; and growing disquiet in foreign nations about the United States being a peaceful explorer of space,

---

*MSC speakers were Paul Purser, Chamberlin, James Rose, Homer W. Dotts, and George MacDougall. Non-NASA visitors were Major Ben J. Loret, Major Earl A. Hoag, and Captain George R. Honold (Air Force), and Bill Nordyke, Donald P. Armstrong, and Mike Weeks (Aerospace Corporation).

which carried the added threat that some countries might expel NASA tracking stations from their territories.[9] After going over these arguments, Boone concluded:

> It is in the national interest that the management of Project Gemini remain with NASA's Manned Spacecraft Center. A change in program management would seriously delay and substantially increase the cost of the manned lunar landing program. Any delay would reduce the chances that the United States will make a manned lunar landing before the Russians do.

A much better choice than giving Gemini to the Air Force would be to enhance the role of the Air Force within the framework that already existed.[10]

Just as surprised by the McNamara proposal as NASA was the Air Force, which shared NASA's distaste for a Gemini takeover, partly because it might jeopardize Dyna-Soar, partly because the costs of a few fully "blue" Gemini flights would far outweigh any foreseeable gains.[11]

NASA's arguments for keeping Gemini seemed convincing enough when presented to top Pentagon officials on 9 January 1963, bolstered as they were by the Air Force's unwillingness to take the program. McNamara and Gilpatric readily agreed not to press for transfer of Gemini. However doubtful the future role of military man-in-space, they thought the Air Force remiss in failing to accept NASA's offer of a larger part in Gemini. That was what McNamara now wanted as a formal pact between the two agencies; and he wanted it soon, before he began to present his case for the coming year's Defense budget to Congress on 21 January. Perhaps as much as $100 million in Defense funds could go to Gemini. McNamara's key idea was a joint management board to run the project and he promised to forward a draft agreement soon.

A jointly managed Project Gemini had no more appeal for NASA than an outright transfer. Boone dismissed the proposed board as "a completely unnecessary organizational appendage"[12] even before he saw the promised draft. It arrived on Saturday, 12 January, and did nothing to soften Boone's judgment. Claiming that "both parties [DOD and NASA] consider that the national interest requires the program to be jointly managed," McNamara proposed an eight-man Gemini Program Steering Board to approve program and funding plans, to safeguard both Defense and NASA experimental objectives, and to resolve schedule and resource conflicts. Although GPO would report to the new board, project management would remain unchanged. Defense intended to pay for its enlarged role with money for current Gemini needs, as well as future board-approved changes.[13]

NASA's top management discussed the plan on Monday afternoon,

14 January, and Boone drafted a reply. McNamara's "joint management," in Boone's view, equaled "rule by committee," which "in this case would be ineffective, uneconomical, and in fact unworkable." Changing Gemini also threatened Apollo and might cost the United States its chance to win the space race. The proposed joint board also violated the Space Act of 1958, certainly in spirit and probably in letter. There seemed to be room enough for the Air Force in the current Gemini setup. If not, a joint planning and review (as opposed to management) board to advise the NASA Administrator ought to serve the purpose. Boone concluded by stating "NASA's strong interest in the Dyna-Soar program," hinting that NASA would endorse the Air Force project if Defense relaxed its demands on Gemini.[14]

NASA's revised version of the Defense draft altered enough words and accents to transform its meaning. Gone was any hint of "joint management." The steering board had become the Gemini Program Planning Board, limited to watching over a program of Gemini experiments. There was no mention of approving program plans or allocating resources. At most, the board could inform the NASA Administrator and the Secretary of Defense of such problems as planning defects or schedule conflicts. NASA repeated, and stressed, its claim to sole control of Gemini. GPO would not report to the board. The Air Force would be restricted to joining "in the development, pilot training, preflight check-out, launch operations and flight operations of the GEMINI program to assist NASA and to meet the DOD objectives," just as it had been doing.[15]

The Defense Department accepted NASA's terms in a series of meetings between spokesmen for the two agencies over the weekend of 19-20 January. Willis H. Shapley, Deputy Chief of the Military Division of the Bureau of the Budget, arranged the meetings and prepared a series of notes designed to clarify the intent of the agreement proper and to distinguish it from some rumored proposals that had surfaced in the press. *Aviation Week and Space Technology,* for example, had reported in its issue of 10 December 1962 that NASA and the Air Force had agreed on a cooperative Gemini/Blue Gemini program: NASA would fund Gemini development and fly the first missions; the Air Force would fly copilots on one or two of the early missions and buy the last four or five Gemini spacecraft for its own flights plus a few extra beyond the twelve NASA had ordered.[16]

Shapley's notes mostly covered management relations between NASA, Defense, and the proposed Gemini Program Planning Board; but they also touched on funding and the domestic and foreign impact of the new arrangements. Gemini was not to be thought of as a joint program, but rather as a program serving common needs, with the Department of Defense paying for the military features, NASA in full charge of the program, and the role of the board strictly advisory. Defense funds were to be used for nothing but the changes geared to military needs; the money was specifically not to be used to speed up the current NASA program nor to make up slippages and overruns. No major change in policy toward the Air Force role in space was intended, and the new agreement was to be presented to the public as the latest in a series of efforts to enhance cooperation and to avoid duplication between NASA and the Pentagon.

Webb signed the revised agreement and sent it, along with a slightly edited version of Shapley's notes, to McNamara on 21 January. The notes were not part of the formal document, but they helped fill out the record of understanding between the two agencies.[17] The new pact was made public the next day. Webb and McNamara "joined in stressing the national character and importance of the Gemini project" and in their determination to see it "utilized in the national interest, and to avoid unnecessary duplication of effort in this area as in all others"—citing the agreements on the management of Cape Canaveral (also announced on 22 January) and on such earlier undertakings as Dyna-Soar and the national launch vehicle program as examples of similar cooperation.[18]

How a seemingly larger Defense role in Gemini might affect international opinion was the subject of still further concern. NASA assured the State Department that Gemini's goals remained unchanged, its peaceful scientific character unaltered. NASA still ran Gemini and planned to make Gemini's scientific data as widely available as Mercury's. The new agreement simply augmented military support of the same kind already known to the manned space flight program. Gemini was still open, NASA still managed it, and its foreign network stations would have no military personnel except medical.[19]

Although the NASA/Defense agreement of 21 January left NASA clearly in charge of Gemini, rumors of an Air Force takeover persisted.[20] Real changes were small. The major innovation was the Gemini Program Planning Board, a strictly advisory body whose planning was to be confined to military experiments for Gemini flights. Its co-chairmen were Seamans for NASA and Brockway McMillan for Defense. McMillan was Assistant Secretary of the Air Force for Research and Development. Holmes and Boone were the other NASA members; and the Department of Defense named General Bernard A. Schriever, Commander of Air Force Systems Command, and Lawrence L. Kavanau, Special Assistant for Space to the Director of Defense Research and Engineering. The group held its first meeting on 28 February 1963 at NASA Headquarters in Washington.[21] The board in this as in later meetings did attend to the place of military experiments in Gemini. But experiments did not remain its only concern, nor did they turn out to be the board's signal contribution to Gemini.

## CHAMBERLIN DEPARTS

The dispute between NASA and the Department of Defense about who was to have the last word in Gemini, whatever might be its long-range impact, agitated only the highest echelons. MSC engineers knew little of what was going on and, in any case, had their hands full with their own problems. Gemini reprogramming had slowed the rate at which money was being spent, but costs still spiraled upwards. Although stretching out the program was bound to offset immediate savings by larger total costs unless parts of the program were chopped out, the size of the increase soon surpassed anything that might have been expected. Meanwhile the revised program suffered from the growing severity of the technical problems that had afflicted it before and during the fall budget crisis. Paraglider testing and Titan II anomalies loomed largest.

Despite some talk about dropping paraglider from Gemini to meet fiscal constraints, paraglider development came through largely unscathed. While other major systems suffered more or less drastic cutbacks, paraglider's budget expanded. By the end of 1962, contract changes and overruns had raised the price of the current phase of paraglider development from four and a half to over seven million dollars.[22]

North American Aviation, the paraglider contractor, was still having problems with flight testing. The success of 23 October 1962, which concluded the test series of a half-scale model launched with its wing already deployed, proved only a respite. The next step was trying to deploy the wing in flight. North American refitted the half-scale test vehicle at its plant in Downey, California, and shipped it back to Edwards Air Force Base for its first flight test, scheduled for 27 November. The all-too-familiar pattern of minor problems, mostly electrical, delayed the flight day by day until 10 December, and then the results were disappointing. The capsule tumbled from the helicopter, fouling the drogue parachute intended to pull the can, in which the wing was stored, away from the paraglider. Wing inflation intensified the tumbling and the emergency drogue parachute ejected too soon. When the capsule spun down past 1600 meters, the minimum recovery altitude, radio command detached the wing and allowed the capsule to descend on its emergency parachute.[23]

The next attempt, on 8 January 1963, after its share of delays, produced even worse results. There was no tumbling, but the storage can was late in separating; so the capsule was falling too fast when the wing started to inflate and its membrane tore. As the capsule fell below 1600 meters, its wing not yet fully deployed, emergency recovery was ordered to no avail. The main parachute remained packaged, and the capsule crashed. Picking through the wreckage, North American in-

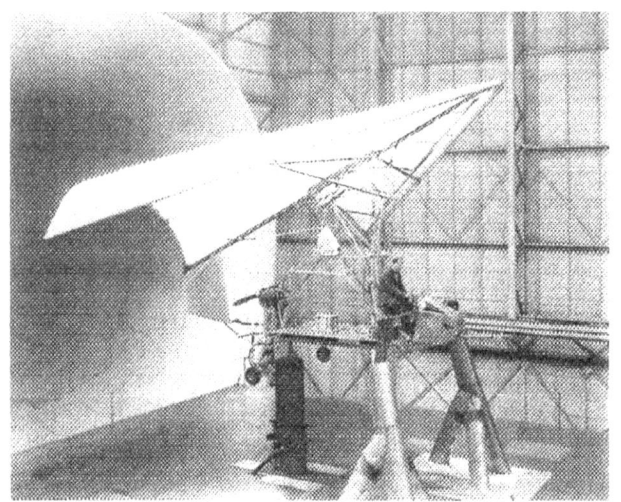

## The Paraglider

*Full-scale paraglider training vehicle is studied for handling characteristics in Langley Research Center wind tunnel.*

*Half-scale paraglider and spacecraft model ready for helicopter tow test.*

*Full-scale paraglider test vehicle at North American Aviation.*

*Artist's conception of a Gemini landing by paraglider.*

124

spectors found that a squib switch in the emergency parachute's electrical system had misfired. That was not the only problem, but it was the most discouraging—the switch was a standard item, much used in the space program and not known to have failed in 30 000 successive firings. GPO warned North American to be sure everything that had gone wrong was corrected before trying again.[24]

A month later, North American reported to the paraglider coordination panel that five distinct failures had been spotted, studied, and fixed. The panel was convinced, but Chamberlin was not. After an extended meeting with George Jeffs, manager of the paraglider program for North American, Chamberlin decided to give the trouble-plagued half-scale flight-test program another chance.[25] Once again, the current crop of troubles had little impact on plans for the next phase of development, which covered the rest of flight testing, pilot training, and paraglider production. Part of Phase III, gearing up for production, was worked out and under way by 22 January. North American's proposals for the rest of the program were ready by the end of the month. GPO approved and, with the concurrence of NASA Headquarters, readied a new contract.[26] But the Office of Manned Space Flight had second thoughts and stopped the procurement action "for the time being."[27] The halt proved to be permanent.

The Gemini paraglider program foundered on North American's third attempt to deploy a half-scale wing in flight. Although the first two flights had been at least partial successes, the third, on 11 March, offered no comfort at all. The storage can failed to separate, so the wing could neither eject nor inflate. When the radioed command to deploy the emergency parachute produced no response, the second half-scale test vehicle joined the first as wreckage.[28] Paraglider testing came to an abrupt halt.

Gemini's other major headache early in 1963, Titan II, posed a far greater threat to the program as a whole. There would still be a Project Gemini without paraglider, but not without Titan II. Despite some hopeful signs, the status of the launch vehicle remained very much in doubt. The central problem was still the lengthwise vibration, or Pogo, that bounced the vehicle while its first-stage engine was burning; but other technical problems began to compete for attention. Efforts to resolve them were coming up against a crucial disparity between Air Force and NASA goals in Titan II development.

The Martin Company's proposed answer to Pogo—a surge-suppression standpipe in the first-stage oxidizer feedline—was installed in the soon to be infamous Missile N-11, the eighth Titan II that the Air Force launched in its missile development program, on 6 December 1962. The supposed cure, far from damping the Pogo effect, raised it to $\pm 5g$, and the violent shaking induced the Stage I engines to shut down too soon.[29] A rueful Robert Gilruth told his fellow members of the Manned Space Flight Management Council that he saw one hope: "the fact that the addition of the surge chamber affected the oscillation problem may indicate that the work is being done in the right place."[30]

The next Titan II, launched on 19 December, carried no standpipes; but increased fuel-tank pressure, which had shown good results on some earlier flights, again reduced the Pogo level. This missile also featured oxidizer feedlines made of aluminum instead of steel, which seemed to have some bearing on the sharply lessened amplitude of oscillation. This was disconcerting, no reason for the effect being readily apparent. The Pogo problem clearly needed more study.[31]

In the tenth flight, on 10 January 1963, Pogo hit a new low of six-tenths the force of gravity ($\pm0.6$g) at the spot on the missile where a manned spacecraft would be located. This was getting close to the level tolerated on Mercury flights, roughly $\pm0.45$g. But Gemini's astronauts were supposed to take a larger part than Mercury's in flying their craft into orbit. NASA's goal for the Titan II remained $\pm0.25$g at most. Nonetheless, despite the still large gap between performance and goal, increased fuel-tank pressure had so reduced "POGO type oscillations" that Gilruth could say, "this now becomes a secondary problem."[32]

He may have been more concerned about another problem than he was optimistic about Pogo. Despite the low Pogo level on the tenth flight, the missile's second-stage thrust was only half what it should have been. On some earlier flights, the failure of second-stage engines to build up to full thrust had been blamed on Pogo. That now appeared doubtful. Another source of unease, and the one Gilruth now tabbed as the major problem, was the threat of unstable combustion in the second-stage engine. Static firing tests during January 1963 showed that the Aerojet-General motors might have trouble reaching a steady burn after the shock of starting.[33]

But this was as yet mostly surmise, and Chamberlin's concern still centered on Pogo, chiefly because he was not at all certain how far the Air Force Ballistic Systems Division (BSD), which was in charge of Titan II missile development, would go to meet Gemini's much stricter demands.[34] His fears were confirmed on 29 January, when BSD's Titan Program Office froze the missile design with respect to devices for cutting vibration levels, since increased pressure in first-stage fuel tanks and aluminum oxidizer feedlines reduced Pogo below specifications for the missile airframe and systems.

This was an answer only for the missile. Tank pressures were nearing structural safety limits, and more pressure could not lower the vibrations much further, anyway. But the level was still too high for Gemini. BSD intended to keep looking for a way to achieve the lower value NASA wanted; but early in March, BSD decided that it could no longer accept the costs and risks of efforts to reduce the oscillations any further.[35]

*Secretary of Defense Robert McNamara (left) in Houston for a briefing on the Gemini program by Robert Gilruth (right) and his staff.*

Chamberlin had no direct line to BSD, his only channel being through SSD. With BSD in charge of missile development and SSD of Gemini launch vehicles, NASA was largely a spectator. Chamberlin could do little more than appeal to SSD to intercede with BSD. Since there was no flight test program for the Gemini booster, the Titan II missile research and development program was the only chance to solve Gemini problems. But BSD was responsible for a weapon system, not a launch vehicle, and was understandably loath to risk the missile for the booster.

During March, therefore, Chamberlin spent a lot of time on the telephone, asking Richard Dineen, in charge of Gemini launch vehicle development for SSD, for help not only with Pogo but on the threatening combustion instability problem. Chamberlin hit hard on his long-standing demand for a rigorous qualification program but now stressed that qualification must be "followed by a suitable number of successful flight tests" to reach the required level of confidence in a booster for manned space flight. He wanted to know what plans Dineen had for making sure that the Air Force test program would meet Gemini's needs, and Dineen promised a report in short order.[36]

Word of Titan II's troubles was slow to reach NASA's upper echelons. When James Marsh, head of the Gemini launch vehicle program at Aerospace Corporation, discussed the current status of the booster at a meeting of the newly formed Gemini Program Planning Board on 7 March, he was far from alarmist. Seamans got the impression that things were well in hand. A detailed redesign of the turbopump impellers in the first-stage engines would take care of the Pogo problem, according to Marsh, and the unstable burning in the second-stage engines was no risk to Gemini.[37]

This view was rudely shattered a week later, when Seamans traveled with Secretary McNamara and a party of Defense officials to Houston for a close look at Gemini. He learned for the first time that MSC was now thinking of two unmanned flights, rather than one, cutting the number of manned missions to ten, the first delayed five months until August 1964. Trouble with Titan II was offered as the main reason for this drastic change in schedule, and combustion instability was cited as potentially a greater problem than Pogo. McNamara assured Seamans and MSC that Titan II would be fixed, but Seamans was still doubtful.[38]

This was only three days after the crash of the second half-scale paraglider test vehicle. The conjunction of the newly revealed impact of Titan II problems and the latest in the series of paraglider mishaps suggested that Project Gemini was in deep technical trouble. To make matters worse, Gemini had new money worries. The reprogramming effort of the last quarter of 1962 had slowed the rate at which Gemini was spending money but at the expense of stretching out the program. In the nature of things, a longer program was liable to cost more overall; when Holmes reported, early in February, that Gemini's total cost would reach $834.1 million, the figure was not too disturbing. That was about $60 million over the lowest estimate in September 1962 but well short of the $925 million that had then appeared to be a possibility.[39]

Just a month later, however, on 8 March 1963, MSC's revised preliminary budget for fiscal year 1964 reached NASA Headquarters, and it was a shock. Gemini's estimated total had shot over the billion-dollar mark. The new figures was nearly twice the cost first approved in December 1961 and almost $200 million higher than the figures Seamans and other NASA officials had been using as the basis for NASA's fiscal year 1964 budget request, most recently in House hearings earlier that week.[40] So large an increase, coming on the heels of what had seemed to be a resolution of Gemini's funding problems, took NASA Headquarters by complete surprise. Chamberlin, as manager of Gemini on the field level, knew what was happening. But, waiting for an opportune moment to break the news, he was overtaken by events.

Unexplained cost increases combined with seemingly critical problems in paraglider and Titan II development to bring Chamberlin's tenure to an abrupt end.[41] On 19 March, Gilruth relieved Chamberlin of his duties as project manager and assigned him to the post of Senior Engineering Advisor to the Director, cutting him off from any direct connection with Gemini. Charles Mathews took over as acting manager. He came to Gemini from the Engineering and Development Directorate, where he had recently added the job of Deputy Assistant Director to his work as Chief of the Spacecraft Technology Division. Mathews was a charter member of Space Task Group, having come

*Gemini program managers were James Chamberlin (top, fourth from left, with astronaut Virgil Grissom, James McDonnell of McDonnell Aircraft, astronaut Walter Schirra, baseball star Stan Musial, and astronaut Scott Carpenter) and Charles Mathews (below left, with Robert Gilruth).*

with Gilruth from Langley's Pilotless Aircraft Research Division. He had headed STG's Flight Operations Division until 17 January 1962, when he moved over to the Engineering and Development Directorate as chief of what was then called the Spacecraft Research Division.[42]

When Chamberlin left Gemini, an era ended. In the large and complex undertakings of modern high technology, one person can seldom be credited with so large a share in the shaping of a project as Chamberlin deserved for Gemini. Much of the ultimate success of the project had its roots in Chamberlin's brilliance as a designer and skill as an engineer, but so did some of the current harvest of troubles. The talented engineer can always see new ways to improve his machines, but the successful manager must keep his eyes on costs and schedules, even if that sometimes means settling for something good enough instead of better.

But perhaps in a deeper sense, Chamberlin can be seen as a victim of the way Gemini was created and funded. Approved as something of an afterthought in the American manned space flight program, absent from NASA long-range budget plans, Gemini began with shaky finances. Crushing time pressure made things worse. Gemini, although in most ways just as sophisticated as Apollo, began later and had to finish its flight program much sooner than the lunar program. As Chamberlin later remarked, "we went ahead as fast as possible with whatever funding could be scrounged.... If Gemini were too late, there would be no need for it, and it would be cancelled." In this setting, technical problems that might otherwise have appeared little more than routine assumed a more ominous guise.

Chamberlin's colleagues in and out of NASA deeply respected him as an engineer and designer but also saw his flaws as a manager and recognized the difficulties of the situation. His sudden and largely unexpected departure was thus not the blow to project morale that it might have been. The shock was also eased by the identity of the man who replaced him. Mathews was well known and widely esteemed. He took over a program that did seem to be in trouble.[43]

## GEMINI REGROUPS

The shaky status of Gemini costs and schedules was the major factor in Chamberlin's ouster, and it was to those matters that Mathews first turned in his new role as acting program manager. An early move was a critical review of the Gemini flight program. This produced one quick decision: an unmanned mission would be flown in place of one of the manned flights; only 10 of the 12 Gemini flights were now to carry crews. This was largely a response to the stubborn problems in Titan II development. The first flight had been planned most recently as a suborbital ballistic shot to test spacecraft heat protection and validate spacecraft structure and systems. With launch vehicle status uncertain, however, this no longer seemed sufficient qualification for manned missions. Another question mark was the spacecraft itself, which did not seem likely to be ready in time.[44]

GPO had a new flight schedule to submit to Manned Space Flight Director Holmes by 11 April. It differed sharply in some key ways from earlier plans. The major change was that the first flight, still due in December 1963, was to be orbital, its primary objective the flight qualification of the booster. The spacecraft would serve chiefly as an instrument carrier, neither separating from the launch vehicle's second stage nor being recovered. Gemini's second flight, postponed from March to July 1964, was now what the first had been—a suborbital ballistic flight intended to prove the spacecraft could withstand high heating rates but also to qualify all launch vehicle and spacecraft systems for manned flights.

The first men to fly in Gemini now had to wait for the third mission, in October 1964, five months later than had been scheduled for the third flight and seven months past the former date for the first manned flight. The mission was not only late, it was much reduced in scope. First planned for a full day, or 18 orbits, the mission now seemed likely to be no more than three orbits, mainly for systems evaluation.[45] The three-orbit limit became official in mid-June 1963. This raised the question of what to do with the package that both of the first two manned spacecraft were supposed to carry into orbit to practice the final stages of rendezvous. Three orbits hardly seemed long enough. By the beginning of July, the rendezvous evaluation pod was cut from the first manned mission.[46]

The pod stayed on the fourth flight and second manned mission, scheduled for seven days in orbit during January 1965, three months after the third. This longer interval between launches was planned for the rest of the program. The two months that had been allowed no longer seemed time enough to check out machines and train crews. Another change in the flight program inserted a rendezvous mission between the two longer flights, so the fifth would be a rendezvous mission and the sixth would remain in orbit 14 days. The two long missions had been back-to-back, but this left little time to absorb the lessons of one such flight before launching another. The last six missions, each about three days long, all focused on rendezvous. The final flight was scheduled for January 1967, nearly two years after the date first approved in December 1961 and more than a year later than expected after reprogramming in late 1962. The new flight plan also reflected the uncertain status of the paraglider landing system, now scheduled only from the seventh flight on. Earlier spacecraft would rely on parachutes, and the first land landing was not expected until October 1965.[47]

NASA Headquarters approved the new Gemini flight plan on 29 April 1963.[48] The lengthened schedule and spaced-out launches eased the pressure on Project Gemini in terms of both time and money. Technical problems and money shortages were the proximate cause of the changes, but throughout 1962 the shape of Gemini had been subtly shifting. Mercury technology proved less easy to transfer to Gemini than expected, partly for technical reasons—the planned coupling of two Mercury environmental control systems to provide for a Gemini crew, for example, went by the board as engineers tried and failed to convert the concept into detail specifications[49]—but mainly because the image of Gemini had altered in the eyes of its makers. "Instead of being merely a transition between Mercury and Apollo," Gilruth told his colleagues in the Management Council on 30 April, "the Gemini program now actually involves the development of an operational spacecraft."[50]

Holmes spelled out what this meant in a lengthy memorandum to Seamans on 3 May. By building into Gemini the most up-to-date technology, rather than merely modified Mercury equipment, "Gemini would have extensive and most useful applications in earth orbital space operations," even, ultimately, "as a resupply vehicle for future space stations." It would also produce a beneficial side effect: the new Gemini promised to be a much greater help to Apollo in such areas as systems development, preflight checkout, and mission training. None of this came cheaply, either in time or money, but Holmes argued it was worth it because "we have a much more valuable and worthwhile Gemini Program than could have been had if we had not taken advantage of our increased knowledge to develop and design the best spacecraft possible within the limits of our present technology."[51]

These were the arguments that NASA spokesmen used to explain the higher costs that Gemini had incurred in the past fiscal year and to defend their budget request for fiscal year 1964 to congressmen growing restive in the face of soaring NASA needs. Gemini, Holmes told the House Subcommittee on Manned Space Flight, was "much more than a big, overgrown Mercury."[52] It had, said Webb, "what I would characterize as the potential for the first workhorse of the Western space world in very much the same way that the DC-3 airplane became a great workhorse of aviation for many, many purposes."[53]

How much of this was merely after-the-fact rationalization may be open to question, but whatever hopes NASA officials might have for using Gemini or helping Apollo depended on solving some urgent problems. Development of the new technology that was to transform Gemini was lagging. The most advanced spacecraft systems—propulsion, escape, and fuel cell—were running into trouble; the paraglider program had faltered; and, worst of all, the Titan II launch vehicle posed a question mark for manned space flight. Maybe Gemini would become a workhorse, and maybe that prospect was good reason to delay the flight program. But the many technical problems, Gemini's new acting manager admitted when interviewed by a leading trade journal, had already wrecked the old schedule.[54]

## ATTACKING PARAGLIDER AND TITAN II PROBLEMS

The most pressing worry when Mathews took charge of the project in mid-March 1963 was what to do about the trouble-plagued paraglider development program. Back-to-back failures, as North American tried to deploy the wing in flight, had destroyed both half-scale test vehicles. GPO had been funding paraglider on an interim basis since February, little money was left, and North American was ready to quit unless it got new directions. With neither time nor money enough to replace the two lost test vehicles, GPO had to work out a new test program with North American, using the hardware still on hand or almost ready—the two full-scale test vehicles slated for deployment tests, the half-scale boilerplate left over from emergency parachute system qualification, and the paraglider trainer that North American was building.[55]

Spokesmen for North American and MSC met in Houston 27-28 March to discuss the options. Telephones in GPO, in the Gemini Procurement Office, and in North American were busy over the next two weeks as the main features of a revised test program were argued, talked out, and settled. The key decision was to divide the flight sequence in half and work through the problems of each phase separately before trying to demonstrate a complete flight from deployment through landing.[56]

Spreading the wing in flight was still the crucial problem, and it was to be tackled with the two full-scale test vehicles. The new test plan, however, was simpler than the old. As the vehicle dropped from a high-flying aircraft, its wing would inflate and deploy to convert its fall into a glide down to 3000 meters. That ended the test sequence. Explosive charges would sever the cables that suspended the test vehicle from the wing, and the now wingless vehicle would descend to Earth beneath a large parachute. The rest of the flight sequence, gliding from 3000 meters to a landing, was to be studied with two tow-test vehicles, modified versions of the paraglider trainer. Towed by a helicopter to the proper altitude and then released, this vehicle would be flown by a pilot down to the California desert. In the final stage of the program, Gemini static articles would be fitted with standard paraglider gear and flown through the complete flight sequence from deployment to landing.[57]

If everything went according to plan, the paraglider landing system could be ready for the seventh Gemini spacecraft. By the time McDonnell started building the tenth spacecraft, paraglider gear could be installed at the proper place on the production line.[58]

On 12 April 1963, Mathews outlined for North American what had to be done at once to put the new program into effect. The company was to stop all work on landing gear for the full-scale test vehicle, since it would now land via parachute, and to forget about trying to convert the half-scale boilerplate into a half-scale test vehicle. Instead, the boilerplate would be used as a tow-test vehicle to work out takeoff techniques needed later for manned flights. North American also had to qualify the new full-scale parachute system, which differed substantially from the emergency system—using three Mercury-type parachutes—that North American had tried hard to qualify, without much success, during the summer and fall of 1962. By the end of April 1963, North American had shifted gears and was working along the lines laid out earlier that month.[59]

The reoriented paraglider program was formalized in a new contract between North American and NASA on 5 May 1963 that also

*133*

closed out the earlier contracts. MSC and the contractor agreed on a year-long program (to May 1964) more tightly focused on the basic design of a workable paraglider system than the old had been, with such matters as flight training and production postponed until the design had been proved.[60] NASA settled the earlier contracts with North American for $7.8 million and negotiated a $20-million price for the new effort that was intended to save paraglider landing for Gemini.[61]

Although doing something about paraglider was the most pressing problem Mathews faced when he took over Gemini, Titan II was the greater concern for the program as a whole. So far, Air Force efforts toward clearing up the troubles had been limited to what was needed to make its missile work. Nothing extra was yet being done to see that Titan II met Gemini's needs, although Bernard Schriever had assured Holmes that any Titan II problems that threatened Gemini would be taken care of.[62] Pogo seemed to Mathews, as it had to Chamberlin, the most urgent, and Mathews, like Chamberlin, insisted that $\pm 0.25$g at the spacecraft was the highest level of vibration that NASA could accept. BSD, however, professed to be content with the g-level of $\pm 0.6$g already achieved, well below earlier levels as high as 5g. That was low enough for the missile, and BSD firmly refused to spend any more of its money to lower it further.[63]

GPO could do little to change BSD's stand, but Schriever, whose command embraced BSD, did have something to say about it. He ordered top officials of both BSD and SSD to his headquarters at Andrews Air Force Base in Maryland on 29 March 1963 to present a status report on Titan II problems related to its role as Gemini launch vehicle. Spokesmen for the major Titan II contractors—Martin, Aerojet, Aerospace, and Space Technology Laboratories—were on hand to discuss their efforts. What Holmes and the other NASA representatives Schriever had invited to the meeting heard was far from reassuring.

Brigadier General John L. McCoy, Director of BSD's Titan System Program Office, led off with an account of the two outstanding problems, longitudinal oscillation and combustion instability. Neither, he stressed, now threatened missile development. Trying to meet Gemini standards by changing any of the missiles still to fly in the development program was too chancy. McCoy's job was to develop a weapon system, which he objected to risking for Gemini.

The contractors argued that the problems were just about solved. Both Aerospace and Martin-Baltimore endorsed the optimistic view of Aerojet-General's chief project engineer for Titan II engines, Alvin L. Feldman. Feldman pointed out that Pogo had already responded to increased fuel-tank pressure, and he saw even more promise in a combination of standpipes in the oxidizer lines and mechanical accumulators in the fuel lines. Unstable burning might be handled by modifying the baffles on the injector that fed propellants to the engine or by starting the flow of propellants with some inert fluid.

A closed-door session limited to NASA and Air Force officials followed this open session. Here Holmes vented his frustration at the parade of numbers, statistics, and percentages on Titan II problems he had heard. The crucial point, he insisted, was that no one knew what caused either Pogo or unstable burning; without that knowledge, the booster could not be judged man-rated. Since the Air Force was now a bigger partner than before in Gemini, Holmes thought that Defense funds ought to pay a share of whatever the price might be to fit the launch vehicle to Gemini. But even if NASA had to pay the whole bill, even if Gemini had to face more delays, Holmes wanted these shortcomings corrected. Lieutenant General Howell M. Estes, Schriever's second-in-command, agreed. They decided on a joint development and test program expressly designed to bring Titan II up to Gemini standards, with Air Force Titan II money to get it started and the question of funding the rest to be referred to the Gemini Program Planning Board.[64]

Just three days later, on 1 April, McCoy was heading a new Titan II/Gemini Coordination Committee,* which, by 5 April, had drawn up a "Joint Titan II/Gemini Development Plan on Missile Oscillation Reduction and Engine Reliability and Improvement." It spelled out the work needed to cut Pogo levels to NASA standards and to reduce the incidence of combustion instability in the second-stage engines. It also outlined an "augmented engine improvement program" to clean up the design of the first- and second-stage engines and to enhance their reliability. McCoy's committee planned to direct the effort, with funds supplied by BSD's Titan System Program Office. The plan to improve and man-rate Titan II had two major restrictions: the weapon-system's flight test program was not to incur undue delays by waiting for Gemini items; and McCoy had the final say on if and when to fly Gemini improvements, with missile program objectives taking precedence.[65]

The Gemini Program Planning Board concurred in the plan a month later, on 6 May, and recommended that the Department of Defense pay for it, starting at once with current Defense emergency funds. This meant $3 million from fiscal year 1963 money and another $17 million from the next year's budget. The Air Force provided half the $3 million by the end of the month, with a firm promise for the balance.[66]

In acting on the Titan II plan, the board was moving beyond its charter, which called for it simply to decide what military experiments

---

*Members were Richard C. Dineen of SSD, James A. Marsh of Aerospace, and James G. Berry, Titan II project director for Space Technology Laboratories.

should be carried on Gemini flights. Its roster of members, however, included Holmes and Schriever, as well as Seamans and McMillan, making it the logical group to coordinate a high-level attack on Titan II's problems. When the board submitted its recommendations to Secretary of Defense McNamara and NASA Administrator Webb on 29 May, no one was surprised that it covered not only experiments but the pursuit "with utmost urgency" of the Titan II improvement plan, using Defense funds and the missile test program.[67] McNamara and Webb endorsed the board's findings. McNamara specifically agreed to pay for the program and directed the Secretary of the Air Force both to fund it and to flight-test the improvements in the missile program. In a memorandum to the board members, Webb stressed

> the urgency we attach to the development of the Gemini Launch Vehicle. It is of the utmost importance that the cause of the present deficiencies in the Titan II be determined and remedial action accomplished as expeditiously as practicable. . .to eliminate the launch vehicle as a potential source of delay in the Gemini schedule.[68]

The delay was already more than potential, as attested by the major role Titan II problems had played in Gemini's new flight program. But further delays loomed ahead as the Titan II missile test program unexpectedly faltered during the spring of 1963 and threatened to undo the improvement plan before it had fairly begun. The 18th flight test of the Titan II missile was launched on 24 May 1963. It was only the 10th fully successful flight and the last for months to come.[69]

The next launch, five days later, produced a particularly disappointing failure. Martin, Aerojet, Aerospace, and Space Technology Laboratories had worked hard to confirm the hypothesis that Pogo during first-stage flight was caused by coupling between the missile structure and its propulsion system, the couple making an unstable closed loop. A study of year-old static-firing data led Sheldon Rubin of Aerospace to believe he had found the missing link in the analytic model; the partial vacuum produced by pumping caused hydraulic resonance in the fuel suction line. If valid, this finding would correct the two major shortcomings of prior analyses, which had failed to predict where oscillations ceased during flight and had wrongly predicted that oxidizer standpipes alone would suppress Pogo. Rubin's corrected model showed why Missile N-11 in December 1962 was less stable than other Titan IIs and how adding fuel accumulators as well as oxidizer standpipes would suppress Pogo. The missile launched on 29 May carried Pogo suppression devices for both oxidizer and fuel to test their combined effect. But, leaking fuel in its engine compartment, the missile burst into flame as it lifted off. Its controls damaged by the fire, the missile pitched over and broke up 52 seconds later. In contrast to Missile N-11, the Pogo devices were absolved from any blame for the failures, but the flight ended too soon to provide any Pogo data and the problem remained unsolved.[70]

This setback was followed by another, on 20 June, in the 20th Titan II flight. This was purely a military test, the missile being launched from a silo at Vandenberg Air Force Base in California. First-stage flight was troublefree, with Pogo levels low enough ($\pm.62$g) to meet Air Force standards. But partial clogging of the tiny holes in the oxidizer injector of the second-stage gas generator caused thrust to fall off shortly after staging to about half the required value. The same thing had happened in two earlier tests; had the missile been carrying a spacecraft, its crew would have been forced to abort the mission.[71]

Back-to-back failures at this stage in the program compelled BSD to suspend Titan II flight testing. Only half the 20 flights so far launched could be called fully successful, and McCoy now faced the task of making good on at least 12 of the 13 flights still left him, to prove that Titan II was ready to join America's strategic deterrent forces. The missile had to come first, and McCoy again ordered a halt to any further attempts to lower Pogo levels as too great a risk to what remained of his test program. Although Major General Ben I. Funk, SSD commander, appealed McCoy's decision to Systems Command Headquarters, the whole question of Gemini-Titan development, and particularly of flight-testing a cure for Pogo, was once more unsettled.

## A CLOUDED FUTURE

In the aftermath of reprogramming, Gemini was buffeted by new crises. An offhand Defense Department bid to take over the program flustered NASA's top echelons briefly, but technical problems began taking on fearsome proportions early in 1963, with paraglider and Titan II looming as the greatest question marks. When the first months of 1963 also revealed that Gemini's money troubles had not been settled, the stage was set for a change of project managers. Charles Mathews replaced James Chamberlin as head of a faltering program. The framework was solid enough, a tribute to Chamberlin's engineering efforts, but costs, schedules, and administration were not. Mathews moved swiftly and smoothly to take these problems in hand. In short order, the status of the program was reviewed; its schedules, budgets, and objectives reassessed; and its revision outlined. By mid-1963, Gemini's managerial worries, both internal and external, had been at least temporarily resolved by a tightened organization, a lengthened schedule, and a modified program. But the major technical problems persisted and even worsened.

With many of the Gemini launch vehicle's parts still short of flight status and with BSD firmly opposed to risking its own program to solve Gemini's problems, the prospect of meeting the December 1963 deadline for the first Gemini launch was dimming. NASA was no longer concerned simply with the status of the vehicle and the effect of specific problems like Pogo and unstable combustion on its chances of being ready in time. Although its promise had been great, Titan II's flight record was so poor that NASA was beginning to wonder whether it belonged in Project Gemini at all.[72]

# VII

# The Darkest Hour

THE easing of Gemini's managerial problems by mid-1963 opened the way for a concerted attack on Gemini's technical problems. Even under new management, however, the last half of the year saw Project Gemini at its lowest ebb. The Gemini spacecraft, the Agena target vehicle, and, most seriously, the Titan II launch vehicle—each raised problems that threatened to overwhelm the program. This was to be Gemini's darkest hour, and it began with another dual flight that raised new fears of a Soviet victory in the race for first space rendezvous. On 14 June, Lieutenant Colonel V. F. Bykovsky orbited aboard *Vostok V*. Cosmonaut Valentina Tereshkova followed two days later in *Vostok VI*. The two passed within five kilometers of each other. Once again, however, there was a crumb of hope in the Vostok's lack of maneuvering capability. It was a faint hope.[1]

## TITAN II IN JEOPARDY

Gemini's biggest question mark in mid-1963 was the launch vehicle. Flight tests of the Titan II missile, suspended in June after two successive failures, had yet to produce results good enough to convince anyone that a booster derived from this missile was a safe bet for Gemini. To make matters worse, Brigadier General John McCoy, director of Titan programs for the Air Force Ballistic Systems Division (BSD), strongly opposed any changes in the missile to meet Gemini standards—and for sound reasons. He could not afford to risk the failure of the missile program for a chance to help Gemini.

As the Titan II program faltered, NASA concerns mounted. The Gemini Program Planning Board persisted in its efforts to resolve the impasse between NASA and BSD. On 28 June, the board asked NASA to state the least it would accept for launch vehicle performance, the Air Force to describe its program in detail. Board co-chairman Robert Seamans, NASA's Associate Administrator, asked MSC Director Robert Gilruth for a precise statement of MSC standards for making Titan II over as the Gemini launch vehicle. The response, on 1 August, was a brief review of "Gemini Launch Vehicle Specifications and Requirements," which pinpointed the three major problem areas that made the Titan II unsafe for manned space flight—longitudinal oscillation (Pogo), dynamic instability of the second-stage engines, and detail design faults of Titan II engines. MSC insisted "that these problems must be satisfactorily solved and the solutions incorporated into the GLV prior to its use in the manned Gemini program."[2]

Every Titan II so far flown had displayed Pogo, although the level had varied, reaching a low of just over one-third the force of gravity ($\pm 0.35$g) in the 17th test flight on 13 May 1963. This potential hazard to pilot safety prompted a survey of available data on human tolerance of such vibration, leading MSC to conclude that Pogo should be completely eliminated, or at least not allowed to exceed $\pm 0.25$g. A test program on the centrifuge at NASA's Ames Research Center in California, completed in July 1963, tended to confirm the validity of this stand; an MSC astronaut test program conducted immediately after the Ames tests provided even stronger support. Higher levels might be tolerable, but 0.25g still seemed a prudent upper limit. MSC preferred an experimental program to trace Pogo to its source and eliminate it but would settle for this bearable limit if proved on Titan II flights before the vehicle flew in Gemini.[3]

The second major problem, combustion instability, had not yet occurred in flight, but Aerojet-General's ground tests had revealed incipient instability during second-stage starting—that is, the initial engine-firing pulse could trigger uneven burning in stage-II engines. In a statistical sense, the engine was stable, since Aerojet-General could show that the instability rate was no more than two percent in ground tests. From a physical viewpoint, however, the engine had to be described as dynamically unstable, and that risk could not be accepted when human lives were at stake. Statistical reliability was not enough for a manned booster. Aerojet-General must develop and prove a dynamically stable engine before the first manned Gemini flight.[4]

The third major area of concern comprised a range of problems, each minor in its own right but significant in the aggregate. Of the 10 full or partial failures in the 20 Titan II test flights to date, Pogo could be blamed for only one, dynamic instability for none at all. The others resulted from small defects—a clogged injector, a failed weld, a broken line. The central problem seemed to be "a real lack of understanding on the part of Aerojet of procedures and responsiveness to problems that must be associated with the development of engines for use in a manned launch vehicle."[5]

When several top-ranking MSC officials visited Aerojet's Sacramento plant in July 1963, they were dismayed at what they saw and concerned about a number of questionable practices in design, manufacturing, and quality control, in general, and several components—turbine idler gears, main fuel valves, turbine seals, and turbine manifolds—in particular. The Air Force Space Systems Division (SSD), NASA's agent for launch vehicles, had already spotted 40 engine parts that could be improved. MSC judged that most of these changes had to be made and the results confirmed in flight before the booster was committed to the first manned Gemini mission.[6]

The Gemini Program Planning Board heard NASA's report on launch vehicle performance standards on 5 August 1963, revised the wording slightly, and accepted it. With this statement as a basis, MSC and SSD were to arrange a formal agreement on the goals of reduced Pogo, a stable second-stage engine, and improved engines. They were also to agree on the programs needed to achieve these goals and the criteria for deciding when the goals had been met.[7]

Although Titan II itself was still a question mark, the managerial logjam that had so far prevented a concerted attack on its shortcomings as a manned booster now appeared to be breaking up. Major General Ben Funk, SSD Commander, told Gilruth on 8 August that Air Force Headquarters had approved the "augmented engine improvement program." Funk agreed that Aerojet's efforts left something to be desired, then outlined a series of steps he had taken to tighten up the firm's work. He had still another piece of good news. The decision to fly no more Pogo fixes on Titan flights had been reversed. The gas generator clogging problem that had marred the Titan II flight of 20 June seemed to have been solved, and the booster would soon be flying again. Missile N-25, scheduled for a September launch, would carry standpipes and accumulators to suppress Pogo.[8]

Aerojet-General began work on the improved engine program in September. That same month also saw a start on the Gemini Stability Improvement Program, or Gemsip, an effort to redesign the injector of the second-stage engine to overcome incipient combustion instability.[9] When the Gemini Program Planning Board met again, on 6 September, MSC and SSD had agreed on the statement of "Gemini Launch Vehicle Specifications and Requirements for Major Titan II Problems" that the board had requested.[10] It fully met NASA's demands. Things seemed to be moving at last.

Titan II, however, had yet to prove itself. Missile problems had already prompted NASA, earlier in 1963, to replace one of Gemini's manned missions with a second unmanned flight. Still unsolved, they now forced NASA to plan yet another unmanned flight. On 12 July, Mathews told MSC's senior staff that GPO was thinking about backing up the first Gemini flight with an extra unmanned flight (making a total of 13 instead of 12) roughly midway between the first two scheduled missions, or about 1 April 1964. The proposed payload was a boilerplate capsule with instrumentation pallets like those in Spacecraft 1.[11]

At a meeting on 5 August, the Gemini Program Planning Board agreed to review the plan. The next day, Mathews wired Walter Burke at McDonnell to begin work on the adapter that would attach capsule to launch vehicle. NASA Headquarters approved the new mission and suggested calling it Gemini 1A, or GT-1A.* Based on data from McDonnell and SSD, the project office figured the cost of the extra flight at around $2 million.[12]

William C. Schneider, Gemini Project Manager at NASA Headquarters, presented NASA's case for the extra flight to the planning board on 6 September. In essence, NASA wanted to guard against a failure of the first mission by planning a contingent mission, identical to GT-1, to fly before the scheduled GT-2. The board concurred, and Mathews wired Richard Dineen, SSD's Gemini launch vehicle overseer, to make sure that the second launch vehicle would be ready in time to meet the date for GT-1A. The new mission was strictly a backup, however, to be flown only if GT-1 failed to meet its objectives. The decision waited on the outcome of the first mission.[13]

For GT-1A, MSC diverted a boilerplate spacecraft being built for flotation tests by a local Houston contractor. Named Boilerplate 1A, it arrived at the Center on 24 September, where the Technical Services Division began the task of making it flightworthy. Regular biweekly panel meetings started early the next month, and the rebuilt boilerplate was ready in mid-November. It left Houston via flatbed truck on 13 December, reaching Cape Canaveral three days later, there to have its wiring and equipment installed; the work in Houston had been limited to the structure. The adapter, built and instrumented by McDonnell, arrived at the Cape 27 January 1964. By then, however, the threat that had called forth the effort had largely dissipated, and little further work was done before GT-1A was formally canceled on 17 February.[14]

That cancellation reflected a striking turnaround in Titan II prospects from their lowest ebb during the summer and fall of 1963. BSD resumed the flight test program on 21 August. Although the flight itself was a success, NASA suffered another setback. This missile was the

*GT, for Gemini-Titan, had become the standard designation for non-rendezvous missions; GTA, for Gemini-Titan-Agena, for rendezvous missions.

first of five planned to carry the Gemini malfunction detection system, crucial for Gemini because it was to provide spacecraft pilots with the data they needed on existing or impending booster problems during launch. BSD had agreed to fly the system "piggyback"—installed, working, and reporting to ground receivers and recorders, but not otherwise acting on the missile. The system flown on 21 August suffered a short circuit 81 seconds after liftoff and provided no further data.[15]

Titan II's next launch, on 23 September, did little to dispel the gloom. A guidance malfunction threw the missile out of its planned trajectory. Since the missile was guided inertially and the Gemini booster used radio guidance, this had no direct bearing on Gemini. That was small consolation, however; Pogo reached $\pm0.75$g, very nearly the worst since the disastrous flight of Missile N-11 in December 1962.[16]

The heart of the matter was foot-dragging by BSD on the question of flying Gemini fixes. Once again, the planning board took a hand. It decided to replace the agreement between MSC and SSD of 6 September with a more authoritative Memorandum of Understanding between the co-chairmen of the board, Seamans of NASA and Brockway McMillan, Under Secretary of the Air Force. The board directed NASA to submit another statement of requirements for the Gemini booster and the Air Force to provide a development plan, complete with costs and schedules, for dealing with Pogo, combustion instability, and engine improvement. The board specifically asked the Air Force for a schedule of all remaining Titan II flights, with a plan for flight-testing changes to reduce or eliminate Pogo and unstable burning.[17]

The meeting of the board took place on 11 October 1963. Four days later, the flight-test question was finally resolved. General Bernard Schriever, a member of the board as well as commander of Air Force Systems Command, called a meeting in Los Angeles of BSD, SSD, and Titan II contractors. Schriever himself firmly supported an active program to clean up launch vehicle problems. Of special concern was whether to follow through with plans to fly Missile N-25 with oxidizer standpipes and fuel accumulators. Aerospace, backed by Space Technology Laboratories, argued strongly for the planned flight, especially since engine ground tests begun in August had confirmed fuel-line resonance as the culprit in the failure of Missile N-11 and shown that fuel accumulators would solve the problem. They carried the day, winning the crucial decision to proceed with the test flight of N-25 as planned. Funk planned to see his BSD counterpart regularly and arranged for meetings between the two project managers, Dineen and McCoy, to make sure that there was no more backsliding.[18]

Later events were to prove that this time the question had, indeed, been settled. Meanwhile, however, only the test flights could show that more determined management was the answer to the technological problems. Titan II was still in trouble, and the weekly status reports that Seamans was getting from the Air Force Systems Command after mid-September reflected a promising beginning but little more.[19] Some thought was even being given to dropping Titan II from the Gemini project altogether. The Propulsion and Vehicle Engineering Laboratory of NASA's Marshall Space Flight Center began to study the desperate expedient of substituting the Saturn I launch vehicle for both Titan II and Atlas.[20]

### PARAGLIDER ON THE WANE

Work on the reoriented paraglider program of May 1963 got off to a quick start. Before the end of the month, North American Aviation was working out techniques for launching a tow-test vehicle from the ground. This preliminary effort, which involved first a car-towed half-scale vehicle and then one towed by helicopter, was designed to show what the paraglider would do during towing and liftoff and to work out proper towing techniques, all this to prepare for that part of the new test program in which a pilot would fly the test vehicle from an altitude of 3000 meters to a landing. NASA's Flight Research Center also conducted a series of tow tests, the whole effort being completed in mid-October 1963.[21]

May 1963 also saw North American begin work on the other phase of the new test program, testing the deployment sequence with the full-scale test vehicle. Since this phase of testing called for the test vehicle to land by parachute, the first step was to qualify a parachute recovery system, one standard Gemini parachute backed up by a second. North American got off to a smooth start. Two drops of a small bomblike test vehicle on 22 May and 3 June showed that the system's two small stabilization parachutes worked. The contractor quickly began testing the full system on a boilerplate test vehicle. A minor malfunction marred the first drop on 24 June, but three good tests followed in July, with only one more needed to prove the system. What was to have been the final drop, on 30 July, brought a crucial setback. Both main and backup parachutes failed, and the boilerplate crashed.[22]

The company wanted to get on to the next phase of testing and argued that the failure could be safely ignored, partly because North American believed it knew how to correct the problem, partly because further tests would require a new boilerplate and mean a delay in the program. The logic was sound enough, but GPO feared that, although the immediate problem might be easily corrected, its root cause—the instability of the vehicle—might produce other, and worse, problems.

*144*

## GEMINI PARACHUTE LANDING SEQUENC

| | |
|---|---|
| 50,000 FEET | HIGH ALTITUDE DROGUE CHUTE DEPLOYED |
| 21,000 FEET | OPEN CABIN VENT VALVE |
| 10,600 FEET | PILOT PARACHU DEPLOYED |
| 9,600 FEET | R&R SECTION SEPARATION |
| 9,000 FEET | MAIN CHUTE DEPLOYMENT |
| 6,700 FEET | TWO POINT SUSPENSION |
| 1,500 FEET | CABIN WATER SEAL CLOSED |
| SEA LEVEL | TOUCHDOWN |
| | JETTISON CHUT |

## The Gemini Parachutes

If Gemini were forced to use parachutes instead of the trouble-plagued paraglider for landing the spacecraft, the landing sites would shift from land to sea. Below, left, is a water landing test in the Salton Sea at El Centro, California. The notable difference from Mercury landings is that the Gemini spacecraft lies in the water horizontally rather than vertically. Since an emergency landing on land could not be ruled out, tests in the California desert (below right) sought impact data on vertical landings.

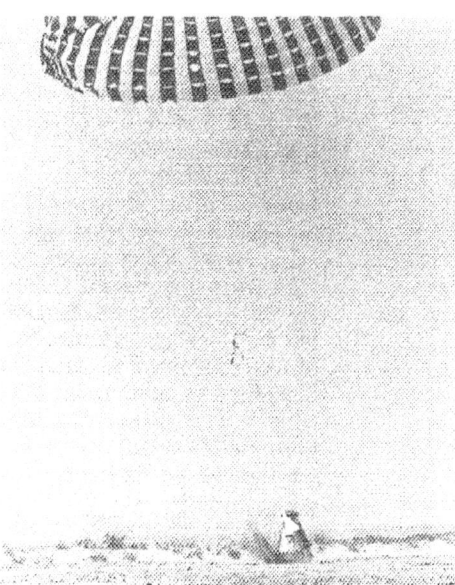

GPO and North American agreed on two further drop tests. McDonnell furnished the new boilerplate, which North American, on the basis of spin-tunnel tests, modified to provide a more stable suspension system. That took time; over three months elapsed before the next drop, on 12 November 1963. Everything worked, and another test three weeks later confirmed the result; the parachute recovery system was at last qualified for full-scale vehicle deployment tests.[23]

Proving the parachute system was not the only source of delay. Design engineering inspections of the full-scale test vehicle on 1 August and the tow-test vehicle on 27 September produced the normal share of required changes. Wind tunnel tests of North American's first full-scale prototype wing at Ames Research Center in October yielded too little data and had to be repeated in early December. So it was late November before the contractor could deliver the first tow-test vehicle to Edwards Air Force Base to begin its manned program and mid-December before the two full-scale vehicles arrived.[24] With almost two thirds of the time available under the new contract exhausted, North American had yet to begin the major flight-testing portion of the program.

By the fall of 1963, the status of paraglider in Gemini was once more in jeopardy—only partly because of North American's troubles. The inflated frame used in the paraglider design was being challenged by advocates of what seemed to be a viable alternative—an all-flexible gliding parachute, the so-called parasail. This device offered a lift-to-drag ratio ranging from 0.9 to 1.2, lower than paraglider's but still enough to provide worthwhile range and control. It was further handicapped by its relatively high rate of descent, which required landing rockets to cushion impact with the ground. But, overall, parasails matched conventional parachutes closely enough to promise a reasonably quick and relatively cheap development of a reliable device for land landing.

The gliding parachute had, in fact, competed with the inflated-frame paraglider design back in 1961, when the choice of a land-landing technique for what was then the Mercury Mark II project was being made. Although rejected for Mark II, the concept persisted as the subject of a modest research and development program at MSC.[25] As paraglider faltered, parasail seemed more attractive. Project Gemini's new manager, Charles Mathews, was more receptive to parasail—or less committed to paraglider—than James Chamberlin had been. Supported by MSC Director Gilruth, Mathews called on GPO for another look at parasail. In April 1963, after the second half-scale test vehicle had crashed but before the future of the paraglider program was decided, he asked McDonnell to study changing Gemini's landing system from paraglider to parasail.[26]

While McDonnell pursued its study, MSC's Flight Operations Division and Systems Evaluation Division continued testing a parasail system and pressing for its adoption. Paraglider still had highly vocal backers, however, who denied that its problems involved anything more than sequential details that would have to be ironed out for any recovery device, even conventional parachutes. Claiming that paraglider development had been known from the first to be a hard task, they objected to dropping it after so much of the work had already been done.[27] The lines were drawn where they had been in 1961: Flight Operations Division and the Engineering and Development Directorate still opposed paraglider; most of the project office and the prospective pilots, supported by Flight Crew Operations, favored it.

When McDonnell finished its study early in September 1963, the issue was carried to NASA Headquarters. The company's informed guess at the cost of a parasail and landing-rocket system for the Gemini spacecraft was $15.7 million, with a good chance to be ready for Spacecraft 7. When the parasail proposal was informally presented to NASA Headquarters on 6 September, it was rejected. Dropping paraglider on the verge of flight testing, leaving nothing to show for all the time, money, and effort already spent, was out of the question. The alternative, going ahead with parasail development as something to fall back on if paraglider failed, was ruled out for lack of funds to support both tasks at once.[28]

Although reprieved, the paraglider program did not come through unscathed. High-level talks between MSC and NASA Headquarters produced still another reorientation of the program.* The paraglider landing system program was stripped of all other objectives, leaving as its only goal proving paraglider's technical feasibility—which meant primarily showing that the wing could be inflated and deployed in flight to achieve a stable glide—with the accent on staying within the $16.1 million budgeted for fiscal year 1964. Until that goal had been met, there was to be no further work on a prototype system for Gemini, much less on production. Gilruth insisted on a clear understanding that paraglider might still fly on Gemini if the flight tests succeeded, that paraglider's future in Gemini had not been foreclosed.[29] The implication of foreclosure was nonetheless there.

Under orders from MSC, North American ceased its efforts to keep the full-scale test vehicle fitted with the latest Gemini equipment. MSC also directed McDonnell to stop all testing related to installing the paraglider, to design parachute versions of all Gemini spacecraft, and to plan on putting paraglider in the last three, the last two, or

---

*Major participants were MSC Director Gilruth, NASA Associate Administrator Seamans, George E. Mueller (who had recently replaced Brainerd Holmes as Deputy Associate Administrator for Manned Space Flight), and George Low (Mueller's Deputy Director for Programs).

only the last spacecraft. Nothing of paraglider was to remain in the spacecraft except the option to put everything back if the flight testing succeeded. Parachutes had, by late 1963, displaced paragliders as the planned means of recovery through the ninth mission. Paraglider landing was still listed for the last three Gemini flights, but some planners, SSD Commander Ben Funk among them, assumed paraglider would not be included in the tenth mission, either, "and probably will not be carried on any of the twelve flights."[30] The very fact of paraglider's doubtful status had already begun to close off any real chance to fly in Gemini, whether it proved itself or not.

A common feature of spacecraft development, and always a matter of concern, seems to be an innate tendency toward weight growth. Gemini was no exception. A complete paraglider landing system weighed almost 360 kilograms more than a conventional parachute recovery system. Once paraglider's place had been questioned, that difference was seen as a bonus and was simply used up. Experiments, for instance, began to encroach on as yet unfilled space allotted to paraglider, especially after January 1964, when the Manned Space Flight Experiments Board was formed. Gemini's planners were beginning to look on paraglider as an extra demand on the payload budget, already pushing the limits set by the booster. If paraglider were to be restored, some other mission objectives would have to give way.[31] In other words, even if North American succeeded in showing that paraglider worked, that could no longer guarantee an attempt to fly the system in Gemini. Everything rested on the outcome of North American's upcoming effort to deploy the wing on the full-scale test vehicle in flight; although success could not ensure a place for paraglider, failure would surely bar it.

### SPACECRAFT SYSTEMS BECOME MORE TROUBLESOME

Work on the systems that made up the Gemini spacecraft was moving along well in early 1963. Design had largely been completed, and developmental tests were starting.[32] In some instances, this revealed unexpectedly hard problems. Three systems, in particular—fuel cell, propulsion, and escape—began to emerge as potentially critical areas. As a group, these systems called for the largest advance beyond existing technology. Each was essential to a major Gemini objective, each was new to the manned space flight program, and each resisted efforts to resolve its problems.

A major innovation in the Gemini spacecraft was the substitution of fuel cells for conventional batteries as the prime source of electrical power during flight. McDonnell had subcontracted the development of this system to General Electric (GE). By the end of 1962, GE had completed facilities at its Direct Energy Conversion Operation in West Lynn, Massachusetts, to produce fuel cells. GE had also surmounted the first serious development problem: leakage of oxygen through the cell's ion-exchange membrane, which proved to be largely the result of mechanically induced stresses rather than an inherent design weakness.[33]

Solving this problem, however, exposed another. With leakage controlled, fuel-cell test units working over longer times showed degraded performance. The cause appeared to be contamination of the membrane by metal ions from the fiber glass wicks that removed water produced by the operation of the cell. Leaks in the tubes that fed hydrogen to the cell were a second source of test failures. Both problems demanded design changes. Dacron cloth replaced fiber glass wicks, and a titanium-palladium alloy supplanted pure titanium tubing, which had proved susceptible to cracking. Slow delivery of both materials, as well as the necessary redesign, began to affect schedules. Dacron produced its own problems: the new wicks touched the membrane, drew off electrolyte, and impaired cell function. Thinner wicks were an easy answer.[34]

The test failures, design changes, and revised production techniques combined to delay the fuel-cell program. GPO began looking for ways to increase the rate of fuel-cell production and to install fuel cells at a later point in spacecraft assembly. A visit to GE in May 1963 convinced both GPO and McDonnell that the current program was unrealistic; schedules allowed too little time for testing and failed to provide for contingencies or troubleshooting.[35] Throughout the spring and summer of 1963, McDonnell and GE kept juggling test and production units, trying to meet ever less tenable schedules, as slippage in the fuel-cell program mounted.[36] These efforts were complicated by further development problems.

The project office was far from certain that fuel cells would be ready on schedule, even when GE began shifting its main effort from engineering and development to making fuel-cell stacks on the production line.[37] On 27 August 1963, GPO asked McDonnell for an engineering evaluation of batteries for electrical power in Spacecraft 3, the first man-carrying ship, scheduled for October 1964; the fuel cells were to remain aboard to be used only on a test load for purposes of flight qualification. When and if proper operation was confirmed, they might then be hooked into the spacecraft main electrical system. McDonnell had a plan for dual installation of batteries and fuel cells ready within a month.[38] Mathews then requested a design study of substituting batteries for fuel cells in all seven spacecraft planned for two-day rendezvous missions.[39]

NASA Headquarters also took action. George E. Mueller, NASA's

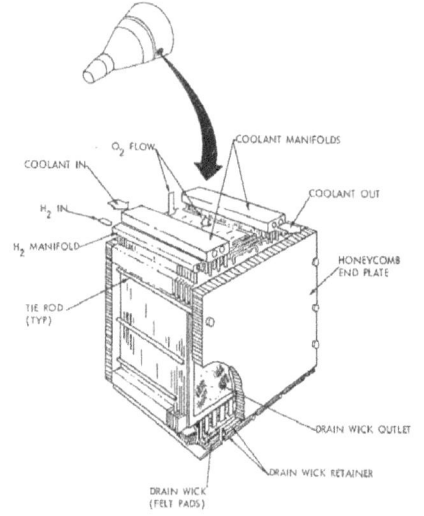

**Gemini Fuel Cells**

ION-EXCHANGE MEMBRANE FUEL CELL

PRINCIPLE OF OPERATION

LOAD

ANODE— +CATHODE

GAS CHAMBERS
ION-EXCHANGE MEMBRANE
CATALYTIC ELECTRC

HYDROGEN IN
AIR OR OXYGEN IN

$4\bar{e}$
$2H_2$
$4H^+$
$4\bar{e}$
$O_2$
$4H^+$
$2H_2O$

$H_2O$
DRAIN

ANODE
$2H_2 \rightarrow 4H^+ + 4\bar{e}$

CHEMICAL REACTIONS
OVERALL
$2H_2 + O_2 \rightarrow 2H_2O$

CATHODE
$4\bar{e} + 4H^+ + O_2 \rightarrow 2H_2O$

COOLANT MANIFOLDS
$O_2$ FLOW
COOLANT IN
COOLANT OUT
$H_2$ IN
$H_2$ MANIFOLD
HONEYCOMB END PLATE
TIE ROD (TYP)
DRAIN WICK OUTLET
DRAIN WICK RETAINER
DRAIN WICK (FELT PADS)

The Gemini fuel cell that supplied electrical power to the spacecraft consisted of three stacks connected in parallel to form a battery section. Each stack was made up of 32 cells between the end plates. At top left is a sketch of a fuel cell stack and its location in spacecraft equipment adapter section. At top right is a schematic of the principle of its operation. At bottom left is a set of three fuel cell stacks assembled without their cover. At bottom right is a fuel cell with cover undergoing test at the Direct Energy Conversion Operation, General Electric, West Lynn, Massachusetts.

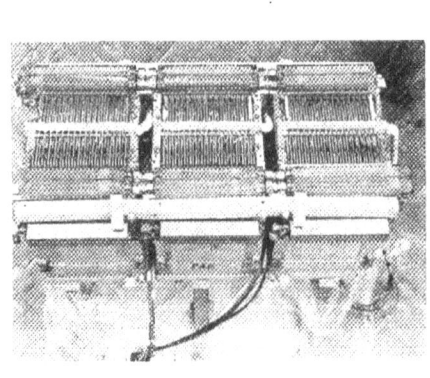

new Deputy Associate Administrator for Manned Space Flight, arranged for three senior engineers from Bell Telephone Laboratories*
to visit the GE plant to assess the status of the fuel-cell program.[40]
Rumors were already circulating that fuel-cell problems might force
NASA to limit all Gemini missions to two days.[41] GE experiments had
shown that Gemini fuel cells had an operating life of 600 hours in
theory, but a number of factors, among them the high operating temperatures imposed by a newly redesigned cooling system, had reduced
that figure to less than 200 hours in practice.[42] Fuel-cell problems were
never conceptual. As a source of electrical power for long-term orbital
missions, no one doubted that cells had a solid edge over batteries.
The rub came in trying to convert that concept into hardware to meet
Gemini specifications—essentially a matter of nuts and bolts, compounded to some extent by managerial shortcomings. This was clearly
pointed up in the findings of the Bell experts, who toured the GE
plant on 29-30 October 1963.

Their key tasks were to spot the development problems that remained and to answer two questions: Could GE solve these problems?
What were the contractor's prospects of meeting Gemini production
schedules? The team pinpointed technical matters of fuel-cell structures, materials, and the like, as exemplified by uneven current distribution because of poor contact between membrane and catalyst or catalyst and rib. The Bell engineers thought that GE could solve these
problems, given enough time. Whether there was time, however, was
something else; the team suggested that NASA might want to think
about a backup program. GE was already six months late. Despite its
stated intent to make up the lost time, GE would be doing well to
maintain the current schedule. The Bell recommendations, like those
put forward a little later by McDonnell in a survey of possible fuel-cell
changes to meet Gemini operational needs, were restricted to narrow
technical considerations.[43]

Fuel-cell production came to a halt on 26 November, as two GE
task groups tried to resolve persistent engineering and manufacturing
problems. Testing of the stacks on hand continued, but GE could build
no new ones until a thorough study had revealed the causes of poor
fuel-cell performance.[44]

Still fearing that fuel cells might not be ready for Spacecraft 3,
Mathews instructed Walter Burke to alter the spacecraft's electrical system to accept either batteries or fuel cells as power sources when the
spacecraft reached Cape Canaveral. By mid-December, convinced that
the fuel-cell system could not be qualified in time, GPO opted to fly

---

*N. Bruce Hannay, Frank J. Biondi, and Upton B. Thomas.

the first manned mission with batteries. But Spacecraft 2 would be fitted with both systems, chiefly to afford a chance to qualify the fuel-cell reactant system. The reactant supply system was a distinct development. The system, subcontracted to AiResearch, stored and fed to the cells the hydrogen and oxygen they ran on.[45]

There was still little reason to believe that fuel-cell problems could be resolved even for later Gemini flights. On 20 January 1964, Mathews asked Burke to begin work on a battery-operated system for Spacecraft 4. Switching from fuel-cell to battery power for these two spacecraft cost Project Gemini almost $600 000.[46] The GE task groups having completed their intensive six-week search for the causes of the problems, a meeting was scheduled in Houston on 27 January 1964, between NASA and its contractors to review fuel-cell status and to decide what to do about it.[47]

Although some missions might have to be curtailed, the Gemini spacecraft could carry men aloft without fuel cells by using conventional batteries. No such easy answer existed for the escape system. Any effort to replace it with something else would not only be difficult but far more costly. In the spring of 1963, some thought the change would be worth whatever it cost. MSC's Flight Operations Division revived a proposal to replace ejection seats with an escape tower, the system used in Project Mercury. Doubtful that the seat could be qualified in time and skeptical of its value as an escape device in any case, chief of Flight Operations Christopher Kraft urged Gilruth to start a backup program to see, at least, if an escape tower could be used for Gemini.[48]

Gemini Project Office, seconded by the astronauts and Flight Crew Operations, still believed that Gemini ejection seats could be made to work. Hard-to-solve problems were only to be expected in the development of so advanced a system.[49] Things were, in fact, starting to look up. Simulated off-the-pad ejection (Sope) tests had been suspended in the fall of 1962 until all system components were ready and the complete escape sequence, including recovery of dummy astronauts, could be demonstrated. The system had also grown more complex; it now included a device—a hybrid of balloon and parachute called a ballute—to prevent an astronaut from spinning during free fall if he had to eject from an altitude much higher than the 2000 meters at which his personal parachute was set to deploy.[50]

When Sope testing resumed on 7 February 1963, the results were disappointing from the standpoint of proving the complete escape sequence—the ballutes failed to inflate and release and the personal parachute did not deploy properly. But, in the view of Kenneth Hecht and his colleagues in GPO who were in charge of escape-system development, the test marked a real breakthrough. They had been convinced that the key problem was dynamic, the relationship between rocket-motor thrust vector and the shifting center of gravity of the

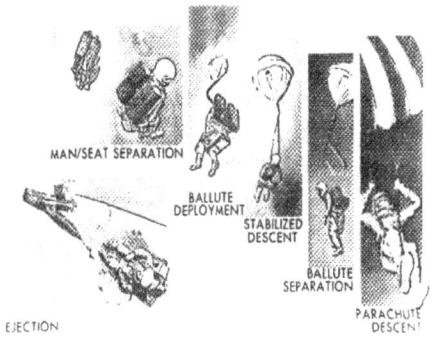

MAN/SEAT SEPARATION

BALLUTE DEPLOYMENT

STABILIZED DESCENT

BALLUTE SEPARATION

PARACHUTE DESCENT

EJECTION

## The Ballute

*When the Gemini ejection seats were used in an emergency during flight, a ballute had to be used to stabilize descent until parachute altitude was reached. This sequence is shown in the sketch above. At right, a jump test of the ballute is being conducted at the Naval Parachute Facility, El Centro, California.*

seat-man combination. Analysis of the data from the test revealed that they had been overlooking a significant factor in their calculations—the tendency of the ejecting mass to tip as a result of its inertia when it left the end of the guide rails. With that factor accounted for, the key problem was solved. "The remaining technical problems," Hecht later recalled, "were in debugging the details of a very complex design."[51]

That, however, was no small order. Measures were taken to ensure that the personal parachute would deploy at the low dynamic pressure associated with off-the-pad aborts. McDonnell and Weber engineers also cleaned up the makeshift additions to seat design that had piled up in the course of development. But the complete escape sequence still had to be proved. All that took time. The new package was given its final checkout on 22 April 1963.[52] Three weeks later, on 15 May, Sope testing was under way again, with heartening results.

*153*

The last four tests in the series of 12, which had begun in July 1962, were almost flawless, only an insignificant failure of part of the test gear marring the final test, on 16 July. The development phase of pad ejection testing was now complete.[53]

Still unfinished, however—indeed, scarcely begun—was a second series of development tests, sled-ejection tests. These were not so novel as the Sope tests, being in common use for all ejection-seat development. They simulated ejection at high dynamic pressures—as might be met in an escape during first-stage booster firing. In the Gemini tests, conducted at the Naval Ordnance Test Station in California, two ejection seats were mounted side by side in a boilerplate spacecraft carried on a rocket-propelled sled running on tracks. Known as the Supersonic Naval Ordnance Research Track, it was, obviously, called "Snort." But the delays met in Sope tests, compounded by the reprogramming of late 1962, slowed the sled program.[54]

This may have been just as well, because the test vehicle was badly damaged in its first run, on 9 November 1962. This was not an ejection-seat test. The test station needed a trial run to confirm its data on sled performance and structural soundness. It got what it wanted, but a rocket motor broke loose and smashed into the boilerplate, starting a fire. Although both boilerplate and sled needed a lot of work, GPO foresaw no delay in the sled-test program itself, since other factors had already required it to be rescheduled, leaving ample time for repairs.[55]

Flawless Sope tests on 15 and 25 May 1963 showed that the new seat design was working and sled tests could begin. A dynamic dual ejection on 20 June was a success, followed by a second good run on 9 August. That turned out to be the last test in 1963. The seat system went through still another redesign, this time to provide for the automatic jettison of backboard and egress kits.[56] A more serious problem, and one that persisted, had little to do with the system itself. Testing was continuously hampered by shortages and slow delivery of parts, particularly the pyrotechnic devices that were crucial to so many of the system's functions.* [57]

Although fuel-cell and escape systems had begun to look troublesome in 1962, the thrusters on which the Gemini spacecraft relied for attitude control and maneuvering in orbit and for control during reentry seemed at first to present no special problems. The subcontractor

---

*The ejection seat was not the only system in Gemini having troubles with pyrotechnics. They seemed to be causing problems throughout the program, so much so that, in August 1963, Charles Mathews established an ad hoc committee to review the Gemini pyrotechnics systems—design, qualification, and functions. Headed by Russell E. Clickner (Mercury), the committee consisted of Joe W. Dodson (Mercury), Roger N. Messier (Technical Services), Chester Vaughan (Systems Evaluation and Development), and Robert Cohen and Percy Miglicco (Gemini). The work of the committee had a widespread influence on Gemini pyrotechnics and associated systems—circuitry, redundancy, system design, logic, and qualification testing.

for both these systems, Rocketdyne Division of North American, focused its research effort on developing an engine of 111 newtons (25 pounds of thrust) able to perform within specification for five minutes of constant burning. McDonnell and Rocketdyne engineers assumed that a thruster design able to meet that standard could also sustain the pulsed, or cyclic, firing that would be called for in practice. They also thought that a working, 111-newton-thruster design need only be scaled up to meet the performance demanded of the 445-newton (100-pound-thrust) maneuvering thrusters. They were wrong on both counts.[58]

Then Rocketdyne began running into trouble in steady-state thruster firing. Early tests of the small thrusters showed they tended to char through their casings and to fall off sharply in performance within little more than a minute of continuous firing. When this problem was fixed early in 1963 by a makeshift strengthening of the throat region of the thruster, which allowed it to attain a full five minutes of firing and more, Chamberlin was cautiously optimistic about having qualified units ready to be installed on time.[59]

That hope suffered a setback when Rocketdyne turned to pulse testing and found that pulsing thrusters burned out their ablative liners far more quickly than identical thrusters firing continuously. Char rates—the speed with which thrust-chamber liners burn up—were one and one half times greater in pulsed firing, and thrusters were failing as their lining material was exhausted and their casings burned through. Such expedients as oxidizer to fuel ratio lowered (from 2.05:1 to 1.3:1) to reduce chamber temperatures and thus char rates, thickened ablative linings, and shortened firing times (for some thrusters) could only alleviate, not solve, the problem. In May 1963, Rocketdyne had neither completed the design of the reentry control thrusters nor fired the attitude thruster through a full pulsed duty cycle. The company had fallen three months behind schedule in delivering the thrusters and other parts of the system to McDonnell for Spacecraft 3, and development testing was equally laggard.

To make matters worse, new tests revealed that the larger maneuvering thrusters could not be simply enlarged versions of the attitude engines. Rocketdyne had, so far, done very little work on the maneuver thrusters, partly because of its focus on the smaller model and partly because it had been slow to provide test hardware and facilities. During April 1963, testing of the larger OAMS thrusters had ceased altogether. The new findings now compelled the company to reactivate that test program at once.[60]

Rocketdyne made one design change after another in an effort to put together a thruster that worked, with no striking success. By July 1963, McDonnell was willing to accept a version of the attitude thruster that could not be ready until Spacecraft 5. Relaxed test require-

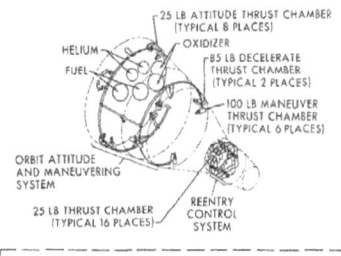

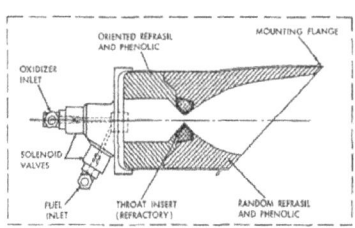

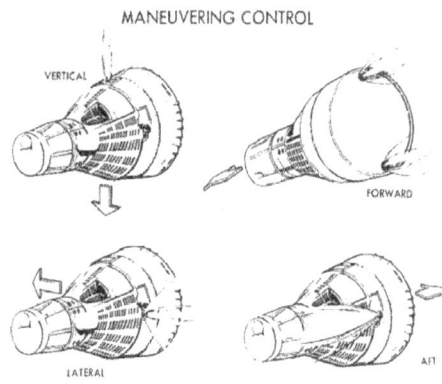

**Gemini Thrusters**

MANEUVERING CONTROL

*The schematic above shows the arrangement of the thrusters on the Gemini spacecraft; the inset shows a cutaway of a thruster. At right are the maneuvers possible with various thruster combinations.*

ments and less stringent performance standards—lower oxidizer to fuel ratios, shorter firing times, and reduced thrust ratings and specific impulse for all engines—helped a little, but grounds for real optimism were slight.[61] As the summer of 1963 drew to a close, no small OAMS thruster had achieved a full mission duty cycle. A few larger OAMS thrusters had, but too few to be sure and with too small a margin of life beyond the duty cycle. The reentry control thrusters looked a little better, largely because of the lesser demands placed on them. They had to function only for a relatively brief time during reentry and could be expected to run dry before burning through.[62]

Even the reentry thrusters, however, hardly inspired confidence. Stabilizing the spacecraft at subsonic speeds during the last phase of reentry, from roughly 15 000 to 3000 meters, had been intended as one function of these motors. (The other, and more important, was to hold the spacecraft in the correct attitude for retrofire to control the angle of reentry and thus to prevent either too steep or too shallow a flight back into Earth's atmosphere.) But, in September 1963, GPO decided that the thruster problems were severe enough to warrant seeking another way to steady the spacecraft. Since the first six Gemini spacecraft were then slated for parachute recovery, GPO decided to add a drogue parachute to the system for this purpose. Development testing of the parachute recovery system had finished in February, and qualification testing was well advanced. Mathews ordered a halt to these tests on 3 September and directed McDonnell to add the drogue.

The first hope, that the new system could be ready for Spacecraft 2, did not survive a close look at the effort required. It was slated instead for Spacecraft 3, the first manned spacecraft; Spacecraft 2 would fly with the non-drogue version.[63]

Rocketdyne, still struggling to meet the 232.5 seconds of pulse operation required of the small attitude thrusters and the 288.5 seconds demanded on the larger maneuvering thrusters, received a jolt in October 1963 from a McDonnell warning that thruster life would have to be doubled or tripled. Astronauts flying simulated missions used the thrusters even more strenuously than they were designed for, and there seemed to be no choice but to widen the margin of performance. Several months elapsed before the new demands were settled at 557 seconds of pulse operation for the small thrusters and 757 seconds for the larger ones. In the meantime, however, thruster testing at Rocketdyne ground to a halt, and the program threatened to founder. No end to development testing was yet in sight, and the start of qualification testing was a long way off. During November and December, Rocketdyne undertook an intense study of the basic features of small ablative rocket engines; McDonnell began work on an alternative design, cooled by radiation rather than ablation; and GPO was thinking seriously about the drastic step of starting qualification tests before development tests were completed.[64]

## A NEW HEADACHE

Despite its key role in Gemini, the Agena target vehicle had received far less attention from GPO during 1962 and early 1963 than other parts of the program, chiefly because time seemed more than ample. Since it was not scheduled into the flight program until the fifth mission, Agena started with seven months more lead time than the spacecraft and Titan II, and that margin more than doubled as a result of the reprogramming crisis of late 1962 and the revised flight schedule of April 1963. By the spring of 1963, although still slated for the fifth mission, Agena's maiden flight was not expected until April 1965, 13 months later than originally planned and trailing the first Gemini mission by almost a year and a half.[65]

That was just as well, because Agena development had moved very slowly. Agena's two propulsion systems, primary and secondary, were subcontracted to Bell Aerosystems Company in Buffalo, New York. The primary system was built around the Bell Model 8247 engine, into which were pumped storable, hypergolic propellants: unsymmetrical dimethyl hydrazine as fuel, inhibited red fuming nitric acid as oxidizer. Its rated thrust was 71 000 newtons (16 000 pounds), and it helped push Agena into orbit (the main boost coming from the Atlas launch vehicle) as well as powering later orbital changes.

The major change in the new engine from the standard model on which it was based was in the starting system. Solid-propellant charges, or "starter cans," in the standard model fed high-speed gas to start the turbine which pumped propellants to the engine. Since these cans could not be reused, the number of times the engine could be restarted was limited by the supply of extra starter cans that could be carried. Gemini required an engine that could start at least five times, and Bell proposed to meet this demand by switching to a liquid-propellant starting system. Liquids were stored in rechargeable pressurized tanks, which fed them to a gas generator where they were converted to gas and transmitted to the turbine. MSC approved the change in September 1962.[66]

Like the primary system, the secondary propulsion system was a modification of a system already in use. Several Agenas had carried an auxiliary propulsion system to permit small adjustments of orbits. Two major changes set off the new model, 8250, from the former system: the new secondary propulsion system was modularized instead of having its parts scattered at various sites in the vehicle, and stainless steel bellows were used in place of Teflon bladders to expel propellants from their storage tanks. The Gemini-Agena secondary system comprised two identical modules, separately mounted but fired in unison. Each module was self-contained, with propellants, pressurized nitrogen to operate the bellows, controls, plumbing, and two thrusters. The larger of the two thrusters, rated at 890 newtons (200 pounds), was intended chiefly for minor orbital adjustments, and the smaller 71-newton (16-pound) thruster for orienting the Agena just before the primary propulsion system fired. MSC had approved the modified secondary propulsion system in August 1962.[67]

Bell had just started its test program when, in the fall of 1962, Gemini's budget crisis struck. While Agena's role in Gemini was under fire, development stopped. But when the smoke lifted, Agena was still very much a part of the program. Contract negotiations between SSD, as NASA's agent, and Lockheed Missiles & Space Company, the prime contractor, began in January 1963.[68] Testing of Agena propulsion systems could now begin. When it did, Gemini confronted a major new problem area.

By April 1963, Bell had completed a development version of the primary propulsion system, test-fired it, and shipped it to the Arnold Engineering Development Center (an Air Force test facility in Tullahoma, Tennessee) for a series of tests to prove that the engine would restart at the pressures and temperatures it would meet in Earth orbit. Tests began on 3 May and continued over the next two months with few surprises, although two problems did emerge. One involved the turbine, which tended to spin too fast. The other trouble spot was the latch-type gas generator valve that controlled the flow of propellants from the start tanks to the gas generator. These valves sometimes opened when they should have stayed closed, failed to open on command, or stuck open. SSD reported to MSC's Atlas/Agena panel that both problems were being closely studied.[69]

Bad luck rocked the program on 15 July, however, when the two problems combined. The valve failed during a test, calling for an emergency shutdown of the engine. A mistake in the choice of shutdown procedures spun the turbine out of control and destroyed the turbopump assembly. That was the end of testing at Tullahoma. Bell planned to finish the series in its own plant in Buffalo, once the problems had been corrected.

The turbine was fairly easy to fix by adding an electronic circuit to monitor its speed and shut it down automatically if it started spinning too fast.[70] But the gas generator valve was not so simply fixed. The failure on 15 July was not its first. A new design was clearly called for. Bell set out to improve its latch-type valve, but how good even an improved version could be was a real question. Bell also went to work on an alternative design, solenoid operated rather than latch-type. Tests over the next few months lent weight to the view that a solenoid valve was not only inherently more reliable but also reduced the complexity of the engine as a whole.[71]

These advantages, and the still unanswered questions about the latch-type valve, swayed a meeting at the Bell plant on 15 November. The participants decided to switch to solenoid gas generator valves in the Gemini-Agena primary propulsion system and forget about latch-type valves. But development had been much delayed. Preliminary flight-rating tests had been scheduled to begin in September. Switching to the new valves would cost four months and postpone the start of these tests until January 1964.[72]

Problems and delays also cost money. Negotiations in January and February of 1963 had set the price (including Bell's fee) of primary system development at $4 771 030. The price tag for solving the turbine problem would be about $300 000. Total costs kept going up, especially after the valve design proved hard to resolve. Toward the end of August, the money actually being spent began to exceed that predicted. By late October, Bell's guess at the cost of completing the program had climbed to $6.177 million, which Lockheed thought was at least $300 000 too low.[73]

Agena's secondary propulsion system developed along the same lines. The new stainless steel bellows produced delays and rising costs. Negotiated cost and fee was $4 395 811; by the time that figure was settled in May, Bell was already asking for an additional $500 000 for the bellows. Scarcely a month later, actual spending was passing predicted expenses as bellows and tanks required still further design work and more testing. In mid-October, Bell's best estimate for the secondary system was $4.63 million, while Lockheed forecast $5.2 million.[74]

Growing engine costs were only part of a trend that brought the Gemini-Agena program to another critical pass in the late summer and fall of 1963. Other program costs were also rising, and the comfortable schedule cushion with which Agena had emerged in the revised program of April had eroded. Shortly before NASA Headquarters sanctioned the revised program, Lockheed estimated the cost for its work at roughly $50.4 million, with $17 million needed for fiscal year 1964.[75] After meetings in May and June to settle details of the new schedule, Lockheed reported its projected total cost as $53.285 million, but SSD had set its sights even higher. NASA's Air Force agents wanted $37.2 million in fiscal-year 1964 funds for Atlas-Agena, with $26 million of that earmarked for Lockheed's Agena contract. GPO protested. Mathews thought that was too much money in view of the stretched-out schedule and wondered if the program could be completed at any reasonable cost with money being spent at that rate. He warned SSD that such spending could not be allowed.[76] When SSD replied on 10 September 1963, current demands were down but the price of the total program was up again, to $57.46 million for Agena and $103.555 million for the entire Atlas-Agena program.[77]

As costs rose, schedules slipped. One source of delay was attempted improvements. The first Agena D programmed for Gemini was AD-13. Meanwhile, however, the Air Force had started a program to improve the standard Agena, the first of which was to be the AD-62 model. The improved version, unlike the earlier model, came equipped with Bell's 8247 engine, which Gemini needed anyway. Since there seemed plenty of time, Lockheed's contract was amended to replace AD-13 with AD-62 as the first Agena for Gemini, at a cost of two months. Another month or more vanished when the Air Force decided to put the restartable Bell engine in AD-71, rather than AD-62, and GPO agreed to take that one. Work on test facilities at Lockheed was slower than expected, adding to the slippage, and development problems in the propulsion systems threatened to delay the program still further.[78]

The Gemini Project Office was less than happy with the course of events, its manager least of all. Mathews was concerned about rising costs, of course, but he was just as concerned with the dearth of information that was reaching him through the filter of SSD. With the Air Force running the Gemini-Agena development program for NASA, Mathews could only plead with his agent to exert more control. Not only was GPO being bypassed in the process that approved changes Lockheed wanted to make, but the project office was not always even told what these changes were. Mathews observed, with good reason, that such decisions as switching from AD-13 to AD-62 (and later AD-71) for the first Gemini-Agena were bound to cause program delays. He urged SSD to think twice about any further changes "considering the deleterious effects that improvements can have."[79]

SSD, however, was not really much better informed than GPO about Lockheed's changes. Mathews' protests about the lax and shallow control SSD imposed on Lockheed highlighted the gulf that divided NASA from the Air Force on the administration of government contracts. The Air Force preferred to accept Lockheed's record in filling past contracts as proof of its competence. The government was, in essence, paying for Lockheed's expertise. Pressing for too many details of funding or technology might hinder progress, cutting into the contractor's flexibility without adding much to its prospects for doing the work. To the Air Force, NASA's demands for detailed technical and financial data seemed at best superfluous, at worst harmful. What NASA wanted, of course, was real control of the program, and that demanded precise and thorough information. Lockheed was merely a case in point. The conflict between NASA and the Air Force over how tight a rein the government needed to exercise spanned the whole range of contract management. For NASA, it was a basic and never-ending problem.[80]

In an effort to bring the Gemini-Agena program into line, Mathews dusted off and sent to Charles Wurster, SSD's chief of Gemini-Agena engineering, a formal statement of work that dated back to July 1963. Such a document was needed, in any case, since there had been no formal work statement since Marshall Space Flight Center had left the picture. The new statement diverged most sharply from the old in the stress it laid on schedules and management. GPO insisted on tight control of all contractors, chiefly by using the system of coordination panels to keep close watch on what was going on. GPO also wanted the last word on any changes, with none to be approved until that office was satisfied that it had every piece of relevant data. So widely did NASA and Air Force viewpoints diverge that it was 18 months and 15 versions of the work statement later, in March 1965, before MSC and SSD finally agreed.[81]

NASA also planned to bring the Aerospace Corporation into the target vehicle program in a role analogous to that it already held in the launch vehicle program, general systems engineering and technical direction. The official end of Mercury in June 1963 had freed a number of experienced engineers for other work. Wurster suggested, and Mathews agreed, that Aerospace had something to contribute to Gemini's Atlas-Agena program, especially in view of the work it had done with Mercury's Atlas launch vehicle. Also in favor of the plan was a chance to impose a degree of technical continuity via Aerospace across all phases of Gemini being carried out under Air Force contracts.[82]

Even if these measures worked, however, they would take time to show any effect. In the meantime, the Gemini Atlas-Agena program was in trouble, with engine development lagging badly, funding and schedules still changing for the worse without much warning. By the end of 1963, most of the time that had seemed so ample in the aftermath of the revised Gemini flight program just eight months before had vanished. The schedule for completing Agena development and for building the first target vehicle now had no slack, and any further problems threatened to delay the first rendezvous launch.[83]

## SILVER LININGS

The last half of 1963 witnessed Project Gemini beset by technical problems that stubbornly resisted solution. No major Gemini system—whether launch vehicle, paraglider, spacecraft, or target vehicle—could confidently be judged ready to fly. These months, in which the approved Project Development Plan of December 1961 had scheduled Gemini's first four flights, became instead a time of troubles; even the revised schedule of April 1963, which called for a first flight before the end of that year, proved beyond reach. And as if to underscore those troubles, the Soviet Union showed that it still held the lead in the space race; 1 November 1963 saw the launch of *Polet I,* a new spacecraft planned "for use in manned orbital rendezvous flight." Although unmanned, it "described complex figures in space" that shifted its first nearly circular orbit to a highly elliptical 1437- by 343-kilometer orbit.[84]

Yet, throughout these months that seem so trying in retrospect, the enthusiastic engineers and technicians, both in government and industry, sustained optimism that transcended the hard facts.[85] Part of that optimism might be chalked up to experience. The pattern of rising costs, sagging schedules, and tough problems was a familiar one at the cutting edge of aerospace technology. Then, too, although the precise nature of Gemini's problems could not have been predicted, they did arise where they were expected—in those systems that demanded the greatest advances beyond current technology. That the escape system, for example, should be hard to develop and qualify scarcely came as a surprise. It had to meet standards far more stringent than had ever been imposed on ejection seats before, and the general nature of the problems to be met could be, and were, foreseen.[86]

Initial schedules and cost estimates tend to be based on the most optimistic assumptions, the completely troublefree development of many complex systems. And these estimates depend on guesswork when new technology is involved—informed and reasoned, to be sure, but guesswork nonetheless. Rightly or wrongly, an organization like NASA assumes that Congress, the source of the money to make things go, prefers fast, cheap programs: the shorter the time and the lower the price, the better a program's chances for support. But there is another, perhaps more weighty, reason for planning optimistically. If time and money are provided for contingencies, then they tend to be used simply because they are there. On the other hand, starting with the strictest limits and yielding further increments of time and money grudgingly may well produce the optimum achievement of the desired goal.[87]

In reality, most of Gemini's troubles in 1963 and later were the product of careful planning and design, credited to the program's first manager, James Chamberlin, that got the project off to such a quick and promising start. This auspicious beginning encouraged NASA to move toward a more ambitious program, to push Gemini closer to its design limits. Problems that might have looked only mildly worrisome in the context of the original Gemini concept took on a more threatening guise when the margin for error had been much reduced.

For a variety of reasons, then, Gemini workers were more confident than a backward look at the difficulties may seem to warrant. But the problems were real; and their gravity should not be downgraded even though, in almost every instance, they responded finally to efforts to resolve them.

# VIII

# Gemini Rising

THE faith that sustained Project Gemini's managers and workers through the dark days of 1963 was not misplaced. Even before the year was over, some of the hardest problems had begun to yield. Gemini's prospects were far brighter by the spring of 1964 than they had been in the fall of 1963. There was still much work to be done, and not every effort at problem-solving was crowned with success. The project that stood on the verge of proving itself in the spring of 1964 was not the same project that had begun two years and more before, nor even the same project that emerged from the budget and managerial crises of late 1962 and early 1963. But most of what its founders had set out to prove had survived, and what had been lost could be balanced with what had been gained.

On 1 November 1963, "Program" replaced "Project" in the title of the office that directed Gemini. This change reflected its responsibility for the program as a whole, and not merely for the spacecraft. Since that had been true from the outset, the new name did no more than underwrite a reality that already existed. MSC Director Robert Gilruth announced it as part of a major reorganization designed to strengthen both Gemini and Apollo now that Mercury was over.* Mercury's man-

---

*Other major elements affected by the reorganization were Flight Operations and Flight Crew Operations Divisions, which emerged as Directorates. Walter Williams went to NASA Headquarters as Deputy Associate Administrator for Manned Space Flight Operations, leaving James C. Elms as sole Deputy Director of MSC. But Elms, who had come to MSC to strengthen its organization, decided his work was done and resigned in January 1964 to return to industry. George Low, NASA Deputy Associate Administrator for Manned Space Flight, was appointed Deputy Director of MSC on 19 January, to take effect on 1 May 1964.

ager, Kenneth Kleinknecht, joined Gemini as deputy manager under Charles Mathews. Kleinknecht brought with him about a third of his former staff.[1]

On the same day, 1 November 1963, an important realignment of NASA Headquarters also went into effect, and for much the same reason: Project Mercury's demise was a chance to reassess the agency's management structure. James Webb, Hugh Dryden, and Robert Seamans had become dissatisfied with the November 1961 reorganization. Headquarters had failed to secure the strong program direction over Apollo that Webb had wanted. When hardware development problems continued to mount, with attendant escalating costs and slipping flight schedules, something very definitely had to be done. Moreover, having a program the size of Apollo, along with all the other programs NASA was pursuing, made it difficult for one man—Seamans in this case—to serve as "general manager" over day-to-day affairs. In 1961, Webb had needed decision makers at the program level, but in 1963 he needed this talent, armed with the proper authority, at the administration level to unify the agency, provide direction to the field centers, and lessen some of the autonomy the latter had held onto so tightly. The major change involved putting the field centers under Headquarters "Associate Administrators" for special activities—George Mueller for Manned Space Flight, Homer Newell for Space Science and Applications, and Raymond L. Bisplinghoff for Advanced Research and Technology—rather than under *the* Associate Administrator as they had been. Mueller, who had replaced Brainerd Holmes as chief of manned space flight, now took charge of both the program and the centers carrying it out—MSC, Marshall Space Flight Center, and Launch Operations Center. Mueller also set up a Gemini Program Office in Washington,* chiefly as a device to oversee Gemini and to bring together in a single group all those in NASA Headquarters whose work related to Gemini. William Schneider had taken over a tiny liaison office of seven people from Colonel Daniel D. McKee earlier in the year. Now he headed a program office seven times that size. Several months would elapse before the effects were felt in Houston.[2] In the meantime, some of Gemini's most severe technical problems were at last beginning to respond to hard work in the field.

TITAN II MAKES THE GRADE, BUT NOT PARAGLIDER

What had been Project Gemini's greatest concern—whether Titan II could function as a booster for manned space flight—was soonest

---

*This was for NASA the beginning of the "five-box" program organization that Mueller demanded. In Headquarters, under Acting Gemini Program Director George Low and his Deputy,

laid to rest. Titan II Missile N-25 was launched 1 November 1963 from the Atlantic Missile Range, the 23d in the series of test flights conducted by the Air Force Ballistic Systems Division (BSD). It furnished the first real proof that Titan II would do for Gemini. Missile N-25 was equipped with the standpipes on its oxidizer lines and mechanical accumulators on its fuel lines that the revised theory had predicted would suppress the severe lengthwise bouncing (Pogo) that threatened Titan II's role as a manned booster. The November flight proved it worked. The devices installed in fuel and oxidizer feedlines reduced Pogo to the lowest level ever in a Titan II flight, only one-ninth the force of gravity ($\pm$0.11g), and for the first time well below the $\pm$0.25g that NASA insisted marked the upper limit for pilot safety.[3]

The Gemini Program Office had no way of forecasting that the next five months were to see the Titan II test flight program produce an unbroken string of successes. But, knowing that standpipe and accumulator had worked on Missile N-25, GPO inferred that the theory behind installing these devices had been confirmed and acted quickly, sure that the Pogo problem had been solved. On 6 November, GPO decided to procure several sets of the suppression devices for Gemini launch vehicles. The soundness of that action was soon confirmed. Titan II launches on 12 December 1963 and 15 January 1964 both carried the oscillation dampers and both met NASA standards. The 15 January flight, added at Aerospace urging, proved the devices effective even with reduced fuel-tank pressures. This was all the more heartening because raised tank pressures had lowered Pogo levels in some earlier missile flights.[4]

While Titan II was proving itself in flight, NASA and the Air Force completed their nearly year-long efforts under the aegis of the Gemini Program Planning Board to fix standards for the Gemini launch vehicle. NASA's final statement, on 15 November 1963, rehearsed its long-stated demands: longitudinal oscillations during powered flight must be no greater than $\pm$0.25g, incipient combustion instability must be eliminated, and all known design shortcomings and anomalies revealed in Titan II ground and flight tests must be corrected. On the same day, BSD and SSD (Space Systems Division) of the Air Force Systems Command issued a plan to prove in flight their program to reduce Pogo and improve engines. These two documents, along with the earlier Air Force plan for cleaning up Titan II problems, answered the board's request of 11 October 1963 for data on

Schneider, were Major Richard C. Henry, Program Control, Acting; Eldon Hall, Systems Engineering; LeRoy E. Day, Test; John A. Edwards, Flight Operations; and Dwight C. Cain, Reliability and Quality.

which to base a formal Memorandum of Understanding between NASA and the Air Force.[5]

What NASA required and how the Air Force planned to respond were discussed for the last time at the board meeting of 3 December. The board accepted the NASA specifications as reasonable, the Air Force plans to resolve the problems and verify the results as technically feasible. Then the co-chairmen of the board, Brockway McMillan for the Department of Defense and Robert Seamans for NASA, signed the formal "Memorandum of Understanding on Certain Design Requirements for the Gemini Launch Vehicle."[6] No further managerial obstacles blocked the way to a man-rated Gemini launch vehicle.

The compound of jurisdictional disputes and technological problems that had made the launch vehicle the single biggest question mark in the Gemini program until late in 1963 vanished almost overnight. By mid-January 1964, Titan II no longer seemed a concern. After the missile's third success with Pogo suppression gear, on 15 January, Seamans was convinced "that the currently completed flight demonstrations of POGO fixes indicated a qualitative understanding of the problem and its solution and provided sufficient confidence to go ahead with the Gemini program." Another sign of the times was the end of the weekly Titan II status reports Seamans had been getting from Air Force Systems Command because, "based on the successful resolution and flight verification of the axial oscillation fix (Pogo) on missiles N-25, -29, and -31, the primary requirement, for which this weekly report was originated, has been satisfied."[7]

Pogo had not, of course, been the only problem, although it was the greatest. Still to be resolved was the potential instability of Titan II's second-stage engine, which Aerojet-General had begun to tackle in October 1963 with Gemsip, the Gemini Stability Improvement Program, focused on working out a new design for the propellant injectors. Gemsip ended 18 months later with complete success, having cost the Air Force about $13 million. NASA spent $1.45 million to install the changes in the last six Gemini launch vehicles. The first six flew with the old-style injectors, which NASA later defended on the somewhat specious grounds that no instability had shown up in a Titan II flight. That was essentially a statistical argument of the kind earlier rejected as a basis for man-rating. NASA found a better reason for going on with the flight program. Aerojet engineers knew that any number of techniques might be used to reduce starting shocks, the major trigger for unstable burning. Very early in Gemsip, they found that a certain minimum pressure in the cartridges that started the motor eased the problem. Temperature conditioning—keeping the start-cartridge temperature above a critical value—proved even more effective. This was the finding that chiefly convinced NASA that Titan II's second-stage engine was safe enough for manned missions, although only Aerojet's redesigned injector finally provided a dynamically stable engine.[8]

NASA's third concern about Titan II had been just how reliable some engine parts were. This was less a matter of design than of the general standards of manufacturing and quality control observed by Aerojet-General. The Air Force, however, saw potentially dangerous weaknesses in design that demanded the development of new parts, an effort that got under way in September 1963 as the Augmented Engine Improvement Program. NASA deemed improved engines nice, but not vital (as damped Pogo and stable second-stage engines were) for Gemini. This was just as well, because the engine improvement program produced small results for the $11 million it cost the Air Force: some minor design shortcomings corrected, welding techniques improved, and better assembly methods adopted. NASA did buy one product of the program for Gemini, redundant shutdown circuitry, at a cost of $1.5 million. But the rest of the hardware developed under the program looked more risky than what it was intended to replace. The Air Force canceled the program in November 1964.[9]

Looking back, NASA officials had nothing but praise for the hard work put in by the Air Force and its contractors to man-rate Titan II for Gemini even while they were trying to prove it as a missile. As George Mueller reported to NASA Administrator James Webb:

> In the broad view of this booster program where a military vehicle, the Titan II, was selected prior to its development and a program of man-rating carried out actually in parallel with the flight test and acceptance of the military versions, we have, I believe, a unique situation. It is unique not only in technical complexity but also in management relations and control. . . . [T]his collaboration between two demanding users has produced an unusually reliable military launch vehicle . . . [and] a man-rated launch vehicle with a remarkable record of success. . . .Configuration management is not a new term but the detailed application of the Air Force to the GLV [Gemini launch vehicle] development is a model of its kind and a significant contribution toward improved management of all major programs, in DOD and in NASA. We have seen major improvements in electrical circuit design, in electrical soldering and welding techniques, in assembly procedures and in test specification.[10]

This picture of a smoothly meshed team moving from success to success, although true enough for the last six months of the program, slighted the obstinate technical and managerial problems that had to be surmounted before the happy outcome was reached.

Even in retrospect, the record of Titan II research and development flights was spotty, especially in view of the high promise that had induced NASA to choose it for Gemini in the first place. Only 22 of the 32 flights that comprised the test program would have succeeded in launching a Gemini mission. Based on Titan II flight tests, in other words, every third or fourth Gemini mission would have been abortive; this does not include the Pogo that rattled missiles during first-stage flight without compromising Air Force test objectives. This picture was, nevertheless, far brighter than it had been in mid-1963—half the 20 tests flown by 20 June would have been failures on Gemini. The concentration of all 10 unsuccessful flights in the earlier part of the program, however, may have held the greatest promise. The unbroken string of 12 nearly flawless flights that concluded the Titan II test program strongly implied that the missile's problems had, in fact, been solved. With Pogo reduced to tolerable levels by techniques that accorded with theoretical analysis, the threat of combustion instability eased by an operational expedient, and a series of successes to show that other troublesome areas had been cleared up, Titan II could be judged man-rated in the early spring of 1964. This judgment seemed amply confirmed by *Gemini-Titan I*, launched 8 April 1964,* the day before the last flight in the missile's research and development test program and well before men were first scheduled to ride the Titan.[11]

The striking vindication of Titan II in the final months of 1963 had no parallel in the paraglider program. Paraglider's only chance to regain a place in Gemini hinged on the outcome of North American's new series of deployment flight tests with the full-scale vehicle. A full-scale wing was to be uncased and inflated in midair, to prove it could support the vehicle in stable gliding and maneuvering under radio control. Each of the planned 20 tests was to end with the wing cut loose at 3000 meters and the test vehicle landing by parachute. The parachute system was qualified on 3 December 1963, clearing the way for flight testing of the full-scale vehicle to begin on 22 January 1964. The first test did nothing to dispel doubts about paraglider; the second test, on 18 February, was also a failure.[12]

That same day, George Mueller told the House Subcommittee on Manned Space Flight that the paraglider "is not presently scheduled on the . . . Gemini spacecraft."

"Will it be used at all in the Gemini program?" one of the Representatives wanted to know.

Mueller replied, "That will depend upon the development status of the paraglider which we will evaluate next spring. It will also depend upon the needs for a paraglider for precise landing of the Gemini spacecraft which we are developing now with the Air Force."

Further probing revealed that paraglider could be ready for the tenth Gemini mission, particularly if the Department of Defense lent

---

*This flight will be discussed in detail in Chapter IX.

its support—this from George Low. But, he added, "we have no money included [in] 1965 or beyond for the paraglider under the assumption we will not go into production."[13]

NASA's public position was that, while land recovery appeared to be both desirable and feasible, it was riskier than water landing. Crew safety, the paramount concern, dictated the proven mode of water landing for all 12 Gemini flights.[14] The risks of land recovery were real enough, needless to say, but they had been just as real in 1961 when NASA decided to adopt land landing as a major Gemini objective. Toward the end of winter in early 1964, however, the means to that end, a paraglider landing system, had yet to achieve a level of performance great enough to rely on. After nearly three years of work, there was still no certain answer to the key paraglider problem—how to unship and inflate the wing from a two-tonne spacecraft plunging downward through the atmosphere. The risk that loomed so large early in 1964 was perhaps not so much land landing as paraglider landing.

Paraglider still had ardent defenders in NASA, and the decision to strike it from Gemini was not yet final.[15] But NASA was ready to drop the paraglider, the more so since the system might still fly in another version of Gemini. In the spring of 1963, under the auspices of the Gemini Program Planning Board, the Air Force had begun laying the groundwork for its own Gemini program, Gemini B/Manned Orbital Laboratory (Gemini B/MOL). The Air Force X-20 orbital glider, still often called by its former name, Dyna-Soar, had been canceled in December 1963, a victim of low priorities and lagging development. Some X-20 funds were diverted to the new MOL program, which projected two men in a modified Gemini spacecraft launched by a Titan III. In orbit, the crew would transfer to a separately launched laboratory for two to four weeks, after which they would return in their spacecraft.[16]

Air Force planning had progressed far enough by January 1964 to require a formal agreement between NASA and the Air Force in the form of a memorandum signed by Seamans for NASA and Harold Brown for the Air Force.[17] Although Gemini B/MOL would not be officially approved until August 1965 and design work was only beginning, NASA saw a chance to save paraglider. On 17 March 1964, George Mueller asked the Air Force for "an expression of the DOD interest in this capability," whether for Gemini B/MOL or any other program. Six weeks later, having concluded that paraglider development had too many problems to warrant putting it in the new program, the Air Force discounted any prospect of joining in paraglider development and threw the problem back to NASA: "Should the NASA qualify and demonstrate the paraglider in the NASA Gemini program, consideration would be given to its application to the Gemini B/MOL."[18] By then, however, it was too late.

*171*

North American's further efforts to fly the full-scale test vehicle produced a string of failures, each distinct in detail but united in a single root cause, "an inability to adequately predict the wing loads of flexible structure[s]." The fifth failure in a row, on 22 April, was the last straw. The next day, William Schneider, NASA Headquarters Gemini chief, informed George Mueller that he planned to transfer what was left of the paraglider program to Flight Research Center and to spend no more Gemini money. A week later, the program office in Houston began cutting back paraglider work and phasing the program out of Gemini. Early in May, GPO and North American agreed to run the rest of the flight-test program with the equipment and money already committed. Paraglider was dead as far as Gemini was concerned, although a public statement of its demise waited until 10 August.[19]

Ironically, North American achieved its first full-scale test vehicle success on 30 April, the day after phasing it out of Gemini began. In fact, the worst was over. Before the end of 1964, North American flew 19 more tests for a total of 25, 5 more than originally planned. By July, the deployment sequence was no longer giving much trouble, although a stable glide after the wing inflated was harder to manage. The last three flights, however, displayed the complete sequence without flaw.[20]

The last full-scale test vehicle flight was on 1 December 1964. Two days later, NASA told North American there would be no more money for flight testing, but equipment on hand might be used, if the company cared to spend its own money. North American seized the chance to complete the other major portion of the May 1963 program—working out landing techniques with a piloted tow-test vehicle. Tow-testing had begun during the summer. On 29 July, a helicopter had towed the vehicle up to a height of a few hundred meters, around the test area, and back to a safe landing. A free flight followed on 7 August, but the vehicle went into a series of uncontrolled turns, forcing the pilot to bail out. North American attacked the problem with dispatch and came up with an altered wing design. On 19 December, a pilot flew the tow-test vehicle through the complete test to a safe landing.[21]

NASA had long since decided to dispense with paraglider for Gemini, however, and that was irrevocable.*[22] The system's shortcom-

---

*Paraglider's partisans in NASA had not lost faith, and the concept itself retained enough of its pristine attractiveness to justify a further effort. During the last half of 1965, North American conducted a research and development program under NASA contract to determine flight and landing characteristics in a series of 12 manned tests, plus a number of associated unmanned flights. More recently, both the Army and Air Force have been interested in developing the system as part of an unmanned cargo delivery system for combat situations.

ings, or at least North American's slowness in coming up with answers, account chiefly for paraglider's failure to survive in Gemini. But the immediate reason for the abrupt action in the last week of April 1964 to kill what remained of the Gemini paraglider may have had more to do with money than with technology.

## MONEY AND MANAGEMENT PROBLEMS AGAIN

Gemini's chronic budget ills were marked from time to time by acute episodes. The crisis of late 1962 had scarcely subsided before the project reeled under a new round of cost increases. By 8 March 1963, the program's total price tag stood at just over $1 billion. NASA's projected budget for fiscal year 1963 had been $232.8 million after the impact of reprogramming had been assessed; actual expenditures topped $289 million. The pattern repeated in fiscal year 1964, with a planned budget of $383.8 million exceeded by $35 million. By 2 March 1964, NASA expected to spend over $1.2 billion on the program.[23] These increases reflected, in part, Gemini's changing scope and the technical problems that somehow proved harder to solve than anyone had expected. They also reflected, perhaps inevitably in so large and complex a program, mistakes, errors of judgment, and mismanagement, though Gemini appears to have suffered less from those ills than other programs of comparable size. Swelling costs were, for whatever reason, evident throughout the program.

NASA and McDonnell had finished negotiating the Gemini spacecraft contract in February 1963, settling on a total cost plus fixed fee of $456 650 062. This figure was not so firm as it then seemed. At the end of 1963, McDonnell estimated total spacecraft costs at upwards of $612 million. Something less than half the difference could be ascribed to approved changes in the program, as exemplified by the $2.7-million price for adding drogue stabilization to the parachute recovery system, though this change was itself prompted by development problems with reentry thrusters. Much of the balance derived from cost overruns on major Gemini subcontracts, with thrusters by Rocketdyne and fuel cells by General Electric the chief culprits. The new year brought no relief. In March 1964, when NASA estimated the total cost of Gemini at $1.2203 billion, the spacecraft accounted for $667.3 million.[24]

Launch vehicle budgets were equally ephemeral. The billion-dollar estimate of March 1963 had included $240 million for the Gemini booster. As the year wore on, Air Force Space Systems Division found the situation "extremely fluid. Costs were constantly increasing and changes were being approved so fast it was difficult to keep track of them. . . . Engine problems were causing late deliveries and increasing costs." When SSD completed its first comprehensive review of the Gemini budget in January 1964, it felt obliged to revise the cost upward to $296 million. Just two months later, after another hard look at launch vehicle costs, SSD claimed to need $324 million. This was the same month, March 1964, when NASA was counting the booster's share of a $1.2-billion Gemini budget as $281 million. Toward the end of the month, Gilruth warned Major General Ben Funk, SSD Commander, that MSC's 1964 booster money had been exhausted. With three months of the fiscal year still to go, the $46.9 million allotted looked as if it would fall $30 million short of expenses. Gilruth was much concerned about funding in the coming two years and asked Funk to take another look at his needs. Funk replied with an estimate of $332 million that included $75.3 million for fiscal year 1965, $8.4 million higher than NASA had planned.[25]

Inexorably rising costs plagued target vehicle as well as launch vehicle development, and for much the same reasons: technical problems compounded by the fact that NASA and the Air Force simply did not agree on how a development program ought to be managed. NASA wanted more control than the Air Force thought wise to impose. NASA efforts to promote its view during late 1963 had availed little, and Mathews' communications with SSD grew more caustic. On 5 February 1964, he scored Bell and Lockheed (and, by implication, SSD) for the sorry job being done on Agena engine development. Costs had "continued to increase even at this late date to a level far beyond that considered reasonable by this office." The excuses offered were, in Mathews' view, worthless:

> The emphasis which BAC [Bell Aerosystems Company] has placed on the fact that the development effort was to be one of minimum cost has apparently led them to a belief that sound technical judgment was no longer required or that minimum cost eliminated its use. The GPO does not consider this argument valid or useful.

The fault was as much Lockheed's as Bell's. Mathews believed that

> the costs quoted by BAC and submitted by LMSC [Lockheed Missiles & Space Company] are excessive or unjustified in many areas. Moreover, these costs have increased and are continuing to increase with apparently little financial hazard to BAC and only after-the-fact recognition by LMSC. . . . GPO must express dissatisfaction with LMSC and BAC management of these programs.[26]

Engine development costs were only part of the problem. The first "firm" budget for the Gemini Atlas-Agena program was ready in September 1963. SSD projected a total cost of $103 million, with Agena's share as $57.5 million. By March 1964, NASA was prepared to spend $137 million for the program, $93 million on Agena alone. The $37 million programmed for Agena in fiscal year 1964 was almost exhausted, although that figure was $2.4 million higher than Lockheed had, in September, claimed to need. Mathews termed the situation "critical" and demanded a complete explanation in writing for the discrepancy between current costs and the September projections. GPO once again saw, in "the contractor's frequent increases in the estimated costs," signs of "a serious need for improvement by the contractor in proper planning and cost control." Mathews warned SSD and Lockheed that "lack of adequate cost control places this program in real jeopardy."[27]

Ironically, at the same time that Mathews was urging SSD to get Lockheed under control, the contractor was finding that it needed still another $2.5 million in 1964 funds, a request that was duly passed along to GPO on 4 April. Lieutenant Colonel Mark E. Rivers, Jr., who had just replaced Major Charles Wurster as chief of Gemini-Agena engineering for SSD, saw signs of sloppy management in the new Lockheed request, which appeared to be based on small changes that had piled up unnoticed over several months.[28]

This, then, was the setting in April 1964 when North American, for the fifth time in a row, failed to deploy the paraglider wing in flight. Mounting costs in all phases of Gemini development had stretched the 1964 budget to the breaking point, and the trend was still upward. Paraglider had been budgeted for $16.4 million in 1964, but that would be the last of the money. Keeping paraglider meant finding new funding or cutting back other parts of the program. In the money budget as in the weight budget, once paraglider's status became doubtful, its place was preempted. Against this confluence of forces—technical, operational, and budgetary—paraglider could not stand.

Whether the target vehicle program could survive was also a question. In late April, budget pressures forced Mathews to discuss with his staff some desperate measures. Paraglider, Atlas-Agena, and even one of the planned Gemini missions were on the chopping block. Once again, however, MSC was able to reprogram funds to save the full 12-flight program and, via Agena, the rendezvous objective, if not paraglider and land landing.[29] One of the factors that may have made Agena's place in Gemini shaky in April 1964 was a new round of technical problems that had cropped up earlier in the month.

Bell's efforts to complete development testing of Gemini-Agena propulsion systems during 1963 had produced spotty results and many delays, which had, in turn, postponed the start of preliminary flight-rating tests of these systems. Scheduled to begin in June 1963, testing of the main engine had been put off until January 1964 but began only on 6 February. Still another two weeks elapsed before the secondary system began its tests on 17 February. Both programs soon ran into trouble.[30]

Main engine testing proceeded with only minor problems through the first week in April. In the following week, however, the test program encountered what proved to be a six-week delay when the test unit's fuel and oxidizer start tanks failed. These tanks were stainless steel canisters with bellows inside them to push the propellants that started the main engine. Visible lengthwise cracks in their outer shells allowed the gas that was supposed to force the propellants from the tanks to escape. The steel in the shells had corroded. Tanks with a new heat-treated steel shell replaced the defective tanks, and testing resumed in May. But the tests, which should have ended in April, ran into late June. Alarmed by the threat of increased cost such a failure implied, GPO demanded a complete written account of the causes and effects, a point of special concern being "indications that subcontractors may have failed to process materials in a manner essential to the proper operation of components being developed."[31]

Agena's secondary propulsion system, like the main engine, started preliminary flight-rating tests smoothly, then ran into trouble early in April. Failure of a propellant valve, however, imposed only a minor delay. A harder problem emerged later in the month during high-temperature firing, when the wall of a thrust chamber burned through after 354 seconds. While well beyond the 200 seconds regarded as the system's longest useful life in orbit, it fell below the specified time of 400 seconds. Bell installed a new thrust chamber and finished the tests—in mid-August instead of the scheduled mid-June. The failure, however, needed to be explained, and that meant more tests. Bell planned a series of six tests over two weeks, beginning early in September. Test-cell problems hampered the work, which did not end until mid-November and then after only four tests. The four were, however, enough to spot the problem—elevated propellant temperatures—and to show that it would not affect the system's performance in orbit.[32]

Bell's slow progress in its test program delayed Lockheed's testing. Because of the scope of changes in propulsion systems required to adapt the standard Agena D for Gemini, Lockheed planned a series of static firings using an Agena skeleton fitted out with propulsion and propellant systems at its Santa Cruz Test Base in California. Lockheed received the propulsion systems from Bell in February and March and had the test assembly at Santa Cruz by the end of March. Checkout problems and Bell's cracked start tanks in April held up the testing. Lockheed returned the main-engine start tanks to Bell, but they were not replaced until mid-May. Other minor problems delayed the first firing until 16 June. Once under way, however, the test program moved quickly to an end on 7 August 1964 with no further mishaps.

Post-test analysis confirmed that the propulsion systems had come through in fine shape.[33]

In the meantime, doubts about the Agena's ability to perform its mission had been growing. On 15 April 1964, SSD suggested flying a non-rendezvous Gemini-Agena mission to bolster confidence. GPO dismissed this scheme but accepted an alternative recommendation that one target vehicle be assigned the role of development test vehicle. This would be helpful for troubleshooting malfunctions and testing changes and would also allow further development testing, should the need arise. The plan was approved in May and the first Gemini-Agena target vehicle, GATV-5001, was to be the test vehicle. AD-71, the first standard Agena D for Project Gemini, had been accepted by the Air Force on 30 April and transferred to the final assembly area at the Lockheed plant, where it was being converted to GATV-5001. Despite its new role, GATV-5001 was expected to remain in flight status until GPO decided otherwise, although GATV-5002 was now tentatively scheduled for the first rendezvous mission. GATV-5001 was not likely to fly unless GPO later opted for a non-rendezvous mission. So GPO canceled one of the eight Atlas boosters then under contract as Agena launch vehicles, saving the program $2.15 million.[34]

## A SET OF BREAKTHROUGHS

The three spacecraft systems that had caused the most trouble in 1963—escape, fuel cell, and thruster—each enjoyed a sharp change of fortune as the year turned. Problems that had resisted the best efforts of NASA and contractor engineers for so many months suddenly yielded. All the answers were not in yet, but by the spring of 1964 the prospect that any of these systems might fail to meet Gemini needs had largely vanished.

Escape system development trials had come to a halt in August 1963 as the system went through another series of design changes and some of its key parts, particularly pyrotechnics, remained hard to get. Active testing resumed on 22 November, with the first in a projected series of about 30 drops of the ballute, which had been added to the crew parachutes for the sake of high-altitude stability. The first 10 tests, which involved both men and dummies and used a ballute 91 centimeters (36 inches) in diameter attached by a single riser, ended on 9 January 1964. In each case, the subject spun too rapidly on the riser.* This was solved by raising the ballute diameter to 122 centimeters (48 inches) and using two-point suspension. Fourteen more drops

---

*The Air Force furnished the human subjects for these tests—Colonel Clyde S. Cherry, Chief Warrant Officer Charles O. Laine (who made the first jump), and Chief Warrant Officer Mitchell B. Kanowski.

over the next few weeks, the last on 5 February, confirmed the changes, and the ballute was ready for its qualification tests.[35]

Only two days later, sled-ejection development trials also came to an end. Testing had resumed with the fourth run, on 16 January, and ended with the fifth, on 7 February. Everything worked in both tests. Since simulated off-the-pad and ballute development tests had already been completed, the successful 7 February test brought the development phase of escape-system testing to a close.[36] Neither fuel cell nor thruster was so far advanced.

Fuel-cell production had stopped in late November 1963, as a pair of GE task groups sought to resolve the system's stubborn engineering and manufacturing problems. Within six weeks they had finished their work, which furnished the basis for turning the program around. Everyone involved in the fuel-cell program gathered in Houston on 27 January 1964 to review development status and decide what to do about it. All agreed that the system needed redesigning. The current PB2 model was to be discontinued; the units already built were to be used for limited testing and to be carried in Spacecraft 2 to gather data and help qualify the reactant supply system. All future cells were to be the new P3 design, and they were to be installed in every spacecraft beginning with the fifth.[37]

Major changes in the new model reflected the narrow technical nature of the problems: dams (or baffles) were added to improve hydrogen distribution; the water collection wick was removed from each cell; and the orifice of the hydrogen feed tube of each unit was restricted so that any stoppage caused by water clogging could be cleared. Other changes included adding Teflon to the electrode to cut the loss of active material from the membrane and an anti-oxidant to the membrane to slow the rate of polystyrene breakdown. Tests had also suggested that the crucial problem of short operating life might respond to reduced temperatures. When further tests confirmed this finding, the coolant supplied to fuel cells was adjusted for lower temperatures.[38]

Although fuel-cell problems were largely technical, GE decided the program could be better managed. It reorganized the Direct Energy Conversion Operation to work solely on the Gemini fuel-cell program. Roy Mushrush, the new manager, had a background as corporate troubleshooter for GE. He arrived on the scene with a blank check on the company's resources for whatever help he needed. Mushrush was seconded by Frank T. O'Brien as Gemini manager. Both men impressed a NASA visitor with their enthusiasm, and morale throughout the plant remained high despite the shakeup.[39]

The fuel-cell program was still a question mark, and no one could be fully certain that the system would be ready in time for Gemini. But in the early spring of 1964, the program's technical and managerial problems seemed to have been taken in hand, and prospects were a good deal brighter than they had been. By the end of May, GE had finished switching to the P3 design and had started a broad test program.[40]

Rocketdyne's thruster development program was also turning a corner. So far, attempts to improve performance had been little more than stopgaps, centered chiefly on cutting the engines' thermal load by dropping the ratio of oxidizer to fuel. But lower working temperatures and longer engine life were being achieved at the expense of combustion efficiency and specific impulse. This was one of three major topics discussed at a review of thruster problems in Houston on 23 December 1963. Rocketdyne was directed to cut the current oxidizer to fuel ratio of 1:1.3 still further, if that could be done without harm to good starting and stable burning.

Study of another expedient was also approved: shifting the side-firing thrusters to align them more closely with the spacecraft center of gravity and so reduce demands on the smaller attitude thrusters in holding spacecraft attitude during lateral moves. Development of this small engine was the least hopeful aspect of thruster work—no one really understood what its design ought to include, and tests produced large and hard-to-explain variations. No attitude thruster had yet shown itself able to fire through a complete mission duty cycle without failure.

A third decision, of greatest impact on the program, grew out of the 23 December review. André Meyer, chief of GPO administration, had been urging a change in the design of the ablation material lining the thrust chamber. A newly developed parallel-laminate material showed promise as an answer to thruster-life problems. Meyer wanted the laminates oriented nearly parallel to the motor housing, instead of perpendicular as before. His efforts to convince both McDonnell and Rocketdyne to make this change had been resisted because of its expense, but now, strongly backed by MSC Director Robert Gilruth, the idea was accepted and an engine to test the concept was ordered built.[41]

The thruster picture brightened perceptibly over the next month. Further tests confirmed that reduced oxidizer-to-fuel ratios prolonged engine life, bringing the maneuvering thrusters within sight of their required mission duty cycles. The performance of the smaller attitude thrusters also improved, though not as much. By mid-January 1964, NASA Headquarters felt sanguine about the prospects for Gemini's big thrusters but saw little hope for so happy an outcome to the development of the smaller thrusters. There was strong support for a study of a radiation cooled engine as a backup.[42]

Meanwhile, Rocketdyne's efforts during the last two months of 1963 to work out the basic problems of small ablative engines had also borne fruit. A search through the files uncovered a research report on the problem of heat flux in small engines and an answer in the technique of "boundary-layer cooling." The injector of a maneuvering thruster was modified to spray about a quarter of its fuel down the walls of the thrust chamber before firing. On 25 January 1964, Rocketdyne tested the engine through its full mission duty cycle without failure, its liner charring only to a depth of little more than a centimeter (one-half inch). A second thruster produced the same results. Since the lining of the flight weight engine was twice that thick, the margin seemed ample. Buoyed by these results, GPO, after a meeting at the McDonnell plant in St. Louis on 13-14 February, ordered McDonnell to have boundary-layer cooling designed into the larger thrusters in time for Spacecraft 5.[43]

The smaller attitude thrusters did not respond as well to boundary-layer cooling, although it helped. A modified injector, combined with an oxidizer-to-fuel ratio of 0.7:1, allowed one small engine to survive a 570-second firing on 15 February with some of its liner intact; in earlier tests with the same ratio but without the injector, the liner had not lasted beyond 380 seconds. Two flight-weight engines with the new injector and lower ratio lasted for 435 and 543 seconds. Another change made these results look even better. Canting the lateral engines to direct the thrust vector closer to spacecraft center of gravity (as suggested at the 23 December meeting) was shown to reduce the thruster life needed to less than 400 seconds.[44]

By mid-March 1964, thruster development and qualification appeared likely to be completed in time, though without much leeway to handle any new problems and with performance that was still marginal. In April, that status was transformed. Thrust chambers lined with laminated ablative material oriented almost parallel (at an angle of only 6 degrees) to the motor housing achieved dramatically better performance. The first modified attitude thruster endured 2100 seconds of burning without failure on 14 April, a fourfold increase over the best prior test. And the next day, a maneuver thruster with boundary-layer cooling and the 6-degree wrap fired for 1960 seconds, the test ending only when fuel was exhausted. Just as striking was the first test of a lateral thruster with the new wrap: 3049 seconds of firing time without failure. George F. MacDougall, Jr., Deputy Manager of Program Control in GPO, reported the results to the MSC senior staff as "a major breakthrough."[45]

Convinced that the answer had been found, GPO lost no time. Within two days after the first tests of the small and large thrusters, McDonnell and Rocketdyne had orders to replace 90-degree with 6-degree wraps in all thrusters and to see that the new thrusters were installed in the orbital attitude and maneuvering systems of all spacecraft beginning with the fifth and in the reentry control systems of all spacecraft as soon as possible. By 1 May, however, Spacecraft 5 looked too early for a complete set of new engines. Instead, all its attitude thrusters would have the modified injector and 6-degree wrap, but only the aft-firing maneuvering engines would feature the new design. The less critical lateral- and radial-firing engines would be the old model. All thruster designs were now frozen, with further testing limited strictly to qualification.[46]

Rocketdyne was by no means home free, but the worst of the spacecraft propulsion systems' technical problems did appear to be over by the spring of 1964. The fuel cell also seemed to be in good shape. Gemini's escape system, already through its development test program, may have looked best of all. As later events were to show, the promise was not quite that easy to fulfill. But none of these three most stubborn systems was slated for the first Gemini spacecraft, which McDonnell had been building in its St. Louis plant.

### TOWARD GEMINI-TITAN 1

The primary objective of the first Gemini mission, as it emerged from the revised flight program of April 1963, was to prove the Titan II able to launch the Gemini spacecraft and put it into orbit within the constraints imposed by manned space flight. To gather and report data were the spacecraft's main functions. Spacecraft 1 was, therefore, unique among the products of the Gemini assembly line in St. Louis in being largely without standard spacecraft systems. For the most part, it carried dummy equipment and ballast to match normal weight, center of gravity, and moment of inertia. Structurally, however, Spacecraft 1 differed from later models in only one important respect. Since mission plans did not call for the spacecraft to be recovered, the heatshield simply completed the structure. Four large holes bored in the ablative material ensured the total destruction of the spacecraft when it plunged back into the atmosphere.

Working equipment was mounted on two special pallets (much like the "crewman simulator" used in Project Mercury) located where the crew would be in later flights. Spacecraft 1 carried two active Gemini systems: a C-band radar transponder and related gear to help ground radar keep track of the spacecraft, and three telemetry transmitters to return data to Earth. Data were to be gathered by a set of special instruments that measured pressure, vibration, acceleration, temperature, and structural loads.[47]

McDonnell began testing Spacecraft 1 on 5 July 1963, with plans to have it at Cape Canaveral by mid-August. The first phase of spacecraft systems tests centered on making sure that each working piece of equipment functioned properly. Many parts did not, bringing testing to a halt on 21 July. The instrumentation pallets had several defects, especially in their electrical circuits and in their response to vibration. Other problems included a transmitter and a radar beacon that had to be returned to their makers to correct out-of-specification performance. With these matters taken care of, testing resumed on 5 August and proceeded smoothly to the end of the first phase on 21 August.[48]

Four days later, McDonnell workmen mated the major spacecraft modules. The now fully assembled vehicle was ready for the second phase of systems tests, checking its overall working and the compatibility between the mated sections. It was now slated to arrive at the Cape on 20 September. During the first half of the month, tests alternated with leftover manufacturing tasks, which slowed things down, but not seriously. All systems performed well during the last half of the month, as the spacecraft was vibrated to simulate a launch, then transferred to the altitude chamber for simulated flight tests under orbital conditions. A complete integrated systems test on 30 September concluded the testing.[49]

A good share of the program office and a sampling of the rest of NASA were on hand the next day to watch Spacecraft 1 as it rolled out of the test area in the McDonnell plant. Throughout the morning, McDonnell experts lectured their NASA guests on the spacecraft, the status of each of its parts, and the results of testing. After lunch, the NASA party retired behind closed doors to ponder the fate of the spacecraft. The McDonnell staff gathered late in the afternoon to hear the decision. Spacecraft 1 had been accepted for shipment to the Cape.[50]

When it arrived on 4 October, it entered a new round of testing. GPO had decided early in the program that Gemini preflight checkout would conform to the Mercury pattern, even though the two-man spacecraft had been designed to render that kind of repeated testing unnecessary. Plans called for the spacecraft to be broken down to its major modules, each of which was retested to the subsystem level. After being put back together again and passing a series of integrated tests culminating in a simulated flight, the spacecraft was to be transferred from the industrial area to the launch complex.[51]

Spacecraft 1, lacking most of Gemini's normal systems, was much easier to check out than later models; by the evening of 12 February 1964, the task was finished. The next step was a formal Preflight Readiness Review of spacecraft status, both physical and functional. Gemini Manager Charles Mathews and a team of engineers from Houston and Cape Kennedy* conducted the review on 18-19 Febru-

---

*President Johnson issued an Executive Order on 29 November 1963, changing the name of the Launch Operations Center to the John F. Kennedy Space Center (KSC) in honor of the late President. The Department of the Interior concurred and Cape Canaveral became Cape Kennedy.

ary, finding nothing that would prevent the spacecraft from being moved to the launch complex nor that seemed likely to delay the launch.[52]

The launch vehicle was not ready for mating, so Spacecraft 1 waited until 3 March before its transfer to complex 19. While the spacecraft waited, minor work continued, especially on the spacecraft shingles. These beryllium shingles were part of the heat protection structure and covered the external surfaces of the two forward modules—the rendezvous and recovery canister and the reentry control system. A fully acceptable fit was not, in fact, achieved until after the spacecraft had been mated to the launch vehicle.[53]

Building and testing the first Gemini launch vehicle was not as easy as getting the spacecraft ready, because GLV-1 had the same role as the later boosters in the program. Just as McDonnell had been building spacecraft despite hard-to-resolve problems in some spacecraft systems, the Baltimore division of Martin-Marietta had been building launch vehicles for Gemini, even during the long months when the Air Force and its contractors were struggling to make Titan II reliable.*

Titan II was built around its propellant tanks, one for fuel and one for oxidizer in both the first and second stages. Martin's Denver division, which held the missile contract, provided the tanks for Gemini boosters as well and shipped the set for GLV-1 to Baltimore in October 1962. After a lengthy series of tests, with special attention to welded joints to be sure they were both strong enough and leakproof, the tanks were ready for formal inspection in mid-February 1963.† Only three passed. The second-stage oxidizer tank was cracked. It was returned to Denver and replaced by the tank intended for GLV-2, which reached Baltimore on 1 March.[54]

By 21 May, the first Gemini launch vehicle was fully assembled and ready to begin testing as a unit. A check for wiring continuity revealed a short circuit in the second stage where a wire's insulation had been cut through by a defective clamp. When inspectors found several other clamps with the same defect, every one of the more than 1500

---

*GLV-1 was already at the Cape on 26 October 1963, a week before the flight of Titan Missile N-25 first promised an answer to the Pogo problem. It was mostly Martin-Denver people who were struggling with missile problems.

†The inspection team, headed by Major Robert Goebel (SSD), included representatives of Martin, NASA, Aerospace, and the Air Force. Coordinating the team's activities was John R. Lovell, GLV-I's "chaperon." A launch vehicle chaperon started his duties at Denver with the building of the tanks, then traveled with the tanks to Baltimore and went through all the testing, keeping complete records of everything that took place and the results. He flew to the Cape with the assembled vehicle and remained with the booster until it was launched, when he returned to Baltimore. Aerojet-General also used the chaperon system, calling its people "guardian engineers." J. W. Gustafson shepherded the first- and second-stage engines from their beginnings in Sacramento, California, to liftoff at the Cape.

wiring-harness clamps in GLV-1 was removed, all wiring inspected, and a new set of clamps installed.[55]

When electrical continuity had been confirmed, the first stage was erected in Martin's new Vertical Test Facility on 2 June, the second a week later. This facility was a tower 50 meters high, adjoined to a three-story blockhouse fitted with test and checkout equipment, or AGE,* matching the AGE at complex 19 in Florida that would later ready GLV-1 for launch. The tower and blockhouse inside the Martin plant were designed to provide test data and to be compared with data gathered during checkout at the Cape.[56]

The first phase of the test program, subsystem functional verification to make sure that each of the vehicle's subsystems was working, began on 10 June. These tests went more slowly than planned. For one thing, the second stage had been late going up, partly because of electrical problems and partly because its engine arrived late. For another, minor troubles cropped up—hydraulic tubing that was not fully cleaned, solder flux that had boiled from a pinhole in a joint and gummed a gyroscope. By the end of June, subsystem testing had fallen about two weeks behind schedule, a source of concern but as yet no threat to the launch planned for December 1963. The functional verification tests lasted until late July, when a review of the data by SSD and the Aerospace Corporation found GLV-1 ready for the next phase of testing.[57]

GLV-1 began combined systems tests on 31 July with a series of tests designed to uncover any interference between the vehicle's several electrical and electronic systems. Five systems failed to meet standards after the first round of testing. Efforts to correct the problems—mainly by adding filters and grounds to Age and airborne circuits—produced results, though slowly. Only after the sixth test, on 5 September, was all interference cleared up. The launch vehicle's last hurdle was a combined systems acceptance test (CSAT), which included a complete launch countdown, simulated engine start, liftoff, and flight, and ended with the simulated injection of the spacecraft into orbit. After several practice runs in conjunction with the electrical-electronic interference testing, Martin conducted the formal CSAT on 6 September, then presented both the data and the vehicle to the Air Force on 11 September for acceptance.[58]

For the next week and a half, the Vehicle Acceptance Team, headed by SSD's Colonel Richard Dineen, met at the Martin plant in Baltimore. SSD, NASA, and Aerospace inspectors explored the vehicle

---

*AGE is one of those acronyms that tend to take on a life of their own. The formal meaning of AGE is aerospace ground equipment, but the acronym was (and is) immeasurably more common in use.

and studied its manufacturing and test records. This detailed inspection disclosed severe contamination of electrical connectors throughout, as well as a broken idler gear in the turbopump. These defects, plus the fact that 42 major components had yet to achieve documented flight status, forced the team to reject GLV-1. Failing to pass this type of inspection on the first try was not unusual, but it meant another long delay before GLV-1 reached the launch site.[59]

SSD and Aerospace members of Dineen's team also conducted a First Article Configuration Inspection (FACI) of GLV-1, with far more encouraging results. FACI had been a standard Air Force procedure since June 1962, a kind of audit of the actual product—as compared to engineering design—to provide a baseline for later products under the same contract. No SSD launch vehicle had ever made the grade on its first try, but GLV-1 did. Such defects as contaminated electrical connectors or broken gears, which barred its acceptance for Gemini, did not reflect discrepancies between design and product.[60]

No sooner was the inspection over than Martin technicians began to set things right. Armed with magnifying glasses, they searched every one of the 350 electrical connectors aboard GLV-1 for traces of contamination and found 180 needing to be cleaned or replaced. All flight control equipment that had produced transient malfunctions during CSAT was removed and analyzed. Defective units were replaced and wiring harnesses reinstalled. At the same time, Martin tried to complete documentation of failure analyses and qualification of flight hardware. This extensive reworking of GLV-1 invalidated most of the earlier test results. Martin's plan for an informal retest of problem areas only was rejected in favor of a full-scale repetition of CSAT. Subsystems testing and a preliminary acceptance test were finished by 2 October.[61]

The second formal acceptance test of GLV-1 ran on 4 October, uncovering little that needed to be corrected. Dineen's team reconvened at Baltimore on 9 October and took only two days to complete its work and decide that GLV-1 could be shipped to the Cape. The team was scarcely enthusiastic about the vehicle. Much work remained to be done on GLV-1, but it could be done at the Cape, and there at least GLV-1 could be helping to check out the launch complex itself.[62]

On 26 October 1963, GLV-1's two stages, each strapped to an eight-wheeled trailer, were towed to the Martin Airport, next to the plant, and rolled through the rear loading door of a huge C-133B cargo aircraft provided by the Military Air Transport Service. A four-hour flight brought the two stages to Florida. Still on their trailers, they were rolled from the aircraft into the hands of Joseph M. Verlander's Martin-Canaveral crew, who towed them to Hangar H to be unpacked, inspected, and fitted with the gear (such as lifting rings) required to erect them. There they remained, under guard, over the

*Titan in Vertical Test Facility, Martin-Baltimore.*

*Titan 2d stage is loaded on C-133B in Baltimore (above right), arrives at Cape Canaveral (right), 26 October 1963.*

*Titan 1st stage is raised by erector (left); 1st and 2d stages are test-fired sequentially (right).*

*Gemini spacecraft is
unloaded at the Cape
from C-133B, 4 October
1963.*

*Spacecraft is mounted
in test and checkout
hangar at the Cape.*

*Spacecraft at launch pad
(below); mated and being
checked in White Room by
Guenther Wendt (right).*

187

weekend. On Monday morning, 28 October, the trailer bearing the first stage reached complex 19.

At the launch complex, the Martin crew trundled the first stage up the long ramp to the launch vehicle erector, which rested on its side parallel to the deck of the test stand. The trailer rolled through the large door (the roof when the erector was standing) and stopped a meter and a half (five feet) from the other end. The crew secured the stage, removed the trailer, and closed the roof-door. A 150-horsepower electric motor then winched the 127-tonne (140-ton) erector upright, a process that took several hours. The trailer-borne second stage arrived at the launch pad a day later. Ordinarily, the next step was mounting the second stage on the first, but GLV-1 was slated for a special static firing test in mid-December, the sequenced compatibility firing of both stages. So stage II was placed in the second-stage erector, a smaller structure used only for checkout or static firings, and the two stages were cabled together. After checking to be sure there was no interference, Verlander's team applied electrical power to the two stages standing side by side on 13 November.[63]

Work at the Cape on GLV-1 was already a week behind schedule. Problems in Baltimore had pushed the launch date from December 1963 to February 1964. Another two-month delay now threatened. Mathews announced himself "greatly concerned with the present situation regarding the Gemini Program at the Atlantic Missile Range." Four distinct groups—SSD, the Air Force's 6555th Aerospace Test Wing (in charge of all Cape launches), Martin-Baltimore, and Martin-Canaveral—were testing and checking out the launch vehicle, with no formal understanding on how responsibilities were to be divided among them. Clarification was not long in coming; but meanwhile matters had become so confused that two distinct Launch Test Directives had surfaced. To make things worse, NASA people at the Cape complained about lack of access to technical data from the contractors. Poorly meshed working groups compounded other problems—a time-consuming review of the official work plan, procurement snags, and, most serious, questions of compatibility between booster and AGE—which extended the planned number of working days to get GLV-1 ready for launch from 86 to 118. By 22 November 1963, Mathews had to tell Seamans that even the already late 28 February 1964 launch date was likely to drop back to 1 April although GPO was working hard to improve the prospect.[64]

In one move to help resolve management problems, Mathews united the several coordination panels that had been dealing with Titan II and related areas into a single Gemini Launch Vehicle Coordination Committee with six standing panels.* All panels were to meet at

---

*Jerome Hammack of GPO was chairman of the Coordination Committee, with Lieutenant

the same time every third week, then report to the parent committee, which would decide what action was to be taken. That should mean no more delays caused by uncertain authority, duplicated effort, or conflicting decisions.[65] Mathews and GPO launch vehicle manager Willis Mitchell also took steps to make good some of the time already lost. The Martin crew switched from two 8-hour to two 12-hour shifts a day. Checkout problems persisted, however, and the scheduled sequenced firing slipped from 20 December 1963 to 3 January 1964. Although a February launch of GLV-1 seemed out of the question, Mathews still hoped to launch by 17 March.[66]

But the problems refused to end. The combined systems test scheduled for 13 December was twice postponed and finally completed on New Year's Eve. Lack of compatibility between the booster and its support systems in complex 19, as well as a faulty turbopump assembly that had to be returned to Aerojet-General, were the major causes of delay. Next was the so-called wet mock simulated flight test, a complete countdown that included filling the propellant tanks; it was voided on 3 January by procedural errors after propellants had already been loaded. The test was called off two and a half hours before the simulated launch, although the count went on until T−30 (30 minutes before launch) to see if any other problems turned up and to give the operations crew some practice. Another try, on 7 January, was a success.

The countdown for sequenced compatibility firing was now set to begin, but a three and a half hour delay was imposed by contaminated oxidizer. Then, during the countdown, a malfunctioning first-stage propellant valve caused the test to be called off 20 minutes before firing. A second try, on 14 January, had to be canceled because unusually cool weather had chilled the engine start cartridges below the 275 kelvins (35°F) specified as the lower limit by Aerojet-General to prevent combustion instability. At last, on 21 January, the third attempt overcame some minor problems and delays to show the whole sequence of fueling, countdown, ignition and shutoff commands, guidance control, and telemetry. First-stage engines fired for 30 seconds and cut off. The second-stage ignited and fired for 30 seconds, halted by radio signal from the ground computer as in real flight. Sequenced compatibility firing proved that the engines delivered the required thrust and gimbaled properly. This static firing, the only one performed on a Gemini launch vehicle, met all prelaunch standards.[67]

---

Colonel Alfred J. Gardner, Chief, Engineering Division, Gemini Launch Vehicle Program, SSD, as associate. Panels were headed by John W. Smith (structures), John J. Turner (systems), Marlowe D. Cassetti (launch guidance and control), Donald Jacobs (abort), Carl Kovitz (test operations), and Richard E. Lindeman (cost, schedules, and contracts). All the panel chiefs were from GPO, except Cassetti, who worked in the Flight Operations Directorate.

*189*

With static firing finally out of the way, the ground crew could now begin getting the booster ready for the spacecraft. That meant putting the second stage on top of the first, which was scheduled for 27 January. But post-firing cleanup found a defective rotor in one of the turbopump assemblies. Shipped to the West Coast for repair, it returned to the Cape on 29 January. Then a missing seal held up its reinstallation until 7 February.

The launch crew did not wait for the new seal; the turbopump assembly could be put back in the second stage after it was erected. On 31 January, they removed the stage from the small erector and secured it in the launch vehicle erector, which was then winched upright. The upper stage was gently lowered onto the first, and the two were bolted together. GLV-1 had assumed its final form. Before the spacecraft could be mated to the booster, there were still subsystem functional verification tests (like those done earlier in Baltimore) to be conducted. Although these tests were supposed to start on 14 February, lack of spare parts and questions about failure analyses imposed another week's delay. Once testing began on 21 February, however, it went smoothly to verify the launch vehicle's readiness for full systems testing by 3 March.

On that day, Spacecraft 1 arrived at the launch complex to be installed in the spacecraft erector support assembly in a controlled-access "white room" atop the launch vehicle erector.[68]

## TIGHTENING LAUNCH SCHEDULES

The revised flight program of April 1963 had projected the first manned mission, Gemini 3, for October 1964. But as 1964 approached, that prospect was dimming. The first Gemini flight was held up by the late delivery and protracted testing of its booster, and Spacecraft 2 was falling behind schedule at the McDonnell plant. Efforts to install spacecraft test and checkout equipment at the launch site in Florida moved slowly enough to suggest that time might be too short there as well. The already certain delay of the first mission, added to the all-too-likely chance that the second would also be late, made the prospects for launching Gemini 3 in 1964 look poor.[69]

At a meeting on 13 November 1963, the Gemini Management Panel* decided that the program's current schedule needed rethinking. The key question was just how much spacecraft and booster testing had to be repeated at the Cape to ensure a successful mission. Two panel members, MSC Gemini Program Manager Charles Mathews and

---

*MSC Director Robert Gilruth had formed the panel in October 1962 to deal with managerial and technical problems. It brought together the heads of the organizations in charge of Gemini—from NASA, the Air Force, and major contractors.

Space Systems Division launch vehicle chief Richard Dineen, set up an *ad hoc* study of work plans and schedules aimed at seeing men in orbit via Gemini before the end of 1964. Mathews reported the findings to the panel at its next meeting, 13 December 1963. Gemini 3 could be launched in November 1964 by cutting down spacecraft testing at the Cape that merely repeated work already performed in St. Louis and by better integrating the entire checkout effort. Launch-vehicle testing was already fairly well meshed between Baltimore and the Cape and needed only to be smoothed out.[70]

Spacecraft checkout procedures were altered sharply "to get a complete working spacecraft out of the McDonnell plant." All testing in St. Louis, along with whatever manufacturing tasks were left after systems testing began, was to be modeled on Cape practice. This meant that the McDonnell test crew had to be retrained. John J. Williams, Assistant Manager for Gemini of MSC Florida Operations,* took a Launch Preparation Group of 200 people, drawn from both NASA and McDonnell, to spend nearly nine months in St. Louis. They throughly revamped the testing process, training the St. Louis crew and actually checking out the second and third Gemini spacecraft. About half the group returned to the Cape with Spacecraft 2 in September 1964, and the rest stayed until Spacecraft 3 was ready in January 1965. The retrained McDonnell crew took over when Spacecraft 4 began systems testing. Basic to the new process was cutting down on repeated testing. Once a subsystem had been tested, it would take its proper place in the spacecraft and stay there. No longer was the spacecraft to be taken apart after it reached the Cape, tested, and put together again. Systems were to be rechecked, of course, but only as part of the complete spacecraft, not as individual pieces.[71]

The booster offered fewer problems in meeting Gemini schedules. Aside from efforts to speed up work on GLV-1, already at the Cape, the only major step was to strike flight readiness firing from the test program planned for the first three launch vehicles. With spacecraft checkout streamlined and booster testing smoothed out, GPO looked forward to getting back in step with the April 1963 schedule, even though the first flight was now going to be about three months late. The eight months that had been allowed between the first two flights was cut to five, with Gemini 2 only a month behind schedule, in August instead of July 1964. By then keeping to the three months between later flights, the first manned mission could be launched in November, a month late, but still in 1964.[72]

---

*On 30 March 1964, Gilruth announced that the Preflight Operations Division had become an autonomous unit known as MSC Florida Operations. Directed by G. Merritt Preston, the group would perform much the same duties as it had in Mercury. The only major change would be the participation in testing at McDonnell.

# IX

# A Taste of Success

WHILE Gemini's first spacecraft and launch vehicle were moving toward their mating on complex 19 at Cape Kennedy, the Gemini Program Office itself was coping with another kind of move. The permanent home of the Manned Spacecraft Center at Clear Lake, though not quite finished, was ready to be occupied. GPO began shifting its desks from the old Veterans Administration building in downtown Houston to the new campus-like setting near Clear Lake on 6 March 1964. Shortly after the transfer had been completed, Program Manager Charles Mathews announced a reorganization of GPO. Major changes reflected the growing stress on schedules and testing as Project Gemini poised on the verge of its first flight. Project Administration changed its name to Program Control.* Scott H. Simpkinson left Mathews' staff to take charge of a new Test Operations Office dealing with reliability and quality assurance as well as test planning and evaluation.† Launch Vehicle Integration became Vehicles and Missions, divided into vehicle development and mission planning offices, plus a

---

*The former chief of project administration, André Meyer, became Mathews' senior assistant; Major Richard C. Henry transferred from the Washington program office to head the new GPO Program Control Office; George MacDougall stayed as second-in-command and acting head of production engineering; Walter Wolhart headed cost engineering; and James E. Bost program engineering.

†W. Harry Douglas came from the Spacecraft Office as deputy manager and acting head of reliability and quality assurance; Charles K. Williams ran test planning; and Victor P. Neshyba, test evaluation.

*Above, the Manned Spacecraft Center, Houston, Texas, as seen in January 1962. The spacecraft center site is to the left of the Jim West mansion seen in the foreground. Right, the site as seen in September 1964 from a different angle. The West mansion is hidden in the trees at center.*

new integration office to keep tabs on spacecraft/launch vehicle and spacecraft/target interfaces.* The Spacecraft Management Office simply changed its name to the Spacecraft Office.† The Houston-based strength of the program office had now reached 117; GPO also maintained representatives at Martin in Baltimore and Lockheed in Sunnyvale, California, as well as resident manager's offices at McDonnell in St. Louis and Kennedy Space Center at the Cape.‡ This was the organization that, with only minor changes, saw Project Gemini through to its end.[1] Before that happy end, however, there was the more immediate matter of Gemini-Titan 1.

## THE FIRST FLIGHT

By 3 March 1964, spacecraft and booster were at last together on launch complex 19 at Cape Kennedy. The series of tests that showed all booster systems were working had just been completed, and the spacecraft had been hung on a tripod in the "white room" atop the launch vehicle erector. This room, with its four levels and 4.5-tonne

---

*Willis Mitchell remained manager; Jerome Hammack became deputy manager and acting head of vehicle development; Wyendell B. Evans, of mission planning; and Lewis R. Fisher, of systems integration.

†Duncan Collins continued as manager and also acting head of electrical and electronics suboffice, with Homer Dotts as his deputy manager and acting chief of the structural and mechanical suboffice. Guidance and control was the province of Richard Carley, and Kenneth Hecht was responsible for escape, landing, and recovery.

‡The Martin-Baltimore representative was Harle Vogel, and the Lockheed-Sunnyvale liaison was A. B. Triche. Wilbur H. Gray was head of the Office of the NASA Resident Manager at McDonnell throughout the program, ably assisted by Andrew Hoboken; the 48-person office focused mainly on engineering and quality control. Walter Kapryan was resident manager at the Cape.

(5-ton) crane to hoist the spacecraft, was sealed off from the outside world and maintained at a constant temperature of 295 kelvins (72°F) and a constant relative humidity of 50 percent, to provide a controlled environment for the spacecraft and the upper stage of the booster. Next to the erector was an umbilical tower 31 meters high. Its seven booms supported 31 cables and lines to spacecraft and booster, feeding electrical power, propellants, and other needs until the moment of launch. Gemini-Titan 1 was scheduled to lift off on 28 March 1964.[2]

A premate systems test on 4 March confirmed the spacecraft ready for mating the next day, when the spacecraft-to-launch-vehicle adapter would be bolted to the booster's upper stage. The effort was delayed briefly when a McDonnell worker dropped his wrench on the dome of the oxidizer tank just below the spacecraft. A plastic sheet protected the dome, but the impact produced a scratch 0.95 centimeter (0.375 inch) long and 0.0038 centimeter (0.0015 inch) deep in the steel surface, just 0.16 centimeter (0.64 inch) thick at the point of impact. The area was burnished to the depth of the scratch and tested to confirm that the metal was still solid.[3]

After the spacecraft and launch vehicle had been mechanically mated, they also had to be connected electrically. But first the booster's status had to be checked in a combined systems test. That was slated for Sunday, 8 March, to be followed by three electronic-electrical interference tests between 9 and 13 March, to make sure there was no serious incompatibility. Minor problems delayed the booster combined systems test until Tuesday, and interference testing did not start until Thursday, 12 March.[4]

The first try at an interference test had to be scrubbed, and that cost another four days. On Monday, 16 March, however, the test went off without any trouble, prompting the crew to run through the second test at once. The attempt went awry through a procedural error. Another try, on Thursday, 19 March, brought bad news. Some amplifiers in the circuits that controlled the booster's tandem actuators (which shifted the engines to alter flight path) showed noisy outputs. A special dry run the next day produced the same problem, and the third interference test had to wait until the trouble was resolved. There was some question about how that was to be done, which was settled on Tuesday 24 March, when Martin troubleshooters pinpointed the problem—in the test equipment. Another test, on Wednesday, confirmed the finding. A conference that evening concluded that the data from the dry run the previous Friday met the intent, if not the precise format, of interference testing. The test equipment was removed that night.[5]

But the tests had taken almost two weeks longer than planned, forcing the launch to be postponed to 7 April 1964. Things now began to move more smoothly. On Friday, 27 March, a combined systems test and simulated flight produced no serious problems.[6] The following Tuesday, 31 March, all the nonflight parts that GLV-1 had carried to the Cape were replaced and Pogo gear installed. GLV-1 was scheduled to have its tanks filled with propellants that night as part of a complete countdown exercise, the wet mock simulated launch.

At 9 p.m., as shift workers were clearing the area for the start of tanking, someone saw smoke pouring from a switch at the pad. A burnt-out transformer and switch motor forced the test to be suspended, since there were no spares on hand and the switch performed a crucial function. It automatically transferred the launch complex to auxiliary power if commercial power failed. Safety demanded that the launch area be deluged with water in case of propellant leak; a power loss would leave that system inoperable for about 30 minutes if the automatic switch were not working. Workmen found a spare transformer at 1:18 Wednesday morning and installed it, but a new motor was harder to locate. One was finally borrowed from the blockhouse since that system could be run by hand.[7] But another day had been lost.

Propellant loading resumed just before 10 Wednesday night and finished four hours later. The countdown began at 5 o'clock Thursday morning, but now came weather trouble. The Cape was under an "atmospheric inversion," a blanket of warm air above cooler air near the ground, which would block the upward dissipation of toxic fumes in case of accident. The count was held from 7 to 8:30, when the inversion started to break up. Ground crews then removed the propellant lines leading to the booster tanks and the count resumed. It followed its normal course until three minutes before launch, T−3, when a minor problem (quickly corrected) required the count to be recycled to T−5. Five minutes later, at half-past noon, the count reached T−0, the moment when the booster's first-stage engine would have ignited in a real launch. The test was a complete success, free of spacecraft problems and marred only by a minor procedural error in the launch vehicle countdown. After a vibration test of GLV-1, the tanks were drained of propellants, a five-hour process finished at midnight.[8]

The Spacecraft Flight Readiness Review Board* convened Friday afternoon, 3 April, in the conference room of the Engineering and Operations Building, headquarters for MSC's Florida Operations. A check of items left open from the preflight review of 18−19 February

---

*The board was headed by Walter Williams and recorded by Lester Stewart; other members were Mathews, F. John Bailey, Jr., Christopher Kraft, Donald K. Slayton, and Merritt Preston from, respectively, the Gemini Program Office, Reliability and Flight Safety, Flight Operations, Flight Crew Operations, and Florida Operations. They evaluated all waivers, deviations, modifications, discrepancies, and work done at the Cape. McDonnell and MSC systems engineers were on hand to answer questions and assist the board.

A TASTE OF SUCCESS

showed that everything had been taken care of except a circuit breaker not yet fully qualified. It was close enough, however, for McDonell to certify it flightworthy, a judgment the board shared. Only two new problems had cropped up since the earlier review, both easily corrected. The board judged all systems ready for flight, pending the outcome of the final systems test, a simulated flight scheduled for 5 April. When the simulated flight went off without a hitch on Sunday, Spacecraft 1 was ready for its mission.[9]

Flight readiness of the launch vehicle was reviewed Saturday afternoon. The Air Force reported two problems, one of which turned out to be nonexistent. The other involved a missing report of the results of an analysis of a failure in the secondary autopilot. The report was still absent on the eve of flight, but a phone call confirmed that the problem had been analyzed. After the simulated flight on Sunday, Walter Williams convened the Mission Review Board. Spokesmen for every group involved in the mission reported everything ready—"all systems 'go.'" At noon, Williams announced that NASA was "proceeding toward a launch not earlier than 11:00 a.m. Wednesday, April 8."[10]

The final decision for launch came on Tuesday morning. At 7:30, 7 April, SSD's Status Review Team for GLV-1 met, took a last look at the launch vehicle, and agreed it was ready to go. That recommendation was passed on to the Flight Safety Review Board at 9:00 a.m. The board approved GLV-1 for flight and committed it to launch, with liftoff set for 11 the next morning.[11]

Preparations for the final countdown were already under way. The first part of the planned 390-minute split countdown started before dawn on Tuesday. That 60-minute segment ended at 5 a.m., when the count was held for 23½ hours to prepare the spacecraft for final countdown, install and hook up pyrotechnics, run some launch vehicle tests, and load propellants. GLV-1's tanks were topped off at 4:10 Wednesday morning, with about 75 people from Martin, the Air Force, Aerojet-General, and Aerospace on hand. Thirty systems experts from McDonnell and MSC arrived at the blockhouse at 4:30. The hold ended right on time, an hour later, and final countdown began at 6 a.m. or T−300. No flaw marred the entire five-hour process.

One second after 11 o'clock Wednesday morning, 8 April 1964, the booster's first-stage engine ignited. Of this one-second discrepancy, a joking Williams later remarked to a roomful of reporters, "There must be something wrong with the range clock." Four seconds later, the 136-tonne (150-ton) vehicle lifted from the pad on that curiously lambent flame so distinctive of Titan II's hypergolic propellants.[12]

Within moments, *Gemini-Titan 1* vanished into the hot Florida sky, beyond reach of human senses but not electronic sensors. Telemetered data flowed back to mission controllers at the Cape, telling

**Gemini I**
**8 April 1964**

Intent launch team in the blockhouse of pad 19, Cape Kennedy
(above left); Chief Test Conductor Paul Donnelly monitors the
final minutes of countdown (center); and the unmanned Gemini-
Titan lifts off, beginning the
flight program of Gemini.

them that the launch was as nearly perfect as it looked. Two and a half minutes after liftoff, the 118 tonnes (130 tons) of propellants in its first stage exhausted after driving *Gemini-Titan 1* 64 kilometers high and 91 kilometers downrange, GLV-1's first-stage engines cut off. The second-stage engine flared into life, and the four bolts that had held the two stages together exploded as they were designed to, cutting the spent first stage loose from the still-accelerating second stage and spacecraft. Five and a half minutes after launch, the second-stage motor stopped, its 27 tonnes (30 tons) of propellants gone. Now 1000 kilometers downrange and 160 kilometers high, coasting at a speed of 7888 meters (25 879 feet) per second, Gemini Spacecraft 1, with the second stage of GLV-1 still attached, was in orbit.[13]

Everything had gone beautifully. Purists might cavil at an excess 7 meters (24 feet) per second launch-vehicle speed that propelled the spacecraft into an orbit reaching out 320 kilometers instead of the programmed 299 kilometers. But they could scarcely deny the handsome achievement of the main goals—proving that the booster could do its job and that combined with the spacecraft its structure was sound. "There's no question these objectives were met," Walter Williams observed to the press shortly after launch.* The nearly flawless performance of the launch vehicle elated its sponsors, prompting one of them, Major General Ben Funk of SSD, to call it "just completely a storybook sort of flight."[14]

The mission of *Gemini-Titan 1* was much shorter than its actual trip. Only the first three orbits were part of the flight plan. When Spacecraft 1 passed over Cape Kennedy for the third time, about 4 hours and 50 minutes after launch, the first Gemini flight came to a formal close. The spacecraft had been expected to orbit Earth for three and a half days. Because of its slightly higher than planned orbit, it actually stayed up for nearly four days. During that time, the Manned Space Flight Network,† a round-the-world system of tracking stations controlled from Goddard Space Flight Center in Maryland, followed the vehicle by radar. On Sunday, 12 April, during its 64th pass, the steadily slowing spacecraft plunged back into the atmosphere, ending its career in flames over the South Atlantic, midway between South America and Africa.[15]

---

*This was Williams' only Gemini launch. On 16 March, this veteran director of all the country's manned space flights resigned from NASA to accept a position as vice president and general manager of Aerospace's Manned Systems Division, to take effect after the first Gemini flight. Williams was replaced as Gemini Operations Director by Kraft, who had become MSC Assistant Director for Flight Operations in the November 1963 reorganization.

†Network stations used for *Gemini-Titan 1* were Kennedy; Grand Bahama Island; San Salvador; Bermuda; Woomera, Australia; Hawaii; Point Arguello, California; White Sands, New Mexico; and Eglin Air Force Base, Florida.

NASA Associate Administrator Robert Seamans commended "the Air Force for its most successful Launch Vehicle Program."[16] So fine a performance of the first mission augured well for those to follow and surely enhanced the prospect that Gemini astronauts would be in orbit before the end of the year. But the glow of accomplishment soon faded before the hard work yet to be done. While the launch vehicle was now qualified for manned missions, the spacecraft was not. Despite the gratifying success of *Gemini-Titan 1*, and some real progress on troublesome spacecraft systems, there was no time to rest on laurels. The target vehicle for Gemini's later missions was still a very large question mark, and Gemini's chronic money woes were far from settled. For all of that, Gemini's future in the spring of 1964 must have looked much brighter than it had only a few months earlier.

## POSTSCRIPTS AND PROSPECTS

So bright, in fact, did the future seem that the long dormant idea of using the Gemini spacecraft for a lunar mission stirred again. George Mueller, NASA's Associate Administrator for Manned Space Flight, had some reason to be concerned about the outlook for Project Apollo in the spring of 1964. Only a few months earlier, plans for manned flights using Saturn I had been canceled, leaving Gemini as the only possible system for manned orbital flights during the next two years or more. Mueller wanted to know if a Gemini lunar mission could be flown. If it could, then a contingency plan was to be prepared for a Gemini flight around the Moon in case Apollo suffered a serious setback. A review of past studies strongly suggested that the idea was feasible and that McDonnell should be asked to conduct a more detailed study.* [17]

But that was not to be. During a tour of the plant in Louisiana where Saturn rockets were built, Wernher von Braun, Director of Marshall Space Flight Center, told a journalist that Gemini might be able to fly around the Moon, but only as "a possible project to salvage this country's prestige if the manned lunar goal proves impossible." Whether this was intended to squelch an Apollo rival, the effect might have been predicted. The same factors that had blocked the idea before still held. NASA had too much invested in Apollo—too much money, time, and prestige—to really think about Gemini to the Moon. Funds, in any case, were tight. On 8 June, Seamans told Mueller there would be no money for study contracts. "Any circumlunar mission

---

*The review was done by William B. Taylor and John L. Hammersmith, of Mueller's Gemini and Advanced Manned Missions offices, respectively.

studies relating to the use of Gemini will be confined to in-house study efforts."* [18]

But that was never more than a side issue. In mid-1964, the first task was still Project Gemini, however attractive the prospects of a more ambitious program might seem. The outstanding performance of *Gemini-Titan 1* and the qualification of the Gemini launch vehicle were most cheering portents. When the Gemini Management Panel met a week after the mission, on 15 April, a comfortable optimism suffused the group. The current work schedule called for the second flight toward the end of August and the third in mid-November, with almost a four-week cushion in each instance to handle unforeseen problems.[19]

This bright outlook darkened in the late summer before a series of natural disasters. First lightning, then hurricanes, conspired to abuse the second Gemini launch vehicle on complex 19 at Cape Kennedy and to delay its flight long past the scheduled time. Even had the weather been perfect, however, McDonnell's difficulties in getting Spacecraft 2 ready to fly might have compromised the schedule.

Late deliveries—notably of thruster systems from Rocketdyne and fuel-cell stacks from General Electric—had slowed construction of the spacecraft during 1963. Parts had failed tests that had to be passed before they could be installed in the spacecraft; modifications meant further delays. Spacecraft 2 could not begin its systems tests until 13 January 1964.[20]

The Spacecraft 2 Design Engineering Inspection (DEI), earlier set for November 1963, had been postponed in the face of these delays until February 1964. MSC formed a permanent DEI board 31 January 1964 to make sure that the spacecraft as a whole and each of its parts would do what they were intended to do—that the spacecraft could, in fact, be expected to achieve its assigned objectives. Normally, the DEI for each spacecraft would fall between the end of manufacturing and the start of systems testing, but the DEI for Spacecraft 2 was a little late. The nine-member board convened at the McDonnell plant on 12 February.† Also present for the two-day meeting were 50 experts from

---

*The in-house studies did continue, culminating in a paper in July 1964 by Calvin C. Guild, enumerating 16 different missions that could be classified as "advanced" (beyond the 12 then scheduled for Gemini) and that used the Gemini spacecraft or techniques derived from the Gemini program. Among them were the demonstration of land landing with either paraglider or parasail, a combined launch in which Gemini would rendezvous with Apollo and check out ship-to-ship communications, a minimum rotating space station experiment to provide experience in artificial gravity for long-duration space travel, space assembly and repair missions, and a lifeboat rescue mission.

†Chairman and vice chairman of the permanent DEI board were to be the head of reliability and flight safety and the manager of the Gemini program. The other five would come from the

(Continued)

GPO and McDonnell, as well as another 50 observers from other MSC offices, NASA Headquarters, and the Air Force. The board looked over the hardware and studied the records to see that each part either matched design specifications or was the subject of a proper waiver. A long list of minor discrepancies ended up as 22 mandatory changes, 4 conditional, and 10 to be studied.[21]

The first phase of spacecraft systems tests went slowly, as problem after problem turned up; troubleshooting them, working out the required changes, and testing the results all took time, adding to the delays. By mid-April 1964, Spacecraft 2 had become the "pacing item" for the second Gemini mission, a dubious honor held by the launch vehicle before the first flight. Getting the spacecraft ready was now the crucial factor in meeting the scheduled launch date.[22] This was not altogether a surprise. Spacecraft 1 had been little more than an instrumented shell, but GLV-1 had been a launch vehicle in every sense of the term. The Martin crews working on GLV-2 were going over ground they had already surveyed, but Spacecraft 2 was the first fully equipped ship to go through the McDonnell plant and its slow progress reflected its novel status.

After the modules of the spacecraft had been mated, the second phase of systems tests began, on 3 July. Further problems hampered testing into the next month.[23] Whatever delay might have resulted, however, became purely academic after mid-August, when Florida weather dealt the first of a series of time-consuming blows to GLV-2.

## GLV-2 AND THE ELEMENTS

While spacecraft testing floundered past snag after snag, GLV-2 had been moving briskly through its test program despite some rough spots. At the outset, the second-stage oxidizer tank was found defective, and a new tank had to be built. Since the first-stage tanks were not yet ready, the delay was inconsequential. Martin-Baltimore received all four tanks from Denver on 12 July 1963. Engines were late in arriving from Aerojet-General, but testing went ahead with nonflight first-stage engines. By the end of January 1964, GLV-2 had completed its horizontal test program. Early the next month it was standing in the Vertical Test Facility; and, after two weeks of modification work, functional verification tests of subsystems began on 21 February.[24]

GLV-2 finished these tests by 13 April, in roughly two thirds the

GPO spacecraft office, three directorates (Engineering and Development, Flight Operations, and Flight Crew Operations), and Florida Operations. Members for the Spacecraft 2 DEI were F. John Bailey, Mathews, Homer Dotts, Aleck C. Bond, John D. Hodge, Virgil I. Grissom, John Williams, and Walter Williams, with Robert T. Everline as recording secretary.

time taken by the first booster. Another week saw it through electrical-electronic interference tests and three preliminary combined systems acceptance tests (CSAT), an effort that had cost GLV-1 over a month. The formal CSAT was run on 22 April with no trouble, and the results were approved by the Vehicle Acceptance Team the following week. The dummy engines still had to be replaced, which took a month. By mid-June, GLV-2 had been inspected and formally accepted for the Gemini program. Since spacecraft work was lagging, the booster's transfer to the Cape was postponed so Martin crews in Baltimore could complete some of the modifications that would otherwise have been made by the Martin-Canaveral team.[25]

Workmen loaded the booster aboard an Air Force C-133B aircraft on 10 July 1964. By noon the next day, both stages had been unloaded and secured. Working a two-shift, five-day week, Martin's Cape crew expected to have GLV-2 ready for Spacecraft 2 by mid-August. Everything proceeded routinely through July and into August, with only minor problems causing small delays. This was of no moment, since the spacecraft was still in St. Louis. Its shipment, scheduled for 1 August, had been postponed for three weeks; it could not now reach complex 19 before the first week in September. The Martin crew nevertheless prepared for the final test of the booster before its mating with the spacecraft and were almost through by 17 August.[26]

But that Monday a severe thunderstorm pounded Cape Kennedy. About half an hour before midnight, lightning struck complex 19. There was no visible damage to the blockhouse, erector, or rocket, but that proved nothing about the status of the electrical and electronic gear. Whether GLV-2 was fit to fly was a real question. NASA labeled the event an "electro-magnetic incident" and demanded a thorough investigation. Inspectors from Martin, Aerospace, and the 6555th Aerospace Test Wing found no signs of any physical damage, but they did locate a number of failed parts, mostly in the ground support equipment. This suggested that the complex had not taken a direct hit but rather had suffered the electromagnetic effects, or induced static charges, of a nearby lightning strike. A test order issued on 20 August set the task: To "re-establish confidence in all [launch vehicle], AGE, ... and Facility Systems, and to determine that all degraded equipment is replaced and appropriate reverification tests are successfully completed." The next day, Gemini manager Mathews flew in from Houston for an "Incident Status Meeting." A three-man steering committee was appointed to oversee the efforts of Air Force, Aeorspace, and Martin work crews.*[27]

---

*The 20 August test order was approved by Martin's Chief Test Conductor and Gemini Project Engineer, Francis X. Carey and William R. Williams. Lieutenant Colonel Stewart V. Spragins, 6555th Aerospace Test Wing, concurred. These three men made up the steering committee.

Two weeks seemed ample to put things back in order. Most sub-systems would have to be retested, and all booster systems, test equipment, and facilities would have to be checked out. Any equipment that might have been affected had to be repaired or replaced. After some consultation, NASA agreed that no airborne units with semiconductors ought to be retained. Once new units were installed, testing could begin again as though the vehicle had just arrived at the Cape.[28]

Before the work was finished, however, Hurricane Cleo belied the forecasts and brushed the Cape on Thursday, 27 August. The Martin crew had time to get the second stage down and under cover, but the first stage remained upright, lashed in place with the erector lowered. Cleo's winds were well below the upper limit that the booster was designed to withstand. With the weather still bad on Friday, the second stage stayed in storage over the weekend. On Monday, the Air Force was getting ready to launch its first Titan IIIA from the next complex, which hampered work on pad 19 for most of the day. By 3 o'clock the next morning, however, the Martin crew had stage II back in place atop the first stage. Further work was delayed by the countdown on the nearby pad, which ended at 10 a.m., Tuesday, when the Titan IIIA blasted off. GLV-2's repeat of subsystems functional verification tests began on Thursday, 3 September.[29]

By then, MSC was just about ready to give up on GLV-2. The Center proposed dropping it from the program and moving each of the other launch vehicles up a notch. GLV-3 would launch Spacecraft 2, and the flight program would lose one mission. The Air Force, strongly seconded by the launch vehicle contractors, urged NASA to stick with GLV-2. A thorough review of the effects of both lightning and hurricane, the measures taken to counter them, and the test results had convinced the Air Force and its contractors that GLV-2 was still as sound as ever. Their case was solid enough to convert the skeptics. An Air Force spokesman concluded: "Based on technical considerations, Martin Marietta Corporation, Aerojet-General Corporation, [and] Aerospace Corporation recommend fly GLV#2. In addition, SSD has reviewed cost and schedule considerations and concludes fly GLV#2." NASA agreed, and the work went on.[30]

Testing had scarcely begun, however, before Nature intervened a third time. Cleo had struck only a glancing blow, but Hurricane Dora was aiming straight for the Cape. As Dora approached on 8 September, Martin workers raced to get both stages of GLV-2 down and safely under cover in a hangar. Wednesday was a day of waiting as Dora passed by. On Thursday, Dora was no longer a threat, but Hurricane Ethel was heading for the Cape and due to arrive by the weekend. GLV-2 stayed under wraps. By Monday, 14 September, the danger was past, and GLV-2 was back in place before the end of the day. The rest of the week was largely given over to replacing semiconductor units and to a thorough inspection of booster and launch complex. Testing resumed after the weekend, on 21 September.[31]

That was the day Spacecraft 2 finally arrived at the Cape. The second phase of systems testing at St. Louis had lasted through August and into September, with frequent interruptions for the receipt and installation of a number of pieces of flight equipment. A simulated flight on 15 September completed testing. A Spacecraft Acceptance Review Board headed by Charles Mathews had already gone over the spacecraft to make sure it was ready for the final simulation.* The board met again on 17 September and decided that Spacecraft 2 was now ready for delivery. It was shipped to Florida the following Monday, 21 September.[32]

GLV-2's misfortunes during August and September 1964 forced NASA to forego its goal of a manned Gemini flight before the end of the year, as a rueful Mathews informed the Gemini Management Panel on 29 September. The second flight was now set for mid-November 1964, the third for the end of January 1965. There seemed no need to alter planned dates for the later Gemini missions, although the schedules would have to be tightened. Once again, Gemini's slowness was highlighted by a Russian first. On 12 October, the Soviet Union orbited *Voskhod I.* The three-man crew flew in a "shirtsleeve" environment (flight coveralls rather than space suits) and all remained in the spacecraft to a land landing (previously only Yuri Gagarin was believed to have stayed with his vehicle until it landed, the others leaving the spacecraft and coming down by parachute).[33]

GLV-2 began an expected two weeks of subsystems tests on 21 September, with the combined systems test that preceded spacecraft mating scheduled for 6 October. Spacecraft 2 should have taken only 11 working days in the hangar area before it joined the booster at the launch complex on 25 October. Once again, however, work on the booster went smoothly, but the spacecraft lagged. GLV-2 completed subsystems tests and the premate test on schedule. In another week the launch vehicle finished electrical-electronic interference tests, the last step before it was ready to receive the spacecraft. While the launch vehicle was being tested, so was the worldwide tracking network. From 9 to 16 October, Goddard and MSC put the tracking stations through their paces.†[34]

---

*Members of the board were Scott Simpkinson (Gemini Test Operations), Duncan Collins (Gemini Spacecraft Manager), Arnold D. Aldrich (Flight Operations Directorate), Philip M. Deans (Engineering and Development Directorate), Robert Everline and Galloway B. Foster, Jr. (Gemini Office of Program Control), Bailey, Slayton, and John Williams.

†For the network test, Kraft, Hodge, Eugene F. Kranz, and Glynn S. Lunney took turns as flight director. The network was not quite the same as for the first Gemini mission: the sites this (Continued)

The spacecraft, however, had yet to arrive at the pad. Work had gone well enough the first week, but trouble cropped up in getting the thrusters ready for a static firing test. After firing, the system had to be flushed and purged, another delay. By 10 October, Spacecraft 2 was already eight days behind schedule; it lost another two days while pyrotechnics were installed. Spacecraft 2 was ten days late when it reached complex 19 on Sunday, 18 October, and settled in the tripod in the white room an hour before noon.[35]

Attempts to run the spacecraft premate systems test brought new problems. As one was solved, another appeared; and it was 27 October before the test was complete. The final step before the spacecraft was joined to the launch vehicle was a premate simulated flight, run in two parts. Despite more than one discrepancy revealed by the test, the spacecraft was mechanically mated to its booster by noon Thursday, 5 November.

After the mating Martin conducted tanking exercises on the launch vehicle to check calibration, to see whether or not the launch crew could load the tanks accurately with the equpment on hand, and to train for launch loading. The Martin crew found some differences between the data gathered from calibration and what they thought they had loaded. This led to a series of tanking exercises throughout the program and set up "a new family of people, called the Wednesday Evening Tanking Society and the Thursday Evening Tanking Society—the WETS and the TETS."[36]

The troubled course of testing and checkout now smoothed. Over the next month, any problems that showed up were handled quickly, as Gemini 2 ticked off the milestones on its way to a 9 December launch: electrical interface integrated validation, 9 November; joint guidance and control test, 12 November; joint combined systems test after electrical mating, 17 November; wet mock simulated launch, 24 November; spacecraft final systems test, 28 November; simulated flight test, 3 December; and launch precount, 7 December.[37]

## SETBACK AND SUCCESS

Loading propellants aboard GLV-2 began in earnest on Tuesday, 8 December, an hour before midnight and finished shortly after three o'clock in the morning. The final countdown started an hour later. It went smoothly, though not quite so smoothly as the first Gemini countdown—there were three holds for a total of 41 minutes. The count

---

time were Cape Kennedy Mission Control; Goddard; Carnarvon, Australia; Hawaii; Canary Islands; Bermuda; Guaymas, Mexico; Corpus Christi, Texas; and two tracking ships—the *Rose Knot* Victor and the *Coastal Sentry* Quebec. Although it was not completely operational, the new Mission Control Center at MSC monitored the exercise.

reached zero at 11:41 Wednesday morning, and the first-stage engines ignited. One second later, a signal from the master operations control set shut down the engine. Flight controllers in the Cape control center observed that the launch vehicle had lost hydraulic pressure in its primary control system and had switched over from primary to secondary guidance and control. Within the blockhouse, technicians began to power down the spacecraft and, at three minutes before noon, Flight Director Christopher Kraft officially canceled the flight.[38]

The proximate cause of the shutdown was the command from the master operations control set, an automatic response to an automatic function—the switchover from primary to secondary flight control during the 3.2 seconds between ignition and liftoff. After the engines ignited, the launch vehicle remained bolted to the stand until thrust built up to 70 percent of maximum. During that time, a switchover in the control system was an automatic shutdown order. The GLV-2 switchover followed automatically when the booster's malfunction detection system sensed the pressure drop in the primary hydraulic system. GLV-2, in other words, spotted its own hydraulic failure, responded by switching over to its secondary system, and then, because it was still on the ground, commanded its engine to shut off.

Having saved itself, GLV-2 stood poised on the pad—a giant question mark. Why had its primary control system failed? The answer was quick in coming. Unexpectedly high pressure in one of the hydraulic lines had burst the aluminum housing of a servovalve, letting the hydraulic fluid leak out. This valve controlled one of the booster's four tandem actuators, the devices that moved the thrust chambers to steer the vehicle in flight. Why the valve housing had failed was a lesson in the folly of unneeded "improvement." At some time during development, someone had decided that the walls of the housing were twice as thick as they needed to be; a third of a centimeter of aluminum was ample to meet design pressures. No one, however, thought to test the actual pressure the housing would have to withstand, nor was any impulse test, as such, included in system qualification. More likely than not, one or another Titan II had suffered the same sort of hard start, but the stouter housings that remained standard in the missile could survive such a pulse while the lighter structural shell in the Gemini booster could not.[39]

When GLV-2 shut down, Spacecraft 2 posed something of a problem. Launch crews knew what to do with a ready-to-go booster, since they dealt with one after the mock launch that was a regular feature of launch vehicle checkout. There was no comparable background for the spacecraft, however, and that led to some hasty improvisation. Aside from its propellants, the spacecraft fairly bristled with pyrotechnic devices, all armed for flight. Should one of them explode, the results might be catastrophic.

Draining the booster of propellants took first priority, so Wednesday had passed and Thursday was well along before the main part of spacecraft "safing" was complete. One particularly ticklish operation remained, however—pulling the pyrotechnics from the isolation valves that barred propellants from the spacecraft thrusters until time to fire. The problem was complicated by the fact that the explosive cartridge was not a replaceable unit, and the whole valve assembly had to come out. But this might allow propellants to reach the thrusters or to spill their highly noxious chemicals over the workers. The makeshift answer was to freeze the propellant lines. After one or two false starts—no one was quite sure how to do the freezing—copper tubing was wrapped around the lines (which were packed in dry ice), liquid nitrogen was run through the tubing, and the whole thing was sprayed with $CO_2$.* That worked, and the valve assemblies were replaced over the weekend.[40]

There was really not much that could be done with the spacecraft over the next few weeks besides making sure it remained in flight status, and nothing much could be done with the launch vehicle until new actuators arrived.† A product of Moog Servocontrols, Inc., the tandem actuators had been taken back to the vendor's plant in East Aurora, New York, for extensive tests. Then the actuators had gone to Martin-Baltimore for further testing. The lightweight servovalves had to be redesigned. Work was further curtailed by the holidays. A messenger reached the Cape with the four new parts on 6 January 1965. They were installed at once and testing resumed, focused mainly on the flight control system. The new round of launch preparations went quickly; by Thursday, 14 January, the last major test was complete. Reviews of spacecraft and launch vehicle gave both a clean bill of health, and launch was set for 9 o'clock Tuesday morning, 19 January.[41]

The countdown began two hours past midnight. It was almost flawless, although it did produce one disappointment. Spacecraft 2 had been slated to carry six fuel-cell stacks of the old model P2B, left over after the design had been updated early in 1964. Despite their known defects, flight testing them with the reactant supply system seemed like a good idea, but only on a "non-interference with flight" basis and with a dummy load, since electrical power would actually be supplied by battery. The six stacks assigned to Spacecraft 2 had behaved erratically

---

*A motor-operated shutoff valve was installed in later spacecraft to make draining the hypergolics a simpler and safer operation.

†During the lull in Cape activity, NASA realigned its field center operations on a noninterference-with-Gemini basis. MSC's Florida Operations was transferred to Kennedy Space Center and renamed the Launch Operations Directorate (with Kurt H. Debus as Director and Merritt Preston as Deputy Director) to "place the responsibility for assembly, checkout, and launch of the total Apollo space vehicle with a single organization."

since they were first installed in St. Louis. When they acted up during the abortive countdown on 9 December and threatened to delay the launch, they were scratched from the mission. Only one stack proved to be still operable; it was activated on 18 December, then shut off and left alone until the next launch attempt. An hour and a half after the countdown started on 19 January, hydrogen intake to the stack was blocked by a stuck valve. Two hours of work left troubleshooters faced with breaking the spacecraft wiring to correct the problem. Since that would have meant a hold in the countdown, the attempt to activate the stack was called off, and the fuel cells were not operated on Gemini 2.[42] Aside from the fuel-cell problem, the countdown produced only the most minor anomalies and one preplanned two-minute hold.

At four minutes after 9 Tuesday morning, Gemini 2 began the last unmanned flight in the Gemini program. GLV-2 hurled the spacecraft 3430 kilometers across the South Atlantic through an arc that peaked 160 kilometers above the ocean's surface. The spacecraft endured the most severe heating Gemini was ever likely to meet as it plunged back into the atmosphere, its heat protection proved, its structural integrity uncompromised, and all systems working. It dropped into the South Atlantic on its parachute about 18 minutes after launch, bobbing in the water for an hour and a half until it was picked up by the U.S. Navy's aircraft carrier *Lake Champlain*.[43]

Some small question marks dotted the mission, but overall it looked quite good. The postflight news conference was a scene of quiet jubilation, with pats on the back for everyone involved. Nothing earth-shaking turned up in the detailed study of the recovered spacecraft—only minor scratches, chars, corrosion from exposure to sea water, just about what might have been expected—nothing that would in any way militate against the forthcoming launch of Gemini 3, the first to carry men aloft.[44]

### DOWN TO THE WIRE

While most eyes had been focused on *Gemini 2* at Cape Kennedy, work on still-to-be-resolved development problems continued elsewhere. Two spacecraft systems indispensable for Gemini's first manned mission—thrusters and ejection seats—remained question marks through most of 1964, and a third—fuel cells—though not slated for Gemini 3, was as yet unqualified. What may have been the largest question of all centered on the Gemini Agena, which throughout 1964 fell further behind schedule.

In April 1964, Rocketdyne seemed at last to have solved its major problems in developing workable thrusters for Gemini, but misgivings persisted. When the Jet Propulsion Laboratory approached Rocketdyne about developing a small engine for the Surveyor spacecraft, Mathews protested. He argued that the company was still a year away from having the Gemini orbital attitude and maneuvering system and reentry control system on a sound footing, and that the main reason the work had improved was the belief that it would get no more NASA small-engine contracts until Gemini work was almost done. Workloads in the California plant were heavy, as shown by the large demands for overtime, and the original $30-million contract had ballooned to over $74 million, of which almost $36 million was an overrun.

Despite the enormous infusion of effort and money, Rocketdyne had failed to maintain schedules and deliveries. Engines for Spacecraft 2, for example, due in February 1963, arrived only in January 1964, and "the delivered products leave much to be desired." Mathews thought it "quite evident that all three interested parties, the Gemini Program Office, the Surveyor Program, and Rocketdyne, will benefit through the selection of a vendor other than Rocketdyne," since the added work could only hamper Gemini without contributing much to Surveyor.[45]

This concern was echoed by manned space flight chief George Mueller;* in a memorandum to his counterpart in the Office of Space Sciences, which had charge of the Surveyor program, he urged that Rocketdyne be denied the contract. MSC Director Gilruth also acted, setting up a special committee to survey Rocketdyne's Gemini program. After hearing some harsh committee findings on 5 August 1964, Rocketdyne's president promised that whatever NASA wanted would be done. Gilruth sent him a long list of recommendations a week later. Some changes were already under way even while the committee was meeting, and more followed, including a reorganization of Rocketdyne's Space Engine Division.[46]

Among the recommendations was a full-scale NASA audit of Rocketdyne's business management practices and Space Engine Division operations. It was a large undertaking, and a report was not ready until April 1965. Its findings revealed a badly managed program. Having "grossly underestimated the magnitude and complexities" of its Gemini subcontract, Rocketdyne had been slow to set up a sound organization. As a result, budgets were poorly controlled "and operations were inefficient," producing "significant cost overruns and delays." Not only had outright overruns very nearly doubled the cost of the program, but, of the 358 engines that should have been delivered by November 1964 under the original contract terms, only 167 had actually been received. Frequent personnel changes at top levels reflected the

---

*Mueller, of course, had an additional concern that did not affect Mathews: Rocketdyne was also the contractor for the Apollo thrusters and was a competitor with Space Technology Laboratories, Inc. (STL) for the lunar module descent engine. In January 1965, STL was awarded the development and production contract.

program's weak management, as did the company's complete inability to provide records showing the reasons for technical problems, what action they prompted, or what impact each problem had on costs and deliveries. The auditors recommended "that Rocketdyne's fee under the Gemini subcontract be adjusted."[47]

When this report was released in the spring of 1965, the worst was already over. Rocketdyne's performance had, in fact, begun to improve markedly in mid-1964, although as late as October Gilruth still thought an alternative source for thrusters might be a good idea. McDonnell received the first long-duration attitude maneuvering thrusters in October 1964, just five months after the new design had been released to production. By the time the audit report was issued, both the attitude and reentry control systems had been fully qualified in their Spacecraft 3 version. How greatly things had changed was shown most clearly when the long-life thrusters, not expected to be ready before Spacecraft 5, were actually installed in Spacecraft 4.[48]

Qualification of the Gemini escape system, like that of the spacecraft rocket systems, was essential before astronauts could be committed to a mission. Rapid progress early in 1964, which saw the development test program concluded, augured well, as did a good start on dynamic proof-testing. A preliminary sled-ejection test on 4 June 1964, to see if hatches and hatch actuators functioned properly under abort conditions, went off without a hitch. Qualification testing began on 1 July with a sled run to simulate conditions of maximum dynamic pressure after an abort during the powered phase of launch vehicle flight. Once again, everything worked.[49]

The same problem that had delayed development testing, one that had little to do with seat design, again brought the test program to a halt. Some of the pyrotechnic devices on which escape-system operation depended failed to arrive. The result was a four-month gap after the July run. In the meantime, NASA had decided to go ahead with a new test series. Sled and tower tests had been the only dynamic simulations planned for the system. Neither, however, could show the system working through its entire sequence as in a high-altitude abort. That became the purpose of a plan to eject the system from a high-flying F-106, worked out at a meeting between NASA, McDonnell, Weber Aircraft (the maker of the system), and the 6511th Test Group at El Centro, California, on 12 June. The first test, intended merely to show that the seat would work with the airplane, was set for September with the F-106 on the ground. Two flights, using production escape systems, were to follow, with the whole series to be finished in a month. Once again, however, lack of pyrotechnics caused delays. Enterprising engineers borrowed some from the ejection seat in North American's paraglider tow test vehicle, enabling them to run the ground test on 15 October. But nothing more could be done for three months.[50]

*Manned Spacecraft Center was visited by a steady stream of program officials from NASA Headquarters in 1964 as the Gemini flight program got into high gear; left, Associate Administrator Robert C. Seamans, Jr., receives a briefing; below, Associate Administrator for Manned Space Flight George E. Mueller (left) reviews the program with Maxime A. Faget and Charles W. Mathews.*

Enough pyrotechnics were on hand for another sled run on 5 November, which revealed a flaw in seat design. An instant after it had been ejected, one of the seats suffered a structural failure of its armrest and side panel that stopped the separation and recovery sequence. Seat and dummy smashed into the ground, strewing wreckage for 140 meters along the track. The hard question now was whether or not the test program had to be revised. The answer was no, provided the reworked seat structure performed well in a test approximating the most severe conditions for which the system was designed. In a sled run on 11 December, it did just that. The system came through with flying colors, bringing that part of the qualification program to an end.[51]

It was perhaps just as well that *Gemini 2* had been so long delayed. By the end of 1964, only one of the four major parts of escape-system qualification had been completed. Still to be conducted were simulated off-the-pad ejection (Sope), personnel parachute, and high-altitude ejection tests. All three resumed in January 1965, when pyrotechnics at last began to arrive.

First to get under way, on 11 January, was parachute testing. Four dummy drops and 12 live jumps from low altitudes over the next month turned up only minor problems. High-altitude testing followed.[52] In the meantime, on 16 January (a year and a half after Sope development tests ended) Sope qualification testing began. Shortage of pyrotechnics had again been the chief culprit in the delay. The first try failed. One seat worked, but the catapult on the right-hand seat fired too soon and exploded when the seat jammed against the still partly closed hatch. Almost a month passed while all hatch actuators were modified and the results checked out. Both the redesigned actuators and the escape system proved themselves in flawless Sope tests on 12 February and 6 March.[53]

High-altitude ejection was the last test program to resume but the first to finish. Nothing went wrong in the first test, an ejection at 4780 meters at mach 0.65 on 28 January. Two weeks later, however, in a test at 12 000 meters at mach 1.7, the aneroid device that was supposed to trigger parachute deployment failed, although everything else worked. That device also failed to deploy the ballute on 17 February, in the first high-altitude live jump, forcing McDonnell and Weber engineers to redesign the aneroid-controlled firing mechanism. Although the aircraft ejection test did not have to be repeated, since being ejected from the F-106 did not cause the failure, the parachute test program did have to be revised. That meant an extra 10 dummy drops and 5 live jumps, which began on 2 March. The final jump, on 13 March, qualified the personnel parachute system and completed the qualification of the Gemini escape system as a whole.[54] And not a moment too soon. The launch of the third Gemini mission, the first to carry a human cargo, was only days away.

The demand for fuel cells was not so pressing in late 1964 as for thrusters and ejection seats, since Spacecraft 3 and 4 were already being converted to battery power as a result of earlier problems. GE's redesigned fuel cell, the P3, had not at first lived up to its promise. Test sections performed erratically, their outputs tending to decay under load and their lives falling far short of requirements. This prompted NASA Headquarters to ask GPO on 10 July to provide a backup battery-power module in case fuel cells were not ready for the fifth Gemini mission. This was a drastic step, since Gemini 5 was slated for seven days; a battery installation to handle so long a mission meant a severe weight penalty and a narrow limit on what might be achieved during the flight. One of the main reasons for putting fuel cells in Gemini had been to ease constraints on such lengthy missions. GPO directed McDonnell to work out with Eagle-Picher, the battery subcontractor, a plan for a backup system.[55]

Early in August, GPO enlarged the scope of the study, asking McDonnell to cover the effects of substituting batteries for fuel cells in all two-day rendezvous missions, of using Agena-supplied power in a combined long-duration and rendezvous mission, and of such changes on the fuel-cell program itself. McDonnell found the feat possible but costly, especially in weight. At a meeting on 14 August, Mathews and Burke decided to provide Spacecraft 5 with a combined system of batteries for the peak loads and fuel cells for basic power needs. If most of the experiments planned for the mission were discarded, Spacecraft 5 would only weigh 30 kilograms more with its battery-augmented system. NASA Headquarters sanctioned the change on 1 October.[56]

The combined system reflected GE's success, finally, in pinpointing the sources of fuel-cell shortcomings. GE engineers found that the life of test stacks declined as electrical load and the temperature of reactants rose. The greater the load—the amperage drawn from the stack—or the higher the inlet temperature, the shorter the stack's life. With a constant load, a change of only 17 kelvins (30°F) in reactant temperature—313 kelvins (103°F) instead of 330 kelvins (133°F)—more than doubled stack life, from 125 to 290 hours. Holding the temperature constant and varying the load produced similar results. With batteries to handle peak loads, a major factor in truncated fuel-cell life might have been countered.[57]

These findings were based only on analysis of prior test data. Now GE revised its test program to see what effect lowered inlet temperatures and reduced loads actually had on test stacks. The results confirmed the premise. Two test units under a steady three-ampere load with reactants at 297 kelvins (75°F) lasted 1100 and 800 hours. Further tests produced equally encouraging results at various levels of load and temperature under normal and abnormal conditions. All difficulties were not yet out of the way, but those that remained were largely matters of detail.[58]

Concern about "the rapidly rising costs of the General Electric fuel cell development program, coupled with the lagging development," persisted for a while; but, significantly, that worry was expressed in a memorandum never sent.[59] The Gemini Program Office in Houston retained some doubts about fuel-cell prospects through the early fall of 1964, urging NASA Headquarters to allow batteries to replace fuel cells in Spacecraft 6 to ensure meeting the prime objective of that mission, rendezvous with an Agena target vehicle. Headquarters demurred until 6 November, but then granted the change.[60]

That decision stood, Spacecraft 6 eventually flying with battery power. In the meantime, however, the response of fuel-cell test units to lower temperatures was so marked during late summer and early fall as to convince both NASA and its contractors that the power system for Spacecraft 5 need not be augmented by batteries. That change was therefore canceled on 18 December 1964. The Gemini fuel cell completed its basic qualification test program in May 1965, three months before it flew in the fifth Gemini mission.[61]

Agena was still further down the line, and its lagging pace showed no signs of speeding up during 1964. Project Gemini received its first Agena D at the end of April 1964, but nearly five months passed before it was converted into GATV-5001, the first Gemini Agena Target Vehicle. Lockheed completed that effort on 24 September and transferred the vehicle to the systems test complex, where cabling it up for preliminary vehicle systems tests began the next day. Not too surprisingly, testing did not run smoothly.

The hardest and most stubborn problems centered in Agena's command and communication (C&C) system—the electronic devices for tracking the vehicle, monitoring its subsystems, and passing commands to the vehicle in orbit. Because of Gemini's unique demand for rendezvous and docking, Lockheed had to design and prove a new C&C system for the Gemini Agena. The new design struck GPO as very good, a judgment confirmed by a special consultant group from Stanford Research Institute, which recommended only minor changes. During testing in October, however, parts of the system started acting up. Troubleshooting got GATV-5001 through its testing, but it seemed all too likely that the C&C system suffered from basic defects in its mechanical and electronic design. The question became, as Mathews later recalled, "Should we live with what we had, or should we back off and completely redesign the configuration?" When the problems persisted, the Air Force insisted on redesign, and Lockheed finally initiated a "Ten Point Plan for C&C Equipment" in February 1965.[62]

In the meantime, GATV-5001 had emerged from its preliminary tests in November 1964 and gone to Lockheed's Santa Cruz Test Base for a round of captive-firing tests. First, however, the target docking adapter had to be installed. This was the unit, built by McDonnell but carried aloft by Lockheed's Agena, to which the spacecraft would attach. When Lockheed workers hoisted the adapter into the test stand and tried to mate it with the Agena, they found it did not fit. After some struggling, they managed to get the two physically hooked together, but the wiring failed to match. The captive firing had to be postponed until January.[63]

The test on 20 January 1965 simulated a full two-week mission. It included repeated firings of both primary and secondary propulsion systems, with operational data transmitted to telemetry stations at the test site and at Lockheed's Sunnyvale plant. The propulsion systems worked well, but the C&C system again had problems. One part, the programmer time accumulator, jumped erratically, picking up almost eight extra weeks. Shipped back to Sunnyvale on 1 February, GATV-5001 lost three weeks while Lockheed tried to fix the capricious timer.

A makeshift fix allowed GATV-5001 to move on to the next phase, electromagnetic and radio-frequency interference tests, while engineers continued their efforts to diagnose and cure the jumping timer. By 23 February, when the interference tests began, GATV-5001 was more than a month behind schedule.[64]

Interference tests ended 9 March, but the vehicle stayed in the anechoic chamber for another week while Lockheed checked out its answer to the erratic timer and to a telemetry synchronization problem that had also cropped up. On 18 March, GATV-5001 moved to the systems test complex for a planned six days of "minor" modifications: filters were to be installed in the command controller (another part of the C&C system) and the forward auxiliary rack (which supported the target docking adapter and housed most of the C&C gear) was to be aligned. These two tasks proved to be more than minor. The first eventually required a complete redesign, the second extensive machining. The result was another lost month. By the end of March, GATV-5001 was 66 days behind schedule.[65]

Final systems testing got under way on 9 April and ended with a simulated flight on 6 May. On 27 May, the Air Force and Aerospace team found GATV-5001 formally unacceptable for Gemini, since FACI (first article configuration inspection) from 10 to 26 May had shown that it was not flightworthy. SSD took the vehicle anyway, but conditionally. Lockheed was expected to correct all defects; some were merely matters of paperwork, but others, like propulsion and C&C systems qualification, were major efforts. GATV-5001 was then flown to the Cape on 29 May, to be used as a development test vehicle.[66]

In the meantime, the first Atlas booster for Gemini had joined the program on 1 December in San Diego. It had then been shipped by truck to Cape Kennedy, a six-day trip. It was erected on complex 14 a week later, to help in checking out the launch pad and ground support equipment. Finished with that by 11 February, the Atlas was moved to a hangar, there to be modified and stored until GATV-5002 arrived.[67]

### A VOTE OF CONFIDENCE

On Tuesday afternoon, just a few hours after the launch of *Gemini 2*, the program received another vote of confidence. Although the second launch had been long delayed, the nature of the delays in no way cast doubts on Gemini itself; NASA and its contractors decided that Gemini missions should be launched at two-month intervals, instead of the three-month cycle then planned.

In September 1964, the Air Force had not only convinced NASA that GLV-2 ought to fly, but also proposed to speed up the program by launching every two months. Although the Vertical Test Facility at

Martin-Baltimore had been designed to handle two launch vehicles at once, only one of these test cells was working. The Air Force suggested opening the second cell to speed up launch vehicle deliveries. SSD Commander Funk assured his Gemini colleagues that the Cape crew could handle launches only 60 days apart.

LeRoy E. Day, Headquarters Gemini Test Director, took charge of a task force to canvass spacecraft, launch vehicle, and target vehicle contractors about the practicality of the plan. A two-month study convinced Day and his group that it could be done. Although NASA's checkout crew at Cape Kennedy expressed a measure of skepticism based on their experiences in Project Mercury and the opening stages of Gemini, the Gemini Program Office had more faith. GPO had, in fact, been thinking of less time between launches when it imposed revised test and checkout procedures in St. Louis and at the Cape early in 1964. When Day presented his findings to Gemini's top echelon on 19 January 1965, they bought the plan and wanted it put into effect by the fifth mission. This vote of confidence in Gemini was founded on a technological judgment, and in that sense it was fully justified. Later events were to show that fitting astronaut training into the shorter schedule was a harder task, although it produced no problems that could not be surmounted.[68]

As 1965 dawned, Project Gemini had cleared most of the hurdles in its path. The past year had seen its last serious development problems overcome. Agena was perhaps not as far along as it should be, but there was plenty of talent at hand to put that in order. The repeated setbacks suffered by GLV-2 could only be seen as acts of God, not defects in technology. That could not be said of its failure on 9 December, but little more than a month of hard work was needed to put matters right. The second Gemini mission, on 19 January 1965, almost matched the first, on 8 April 1964, in the quality of performance. Gemini's spacecraft and launch vehicle had been proved. All that remained, the last hurdle, was sending men aloft. Although the publicly scheduled date for Gemini 3 was the second quarter of 1965, Charles Mathews told the Gemini Management Panel shortly after the flight of *Gemini 2* that late March looked like a good bet.[69]

# X

# The Last Hurdle

ON 13 April 1964, the Monday after the flight of *Gemini-Titan 1,* the men and women of the press gathered in the auditorium at the Manned Spacecraft Center to learn who would be the first to fly the Gemini spacecraft. Robert Gilruth, Director of the Manned Space-craft Center, introduced the four astronauts assigned to Gemini 3, the prime and the backup crews. Commander of the first team was Virgil I. Grissom—"Gus." His crewmate was John W. Young. Backing up the mission were Walter M. Schirra, Jr., and Thomas P. Stafford.[1]

The stocky, crew-cut Grissom, an Air Force major,* was an old-timer in NASA's manned space flight program, one of the original seven Mercury astronauts picked five years earlier. He already had a quarter of an hour of spacecraft flying time as passenger on the subor-bital flight of *Liberty Bell* 7 in July 1961, Project Mercury's second manned mission, and would therefore be the world's first two-time space flyer. Young, his crewmate, was a younger man and a newer as-tronaut; a Navy lieutenant commander, he had been one of the nine pilots selected for the space program in September 1962. Schirra, like Grissom, was one of the Mercury seven. Born in 1923, he became the old man of the astronauts corps when John Glenn resigned early in 1964. In October 1962, Schirra had ridden *Sigma* 7 (the fifth manned Mercury spacecraft) through six orbits in the penultimate Mercury

---

*Grissom, a captain in the Air Force when he joined the astronaut ranks, had been promoted to major in July 1962, one year after his Mercury flight.

mission. Stafford, Schirra's copilot in the backup crew, was an Air Force major who became an astronaut at the same time as Young.*[2]

Gilruth voiced NASA's "high hopes of flying by the end of the year," 1964,[3] leading America back into space after an 18-month hiatus. Those hopes foundered in the storms that lashed Cape Kennedy during the summer. When the launch vehicle for Gemini 2, after passing so smoothly through test and checkout, betrayed the mission in December, even Gemini's unmanned prelude remained unfinished at year's end. But the opening quarter of 1965 saw the success of *Gemini 2* in January and then, scarcely two months later, Grissom and Young in orbit aboard "Molly Brown." With that, Project Gemini had clearly advanced a long step beyond Mercury and opened a new era in manned space flight.

## THE MEN FOR GEMINI 3

Within a week after they had been publicly assigned to the mission, the Gemini 3 astronauts were busy training for it. All astronauts were in training from the time they joined NASA, but for Grissom and Young, Schirra and Stafford, the focus now shifted to a specific mission. Their first assignment was the Gemini mission simulator at the McDonnell plant in St. Louis. This training complex included a flight simulator that matched the inside of a Gemini spacecraft and provided its riders with almost all the sights, noises, and shakings they should meet in a real flight, from prelaunch to postlanding. Because astronauts varied in size† and missions differed in goals and onboard tasks, no two spacecraft were identical, and the mission simulators had to be altered and updated for each flight. But the simulator in St. Louis had not yet been engineered to an exact replica of Spacecraft 3, so the 36 hours that Grissom and Young spent in it over the next two months, as well as the 34 that Schirra and Stafford flew, were devoted mainly to learning general systems and operations.[4]

---

*The others who became astronauts with Stafford and Young were Neil A. Armstrong, Frank Borman, Charles Conrad, Jr., James A. Lovell, Jr., James A. McDivitt, Elliot M. See, Jr., and Edward H. White II. They were introduced to the public on 17 September 1962.

†In January 1963, shortly after the second group of astronauts was selected, the pilots were given specialty assignments in the MSC programs. Grissom, one of the smaller astronauts, was assigned to the Gemini spacecraft. Because of this and his Mercury experience, he was very close to the McDonnell engineers and technicians—so close, in fact that the cockpits of the first three spacecraft were designed around him, giving him the best view of the instrument panel and out the window. The spacecraft was familiarly dubbed the "GUSMOBILE." Although Young was only two inches taller, his seat had to be compressed so he could fit into it. Stafford had to have adjustments made on both the seat and hatch to accommodate his six-foot frame. By July 1963, the program office had discovered that 14 of the 16 astronauts could not be fitted into the cabin as designed, and all later cockpits had to be modified.

**Gemini Astronaut Program**

*Above, centrifuge run, Ames Research Center.*
*Right, water egress training, Galveston Bay.*
*Left, weightless in KC-135, Wright Air Force Base.*
*Right, desert survival training, Stead Air Force Base.*

*Right, dynamic crew procedures simulator, Ling-Temco-Vought.*
*Left, Gemini mission simulator, MSC; below, same simulator at McDonnell.*
*Below right, suiting up for simulated flight, Cape Kennedy.*
*Below center, Young and Grissom simulate Gemini 3 mission, Cape Kennedy*
*Below left, Young and Grissom practice use of launch pad abort rescue vehicle, Cape Kennedy.*

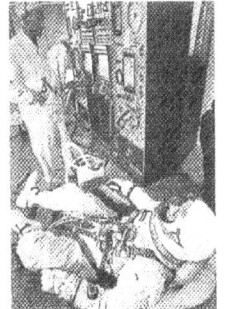

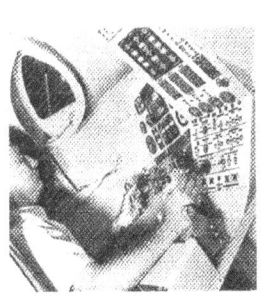

221

On 10 July 1964, McDonnell workmen began taking the simulator apart to ship it to Houston, there to be set up to match Spacecraft 3. The second Gemini mission simulator was already at the Cape, although not yet updated for Gemini 3. That was supposed to have been done by mid-July, but it was not finished until October. Final checkout took the better part of a month, and the Gemini 3 crews could not begin flying simulations in Florida before 9 November.[5]

But no such hangup ever left the astronauts with time on their hands. On 10 and 11 May, all four were in St. Louis to review a mockup of the cockpit. In the months that followed, they kept a close eye on their ship, watching as it passed through its series of tests and inspections in the McDonnell plant. They also joined in the testing itself. During the second phase of systems tests in October and November, Grissom and Young spent more than 14 hours in the cockpit, 9 of them while the spacecraft was undergoing altitude chamber tests. Schirra and Stafford were not far behind, with 8 cockpit hours.[6]

During July and August, the four Gemini 3 pilots (and all their fellows) were in Dallas for a training program on the moving-base abort simulator created by Ling-Temco-Vought, Inc. This device projected the Gemini 3 launch profile in striking detail, complete with such cues as noise, vibration, and a wide range of motions that might be caused by one launch anomaly or another. The trainees also learned how to deal with any number of booster or spacecraft systems malfunctions.[7]

Throughout their training, the prospective spacemen also kept their more mundane flying skills intact. Each managed to average 25 hours a month in the cockpit of an Air Force jet. They also put in more than 200 hours apiece in innumerable briefings, three of them formal affairs that lasted two days each at Houston, St. Louis, and Cape Kennedy, the others an ongoing series of informal systems familiarizations that were part of each training activity. Periodic reviews of mission plans, physical examinations, fittings for flight suits, sessions on experiments to be carried on the spacecraft and on biomedical aspects of the mission, and any number of other operational matters helped fill the hours to overflowing.[8]

In October 1964, the Gemini 3 crews tackled still another aspect of training, practice in getting out of their spacecraft after it landed. The three-part program began with a review of egress procedures in the Gemini mockup at the McDonnell plant, then moved to the flotation tank at Ellington Air Force Base, just up the road from the Manned Spacecraft Center. The tank was a king-size swimming pool, where the crews rehearsed (both with and without space suits) climbing in and out of a boilerplate spacecraft that was either floating or submerged.[9] Grissom and Young completed the third phase of this training in emergency egress from a floating spacecraft during February 1965.

They rode a boat out into the Gulf of Mexico, where a model space-craft was dumped into the water. Then, fully suited, they went through the postlanding checklist and practiced getting out of the spacecraft and into their one-man liferafts. The crews also took refresher courses in parachute landing that month.[10]

During November and December 1964, the four crewmen spent part of their time in Johnsville, Pennsylvania, at the Naval Air Development Center, the site of a man-rated centrifuge run by the Aviation Medical Acceleration Laboratory. The first phase of centrifuge training had taken place in July and August 1963, when Gemini controls and displays had been evaluated and all the astronauts had been spun through acceleration profiles for launch and reentry. For pilots not yet assigned to a mission, the second phase simply provided more of the same. But for the crews of Gemini 3 and Gemini 4,* it was an important part of mission training. They worked in pressure suits, and the others trained in shirtsleeves. Grissom rode the centrifuge for 9½ hours, Young for 11 hours; Schirra and Stafford spent only a little less time in the centrifuge than the prime crew.[11]

When the mission simulator at Cape Kennedy had been updated to match Spacecraft 3, both crews began working in it off and on for the next four months. During that time, Grissom put in more than 77 hours flying his mission on the ground, rehearsing every phase of his planned flight again and again, not only when everything went right but also when something went wrong.† Young put in even more time than Grissom, over 85 hours, in the Cape simulator. Schirra managed to get in 43 hours, Stafford 54.[12] In January 1965, Grissom and his fellow crewmen were back in Dallas for more work on the abort simulator, this time focused on how best to deal with each type of booster or spacecraft malfunction. By the time this training was over, Grissom had run through 225 aborts and Young 154; Schirra and Stafford each totaled only slightly less than Young.[13]

When Spacecraft 3 arrived at complex 19, the crewmen resumed their active role in spacecraft testing. Sandwiching this exercise between trips to Houston for egress and parachute training, Grissom and Young still managed to spend almost 19 hours in the cockpit, beginning with the premate flight test on 14 February and ending with the

*On 29 July 1964, James A. McDivitt and Edward H. White II had been introduced to the press as the prime crew for Gemini 4. Frank Borman and James Lovell were announced as the backup crew.

†The following figures suggest how thoroughly NASA tried to prepare a pilot for his mission. Grissom flew 20 normal and 46 aborted launches; 13 normal speed, 5 overspeed, and 4 underspeed insertions into orbit; 8 platform alignments; 9 runthroughs of the flight plan; 107 retrofires; and 64 reentries. He experienced 51 simulated failures of the booster and 211 systems malfunctions: 57 sequential, 34 electrical and communications, 17 attitude control and maneuver electronics, 30 orbital attitude and maneuver, 16 reentry control, 36 guidance and control, and 21 environmental control.

final simulated flight on 18 March. Schirra and Stafford got in more than 14 hours of cockpit time. Altogether, the prime crew had logged 33 hours in their spacecraft before the final launch countdown began, and the backup crew had spent 22 hours.[14]

Nine months of grueling work were ready to pay off. By February 1965, Grissom was sure that "We're ready to go." NASA agreed. Rumors already put Gemini's first manned flight earlier than the officially announced April or May. And NASA Administrator James Webb, speaking at Nebraska Wesleyan University in Lincoln, hinted that the launch might come in late March.[15] The men were ready, and the machines very nearly so.

## THE MACHINES FOR GEMINI 3

McDonnell finished building Spacecraft 3 in December 1963 and moved it from the production floor to the white room in the St. Louis plant. Engineering changes and equipment installation filled the next six months. Despite some NASA worries about tight schedules, the spacecraft was ready to begin the first phase of systems testing by the end of May 1964, directed, like Spacecraft 2, by the Launch Preparations Group from the Cape. The Development Engineering Inspection (DEI), the first of the periodic reviews to make sure that McDonnell was giving NASA just what it wanted, was held on 9 and 10 June. This first review was chiefly a close look at the modules to be tested, to see that they matched specifications and were actually ready to begin testing. The DEI produced its share of changes, but nothing stood in the way of getting on with the tests.*[16]

While Spacecraft 3 was moving through the McDonnell plant, Gemini Program Manager Charles Mathews took a step that showed the program had entered a new phase. During July, he set up a Gemini Configuration Control Board to be, as he later informed McDonnell, the "one official route for all configuration change action to provide continuity and coordination." Each Monday morning, Mathews met with the heads of the Gemini Offices of Program Control, Spacecraft, Vehicles and Missions, and Test Operations to review all proposed changes and to pass on them—and every change now had to

---

*On 8 June 1964, George Low, MSC's new Deputy Director, made a change in the permanent DEI board established by his predecessor, James Elms. Low himself, instead of John Bailey (Chief, Reliability and Flight Safety) would be chairman. Members of the Spacecraft 3 DEI were Low, Charles Mathews, Duncan Collins (Gemini Spacecraft Office), Bailey, Max Faget (Director, Engineering and Development), Christopher Kraft (Director, Flight Operations), Grissom (in a dual role as astronaut and representative of Flight Crew Operations), John Williams (Florida Operations), and Robert Everline (Gemini) as recording secretary. The board reviewed 45 requests for changes—the board agreed that 17 were mandatory, 6 possible after further study, 16 unnecessary, and 6 undesirable.

be formally presented and justified. When the board met for the first time,* on 27 July, the development era of Gemini had clearly ended. From then on, the main concerns of the program were production and operations.[17]

July also saw McDonnell present NASA with its plan for converting the Gemini contract from fixed fee to incentive fee. This was a direct McDonnell response to a NASA request based on a clause in the contract negotiated in 1963. The idea was to give the company a chance to earn greater profits by cutting costs, meeting schedules, and delivering an outstanding product, but to receive less money if it failed in any of the three areas. With development almost complete, such a plan became feasible. Mathews had appointed a Gemini Incentive Task Group on 2 March 1964, naming as its chairman Kenneth Kleinknecht, his deputy and former Mercury manager.†[18] The formal Request for Proposal was ready for McDonnell by 19 May, after a review by NASA Headquarters. Walter Burke, McDonnell Vice President and General Manager for Spacecraft and Missiles, arrived in Houston on 7 July with a group of colleagues to address a large NASA gathering on his company's ideas.‡[19]

During the spring of 1964, the Air Force Space Systems Division (SSD) had also been working out incentives with its major Gemini contractors, Martin and Aerojet-General for the launch vehicle and Lockheed for the target vehicle. NASA kept close tabs on the progress and drew on SSD experience for the McDonnell proposal. Martin's contract was converted on 10 June and Aerojet-General's on 17 June; Lockheed negotiations were completed early in August.[20] MSC's talks with McDonnell lasted through the fall of 1964, the last details being settled on 18 December, and NASA Headquarters approved the plan on 28 January 1965. It called for a total cost of $712 301 640 for the spacecraft, plus a fee that might range from $28 075 581 to $55 775 581 as the company's performance ranged from poor to good.[21]

Contract changes notwithstanding, McDonnell had completed its tests of Spacecraft 3 modules on 12 September 1964, and was ready to

---

*Members (and alternates) were Mathews, chairman (Kleinknecht), Duncan Collins (Homer Dotts), Willis Mitchell (Jerome Hammack), Scott Simpkinson (Harry Douglas), Richard Henry (George MacDougall), and Stephen D. Armstrong (James I. Brownlee).

†Kleinknecht's team: John B. Alldredge, Leroy E. Kroeker, and Charles D. Heald (from MSC procurement); John E. Roberts, Gregory P. McIntosh, Walter Wolhart, and George MacDougall (GPO); Earle B. Young (MSC Resources Management), and Richard Henry (NASA Headquarters, who later transferred to MSC GPO). Available on an as-needed basis were William A. Summerfelt (incentive approach, schedule, and program planning), Joseph Fernandez (cost), Anthony L. Liccardi (configuration control and specifications), Richard A. Schmidt (incentive management), and Sidney A. Cariski (contracts and procurement), all from NASA Headquarters.

‡Burke was assisted by several key McDonnell Gemini figures, among them A. E. Smith, Harry W. Oldeg, J. M. Gardner, Jr., and Frank Morgan.

mate them. On 21 September, Scott H. Simpkinson, chief of Gemini Test Operations, arrived in St. Louis at the head of 22 engineers from GPO and other MSC elements to join the Launch Preparation Group and MSC's resident McDonnell office for the second major review of Spacecraft 3, the Module Test Review.* Twelve teams under the review board took a careful look at results from the first phase of testing, just completed, and reported their findings to the board, which announced the next day that the modules of Spacecraft 3 were indeed ready to be mated and that the second phase of systems testing might begin.[22]

Spacecraft 3's third major review began on 3 December as the first half of a two-part Spacecraft Acceptance Review (SAR).† The spacecraft had completed all systems tests except its last, the simulated flight. After its review of the test results, the acceptance board allowed McDonnell to proceed with the flight simulation. When this test was finished on 21 December, the board met for the second part of its task, a study of all test results, documentation, and overall spacecraft status. Three days after the simulated flight, on Christmas Eve, the board had "determined that Spacecraft 3 is acceptable for delivery."[23]

After the holidays, the spacecraft was loaded aboard a C-124, which delivered it to Cape Kennedy early Monday evening, 4 January 1965. The concept that a fully checked out and integrated spacecraft was being delivered had by then been largely accepted. Work in the industrial area at the Cape, from the time the craft arrived until it was transferred to the launch complex, centered on putting it in shape to fly by clearing up manufacturing shortages and installing seats and pyrotechnics, rather than by testing, with two major exceptions.

Because this was the first man-bearing Gemini spacecraft, it was the subject of a special communications test at the Merritt Island Launch Area radar range. The spacecraft communications systems were checked out in a radio-frequency environment that matched as closely as possible the conditions they would meet in orbit. Testing of the spacecraft propulsion systems was the other exception. Spacecraft 3 went through a complete end-to-end propulsion systems verification test program, including static firing (as had its predecessor), partly to check out procedures and gear, partly to build some confidence in sys-

---

*Members of Simpkinson's review board were Homer Dotts (Deputy Spacecraft Manager), Wilbur Gray (GPO Resident Manager), Charles Williams (Spacecraft 3 engineer), Walter Kapryan (Cape Manager, GPO), Grissom (for Flight Crew Support Office), and Everline, coordinator and recorder.

†The Spacecraft 3 SAR board consisted of Homer Dotts, chairman, Andrew Hobokan (Deputy Resident Manager), Phillip Deans (Engineering and Development Directorate), John Williams, Grissom, Melvin F. Brooks (Flight Operations), Norbert B. Vaughn (Reliability and Quality Assurance), and Don R. Coryell (Gemini), coordinator and recorder.

tems whose development had been fraught with problems and which were not yet fully qualified. Even with these two special tasks, however, Spacecraft 3 was ready to move to the launch pad a month after it arrived at the Cape.[24]

The launch vehicle for Gemini 3 had been late reaching the Cape through no fault of its own; the long delay in launching *Gemini 2* had left it with no place to go. GLV-3 had, in fact, been built and tested in Baltimore with admirable dispatch. Completed early in June 1964, the vehicle had passed its horizontal tests and finished its checkout in the Vertical Test Facility by the last day of July. Another three weeks saw it through its combined systems acceptance test and review by the Vehicle Acceptance Team. When the team approved GLV-3 on 21 August, GLV-2 was still sitting on the launch pad in Florida, so GPO decided to have the Martin crew in Baltimore install the engineering changes on GLV-3 that were to have been done at the Cape. After looking over these changes, the acceptance team ordered a second combined systems test. The test rerun and the results approved, on 9 October the team once again accepted GLV-3. Martin-Baltimore formally turned it over to the Air Force on 27 October. Since Gemini 2 was still unlaunched, the Baltimore crew installed another set of modifications that had been slated for the Cape, finishing in mid-January.[25]

Now there was room at the Cape for GLV-3, but the Air Force could no longer spare the C-133B that had carried the first two launch vehicles to Florida. A converted Boeing 377 Stratocruiser, nicknamed "Pregnant Guppy," had to serve instead, although it could not hold both stages at once. It flew the second stage down on 21 January, went back to Baltimore to pick up the first stage, and returned to the Cape on 23 January. Two days later, GLV-3 was standing on the launch pad waiting for the spacecraft, which joined it on 5 February. The pace then slowed somewhat, as premate tests of the spacecraft proved troublesome. Nevertheless, spacecraft and launch vehicle were mechanically mated on 17 February, less than a month after the launch of *Gemini 2*. Another month was ample time to complete systems testing, and the simulated flight test on 18 March concluded the task of checking out the machines for Gemini 3.[26]

PLANS FOR GEMINI 3

The precise scope of the third Gemini mission remained uncertain until very nearly the eve of flight. That its primary purpose, as spelled out in the "GT-3 Mission Directive," was

> to demonstrate and evaluate the capabilities of the spacecraft and launch vehicle system, and the procedures necessary for the support of future long-duration and rendezvous missions[27]

had been settled by the rescheduling decisions of April 1963. Gemini 3, in other words, was to show that Project Gemini was ready to meet its major goals. But just how that was to be done was not clearly defined until early 1965.

Such key questions as how long the mission was to be and how its specific objectives were to be met were much discussed. NASA Headquarters had tentatively approved the three-orbit flight suggested by the program office in April 1963. This seemed too short a mission, however, to use the rendezvous evaluation pod (REP), long planned to check out spacecraft radar and maneuvering systems. If the mission could not be lengthened, some other means must be found "to demonstrate and evaluate . . . the procedures necessary for the support of future . . . rendezvous missions." Equally unclear was how so short a flight could do much to prepare for future long-duration missions.[28]

MSC's Flight Operations Division did prepare a tentative mission plan in October 1963 that outlined possible use of the pod during the second orbit of a three-orbit mission. But the matter was settled when, on 4 January 1965, NASA Headquarters decided to strike the pod from Gemini 3.[29] The question of mission duration surfaced again late in the summer of 1964. Word leaked to the press that Grissom and Young, backed by the Astronaut Activities Office, were pressing for an open-ended mission; that is, leaving it up to the crew to decide how many orbits to try for after Spacecraft 3 was in space. GPO was averse to the idea, since the tracking network was then geographically limited and could only fully cover three orbits. Going beyond that on the first flight might be risky. NASA Headquarters again stepped in and squelched the idea. When a reporter asked Grissom what he thought about the decision, the answer was a curt, "We can do all the testing of the spacecraft we need in three trips."[30]

One of the first-order objectives for Gemini 3—one that had to be achieved for the mission to be judged a success and any threat to which was cause enough to hold or cancel the flight—was to "demonstrate and evaluate the capability to maneuver the spacecraft in orbit using the orbital attitude and maneuver system (OAMS)." Early planning thus called for several OAMS firings.[31] The reason for these firings suddenly expanded in January 1965. NASA Headquarters sent Flight Operations in Houston a set of preliminary data, with orders to revise the flight plan to protect the Gemini 3 crew against the danger that Martin Caidin, in his space thriller *Marooned*, had posed: the failure of spacecraft retrorockets to work, stranding the crew in space. Headquarters proposed three OAMS maneuvers to place the spacecraft in a "fail safe" orbit, one from which it would reenter whether the retrorockets fired or not. Actually, Gemini orbits were too low to be permanent, so spacecraft reentry was inevitable. What the fail-safe maneuvers were designed to achieve was the spacecraft's return promptly enough to ensure that the crew survived. Coming as it did less than three months before the planned launch, the new demand threw mission planning into turmoil. But the response was rapid. A revised tentative flight plan was ready in little more than a month, and the final plan followed on 4 March.[32]

The new plan called for firing the aft thrusters to free the space-craft from the second stage of the launch vehicle, adding about 3 me-ters per second to its speed and putting it into an elliptical orbit with a perigee of 122 kilometers and an apogee of 182 kilometers. Just be-fore first perigee, about an hour and a half into the flight and over Texas, a burst from the forward thrusters would cut 20 meters per second from spacecraft velocity and convert its orbit to a near circular 122 by 130 kilometers. During the second pass over the Indian Ocean, some 2 hours and 20 minutes into the mission, would come a series of out-of-plane burns totaling 4 meters per second, a part of the former flight plan to check out the OAMS, with no bearing on the fail-safe plan. Finally, over Hawaii on the third time around, there was a pre-retrofire burn to reduce speed by 28 meters per second, putting the spacecraft into an elliptical reentry orbit with a perigee of 63 kilome-ters.[33]

Another relative latecomer to Gemini 3 was a set of experiments. Although Project Mercury had included some in-orbit experiments, no one seems to have given much thought to Gemini in that context until Mercury ended in mid-1963. That summer, the Headquarters Office of Space Sciences began looking for proposals. It joined with the Office of Manned Space Flight in setting up a Panel on In-Flight Scientif-ic Experiments, or POISE, to pass on the merits of proposed experi-ments. A Manned Space Flight Experiments Board was chartered in January 1964 to decide which experiments would go on which mission, Apollo as well as Gemini.[34]

MSC had earlier formed its own experiments panel, which met for the last time on 16 January to pass on its advice about experiments for the first two manned Gemini missions to the NASA Headquarters group that had superseded it. Noting that Spacecraft 3 had already been built and that the shortness of the planned mission sharply limit-ed any active participation by the crew, the panel stressed the need to find experiments that would largely conduct themselves and were nearly complete in terms of planning, design, and hardware. The panel members believed, although GPO did not, that two experiments left over from the proposed but never flown Mercury-Atlas 10 met these stringent criteria: one intended to explore the combined effects of ra-diation and low gravity on cells, the other to study cell growth at zero gravity. Both were approved by the Headquarters board when it met in Washington the following month.[35]

The first experiment had been prompted by signs of radiation damage to cells after earlier flights, the biological effects being in some cases greater than might have been predicted from the length of exposure; this was a matter of special concern in light of plans for long-duration manned space flight. Either (or both) of two reasons might explain this anomaly: unknown biological effects produced by the "heavy primaries" component of radiation, blocked from Earth's surface by the atmosphere and hence inaccessible to terrestrial laboratories, or the interaction of radiation with some aspect of the space flight environment, such as prolonged weightlessness. Experiment S-4 was designed to furnish a basis for weighing these alternatives.

Human blood samples were to be exposed to a known quantity and quality of radiation (both in the spacecraft and on the ground) during the zero-gravity phase of the mission. The frequency of various chromosomal aberrations in both samples could then be compared. To be mounted on the right-hand hatch, the experiment was wholly self-contained in a half-kilogram (one-pound) hermetically sealed aluminum box that held the blood samples, a radiation source, and instrumentation. The copilot had only to twist the handle and push it in to start the irradiation of the blood samples. Twenty minutes later he would twist the handle in the opposite direction and pull it out to stop the experiment. Word of these actions relayed to the ground would allow them to be duplicated.*[36]

The second experiment was designed to explore the possibility that cells might be directly affected by low gravity—that long-term weightlessness might produce changes with important implications for prolonged space flight. Because the effects were easier to detect in simple cell systems than in complex organisms and because theory argued that effects would appear only in cells upward of one micron across, the eggs of a sea urchin were selected as the experimental material. The eggs were to be fertilized at the start of the experiment, and the possible changes brought about by low gravity observed at several stages of the development.

The cell growth experiment was also self-contained, a sealed 2/3-kilogram (1½-pound) cylinder, to be mounted on the left-hand hatch and worked by the command pilot. The handle had to be turned five times—once half an hour before flight to fertilize the eggs, then four times in flight to fix the dividing cells at specific stages of growth in successive samples. Each time the handle was turned, the fact was relayed to the laboratory, where the action would be duplicated on an identical package. Results from the simultaneous experiments would be compared later.†[37]

---

*Michael A. Bender, U.S. Atomic Energy Commission, was principal investigator on the radiation experiment.

†Principal investigator for the cell-growth investigation was Richard S. Young, Ames Research Center.

A third experiment found its way into Gemini 3 by a more round-about path. Spacecraft falling back into the atmosphere are sheathed in an ionized plasma that blocks all radio communication, a source of much concern in at least two Mercury missions. In the first manned orbital flight, with John Glenn in *Friendship 7*, the five-minute black-out followed a signal that the capsule's heatshield was unlatched. Although the signal was wrong, Mercury control spent an agonizing five minutes until the radio link was restored. Then in the very next flight, M. Scott Carpenter's *Aurora 7* overshot its planned landing point by 400 kilometers because the capsule was misaligned at retrofire. In either case, communications with the reentering spacecraft would have made many hearts beat more calmly.[38]

A reentry communications experiment had been proposed and accepted for Mercury-Atlas 10, but when the program ended with that mission unflown, it was suggested for Gemini. Tentatively assigned to Spacecraft 3 in March 1964, the experiment failed to win a firm place for months, largely because of its half-million-dollar price tag. In July, however, the Office of Advanced Research and Technology in NASA Headquarters agreed to share the cost, and the experiment had its place in the mission confirmed.[39]

Research had shown that, for small objects, adding fluid to the ionized plasma during the reentry blackout could restore communications by lowering the plasma's frequency enough to allow UHF radio transmission to get through. Whether the same technique would work for an object as large as the Gemini spacecraft was now to be tested. A water expulsion system would be installed on the inside surface of one of the landing-gear doors, relics of the days when landing skids were to be used with its paraglider wing. The experiment was fully self-contained except for a starting switch inside the cabin to be thrown by the copilot when the spacecraft had fallen to about 90 000 meters. At that point, the plasma sheath would surround the spacecraft, blacking out communications. Water would be automatically injected into the plasma in timed pulses for the next two and a half minutes, while ground stations monitored and recorded UHF radio reception.*[40]

## Maneuvers of "Molly Brown"

During the first two days of March 1965, the Office of Manned Space Flight held a Design Certification Review in Washington. The review board† asked for, and got, formal pledges from the top execu-

---

*The experiment had originally been proposed for Mercury by William F. Cuddihy of Langley Research Center. His colleague, Lyle C. Schroeder, later took over as principal investigator for the Gemini experiment.

†The board consisted of Mueller, Gilruth, Kurt Debus (Director, Kennedy Space Center), Wernher von Braun (Director, Marshall Space Flight Center), Major General Osmond J. Ritland (Air Force Systems Command), and Major General Ben Funk (Commander, SSD).

tives of all major Gemini contractors that their products were ready for manned space flight, barring something unforeseen turning up during what remained of Cape checkout. A week later, the spacecraft Flight Readiness Review revealed only minor and quickly corrected problems. The launch vehicle passed its final test, simulated flight, on 18 March and its Flight Readiness Review on Saturday morning, 20 March. When the Mission Review Board* met that afternoon, weather was the only thing that might delay the mission. Early Monday morning, the launch vehicle contractors confirmed that GLV-3 was ready to go; at 9 a.m., the Flight Safety Review Board committed the booster to launch.[41]

Martin's pad crew started loading oxidizer aboard GLV-3 at 6:22 that evening, 22 March, and five hours later all tanks were full. The final countdown began at 2 o'clock Tuesday morning, under overcast skies. Included in the countdown were static firings of both spacecraft rocket systems. This had been a matter of dispute between the astronauts and the program office. They agreed on plans to fire one ring of the reentry control system but not on OAMS firing. GPO, backed by the Preflight Operations Division, preferred to fire only the lateral thrusters, but the pilots wanted to fire the aft thrusters too. The matter was settled in May 1964, when NASA Deputy Director for Gemini William Schneider decided both would be fired. Although he knew that the extra test time might affect the launch, he believed "that this will save time in the long run and will increase the confidence in flying a successful mission."[42]

Grissom and Young, who had reviewed their flight plan and gone to bed about 9 o'clock the night before, were awakened shortly before 5 a.m. After steak and eggs, a launch-day breakfast tradition inherited from Mercury, they were driven from their Merritt Island quarters to pad 16, site of the preflight ready room. They arrived about 6 and had their suits on about 45 minutes later. Shortly after 7, a van bore them to pad 19. They mounted the elevator for the 11th level, where their spacecraft awaited them. At 7:30, they were inside with the hatches sealed. Because the so-far flawless countdown had moved faster than expected, they were about 20 minutes ahead of schedule. Young later complained about this extra time spent flat on his back and fully suited; the planned wait was bad enough.

Weather was still the big question mark, the overcast not having

---

*With Williams gone, Kraft became chairman of the spacecraft and launch vehicle Flight Readiness Review Boards and the Mission Review Board. Everline was coordinator and recorder for the spacecraft and mission reviews and James B. Jackson for the launch vehicle board. There were two new members—William Schneider from Headquarters and Max Faget. The other members were the same as for the first two missions: Mathews, Bailey, Slayton, and Merritt Preston.

lifted as expected. Grissom and Young had been in the spacecraft less than an hour when the count was halted, just 35 minutes before launch, because the first-stage oxidizer line had sprung a leak. A handy wrench applied to a poorly seated nut solved the problem, but the count was held for 24 minutes to make certain the leak had stopped. By the time the countdown resumed, the clouds over the Cape had begun to scatter. Thirty-five minutes later, at 9:24 Tuesday morning, 23 March 1965, the sky was almost clear when the engines of GLV-3 burst into life. With a "You're on your way, Molly Brown," from CapCom (capsule communicator) L. Gordon Cooper, Jr., the third flight of Gemini, the first to which men entrusted themselves, began.[43]

Officially the flight of *Gemini 3*, unofficially it was the voyage of "Molly Brown." During Project Mercury, each pilot had named his own spacecraft, although Cooper had some trouble selling NASA on *Faith* 7 for the last spacecraft in the program. Grissom and Young now had the same difficulty with "Molly Brown." Grissom had lost his first ship, *Liberty Bell* 7, which sank after a faulty circuit blew the hatch before help arrived. "Molly Brown," the "unsinkable" heroine of a Broadway stage hit, seemed to Grissom the logical choice for his second space command. NASA's upper echelons thought the name lacking in dignity; but since Grissom's second choice was "Titanic," they grudgingly consented, and the name remained "Molly Brown," though only quasi-officially. Later spacecraft were officially referred to by a Roman numeral, although a few had nicknames as well.[44]

"Molly Brown" lifted off so smoothly that neither Grissom nor Young felt anything. Their real cues were seeing the mission clock on the instrument panel start running and hearing Cooper announce it from mission control. There was less noise than they had heard on the moving-base simulator in Dallas. When the first-stage engine cut off two and a half minutes later, acceleration plunged from six gravities to one. The second-stage engine ignited, bathing the spacecraft in a flash of orange-yellow light that disconcerted Young for the moment it took him to realize that this was a normal product of fire-in-the-hole staging—that is, second-stage ignition before, instead of after, separation. The launch vehicle had slightly exceeded its predicted thrust, but a warning from Cooper prepared the pilots for the larger than expected pitchdown when the second stage took over the steering. Young, who had never been in space before, was entranced by his view of Earth's horizon and the sense of rapid motion as second-stage thrust built up.[45]

Five and a half minutes after launch, the second-stage engine shut down. The pop of the pyrotechnics that severed spacecraft from launch vehicle sounded like the bark of howitzers to Young. Grissom fired the aft thrusters to kick the spacecraft into orbit. He lost track of

**Gemini 3**
**23 March 1965**

Left above, astronauts Young, Slayton, Grissom, and Stafford check map of Gemini 3's three orbits; right, above, press site, Cape Kennedy; left, Young and Grissom arrive at pad 19; right, artist's rendition of corona-like effect caused by exhaust gases during 2d-stage ignition.

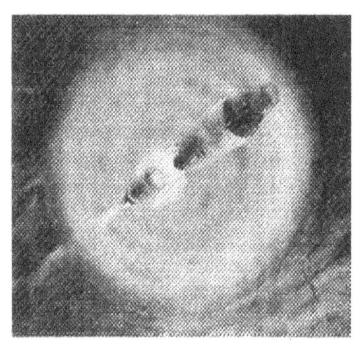

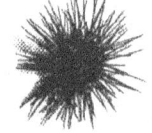 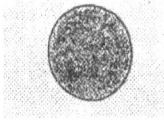

 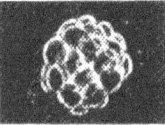

Control room, Cape Kennedy.

Gemini 3 weightlessness experiments; upper left, spiny black sea urchin; upper right, single egg from sea urchin; lower left, cell dividing in two; lower right, cell division into many cells.

Intrepid recovers spacecraft.

White House ceremony in which Grissom receives NASA Distinguished Service Medal: left to right, Vice President Humphrey, Grissom, President Johnson, and Administrator Webb.

the time and fired too long, ending up with his incremental velocity indicator showing a slight overspeed. But he wound up with an orbit of 122 by 175 kilometers, very close to the intended 122 by 182 kilometers. *Gemini 3* was off to a good start—to an almost troublefree flight that closely matched the planned mission.[46]

The match was not perfect. About 20 minutes into the first orbit, just after "Molly Brown" passed beyond range of the mid-Atlantic Canary Island tracking station, the oxygen pressure gauge in the environmental control system reported an abrupt drop. Young, assigned to watch this gauge, naturally assumed that something was wrong with the system. But a quick glance showed odd readings on several other meters and suggested that the real trouble might be in the instrument power supply. Young switched from the primary to the secondary electrical converter to power the dials, and the problem vanished. The whole episode, from Young's first notice of the anomalous reading to his shift from primary to secondary power, took 45 seconds, one clear payoff from intense preflight training.[47]

Grissom's attempt to run the cell-growth experiment was a failure—perhaps, as he remarked later, because he had "too much adrenalin pumping" and twisted the handle too hard. Whatever the reason, the handle broke, ruining the experiment. The radiation experiment gave Young some trouble, but he managed to complete his task. Results were suggestive but inconclusive. Exposed to nearly identical doses of radiation, the inflight blood samples showed more damage than the control samples on the ground. While the effect was small, it did point to interaction between radiation and some aspect of space flight, though just which aspect and how it acted could not be answered. Both Grissom and Young believed that most of the trouble with the experiments stemmed from differences between the packages they flew with and those they had trained with. But they also admitted that they "were not quite as fascinated by sea urchins . . . as . . . by the chance to carry out some real 'firsts' in space flight."[48]

And the *Gemini 3* crew did chalk up at least one historic first by maneuvering in orbit. The first OAMS burn came an hour and a half after launch and lasted a carefully timed 75 seconds, cutting spacecraft speed by 15 meters per second and dropping it into a nearly circular orbit. Three quarters of an hour later, during the second revolution, Grissom fired the system again, this time to test the ship's translational capability and shift the plane of its orbit by one-fiftieth of a degree. During the third pass, Grissom completed the fail-safe plan with a two and a half minute burn that dropped the spacecraft's perigee to 72 kilometers and ensured reentry even if the retrorockets failed to work.[49]

They did work, however. As the three-orbit mission neared its close, Grissom and Young ran through the retrofire checklist. With everything ready, the pilot fired the pyrotechnics that separated the adapter from the reentry module, giving the two spacemen their biggest jolt so far. He then armed the automatic retrofire switch. One after the other, the four rockets exploded into life and burned themselves out. Another set of pyrotechnics cut loose the expended package as "Molly Brown" arced back toward the planet she had left four and a half hours before.[50]

Reentry produced some surprises. At the outset, it matched the simulations both men had been through in training, even to the color and pattern of the plasma sheath that surrounded the spacecraft. Young threw the switch to start the reentry communications experiments just over a minute after the plasma had formed and communications had blacked out. The results were encouraging; at high rates of water flow, both UHF and C-band signals from the spacecraft were picked up by ground stations.[51]

But "Molly Brown" seemed to be off course. The initial computer reading showed that she would miss her planned landing point by more than 69 kilometers, and Grissom's best efforts to reduce that gap were fruitless. Theoretically, the Gemini spacecraft had enough lift to be piloted to a relatively precise landing, but its real lift fell far short of what had been predicted from wind tunnel tests. As a result, *Gemini 3* was about 84 kilometers short of the intended splashdown point. Before they touched down, however, the astronauts suffered another jolt when the spacecraft assumed its landing attitude. After the main parachute deployed, the spacecraft hung from it vertically, with its nose suspended at a single point. Before landing, throwing a cabin switch shifted the spacecraft to a two-point suspension with its front end forward and some 35 degrees above the horizontal. When Grissom hit the landing attitude switch, "Molly Brown" literally dropped into place, pitching both men into the windshield, breaking Grissom's faceplate, and scratching Young's.[52]

The jolt when they hit the water a few minutes later was mild by comparison. Although Gemini was designed to float, all Grissom saw out his window was water. He realized that the still attached parachute was being dragged by the wind, tugging the nose of the spacecraft down. With memories of the ill-fated *Liberty Bell 7* momentarily staying his hand, Grissom released the chute and "Molly Brown" bobbed to the surface, having shown herself fully watertight. The mission plan called for the crew to remain on board until the spacecraft was picked up, a short wait if the recovery ship, the aircraft carrier *Intrepid*, was only about eight kilometers away, as Grissom and Young had last heard before they splashed down. When they learned that the real distance was closer to 110 kilometers, Grissom asked for a helicopter to pick them up and take them to the carrier. Still thinking of *Liberty Bell 7*, however, he refused to crack a hatch until Navy swimmers had attached a flotation collar to "Molly Brown." This spacecraft was not going to sink, but the crew endured a long 30 minutes as the sealed spacecraft grew hotter inside while it pitched and tossed on the long Atlantic swells. "That was no boat," recalled Young. Heat and motion took their toll of Grissom, although Young managed to keep his breakfast down. Once the collar was in place and a swimmer opened a hatch, the two men lost no time in getting out and putting on the "horse collar" hoists that lifted them to the helicopter.[53]

Medical examinations and debriefings began as soon as the two astronauts were in the helicopter and went on for several days. A brief stir ensued when Grissom and Young had little to say to scientists about their observations, mainly astronomical, while in orbit. Other questions were raised about the failure of the cell-growth experiment, but most of the fault could be ascribed to a poorly designed package that was installed in the spacecraft barely a week before flight—a matter of "too little, too late." In any case, the brief mission had centered on engineering evaluation of the spacecraft, with a full schedule that left little time for extra work.

Something of a storm later blew up when the press got wind of Grissom's having eaten part of a corned beef sandwich during the flight. Schirra had bought it at "Wolfie's" on North Atlantic Avenue in Cocoa Beach and given it to Young, who smuggled it on board the spacecraft. When it was time for the crew to eat the space food they carried, Young brought out the sandwich and handed it to Grissom, who ate only a few bites as he wanted no crumbs floating around the cabin. When the news got to Congress, the lawmakers were upset. What was not made clear, apparently, to either the legislators or the press was that the official food was only there for evaluation of its taste, convenience, and reconstitution properties and had nothing to do with any scientific or medical objectives of the mission. No one expected to learn very much about the effects of space food on so short a flight. The fracas did, however, produce some new and more stringent rules about what the astronauts might take with them on future missions.[54]

Despite its minor problems, *Gemini 3* was a complete success as far as its major objectives were concerned. There could be no doubt that Gemini was ready for its role in the manned space flight program. The time of testing was over.[55]

# XI

# Pillars of Confidence

ALTHOUGH the revised Gemini flight plan of April 1963 remained the basic framework of program operations through 1965 and 1966, it proved to be, at least in some respects, still too optimistic. Lagging fuel-cell development forced the Gemini Program Office in August 1964 to settle for four days, rather than seven, as the goal for Gemini IV* and also to delete the practice rendezvous with the evaluation pod from that mission. Gemini V had been slated as the first in which spacecraft would rendezvous with Agena target, but that goal, too, had to be deferred.

If some aims had to be postponed, however, they were balanced by some worthwhile gains. Extravehicular activity (EVA) emerged as a new feature of Gemini IV, and Gemini V expanded to an eight-day mission that included practice with the rendezvous evaluation pod. The new Mission Control Center in Houston assumed flight control duties for Gemini IV,[1] taking over that job from the former control center at Cape Kennedy. Only two months were to elapse between Gemini IV and V, a sign of the progress that NASA was making toward putting space flight on something like a routine basis. Perhaps most important, these two missions set Project Gemini firmly on the path to reaching its major objectives, sweeping aside fears that astronauts might not be able to survive long periods of weightlessness in space and holding out the promise that rendezvous could soon be achieved.

NASA announced the crews for Gemini IV on 27 July 1964, and two days later James A. McDivitt and Edward H. White II, along with

---

*With Gemini IV, NASA changed to Roman numerals for Gemini mission designations. The text will hereafter use Roman numerals for all Gemini missions.

their backups, Frank Borman and James A. Lovell, Jr., talked with reporters in Houston. McDivitt and White, aged 35 and 34, had known each other since college and had been in the same class at the Air Force test pilot school. Borman and Lovell, both 36, first met when they were undergoing testing by NASA. Borman was an Air Force officer and Lovell was in the Navy. All four men were second generation astronauts, part of the group selected by NASA in September 1962.[2]

Their first task after the announcement was to review the status of the spacecraft and booster assigned to their mission. Spacecraft 4 was still being built in St. Louis, with some problems caused by a shortage of parts.[3] In Baltimore, GLV-4 was also in the process of being assembled.[4] After that quick look, the crewmen spent the next five weeks cleaning up work left over from their former assignments. Mission training had to wait until the end of November, when Gemini Simulator 2 became operational in Houston.[5]

Meanwhile, McDivitt and his crewmates, knowing that EVA might be included in Gemini IV, seized every chance to press the case for making it part of their mission. This persistence won NASA management's consent to provide the special space suits that EVA required. The astronauts were not merely chauffeurs; their role in the program went far beyond that of the normal test pilot in determining what was to be done and when. Without the strong pressure from the Gemini IV crewmen, the G4C suit might have been too far down the line to have permitted NASA's late decision to include EVA in the fourth mission.[6] That decision was not, however, quite so late as it appeared.

When Cosmonaut Aleksey A. Leonov walked in space on 18 March 1965, during the *Voskhod II* mission, he revived press complaints that America lagged in the space race and raised fears that a year might pass before a Gemini astronaut matched the Russian's feat. When, a little more than two months later, NASA announced that White would step into space on the next Gemini flight and use a "zip gun" to propel himself, most space watchers merely assumed that NASA was still trying to keep up with its Soviet rival.[7] This may have been true as far as timing was concerned; but EVA had been a part of Gemini thinking almost from the beginning, and studies had begun as early as 1962.[8] The road from study to a place in the flight plan, however, was a rocky one.

Even the public linking of EVA with Gemini IV preceded *Voskhod II* by nearly eight months. At the same press conference in July 1964 where the Gemini IV crewmen took their bow, Gemini Deputy Manager Kenneth Kleinknecht had said one of the crew might open the hatch and stick his head outside during the mission. McDivitt was surprised at how little notice newsmen took of Kleinknecht's statement.[9] At that point, it was still far from certain that even a simple hatch opening would be permitted in Gemini IV. The key questions involved equipment and training.

Gemini IV first appeared as the program's lead-off EVA mission in a "Program Plan for Gemini Extravehicular Operation," during January 1964. Management response was cool, largely because equipment development was only beginning.[10] During the next few months, however, matters improved. The AiResearch Manufacturing Company was awarded a contract for the extravehicular chestpack, the David Clark Company was sent specifications for the extravehicular suit, and McDonnell was authorized to begin an EVA design that was eventually applied to Spacecraft 6.[11]

After Kleinknecht's largely ignored statement in July on standup EVA plans for Gemini IV, the issue continued to be debated within NASA. MSC's Engineering and Development Directorate, and its Crew Systems Division, in particular, opposed any EVA in Gemini missions until crews faced some realistic simulations on the ground.[12] The scheduled altitude chamber tests of Spacecraft 3 in November 1964 offered a good chance to meet that demand. Gus Grissom and John Young wanted to depressurize the cabin during their training for Gemini III and open the hatch at a simulated altitude of 46 000 meters. Selling this idea to McDonnell was not easy. McDonnell, as Young later remarked, "certainly didn't want to take the chance of bagging a couple of astronauts in the altitude chamber," and NASA was none too happy about "putting guys in vacuums with nothing between them but that little old lady from Worcester, Massachusetts [the seamstress at the David Clark Company], and her glue pot and that suit."[13]

Kleinknecht argued that "if we can't do it in the altitude chamber, then we haven't any business doing it 100 miles [160 kilometers] in space." GPO told McDonnell to "include at least one complete depressurization, hatch opening and closing, and repressurization cycle at 40 000 feet [12 000 meters] altitude conditions in each spacecraft manned altitude chamber test commencing with spacecraft 3." The first try at EVA practice left something to be desired, Young recalled, when "we opened the hatch and [then] we couldn't close it." But the three-orbit Gemini III mission was really too short for EVA anyway, and GPO focused its efforts on Gemini IV.[14]

Plans were firmer by the start of 1965, and the Gemini IV crews began training for EVA.[15] Nevertheless, the decision of whether to include EVA in the mission was far from settled, either at MSC or NASA Headquarters. MSC Director Gilruth did approve altitude chamber tests for the crew, but only on 12 March 1965, less than a week before Leonov's space walk.[16] That feat spurred new efforts to get extravehicular activity into an early Gemini mission. With the flight of Gemini III just a week away, that meant Gemini IV. During that

week between *Voskhod II* and Gemini III, Gilruth and Deputy Director George Low had their first look at a "hand held maneuvering
unit," which had been designed and built without fanfare in MSC's
Crew Systems Division. That device, along with a display of the progress with other EVA equipment, brought the Center's top management
solidly behind trying for EVA in its second manned mission.[17]

The hardware still needed to be qualified. Gilruth gave the job to
Crew Systems with a warning to keep the work as quiet as possible,
perhaps to avoid any appearance of too-hasty reflex to Russian accomplishments. A model spacecraft was quickly installed in MSC's 6-meter
vacuum chamber, and preliminary testing was begun.[18] By the end of
April, the vacuum chamber was ready for full-scale EVA simulation,
and Flight Operations people had come into the picture to begin working out techniques for handling EVA as a flight control matter.[19]

But NASA Headquarters had yet to be won over. Manned space
flight chief George Mueller learned about the MSC plans when he visited Houston on 3 April; his response was lukewarm, perhaps because
of the still unqualified status of the hardware. Although he offered no
encouragement, Mueller was not inclined to order a halt, and MSC
went ahead with its plans. On 14 May, when Gilruth arranged an EVA
demonstration for Associate Administrator Robert Seamans, he won a
high-ranking ally. Seamans promised to discuss MSC's new venture
with Administrator James Webb and his deputy, Hugh Dryden.[20]

The next day, Mathews and three of his men were in Washington
for another attempt to convince Mueller that EVA belonged on Gemini IV. Mueller's crucial question was how EVA, not officially scheduled
until Gemini VI, could be moved up two flights; the answer was simply
that everything was ready: all EVA gear was qualified, or nearly so,
and the crew was trained. After he got back to Houston, Mathews
called Mueller on 19 May to report that the last piece of EVA equipment was now flightready.[21]

Seamans, as he had promised, did describe the EVA plan for
Gemini IV to Webb and Dryden. Webb liked it, but Dryden objected
strongly; he thought it smacked too much of a reaction to what the
Russians had done. At Webb's request, Seamans drew up a brief stating
the reasons for putting EVA on the current Gemini mission, which
concluded: "The hardware for extravehicular activity is flight qualified
and the astronauts are trained for this operation. Since extravehicular
activity is a primary goal for the Gemini program, it is recommended
that this activity should be included in Gemini IV." Webb gave the
paper to Dryden. On 25 May, Dryden called Seamans to his office and,
without saying a word, handed him a document. It was the case Seamans had made for EVA; scribbled on one corner was "Approved,
after discussing w. Dryden [signed], J. E. Webb, 5-25-65."[22]

There was still a question about how and when to make public the plans for EVA. MSC opinion was divided. Some favored breaking the news after the fact, some while EVA was in progress, and others at the premission press conference 24 hours before launch. In April, MSC decided to announce it at the press briefing, if it were approved. Seamans, however, rejected that scheme as incompatible with NASA's historic policy of openness on plans for manned launches and ordered EVA material to be included in the press kit for Gemini IV. When the kit appeared on 21 May it contained a one-page discussion of "Possible Extravehicular Activity." On 25 May, the same day EVA was approved, the press was informed that White would leave his spacecraft and walk in space.[23]

One reason for Mueller's resistance to EVA was a plan to combine it with rendezvous. Gemini IV was scheduled to rendezvous with the second stage of its booster in orbit, and White could then use his zip gun to propel himself over near the floating stage. This idea was also a latecomer to Gemini. The rendezvous evaluation pod scheduled for the fourth mission had been forced out in January 1964, when problems with the radar design made it unlikely that that crucial equipment would be available in time.[24] A bit of joking by Gordon Cooper over the communications link to Grissom in *Gemini III* on 23 March 1965, suggested another kind of practice rendezvous.

> Cooper: I have a time for when you'll be nearest the booster [second stage]. Would you like to have that so that you can look for it?

> Grissom: Roger.

> Cooper: Roger. 02 plus zero eight plus five two will be dead ahead at an elevation of plus eight zero degrees at one niner miles. This will be just prior to darkness. It should be very bright. Proceed to see if you can rendezvous.[25]

Gilruth and Low overheard the exchange and thought it sounded like a pretty good idea. Low checked with GPO and Crew Systems and got an enthusiastic response. With Gilruth's wholehearted support, in May 1965 stationkeeping joined EVA as part of the Gemini IV flight plan. The spacecraft would match velocities with the orbiting second stage a relatively short distance away in the same orbital plane and maintain that position for a time. Grissom had maneuvered "Molly Brown," but he had no target. Closing in on a specific object (or point) in space was much more ambitious, especially since McDivitt and White would have to depend on their eyes to track the target, since the rendezvous radar was still unavailable. Martin did install flashing lights on the GLV-4 second stage to help the crew find it.[26] McDivitt and White had still another handicap. There was simply no way for them to train on the ground for stationkeeping—neither the Cape nor the Houston simulator was designed for this task. McDonnell came through by rigging equipment to provide a simulated view of the target against a star background. McDivitt and Borman spent half a day in St. Louis practicing optical rendezvous, but it was makeshift at best.[27]

One other major problem confronted Gemini IV planners, the physiological consequences of a prolonged stay in orbit and of EVA. Charles A. Berry, medical director of the Gemini program, was troubled by the leap of faith implied by the Gemini flight schedule of April 1963, which followed the three-orbit Gemini III with the seven-day Gemini IV. He wanted the length of the mission reduced by half, and trouble with fuel-cell development might come to his aid. If batteries had to be used, the mission could not last more than four days. In August 1964, Mathews reported to NASA Headquarters that Gemini IV would be a four-day mission, not only for medical reasons but also because the fuel cell would have to be replaced by batteries.[28]

Berry was not happy even with a four-day mission. Cardiovascular problems had cropped up in the last two Mercury missions, and every physiologist he met made the same comment about Gemini IV, or so it later seemed:"[Don't you] really know that these guys [are] going to stand up and pass out and might, indeed, die from this flight? "[29] The astronauts would be subjected to much the same kind of physiological strain as that imposed by prolonged bedrest followed by vigorous activity. After their bodies had been deconditioned by days of weightless flight, they had to face high reentry g forces, which might well cause them to faint. If an astronaut fainted during or after landing, he would be held upright by his harness, forcing a perhaps already overtaxed heart to work even harder pumping blood to his head. But astronauts were not bed patients; besides using their muscles for flight tasks they would have been exercising with a bungee cord, a device adapted from the nylon strap and handle of a spear gun that required a force of 300 newtons (70 pounds) to extend it 30 centimeters (12 inches).[30]

EVA added still another medical concern, the disorientation and motion sickness that might overtake a floating astronaut unable to distinguish "up" from "down." Leonov, according to Russian reports early in May 1965, had trouble with his vision and orientation "when he didn't see the spacecraft." Berry, McDivitt, and White studied a filmed interview, with scenes of the space walk, which clearly showed Leonov using numerous reference points—the Sun, the spacecraft, Earth—to maintain orientation. That seemed to be the best answer, the astronaut making sure he knew where he was at all times in relation to the spacecraft.[31]

From a medical veiwpoint, then, some degree of tension marked the approach of the Gemini IV mission. This was, after all, the first four-day flight by Americans, and the Russians were airing their fears of disorientation and physiological dangers at numerous medical conferences. But the crew was trained, and everything that could be foreseen had been considered. There was nothing to do now but wait to see what happened.[32]

## FOUR DAYS AND A "WALK"

About 12 hours before Gemini IV's* scheduled liftoff on 3 June 1965, the Martin crew started fueling the booster and calibrating its propellant loads. Borman and Lovell, the backup crew, flipped spacecraft switches, tested communications circuits, and handled other chores to relieve the prime crew. McDivitt and White had gone to bed at 8:30 the night before. Awakened at 4:10 a.m., they were given a brief physical examination. The astronauts left their Merritt Island quarters after breakfast and boarded a van for the ride to the pad 16 suit-up area, where they were helped into their suits while breathing pure oxygen to get the nitrogen out of their systems and thus prevent aeroembolism, or the bends.[33]

McDivitt and White arrived at pad 19 at 7:07 a.m., rode up in the elevator, and climbed into their spacecraft at $T-100$ minutes. Getting in was relatively easy, but even so White's faceplate fogged. He started his suit fan and cleared up the moisture.[34]

Thirty-five minutes before the scheduled launch, while the erector was being lowered, it stuck at a 12-degree angle from the booster. Raised to its full height, then lowered again, the erector still stuck. After more than an hour, technicians found a connector incorrectly installed in a junction box, replaced it properly, and gave the signal to lower the erector. This time it worked. Space travel was becoming operational. This hold, lasting 1 hour 16 minutes was the only delay for Gemini IV. On Mercury-Redstone 4, the second manned launch in that program, Grissom's *Liberty Bell 7* was scrubbed twice and was plagued by six holds that totaled 4 hours 1 minute.[35]

At 10:16 a.m., Thursday, 3 June 1965, millions of people throughout the world looked and listened while *Gemini IV* lunged spaceward. Television coverage of the launch for the first time had an international audience, as the scene was broadcast to 12 European nations via *Early Bird* satellite.[36] Heightened by the prospect of EVA and the first use of the new Mission Control Center in Houston, interest in *Gemini IV* reached levels never again matched in the program. The Manned Spacecraft Center faced a major challenge in the number of reporters who wanted to cover the story from Houston. Although

---

*The Gemini IV spacecraft had no name, official or otherwise (such as "Molly Brown"), nor did its pilots wear a distinctive patch on their suits, as did all later Gemini crews. A few of the newsmen called the ship "Little Eva," to symbolize the extravehicular activity.

245

MSC's Building 1 auditorium had been "designed to house all large events covered by the news and television services," its 800 seats fell short of the space that would be needed to accommodate the 1100 requests for accreditation NASA had received. To meet the demand, MSC leased one of the new buildings springing up across the highway from the Center for local offices of aerospace companies—and that move came under fire from the local press when its cost was revealed: besides the $96 165 yearly rent, MSC spent $166 000 for modifications, $8000 for television monitors, and $6600 for 610 chairs.[37] But "Building 6," housing the NASA Gemini News Center, served its purpose well as the base for 1068 newspaper, magazine, radio, and television representatives, as well as 60 public relations people from industry.[38] It opened on 25 May, somewhat earlier than the "launch minus five days" that had been customary for news centers during Mercury.

In the spacecraft, McDivitt and White had no doubts about liftoff, as they felt their vehicle pick up speed. There was very little noise. The hush was broken only when the launch vehicle bounced like a pogo stick for a few seconds. Then everything smoothed into near silence again. Pyrotechnics shattered the illusion of quiet at stage 1 and, later, at stage 2 separation. The spacecraft entered an elliptical orbit of 163 kilometers at the low point (perigee) and 282 kilometers at the high point (apogee).[39]

As *Gemini IV* separated from its booster, McDivitt turned the spacecraft around to look for the trailing vehicle. White saw the rocket venting, with propellant streaming from its nozzle. How far was it, and where was it going? McDivitt estimated the distance as 120 meters; White guessed it was closer to 75 meters.

McDivitt braked the spacecraft, aimed it, and thrusted toward the target. After two bursts from his thrusters, the booster seemed to move away and downward. A few minutes later, McDivitt pitched the spacecraft nose down and the crew again saw the rocket, which seemed to be traveling on a different track. He thrusted toward it—no success—and stopped. McDivitt repeated this sequence several times with the same luck.[40]

As night approached McDivitt spotted the booster's flashing lights. He estimated that the distance to the target had stretched to perhaps 600 meters. He knew he had to catch the booster quickly if they were going to stationkeep and do extravehicular activity as planned. For a while, *Gemini IV* seemed to hold its own and even to close with the other vehicle. McDivitt thought they got to within 60 meters, but White estimated it at 200 to 300 meters. The target's running lights soon grew dim in the gray streaks of dawn and vanished with the sunrise. When the target hove into view about three to five kilometers away, McDivitt again tried to close the distance. Additional thrusting did not seem to bring it any closer. Well aware that he was a pioneer

**Gemini IV
3 June 1966**

*Above left, astronauts White and McDivitt train in celestial navigation, Morehead Planetarium, Chapel Hill, N.C.; above center, White practices EVA in pressure chamber at McDonnell; above right, launch tower is stuck, delaying launch.*

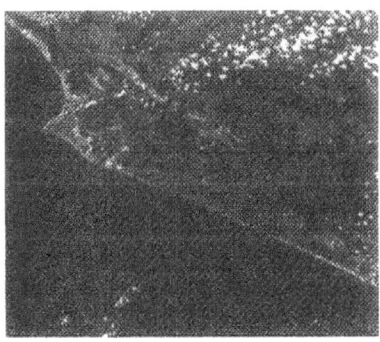

*Above left, new control room at Houston, used for the first time during Gemini IV; center right, White, maneuvering gun in right hand, performs first U.S. EVA; above, Cape Kennedy snapped from orbital altitude; above right, artist Franklin McMahon's sketch of Gemini IV recovery; right, White and McDivitt in helicopter after landing.*

in orbital rendezvous and that choosing the right maneuvers might not be as easy as it seemed, McDivitt had previously asked Mission Director Kraft which was more important, rendezvous or EVA. The space walk, said Kraft. McDivitt knew he had to stop spending fuel chasing the elusive target by the "eyeball" method.

As GPO engineer André Meyer later remarked, "There is a good explanation [for] what went wrong with rendezvous." The crew, like everyone else at MSC, "just didn't understand or reason out the orbital mechanics involved. As a result, we all got a whole lot smarter and really perfected rendezvous maneuvers, which Apollo now uses." Catching a target in orbit is a game played in a different ball park than chasing something down on Earth's essentially two-dimensional surface. Speed and motion in orbit do not conform to Earth-based habit, except at very close ranges. To catch something on the ground, one simply moves as quickly as possible in a straight line to the place where the object will be at the right time. As *Gemini IV* showed, that will not work in orbit. Adding speed also raises altitude, moving the spacecraft into a higher orbit than its target. The paradoxical result is that the faster moving spacecraft has actually slowed relative to the target, since its orbital period, which is a direct function of its distance from the center of gravity, has also increased. As the *Gemini IV* crew observed, the target seemed to gradually pull in front of and away from the spacecraft. The proper technique is for the spacecraft to reduce its speed, dropping to a lower and thus shorter orbit, which will allow it to gain on the target. At the correct moment, a burst of speed lifts the spacecraft to the target's orbit close enough to the target to eliminate virtually all relative motion between them. Now on station, the paradoxical effects vanish, and the spacecraft can approach the target directly. *Gemini IV's* problem was compounded by its limited fuel supply; the Spacecraft 4 tanks were only half the size of later models, and the fuel had to be conserved for the fail-safe maneuvers. When McDivitt and White broke off their futile chase, they had exhausted nearly half their load of propellants.[41]

White had been too busy helping his partner to give much thought to getting ready for EVA. Now that the rendezvous attempt was over, White put the zip gun together, while McDivitt read off a list of things for him to do. White pulled out the umbilical package and mounted suit connectors for the tether and the emergency oxygen chestpack. With 20 minutes still to go before cabin depressurization, the commander noticed that his copilot already looked tired and hot. McDivitt told the Kano tracking station that EVA would be postponed until the third revolution—he wanted White to rest.

While they relaxed, the crewmen talked with Grissom, the Houston CapCom, about the synoptic view of the Gulf of Mexico and all of Florida, including the Cape and its launch complexes. After a 15-minute break, McDivitt picked up the list and White began checking suit hose locks and suit integrity. The flight planners had certainly not foreseen how much time getting ready for EVA would take, McDivitt thought.[42]

Over the Indian Ocean, White was ready for EVA at last—hoses hooked up, umbilical ready, gun in hand, and chestpack in place—and they again rested and chatted. Nearing Carnarvon, Australia, they began to depressurize the cabin. Then a mechanical problem arose— the door would not unlatch because a spring had failed to compress. After much yanking and poking around the hatch ratchet, the door suddenly cracked open. White found the hatch as hard to push up in zero g as it had been on the ground.[43]

Once he had it opened, White rose slowly through the hatch and installed a camera to record his movements as he swam in space, with the zip gun, tethered to his right arm, floating freely by his side. White triggered a burst from the gun, rose above the hatch, and, without imparting any motion to the spacecraft, propelled himself away. Experimenting with the double-barreled device, he traveled about 5 meters but found himself higher above the spacecraft than he intended. He wanted to go over to McDivitt's window. Short bursts of the gun worked well; in fact, it responded throughout much as it had in ground training on an air-bearing table, at least in pitch and yaw. White was less sure about roll, which he thought would be harder to control without using too much fuel. Floating freely, he felt a tendency to pitch, roll, and yaw, all at once. He knew the gun could correct this, but he was concerned about the fuel it would take. Instead, he tugged on the tether and pulled himself aft and high atop the spacecraft adapter. White saw the thrusters firing, expelling plumes of flaming gas, as McDivitt steadied the spacecraft. White propelled himself away from the danger—across the top of the spacecraft and out beyond its nose. He used the gun for two pitchovers and two body turns, each time stopping easily. Then the compressed oxygen fuel bottle was empty—how he wished it had been bigger![44]

There was the usual brief loss of communications between Hawaii and Guaymas, Mexico. While White was using the zip gun over the Pacific, Mission Control was unaware of how he was making out. After the voice circuit was restored, radio listeners had a chance to hear an American human satellite broadcast his views of the spectacle of Earth. White told McDivitt and the world how beautiful it all was, of the pictures he was taking, and how well he was feeling—no vertigo or disorientation whatever. And when McDivitt had to tell him it was time to come back inside, Mission Control and the whole world heard him sigh, "It's the saddest moment of my life."

While he was floating freely, White had paid no attention to the time; and, since they were on the internal spacecraft communications link, Flight Control could not break in on them. Finally, after 15 minutes 40 seconds, McDivitt broke off to ask the ground if they wanted anything. "Yes," Kraft chuckled, "Tell him to get back in." After he passed this on to White, McDivitt heard boots thumping atop the spacecraft. White came back to the hatch as *Gemini IV* was passing over the Atlantic, dismounted the camera and removed electrical connections, and handed all these items to McDivitt along with the gun. McDivitt then helped White get settled, pulling on his legs and guiding his feet into the footwells.[45]

White closed the hatch and reached for the handle to lock it. When it failed to catch, he knew it was going to be as hard to close as it had been to open. Pushing on the handle lifted White out of his seat, so McDivitt pulled on him to give him some leverage. Finally White felt a little torque in the handle and yelled for McDivitt to yank harder. The door was latched.

White sat back, physically exhausted, sweat streaming into his eyes and fogging his faceplate. McDivitt also felt tired, so they rested before extending a radio antenna to find a ground-based voice and tell Earth all was well. Carnarvon answered them. The crew of *Gemini IV* had almost circled the globe in an unpressurized spacecraft.[46]

While White relaxed, McDivitt began powering down some of the spacecraft systems to save electrical power and control fuel, intending to drift for the next two and a half days. Seven and a half hours after liftoff, White went to sleep. He and McDivitt had intended to sleep alternate periods of four hours each, but this was hard to do. The constant crackle of radioed information and orders and the occasional automatic thruster firings kept them awake. Whoever was on duty frequently bumped the sleeper in this uncommonly small bedroom.[47]

*Gemini IV* was the first of the program's longer missions, and it imposed a set of new demands on ground control, which moved for the first time into a three-shift operation. Kraft acted as both Mission Director for the entire flight and Flight Director for the first shift. Eugene F. Kranz directed the second shift and John Hodge the third. Kraft's shift focused its efforts on helping McDivitt and White carry out the flight plan. The second shift concerned itself mainly with keeping track of systems performance and the use of such consumable stores as oxygen and fuel. Realtime flight planning was the special province of Hodge's shift. The basic framework of the flight plan was set before launch; but on the basis of what had already been achieved, how systems were working, and what stocks of fuel and other consumables remained, the third shift was ready by morning with specific instructions for the crew on tasks to be done or eliminated during the day ahead.[48]

Backing up the flight control teams were a number of systems experts who stood by in the staff support rooms of the new Mission

Control Center. They included not only NASA specialists but also contractor people, some of whom were assigned full-time to Houston while missions were in progress. At their home plants, other teams maintained systems under simulated flight conditions to provide quick answers to flight problems. Technical monitors and principal investigators were also on hand in the Mission Control Center for the Gemini experiment program, now more methodically handled by a new Experiments Program Office under Robert Piland in the Engineering and Development Directorate. *Gemini IV's* 11 experiments made it the first American mission to bear some resemblance to the manned space laboratory that had long been a staple of space flight thought.

*Gemini IV* was also the first mission to employ systematic methods to gather, evaluate, and publish information quickly, another demand imposed by longer flights and shorter intervals between missions. Willis Mitchell and Scott Simpkinson of GPO headed the 150-person Mission Evaluation Team that began work at liftoff and kept working through postflight inspection and mission evaluation.[49] *Gemini IV* served as training ground for pilots, flight controllers, and evaluators alike, setting the style for later Gemini missions, as well as for future Apollo flights.

Meanwhile, McDivitt and White drifted through space, watching systems, making observations, and doing experiments. A rigid constraint on fuel usage hampered most of these activities, although several of *Gemini IV's* 11 experiments were largely unaffected.*

Five dosimeters checked radiation in the spacecraft (experiment D-8), especially while *Gemini IV* was passing through the South Atlantic Anomaly (an intense pocket of the ionosphere), where radiation levels were considerably higher than in all other regions. In the Simple Navigation experiment (D-9), the pilots used a handheld sextant in an attempt to get celestial navigation readings, to judge sextant operation and navigational accuracy. McDivitt and White agreed that the sextant might be useful for Apollo.[50]

McDivitt and White had good fortune in the Synoptic Terrain (S-5) and Synoptic Weather (S-6) photography. The 70-millimeter Hasselblad camera worked well and, tourist-like, they tried to capture the view. They were especially smitten with the Nile River area—one saw Cairo, the other Alexandria—and White remarked that a landmark near a body of water was easier to spot. On one occasion, they snapped pictures from the Pacific Coast to Texas, showing good geological detail. They performed like professionals in getting pictures of weather phenomena. Unmanned Tiros weather satellites provide coverage from 640 kilometers, but *Gemini IV* gave the meteorologists a closer look, without a mosaic patchwork, at cellular cloud patterns,

---

*For descriptions, objectives, and results of all Gemini experiments, see Appendix D.

cloud layers in tropical disturbances, lines of cumulus clouds over the ocean, and thunderstorm areas.[51]

The crew used the bungee exerciser (M-3) more than had been planned, but White later said that his desire to do strenuous work dwindled during the flight; although, as McDivitt suggested, this might have been caused by lack of sleep. Both agreed that a systematic exercise program would be needed for long missions. Sensors attached to the pilots' bodies, in the Inflight Phonocardiogram experiment (M-4), gathered data on heartbeat rates, especially during liftoff, EVA, and reentry. As might be expected, their heartbeats were essentially normal except during these periods. The bone demineralization experiment (M-6) did show a greater mass loss in the small finger and heel than that experienced by Earthbound, bedrested patients.[52]

One engineering experiment—Electrostatic Charge (MSC-1)—gave higher readings than expected. Investigation later determined that thruster and water boiler operation produced some moisture, resulting in a high electrical charge, which dissipated very quickly. Concerns that docking in space might generate a harmful jolt were laid to rest. The Proton-Electron Spectrometer (MSC-2) and Tri-Axis Magnetometer (MSC-3), complementary radiation studies, provided useful data about Earth's radiation environment and the magnitude and direction of local geomagnetic fields. Photographing the red-blue Earth limb was the final engineering experiment (MSC-10), designed to help train Apollo astronauts in making navigational fixes.[53]

After 48 revolutions, covering 75 hours of flight, the spacecraft computer was updated during a stateside pass. Told to turn the computer off, McDivitt flipped the switch and discovered that he could not. On the ground, efforts to solve the problem began at once. For the next few revolutions, the crew received instructions for trying different switch positions, but the computer finally quit entirely. Now they would have to resort to a rolling Mercury-type reentry, rather than the lifting bank angle the computer was supposed to help them achieve.[54]

In revolution 62, at 97 hours 28 minutes, they fired their maneuvering thrusters in the proper retroattitude for 2 minutes 41 seconds. Afterward they jettisoned the equipment adapter. Bang! bang! bang! bang! went the retrorockets. White watched the brown, dusty Texas plains pass in review and then released the retroadapter. *Gemini IV* was returning to Earth.[55]

At 120 000 meters, McDivitt started the rolling reentry. As the spacecraft rotated, the crew saw the adapter, trailing them, turn into an orange mushroom as it burned. Without the computer, McDivitt and White suspected, they would land short of the planned Atlantic landing point. The spacecraft was getting some lift, but they were sure it would not be enough. McDivitt and White welcomed the increasing g-rates. White noticed no dimming of vision and no shortness of breath. They talked, watched their instruments, and enjoyed the scenery.[56]

At 27 000 meters, McDivitt slowed the roll rate and stopped it completely at 12 000 meters. Shortly, he punched out the drogue parachute. When it deployed, the spacecraft gyrated instead of stabilizing. At 3230 meters, the main parachute deployed and unfurled with a comforting shock, and then they braced themselves for the 1500-meter, two-point suspension mark. When the spacecraft assumed its new position, the crew lurched forward, then backward, but neither knocked their helmets against anything. The splashdown—at 97 hours 56 minutes 12 seconds after launch—was rough, slamming them against the water. But they were down and safe, so far.[57]

*Gemini IV* missed its mark by 80 kilometers; but several of the recovery ships had begun moving toward its landing site, and one helicopter crew watched the spacecraft descend to the ocean. Within a few minutes, swimmers jumped into the water and attached a flotation collar. Then the pilots were hoisted into the helicopter. Fifty-seven minutes after touchdown, the crew stepped onto a triumphal red carpet on the deck of the aircraft carrier *Wasp* to be greeted by the ship's crew.[58]

During the helicopter ride, an MSC physician reported that the crew seemed to be in good shape. Nevertheless, everyone wondered about their physical condition after being weightless so long. A NASA information specialist, who had seen Cooper stagger after his Mercury flight, was surprised to see White do a jig-step. A colleague commented, "The air of tension [immediately] dissipated." Berry and his medical team met the crew aboard the *Wasp*. Medical examinations over the next 66 hours revealed no major problems. In fact, on the day after the landing, on his way to the ship's medical ward, White noticed some Marines and midshipmen having a tug-of-war. He joined the midshipmen for 15 minutes. Although his team lost, White certainly appeared strong and healthy. Later McDivitt and White inscribed a picture of themselves walking across the red carpet, "The day the straw men fell down." Berry agreed, as he found his patients fatigued but showing no sign of faintness. Although the loss of bone mass in the heel and little finger was not surprising, physicians were startled to find a loss in the volume of plasma—circulating blood. Both lost weight, as have all American astronauts—McDivitt, two kilograms (four and a half pounds); White four kilograms (eight and a half pounds). But they paved the way for an even longer mission.[59]

*Gemini IV* roused great excitement, with all its daily activities heralded in newspapers around the world. Its deeds shunted aside dark clouds that loomed on 7 June (the day that McDivitt and White returned from space), when the U.S. Military Command in South Vietnam announced that its troops would fight alongside Vietnamese forces. President Johnson came to Houston to congratulate them; a million Chicagoans showered them with ticker tape; and Administrator Webb sent them, at the request of the President, to the Paris International Air Show, where they met Cosmonaut Yuri Gagarin, the first space traveler.[60]

## PRELUDE TO GEMINI V

Although *Gemini IV* chalked up a success, rendezvous remained a question mark. Seamans asked Langley Research Center to study orbital mechanics, especially the complex decisions on attitude and velocity changes and probable fuel usage both with and without computers. Langley engineers reviewed the *Gemini IV* mission results and concluded that the fuel allotted seemed ample for stationkeeping but that the crew had simply not been adequately trained for the job. As Paul Purser later remarked, "no one was 'adequately trained' in that the differences between motions on earth and motions in orbit were not intuitively realized or 'second-nature' to anyone."[61]

Another postflight concern was the computer failure. IBM, the subcontractor, was unable to duplicate the failure on a test computer, and the *Gemini IV* computer itself worked perfectly through 500 tests in St. Louis. Since the trouble remained a mystery, IBM modified the Gemini V computer with a manual switch that allowed areas that might have caused the problem to be bypassed.[62]

A number of other questions also had to be answered for Gemini V. Should a fail-safe reentry be flown? Should there be an EVA? What type of suits should the crew wear? Could the crew be trained soon enough to shorten the launch intervals from three to two months? Could the scientists get their experiments ready in time for them to be integrated into crew training?

Fail-safe orbits had been planned for all manned Gemini flights. Missions not slated for rendezvous would use spacecraft thrusters to bring the vehicle into the atmosphere. Other flights would depend on the Agena to push the spacecraft into the atmospheric fringes. NASA Headquarters had imposed this precaution on *Gemini III,* whose crew later had little to say about it. But *Gemini IV's* McDivitt and White lambasted it; saving fuel for the fail-safe reentry had forced them to limit both operations and experiments. With Gemini V slated for 8 days and 17 experiments, Houston wanted to scrub the maneuver. Since the retrorockets had fired as advertised, even after soaking four days in space, Mueller agreed.[63]

White's successful EVA was going to be a hard act to follow. There was little to be gained from merely repeating it, but the environmental system was not ready for anything more advanced. And there were other reasons for skipping EVA in the next several missions. McDivitt and White had trouble stowing things away before reentry; the 8- and 14-day missions coming up would produce even more garbage. As for Gemini VI, that crew wanted to stress only rendezvous and docking. Then, too, the Gemini V pilots had been campaigning vigorously for more comfort in orbit—wearing their helmets, goggles, and oxygen masks but not their suits. They lost that battle and later wore the G4C extravehicular suits that had been bought for them before the decision to fly EVA on *Gemini IV*. With no reason for repeating the standup EVA, Mueller and William Schneider decided there would be no EVA on the next three missions.[64]

Shortening the intervals between missions was part of the problem in getting the crew ready to fly. In September 1964, when plans for speeding up the flight schedule were first being studied, flight operations and crew training had emerged as the most likely stumbling blocks. When the study was completed and accepted in January 1965, Gemini V still did not have a crew and training time was getting short. Cooper and Conrad were finally named on 8 February, with Neil A. Armstrong and Elliot M. See, Jr., as backups.[65] Now there were 12 men (crews for missions 3, 4, and 5) lining up for the trainers and simulators. By the end of June, the Gemini V training program was in trouble. That was eased somewhat when the Houston simulator, which had been used chiefly to familiarize new crews with Gemini systems in general, was refitted more specifically for Gemini V.[66]

One of Gemini V's chief objectives, the practice rendezvous with the evaluation pod, became more urgent after the doubts raised in *Gemini IV*. Cooper and Conrad devoted a large part of their training time to preparing for this exercise, which now seemed a crucial prelude to Gemini VI. It was being planned to simulate, as closely as possible, the terminal phase of a rendezvous with an Agena.

Another requirement for the first rendezvous flight that Cooper and Conrad rehearsed was a simultaneous launch countdown, which involved their Titan II and spacecraft on pad 19 and an Atlas-Agena on pad 14, to give the launch crew and flight controllers some experience in launching two vehicles at precise times. On 22 July, the Gemini V crew went through the motions of a double launch, including five holds—for propellant tanking, a faulty command panel switch, spacecraft problems, erratic range sequencer performance, and spurious pulses received at Lockheed's ground stations. The demonstration lasted 867 minutes instead of the scheduled 505 minutes, but it did give the needed practice. When the test ended, the lowered erector could not be raised. The crew had to be rescued with the "cherry picker," a cabin on the tip of a crane that had been used in Mercury and that Cooper had insisted be included in the Gemini program. Riding it down gave him a sense of vindication.[67]

Although Cooper and Conrad were putting in some very long

days, the scheduled launch of 9 August was simply too soon. Astronaut Chief Donald K. Slayton flew to Washington to try to argue Mueller into delaying the date. On 21 July, Mueller reluctantly agreed to postpone the launch until 19 August.[68]

The usual reviews started on 29 July with the spacecraft readiness review, followed by launch vehicle readiness, 16 August; mission, 17 August; and flight safety, 18 August. On 19 August, Everett E. Christensen of NASA Headquarters assumed the role of mission director.[69]

Although thunderstorms threatened that morning, the operations crew decided to push on and launch, if possible. But the predicted storm welled over the pad area, and—shades of *Gemini 2!*—a lightning strike near the power facilities caused the spacecraft computer to waver. Finally, the erector was raised and the crew was helped out of the craft. Propellants were drained, pyrotechnics removed or defused, and a 48-hour recycle begun.[70]

## THE COVERED WAGON

Although NASA Headquarters refused to allow nicknames for Gemini spacecraft, Cooper was not so easily put off. Conrad's father-in-law had whittled a model covered wagon, which inspired Cooper with the idea of a patch using that motif and the motto: "Eight days or bust." A personal appeal to NASA Administrator Webb led, after much discussion, to approval of the "Cooper patch." But Webb heartily disliked the motto—if the mission did not go the full eight days, for whatever reason, many would say it had "busted"—and turned it down.[71]

On Saturday, 21 August, Guenter F. Wendt, the McDonnell pad leader, hustled Cooper and Conrad into their couches. Precisely at 9:00 a.m., they felt the modified Titan II start them on a far longer journey than any made by a bygone, continent-crossing covered wagon. The start was smooth enough but then came the bumps of Pogo.* A few seconds before staging, the bouncing stopped. *Gemini V* cut loose from the booster's second stage at 163 kilometers altitude, with an orbital apogee of 349 kilometers.[72]

Because of the mission's length, the supply of oxygen and hydrogen for the fuel cell was a concern. Cooper intended to operate the cells at the lowest possible pressure. But Conrad suddenly noticed that the pressure had dipped too low. Flight Control told him to switch on the oxygen heater to raise the pressure. To his surprise, the needle

---

*Pogo oscillations reached +0.38g during stage I flight, exceeding the permitted +0.25g for a total of about 13 seconds. Within three days after the launch, analysis of flight data showed that the oxidizer standpipes had been charged with only 10 percent of the required volume of nitrogen. The fault was quickly traced to prelaunch procedures, which were corrected. This was the only Pogo anomaly to mar a Gemini mission.

continued to drop. At 2 hours 13 minutes, Cooper yawed the space-craft 90 degrees and ejected the rendezvous pod.[73]

Cooper turned the spacecraft to the rear, flipped on the radar, and got an immediate signal. The radar scale showed the pod moving off at a relative speed of two meters per second. Conrad had expected it to drift away and trail behind the spacecraft, but to his astonishment it went out to the side. Finally it started to follow them as they thought it should.

The heater had still not raised the pressure in the cells. *Gemini V* was out of communications range, so Cooper had to make a decision without help from the ground stations, as the pressure had fallen below 138 newtons per square centimeter (200 pounds per square inch). Never having seen a fuel cell working at a pressure that low, he was afraid it might stop entirely, and he reluctantly elected to power down. Without electrical power, rendezvous with the pod was out of the question. *Gemini V's* crew now wondered if, as Administrator Webb had feared, the mission had "busted." Would Mission Director Christensen continue the flight or have them come home?[74]

Flight Director Kraft now had his first major problem at the new Mission Control Center. He knew the spacecraft had enough battery power for reentry even if the fuel cell failed completely, but he needed to know if there would be time enough to reach a good reentry zone, such as the mid-Pacific near Hawaii on the sixth revolution. While Kraft waited for an answer, the fuel cell pressure dropped to 83 newtons (120 pounds). McDonnell set up a test in St. Louis to find out the lowest working pressure for a fuel cell. During the fourth revolution, the oxygen pressure stabilized at 49 newtons (71 pounds). About this time, Kraft was assured that the batteries were good for 13 hours. Mission Control Center learned that the low-pressure tests in St. Louis were going well. With these facts in hand, Kraft decided Cooper and Conrad could fly for at least one day.[75]

Kranz and his crew then came on duty. While he and his problem solvers wrestled with the heater, Edwin E. Aldrin worked with a Mission Planning and Analysis Division team to design maneuvers for some sort of practice rendezvous—now that the pod was out of the picture—just in case the electrical supply should be salvaged. Kranz's team thought it would be safe to go ahead and operate the cells. When Hodge arrived, the three Flight Directors agreed to tell Cooper to turn the electricity back on. They were relieved when the pressure remained stable as the stacks were brought back on the line. Hodge's flight planners gave the crew some experiments and systems checks to perform, which required more and more power.[76]

Thinking they might have to land early, the crew had begun to put things away. Now that they were back in business, the cabin was soon full of loose gear again. Then it was time for some rest.[77] It had been a long, cliff-hanging first day for Cooper and Conrad in their "Covered Wagon."

While *Gemini V* drifted, the cabin got cold. The crew turned the airflow on low but continued to shiver. This was different from Mercury flights, where the capsule had tended to overheat. The suit coolant circuit seemed cold, too, so they took the hoses off and stopped the flow inside the suits. As the spacecraft tumbled through space, the sight of the stars spinning around outside the window bothered them until Cooper covered the windows and blocked out the view.[78]

Cooper and Conrad had no better luck sleeping than McDivitt and White. At first they tried sleeping alternately, but the dozer was soon disturbed by the ground calling, "Gemini 5, Gemini 5, Gemini 5." As long as one of them was awake, there would be radio transmissions, and they decided this sleep schedule would not work. So they tried, not altogether successfully, to sleep, eat, and work together.[79]

Cooper and Conrad considered the third day the high point of the flight. They worked steadily on experiments and did a series of maneuvers for a "phantom rendezvous." Setting up their calculations on the assumption that they were tracking an Agena in a different orbit than the spacecraft, the flight controllers would pass information to the crew, just as though the target vehicle really existed. Using both ground and spacecraft computations, Cooper would then maneuver *Gemini V* to a rendezvous with this moving point in space, giving him a chance to check out the complete maneuvering system. Such precise moves were new to manned space flight, but Cooper came through like a champion, bringing his spacecraft to the exact position Kraft had asked for. Doubts about being able to accomplish rendezvous faded, and the mission planners were confident and ready for Gemini VI.[80]

The crew powered the electrical systems down again and resigned themselves to drifting in space, performing experiments when possible. Since the inertial guidance platform was not working, they had little success, although they did some experiments, performed radar tests, and made vision tests. They saw smoke at Laredo, Texas, for example, but did not see a checkerboard pattern that had been laid out for them on a field. In the evening, Cooper asked for some uninterrupted sleep and got it.

Cooper slept seven hours and Conrad five, so their work day began at a more normal time. It was to be the last busy shift. First, they saw a rocket sled test as they flew over Holloman Air Force Base, New Mexico. Over Vandenberg, on the next pass, they sighted the contrail of a chase plane just before they glimpsed the ignition of a Minuteman missile. In the Atlantic, they observed their prime recovery carrier, *Lake Champlain*, with a destroyer astern. But, down below in Mission Control Center, a new problem was causing fresh worries.

Since there was no way to dump the fuel cell's product water overboard, its storage tank had been partitioned by a bladder wall; one side held drinking water, the other stored the acidic liquid. As the crew drank, more room for the fuel-cell discharge was provided. But the cells were producing 20 percent more fluid than had been foreseen. When an analysis by Kranz' team disclosed that, even at the high rate of production, there would be some room left at the end of the mission, everyone sighed in relief. Then still another problem arose to plague the mission controllers.

Late in the fifth day, the orbital attitude and maneuvering system (OAMS) grew sluggish, and one thruster quit. Kraft canceled all experiments that required fuel, and the crew turned off the electrical system to help reduce the water buildup. Although several possible solutions to the thruster problem were worked out, none was successful. So Cooper and Conrad again drifted through their rest and sleep period, awakening only to find that the whole OAMS had become erratic. Two thrusters had now stopped. The spacecraft drifted for the rest of the mission, with Cooper only turning on the system occasionally to stop excessive tumbling. When things had been working right, the crew had been busy. Now Conrad mentally kicked himself for not bringing a book.[81]

Despite all the problems, the crew did a creditable job on the experiments. Only one of the 17 had to be scrubbed—D-2, Nearby Object Photography—since it depended on rendezvous with the pod. Two complementary Department of Defense experiments were successful. Experiment D-1, Basic Object Photography, proved that the crew could acquire, track, and photograph celestial bodies. Weather conditions somewhat hampered D-6, Surface Photography, but Cooper and Conrad did obtain photographs of Merritt Island, Florida; Tampico, Mexico; Rocas Island, Brazil; and Love Field, Dallas, Texas.

Defense experiments D-4/D-7, Celestial Radiometry and Space Object Photography, were combined to make irradiance measurements on celestial and terrestrial backgrounds and on rocket plumes. The final defense experiment—S-8/D-13, Visual Acuity/Astronaut Visibility—combined use of an inflight vision tester and the observation of rectangular marks in fields near Laredo, Texas, and Carnarvon, Australia. Weather and operational problems made ground observations difficult—they never were able to see the Carnarvon field, but the Laredo pattern was partially read in the 48th revolution. The tester showed that the crew's vision did not change during the eight-day flight.[82]

*Gemini V* carried the same medical experiments as *Gemini IV*, plus M-1, Cardiovascular Conditioning, and M-9, Human Otolith Function, to see if the ability to perceive the horizontal deteriorated during flight. Postflight responses were not significantly different from those reported before the mission. Conrad wore inflatable leg cuffs for

M-1. When activated, the cuffs pressurized automatically for two minutes out of six. They could be run continuously throughout the flight or be turned off. Conrad had some problems with the equipment but he felt the cuffs might be useful for extremely long missions. His pulse rate returned to normal faster than Cooper's after the flight, and he lost four percent less plasma volume. But this could not be conclusively traced to the use of the cuffs, since individual responses differ. Principal investigator Pauline Beery Mack found that both had lost more calcium than the *Gemini IV* crew, but she was unwilling to predict a trend since "a form of physiological adaptation may occur in longer space flight."[83]

Cooper obtained the first photographs of the light of the moonless sky (zodiacal light and the gegenschein), experiment S-1. He made a series of stepped exposures and took two pictures of the gegenschein, a faint nebulous light opposite the Sun. Like their predecessors, Cooper and Conrad took synoptic terrain and weather photographs. Pictures of the Zagros Mountains showed more detail than the official Geologic Map of Iran. The crew also provided pictorial cloud studies, including tropical storm Doreen. S-7, Cloud-Top Spectrometer, the other science experiment, proved the feasibility of making cloud altitude measurements from spacecraft.[84]

During the mission, Hurricane Betsy moved relentlessly toward the planned landing area. The landing area sea-state contraints for Gemini were considerably relaxed from those of Mercury. For Mercury, the limits were winds no more than 34 kilometers per hour (18 knots), waves no more than one and a half meters (five feet); for Gemini, winds up to 47 kilometers (25 knots) and waves up to two and a half meters (eight feet) were acceptable. Weather for Mercury in all of the recovery areas—primary, secondary, or contingency—had to be good. No such restraints were ever placed against Gemini—but it certainly could not be expected to touch down in a hurricane area. The Weather Bureau recommended that *Gemini V* be brought down early to avoid landing too near the storm. Kranz agreed in plenty of time for the *Lake Champlain* to reach the new recovery zone.[85]

Because of the erratic, and sometimes inoperable, OAMS, Kraft allowed the crew to use one of the two rings of the reentry control system to position the spacecraft properly more than one revolution before coming back to Earth. During the 120th pass, Cooper told McDivitt (CapCom in Houston for reentry) that *Gemini V* was ready for retrofire.[86]

In the darkness near Hawaii, on the morning of 29 August, at 190 hours 27 minutes 43 seconds, the first retrorocket went off, followed by the second and third. After what seemed like an eternity, the fourth fired. Cooper peeked out the window and felt as if he were sitting "in the middle of a fire." With the control system thrusters spew-

## *Gemini V*
## 21 August 1965

Above left, rendezvous evaluation pod nestled in the adapter; 5-meter segment of Gemini V booster being recovered from the Atlantic—first large section to be recovered.

Above, Gilruth, Seamans, and Webb are interviewed on ailing fuel cell; right, Gemini V photograph of the lakes and mountains of Tibet.

Left, control room congratulations after landing (Hodge, Mueller, and Kraft); below right, astronauts Conrad and Cooper on world tour, receive keys to Addis Ababa from Lord Mayor Zewde Gebre Hiwot; below, cosmonauts Belyayev and Leonov meet astronaut Conrad in Athens conference of the International Astronautical Federation.

ing flame in front and the retrorockets firing behind, a nighttime reentry had to rely strictly on instruments, Cooper discovered. There was absolutely no way of seeing the horizon or a landmark. He and Conrad stayed on instruments until they had passed over the Mississippi in the morning light.[87]

Cooper held the spacecraft at full lift until it reached the 120 000-meter altitude and then rolled it to a planned bank angle of 53 degrees. The reentry gauge soon indicated that they were high—there might be an overshoot of the landing point. Cooper, responding to the instrument, slewed to 90 degrees left instead of 53 to create more drag and reduce the landing error. The g-loads quickly shot from 2½ to 7½.[88]

At 20 000 meters, Cooper punched the drogue parachute button. *Gemini V*, unlike *Gemini IV*, did not oscillate—it was completely stable on the drogue. Cooper then cut in the second control ring thrusters to discard the fuel as the spacecraft came straight down. He and Conrad watched the main parachute as it unfurled and felt the expected jolt at two-point suspension. In contrast to the McDivitt-White landing, impact was very, very soft.

*Gemini V* landed 190 hours 55 minutes 14 seconds after launch, 130 kilometers short of the planned landing point. The computer had worked as it should in this case—the error had been human. Earth's rotation rate is 360.98 degrees per day. But, in programming the computer, someone had left off the two decimal-place numbers and fed the machine just the 360 degrees. Cooper's efforts to compensate for what he recognized as an erroneous reading had brought them down closer to the ship than they would otherwise have been.

The short landing caused no problems for the U.S. Navy recovery forces. A helicopter soon arrived over the spacecraft and three swimmers dropped into the water. Cooper and Conrad were very comfortable. With a calm sea, Cooper wanted to stay with the spacecraft on this pleasant summer morning (about 8:30, Cape time) until he learned that the carrier was still 120 kilometers away. Then he and Conrad rode the helicopter to the *Lake Champlain*.[89]

The admiral welcomed them aboard ship. Asked what they had been thinking about when it looked as though the fuel cell heater problem might cause the mission to end early, Conrad pointed out a picture he had drawn between the spacecraft seats of a covered wagon halfway over a cliff.[90]

Although the crew's worries were over, Berry's were not. His postflight concern was the trend in plasma volume and calcium losses, which were increasing on these longer missions. He was aware that the crew had been forced to drift through space the last three days, with little to do; but they should have exercised more. Two days later, to Berry's relief, both were physiologically almost back to normal.[91]

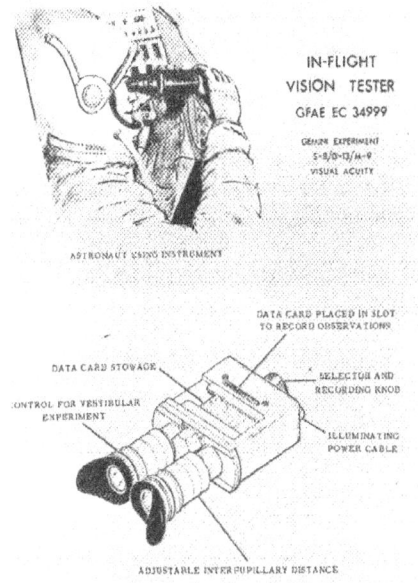

IN-FLIGHT
VISION TESTER
GFAE EC 34999

GEMINI EXPERIMENT
S-8/D-13/M-9
VISUAL ACUITY

ASTRONAUT USING INSTRUMENT

DATA CARD PLACED IN SLOT
TO RECORD OBSERVATIONS

DATA CARD STOVAGE

CONTROL FOR VESTIBULAR
EXPERIMENT

SELECTOR AND
RECORDING KNOB

ILLUMINATING
POWER CABLE

ADJUSTABLE INTER PUPILLARY DISTANCE

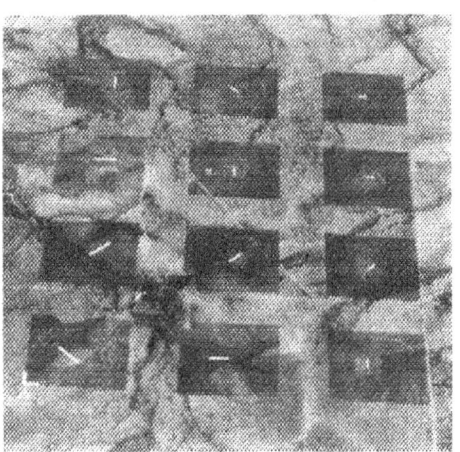

*Two visual acuity experiments on Gemini V: left, the vision tester; below, ground-marker test near Laredo, Texas.*

A safe landing and healthy crew after an eight-day space voyage increased NASA's confidence in achieving its lunar-landing goal during the sixties. In a span of only three months in 1965 and after just two long-duration flights, medical fears of weightlessness began to subside. Hugh Dryden reflected this optimism in his report for the President:

> The primary objective of the Gemini V mission to demonstrate man's ability to function in the space environment for 8 days and to qualify the spacecraft systems under these conditions was met. This milestone duplicated the period required for the manned lunar exploration mission.

> Gemini V also demonstrated the capability of man to withstand prolonged periods of weightlessness. The adaptability of the human body was indicated by the performance of the astronauts. For example, their heartbeat rates gradually dropped to a level significantly lower than their preflight normal rates, but by the fourth day, adapted to the weightless condition and leveled off. Upon return to Earth, the heartbeat rates were slightly higher than normal, as expected, but returned to normal rates during the second day. This has assured us of man's capability to travel to the Moon and return.[92]

Postflight activities for Cooper and Conrad included a six-nation goodwill tour assigned to them by President Johnson. During the trip, they attended the International Astronautical Federation Congress in

Athens, where they talked with the crew of *Voskhod II*, Russian Cosmonauts Aleksey Leonov and Pavel Belyayev.[93]

NASA now turned to plans for the rendezvous and docking mission and for the final long-duration flight, both scheduled before the end of the year. The goal of five manned flights in a single year seemed phenomenal, compared with the experience of Project Mercury. But *Gemini IV* and *Gemini V* had indeed proved to be pillars of confidence, a solid base from which to build.

# XII

# "Spirit of '76"

IN November 1964, halfway through the 11 months of training for the first manned Gemini flight, Donald Slayton confidentially told Walter Schirra that the *Gemini III* backup crew would pilot the first rendezvous mission, Gemini VI. The following February, Schirra let newsmen know that he and Thomas Stafford would be the first Americans to rendezvous and dock in space. Two months later, Public Affairs Officer Paul P. Haney jested, "The purpose of this [news conference] is to reveal one of the best-kept secrets in NASA history—the identification of the prime crew on GT-6." Gus Grissom and John Young received the backup assignments, reversing the crew roles for *Gemini III*.[1]

For the seventh Gemini mission, NASA had scheduled its longest flight of the program—14 days. Crew selection was beginning to follow a leap-frog pattern; that is, the backup crew for one mission became the prime crew for a later flight.* On 1 July 1965, NASA picked the *Gemini IV* backup crew, Frank Borman and James Lovell to fly Gemini VII, with Edward White and Michael Collins as alternates. Collins was the first member of the third astronaut class (selected in October 1963) to be named to a flight.†[2]

---

*The pattern was sometimes broken. Neither Grissom nor White flew a second Gemini mission; David R. Scott joined Neil Armstrong on *Gemini VIII*; and Elliot See (Armstrong's fellow backup pilot on *Gemini V*) and Charles A. Bassett II received the prime assignments for *Gemini IX*.

†Others in the third group were Edwin E. Aldrin, Jr., William A. Anders, Charles Bassett, Alan L. Bean, Eugene A. Cernan, Roger B. Chaffee, R. Walter Cunningham, Donn F. Eisele, Theodore C. Freeman, Richard F. Gordon, Jr., Russell L. Schweickart, David Scott, and Clifton C. Williams. This brought the strength of the corps to 30.

NASA officials had been wrestling with the order of flights—which mission was to carry out what major program objectives—for a long time. Slow progress on some systems had forced a shuffling of tasks. A prime example was the pitfall-strewn route of the Agena target vehicle to the launch pad, which affected schedules for both rendezvous and long duration. When Charles Mathews took over the Gemini program, the target vehicle was in real trouble. Thus, flight schedules were changed to fly an Agena mission before the Gemini endurance test. Then, if anything happened, there would be time to work on the vehicle before the next rendezvous flight.[3]

Although Gemini Agena Target Vehicle (GATV) 5001 had been shipped to Cape Kennedy in May 1965, it was a test vehicle and unqualified for flight. In August, NASA officially assigned GATV 5002 to the first rendezvous mission. It was of better production quality than 5001, but NASA officials still doubted that its main engine could be trusted for docked maneuvers with a manned spacecraft. The Gemini Program Office firmly opposed firing the big engine. This was an old argument. Schirra, in particular, chafed at the limitation and tried hard to get it lifted. When that failed, he was willing to settle for a chance to try out the smaller secondary engines. For a while, Schirra thought he had won his point, but no reference to out-of-plane docked maneuvers appeared in the final flight plan.[4]

Rendezvous techniques remained largely in the realm of theory. When training for Gemini VI began in the spring of 1965, little had yet been done toward planning crew procedures for making the final maneuvers. Dean F. Grimm of MSC's Flight Crew Support Division joined forces with Astronaut Edwin Aldrin, who had studied the pilot's role in rendezvous for his doctoral dissertation at the Massachusetts Institute of Technology.

In 1963 and 1964, Aldrin worked hard at selling the project office and flight operations on a concentric rendezvous. The target would be launched in a circular orbit 298 kilometers high, the spacecraft in a lower elliptical orbit. Since the spacecraft was closer to Earth, it took less time to circle the globe and could catch up for rendezvous. Aldrin and Grimm worked out the trajectories and maneuvers that would allow the spacecraft to intercept the target.[5]

A two-week review in April 1965 convinced Grimm and Aldrin that MSC's plans for an active human role in rendezvous were in poor shape. Most work seemed to stress a closed-loop concept that relied more on machines than on men. Radar and computer would make rendezvous nearly automatic. Of course, if either failed, so did the mission. Aldrin and Grimm believed the pilots should have options if the equipment malfunctioned. Grimm went to St. Louis and persuaded McDonnell to rig a device that could simulate trajectories, orbital insertion, and spacecraft-target rendezvous.* A computer allowed flight profiles to be set up that varied the series of maneuvers leading to target interception. Crewmen learned what to do if any piece of equipment failed, and they profited from merely going through the motions as they tried to decide which procedures were useful and valid. Schirra and Stafford rejected, for example, an early concept for doing rendezvous with the spacecraft inverted—head toward Earth—using the inertial guidance system to judge spacecraft attitude. They both disliked this method because they lost their sense of direction. Overall, the prime crew participated in 50 complete rendezvous simulations. As Schirra and Stafford trained on the simulator, they took notes and discussed with Aldrin and the others the best procedures to use. These were then incorporated into charts that would be carried in flight.[6]

Aside from concerted efforts to qualify the Agena and to pull together rendezvous plans, Gemini VI preparations were fairly routine. Measures taken to shorten the launch intervals were surprisingly successful, and the 25 October launch date was not hard to meet. In April 1965, GLV-6 became the first Gemini launch vehicle to be erected in the new west cell of the Vertical Test Facility at Martin-Baltimore; tests on GLV-5 were still in progress on the old stand (now called the east cell). GLV-6 reached the Cape early in August and went into storage until *Gemini V* was launched. Spacecraft 6 arrived in Florida about the same time, but it did not go into storage. Instead, it was hoisted atop a timber tower for electronic compatibility tests with GATV 5002, because the target vehicle's command and communications system had just undergone major modifications. Originally intended as a one-time exercise for the first Agena, these tests became a major part of prelaunch checkout for all Gemini-Agena missions. When they were finished, the test operations group was confident that the Agena would respond reliably to all spacecraft and ground control commands.[7]

*Gemini VI* was the last of the program's battery-powered spacecraft, which limited the flight to two days at most. Schirra, in fact, thought the power would be pretty thin for even this amount of time. When the mission directive neared its final version by the end of September, it provided that the "mission may be cut to one day if all objectives are completed." The crew, in other words, could come home as soon as they completed rendezvous and docking with the Agena; everything else was secondary, even experiments. There were seven of these: two depended upon rendezvous with the Agena, one was medi-

---

*Grimm and Aldrin had help in setting up rendezvous procedures: at MSC, Branch Chiefs Paul C. Kramer (Crew Safety and Procedures) and Edgar C. Lineberry (Rendezvous Analysis); at McDonnell, Charles A. Jacobson, Marvin R. Czarnik, William Murphy, Walter Haufler, and William E. Hayes. Gordon Cooper and Charles Conrad, the *Gemini V* crew, acted as engineering test pilots until the Gemini VI crews could take over.

cal, three were photographic experiments as carried on all flights and used crew time only when it did not interfere with the major task—rendezvous—and one was passive. "On my mission, we couldn't afford to play with experiments," Schirra later said, "rendezvous [was] significant enough."[8]

## "NO JOY, NO JOY"

On launch day—25 October—at pad 14, a team from General Dynamics Corporation conducted the countdown of the Atlas launch vehicle capped by the slender Agena. Although this would be its maiden voyage in Project Gemini, Agena was a veteran, in one model or another, of more than 140 flights since 1959. The countdown, presided over by NASA Mission Director William Schneider,* proceeded simultaneously for the Atlas, Agena, modified Titan II, and spacecraft.[9] Noticeably absent were the delays that had plagued Mercury launches. Fifteen minutes before the Atlas-Agena was to leave its moorings, Schirra and Stafford climbed into the spacecraft and settled into their couches.[10]

At 10 o'clock, General Dynamics launch chief Thomas J. O'Malley pushed the button that sent the Atlas-Agena skyward. Signs that something was wrong appeared minutes later when the target cut loose from the booster. The Agena seemed to be wobbling, even as its attitude control system labored to keep it stable. The small secondary engines ignited and the gas generator valve opened to fire the main engine and boost the Agena to orbit. A telemetry signal in the Mission Control Center showed that the big engine had started exactly on time.[11] But that was the last good news. In Houston, Schneider, who thought Agenas always flew, was astounded to learn there was a problem. In fact, Air Force radar was tracking what seemed to be five pieces of the target vehicle.[12]

In the meantime, Public Affairs Officer Paul Haney, trying to keep the public informed, had little or nothing to report. Ten minutes after liftoff, he could only repeat that no telemetry signals were coming into the stations along the flight control network and that, over on pad 19, Schirra and Stafford were continuing their preparations for flight. After 50 minutes, the last flicker of hope gone, Haney told his listeners, "We have had a conversation with [the Carnarvon tracking station] . . . and their report keeps coming back—No joy—No joy." The mission was scrubbed.[13]

Actually, only six minutes after launch, a deadening sense of fail-

---

*Schneider, newly named Deputy Director for Mission Operations and Gemini Mission Director in NASA Headquarters, retained that position throughout the remainder of the program. LeRoy Day replaced Schneider as Deputy Director of the Gemini program in Washington.

ure was spreading among those closely connected with the target vehicle's development. Jerome B. Hammack, who kept tabs on the Agena
for GPO, was in the pad 14 blockhouse, listening to the flight controllers' comments. He was soon convinced that there was deep trouble.
The Air Force officer in charge of Atlas-Agena launches, Colonel L. E.
Allen, thought the Agena had probably exploded. The two men headed for the Lockheed hangar, where others also gathered for the wake.
Hasty study of partial telemetry data threw little light on the cause of
the disaster, but newsmen were clamoring for a press conference.
NASA and Air Force officials told reporters that they did not know
exactly what had caused the failure, but that ten days might be enough
time to decide what to do to keep it from happening again.[14]

## THE VII/VI-A DECISION

The gloom that descended upon Gemini was quickly pierced by a
ray of hope. While the futile countdown for the spacecraft launch was
still under way, Frank Borman rushed from the outside viewing stand
to the Cape Kennedy Launch Control Center to find out what had
happened. He found himself standing with Gemini VII crewmate
James Lovell near two McDonnell officials, spacecraft chief Walter
Burke and his deputy, John Yardley. The astronauts heard Burke ask
Yardley, "Why couldn't we launch a Gemini as a target instead of an
Agena?" Yardley recalled that the Martin Company had proposed a
rapid-fire launch demonstration some months before. He asked Raymond Hill, now in charge of McDonnell work at the Cape, what he
remembered about the study. Hill briefly outlined the plan, and all
three began discussing how it could be adapted to carry out Burke's
idea.

Borman listened with growing excitement as the McDonnell idea
jelled. What he heard made sense, with one exception. When Burke
began to sketch on the back of an envelope how an inflatable cone
could be attached to Spacecraft 7 to permit docking, Borman drew the
line; he disliked the thought of *anything* nuzzling into the equipment
housed in his spacecraft's adapter section. Burke and Yardley found
NASA manned space flight chief George Mueller and Gemini Manager
Charles Mathews and tried out their scheme on them. Neither NASA
official gave it much of a chance. The two McDonnell engineers left
the building to see if they could sell their concept elsewhere.[15]

Burke's brainstorm was built on more than just a vaguely recalled
Martin proposal. Shortening the launch intervals to two months had
proven that hardware could be put into the pipeline faster than in the
past. But if Gemini VII were to be the target for Gemini VI, the two
vehicles would have to be launched less than two weeks apart. Mueller
and Mathews simply refused to believe that it could be done. Ironical-

Left, the vehicles for the Gemini VI rendezvous mission: Titan with Gemini Spacecraft 6 on top on pad 19, and the Atlas capped by Agena target vehicle 5002 on pad 14. Below left, the rendezvous simulator at McDonnell, which reproduced the last 30 meters; center, rendezvous simulator at Langley which reproduced the last 65 meters; right, docking test at Cape Kennedy to determine whether the particular Agena and the particular Gemini spacecraft were compatible for docking.

A grim Mission Control waits for final confirmation of the Agena launch failure, which would force a scrub of the Gemini VI mission: foreground, left to right: Donald Slayton, John Clark, GSFC; standing, left to right, Elliott See, Glynn Lunney, Charles Bassett, Howard Tindall, Robert Gilruth (arm on rear console), George Low, Edwin Aldrin; seated, center, John Hodge, Christopher Kraft, and James Elms.

ly, they were the prime movers in urging shorter schedules; but Burke's idea far exceeded their expectations.[16]

In September 1964, Mueller had asked Schneider if he thought activating a second launch complex would help to shorten the time between launches. Schneider's first reaction was no. But, in February 1965, he had his office study the value of launching two Gemini spacecraft either simultaneously or in quick succession. Eldon W. Hall, Schneider's Systems Engineering Director, reported that having two crews in orbit at the same time and trading pilots in mid-space would have public appeal. Other advantages might be using an unmanned Gemini for a space rescue or completing a rendezvous mission if a spacecraft failed to launch. But none of these things was worth the cost of a second pad and spacecraft modifications. In summary, Hall said, "It might be nice, but there is no overwhelming necessity."[17]

Mueller seized every chance to push for shorter launch schedules and new objectives to wring added experience from the Gemini program, especially for Apollo. In Houston, Mathews kept his staff on the lookout for new ideas for the missions. He had helped Hall with the report and agreed that the expense would be too great. Mathews did, however, arrange to procure spare parts for pad 19 so it could be swiftly restored after a launch.[18]

Because of the daily contact between NASA, the Air Force, and contractors, ideas for speeding up the program flowed freely at the Cape. One of these—a rapid turnaround of the launch vehicle—was the result of collaboration between Joseph Verlander, Martin chief at Kennedy, and Colonel John Albert, Chief, Gemini Launch Vehicle Division, 6555th Aerospace Test Wing. They proposed getting a fully checked Gemini Titan ready for launch and then parking it somewhere while a second launch vehicle was prepared for flight. One problem was how to move the first booster, since the engine contractor, Aerojet-General, insisted that the vehicle had to remain upright once it had been erected and checked out. The answer to that was a Sikorsky S-64 Skycrane, a helicopter powerful enough to lift and carry the upright Titan II. It was really quite a simple plan, though carrying it out might involve a lot of complexities. After a booster and spacecraft had been checked out in the usual manner, the spacecraft would be transferred to bonded storage and the launch vehicle would be hauled by helicopter to nearby pad 20, which was not in use at the time. Then a second booster and payload would be readied on pad 19 and launched. The stored and parked vehicles would be immediately returned to the pad and launched in five to seven days.

No one seemed interested in the Gemini "rapid fire," or "salvo," proposal except its creators. When Verlander told O.E. Tibbs about it, the Martin vice president frowned on the idea of using the Skycrane helicopter. Albert outlined the plan to SSD Commander Ben Funk and

SSD Director of Gemini Launch Vehicles Richard Dineen but roused only mild interest. Burke and Yardley listened politely but did not seem impressed. Mathews told Verlander and Albert frankly that there was no place in the Gemini program for such an unorthodox suggestion. In August 1965, Albert took the scheme to *Gemini V* Mission Director Everett E. Christensen, but he received no encouragement there, either. This lack of enthusiasm was daunting, and the Martin plan seemed destined for limbo.[19]

Two months later, however, in the aftermath of an exploded Agena, the idea looked better, at least to Burke and Yardley. But they got no warmer reception than Verlander and Albert. Failing to sway Mueller and Mathews, they left the Launch Control Center for the Manned Spacecraft Operations Building, where an impromptu meeting on what to do next was in progress. Here they again urged their scheme, but, as Merritt Preston, the Kennedy launch operations manager, later said, "Poor Yardley and Burke were pounding a stone wall . . . they got the coldest shoulder I ever saw."

People at the meeting were more interested in the possibility of switching the 3670-kilogram Spacecraft 7 with the 3553-kilogram Spacecraft 6. Albert and others—among them some of the very men who had fathered the rapid turnaround plan—favored the proposed exchange. Having been rebuffed earlier, they now thought more conservatively. They reasoned that some of the time and work invested in Gemini VI launch preparations might be retrieved by using the booster already on the pad and checked out to launch the long-duration spacecraft. Burke and Yardley, on the other hand, pushed for removal of both the spacecraft and the booster, hoping to buy time for their proposal to be given further consideration. But the NASA, Air Force, and industry launch teams wanted to wait and see if GLV-6 had enough power to lift the heavier Spacecraft 7 into orbit.[20]

Mueller called NASA Administrator James Webb in Washington shortly after the Agena explosion and told him about the idea of exchanging spacecraft. Webb discussed it the next morning with his chief associates—Deputy Administrator Hugh Dryden, Associate Administrator Robert Seamans, Associate Deputy Administrator Willis Shapley, and Mueller, now back from the Cape. If the switch could be made, the earliest launch date would be 3 December. If GLV-6 were not powerful enough to lift Spacecraft 7 into orbit, then the launch would take place on 8 December. Gemini VI, postponed to February or early March, would still fly before Gemini VIII. There was no mention of the Burke-Yardley proposal.[21]

Having made little headway at the Cape and with the spacecraft exchange plan gaining support, Burke and Yardley had headed for Houston to broach their idea to MSC Director Robert Gilruth. On

Tuesday morning, 26 October, while Webb and his colleagues were talking about exchanging the two spacecraft, Gilruth listened to Burke, smiled, and said, "Walter, you know things aren't like that in real life." Burke shot back, "Tell me what's wrong with it." Gilruth could come up with no convincing obstacle. He called George Low in to help him nitpick. The Deputy Director was intrigued by Burke's scheme. His only real doubt was whether the tracking network could handle two manned spacecraft at the same time. But that was a question for Flight Operations Director Christopher Kraft.

In the meantime, Mathews had arrived in Gilruth's office. He was no more able than Gilruth or Low to think of any insurmountable barrier to the plan. Gilruth asked Kraft to join them and show them the operational roadblocks that must be there. Taken aback, Kraft first said, "You're out of your minds. It can't be done." After thinking a few moments, though, he was not so sure. He called Sigurd A. Sjoberg, his deputy, to set up a meeting with his flight operations experts for 1:30 that afternoon. Flight Crew Operations chief Slayton was the next to hear the news, and he, in turn, sounded out the pilots for their reaction. Schirra and Stafford greeted the prospect with enthusiasm.[22]

In Florida, hopes for switching the spacecraft faded when an analysis of GLV-6 showed that it lacked the power to orbit the Borman-Lovell spacecraft. At a meeting in the office of John Williams, Director of Spacecraft Operations, the Cape leaders were now forced to consider the Burke-Yardley suggestion they had scorned before. As they tinkered with a tentative work schedule for a nine-day pad checkout, they began to see glimmers of light. Merritt Preston telephoned Mathews in Houston and LeRoy Day in Washington and told them it might work, after all, as far as the machines were concerned. Day found that Mathews was now in favor of the plan. Hardware was apparently no obstacle, but tracking and control operations were still a question mark.[23]

Kraft came back from lunch with Low and outlined the gist of the proposal to his staff. The men in John Hodge's Flight Control Division found it "a hell of a great challenge and to a man they wanted to press on as soon as possible." One of them suddenly said, "Why don't we handle it as if one of the spacecraft were a Mercury-type and the other a Gemini-type spacecraft?" Mercury controllers at the tracking stations observed data on their consoles, summarized it, and forwarded the result by teletype to Mercury Control Center. Gemini VII could be handled that way while it served as a passive target for Gemini VI. For Gemini missions, the stations were fitted with computer communications processors. As the spacecraft passed overhead, the processors interrogated the appropriate systems for specific data, which were automatically transmitted to Mission Control. Gemini VI, the active partner in the rendezvous, would be controlled by the more sophisticated system. With this as a basis, an operational mode was laid out.

After Gemini VII lifted off, flight control would be carried out in the normal manner while the pad was being prepared for the second launch. Once the flight controllers were sure the orbiting spacecraft was operating properly, Mission Control would concentrate on Schirra and Stafford in their spacecraft, and the tracking network would watch Gemini VII, record data, and send information by teletype to the Houston controllers. This mode would continue until the complicated rendezvous mission ended and Gemini VI-A (so called to distinguish it from the originally planned mission whose objective had been rendezvous with Agena) returned to Earth. Then Gemini VII would become the focus of communications again. Kraft was soon convinced that the operation could be carried out safely. He told his Mission Planning and Analysis Division to set up the flight plan so the second launch could take place as soon as the pad was ready.[24]

At 3 o'clock that afternoon, Kraft told Gilruth that he was ready to talk, and he sounded excited. An hour later, Gilruth, Low, Mathews, Slayton, Burke, and Yardley heard what Kraft had to say. They talked about it for an hour, then Gilruth called Mueller, who liked the dual control idea but wanted to sleep on it. Burke and Yardley left for St. Louis with a promise from Gilruth to let them know within 24 hours what Headquarters decided.[25]

But the news was beginning to leak out. James C. Elms, Mueller's deputy, heard from Washington reporters that there were rumors that NASA was going to fly two manned spacecraft at the same time. He phoned Houston to ask Low what was going on. When Low had told him about the plan, they decided to warn Mueller about the danger of news leaks. Realizing that speed was now vital, Mueller called Seamans at home. It was too late to do anything that evening, and Seamans asked Mueller to come over the first thing in the morning to discuss the subject. Although Seamans was very interested in what he heard on Wednesday morning, he told Mueller to keep it quiet until he could pass it along to Webb and Dryden.[26]

That afternoon, 27 October, Webb, Dryden, Seamans, and Shapley met to discuss the Burke-Yardley proposal. Because Dryden had been concerned about adding extravehicular activity to *Gemini IV* at the last minute, Seamans felt he had to play devil's advocate. Even before Seamans finished, Webb was intrigued. Believing himself to be less conservative than Seamans about novel ideas, however, Webb telephoned Mueller and asked him point-blank if it would work. Mueller asked him to wait while he doublechecked with Gilruth in Houston.

Mueller told Gilruth that Webb liked the idea and thought it important enough for the President to announce. Mueller warned the

MSC Director that there could be no hedging. Once President Johnson made the plan public, the nation would be committed. How, Mueller wanted to know, did Gilruth feel about the proposal after thinking it over for 24 hours? Affirming that it still looked good to him, Gilruth nevertheless asked for half an hour to count the votes. Mueller gave him 15 minutes. Gilruth and Low polled Kraft, Slayton, Mathews, and Preston, stressing what a presidential announcement implied. When the affirmative ballot was unanimous, Gilruth called Mueller, who notified Webb that he had a deal.[27]

Webb then tried to reach presidential aide Jack Valenti but talked with Joseph Laitin, an assistant, instead. Laitin asked the Administrator to send the proposal to the White House so it could be forwarded to the President who was at his ranch near Austin, Texas. Webb drafted a memorandum for the Chief Executive, while Julian Scheer, NASA Assistant Administrator for Public Affairs, composed a press release.[28] The NASA chief informed the President that, barring serious pad damage after the launch of Gemini VII, Gemini VI-A could be flown in time for Schirra and Stafford to rendezvous with Borman and Lovell. Webb told President Johnson, "I believe it will be encouraging to you . . . to learn that we have gained enough strength in . . . the Gemini program to consider . . . such a quick turnaround."[29]

On Thursday, 28 October, a press conference was held at the Texas White House to announce the Gemini VII/VI-A rendezvous mission. That a plan of such scope could be suggested, discussed, approved, and announced in scarcely three days was a sign of the managerial and technical trust that Gemini had already come to inspire. William D. Moyers, the President's Press Secretary, told the news media about the plan and answered questions from reporters. Moyers said the mission was targeted for January; back at MSC, however, everyone from Gilruth on down was working toward an early December flight.[30]

At Cape Kennedy, normal methods now had to be suspended. From the hardware standpoint, success depended upon the performance of the launch preparation teams. Members of the NASA, Air Force, and Aerospace teams met and agreed on the best way to implement the plan. In this emergency situation, Aerojet-General engineers came through with procedures for handling the vehicle in a horizontal position, even though they had said earlier it must not be done. The Air Force's 6555th Aerospace Test Wing took GLV-6 down, one stage at a time, and placed it in bonded storage under plastic cover. On 29 October, the team erected GLV-7 on pad 19.[31] Spacecraft work began when the McDonnell Cape team was rounded up to hear about the new mission. "Oh, man, you are crazy!" was the first reaction of pad leader Guenter Wendt when he saw the "S/C #6 Pad Schedule," which listed tasks for nine hectic days after the Gemini VII launch. But he,

275

like everyone else, tackled the challenge enthusiastically. While these exact schedule details were being pinned down, Spacecraft 6 was secured in a building on Merritt Island.[32]

Crew training presented no serious problems. Schirra and Stafford were honed and ready to go. They stepped aside while Borman and Lovell flew the simulator, taking only occasional sessions to keep sharp. Rendezvous plans remained unchanged. But Gemini VII's flight plan was altered to circularize the orbit, so Spacecraft 7 would travel in the same path that the Agena would have used.

Although Kraft's group had a workable concept for flight control, the operations experts still had a lot of work ahead setting up simultaneous controls for two manned spacecraft. Goddard Space Flight Center, in charge of the tracking network, began altering station layouts to allow voice communications with Gemini VII and VI-A at the same time. Equipment at Goddard was also adjusted to ensure that computer programs for two manned spacecraft could be prepared.[33]

Schirra and Stafford wanted to add extravehicular activity to the flight plans—perhaps Stafford could change places with Lovell in a demonstration of space rescue—but they met a pronounced rebuff. Borman's goal was a 14-day mission. He wanted nothing to do with any proposal that might threaten it. "Wally could have had all the EVA he wanted," Borman later said, "but I wasn't going to open the hatch." There were real hazards in trying to exchange pilots in midspace, since the life support hoses would have to be detached and reconnected in a vacuum, leaving the pilots with only the backup system to depend on as they traveled between the two spacecraft. It might have looked great in the headlines, Borman added, "but one little slip could have lost the farm."

Schirra and Stafford did not give up and turned to Low for help. The Deputy Director learned that Stafford, one of the taller astronauts, sometimes had trouble getting out of and back into the spacecraft in zero-g tests. Even the barest chance that this might happen during the mission made the whole idea seem too risky to Low, but he passed the crew's wishes on to NASA Headquarters. The consensus in the executive offices was that there should be no EVA on Gemini VII-VI-A. Ironically, Spacecraft 6 was the first vehicle to be specifically designed for EVA. Schirra had worked hard to get it out earlier, so he and Stafford could focus on rendezvous. He had done too good a job. As he later remarked, "I wrestled that out of there so well that I couldn't get it back in when we had the delay."[34]

## SUITCASE FOR A FORTNIGHT

Frank Borman and James Lovell had put in long hours getting ready to spend two weeks in space. Working directly with the *Gemini*

*IV* pilots and talking with the crew of *Gemini V*, Borman and Lovell learned much about what to take with them and how to prepare themselves physically and psychologically. They already knew the spacecraft systems, but they needed to figure out how to live in such confined quarters for so long and still perform useful work. As successful as the preceding missions had been, they still wondered if six extra days could be safely added to the flight. Edward White and James McDivitt had been fatigued; Cooper and Conrad tired and bored. Both crews stressed the impossibility of sleeping alternately. Borman and Lovell resolved to sleep and work together.

The astronauts and mission planners had learned another lesson from *Gemini IV* and *V*. Prescribing tasks for assigned times during a flight was useless. So Borman and Lovell would take off with what was, in essence, a flight plan outline. Experiments and other tasks would be carried out only when the flight controllers and crew could fit the job to the opportunity. The only prescheduled tasks fell between launch and stationkeeping, the first four hours of a 330-hour mission.

Another innovation that the crew welcomed was adjusting the sleep-eat-work-relax cycle to their more normal, Earthbound habits. Borman and Lovell had two work periods each day, coinciding with morning and afternoon in the United States Central Standard Time zone. This schedule also fitted the specialized activities of the three flight controller shifts—to execute the flight plan, to analyze systems performance and the supply of consumables, and to keep up with what had been done and plan the next segment of activities.[35]

Stowage of food and gear was a special problem on a two-week flight. Unfinished meals and food wrappers could quickly clutter up the spacecraft, as Cooper and Conrad had learned in the eight-day mission. Extra storage space in the small cabin had to be found before the 14-day trip. GPO Deputy Manager Kenneth Kleinknecht went with Borman and Lovell to St. Louis, where Spacecraft 7 was going through its test phases, to help them hunt for more space. The search for an extra garbage dump was successful: waste paper from their meals could go behind Borman's seat for the first seven days and behind Lovell's for the next seven. After working out procedures, the crew practiced stowing for launch, orbit, and reentry, until they were sure they knew where to put every scrap of paper.[36]

Tailoring flight and stowage plans for a 14-day mission was important, but even more significant was a newly tailored space suit to make Gemini VII more livable. In early June 1965, McDonnell started a test program to see if astronauts could ride almost suitless in space. Gordon Cooper and Elliot See, wearing standard Air Force flight suits (with medical monitoring plugs, helmets wired for Gemini communications fittings, and oxygen masks connected to emergency bottles), flew in the altitude (vacuum) chamber in St. Louis to simulated heights of 36 000 meters. Both astronauts were elated over the results, but McDonnell personnel were uneasy—in actual flight, the cabin temperature might go too high. At an MSC-McDonnell management meeting the next month, McDonnell was asked to study another possibility. James V. Correale of the Crew Systems Division had suggested using a lightweight pressure garment similar in operation to a G3C intravehicular suit. Although this soft suit would not allow pilots to complete a mission if the cabin lost oxygen presure, it would provide them enough margin of safety to get to a recovery area.

Test results at McDonnell showed that the spacecraft environmental system actually operated more efficiently with suits off, but NASA and McDonnell engineers did not like the idea of the crew being so vulnerable. The best way to extinguish a fire in space, for example, was by cabin depressurization, which was out of the question if the men were suitless. And they needed protection if they had to use the ejection seats. Therefore, NASA officials snapped quickly at Correale's idea for a lightweight suit. This decision—in August 1965—was too late to benefit the crew of *Gemini V*, but there was enough time to get the suit ready for Gemini VII.

To produce a more comfortable suit, the David Clark Company removed as much corsetry as possible from the 10.7-kilogram (23-1/2-pound) Gemini pressure suit. The suit was designed to be removed during flight without requiring too much energy or space. A soft cloth hood—which used zippers, as opposed to a neck ring, for fastening to the torso portion—replaced the fiber glass shell helmet. The contractor, working with MSC's Crew Systems Division, managed to cut suit weight by a third, but the 7.3-kilogram (16-pound) suit was still somewhat heavy. In evaluation and training sessions, however, Borman and Lovell found the new garment handy. The soft hood could be zipped open, and the complete suit could be removed and laid on the side of the seats, without having to be stowed away.

If the spacecraft systems were performing properly, the crew would take the suits off after the second day in space. The garments would then be worn only for such critical phases of the mission as rendezvous, reentry, and landing. Use of the lightweight suit, designated G5C, was approved in August; by November, qualification was completed.[37]

Gemini VII carried more experiments than any other flight in the program. Because it was the last long-duration mission, its medical experiments were particularly important in assessing man's capabilities for the lunar landing program. Of 20 experiments, eight were medical, a higher ratio than in any other Gemini flight (see Appendix D).[38]

Two of the medical experiments—calcium balance study and in-flight sleep analysis—were better suited to a clinic than to a small space-craft cabin and were viewed with something less than enthusiasm by the crew. Even the name of the "Inflight Electroencephalogram" (EEG) experiment made the astronauts a little nervous. Although it was merely a study of sleeping habits in Gemini, the EEG was normally used to diagnose subtle disturbances such as incipient epilepsy and brain tumors. But some specialists believed brain wave recording could offer more information, and the astronauts were understandably wary of how the results might be interpreted. Changing the name to "Inflight Sleep Analysis" solved only half their problem. Since normal hair growth would dislodge the scalp sensors after 48 hours, the information had to be gathered at the worst possible time—the first night, when most people have difficulty sleeping in a new environment, anyway.[39] Borman and Lovell also turned a jaundiced eye on the calcium-balance study. It was a nuisance because they had to keep a complete record of body intake and wastes for 9 days before the flight, 14 days during it, and 4 days afterward. Before and after the mission, a nutritionist from the National Institutes of Health limited the items they could eat and drink and weighed out their meals in grams. Almost a month of this regimen did not appeal to the crew.[40]

The only other medical experiment making its space flight debut was "Bioassays of Body Fluids."* Its purpose was to study the effect of space flight on body fluid chemistries that might be affected by physical and mental stresses. The experimenters hoped to draw some conclusions about the physiological costs of space flight by analysis of urine samples.[41]

In categories other than medical—scientific, technological, and defense—only three experiments were being flown for the first time. The other nine were repeated from Gemini IV and V. Two of the new experiments were technological: an in-flight laser transmitter to be aimed at a laser beacon at the White Sands Test Facility, New Mexico, to establish optical communications from space; and landmark contrast measurements of selected areas around the world (primarily coastlines), which might be useful to Apollo for guidance and navigation. The third was a Defense experiment to determine the value of star occultation measurements for spacecraft navigation.[42]

The Gemini VII/VI-A decision made Borman's and Lovell's flight more than an endurance test. It changed the amount of fuel they could spend for experiments and stationkeeping with the booster and forced modifications to turn their spacecraft into a target vehicle. Over

---

*This experiment had been part of the Gemini VI mission until the flight was canceled on 25 October 1965.

an early-November weekend, target acquisition and orientation lights, a radar transponder, a spiral antenna, and a voltage booster were installed on Spacecraft 7.[43]

## TWO WEEKS IN A SPACECRAFT

Four years earlier the chimpanzee Enos had barely completed two circuits of Earth. Now Borman and Lovell were ready to try for more than 200 during two weeks in space. On 4 December 1965, they entered the spacecraft and settled in their couches. The minutes to launch ticked off, with the astronauts checking systems, listening over the communications circuits, and waiting to hear the erector go clanking downward. Promptly at 2:30 p.m., the booster rose from the pad. There was no doubt about it, Lovell said, the triple cues of CapCom Elliot See's countdown, the vibration of the launch vehicle, and the noise of the engines all told him he was going someplace.[44]

"We're on our way, Frank!" Lovell shouted. As the launch vehicle boosted the spacecraft skyward, the booster rolled toward its programmed launch azimuth of 83.6 degrees. With only minor deviations in its powered phase, *Gemini VII* slid smoothly into its planned 160-kilometer keyhole.[45]

Shortly after the spacecraft cut loose from its booster only a little over six minutes from liftoff, Borman wheeled *Gemini VII* around to find the launch vehicle. Two seconds of thrust had been enough for the separation maneuver and now he fired for five seconds to get into position for stationkeeping. The afternoon Sun glared through the windows but in less than 30 seconds he saw the booster. Fuel spewed from a broken line, first forming globules and then crystallizing into cascades of flakes. The Titan II bounced and jumped about the sky. Occasionally eclipsing the view of the Sun, the venting fuel created a brilliant and beautiful contrast. For 15 minutes, the crew took turns at formation flying and picture taking. Stationkeeping was easy, but chasing the tumbling second stage was costing more fuel than Borman liked. And at 15 meters, he was too close to such unpredictable motion, anyway. He fired the spacecraft thrusters to move away.[46]

Half an hour into the flight, experiments began. Cardiovascular conditioning cuffs were snapped on Lovell's legs, where they started pulsing. The booster was still in sight, its lights flashing and billions of particles around it. Borman and Lovell saw some unidentifiable objects in orbit five to six kilometers away. About 7:00 p.m., they turned from sightseeing to housekeeping, and at 9:30 they ate their first meal in space. Intermittently, air-to-ground communications dealt with an irksome fuel cell warning light, which blinked on and off. As night fell below, noise from the ground became less frequent, giving the crew a chance to catnap. Borman's suit was warmer then he had expected; he had to turn the control knob to the coldest setting.

After breakfast, at 9:00 a.m., CapCom See told the crew it was time to go to work. Systems reports were run through, their physical well-being was discussed, and the day's experiment load was assigned. See passed on Mission Control's analysis of the fuel cell warning light and news of more mundane events: the theme song of the men aboard the aircraft carier *Wasp* ("I'll Be Home for Christmas"), football scores, and a collision between two airliners over New York. Borman retorted, "It looks like it's safer up here than down there." "We're not down yet, buddy!" Lovell reminded him.[47]

Some 45 hours into the flight, Lovell began doffing his suit, a simple action that took more than an hour in such crowded quarters. At that point, both astronauts had stuffy noses and burning eyes. Borman complained that he was too warm. After Lovell had removed his suit, however, the general cabin environment improved.[48] A debate about suits on or suits off during flight that had started before the launch of *Gemini VII* continued for nearly six days into the mission.

Both astronauts had planned to remove their suits after a two-day check of the environmental system. That changed when Mueller got wind of it. He objected strongly and so did Seamans, who agreed that one crewman should be suited at all times. Either pilot could take his suit off for up to 24 hours, but during launch, rendezvous, and reentry, both were to be suited.[49]

Borman made frequent comments about Lovell's comfort and his own distress. As the hours passed, the rationale of one suit off and one on became ever less persuasive. Even sitting with his suit completely unzipped and his gloves off, Borman sweated while Lovell remained dry. Lovell's first 24 hours unsuited passed, and he elected to sleep suitless a second night. Borman agreed, despite his own discomfort, because Lovell, the larger of the two men, had more trouble getting the suit off and on in the confines of the cabin than he did. Lovell did don some special lightweight flight coveralls but took them off after 15 minutes—it was just too hot.

One hundred hours into the flight, Borman asked the flight controller on the *Coastal Sentry* Quebec to talk to Kraft about taking off his suit. Because he knew of Mueller's opposition, he cautioned CapCom Eugene A. Cernan, on the next pass over Houston, to discuss his request with Slayton first and not to present it to Kraft as an emergency. Cernan agreed.

Meanwhile, the controllers tried to get Lovell to put his suit on and Borman to take his off, so the surgeons could check the effects on both pilots of the suited and suitless conditions. The crewmen wanted to wait until the rendezvous with Gemini VI-A had been completed, but Kraft insisted. After 146 hours of flight, Borman finally agreed. Two hours later, it was his turn to sit in suitless comfort as Lovell sweltered.[50]

The suit question was also working its way up the NASA chain of command, as the daily mission evaluation reports became tinged with concern about how alert the crew would be for the coming rendezvous. When Borman made his request through Cernan, Mission Director Schneider relayed it to George Mueller in Washington. Mueller asked MSC Medical Director Charles Berry (who was also chief flight surgeon during the missions) for a comparative analysis of the two astronauts. Already aware that Gilruth favored suits off, Mueller asked for a poll of the other members of the Gemini Design Certification Board.

Kennedy Director Kurt H. Debus, Marshall Director Wernher von Braun, and SSD Commander Ben Funk all agreed that the reasons for being unsuited outweighed those for being suited. Berry reported that the blood pressure and pulse rates were closer to normal with suits off. The pilots got their wish, and debate ended.[51]

Despite Frank Borman's discomfort, spacecraft operations proceeded efficiently. The crew conducted experiments, evaluated spacecraft systems, and worked, slept, ate, exercised, and rested. Good humor and good spirits prevailed, bolstered by family reports, the daily See-Haney newscasts, and the preparations for sending *Gemini VII* some visitors—the VI-A crew. Borman expressed some concern about the fuel needed to get into position for the meeting, but four orbital adjustment maneuvers worked well.[52] In a nearly circular orbit of 300 kilometers, the spacecraft's orbital lifetime was now theoretically over 100 days.[53] The friendly target was ready.

## Go Back to "Go"

While Gemini VII had rested on pad 19 awaiting launch, welders and repairmen had stood by. Borman and Lovell had barely started their booster-chasing exercise when Elliot See told them that pad cleanup had begun. The normal feeling of anticlimax after a launch was absent. If anything, spirits may have seemed too high. "Everybody was so excited you'd think they were going to launch the next day," John Albert recalled. The Martin crew found minimal damage to pad 19. Workmen wasted little time on normal painting or cleaning. Their objective was to replace critical instrumentation.

The launch team got GLV-6 up and the spacecraft mated to it in one day, complete with standard procedures, tests, and reviews. In addition, *VII's* radar transponder was interrogated as Borman and Lovell passed over Cape Kennedy to ensure that it would answer VI-A's radar transmissions.[54]

After 56 hours of the Borman-Lovell mission, rapid progress in getting Gemini VI-A ready fostered hopes that it might fly on the eighth instead of the planned ninth day. A computer problem dampened these hopes briefly, but, with a new part installed, the final simulated flight test started and ended without problems. On 9 December, Mathews and Funk were convinced that the launch could be made a day early.[55]

On Sunday, 12 December, Astronauts Schirra and Stafford moved through the doors and into the couches of Spacecraft 6 for a second time. After a troublefree countdown, precisely at 9:54 a.m., their Gemini launch vehicle roared into action. The roar was quickly strangled. *Gemini II's* "hold-kill" seemed to be repeating, but this time more critically—there were two men strapped atop this sputtering rocket. At 1.2 seconds, an electrical tail plug dropped from the base of the booster and activated an airborne programmer—a clock in the cockpit that was not supposed to start until the vehicle had lifted off. Because there had been no upward movement, the valves closed to prevent fuel from gushing into the launch vehicle's engines. The malfunction detection system had sensed something wrong and had stopped the engines.[56]

One of the most suspense-filled moments in the whole Gemini program followed. If ever there were a time to use the spacecraft ejection seats to get away from a cocked and dangerous rocket, this seemed to be it.

Kenneth Hecht, chief of the Gemini Escape, Landing, and Recovery Office and long-time ejection seat specialist, was surprised when the crew did not eject, as they should have if ground rules had been strictly followed. If the clock were right, then the vehicle had left the ground. Had it climbed only a few centimeters, the engine shutdown would have brought 136 tonnes (150 tons) of propellants encased in a fragile metal shell crashing back to Earth. There could be no escape from the ensuing holocaust. But neither Schirra nor Stafford had sensed motion cues; and Schirra, who as command pilot would have been the one to pull the "D-ring" for ejection, decided not to, despite the ticking clock.

At the moment of crisis, the veteran test pilot remained calm. With no trace of emotion in his voice, Schirra reported, "Fuel pressure is lowering." Francis X. Carey, the Martin launch vehicle test conductor, was just as matter of fact over the radio circuit to the spacecraft. Just a hint of panic might have caused Schirra or Stafford to pull the D-ring. Schirra relied, with icy nerves, on his own senses. He knew GLV-6 had not moved, and he knew the clock was wrong.[57]

When the smoke had cleared and it appeared that the booster was not going to explode after all, up went the erector. Guenter Wendt and his McDonnell team hastened back to the white room they had so recently left. After checking on the cabin pressure and making sure that the crew had safetied the seat pyrotechnics, Wendt opened the hatches and helped the astronauts, their faces etched with disappointment, out of the spacecraft.[58]

Seamans had been listening in at NASA Headquarters in Washington. Once sure that the crew was safe, he went home. A call from Administrator Webb soon brought word that President Johnson was greatly disturbed by the failure. All was not lost, Seamans told Webb. *Gemini VII* still had six days in orbit—time enough, he hoped, to find the source of trouble and launch VI-A for the rendezvous.[59]

The Martin and Air Force teams began recycling the booster for a launch to take place four days later. So far as they knew, the only thing wrong was a tail plug that had fallen out prematurely. A check through the records left no question that the plug had been properly twisted into its detents. But testing revealed that some plugs did not fit as tightly as others and pulled out more easily. (The harder-to-remove plugs, with a safety wire added, became standard for Gemini.)[60]

As expected, reporters clamored for details about the engine shutdown. Merritt Preston was picked to tell them what NASA knew and what it planned to do. Known to the press as a spacecraft expert, Preston could not be expected to know all the technical details about the launch vehicle and would be saved from having to guess. Although he winced at being placed on the firing line, his explanations at a news conference were well received and he was not pressed for answers. Reporters shared with Gemini officials the belief that it was just a case of a plug pulling out. The malfunction detection system had worked as it should, the crew had remained cool. There seemed every reason to believe that the launch could take place in four days.[61]

Aerospace engineers routinely examined the launch vehicle engine thrust-trace data. The firing trace looked normal at the beginning, but some strange squiggles farther along on the graph suggested that thrust had decayed *before* the plug dropped out. A call to John Albert caught him as he was leaving for a meeting to discuss plans for the launch turnaround. He detoured to get a copy of the graph, which he took to the meeting. A telephone call was immediately placed to the Aerojet-General plant in Sacramento. A detailed analysis tentatively spotted the problem in the vicinity of the gas generator. But the trouble itself needed to be pinned down. By 7 o'clock that evening—12 December—the Cape Aerojet engineers were searching the engine, piece by piece. All through the night they worked, but to no avail. When Charles Mathews came by at 9:00 the next morning, their haggard and worried faces told him there had been no success. Just as he was asking what Aerojet intended to do now, an excited engineer came running in, shouting that he had the answer—a dust cover that had accidentally been left in the engine. Months before, in the Martin Baltimore plant, the gas generator had been removed for cleaning. When the check valve at the oxidizer inlet was taken off, Martin technicians put a plastic cover in the gas generator port to keep dirt out. Later the dust cap was overlooked when the unit was reinstalled. The relatively inaccessible location of the check valve—on top of the engine just under the tankage where it could not be seen and all work had to be done using mirrors and touch—effectively prevented the errant cap from being discovered.[62]

Once the trouble was found, the gas generator was cleaned and replaced in GLV-6 on 13 December. It had suffered no damage, but a question still lingered: Could VI-A be launched in time to rendezvous with VII? At the time of the hangfire, recycling was expected to take four days, but within five hours of the failure, Elliot See told the *Gemini VII* crew that launch was targeted for the third day—15 December[63]—with a mighty effort to reduce the 96-hour recycle to 72 hours. It succeeded.[64] The friendly target was still waiting patiently upstairs.

One question remained unanswered and unanswerable. When Schirra refused to pull the D-ring that would have ejected the Gemini VI-A crew, was that a decision he alone would have made, or was that an indication that none of the astronauts would have used the seats?* The feelings expressed by the only Gemini pilots who faced that decision leave a measure of doubt.

Stafford's concern was the enormous acceleration—more than 20 g's—an off-the-pad abort required to throw the seat in a stable trajectory far enough from the booster to do any good. Even a mentally prepared astronaut might suffer severe injury. At best, Stafford believed, he would have been walking around for months with a crick in his back, like those who had ejected in similar high-impulse Martin-Baker seats. Of course, he would also be alive. And Schirra remarked, "If that booster was about to blow ... if we really had a liftoff and settled back on the pad, there was no choice. It's ... death or the ejection seat."[65]

## THE VISITORS

On 15 December 1965, the mood of those working on the rendezvous mission—planners, pilots, and ground crew—was one of high anticipation. If on this third attempt Gemini VI-A would cooperate and

---

*Early in the program, some thought was given to training Gemini crews on an ejection seat catapult at the Navy's aircrew training laboratory in Philadelphia. When a Navy test subject tried the facility and reported that it was no worse than being catapulted in a plane off a carrier, MSC officials decided it was not worth the effort. Warren J. North, Chief of the Flight Crew Support Division, said that "generally speaking, the flight crews were all in favor of the ejection seats," in spite of the extremely high g-forces.

go into orbit, a truly significant world space "first"—rendezvous—might be chalked up. Russian endurance records had now been shattered in two successive American manned space missions, but achieving rendezvous would be navigationally significant to the Apollo program as well as important one-upmanship. Having a friendly target to approach, one that could point its transponder and talk back as *Gemini VI-A* called out its course and speed, created an atmosphere of confidence.[66]

At 8:37 a.m. *Gemini VI-A* rose from its pad. As if forcing it to move by will power alone, Schirra urged, "for the third time, go." A moment of wonder followed, as the launch vehicle seemed to shimmy. This shaking may have been only an impression; because of their recent experience, both pilots were highly attuned to movement and sound. At engine cutoff, Stafford checked the computer and got a reading of 7830 meters per second. This told them they were on their way. Borman and Lovell in *Gemini VII*, passing near the Cape Kennedy area, saw nothing except clouds; but they soon learned from the Canary Islands communicator that the orbital parameters of *VI-A* were 161 by 259 kilometers. A few minutes later, as they flew over Tananarive, Malagasy Republic, they saw *VI-A's* contrail and got a brief glimpse of the visitors' spacecraft. They put on their suits and waited for company to arrive.[67]

The rendezvous profile—dubbed "M equals 4" by the mission planners for convenience (the "M" had no special meaning)—scheduled the catchup to *VII* during the fourth revolution of *VI-A*. Schirra and Stafford faced six hours of maneuvering to reach Borman and Lovell.[68]

At insertion, the chase vehicle trailed its target by 1992 kilometers. The *VI-A* crew aligned the inertial platform to position their spacecraft for a height adjustment. Over New Orleans, after 94 minutes in space, Schirra ignited the thrusters to speed up by 4 meters per second. The perigee remained the same, but the acceleration kicked the apogee up to 272 kilometers. *Gemini VI-A*, being nearer to Earth and so moving faster, now lagged only 1175 kilometers behind *Gemini VII*.[69]

Near Carnarvon, at 2 hours 18 minutes ground elapsed time, Schirra began a phase adjustment. This had a twofold purpose: to reduce the distance to the target and to raise the chase vehicle's perigee to 224 kilometers. He pressed the button to add 19 meters per second to his velocity. Over the Pacific less than half an hour later, Schirra turned his spacecraft 90 degrees to the right (southward) and ignited the thrusters to push *Gemini VI-A* into the same plane as *Gemini VII*. Now the distance between the two vehicles had narrowed to 483 kilometers.[70]

Three hours 15 minutes into the mission, Elliot See told Schirra that radar contact should soon be possible with *Gemini VII*. The *VI-A* crew got a flickering radar signal, then a solid lock-on at 434 kilometers range. Over Carnarvon, at 3 hours 47 minutes, the aft thrusters fired for 54 seconds to add 13 meters per second to *Gemini VI*'s speed. The result was almost a circle, measuring 270 by 274 kilometers. In slant range distance, the two spacecraft were now 319 kilometers apart and closing slowly.[71]

Schirra and Stafford placed *Gemini VI-A* in the computer (or automatic) rendezvous mode at 3 hours 51 minutes into the flight. While the lower orbiting vehicle gained slowly on its target, Schirra dimmed the lights on his side of the spacecraft to improve outside visibility. At 5 hours 4 minutes, he exclaimed, "My gosh, there is a real bright star out there. That must be Sirius." The "star" was *Gemini VII*, reflecting the Sun's rays from 100 kilometers away.

Gradual catchup of the target vehicle lasted until 5 hours 16 minutes; Schirra prepared to make the last rendezvous maneuvers. The two ships were now close enough to allow Spacecraft 6 to thrust directly toward Spacecraft 7. He fired the thrusters and closed on *Gemini VII* at a rate of better than three kilometers every minute and a half.[72] Schirra and Stafford briefly lost sight of *Gemini VII* when it passed into darkness but soon picked up the target's running lights.[73]

Schirra made two midcourse corrections spaced 12 minutes apart (at 5 hours 32 minutes and 5 hours 44 minutes). Six minutes later, at a range of 900 meters from his target, Schirra began braking his spacecraft by firing the forward thrusters. Soon he had no difficulty seeing *Gemini VII*. Fittingly, in the terminal stage of rendezvous, the *VI-A* astronauts saw the stars Castor and Pollux in the Gemini (Twin) constellation aligned with their sister ship. Then Spacecraft 7 flashed into the sunlight—almost too bright to look at. From a distance of 200 meters, it resembled a carbon arc light. Following the braking and translation maneuver, *VI-A* coasted until the two vehicles were 40 meters apart, with no relative motion between them. The world's first manned space rendezvous was now a fact. In Mission Control, the cheering throng of flight controllers waved small American flags, while Kraft, Gilruth, and others of the jubilant crowd lit cigars and beamed upon this best of all possible worlds. At 2:33 p.m., 15 December 1965, *Gemini VI-A* had rendezvoused with *Gemini VII*.[74]

When Russian *Vostok III* flew within five kilometers of *Vostok IV* on 12 August 1962, some people believed, with the help of *Pravda* news dispatches, that rendezvous had been accomplished. The two spacecraft, however, were in different orbital planes; nor could they maneuver to stop relative motion between them. In simple terms, it was good shooting from the pad, but the result was the same as if two bullets had passed in the middle of a battlefield. Schirra knew what a real rendezvous in orbit was:

> Somebody said. . . when you come to within three miles [five kilometers], you've rendezvoused. If anybody thinks they've pulled a rendezvous off at three miles, have fun! This is when we started doing our work. I don't think rendezvous is over until you are stopped—completely stopped—with no relative motion between the two vehicles, at a range of approximately 120 feet [40 meters]. That's rendezvous! From there on, it's stationkeeping. That's when you can go back and play the game of driving a car or driving an airplane or pushing a skateboard—it's about that simple.[75]

Borman and Lovell had been fascinated by the fireworks of *VI-A's* thrusters during braking and startled by the 12-meter tongue of flame. As Schirra and Stafford neared, there was a second surprise. Borman said, "You've got a lot of stuff all around the back end of you." Minutes later, during stationkeeping, Schirra told Borman, "So do you." Cords and stringers three to five meters long streamed and flapped behind both spacecraft.[76]

Rendezvous maneuvers had cost *VI-A* only 51 kilograms (113 pounds) of fuel. Schirra still had 62 percent left in his tanks. It had been easy, he said, and there was plenty of fuel for stationkeeping, flyarounds, formation flying, and parking the spacecraft in specific relative positions. Borman and Lovell were not so wealthy; Flight Control told them to stop maneuvers when the *VII* tanks dropped to an 11 percent supply.

For more than three Earth revolutions, the two spacecraft stayed at ranges of from 0.30 meters to 90 meters. *VI-A* approached *VII* to examine the stringers on one occasion. On another, they flew nose to nose. Schirra and Stafford swapped the controls back and forth because the Sun streamed so brightly through first one window and then the other. When it was time for Borman and Lovell to perform an experiment, Schirra and Stafford moved out 12 meters and parked. For some 20 minutes, in one instance, neither bothered to touch the steering handle, as the spacecraft remained stable in relation to its sister ship. On the first night pass, the two spacecraft faced each other at distances ranging from 6 to 18 meters. Schirra had worried about visibility during darkness, but it turned out to be excellent—docking light, handheld penlight, and even *VII's* cabin lights were clearly visible to him.

Using what Schirra called his eyeball ranging system, the *VI-A* crew did an in-plane flyaround of *VII*, roving out to 90 meters. Believing this was too far away to be called stationkeeping, Schirra hurriedly brought *VI-A* within 30 meters. The astronauts were highly impressed with their ability to control the spacecraft. Velocity inputs as low as 0.03 meter (0.10 foot) per second provided very precise maneuvering. Because of this fine control, he and Stafford concluded that nuzzling into and docking with a target vehicle would be no problem.

As the pilots' bedtime approached, Schirra flipped the spacecraft blunt-end forward and fired his thrusters to impart a small separation speed. Eventually, the crews settled down 16 kilometers apart. Borman, who frequently caught sight of *Gemini VI-A* in the distance, remarked to the *Rose Knot* Victor tracking ship communicator, "We have company tonight."[77]

After launch, rendezvous, and stationkeeping, Schirra and Stafford were utterly exhausted and hungry. They ate a good meal and went to sleep. When Schirra awakened with stuffy head and runny nose, he was glad that the mission was flexible, with the option of landing after only one day of flight if everything had been done. He and Stafford had achieved all their mission objectives, and the flight controllers would not be able to give too much more attention to *Gemini VI-A*, anyway. *Gemini VII's* fuel cell needed help, and Borman, Lovell, and Mission Control had to focus on its problems if the mission were to be able to last 14 days.[78]

But Stafford caught everybody's attention for a few minutes. In an excited tone he reported:

> Gemini VII, this is Gemini VI. We have an object, looks like a satellite going from north to south, probably in polar orbit. . . . Looks like he might be going to reenter soon. Stand by one. . . . You just might let me try to pick up that thing.

Over "one," the communications circuit, came the strains of the pilots playing "Jingle Bells."* The spirit of Christmas glowed—*Gemini VII* was about to begin its 12th day and *VI-A*, having demonstrated rendezvous in fine fashion, was going home.[79]

Schirra said, "Really a good job, Frank and Jim. We'll see you on the beach." He then flipped *VI-A* blunt-end forward and jettisoned the equipment section; retrofire followed automatically.[80]

Schirra placed the spacecraft in an inverted (heads down) attitude to see Earth's horizon. Nearing the 100 000 meter fringe of the atmosphere, Schirra set the bank angle at 55 degrees left and held it until computer guidance took over at 85 000 meters. The spacecraft threatened to overshoot its planned landing point. This had to be countered by banking first left, then right. Since the Gemini spacecraft obtained

---

*Michael Kapp, producer of the Bill Dana "Jose Jimenez in Orbit" record album in the early sixties, had given Schirra a small four-hole harmonica on 8 December 1965. (Kapp also provided many of the music tapes that were broadcast to the Gemini crews from the Mission Control Center.) Stafford, the other half of the two-man space band, jingled small bells. Frances Slaughter, of the Cape Flight Crew Operations Office, had fastened them to his boots before a training simulation, for a joke, and he took the bells on the flight to provide the rhythm section. It had been Schirra who furnished the corned beef sandwich that had created such a furor for the *Gemini III* crew. Asked some time after his flight why he "didn't get too much static for the harmonica," Schirra replied, "I think the timing was pretty good on that."

**Gemini VII/VI-A
4 December 1965**

*Left, scorched pad 19 after Gemini VII launch; center, first stage of Gemini VI-A's Titan launch vehicle is erected on pad 19; right, Gemini VI-A roars away.*

*World's first space rendezvous.*

The Wasp *welcomes* Gemini VI *(left), then* Gemini VII's *astronauts Lovell and Borman.*

its greatest lift flying straight ahead, banking cut lift and shortened range.

The crew turned the computer off at 24 000 meters, deployed the drogue parachute at 14 000 meters, and punched out the main parachute at 3200 meters. *Gemini VI-A* landed about 13 kilometers from its planned impact point, recording the first successfully controlled reentry.[81] For another first, they did it in full view of live television beamed from the *Wasp* via satellite transmission. As on his Mercury flight, Schirra elected to remain aboard his spacecraft while it was hauled onto the carrier deck. Thus, on 16 December 1965, after 16 revolutions (and 25 hours, 15 minutes, 58 seconds), the world's first manned spaceflight rendezvous mission became a matter of record.*[82]

## THE HOME STRETCH

After the guests had departed, Borman and Lovell realized that their incentive had gone with them. Events such as stationkeeping, experiments, getting out of their suits, and waiting for the *VI-A* visitors had sustained their enthusiasm. Even then, the novelty of space flight had worn thin, and their thoughts had strayed homeward. With *VI-A* gone and almost three days left, the mission began to drag. Beyond all doubt, 14 days inside this spacecraft was "a long haul in that short frame." While in drifting flight, Borman read some of Mark Twain's *Roughing It*, Lovell part of *Drums along the Mohawk* by Walter D. Edmonds—both selected partly because they had nothing to do with the space program.[83]

During a mission as long as *Gemini VII*, impressions only indirectly connected with the flight naturally came to mind. Lovell indulged in a disquisiton on legs, which were

> affected the most by zero g because you don't realize how much exercise you do every day. Just combating Earth's gravity, you do quite a bit; and the legs are designed to do most of that work for you. They get you around—they walk—they lift up your body. Suddenly, for two weeks, this gravity is taken away. The legs don't have a job any more—they're just there. [A man without legs] for Gemini would have been perfect because you could utilize that space for something else. Everything except for maybe EVA. But in that spacecraft, we didn't use the legs for anything.[84]

A few minutes after Schirra had played his spirited rendition of "Jin-

---

*The National Aeronautic Association, representing the Fédération Aéronautique Internationale, certified *Gemini VII/VI-A* for four manned space flight achievements: longest distance in orbit, longest duration in orbit, distance in group flight, and duration in group flight.

gle Bells," Borman and Lovell took off their suits. They might as well be comfortable. Then they had to see about a thruster problem that had greeted them upon awakening. When Borman tried to fire thrust chambers 3 and 4, only whitish, unburnt fuel streamed out. The pitch thrusters stopped the spacecraft from yawing and thrusters 11 and 12 also helped, although they were a little too strong in control. One of the non-working thrusters was tested after the flight. The laminate in the thrust chamber was found to be the old-style 90-degree layup, instead of the new 6-degree design that had solved the burnout problem.[85]

But the thrusters were merely annoying; the fuel cell was a greater concern. Despite the warning light during the first revolution, the cell had provided enough electrical power for the spacecraft to operate normally for 126 hours. The ground analysis team, with an operating model set up in St. Louis, had helped keep it going, but power output was only partial by the end of the 12th day. The next day, the fuel cell threatened to quit completely as the warning light burned continuously. *Gemini VII* might have to end early with a landing in the Pacific Ocean, much as the crew disliked the idea of missing the 14-day goal. Test results in St. Louis, however, showed that the electrical system would carry them all the way. Relieved, Borman slept better than he had on any other night in space.[86]

Borman and Lovell finished their packing on the last day. Asked about their baggage, Borman said the cockpit was clean, he and Lovell were wearing their suits, and they were all set to go home.

Before the retrorockets fired, the ground stations kept the crew busy for two hours on the reentry checklist. Flight surgeon Berry reminded them to elevate their feet and pump their legs. Borman broke in to say that he and Lovell wanted to get out of the spacecraft as soon as possible. They had no desire to wait around to be stylishly hoisted aboard a carrier. As they started their last revolution—number 206— the tracking stations along the circuit bade them goodbye. The music being broadcast included the tune "Going Back to Houston."[87]

With retrofire approaching in the darkness near Canton Island in the Pacific, the crew wondered—as do all astronauts—whether the rockets would fire. Lovell described his emotions graphically.

> Retrofire has a unique apprehension in the fact that both of us are aviators and we understand the apprehension in flying. If you have an accident in an airplane, something's going to happen.... You hit something, or it blows up—you're coming down. Now, in liftoff and reentry, a space vehicle is like an airplane. Something's happening. But if the rockets fail to retro, if they fail to go off, nothing's going to happen. You just sit up there and that's it. Nothing happens at all. So that's the unique type of apprehension, because you know that you've gotten rid of the adapter, you know that you're going to have 24 hours

of oxygen, 10 hours of batteries, and very little water. So you play all
sorts of tricks to get those retros to fire.[88]

The first retrorocket fired automatically and on time. The next
two rockets followed in quick succession and, after a pause, the fourth
fired. As the firings jolted them, Lovell said, with relief, "That's one
big hurdle over with, tiger!" Borman answered, "You're right, ace."

From Houston, CapCom See told them to fly a 35-degree left
bank until computer guidance cut in. A surprised Lovell reminded
Borman that 53 degrees had been planned. Borman questioned See,
who confirmed the 35-degree bank.* By that time, however, the computer "had come in on the line . . . it was actually commanding the
spacecraft," with Borman banking to right and left, following the needles. As Lovell later said, "You have no control over how close you're
going to get to the target. Your only control is how good that computer is doing, or how good your c.g. [center of gravity] was when you set
up the computer and the retrofire time. . . ."[89]

Borman rolled *Gemini VII* head down to use the horizon as a
guide for keeping the proper spacecraft attitude. He could see nothing
from his window, however, and had to depend entirely on his instruments and on Lovell, who finally saw the horizon after about six and a
half minutes and began calling out adjustments. Borman concluded
that reentering was definitely a two-man job for Gemini; there was no
way to follow the needles on an instrument panel and watch the horizon at the same time.[90]

Because they had been weightless for so long, the onset of the g
forces "felt like a ton." During the long glide, which did not have a
sharp angle of descent, g forces never rose higher than 3.9 (contrasted
with an average of 7.7 for the Mercury-Atlas orbital flights). But the
higher g did not bother them too much, since they were very busy
trying to get as close to that carrier as possible.

The reentry control system worked well, holding *Gemini VII*
steady until the drogue parachute came out. The spacecraft rocked 20
degrees to either side, giving the crew a shaking. On the way down,
Lovell opened the snorkel; smoke and an acrid smell filled his hood,
causing his eyes to water. But even his smarting eyes were glad to see
the main parachute deploy. Little did the crew care that they hit the
water with a heavy thud. Borman's thoughts were elsewhere; he was
trying to spot the recovery helicopter. When he did not see any aircraft, he remarked, "Shoot! We must have missed it more than Wally
did." The two command pilots had a small bet on who would land
closer to the target. But Borman was not sure when he began to talk

---

*In the postflight report, Scott Simpkinson's evaluators noted that the flight controllers had
been wrong and had given Borman an erroneous bank angle.

with "Air Boss," pilot of one of the helicopters in the area of the spacecraft's descent; maybe they were near the aiming point, after all.[91]

On 18 December 1965, after 330 hours, 35 minutes, 01 second, *Gemini VII* came to rest on what Lovell called the good old aqua firma, missing the target by 11.8 kilometers.* Mission objectives had been achieved in fine fashion. Provided the crew came through in good physical condition, it could be assumed that an Apollo team could fly safely to the Moon and back.

Borman felt a little dizzy, Lovell not at all. Borman suggested that they get out of their suits, as it was warm in the spacecraft, but the effort was just too great. They turned on the oxygen repressurization valve and were soon comfortable. The pararescuemen were already working on the flotation collar, and the recovery helicopters were hovering nearby. Half an hour after landing, Borman and Lovell were greeted aboard the *Wasp*,[92] the second spaceship crew the carrier's crew had snared in a few days.

When the returning spacemen came onto the deck of the carrier, they were tired but happy. They walked slightly stooped and a little gimpy-legged, partly because of their pressure suits and the ship's roll, but mostly because they were just plain weary. Perhaps even more remarkable than being able to walk across the deck without stumbling was the fact that the crew had been able to get into the "horse collar" to be hoisted into the helicopter. After being weightless for 14 days, this was a severe physical test. Berry was jubilant over the medical results of *Gemini VII*:

> The most miraculous thing was when they could get out of the spacecraft and not flop on their faces; and they could go up into the helicopter and get out on the carrier deck and walk pretty well. They were in better physiologic shape than the V crew. Initially, their tilt-table responses were not as bad and did not last as long. It looked more like four-day responses, by far, than eight-day. The calcium loss was the same way. Amazingly, they maintained their total blood volume. They didn't get any decrease, but they did it in a peculiar way. They lost the red-cell mass still, but they replaced the plasma—they put more fluid in. Apparently, there had been enough time for an adaptive phenomenon to take place.[93]

When the detailed examination started, the physicians found that Lovell, who had worn the cardiovascular cuffs, had less blood pooling in his legs than Borman. After a good night's sleep aboard ship, both

---

*Flight Control had told Borman of the procedures Schirra had used in flying the first computer controlled reentry. Since he was anxious to win his bet, the *Gemini VII* commander was glad to have the benefit of the *Gemini VI-A* commander's experience.

men looked rested and said they were.[94] They had made the long haul in that short frame in fine style.

Christmas week of 1965 was perhaps the high-water mark of manned space flight to that time. The string of successes had an unlooked-for effect, however—manned space flight became almost commonplace, the novelty had all but gone. Who did what and when tended to blur. Any single event, such as America's first suborbital flight or first orbital mission, became hard to recall. Perhaps more than it intended, NASA had achieved the program goal implied in the Project Development Plan of December 1961: to put space flight on something like a routine basis.[95] The routine loses news value, and score cards on Russia versus America in the space race vanished when the lead clearly passed from East to West.

Gilruth may have best summed up the bright look of things at the postrecovery conference on 18 December, when he said:

It has been a fabulous year for manned space flight. . . . I guess you all realize that this year, since March, we have put 10 men in orbit and brought them back. And we have accomplished the major part of the Gemini space objectives at this point in the program. The long duration, which was a major objective, some of us didn't really think you could go 8 or 14 days in that spacecraft . . . we have seen the men return in good shape with all their tasks done. . . . We have seen EVA this year in Gemini, and we have seen rendezvous. We have seen controlled reentry demonstrated, the controlled reentry technique that is so important to Apollo, and we have seen accomplished a whole raft of scientific experiments.[96]

NASA faced the new year with an equal number of manned Gemini flights still to be flown, and it expected to do this with an unbroken chain of successes. Morale was high, as many program objectives had been stamped "Achieved." Postflight celebrations were carried across the seas when President Johnson asked Borman and Schirra to make an eight-nation, good-will tour of the Far East. Meanwhile, engineers at the Manned Spacecraft Center prepared for a "Gemini Midprogram Conference," to discuss the results of the first seven Gemini missions, as they had done for the Mercury program in the Summary Conference held in Houston in October 1963.[97]

# XIII

# Agena on Trial

MANNED space flight and NASA faced the new year of 1966 in an ambiguous position. High achievement had marked 1965, capped by the exciting and important "76" mission at the very end of the year. But the key to more sophisticated missions, the Agena, was in serious technical trouble. Only with Agena could Gemini hope to realize a range of still-to-be-attained goals—docking, re-rendezvous, rendezvous with two separate targets during a single mission, and high-altitude flight—goals that would be indispensable to Apollo, the program to land men on the Moon. But many doubted that Agena could be ready in time to meet Gemini's tight launch schedules. Year's end saw Agena's career in manned space flight once again called into question—and this time a substitute target had already been approved for development.

Agena, though most critical, was not the only problem. Extravehicular activity (canceled in the three previous missions) was supposedly ready to enter a more advanced stage. Unexpected development troubles demanded a last-minute effort (reminiscent of *Gemini IV*) to qualify equipment. Edward White had succeeded in his "space walk," but NASA faced a tougher EVA task—testing the Air Force's Astronaut Maneuvering Unit (AMU), a far more complex personal propulsion system than White had used. Step-by-step progress having been skipped, the EVA set for Gemini VIII in mid-March had to bridge the gap.

At the beginning of 1966, then, the Gemini program had met with success in seven straight missions, five with crews aboard. Not all its goals had been attained, but many had. Now the Apollo program neared its operational stage. Might NASA halt Gemini to concentrate on Apollo? Administrator James Webb had used similar reasoning to conclude Project Mercury earlier than many desired.

Although George Mueller, Associate Administrator for Manned Space Flight, knew of no move afoot to close down Gemini, he foresaw that many engineers in Houston might worry that they were nearly out of jobs. To assuage their misgivings, in December 1965 he made a case for flying all 12 Gemini missions. Even a cursory glance at the program's aims, Mueller said, showed healthy returns for nearly every item. While medical fears had been erased by the outcome of 14 days in space, NASA still needed to perfect techniques for rendezvous and extravehicular activities. Then, too, an experienced cadre of flight crews was essential, not only for flying missions but for astronaut and flight control training as well. LeRoy Day, Mueller's Deputy Director for Gemini, passed this reassurance on to Gemini Program Manager Charles Mathews in Houston.[1] That potential morale threat allayed, the engineers could focus on such technical problems as making Agena work.

### SICKNESS TO HEALTH

Agena's woes were by now chronic. The Gemini Agena target vehicle (GATV) was pacing the program by mid-1965, prompting GPO to consider removing the first production model, GATV 5001, from its job as a test vehicle so it could be used in Gemini VIII.[2] All such plans went up in smoke with the explosion of GATV 5002, which ignited the most demanding piece of engineering detective work in the entire Gemini program. Efforts to cure Agena's ailments spanned more than four months, much of it on a three-shifts-a-day, seven-days-a-week schedule.[3]

An hour after the Agena failure of 25 October 1965, Mission Director William Schneider had left Houston for Florida, where Colonel John B. Hudson, SSD Deputy Commander for Launch Vehicles, had called a meeting of a subpanel of the Agena Flight Safety Review Board for the 26th.* The subpanel members had learned enough from telemetry data to list the tasks to be done: find out why the Agena had failed and what the fixes would entail for design, performance, and schedule; decide if it would be possible to use GATV 5001 and how long it would take to get it ready for launch; and begin cutting red tape that might slow the work.[4]

---

*Present at the meeting were Schneider, Jerome Hammack, Alfred Gardner, Scott Simpkinson (GPO Manager of Test Operations), John A. Edwards (NASA Director of Gemini Flight Operations), Merritt Preston (KSC Director of Launch Operations), Ernst R. Letsch (Gemini Launch Systems Directorate, Aerospace Corporation), and Lieutenant Colonel L. E. Allen (Commander, SLV-3 Division, 6555th Aerospace Test Wing).

Lawrence A. Smith, Gemini Manager for Lockheed, had already sent the taped record of telemetry signals to the plant in Sunnyvale, California, where W. R. Abbott took charge of the failure-search team. Most likely causes of the disaster were a "hard-start" backfire or an electrical short; Abbott's group soon narrowed its search to the engine as the more probable source of trouble. After reporting to Major General Ben Funk's full Agena Flight Safety Review Board, Hudson took his subpanel to Sunnyvale on 1 November; they agreed with Abbott's analysis that a hard start (similar to an automobile engine backfire) had been the cause and that it had resulted from fuel being injected into the firing chamber before oxidizer.[5]

The problem was rooted in NASA's original specification for a Gemini target vehicle able to start and stop its main engine five times during a mission, in contrast to the Standard Agena's two-start engine. This 150 percent increase in demands on the engine at once raised the problem of fuel and oxidizer economy. In the two-start engine, the oxidizer began flowing first, while a pressure switch restricted fuel flow until a given amount of oxidizer had reached the firing chamber. This was known to enhance the engine's starting characteristics, but it was also wasteful. Oxidizer leaked through before engine firing, and some continued to flow after shutdown; the oxidizer would be gone long before the fuel ran out. So Lockheed accepted a proposal by the engine subcontractor, Bell Aerosystems Company, to remove the pressure switch and thus allow fuel to enter the chamber first.[6]

Abbott concluded that in space the presence of fuel in the thrust chamber (perhaps in considerable quantity) had caused the engine to backfire when the oxidizer reached the chamber, causing an explosion. When Funk's review board met in Los Angeles on 3 November to make tentative plans for an engine requalification program, Abbott presented his findings, which were discussed the next day.[7]

But Abbott's and Hudson's groups were not the only ones working on the problem. At NASA Headquarters, Associate Administrator Robert Seamans told George Mueller to form a NASA review board to look into all aspects of the failure, both technical and managerial. Mueller appointed MSC Director Robert Gilruth co-chairman of a Gemini Agena Target Vehicle Review Board and asked Air Force Major General Osmond Ritland to serve with Gilruth.*[8]

And down at Cape Kennedy, Lockheed's Wulfgang C. Noeggerath was working with MSC engineer Horace E. Whitacre to pinpoint the cause of the failure. Unsure that the two of them could explore the matter in the depth needed, Whitacre suggested that Lockheed spon-

---

*Board members were Seymour Himmel (Lewis), George Detko (Marshall), Colonel William C. Nielsen (SSD), Colonel Quenten A. Riepe (6595th Aerospace Test Wing), Morton Goldman (Aerospace), John Bailey (MSC), and Robert H. Gray (KSC).

sor a symposium of rocket experts from around the nation. Noeggerath convinced his superiors that it was a good idea.[9]

The two-day symposium began on 12 November, with 19 scientists and engineers in attendance.[*][10] Noeggerath and Whitacre told the visiting experts that the most likely cause of the Agena explosion had been a premature engine shutdown. Engine firing had produced severe oscillations and mechanical damage. Temperature decreases had indicated fuel spillage. When electrical circuitry failed, the engine stopped, but a valve that controlled tank pressure as fuel was being used remained open. As fuel stopped flowing, pressure built up in the tanks, which ruptured and destroyed the vehicle—a planned flight safety precaution. Whitacre and Noegerrath also reported that the engine had not been tested at simulated altitudes higher than 34 000 meters, since no one believed that the environment above that level made any difference for engine firings.

Although Abbott's backfire theory accounted for the oscillations that had triggered the explosion, not everyone agreed that a single cause was enough to explain what happened. But the symposium could come up with nothing better. On 15 November, it recommended to SSD that engines should be modified so oxidizer entered the chamber first and should be tested at simulated altitudes closer to where Agena would be working—above 76 000 meters.[11]

Funk now formed a "super tiger" team of three senior engineers[†] to review everything that had been found about the explosion and to suggest some answers to the NASA review board. The three agreed with oxidizer starting and with firings at simulated altitudes above 76 000 meters. They also wanted Bell Aerosystems to conduct ground ignition tests for data on engine-firing characteristics. The super tigers presented these recommendations at a meeting in Houston on 20 November,[††] then to the Gilruth-Ritland review board, which approved them. Lockheed announced the formation of a Project Surefire Engine

---

[*]Symposium attendees: S. M. King and D. D. Thomas (Aerospace), E. G. Haberman (Air Force Rocket Propulsion Laboratory), Charles E. Feiler (Lewis), Henry O. Pohl and Whitacre (MSC), D. D. Evans and J. H. Rupe (Jet Propulsion Laboratory), F. D. Sullivan and D. M. Wyckoff (Aerojet-General Corporation), T. F. Reinhardt and Craig M. Schmidt (Bell Aerosystems), Jack R. Hahn and R. S. Levine (Rocketdyne Division, North American Aviation, Inc.), J. J. Kappl (The Marquardt Corporation), R. F. Sawyer (Princeton University), and J. L. Grubbs, Jerome Salzman, and Noeggerath (LMSC).

[†]The super tiger team consisted of Bernard A. Hohman (Group Director, Gemini Launch Systems Directorate, Aerospace), Colonel John Hudson, and L. Eugene Root (president of Lockheed Missiles & Space Company).

[††]Mathews presided at the meeting, which included Gardner, Smith, Letsch, Schneider, Bailey, Hammack, Colonel Jean A. Jack (Deputy Chief of Staff, Test, Arnold), W. von Lunkhuysen and Frederick A. Boorady (Bell Aerosystems), L. T. Barnes (ARO, Inc., Arnold contractor), George Low (Deputy Director, MSC), Joseph F. Shea (Apollo Program Manager, MSC), Willis Mitchell (Vehicle and Missions Manager, GPO), and Richard K. McSheehy (MSC Special Assistant for Apollo Support, Propulsion and Power Division).

Development Task force to carry out the program. This did not end the analysis of the trouble. Reports and recommendations from other NASA centers continued to come to Gilruth until 9 March 1966, one week before the Gemini VIII flight.[12]

## AN ALTERNATIVE TARGET

While Agena's sponsors labored to nurse it back to health in time for the Gemini VIII mission, McDonnell engineers had been thinking of other ways to achieve rendezvous and docking. During launch preparations for the *Gemini VII/VI-A* mission, McDonnell's Gemini Program Technical Director, John Yardley, invited several NASA officials to his motel room in Cocoa Beach, Florida. He outlined a plan for making a poor man's target by bolting a target docking adapter to the rendezvous and recovery section of a spacecraft and fitting it to the Atlas booster. An enthusiastic Mueller told Mathews to prepare a defense of the concept for Seamans, who would have to approve it. To avoid any hint that a new development program was in the offing, they decided to call it simply an "ATDA," for augmented target docking adapter, which accurately reflected its status as a rearrangement of already developed and qualified hardware.[13]

An immediate question was whether the Atlas launch vehicle could handle the proposed ATDA; it was much lighter in weight than the Agena but lacked an engine to boost it to orbit. A call to General Dynamics in San Diego posed the weight question (without disclosing the as yet unapproved plan) and received an encouraging response.[14]

By 5 December 1965, Mathews had the case for the alternate vehicle ready. While Day filled in Seamans' staff, Mueller described the plans to the Associate Administrator himself, who approved it. Four days later a statement of work for the ATDA was ready and McDonnell began building the substitute Gemini target.[15] This target-adapter became something of a sword of Damocles over Lockheed, a weapon that GPO was willing to use at more than one level. Jerome Hammack spurred Lockheed's efforts by sending Smith a picture of the alternate vehicle (often called the "glob"), and Mathews asked Flight Crew Operations for an alternative flight plan, eliminating all Agena maneuvers from Gemini VIII.[16]

Project Surefire was already running into trouble. The crucial simulated high-altitude tests of the modified engine could only be run at the Air Force's Arnold Engineering Development Center in Tennessee, but it was booked solid. Time was running out for Gemini VIII, scheduled for mid-March 1966. John Hudson flew to Vandenberg Air Force Base, where he persuaded General Bernard Schriever, Air Force Systems Commander, to sign a letter moving Agena to the head of the line at Arnold.[17]

*301*

*Competitors for the role of target vehicle for Gemini's first docking mission: above, Agena 5003 at the Lockheed plant at Sunnyvale; right, mockup of McDonnell's ATDA with mockup of the Gemini spacecraft.*

The Agena test program also got a priority from NASA, when Mueller decided that Apollo lunar module engine tests at Arnold could be slipped. On 17 December, after Bell had completed the Project Surefire modifications, the Air Force accepted the main engine for GATV-5003. Bell had already begun the series of 48 sea-level firings that the super tiger team had recommended.[18]

Setting up the test program, however, was only part of the problem. Another source of delay loomed in Mueller's demand that GATV-5003, which had arrived at the Cape on 18 January, undergo a static firing before it was committed to Gemini VIII. A worried Agena team, fearful that this demand would so delay their efforts that the outcome of the Arnold tests would be meaningless, met late into the night of 4 February 1966, the eve of another meeting of the Flight Safety Review Board, to discuss means of getting Mueller to postpone his decision.

The next day, Lockheed's Smith and Aerospace's Hohmann told Mueller that a static firing at this point was just not worth the delay it would entail. Mueller quizzed them closely and asked for a written report on the pros and cons of static firing as well as on everything that had been done to ensure that Agena would not fail again. The pressure was still on, but Mueller had at least not closed the door on Agena—it still had a chance to compete with the ATDA for Gemini VIII.[19] February 14 became the deadline for making the choice, while GPO kept working on both the ATDA and Agena.

Late in January, GPO engineers went to St. Louis to conduct a design review of the ATDA, and Gemini procurement received word to put through the final papers for its purchase.[20] ATDA development was quick because its parts had already qualified for space flight and good luck held its cost down. A spacecraft rendezvous and recovery system fished from the sea for postflight examination after the *Gemini VI-A* mission could be used in building the ATDA. McDonnell put it together by 1 February, and NASA conducted the acceptance review the next day.[21] The stand-in was ready to assume the starring role.

Agena was clearly trailing its rival, but its sponsors hoped to regain lost ground when the second act of the test program began at Arnold on 7 February.[22] Meanwhile, Hohmann and Smith had sent their written reports on static firing to Mueller. In Hohmann's view, static firing was mainly useful for training launch crews, not for proving rockets. He pointed out that Mercury-Atlas 1 had failed at launch, even though it had been static fired, and that static firing would not have disclosed GATV 5002's problems. Smith stressed the penalties in money and time.[23] A quick poll of opinion from NASA Headquarters and the manned space flight centers supported the Hohmann-Smith viewpoint,[24] and Mueller dropped the notion of static-firing the Agena.

That was a plus for Agena's prospects, but the test program at Arnold produced less happy results. After the first six tests, problems with mismatched hardware had already compelled GPO to direct McDonnell to speed up its ATDA testing.[25] The seventh test, on 12 February 1966, was nearly fatal. Fuel lines contaminated by alcohol and water caused a hard start that badly damaged the engine. Fortunately, Bell had just about finished its series of sea-level tests and could send that engine to replace the damaged one. As Agena's time seemed to be running out, its proponents literally worked around the clock, juggling, cajoling, scheming, begging, and snarling when necessary, to reach what had begun to seem an impossible goal. More than once, Day and Mathews pleaded with Mueller to keep the Agena. Finally he gave them one week to return the vehicle to good health in time for the review board meetings to be held in Washington 6 and 7 March. Day later recalled his feeling, all during Project Surefire, that Mueller was just putting pressure on MSC and Lockheed and never really intended to cancel the Agena. Mueller did object to the cost of the modifications. He was not willing to brook what appeared to be a $15 million overrun and so was studying proposals to cut Agena out of the program, using the ATDA or two spacecraft for rendezvous.[26]

On 1 March, the new test series began at Arnold. By the end of the fourth day, 22 firings at simulated altitudes of from 83 800 to 114 300 meters had proved the success of the modifications. Meeting as planned, the Design Certification and Air Force Flight Safety Review Boards approved the modified Agena for flight.[27] The Agena had been requalified just in time to fly on Gemini VIII, for its rival, too, was ready, having also completed its test program on 4 March. Now the ATDA went into storage at Cape Kennedy, to be called out if the Agena again faltered.[28]

## EXTRAVEHICULAR ACTIVITY PREPARATIONS

While one group in the manned space flight program struggled to restore the Agena, another faced qualification problems, on a smaller scale, with extravehicular flight equipment. White's spectacular venture into space during *Gemini IV* was backed by a comparatively simple technology. That first step into space had merely required an astronaut to leave the cabin and see what he could do. White did it in grand style, making up his flight plan as he went along. His successor would not be so free as he attempted such specific tasks as retrieving experiment packages. So easy a first venture misled planners into thinking that EVA would present few, if any, problems. No one really worried when space walks were deleted from the missions between *Gemini IV* and VIII, even though the Astronaut Maneuvering Unit was still scheduled for Gemini IX.

The Air Force, however, was disconcerted. Colonel Daniel McKee, head of the Air Force Systems Command Field Office at MSC, complained about being kept in the dark about the plans for White's exercise. His office should have been involved in the orderly planning that was to lead to the use of the AMU, since at about $12 million that was the single most expensive Defense experiment to be carried on Gemini. The AMU was designed to make an extravehicular astronaut independent of spacecraft systems. A boxlike backpack with sidearm controllers, it consisted mainly of three beams and supporting shelves on which such parts as tanks to store the hydrogen-peroxide propulsion system and the life-support oxygen supply were mounted. Because the spacecraft was so small, the AMU was housed back in the adapter section. The astronaut would go out the hatch tied to a tether, make his way to the rear of the spacecraft, and strap himself into the AMU. This special propulsion system weighed about 76 kilograms (168 pounds), which was no burden at all in the weightlessness of space.[29]

On 20 September 1965, NASA had named the crew for Gemini VIII. Neil Armstrong, a civilian test pilot with long experience in the X-15 rocket research aircraft program, was the command pilot, as he had been in the backup crew for *Gemini V.* His fellow crewman, David Scott, was new to the Gemini program. For the backup crew for Gemini VIII, Navy Lieutenant Commanders Charles Conrad and Richard F. Gordon, Jr., the pattern was similar. Conrad had been pilot on *Gemini V,* and Gordon was newly assigned to Gemini.[30]

Some three weeks before the crew was announced, McDonnell held a briefing on the extravehicular gear Gemini VIII would carry. It comprised two major units: an Extravehicular Life Support System, called the ELSS by the engineers who worked on it, and an Extravehicular Support Package, known as the ESP or, more commonly, the backpack. The life support system was a chestpack designed to do just that—feed vital oxygen to the spaceborne astronaut from the spacecraft's supply, from a primary source in the backpack, and from its own emergency supply. The backpack did more. Designed like the AMU to fit into the spacecraft adapter section, it carried, aside from its own oxygen supply, a radio and 8 kilograms (18 pounds) of propellant for a zip-gun maneuvering unit. The backpack was connected to the spacecraft systems by an 8-meter oxygen-hose tether. Once the astronaut had switched over from the spacecraft to the backpack oxygen supply, he could add a lightweight 23-meter tether to the shorter hose and, theoretically at least, maneuver as far as 30 meters from the spacecraft.[31] Armstrong attended the St. Louis meeting and asked for help in the training program. The crews would need a realistic spacecraft adapter with which to practice donning the backpack. Armstrong also wanted the pilot to leave the spacecraft in the altitude chamber and test the combined backpack-chestpack.[32]

As soon as he was assigned to the mission, Scott began concentrating on the extravehicular exercise, eventually going through over 300 airplane zero g parabolas and more than 20 hours on an air-bearing table. The astronauts practiced EVA maneuvers, supported by an air cushion of 0.0254 millimeter (0.001 inch), on a table roughly 6 by 7 meters. They used a zip gun to move from one place to another, which gave them some idea of what it would be like to start and stop in space. This strenuous training raised some questions.[33]

Scott's zip gun had about 15 times more propellant than White's and used Freon instead of oxygen as fuel, further multiplying the gun's total impulse, since Freon has a density about three times greater than oxygen. How oxygen acted in vacuum was fairly well known, but Scott worried about how Freon would behave. One problem soon showed up: at low temperatures, the Freon caused the zip gun's poppet valve to stick open when triggered, and the escaping gas threatened to tumble the astronaut in space. New seals solved the problem and two new shutoff valves added a safety factor.[34]

By December, Scott and Armstrong were voicing a number of doubts about the equipment, ranging from nitpicks to serious complaints. One that fell into the latter class was the threat of an oxygen ejector in the chestpack freezing and blocking the flow of oxygen from both the spacecraft and the emergency supply in the chestpack. The life support system had been icing up during tests. Although test conditions were more severe than Scott would meet in space, he could hardly be expected to ignore the warning. The designers obliged by installing 20-watt heaters near the ejector.[35]

Another problem was the tangle of umbilicals, tethers, and jumper cables that made donning the chestpack inside the spacecraft so difficult. During early tests, Scott found his movements restricted and his vision nearly blocked by his pressure suit while he was trying to connect everything. Late in December 1965, however, Scott satisfied himself that he could strap on the unit, hook it up, and open the hatch in the McDonnell altitude chamber at a simulated 46 000 meters. Scott went through a full-dress rehearsal in the last few weeks before flight, in the MSC 6-meter vacuum chamber, putting on the chestpack inside the spacecraft, going outside, and then donning the backpack housed in the adapter.[36]

## MISSION PLANNING COMES OF AGE

Technical problems in qualifying the Agena and the extravehicular equipment commanded center stage in Gemini VIII preparations, adding to the already heavy burdens of planning the mission. Project Gemini was entering a more advanced phase, as both spacecraft and target faced missions of growing complexity that would test their capabilities to the limit. Program leaders had to balance their concern for reaching the program goals against the dangers of trying too much too soon. A persistent problem such as the Agena presented could not but raise doubts and cause second thoughts about going forward with some as yet untried operation. Even under the best of circumstances, trying to foresee and counter everything that might go wrong with four major dynamic systems—spacecraft, booster, Atlas, and Agena—made mission planning an arduous task. With major technical difficulties further clouding the issue for Gemini VIII, plans changed quickly and often.

In the summer of 1965, MSC's Mission Planning and Analysis Division had started tailoring a plan for Gemini VIII, the first results of which were discussed on 26 and 27 August. Among the alternative modes of rendezvous being considered was a rendezvous sooner than the fourth revolution of the spacecraft—the "standard" rendezvous that had been scheduled for *Gemini VI*. Despite doubts that the flight control team could support any rendezvous earlier than that, the scheme called "M equals 2" (rendezvous in the second revolution) being studied by the mission planners was worth thinking about. Another subject was a proposed phantom rendezvous with an imaginary target, requiring a thrust from the Agena's main engine of at least 150 meters (500 feet) per second, to take place shortly after the first sleep period. The pilot would then exit the spacecraft for more than two hours of extravehicular activity—that is, floating freely around the world! After that, the spacecraft would undock and withdraw from the Agena, to return later for a second rendezvous. Finally, Gemini VIII's Agena would be left in orbit as a passive target for Gemini IX.

No sooner had the Gemini VIII plan been committed to paper than caution flags were raised. One issue was an old one that the earlier crews had fought—sleeping alternately. Lockheed recommended that one astronaut remain awake whenever the spacecraft and Agena were docked. Mathews consulted with Whitacre, then denied the request on the ground that sleep at this time (after launch, rendezvous, and docking, and before EVA) was necessary for both men. Besides, Whitacre's analysis showed that the tracking network could cope with almost anything that might go wrong. Another question was time. Fuel cell development problems had imposed a limit of two days on rendezvous flights. Could so elaborate a plan be carried out in such a short time? Maybe the time could be expanded. Since the fuel cell's troubles seemed headed toward resolution, McDonnell was asked to see if later rendezvous missions could be extended to three days.[37]

Meetings continued throughout the fall of 1965, as spokesmen for NASA Headquarters, McDonnell, and MSC began to stress re-rendezvous, which they thought might be good training for Apollo. Discussions on firing the two Agena propulsion systems remained inconclusive, as did talk about flying for three days. When a McDonnell study indicated that the fuel cells could support a 72-hour flight, if all supplies were carefully husbanded, that question appeared to be settled.[38] Firing the Agena main engine while docked with the spacecraft, however, was finally rejected for the same reason that it had been on *Gemini VI*—it was not yet deemed safe enough. That meant the phantom rendezvous was out.[39]

Toward the end of February 1966, with problems seemingly well in hand, a "final" version of the flight plan appeared. Like the *Gemini VII* plan, this was more an outline than a precise schedule of events. Crew and flight controllers had a range of options to deal flexibly with circumstances as they arose in the course of the mission.[40]

Operations planning was being paralleled by experiments planning. By November 1965, the Manned Space Flight Experiments Board at NASA Headquarters had approved eight tasks for Gemini VIII. Eventually, ten experiments were approved for the mission,* three of them requiring extravehicular activity. Of these, two were scientifically oriented—S-9, Nuclear Emulsion, to expose an experimental package to radioactivity in space (especially in the South Atlantic Anomaly), and S-10, Agena Micrometeorite Collection, to study the micrometeorite content of the upper atmosphere. In the third, proposed by the Department of Defense, Scott would use a power wrench for weightless work. He would go to the adapter area, pull out a box containing a torqueless motor-driven wrench, use the tool to take five nuts off a special plate, and then rebolt the plate to the box. This simple task—with and without knee tethers—would be compared with doing the same thing on the ground to show the differences in work-

---

*See Appendix D.

ing in one gravity and in weightlessness. Scott and George C. Franklin of the Flight Crew Support Division decided to augment this experiment by adapting a cheap standard socket wrench to fit the nuts and the pressurized glove. They believed that comparing the muscle-powered and the electrically-operated tools would say something useful about energy usage in space.[41]

Mission plans and flight schedules were inseparable, and Apollo again began to intrude. Apollo mission 201 was planned for February 1966; if there were any delays, it would slip into March. The problem was not with launch pads nor, in most cases, with people. A tracking ship, the *Rose Knot* Victor, was the source of conflict. For Apollo 201, a suborbital flight, the ship would be sailing the Atlantic Ocean. But its station for Gemini VIII was in the Pacific. Mueller ruled that the Gemini flight had priority; but Apollo 201 flew as scheduled on 26 February, giving the slow-moving *Rose Knot* time enough to keep its date with Gemini VIII.[42]

Flight Control also shifted for Gemini VIII. Christropher Kraft, who had directed flights for Mercury and all Gemini missions through *VII/VI-A*, had to leave Gemini to begin planning for lunar landing missions as Apollo neared operational status, although he expected to keep an eye out for Gemini lessons that might be of use to Apollo. Kraft's move left Gemini Mission Control short of experienced flight directors. His successor, John Hodge, who headed the Flight Control Division, divided flight direction into 12-hour shifts with Eugene Kranz, Chief of the Flight Control Operations Branch. Clifford E. Charlesworth, flight dynamics officer on past Gemini missions, began training as a flight director.* [43]

In the two weeks before the scheduled launch, equipment problems remained a threat. The extravehicular gear, in particular, was still in trouble, with lines icing and valves cracking. Then, at Cape Kennedy, the spacecraft environmental control system began acting up; and, over on pad 14, Atlas fueling ran into some difficulties. These last two problems did cause a day's delay, from 15 to 16 March. Then everything was ready to go.[44]

## THE WHIRLIGIG

On 16 March 1966, five months after Walter Schirra and Thomas Stafford had been left at the starting gate in NASA's first attempt to launch two vehicles toward rendezvous on the same day, NASA tried again. This time nothing marred the countdown of the Atlas-Agena or the Gemini space vehicle.[45]

---

*See Appendix F for Mission Control Center position descriptions and responsibilities.

The target launch vehicle lifted from pad 14 at 10 o'clock in the morning. Its trajectory was at first low and to the right (south) of the intended flight path. The sustainer engine rammed the target back on track. In a little more than five minutes, the Atlas had done its job. Now it was Agena's turn. After a short coast, its secondary propulsion system burst into life. The crucial test for the Agena came with the firing of its main engine, and the engineers crossed their fingers and held their breaths. But it worked. The engine ignited and carried the target into a 298-kilometer circular orbit.[46] Planners had wondered if the Agena could so position itself that astronauts could catch it. The answer was yes!

With one up and one to go, attention turned to pad 19. Fourteen minutes before the Atlas-Agena lifted, Armstrong and Scott slid through the spacecraft hatches into their couches. As the flight-preparation crew helped harness Scott to his parachute, they found one of its catches full of glue. Backup command pilot Conrad and McDonnell pad leader Guenter Wendt began digging it out. Just a little thing like that, Scott thought, "might have cost us a launch," but he could not help smiling as he watched Conrad sweat over the job. The catch came unglued and Gordon, the backup pilot, tried the fitting a few times to prove to Scott that it was working. Learning of the Agena's nearly perfect orbit, Armstrong said, "Beautiful, we will take that one."[47]

Given the Agena's orbital parameters, the Gemini launch vehicle should lift off at 10:40:59 a.m. The powerful engines of Titan II throbbed into life exactly on time, and Armstrong and Scott felt the hold-down bolts shear for breakaway. GLV-8 started off a little low, as had Atlas, but soon straightened to boost the 3788-kilogram (8351-pound) spacecraft into an elliptical orbit 160 by 272 kilometers.[48]

After the first hurdle had been vaulted, the next challenge was catching the target. Procedures were much the same as those for *Gemini VI-A*, although this time there was no friendly target to point its attached transponder toward the spacecraft's radar. Armstrong and Scott began the chase 1963 kilometers behind the Agena.

Thirty-four minutes into the flight, the Sun set and, in the engulfing darkness, the crew could see brilliant fires streaming from their spacecraft's thrusters. As the radiator in the adapter expelled water, the thrusters fired to compensate for a sideward turn. The Carnarvon, Australia, tracking station told them the radiator was not much of a problem and passed to them the Flight Director's "go" for a day's flight.[49]

Over the Pacific, the two astronauts had some time to sightsee. Molokai, Maui, and Hawaii hove clearly into view. Armstrong tried to see Kauai and Oahu, but cloud banks obscured them. Minutes later, Scott said to his partner, "We're going over Baja California now. Can you see it?" But Armstrong had his eyes on the Los Angeles ship basin in the other direction, and his response was, "Oh, look at all those ships!" Armstrong then spotted the Rogers Dry Lake bed. He looked for, but was not certain he found, Edwards Air Force Base, where he had spent seven years piloting experimental airplanes. Over Texas, both men wanted to see if they could spot their homes, but work preempted this scenic interlude. At the low point of their first circuit of Earth, Armstrong aligned the inertial platform for a height adjustment maneuver. At 1:34 hours elapsed time, he touched off a five-second burst of the thrusters for a small retrograde change in velocity, to lower the apogee slightly. Armstrong noticed a problem in cutting off residual thrust. This resulted in varying computer readings and made it difficult to tell the exact deceleration obtained.[50]

On their mission, Schirra and Stafford had been so preoccupied that they had not taken time to eat, which left them hungry, as well as tired, when they caught up to Borman and Lovell. Scott and Armstrong knew they would be very busy all three days of their mission, so each grabbed a package of food and started preparing a meal, which seemed to take longer than they thought it would. When they had to stop and align the platform for a maneuver to raise the perigee, they placed the food packages against the spacecraft ceiling. Weightlessness was handy.[51]

Nearing second apogee (2:18:25 hours), Armstrong fired the thrusters to add 15 meters per second to their speed. Again, tail-off residuals made it hard to get a computer reading.[52] After this maneuver, Armstrong and Scott pulled their food from the ceiling. Although Armstrong's chicken and gravy casserole had been mixed with water for half an hour, it was still dry in spots and not much like home cooking. But he finished it and washed it down with fruit juice to keep from dehydrating. Then he tried a package of brownies, which were stuck together and crumbly. They were hard to eat without scattering weightless scraps all over the cabin.[53]

The next maneuver was designed to push the spacecraft into the target's orbital plane. Armstrong yawed *Gemini VIII*'s nose 90 degrees south of the flight path. Over the Pacific Ocean, 25 minutes before completing the second revolution (2:45:50 hours), the command pilot punched the aft thrusters to produce a horizontal velocity change of 8 meters per second. He waited for the ground controllers to tell him if any adjustment was needed. Hearing nothing, he assumed his thrusting had been correct. Over the Guaymas, Mexico, tracking station, Lovell, the Houston CapCom, suddenly cut in on the remote site line to order him to add 0.6 meter per second to his speed. With only a minute to get ready, there was little time to turn the spacecraft and no time to align the platform. "It was ... a pretty quick loose burn ... without much preparation," Scott said.[54]

Armstrong and Scott then began the rendezvous radar test. They did not expect to get radar contact as quickly as Schirra and Stafford had, but the Westinghouse development team had promised target acquistion at 343 kilometers. The radar locked on solidly at 332 kilometers, which was good enough.[55]

Over the Tananarive tracking station, 3:48:10 hours after launch, Armstrong nosed the spacecraft down 20 degrees and applied the aft thrusters for an in-plane (with the target) velocity change of 18 meters per second. This gave them a nearly circular orbit close to 28 kilometers below that of the target. The spacecraft was now in position to start the terminal phase of rendezvous.[56]

The crew sighted a shining object 140 kilometers ahead, which must be the Agena. After closing to a range of 102 kilometers, all doubts were erased—the target gleamed in the sunlight. Scott switched the computer from the catchup to the rendezvous mode and watched the distance dwindle on the slide, automatically. Just before sunset, the Agena suddenly disappeared, but at twilight its acquisition lights blinked into view.[57]

When the Agena was at the proper angle (10 degrees) above them, Armstrong aligned the inertial platform for the translation maneuver. Then he pitched *Gemini VIII's* nose up 31.3 degrees and canted the vehicle 16.8 degrees to the left. At 5:14:56 hours, ground elapsed time, the command pilot fired his aft thrusters, later making two small corrections. High over the *Coastal Sentry* Quebec tracking ship, stationed near Antigua Island, at 5:43:09 hours, he braked the spacecraft. Since he could see the Agena, Armstrong judged his braking action by eye as Scott called out radar range and range rate. At a distance of 46 meters, relative velocity between the two vehicles had been canceled. The second rendezvous in the Gemini program had been achieved.[58]

For 36 minutes after rendezvous, Armstrong's delicate maneuvering kept his spacecraft on station with the target vehicle. As the command pilot drove, Scott inspected the Agena—checking antennas, docking lights, and the like. Finding it hard to see all of the target's instrument panel displays near the docking cone, he used the telescopic sight of a hand-held sextant. But a really good look would have to wait until they were docked, when these instruments would become a second dashboard. Meanwhile, Armstrong studied the general appearance of the Agena. It seemed stable, and he nudged the spacecraft to within a meter (about three feet) of the target. Then, at 6:32:42, Keith K. Kundel, CapCom on the *Rose Knot* Victor, radioed, "Go ahead and dock."[59]

Armstrong eased *Gemini VIII* toward the target at a barely perceptible rate of 8 centimeters (3 inches) per second. "About two feet [60 centimeters] out," he told the *Rose Knot* Victor. In a matter of seconds, Armstrong gleefully reported, "Flight, we are docked! It's

. . .really a smoothie—no noticeable oscillations at all." For a moment, the flight controllers in Houston could not realize that they had really done it. Then pandemonium broke loose, with back slaps, hand shakes, cheers, and tremendous grins.[60]

Because there had been some difficulty in verifying the Agena's uplinked* stored program commands for the planned docked yaw maneuver and in loading the target's velocity meter, the flight controllers suspected that Agena's attitude control system might be misbehaving. In fact, Lovell, on the remote link through Tanarive just before the spacecraft passed out of communications range, told the crew, "If you run into trouble and the attitude control system in the Agena goes wild, just . . .turn it off and take control with the spacecraft." With this warning ringing in their ears, Armstrong and Scott began their docked chores.[61]

The Agena was designed to obey orders from the spacecraft, as well as from ground control. Scott commanded the target's attitude control system to turn the vehicle combination 90 degrees to the right. It took five seconds less than the full minute expected. Scott next dialed an order to start the Agena's tape recorder and looked over toward Armstrong. As he did, his gaze skimmed the control panel in the spacecraft. Something had to be wrong—*Gemini VIII* should be in level flight, but the "ball" indicator showed a 30-degree roll. He knew there was no use checking the horizon out the window, as they were passing through Earth's shadow. There would be no help from the ground tracking stations either; they were still out of communications range.[62]

"Neil, we're in a bank," Scott said. He thought perhaps his spacecraft attitude ball had tumbled, but Armstrong's indicator showed an identical mark. The command pilot managed, with bursts from the OAMS, to stop the motion temporarily, but it soon started again. Their immediate reaction was to blame the Agena. As soon as the vehicles were steady enough, Scott commanded the target to turn off its attitude control system, as the communicator had instructed. For four minutes, the two craft steadied and straightened up; the trouble seemed to be over. Armstrong started maneuvering to get the docked vehicles into the correct horizontal position; suddenly they began to roll again, faster and faster. "What's the problem now?" the pilots wondered. They were supposed to do a small test to find out what stress and strain the linkage between the two vehicles could tolerate. That issue was now academic; the immediate question was whether it could stand up under these wild gyrations.

---

*"Uplink" was a term used by flight controllers to denote information telemetered from the tracking network to the spacecraft and Agena. "Downlink" meant the opposite—from space to the ground.

**Gemini VIII
16 March 1966**

*Mission review meeting at Cape Kennedy prior to Gemini VIII: left to right, Charles Mathews, Gemini program manager at Houston; David Scott, Gemini VIII pilot; and William Schneider, Gemini mission director, NASA Headquarters.*

*Rendezvous: 65 meters.*

*Station-keeping: 14 meters.*

*Docking with Agena.*

While Armstrong struggled with the controls, Scott photographed the interaction between the two vehicles out of his spacecraft window. The command pilot soon reported that the OAMS propellant had dropped to 30 percent, a strong clue that a spacecraft thruster might be causing the trouble. While Armstrong fought the controls, Scott cycled the target vehicle switches off and on and off again. Then Armstrong jiggled the spacecraft switches as well, to see if he could isolate the problem. Nothing they did seemed to have any effect.[63]

The crew realized that they would have to break away from the Agena to analyze the situation. Past simulation training gave them no clues to what was happening or how to handle it. Scott transferred control of the Agena to the ground stations (which had been locked out to prevent spurious signals), and Armstrong labored to steady the vehicles enough to divorce them. "Go," Armstrong said, and Scott hit the undocking button. Armstrong gave the thrusters a long hard burst, and the spacecraft pulled straight back.[64]

Almost immediately, suspicion about a spacecraft control problem became an established fact as the spacecraft rolled even faster. "And then we really took off," Armstrong and Scott later reported. *Gemini VIII* soon came into acquisition range of the *Coastal Sentry* Quebec. James R. Fucci, CapCom aboard the ship, was concerned and perplexed. He could not get a solid electronic lock-on, but a blinking light signal indicated that the craft had undocked. Unaware that the spacecraft was rolling, so the antennas could not remain in position, he put in a call to the crew to try to find out about these strange signs he saw on his console.

Fucci: *Gemini VIII,* CSQ Cap Com. Com check. How do you read?

Scott: We have serious problems here ... we're tumbling end over end up here. We're disengaged from the Agena.

Fucci: Okay. We got your SPACECRAFT FREE indication here.... What seems to be the problem?

Armstrong: We're rolling up and we can't turn anything off. Continuously increasing in a left roll.

Fucci: Roger. [37 seconds later] *Gemini VIII.* CSQ.

Armstrong: Stand by.

Scott: We have a violent left roll here at the present time and we can't turn the RCS's off, and we can't fire it, and we certainly have a roll ... stuck hand control.[65]

After backing away from the Agena, the spacecraft had started to whirl at a dizzying rate of one revolution per second. Armstrong suspected that the maneuvering thrusters were about finished. He and Scott were also having trouble seeing the overhead panel dials; their physiological limits seemed near. They were dizzy, and their vision was blurred. Something had to be done. "All that we've got left is the reentry control system," Armstrong said. "Press on," Scott responded. The two men began to throw switches to cut out the OAMS and cut in the reentry control system. Armstrong tried his hand controller—nothing. Scott tried his—still nothing. They started switching circuitry again—maybe something had been set in the wrong position.

The hand controllers responded!

Armstrong steadied the motion and then turned off one ring of the reentry control system to conserve fuel. He then carefully reactivated the maneuver thrusters; now they were able to tell that No. 8 had "failed on"—that is, it had stuck open![66]

Using the reentry control thrusters meant that the *Gemini VIII* mission would have to come to an end as soon as possible. That was a mission rule. True, the spacecraft was operating in a backup mode—but it was the prime mode for reentry. If these thrusters developed leaks, the crew would have absolutely no means of getting the spacecraft into position for the critical retrofire that would return them to Earth. Attitude control before and after reentry was essential to reenter the atmosphere safely. Here was a case where the fail-safe maneuvers that Headquarters had insisted on early in the program were impossible—there was virtually no maneuverability left in the orbital thrusters. Armstrong and Scott also remembered, wistfully, that Kraft, the flight controllers, and engineers had nursed other missions to completion. Could the same be done for them now? This was but a fleeting hope, as the Hawaiian tracking station communicator told them to get their spacecraft into position for reentry.[67]

*Gemini VIII*'s problems were certainly the most frustrating of any Gemini had yet encountered. The flight control team's ability to respond to real problems on previous missions, keeping spacecraft flying to wring all useful data from failures as well as successes, had bolstered confidence in the program and promoted "real-time" planning. But *Gemini VIII*'s failure had forced the astronauts to resort to a last-ditch mode for attitude control before the ground crews had a chance to provide the options that might have allowed the flight to go on.

John Hodge, in his first trial as chief flight director, now had only one choice left—which contingency recovery landing area would be best? If he waited much longer, it would take a full day (or 15 revolutions) for the crew to reach a splashdown point from which they could be quickly recovered. Since the orbital track had precessed westward, landing during the sixth or seventh orbits would have to take place in the Pacific Ocean. When the Landing and Recovery Division recommended a touchdown in the seventh circuit, Hodge agreed.

Kranz had dropped by to listen to the spacecraft and target docking. Since Hodge had been at the flight director's console for 11 hours,

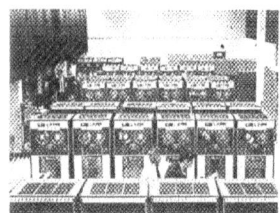

*The windowless Mission Control building in Houston (below, foreground) was the scene of tense activity as it planned an emergency reentry for Gemini VIII. The real-time computer complex (left) spun off millions of calculations to its control room (right).*

## Troubled *Gemini VIII* and Mission Control

*Consoles (above) reported on the capabilities of the recovery force and on the weather; left, the recommendation: shift the landing from the Atlantic to the Western Pacific near Okinawa (William Schneider, John Hodge, Robert Thompson, Christopher Kraft); right, Kraft fields reporters' questions.*

he and Kranz decided that the second shift should report for duty immediately, catch up on all information, and direct the final phases of the mission. Had the flight continued for three days, reentry would have taken place on Kranz' shift, anyway, and he and his men had more practice in recovery procedures than Hodge and his group.[68]

The engineers who had worked so hard on the Agena's problems found their situation just as exasperating as that of the flight controllers. After the docking, Smith, Harold W. Nolan, and others from Lockheed had retired to nearby motel rooms to celebrate the momentous event. Very shortly, Smith called Nolan, saying, "We've got trouble!" Nolan switched on his television, only to hear newscasters reporting that the Agena was at fault. Smith's motel room became the initial Lockheed failure-analysis command post, the first guess being that the target's attitude control system had failed.

Many other engineers and program officials also heard about the spinning spacecraft while out of touch with minute-by-minute developments. Mueller, for instance, had remained at Cape Kennedy only through launch and the early phases of the mission. Then he took off for Washington to attend the annual Robert H. Goddard Memorial Dinner, sponsored by the National Space Club. The pilot of the NASA aircraft heard what was happening over the plane's radio and informed Mueller. They returned to Florida, where Merritt Preston met Mueller's party with a motorcycle escort, the prelude to a hair-raising ride to the old Mercury Control Center in time for spacecraft retrofire.[69]

Most of NASA's leaders at Headquarters had, in fact, already headed for the Goddard dinner—the prestigious social event of the year for the space community. At the opening reception, Deputy Administrator Seamans* was called to the telephone to learn of *Gemini VIII*'s plight. He immediately phoned Houston Flight Control and learned that the spacecraft spinning had been stopped. When he told the chairman of the dinner about the trouble, Seamans was asked to make a brief announcement: he said the flight would have to be aborted, but the crew seemed in no immediate danger. Vice President Hubert H. Humphrey, the principal speaker, asked to be told as soon as the crew had been successfully recovered. Before he had finished his address, Humphrey was able to inform his listeners that Armstrong and Scott had landed safely. Seamans vowed that never again would he be caught in a public position during the critical phase of any succeeding flight. He needed privacy and better communications with the Control Center.[70]

As a rule, McDonnell (the spacecraft contractor) sent several of its

---

*On 21 December 1965, Seamans had been sworn in as Deputy Administrator of NASA, replacing Hugh Dryden, who died on 2 December.

experts from the Cape to Houston after launch and first orbit to be available as troubleshooters. On 16 March 1966, a NASA Gulfstream left Florida for Texas with about 14 passengers, among them several high-ranking McDonnell engineers. Over New Orleans, the pilot cut in a commercial radio broadcast to the cabin. The announcer was talking about an imminent recovery in the Pacific. This was all the startled passengers heard, since the news announcement ended there. Something had obviously gone wrong, but what was it? There was nothing to do but wait until they got to Houston.

Raymond Hill, McDonnell's Gemini manager at the Cape, recalled that his company's policy changed radically after "I . . . was caught with my pants down." In the future, senior McDonnell officials—Hill, Walter Burke, John Yardley, and Robert Lindley—would not be in transit at the same time during a flight. Hill stayed at the Cape, Burke went to Houston for the first day of the flight and then back to St. Louis, and Yardley and Lindley went to Houston and stayed until the mission was over. McDonnell specialists, who had previously remained in St. Louis to handle test set-ups and to answer questions by telephone and teletype, were shifted, along with their subcontractor counterparts, to Houston to work directly with GPO systems engineers during the mission.[71]

Meanwhile, Navy recovery forces in the Pacific were swinging into action. A destroyer, the U.S.S. *Leonard F. Mason*, steamed at flank speed toward the expected landing point 800 kilometers east of Okinawa and 1000 kilometers south of Yokosuka, Japan.[72]

With *Gemini VIII* now flying over the southern latitudes, Kranz had only three tracking stations in position to keep in touch with the crew*—*Coastal Sentry* Quebec, *Rose Knot* Victor, and Hawaii.[73] The spacecraft was in darkness over the Congo when Kranz's Houston flight controllers began the final countdown for retrofire. Through the remote stations, Scott reported, "Props OFF," and Armstrong said, "Hang in there." Seconds later, Scott said, "Okay. Four retros fired in AUTO RETROFIRE. . . ."

Armstrong was worried that he and Scott might land in some remote wilderness where they would be hard to find. He later said he had been thinking of the steamship *Andrea Doria*, which had gone down in the Atlantic on 26 July 1956. Although the liner's radios were operating, it had taken the rescue vessels a day and a half to find the sinking ship. He wanted Scott to doublecheck his every move—"I keep thinking there's something we've forgotten about," he said, "but I

---

*Because the orbital track of a spacecraft during a day's flight ranges from 30 degrees north latitude to 30 degrees south, the maximum number of tracking stations were available during only 3 of the 15 revolutions in that 24-hour time period.

don't know what it is." Scott answered reassuringly, "We've done everything, as far as I know." Over China, *Gemini VIII* slipped down into
the fringes of the atmosphere.[74]

Everything clicked off properly during descent. As they neared a
landing, Armstrong asked his partner, "Do you [see] water out there?"
Looking into the first faint light of dawn, Scott replied, "All I see is a
haze." Then his voice quickened, "Oh, yes, there's water! It's water!"
Less than two minutes later, Scott yelled, "LANDING—SAFE." The
flight had lasted 10 hours 41 minutes 26 seconds.[75]

The crew went quickly through the postlanding checklist, putting
switches and valves in their correct positions. Then antennas were extended so they could communicate with the recovery forces. "Naha
RESCUE 1, Naha SEARCH 1," Scott called, but no answer came.
They were not very worried, however, as Houston Flight Control had
told them the rescue planes would get to them shortly and the *Mason*
should reach them in three hours. This meant their landing had been
very close to the contingency touchdown point.[76]

Several aircraft, including two HC-54 Rescuemasters—one from
Naha Air Base, Okinawa, and the other from Tachikawa Air Base,
Japan—had raced to fetch the crew. The HC-54 from Naha got there
first. Suddenly the pilot shouted, "I got it!" He had seen the spacecraft, with its main parachute in full bloom, drifting to the ocean's surface. Three pararescuemen were equipped and ready to jump.
Armstrong and Scott saw one of the three as he parachuted down.
Because of the waves, the frogmen had trouble hooking the flotation
collar to the spacecraft. The rough sea also made them queasy, a feeling shared by the astronauts. But the swimmers persisted and secured
the collar within 45 minutes of spacecraft landing. Demonstrating excellent cooperation with NASA and careful planning, the Department
of Defense recovery forces had reacted to the emergency landing as
though it were normal. Armstrong and Scott had few complaints about
recovery in this remote area.[77]

Three hours later, as promised, the *Mason* pulled alongside and
fastened a line to the spacecraft. Climbing the Jacob's ladder in sea
swells of 4 to 5 meters was hard, but they made it. On deck, the tired
astronauts managed smiles and greetings for the welcoming sailors. Still
feeling nauseated, the *Gemini VIII* crew headed immediately for sick
bay. Medical personnel helped them strip off their pressure suits.
Their undergarments were soaked with sweat. They were thirsty, but
clinical examination showed minimal dehydration. The *Mason* reached
Okinawa the next day, and the two astronauts flew on to Hawaii, then
home.[78]

Once the manned phase of the *Gemini VIII* mission was over,
Hodge and Kranz turned back to the target vehicle. Because Scott had
the foresight to pass the control of the target back to the ground,

there was a chance to put the Agena through its paces and see how it reacted to commands. There was still hope that the Agena for *Gemini VIII* might be used as a passive target for Gemini IX or X.

### AGENA'S SOLO

After the undocking, the Agena had stabilized quickly. In the 15th revolution over the Carnarvon station more than 21 hours after launch, Flight Control commanded the main engine to fire twice, to place the Agena in a circle 407 kilometers above Earth. The first burn produced its half of the goal, but the second did not. Instead Flight Control found the parameters were now 407 by 626 kilometers.[79]

Melvin F. Brooks, the Agena systems monitor in Flight Control, immediately began conferring with the Lockheed engineers to figure out what had happened. They suspected that the vehicle's center of gravity had been miscalculated. How could they command the vehicle to offset this? On the next main engine burn, the center-of-gravity compensation attempt failed. Brooks and Lockheed engineers huddled again. What could be wrong now? They finally agreed that there also seemed to be trouble in the yaw hydraulics, allowing the engine to gimbal more than it should. The target's orbit now measured 211 by 476 kilometers.[80]

If this Agena were to become Gemini IX's or X's passive target, there were two major problems to contend with, and Flight Control had to decide what to do about them. There was definitely too much fuel aboard* and the orbit was still too high. Hodge and his controllers decided not to try any more plane-change maneuvers; they would simply try to get the vehicle to the altitude they wanted. The next firing, a retrograde maneuver, convinced them that they had the hang of operating the vehicle. So Flight Control concentrated on reducing the fuel supply in both the primary and secondary tanks.[81]

In all, ten maneuvers were made using the two propulsion systems, sometimes with both firing at once. This was considerably more than the five starts required by the contract. The Agena's command and communications system had accepted a total of 5439 commands (45 from the *Gemini VIII* spacecraft). Lockheed's contract had only called for 1000.[82]

Just before the *Gemini VIII*-Agena docking, Scott had commented that he "bet those Lockheed guys are just jumping up and down." And so they had been. Jubilation died quickly when the news came that the spacecraft was in trouble. Agena's solo maneuvers wiped away any suspicions of wrongdong on its part. Somebody else must shoulder

---

*The Agena's electrical system would be dead before a return visit by a spacecraft; with no way to control the target, a load of fuel was a hazard during any rendezvous attempt.

the blame for *Gemini VIII*'s early landing. Why had thruster No. 8 failed in the open position?

From its landing spot in the Pacific Ocean, the spacecraft had been hauled back to its place of birth—the McDonnell plant in St. Louis—so the engineers could analyze its problems. Set up in a controlled laboratory where the investigations could proceed unmolested, the spacecraft was checked over completely for more than a month. Only the most probable cause of the trouble could be identified. Scott Simpkinson's evaluation team decided that:

> The valves on thruster 8 opening unintentionally was probably caused by an electrical short, ... there were several locations in the spacecraft at which the fault could have occurred.

To prevent a recurrence of the thruster problem, McDonnell changed the attitude control circuit switch so that when it was in the "off" position no power could go to the thrusters. Formerly, turning off the power to the electronics packages did not stop power going to the thrusters. They could still fire.[83]

Thus, the *Gemini VIII* mission ended on a dissonant chord—high success (the first space docking), undeniable failure (abbreviation of the mission), and much relief (safe recovery of the astronauts from a dangerous situation). The timing of the failure was especially frustrating. Being out of communications left the flight controllers and engineers helpless. Time after time in later interviews they repeated: if that spacecraft had just been over a ground station, telemetry would have told them that thruster No. 8 was firing continuously; they could have told the crew what to do before the reentry control system was activated and it was too late. Although the Gemini team was chagrined that the crew had been forced to land early, the knowledge that docking could be achieved with relative ease somewhat assuaged their anguish. Moreover, the Agena solo had demonstrated the target vehicle could help fly more elegant missions. There would be no pause in the program.[84]

Press on to Gemini IX!

# XIV

# Charting New Space Lanes

I N October 1965, Elliot See and Charles A. Bassett II learned that they would fly Gemini IX. Chief Astronaut Donald Slayton also told them that their backups would be Thomas Stafford and Eugene Cernan.[1] Stafford was, at that time, copilot for *Gemini VI*. When that mission failed to go and plans brewed for *VI-A* to rendezvous with *VII*, See, Bassett, and Cernan wondered whether Stafford could finish in time to get ready for IX.

But they could not wait for him; the three men started training in November, sandwiching their simulations between those of other crews. They followed Spacecraft 9 through its building and testing, familiarized themselves with Gemini systems, and helped shape a tentative flight plan. Bassett and Cernan focused on extravehicular activities because one of them would go outside the spacecraft and ride the Air Force's Astronaut Maneuvering Unit (AMU).

The trio interrupted their routine early in December to work as communicators in the Houston Mission Control Center during the *VII/VI-A* mission. They then returned to flight training. Stafford, however, had to go through his postmission debriefing before he joined them in February 1966.[2]

## TRAGEDY

One bright winter morning, the last day of February 1966, the Gemini IX foursome checked into Ellington Air Force Base, Texas, for flight clearance to St. Louis in two dual-seat T-38 jet aircraft. They

*323*

planned to spend several days practicing on the rendezvous simulator at the McDonnell plant.

At Ellington, the four fliers learned that weather in St. Louis was gloomy: 180-meter overcast, visibility 3 kilometers, rain, and fog, with little change expected. Instrument flight rules would be required. See called the St. Louis air traffic controllers, saying he would see them in a couple of hours. He and Cernan discussed the different runways at Lambert Field in St. Louis. See then climbed into the front seat of one T-38, with Bassett easing into the back seat. Stafford and Cernan got into the other plane. They took off from Ellington at 7:35 a.m. See and Bassett led, with Stafford and Cernan flying wing position.[3]

Reaching St. Louis just before 9 o'clock, See radioed the Lambert Field control tower and learned that the overcast had lifted to 240 meters since his earlier call, but the visibility had dropped to 2.4 kilometers. Light snow flurries now mixed with the rain and fog. As the aircraft descended through the overcast, the pilots found themselves too far down the runway to land. See elected to keep the field in sight and he circled to the left underneath the cloud cover. Stafford followed a missed approach procedure and climbed straight ahead into the soup to 600 meters, intending to make another instrument approach. He landed safely on his next attempt.[4]

Meanwhile, See had continued his left turn. The aircraft angled toward McDonnell Building 101, where technicians were working on the very spacecraft See and Bassett were scheduled to fly. Apparently recognizing that his sink rate was too high, See cut in his afterburners and attempted a sharp right turn; but it was too late. The aircraft struck the roof of the building and crashed into a courtyard. Both pilots were killed.[5]

NASA named a seven-man board to investigate the accident. Led by Astronaut Alan B. Shepard, Jr.,* the board looked into all aspects of the tragedy—aircraft maintenance, pilot experience, medical histories, and weather conditions. Shepard's group listened to testimony from everyone who had anything to say, sifted the wreckage for clues, and drew conclusions. They found nothing wrong with the aircraft; it had functioned properly to the moment of impact. Within the past six months, See and Bassett had renewed their instrument flying certificates. Before and during the flight, both men had been in good physical and mental condition, as attested by medical examinations and by reported pre- and in-flight conversations. Furthermore, See was reputed to be an excellent test pilot. Careful, judicious, and technically competent, he should never have crashed at all. Weather appeared to have been the major contributing cause, and pilot error prompted by a desire not to lose sight of the field had carried them too low.[6]

*The other members of the investigating team were Alan Bean, Joseph S. Algranti, Harold E. Ream, John M. Kanak, Dick M. Lucas, and John F. Zieglschmid.

On Wednesday, 2 March 1966, Spacecraft No. 9, on its way to the flight dock for shipment to Cape Kennedy, passed an American flag flying at half-mast at the McDonnell plant. The next day, Elliot See and Charles Bassett, attended by their fellow astronauts, were buried in Arlington National Cemetery across the Potomac from the Nation's capital.[7]

NASA assigned the Gemini IX prime crew positions to Stafford and Cernan, marking the first time in the agency's manned space flight history that a backup crew had taken over a mission.* On 21 March James Lovell and Edwin Aldrin were given the backup duties. There would be no delay in the launch schedule.[8]

### THE WHAT AND HOW DEBATES

Problems in getting ready for Gemini launches were causing fewer delays by the spring of 1966 than they had earlier. Vehicles were getting to Cape Kennedy for storage about a month before they were needed on the launch pad. The NASA-Air Force-industry launch teams had gained plenty of experience in reacting quickly to Gemini hardware problems. Merritt Preston, one of NASA's leaders at the Cape, said later, "Habitually we got in trouble on Gemini, but it never got to us because we could always fix it."[9] Spacecraft 8's thruster failure turned out to be a blessing in disguise. As the Cape workmen combed the adapter area around the thrusters on Spacecraft 9, they found a number of likely causes for the malfunction, which they attended to on the spot. Meanwhile, in St. Louis, engineers were exploring ways of dealing with the electrical short in the thruster circuit. GPO and McDonnell decided on a master switch that would cut off all power to the thrusters simultaneously. In case of trouble, the crew could check the system, circuit breaker by circuit breaker, until a short was found. The Cape team installed this switch on Spacecraft 9 with no effect on the launch schedule.[10]

For Gemini IX, the three major questions centered on working procedures rather than technology: tethered versus untethered extravehicular activity, rendezvous in the third spacecraft orbit, and radar versus optical tracking from the spacecraft.

Work on the Astronaut Maneuvering Unit by Chance Vought and the Air Propulsion Laboratory at Wright-Patterson Air Force Base in Ohio set the stage for the tether debates. The manually operated unit was powered by a hot gas, hydrogen peroxide. In a number of tests, the device showed it would be useful to an astronaut in controlling his

---

*During Mercury, when Donald Slayton was replaced as prime pilot on the Mercury-Atlas (MA) 7 mission because of a heart anomaly, his backup pilot, Walter Schirra, did not get the assignment. Scott Carpenter, who had been the alternate on John Glenn's MA-6 flight, flew the mission.

attitude and keeping himself stable while he maneuvered in space. When, early in 1963, the Air Force was given a chance to place experiments in the Gemini spacecraft, the AMU was an obvious choice. It could help pilots working in space on many tasks that the Air Force was particularly interested in—maintenance, repair, resupply, crew transfer, rescue, satellite inspection, and assembly of structures. Since none of these was as yet a primary or secondary objective to NASA, the unit would fly in Gemini merely to confirm what it could do.[11] The tether entered the picture as a safety factor.

At first, the Air Force had in mind a 60-meter tether. But studies suggested that an astronaut might get tangled up in a weightless tether. Although this might be countered by a reel mechanism that would keep the line taut, the real question soon became whether a tether was needed at all. Could redundant or alternate systems offer the same safety provided by tying an astronaut to an orbiting spacecraft? The Air Force thought they could, and some in NASA agreed. Tether development was canceled.[12] Colonel Daniel McKee, head of the Air Force field office in Houston, pointed out that contractors, when they knew the propulsion system would be flown by astronauts not tied to the spacecraft, would be compelled to make highly reliable systems. After all, no one wanted an astronaut floating off into space. But that possibility was exactly what NASA was worried about. Warren J. North, Chief of MSC's Flight Crew Support Division, held that tethers were a spaceman's best friend, "especially if you have oxygen in them."[13]

The dispute persisted, sometimes heatedly. An MSC and Air Force meeting in July 1965, to consider "EVA possibilities for Gemini 8," included "EVA without tether." But NASA Headquarters soon made its official position quite clear. William Schneider, Deputy Director of Mission Operations, wired MSC Gemini Manager Charles Mathews that "EVA shall be based on the use of a tether on Gemini flights thru Gemini 12."[14]

McKee was not so easily discouraged. In February 1966, he was still debating the issue. McKee wanted the matter left open until Gemini XII, when the maneuvering unit was scheduled for its second flight. He prepared a position paper, pointing out that all critical systems on the AMU were backed up and that its test programs had been oriented toward free flight, because this was the unit's ultimate purpose.[15] MSC Director Robert Gilruth forwarded McKee's case to George Mueller, chief of NASA's manned space flight programs, who was still not convinced. Mueller insisted that all Gemini astronauts would be tethered, but even this experience might be helpful to the Air Force in future untethered flights. A new NASA position paper described spacecraft maneuvers that would maintain tether slackness to simulate free space activity. Although "prudence dictates that a tether be used at all times during Gemini extravehicular activity," the door might still be open to untethered flights, "in the event that an operational requirement is identified which cannot be met in [any other] way."[16]

The spinning flight of *Gemini VIII* on 16 March gave the Air Force a chance to push that door open: what might have happened had David Scott been outside and fastened to the spacecraft when it went out of control? He could have been wrapped up like a broken window shade. The Air Force suggested adding a safety disconnect device, at least, as long as NASA persisted in a tether, so a crewman could free himself if something like that happened again.

NASA officials, too, had been thinking about the plight of a crewman caught outside a whirling spacecraft. Scott said that he could have spotted the thruster problem and gotten back into the spacecraft to help Armstrong deal with it. But many in the Office of Manned Space Flight were convinced that, if spacecraft troubles arose when a pilot was outside, the best thing for him to do was to get back inside as quickly as he could. There were too many hazards connected with troubleshooting for him to try diagnosing any problem, let alone using a disconnect to discard the security of a lifeline. That ended the active debate,[17] but there were still some who thought it was a good idea, one that ought to be tried in future programs.[18]

The second major issue on the Gemini IX mission—when to rendezvous with the target vehicle—was not so hotly pursued. Planners for *Gemini VI*, considering possible sources of trouble, had concluded that rendezvous should take place no sooner than the fourth orbit. This was a well researched procedure, which Walter Schirra and Stafford had demonstrated in high style. But some engineers in the Apollo Spacecraft Program Office wanted to tamper with success. Rendezvous in the first, or at least by the third, spacecraft revolution would more closely approximate lunar orbit rendezvous.[19] In September 1965, mission planners began working on a tentative $M=3$ rendezvous (in the third spacecraft orbit) for Gemini IX and X. For the rest of the year, they worked on this new rendezvous scheme.[20]

NASA, Air Force, and industry representatives met in Houston on 20 January 1966 to review the results of these labors. After the spacecraft had separated from the launch vehicle, the first maneuver—"IVAR" for the unwieldly "insertion velocity adjust routine"—would reduce orbital insertion errors. The crew would use the inertial guidance system to raise or lower spacecraft trajectory immediately. At the apogee of the first circuit, the crew would perform a "phase adjust," to establish the proper phase relation between the spacecraft and the Agena. One and a half orbits later came another change, this time a triple play, to correct phase, height, and out-of-plane errors. The final maneuver was to circularize the flight path two and a quarter revolutions after insertion. This would place the spacecraft about 28 kilometers below the target and ready to start firings to catch it. The remaining maneuvers were similar to those required for a fourth-orbit rendezvous.[21]

No one doubted that this sequence would work but some saw no reason for an M=3 at all. Two camps formed. One group insisted that it closely approximated lunar orbit rendezvous; the other maintained that the kinship was so slight that it was not worth doing. The second group also contended that ground tracking and ground computer capabilities for this approach were not as good as they were for rendezvous in the fourth revolution. Schneider believed that the third-circuit concept would be useful to Apollo operations. Mueller agreed with him, and that settled the issue.[22]

The third Gemini IX debate, radar versus optical tracking, grew from a type of rendezvous clearly applicable to Apollo. This matter first came up when several engineers, looking for ways to keep the spacecraft from getting too heavy, wanted to pull the radar out of both Apollo vehicles. The command module lost its radar in February 1965 when the ASPO Configuration Control Board ruled that the astronaut aboard the mother ship could use an optical sight to help rendezvous with the radar-and-flashing-light equipped lunar module. Later that year, with weight reduction becoming even more pressing, the lunar module's radar was the candidate for removal. This meant that during lunar operations—whether on takeoff from the Moon or at any time the two vehicles were apart—rendezvous of the two ships would depend entirely on astronaut eyes, optical sights, flashing lights, and computers. This was too much for the men who had to fly the machines; they did not entirely trust their eyes or the suggested equipment. They wanted the help of electronic radar signals on one vehicle bouncing back from the transponder of the other. At least, they said, the radar should remain on the lunar module.[23]

Stafford and Cernan did agree to include a test on Gemini IX to compare optics and radar by performing a rendezvous from above the target vehicle. In this exercise, the Agena would be over the Sahara Desert, which would simulate the lunar surface, and the crew would try to fly down to it, using both radar and optics.[24]

## PREPARATIONS FOR GEMINI IX

When Stafford and Cernan returned to training in mid-March 1966, after the See-Bassett accident investigation, the command pilot spent little time on the spacecraft systems. After all, he had put in more than 300 hours in the spacecraft simulator in the past two years. He concentrated instead on flight planning, which was more complicated for this mission than either of the two he had worked on before. It was also subject to more changes. Cernan and Aldrin, on the other hand, had to focus on extravehicular training, which was dominated by the scheduled use of the maneuvering unit.[25]

Working up the flight plan, with its heavy emphasis on rendezvous and extravehicular activity, began in 1965 and lasted until Gemini IX was launched. By January 1966, three types of rendezvous had been included: third spacecraft orbit, from above the target vehicle, and a very high altitude maneuver to reach an imaginary (or phantom) target. The phantom rendezvous (which depended on the Agena's propulsion system) was soon canceled by the planners, both because they still did not completely trust the target vehicle's engines and because they did not want to expose the crew to too much radiation.[26]

Gemini IX soon picked up a third rendezvous, anyway, one that *Gemini VIII* missed doing—re-rendezvous from an equiperiod orbit. The spacecraft thrusters were used for an upward velocity change to separate it from the target. If the firing were precise and all conditions were right, the spacecraft and Agena would automatically rendezvous at the end of an orbit, because the more elliptical spacecraft orbital path would intersect the circular orbit of the target at the proper point. Theoretically, the closing maneuvers should involve only braking the spacecraft to reachieve stationkeeping (alias re-rendezvous) with the target.

Stafford was beginning to worry about doing three rendezvous; his spacecraft was the last to have the smaller tanks—150 kilograms as opposed to (on Spacecraft 10) 208 kilograms of maneuvering fuel. But the equiperiod rendezvous was designed as a fuel-cheap way to evaluate maneuvers and lighting conditions for a dual rendezvous with a passive target scheduled for Gemini X. And Mathews decided that the lunar module abort rendezvous could remain in the flight plan for Gemini IX, but it would have a lower priority and would be contingent on fuel and time.[27]

So rendezvous was the first major objective on Gemini IX, and preparing for the different types produced its share of headaches. But the second most important activity, extravehicular work with the AMU, was a bigger source of trouble.[28]

The AMU had been ticketed for at least two flights from the start. This backpack, with its oxygen supply and radio, was powered by hydrogen peroxide, a relatively unstable chemical. Several MSC engineers were unhappy about using it. Warren North was one of them; North also worried about the high-temperature jet hitting the astronaut's space suit. Cernan's personalized jet-pack weighed 76 kilograms and its 10.2 newton (2.3-pound) thrusters operated in pairs—forward and back, up and down, but not from side to side. This caused another worry.[29] But Aldrin, on a training trip to California, suddenly got an idea. He tested it on his next trip to the Ling-Temco-Vought (formerly Chance Vought) plant in Dallas. After he mounted the training machine, a burst from the two aft thrusters sent him across the air-bearing table toward his target. A brief nudge from the small control jets at one shoulder and knee turned him to the side. He could now use his forward- or backward-firing thrusters to move sideways with respect to his path toward the goal.[30] North's fears that the heat of the AMU thrusters might damage the pressure suit proved valid, and its insulation had to be changed. The Mylar insulation was replaced by 11 layers of aluminized H-film (a thin sheet of polyamide with a coating of aluminum on one side).[31]

The spacecraft also needed some rework to fit it for extravehicular tasks. At NASA's request, McDonnell bonded 80 Velcro hook patches to the surface of the spacecraft. Then Velcro pads, which would cling to the patches on the spacecraft, were added to Cernan's gloves to help hold him in place as he moved about. With body position so important in checking out and donning the AMU, two handholds and a footbar were installed as restraints. Velcro pile on the footbar would mate with Velcro patches on Cernan's boots. During zero-g flights, he found this was not enough. After stirrups were added, he and Aldrin had no difficulty in checking out the unit in further practice flights.[32]

## ATTEMPTED LAUNCHES

Everything was ready for Gemini IX on 17 May 1966. In the Mission Control Center, Eugene Kranz assumed his duties as flight director, presiding over a three-shift operation. The other two flight directors were Glynn S. Lunney and Clifford Charlesworth. Only 200 newsmen were on hand, compared to the thousand or more who had covered *Gemini IV* the year before.[33] Gemini was becoming more routine, hence less newsworthy.

After a smooth countdown, Atlas launch vehicle 5303 rose from pad 14 at 10:12 a.m. For two minutes the rocket's three engines rammed Agena 5004 skyward. Only ten seconds before the two outboard engines were supposed to stop, however, one of them gimbaled and locked in a hardover pitchdown position. The whole combination—Atlas and Agena—flipped over into a nosedive and headed like a runaway torpedo back toward Cape Kennedy.[34]

Shortly after the booster engines stopped firing, the guidance control officer reported he had lost touch with the launch vehicle. Richard W. Keehn, General Dynamics program manager for the Gemini Atlas, was alarmed and puzzled. Telemetry showed that the sustainer engine had cut off, and a signal that the Agena had separated from its launch vehicle followed. Agena signals kept coming until 456 seconds after launch—then there was silence. Keehn raced over to Hangar J, the General Dynamics data station, where the telemetry tapes pointed to an Atlas engine problem. But television reports implied that the target vehicle was in trouble again, and Lockheed officials winced whenever they heard someone speak of the "Agena bird"; this was ironic in the light of the problems and delays caused by Atlas in the Mercury program and the success of Agena in Project Surefire and *Gemini VIII*. Meanwhile, the Gemini IX Atlas and Agena had plunged into the Atlantic Ocean 198 kilometers from where they had started.[35]

As contractors worried about technical problems, NASA again faced the necessity for a quick recovery plan when a target vehicle failed to reach orbit. This time, however, the agency had something in the hangar, an alternate vehicle—the ATDA. After the Agena exploded in October 1965, NASA had ordered General Dynamics/Convair to be prepared to furnish a backup Atlas within 14 days of another such catastrophe.[36] And in April 1966, just a month before the attempted launch of Gemini IX, Schneider had reminded Preston that he would have to be ready to launch the alternate target in a hurry if the Agena again failed to keep its orbital appointment. Now it had. On 18 May, Mathews wired Colonel John Hudson, Deputy Commander for Launch Vehicles, Air Force Space Systems Division, to prepare Atlas 5304 for launch on 31 May in a mission now called Gemini IX-A.[37]

With what had been the backup plan now in effect, the next question was what to do if the ATDA, too, failed. At a staff meeting on 18 May, Mathews announced that Gemini IX-A would be launched anyway, to rendezvous with the *Gemini VIII* Agena, still in orbit. McDonnell, in any case, was confident of the ATDA. When Mathews asked, in a management meeting in St. Louis the next day, "Does anyone have any reservations about flying the ATDA?" the answer was no.[38] That was just as well, because the motion of a rendezvous with the old Agena soon had to be abandoned. Its orbit had not decayed to the expected extent, and it was still sailing around Earth 402 kilometers up. Without the help of Agena, high-altitude flight might take too much spacecraft fuel and leave the crew stranded with no way to get to the lower orbit needed for retrofire.[39] Deputy Administrator Robert Seamans and Mueller agreed with Mathews that rendezvous with Agena 8 was too risky, but Gemini IX-A would still fly, even if the substitute target did not make it. Extravehicular activity with the AMU was a much needed venture in its own right.[40]

Long before these decisions were made, the Atlas contractors were frantically busy. Keehn had bundled up the telemetry tapes and headed for San Diego, where study of the data plus some tests located the trouble in the electrical wiring.[41] Within a week, Keehn and his group pinpointed the cause of the failure: a pinched wire in the autopilot that produced a short circuit. This meant some extra work on the electrical connectors, and General Dynamics asked NASA for an extra day to complete the task and prepare Atlas 5304 for launch. The agency set 1 June as the new date.[42]

Although General Dynamics had accepted the blame for the mission failure, Lockheed was worried about telemetry signals that indicated a problem with an Agena inverter. A nagging question persisted. Could the target vehicle have gone into orbit if the Atlas had worked? This inverter provided power to both the gyroscope and the sequence timer. To Lockheed's relief, a series of row cameras located at Melbourne Beach, Florida, got pictures of the Atlas' outside loop. They showed that the Agena passed through ionized gases from the booster's exhaust, which caused an electrical short and failure of the inverter.[43]

On 1 June 1966, men and machines were again gathered at the Cape Kennedy launch site, this time to try to send the alternate target vehicle and Gemini IX-A into coordinated orbital flight. At the appointed time, 10:00 a.m., the Atlas rose from pad 14. After a six-minute boosted phase, it tossed the ATDA into a nearly perfect 298-kilometer orbit. Just one thing marred the picture: telemetry signals suggested that the launch shroud covering the docking port had only partially opened and had failed to jettison.

Concurrently, over on pad 19, Stafford and Cernan were going through their countdown to launch. When the count reached the three-minute mark, a hold was called so the spacecraft could be launched precisely on time for the best catchup trajectory with its target. Almost immediately after the count resumed, problems developed in the Cape ground launch control equipment when it tried to send the spacecraft refined information on the exact launch azimuth. The launch window (only 40 seconds long) closed, and Mission Director Schneider delayed the flight for 48 hours. For the second time, Stafford and Cernan had to take the elevator down. Stafford later said, "Frank [Borman] and Jim [Lovell] may have more flight time, but nobody had more pad time in Gemini than I did!" By the time Gemini IX-A lifted off, he had been in the two spacecraft (6 and 9) ready for launch a total of six times.[44]

## AN ANGRY ALLIGATOR

Stafford and Cernan met with no untoward incidents on 3 June. The flight began precisely at 8:39:50 a.m. Stafford watched the instruments more closely than had his predecessors, since he had this new IVAR (insertion error correction) to handle in starting the rendezvous sequence. Six minutes after launch, CapCom Neil Armstrong said, "You are go for IVAR." Seconds later, the command pilot fired the spacecraft thrusters in the chase toward the target vehicle 1060 kilometers ahead.[45]

By the time Stafford and Cernan arrived over the Canary Islands—only 17 minutes after launch—the computers had ground out the figures. Armstrong gave the crew the data for the phase adjustment near the first apogee. At 49 minutes into the flight, the thrusters added 22.7 meters per second to spacecraft speed to raise its perigee from 160 to 232 kilometers. "I felt that one, Tom!" Cernan exclaimed.[46]

During the hour before the triple play—to correct phase, height, and out-of-plane errors—the crew checked systems, went through stowage lists, took off gloves and helmets, and got cameras ready for the rendezvous. To circularize the flight path, at 2:24 hours elapsed flight time Stafford pitched the nose of the spacecraft down 40 degrees and turned it three degrees to the left of its flight path. Fifty-one seconds later, he fired the aft thrusters to add 16.2 meters per second to the vehicle's speed. The orbit now measured 274 by 276 kilometers—22 kilometers below and 201 kilometers behind the target vehicle and closing with it at 38 meters per second.[47]

Over Tananarive, 12 minutes before Stafford had fired the thrusters, the crew got some flickers of a radar contact with their target. A range reading of 240 kilometers between the vehicles showed on the scale. George Towner and the other Westinghouse radar builders were relieved; they had worried about acquisition of a target that would wig, wag, and wobble. The Agena was a stabilized vehicle; the ATDA was not, and its radar reflectivity changed with its continually changing attitude. Within 222 kilometers, however, electronic lockon was relatively good.[48]

At 3:20 hours, the crew caught sight of their goal 93 kilometers away. For some time, it flitted in and out of view on an optical sight. At 56 kilometers, it became quite clear and remained visible from then on. As he drew nearer, Stafford reported seeing flashing acquisition lights. Thinking for a moment that the shroud had jettisoned after all, he said, "All right. We're in business." Surely they could not have seen the running lights so clearly if the shroud were still attached. While making minor corrections, he was glad that he could see the little "shiners" so well, because moonlight, streaming through his window, almost blinded him. The Moon soon became an asset, however, as its rays reflected off the ATDA.[49]

Stafford began slowing his spacecraft at 4:06 hours. During the closure period, he peered out the window, trying to see if the shroud was there or not. Then he exclaimed, "Look at that moose!" As the distance dwindled, he knew that he had been indulging in wishful thinking—"The shroud is half open on that thing!" Seconds later, Cernan remarked, "You could almost knock it off!" When the final braking was completed, the two vehicles were only 30 meters apart and in position for stationkeeping. But it did not seem likely that the spacecraft nose could slip into the mouth of the "moose" and dock.[50]

The crew described the shroud in detail and wondered out loud what could be done to salvage the situation. One of Stafford's remarks—graphic and memorable—became the trademark of the entire mission. His animal analogy switched to reptilian when he said, "It looks like an angry alligator out here rotating around." He itched to nudge it with his spacecraft docking bar to open its yawning jaws, but Flight Director Kranz told him to control the urge.

Perhaps the most significant aspect of this incident was the close examination of an unstable body while discussing it over the air-to-ground circuit. Stafford stayed 9 to 12 meters from the target but moved to a ticklish position only centimeters away in daylight. As the ATDA rotated slowly, he rolled his spacecraft upside down to parallel the movements of this weird looking machine. His performance met, in effect, one of the Defense Department's objectives for the AMU—finding and inspecting unidentified satellites. Stafford said he could plainly see that the explosive bolts had fired but that two neatly taped lanyards held the clam shell partially in place. These lanyard wires had high tensile strength, he was assured from the ground, so it might not be wise to nudge its jaws.[51]

Schneider called James McDivitt and Scott, who were in Los Angeles, and asked them to go to the Douglas plant and look at a duplicate target vehicle shroud to see if the wires could be cut or the shroud removed in any way during orbital flight. The astronauts soon reported that the wires could be clipped, but there were many sharp edges that might tear the astronaut's suit as he worked. In the meantime, ground controllers sent signals to the target to tighten and relax the docking cone, hoping that might free the shroud. But it remained in place—there would definitely be no docking on *Gemini IX-A*.[52]

The shroud episode was embarrassing, and another investigation began immediately. The solution was simple, if one recalls the old saw about too many cooks spoiling the broth. Douglas built the shroud that Lockheed, in turn, fitted to the Agena. The ATDA, however, was built by McDonnell. Before McDonnell technicians made the final installation on the ATDA at the Cape, a Douglas engineer supervised a practice run, with the exception of the final part—the lanyards that operated the electrical disconnect to the explosive bolts. For safety's sake, these were not hooked up. Before the mission, the Douglas engineer went home to his pregnant wife. On launch day, the McDonnell crew followed procedures published by Lockheed, which had been copied from Douglas documents. The instructions said, "See blueprint," but the Lockheed drawing was not used. The Douglas technician who normally hooked up the lanyards knew what to do with the loose ends, even without the blueprint. But he was not there, and the strangers fixing the ATDA's shroud looked at the dangling straps, wondered what to do with them, then taped them carefully down. In orbit, Stafford photographed their neat handiwork.

As Scott Simpkinson, GPO Manager of Test Operations, later said,

**Gemini IX**
**3 June 1966**

*Above left, astronauts Aldrin and Stafford practice docking their Gemini spacecraft with the Agena; above, Mission Control watches on TV as the Atlas-Agena is launched, only to drop into the Atlantic moments later; above right, the backup, ATDA, is readied for launch.*

*Above, the "angry alligator," with clamshell doors hanging ajar, thwarting docking; below left, Cernan on a spacewalk; below right, two ships glad to see each other.*

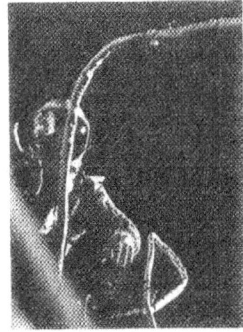

three good lessons were learned from this mistake: (1) simulate processes completely, (2) keep experienced people on the job, and (3) follow written procedures exactly.[53]

*Gemini IX-A* now began its equiperiod rendezvous. Five hours after launch, Stafford nosed the spacecraft down 90 degrees and fired the forward thrusters for 35 seconds to increase his speed by 6 meters per second. The crew quickly found that the target was disappearing below them. Later, in the darkness, they plotted their position with a sextant and checked the result against a preplanned chart solution. Mission planning had been right; all that was necessary to complete the rendezvous was to slow the spacecraft down. At 6:15 hours, Stafford began a series of four maneuvers to bring the spacecraft back to stationkeeping alongside the target. The second of the three rendezvous exercises was easy.[54]

Less than an hour after *Gemini IX-A* returned to its target (6:36 hours elapsed time), the crew got ready to leave again, for the third planned rendezvous.[55] At 7:15 hours, Stafford fired the aft thrusters to decrease the spacecraft speed by 1.1 meters per second and widen the distance between the two satellites.

Stafford and Cernan could now relax a little. It had been an exhausting day. Still wanting to snap the alligator's jaws off, they chatted with ground controllers about the shroud. Then they checked spacecraft systems, ate, and tried to sleep. Cabin noises and lights made sleeping difficult, however, and they only dozed for 40 minutes or so at a time; their scheduled eight hours of slumber were fitful, at best.[56]

The next day—4 June—Spacecraft 9 led its target by 111 kilometers. That retrograde maneuver (against the direction of the flight path) had lowered the orbit of the spacecraft (it now measured 289 by 296 kilometers) and the target traveled a nearly constant 298 kilometers above the planet. Thus the spacecraft, being nearer Earth, illustrated the paradox of slowing down to go faster, relative to the surface of the world, than the object flying overhead. The stage was set for Stafford and Cernan to do a rendezvous from above; but they first had to accelerate the spacecraft in the direction of the flight path so it would leap to a higher altitude than the target. Automatically, then, the lower flying target would reduce the spacecraft's 110-kilometer lead. To rendezvous, the crew only had to cancel out altitude and velocity vectors that had placed their vehicle above and ahead of its objective.[57]

A phase adjustment at 18:23 hours was followed a little more than 30 minutes later by a height adjustment. Another burst from the thrusters put the spacecraft into an orbit measuring 307 by 309 kilometers. The slant range to the target, which had stretched to 155 kilometers, began to shorten. Within 15 minutes, Stafford reported that the vehicles were only 100 kilometers apart. Forty minutes later, Cernan called out a 37-kilometer mark. At 21:02, the distance was 28.6 kilometers. Stafford pointed the nose of his spacecraft down 19 degrees and yawed it to the left 180 degrees, aiming at the other vehicle, which was still below and behind him.[58]

Over the Atlantic Ocean, then the Sahara Desert, on past the African continent, Stafford and Cernan had trouble spotting the target, but the electronic eye of the radar did not. When they were 37 kilometers away, they had seen the vehicle reflected brightly in the moonlight and, later, in the sunlight. As the Sun rose, however, they lost sight of it completely. The range had closed to less than six kilometers before Stafford saw what looked to him "like a pencil dot on a sheet of paper." Without the radar, he said, they would "have blown that rendezvous." But at 21 hours and 42 minutes after launch, *IX-A* and the target were again side by side. Three types of rendezvous had been completed in less than 24 hours.[59]

At the end of the third rendezvous, the Carnarvon, Australia, flight controller told Cernan that Flight Director Charlesworth wanted the crew to start getting ready for EVA. Stafford had begun to worry about the amount of fuel that would be consumed if he continued stationkeeping with the target. Unless the flight controllers thought Cernan might actually do something about the shroud, the command pilot wanted to get out of the vicinity of the ATDA before the pilot got out of the spacecraft. The crew was also pretty tired. As they approached Houston, Armstrong told Stafford to postpone EVA until the third day and to leave the ATDA. Stafford accelerated the spacecraft by one meter per second and moved away forever from the angry alligator.[60]

On 5 June, at 5:30 a.m., nearly 45 hours and 30 minutes into the mission, the crew began preparations for Cernan to emerge from the spacecraft. In the cramped cabin, they worked, rested, and worked again, finishing ten minutes before sunset. Near sunrise, Cernan cracked his hatch. It took more effort than he expected, but he soon stood in the opening, looking out at infinity and waiting for the first signs of daylight. Cernan had no feeling of disorientation nor any sensation of being lost in the dark of space. He heaved out a litter bag, the start of an exercise scheduled to last 167 minutes, during which the pilot would stand, walk, float, or ride nearly twice around the world.[61]

Once outside the spacecraft, Cernan did some simple experiments to get the feeling of working in space. He was startled to find that everything took longer than he had assumed it would from his experience in simulations. Cernan said he really had no idea how to work in slow motion at orbital speeds. Every movement of an arm or leg in free space exacted a reaction from his body. Minute forces that would scarcely be noticed in Earth's gravity upset his equilibrium in space. He had only to twitch his fingers to set his body in motion. On *Gemini IV*,

White had commented on the need for handholds. Now Cernan found that even those installed on Spacecraft 9 were inadequate and that the Velcro was not strong enough to keep his body in position as he edged back toward the adapter. He had to fight the limited mobility of his space suit, and the effort taxed his strength. He constantly referred to the umbilical as the "snake." When he let it out to any distance, it was hard to control.[62]

When he finally reached the adapter, some lights that had been installed especially to help him see were not burning. He asked Stafford to turn them on, but only one lit up. Moving around the adapter was no easier than moving around the rest of the spacecraft. Still, he began preparing the maneuvering unit for flight. He attached penlights; opened and checked the nitrogen and oxygen shutoff valves; positioned the sidearm controllers, umbilicals, and restraint harness; attached the AMU tether; turned on the unit's electrical power; and changed over to the electrical umbilical. Everything, just everything, took much longer than he had expected. He kept floating out of control; he simply could not maintain body position. The few footbars, stirrups, and handbars were insufficient for any task that required leverage.

Ten minutes after sunset, Cernan's faceplate began to fog,* so he rested. But here there could be no such thing as complete relaxation because of the tendency to drift away. He went back to work, but his visor soon fogged again. After the next sunrise, the moisture lessened. As soon as he moved about, it returned. Strangely, he felt neither hot nor cold†—his only problems were this fogged visor and tasks that had to be done with one hand when he really needed two.

When 80 percent of his work was finished, Cernan again had to stop and rest. Like a mountain climber with a backpack, he sat down in the maneuvering unit and found his most peaceful moment in this strange environment. Body molded to the seat, feet against a footbar, and arms atop the handbars, he enjoyed a taste of comfort for the first time since he started this stroll outside. The flight passed into darkness, but by the light in the adapter Cernan could tell just how occluded his faceplate had become.

He began to wonder whether to go on with EVA. Mentally, he ticked off the checklist items that remained: strap in, change to the AMU oxygen lead, start breathing oxygen from the unit's supply, and free his personal transportation from the spacecraft adapter. Cernan

---

*After the mission the fogging problem was duplicated in altitude chamber tests, using the Spacecraft 9 life support system and Cernan's space suit. When a small area of the faceplate was treated with an anti-fog solution, that spot remained clear. As a result, future Gemini crews carried an anti-fog solution to be applied immediately before EVA.

†At one period, Cernan's back did feel hot. Later analysis showed that some of the insulation of the suit had separated, allowing the Sun's rays to penetrate.

knew, from repeated experience in zero-g training flights, that he could do these tasks blindfolded. But then what? he thought. "So you make the connections ... if you can't see, you can't very well go out there and fly because you don't know what to expect." And if he flew the maneuvering unit, anyway? He could finish putting it on, he knew, because he was restrained in the adapter. But when the time came to take it off, he would be standing in free space. Could he take it off with one hand, while holding onto the spacecraft with the other? Would it be wise to try that when he couldn't see? Much better to end the exercise now, he thought. So he and Stafford decided to cancel the rest of the EVA, and Mission Control agreed.

Carefully, Cernan eased himself out of his comfortable seat, leaving his sun visor up to see if that might help defog his faceplate. At sunrise, he detached the AMU's electrical umbilical and connected his spacecraft lifeline. Still almost blind, he groped his way out of the adapter and back along the spacecraft to the cockpit. He slid into the hatch and stood there a few moments. Stafford held on to Cernan's legs so he could rest. Slowly his faceplate began to clear in the center, giving him a narrow range of vision. He tried to retrieve an externally mounted mirror that the command pilot had used to watch what was going on behind the cockpit. As Cernan wrestled with the mirror, his suit's cooling system became overtaxed, causing him to get extremely hot for the first time. His faceplate again fogged up completely. Stafford helped Cernan in and, together, they closed the hatch and started pressurizing the cabin. With their helmets almost touching, Stafford still could not see Cernan through the faceplate. The extravehicular exercise had lasted for 128 minutes instead of the planned 167; fogging had started 63 minutes after hatch opening.[63]

Two major aims of *Gemini IX-A* were rendezvous and extravehicular activity; the third was experiments.* Stafford and Cernan gave closer attention over a sustained period of time to the assigned experiments than had any Gemini crew before. When the space walk was postponed to the third day, the astronauts spent most of the second day on experiments and rest. About the only conversation they would tolerate from the ground was about their workload. On several occasions, when flight controllers forgot, they were reminded that the crew was busy. "My mistake for contacting you," came the response.[64]

Stafford and Cernan carried out M-5, bioassay of body fluids (the only medical experiment), which required wastes to be collected and labeled in laboratory fashion. Like other Gemini crews, Stafford and Cernan disliked this complex and messy task, nor did they enjoy the blood sampling they had to endure before and after the mission.

*See Appendix D.

Stafford equated the physical effort for M-5 to that required for doing a rendezvous and a half.[65]

The Department of Defense sponsored one experiment in addition to the Astronaut Maneuvering Unit—D-14, UHF/VHF polarization—to measure the inconsistencies of the electron field along the spacecraft orbital path and to study structures and variations of the lower ionospheric region. Stafford and Cernan operated the D-14 transmitter five times over Hawaii and once over Antigua during five successive revolutions. Everything worked well, but the number of measurements was limited because the antenna was poorly located. Later, when he was struggling outside, Cernan accidentally broke off the D-14 antenna.[66]

The four remaining experiments were scientific. Two of these involved micrometeorite collection. S-10 was a package mounted on the ATDA for Cernan to pick off during his space walk. This he could not do, but the astronauts did manage to photograph the package. The pictures showed that the device was in excellent condition. The second experiment of this type, S-12, was attached to the spacecraft and operated by the astronauts by remote control. While Cernan was in the adapter, he heard Stafford close and lock the box. Cernan retrieved the package and stowed it in the spacecraft.[67]

Cameras were the principal instruments used in the last two experiments—S-1, zodiacal light photography, and S-11, airglow-horizon photography. Stafford and Cernan took S-11 pictures on three successive night passes, between the 29th and 33rd hours of flight. They got 45 good photographs, under very trying circumstances. The tendency to float upward in zero gravity made pointing the camera and taking the pictures no easy task.

Zodiacal light photography had been scheduled for the space walk. A fogged faceplate, however, was no help in aiming a camera. The pictures had to be taken from inside the spacecraft after Cernan had returned to the more restful confines of his couch. Cernan had to hold the camera against his chest while pointing it out the window at the targets and calling out directions to Stafford for aligning the spacecraft. He obtained 17 good photographs.[68]

On 6 June, during the 45th revolution, they got ready to come home. *Gemini IX-A* touched down 0.70 kilometers from the planned impact point in the Atlantic Ocean, 72 hours, 20 minutes, and 50 seconds after launch. After scanning the panels in the spacecraft and flipping some switches, the crewmen opened both hatches, relaxed, and watched the gently rolling sea. They were close enough to raise their arms and thumb a ride on the *Wasp*. Stafford and Cernan stayed in their spacecraft until it was hoisted onto the ship's deck. After the usual hullabaloo had subsided, Cernan told anyone who would listen to him that extravehicular activity was not easy, not nearly as easy as people believed. And he seemed bitterly disappointed that he had been unable to fly the Air Force's maneuvering unit.[69]

To the public, the frustrations of *Gemini IX-A*—the formidable shroud and the fogged faceplate—overshadowed its accomplishments. Flying formation with and examining an unstable body had been a useful experience. Of even more significance were the advanced rendezvous maneuvers, proving that the flight controllers and crews could handle sophisticated rendezvous techniques that might be applicable to Apollo. Had *Gemini IX-A* been *VIII*, the results might have been viewed differently—as just part of the learning process. But docking, a primary objective, had not been achieved; and extravehicular activity had not succeeded in evaluating the maneuvering unit. Some engineers in MSC Crew Systems Division thought too much was being tried too soon—the simpler maneuvering unit planned for *Gemini VIII* would have been the logical second step in mastering EVA. As it turned out, the cliche to "watch out for that second step" would have made a good motto, but the step was greater than anyone had yet realized.[70]

## LAYING NEW TRACKS

Immediately after *Gemini IX-A*, Deputy Administrator Seamans expressed his dissatisfaction with results and the way missions were being handled. Although the flight, ground, and operations crews performed well in what they did, the achievements fell too far short of mission objectives. Seamans wanted a mission review board set up. He ticked off several items for such a group to study: corrective measures for the Atlas-Agena failure, the guidance update problem that delayed the launch two days, the shroud incident, and the suit environmental control difficulties. He also wanted the board to make sure that objectives and alternatives were carefully selected well in advance of launch.[71] Mueller established the Gemini Mission Review Board, with his deputy, James C. Elms, as chairman.*[72]

The board first laid out ground rules for drafting recommendations for each of the remaining Gemini missions. Benefits for Apollo and for science and technology were weighed against risks to crew safety. Mission planning policies were examined—was too much being programmed or too little?[73] With Gemini X scheduled for 18 July, planning for that flight was nearly firm. The board did measure mission objectives against the new ground rules, but there was neither time nor opportunity for more than minor changes.[74]

Gemini X, like *VIII* and *IX*, was a complex flight with multiple

---

*Members were Edgar M. Cortright (NASA Deputy Associate Administrator for Science and Applications), Major General Vincent G. Huston (Commander, Air Force Eastern Test Range), and MSC GPO Manager Mathews.

objectives. Among these was a dual rendezvous involving two Agenas—one launched for the mission, the other a passive target left over from *Gemini VIII*. Using the target's main engine to propel the docked Agena/spacecraft combination to high altitudes had been hotly debated on two previous missions. When the Atlas dropped into the Atlantic Ocean on 17 May 1966, the time for discussion was past. Since neither *Gemini VIII* nor *IX-A* had provided the hoped-for experience of firing the Agena's main engine while it was docked to a spacecraft, a decision had to be made promptly. There were only three flights left in the program. Nor would there be any preliminary, low-level practice first. The next day, Mathews told his staff that Gemini X would dock with Agena 10 and together they would climb to Agena 8.[75]

On 24 January 1966, John Young and Michael Collins were named to fly Gemini X.* When Young first heard about the dual rendezvous plan, he thought, "they must be out of their minds." The astronaut had two worries. Could he slow down the linked vehicles and stop them in time to keep from crashing into the second Agena? *VIII*'s Agena, having run out of electrical power, was dead, with no radar transponder or other apparatus to help in the search. Could he even find the old Agena, using only optical equipment? Young recalled, "We hadn't worked on any of these procedures. The problem with an optical rendezvous is that you can't tell how far away you are from the target. With the kind of velocities we were talking about, you couldn't really tell at certain ranges whether you were opening or closing."[76]

Young also remarked, "We didn't have an EVA program," but that soon changed. Collins would do experiments, retrieve packages from both the spacecraft and the passive target, test a zip gun, and visit an unstabilized vehicle. The backpack was dropped for missions X and XI and replaced by a 15-meter umbilical to supply oxygen and electrical support.[77]

Deciding what to do was only the beginning; how to do it was the bigger challenge. The second part of the double rendezvous (with the passive Agena) was particularly tricky. Agena 8, like all Earth-orbital vehicles, had been precessing above and below the equator on its orbital path. With no help from the dead target possible, the Gemini X Agena and spacecraft would have to be launched at very precise times. Suppose circumstances delayed the launches? It had happened before—more often than not! The mission planners would have to come up with a new set of numbers in a hurry. With events so closely relat-

---

*Lovell and Aldrin were selected as backup command pilot and pilot, respectively. On 21 March 1966, after the deaths of See and Bassett, they were moved into the backup positions for Gemini IX-A. Bean and Clifton Williams then became the alternate crew for Gemini X. At the same time, NASA also announced the crews for the first Apollo manned flight: Gus Grissom, Edward White, and Roger Chaffee (prime); James McDivitt, David Scott, and Russell Schweickart (backup).

ed, delay or failure at any point threatened all aims of the flight.

While shaping the Gemini X mission for the dual rendezvous, the planners decided to give the crew some helpful experience in onboard navigation, using optical equipment, charts, and the spacecraft computer. The crew would join its first target in the fourth orbit. Mission sequence was the next consideration. When should the dual rendezvous take place—the second day or the third day? Mission planners eventually decided that the second day should be devoted to experiments, the third to chasing the passive target. This, in itself, appeared to create a conflict of aims; although Agena 10 was needed to carry the spacecraft to the second target, many of the planned experiments could not be performed while the vehicles were docked.

About 50 people kicked this problem around at a trajectories and orbits meeting on 28 April 1966. Obviously, the launch dates would have to be jockeyed to get the best phase relationship between the spacecraft and target for both the dual rendezvous and the experiments.[78]

Even assuming that both launches went as planned, shaping the second rendezvous was an exacting task. The North American Air Defense (NORAD) Command, at Colorado Springs, had kept track of Agena 8's whereabouts ever since it ran out of electrical power. To begin the rendezvous, the docked Gemini X/Agena 10 combination should first go into a large elliptical orbit, 298 kilometers at perigee and 752 kilometers at apogee. After six revolutions to judge phase relationships, Agena 10 would then maneuver down to an approximately 398-kilometer circular orbit near Agena 8's space lane, as reported by NORAD.

The high altitude aspect of the flight raised its usual qualms. Although the Gemini Program Office no longer resisted the use of the big Agena engine while the vehicles were docked, McDonnell did not like the idea of the vehicles passing through so many high orbits, which might affect a safe emergency reentry if the retrorockets did not perform as needed. There was also the South Atlantic radiation zone to be considered. In a trajectories and orbits meeting at the end of June 1966, the maximum acceptable altitude for the dual rendezvous was set at 298 by 1065 kilometers, based on radiation constraints and actual radiation levels measured in 1964. But the decision to use Agena for docked maneuvers had already been made, and any misgivings had to be laid aside. After careful study, the planners concluded that an emergency reentry from an elliptical orbit with a perigee of 298 kilometers could be made even if only three out of the four retrorockets fired. Finally, they plotted the spacecraft's orbital track with great care, to avoid the heavy radiation patches.[79]

With the memory of past flights still fresh—when no one had been sure what target, if any, would be waiting—they made alternate and contingency plans for Gemini X. If the target vehicle for this flight did not reach orbit, the mission would be renamed X-A, and the spacecraft would be launched into a 162- by 385-kilometer orbit to rendezvous with the Agena 8 on the 16th revolution. The alternate plans also covered experiments, extravehicular activity, and systems tests.[80]

## THE SWITCH ENGINE

After the premission review, the traditional meal, and the ritualistic suiting up, Young and Collins left the crew quarters on 18 July 1966 for pad 19—to begin the most complex manned flight so far. They had been awakened at noon for a 5:20 p.m. takeoff, when a 35-second window offered the best chance for rendezvous with the two Agenas. The Atlas lifted its payload toward space at 3:39 p.m., just two seconds late.* One hundred minutes later, the Gemini launch vehicle boosted the spacecraft skyward exactly on time. Except for a slight shaking and a buzzing in their ears, Young and Collins had a nice ride to start chasing their first target.[81]

At entry into orbit, *Gemini* X trailed its Agena by 1800 kilometers. Flight Director Lunney told the crew they were all set for a fourth-orbit rendezvous. Collins unstowed a Kollsman sextant to sight on selected stars for an attempt at optical navigation. Young pointed the spacecraft while his crewmate tried to find the horizon. Collins realized that he was using the wrong reference when he saw stars below the line. He had been mistaking the airglow, a band of radiant light from the upper atmosphere, for the horizon. Even after he corrected this, Collins could not get the lens of the sextant to work properly, as the optical image of the stars did not agree with what he had been taught. He laid the Kollsman aside and tried an Ilon instrument, but that was little help as the Ilon had a severely limited field of view.[82]

Young and Collins checked their figures with Lunney, who had been watching their activities carefully through telemetry. When the trio found that the numbers did not agree with those of the ground computers, Gordon Cooper, the Houston CapCom, passed the word that the crew would have to use the ground computations. Young then fired the thrusters to adjust their orbit to 265 by 272 kilometers. When he aligned the platform for the terminal phase, the command pilot did not realize that the spacecraft was turned slightly. As he thrusted toward the target, Young needed two large midcourse corrections. The spacecraft path toward the Agena was not lined up properly. So he had to stop thrusting briefly and take off on a new tack. The final translational maneuvers to reach the Agena cost nearly 181 kilograms

---

*This was the 299th Atlas launch—the 100th for NASA.

of fuel, or three times more than any earlier mission.[83] Five hours and 52 minutes after launch, Young reported a rigid dock.[84]

Because too much fuel had been used, Lunney decided to omit docking practice—backing away and returning to the target's cone. Young and Collins wondered if the second rendezvous might also be canceled, but, some six and a half hours into the mission, the ground controllers started giving the crew the data they would need for the burn. Then, an hour later, the CapCom at Hawaii cleared them to try for second rendezvous.

The Agena main engine roared into life exactly on time. For 80 seconds, the target vehicle thrust the spacecraft upward, adding 129 meters per second to their speed. The crew, at the moment flying backward, had little to say about their reactions to a negative one-g force (a shove to the front of the body—"eyeballs out"—rather than a push on their backsides—"eyeballs in"—as during launch). They were thrown forward from the seats against the body straps. Young later described the first ride on a space switch engine:

> At first, the sensation I got was that there was a pop [in front of our eyes], then there was a big explosion and a clang. We were thrown forward in the seats. We had our shoulder harnesses fastened. Fire and sparks started coming out of the back end of that rascal. The light was something fierce, and the acceleration was pretty good. The vehicle yawed off—I don't remember whether it was to the right or to the left—but it was the kind of response that the Lockheed people had predicted we would get.... The shutdown on the PPS [primary propulsion system] was just unbelievable. It was a quick jolt ... and the tailoff ... I never saw anything like that before, sparks and fire and smoke and lights.[85]

*Gemini X* reached an orbit that measured 763 kilometers at the top and 294 kilometers at the bottom. The Agena had pushed the spacecraft more than 463 kilometers above its initial apogee. Young and Collins now viewed Earth from a higher elevation than any human beings ever had. Instead of gazing at the planet in wonderment, however, they confined their attention to their own little, artificial world. They watched spacecraft systems and kept an eye on the radiation dosage readings (which were within tolerable limits). During his technical debriefing, Young only reported, "We took some pictures at apogee.... I don't know where it was, but it shows the curvature of the earth.... We took some pictures coming down hill. I think it was the Red Sea area." Thus, in rating one impression over the other—record high altitude versus Agena ignition—Young and Collins were more affected by the firing of the switch engine than they were by the unique vantage point they had reached. This lack of awe at their record height was caused, at least in part, by the fact that the switch engine blocked much of the downward view.[86]

*Above, erector on pad 19 being lowered during a range frequency test; Gemini X still wears protective covering. Right, engineer practices retrieving an experiment package from Agena, using the EVA simulator. Collins was scheduled to perform this task during Gemini X.*

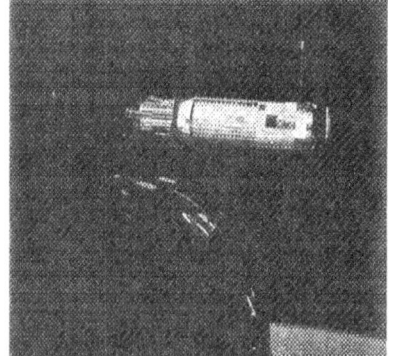

*Rendezvous with Agena.*

*Gemini X lines up for docking with Agena; Agena then propelled Gemini X out to 752 kilometers altitude and to rendezvous with Gemini VIII's Agena.*

Nine hours into the flight, the pilots bedded down, sleeping fitfully. Both were still wondering if the second rendezvous would be done. Besides, neither was "really bone-tired," Collins said. Charlesworth's shift in Mission Control was busy that night, reviewing alternate plans for adapting the mission to fulfill its objectives.

When Young and Collins opened for business after 18 hours of flight, their spirits lifted as the CapCom at Carnarvon gave them the numbers for the next target vehicle firing. With the Agena/spacecraft combination faced about so the main engine would fire directly into the flight path, Young made a 78-second burn to reduce the velocity by 105 meters per second and lower the apogee to 382 kilometers. The pilots were again pressed forward in their seats, but this time they were impressed more by the firepower of the Agena than by its fireworks. "It may be only 1 g, but it's the biggest 1 g we ever saw! That thing really lights into you," Young commented.[87]

Like rendezvous maneuvers in the past, the next Agena burn (and the final one with the main engine) aimed at circularizing the orbit. At 22:37 hours, the target drove the spacecraft along the flight path to add 25 meters per second to the speed. This brought the low point of the orbit up to 377.6 kilometers—only 17 kilometers below Agena 8.[88]

Although rendezvous and docked maneuvers with the Agena were the high point of the first day, the crew also spent a good part of that time on the 14 experiments they carried.* Twenty minutes after launch, the crew turned on a switch to start the tri-axis magnetometer (MSC-3). This device was used, as it had been in other flights, to measure the radiation levels in the South Atlantic Anomaly. Two other experiments were also devoted to radiation—MSC-6, beta spectrometer (mounted in the adapter to measure potential radiation doses for future missions), and MSC-7, bremsstrahlung spectrometer (installed in the cabin to detect radiation flux as a function of energy when the spacecraft passed through the South Atlantic Anomaly).[89]

Some of the experiments had to be done outside the spacecraft. Before the third Agena burn, Collins got ready for his first exposure to outer space, a standup EVA. Preparations went well and the hatch opened easily. At sunset, Collins stood in his seat, setting up a 70-mm general-purpose camera for S-13, a photographic study of stellar ultraviolet radiation. Collins aimed the camera at the southern Milky Way, scanning from Beta Crucis to Gamma Velorum, and exposed 22 frames. The entire night pass was devoted to this task. Young helped

---

*Originally scheduled to do 16 experiments, *Gemini X* lost MSC-5 (lunar ultraviolet spectral reflectance), which was to determine the ultraviolet spectral reflectance of the lunar surface and aid in designing equipment to protect Apollo astronauts from sunburn and eye damage. Because the Moon was out of phase, this chore was deleted before the flight. M-5, bioassays of body fluids had been the bane of all crews from *Gemini VII* through *IX-A*. Mathews had tried in vain to get it out of the earlier missions. This time he succeeded—its cancellation on 12 July 1966 marked the end of medical experiments in the Gemini program.

Collins identify the stars, at the same time controlling the spacecraft and target vehicle combination. With the beginning of daylight, Collins began MSC-8, color patch photography, to see if film could accurately reproduce colors in space. The pilot did not complete this assignment, however, as his eyes began to fill with tears. Young had the same problem. They suspected at first that the anti-fog compound inside their faceplates was irritating their eyes. They closed the hatch at 24:13 hours, about 6 minutes early.[90]

They had noticed a strange odor that they thought might have been the lithium hydroxide used in the environmental control system, but ground engineers finally concluded that their smarting eyes were caused by having both suit fans on at once. They turned one fan off and, at 30 hours elapsed time, began a second sleep period. Bone-tired this time, they rested well.[91]

Young and Collins awakened to a "morning" of increased activity. In addition to normal systems check, the ground network also reminded them of the experiments expected this day—the S-26 ion wake measurement, to study the ion and electron structure of the spacecraft's wake (after it undocked from the Agena), S-5 synoptic terrain, and S-6 synoptic weather photography. The pilots also had to work in two maneuvers to help them catch up with Agena 8.

Their Agena switch engine had accomplished its task, and more. After being hooked to it for 39 hours, however, they were getting a little tired of looking at it. Young said that watching the Agena out his window was

> just like backing down the railroad [track] in a diesel engine looking at a big boxcar in front of you.... The big drawback of having the Agena up there is that you can't see the outside world. The view out of the window with the Agena on there is just practically zilch.[92]

On freeing themselves from their Agena, the crewmen began preparing for Collins' exit from the spacecraft. Young now needed to make the final maneuvers to get the spacecraft close enough to the Agena 8 for Collins to reach it. Collins connected the 15-meter umbilical to his suit and then fastened it out of the way until time to use it.

"45:38. First sighting of Gemini VIII," Young said. "At this minute it's blurry." After the distance between the two vehicles had been calculated, the Houston CapCom (on the remote line through the Canton station) informed Young, "Your range, Gemini X, is 95 [nautical] miles [176 kilometers]." The crew then learned that what they had been looking at was their own Agena just 5.5 kilometers away. Houston offered consolation, "95 miles is a pretty long range," and Young answered, "You have to have real good eyesight for that." They didn't see the *Gemini VIII* Agena until it was 30 to 37 kilometers from them, looking to Young like "a dim star-like dot until the sun rose above the spacecraft nose." NORAD's constant care had paid off. They found Agena 8 just where it was supposed to be.[93]

At 47:26 hours Young started the final closure, with Collins computing the figures for two midcourse corrections. The crew found the old Agena pretty stable, and Young moved in to stationkeep about 3 meters above it. In less than 30 minutes, he told the Houston CapCom that they were going down for a closer look at the micrometeorite collection package. Back in Mission Control Center, fuel usage during stationkeeping was being very closely watched. When it proved to be reasonable, *Gemini* X received a go for the next extravehicular exercise. "Glad you said that," Young answered, "because Mike's going outside right now."[94]

Collins emerged from the spacecraft at dawn. Like Cernan on *Gemini IX-A*, he found that all tasks took longer than he expected. But he picked off the package from the spacecraft exterior. Next, he moved to the adapter to attach his zip gun to the nitrogen fuel supply. Back in the cockpit area once again, he held on while Young moved the spacecraft to within two meters of the Agena.

Collins pushed off from the spacecraft, floated freely in space, and grasped the outer lip of the docking cone on the target. As he clutched at the experiment package, he wished for handholds—or more hands. Cernan had warned him that it would be hard, and it was. He soon lost his grip on the smooth lip and drifted away from the package and from the Agena. He had to decide quickly whether to pull on the umbilical, coiling about like a snake, or to use the hand-held gun. Being about 5 meters away from the spacecraft, Collins chose the gun. It worked, and he propelled himself first to the spacecraft and then back to the Agena, using a series of squirts to get to the package. This time he clung to wire bundles and struts behind the adapter cone and grasped the S-10 experiment. Collins was supposed to attach a replacement device in its place, but he abandoned this idea, fearing he would lose the one he had picked up. Using the umbilical, he pulled himself hand over hand back to the cockpit and gave the S-10 package to Young.

So far, the umbilical had been snubbed so it would extend only 6 meters. The pilot now unsnapped the buckle that released the remaining 9 meters, intending to evaluate the gun. But the gun play stopped before it started. The Hawaii CapCom told Young, "We don't want you to use any more fuel [for stationkeeping]." Young replied, "Well, then, he'd better get back in." To Collins he said, "Come on back in the house."[95]

Getting back into the spacecraft was surprisingly difficult. Collins had gotten himself tangled in the umbilical. Since the pressurized suit made it difficult to see or feel just where the line had wrapped itself about him, he had to wait while Young helped unwind him and got him back into the seat. But fuel remained the big question. Houston called them, "just ... to confirm that you're not using any fuel." Young replied, "We've got everything shut off."

More was shut off than he realized. He soon discovered that the radio transmitter had also been turned off. By this time, Collins was back in his seat. Young reported that hatch closing had been easy. With the long lifeline coiling all over the cabin, Young thought it made "the snakehouse at the zoo look like a Sunday school picnic." A little over an hour later, the crew reopened the hatch and tossed out the chestpack and umbilical. This operation only took three minutes. McDonnell had done an excellent job on this righthand hatch.[96]

Because of the time spent struggling with the umbilical, Collins and Young had to hurry to get set up for an important manuever that would make the point of reentry more precise. They carried out an orbit-shaping activity exactly on time, at 51:38 hours. This retrograde firing, of 30 meters per second, brought the spacecraft perigee down 106 kilometers, making the orbital parameters safe for reentry. After another round of experiments—this time synoptic terrain and weather photographs taken as the spacecraft drifted through space—the crew began their third sleep period.[97]

On awakening (about 63 hours into the flight) on homecoming day, Young and Collins spent more time on experiments and did their packing. Then, 70 hours and 10 minutes after liftoff, the crew felt the first retrorocket ignite as they passed over the Canton Island tracking station during their 43rd revolution. Reentry went remarkably well, with Young steering bank angles by computer solutions. Landing in the western Atlantic at 70:46 hours (4:07 p.m., 21 July 1966) was only 5.4 kilometers from the aiming point. The crew of the primary vessel, the *Guadalcanal*, watched the spacecraft hit the water. Once the swimmers had attached the flotation collar and positioned the raft, Young and Collins climbed out. They were lifted by helicopter to the deck of the recovery ship.[98]

With that part of the mission completed, the flight controllers put the *Gemini* X Agena through its paces. Over a 12-hour period, the main engine was fired twice and the small engine once. Since the first maneuver was intended to study temperatures at higher altitudes, the controllers sent the Agena up to a 1390- by 385-kilometer orbit. They watched it for almost seven hours and found that the temperatures varied little from those at lower orbits. The vehicle was then returned to a circular orbit (352 kilometers) that would make it available as an alternate target for later flights.[99]

*Gemini* IX-A and X had successfully grappled with some of the specific needs of the Apollo program, acquiring operational experience while fostering healthy debates between the two programs on procedures and equipment. Perhaps the greatest benefit to Apollo was the demonstration and practice of several types of rendezvous. Each provided a storehouse of information. In addition, the orbit-shaping maneuvers to the higher altitudes established that the trapped-radiation hazards could be avoided on trips into deep space. Then, too, the very fact that one spaceborne vehicle could meet another, latch onto it, and use it as a kind of space tug offered many possibilities for such space flight concepts as shuttles, space stations, and space laboratories.

There had been problems, but missions *IX-A* and *X* had logged a combined total of three hours and 41 minutes open-hatch experience. Although the extravehicular hiatus between the fourth and ninth flights adversely affected both equipment and operational development, Cernan and Collins had shown that tasks outside the spacecraft were feasible. They found that all chores took longer than foreseen and that body positioning was difficult. During technical debriefings, each extravehicular pilot had pointed out the need for more and better restraints and handholds. These aids were being developed. Overall, perhaps, extravehicular activity remained Gemini's greatest problem. It was and is dangerous, difficult, and deceptive, despite its delights.

The ninth and tenth flights also took several steps forward in experiment performance. Despite operational constraints, usually brought on by limited fuel resources, each situation had been modified to wring the utmost from specific experiments. More and more, principal investigators were being brought in to help with modifications and to assist in rescheduling their tasks for later in the missions, if necessary. These realtime flight changes could not have been carried out in an unmanned flight and would not have been done in an earlier Gemini mission. So, in *Gemini IX-A* and *X*, the experiments program began to achieve maturity.

By the end of *Gemini X*, many of the men and women who had worked full time on the program had begun to have a strong feeling of anticlimax and to wonder about their next jobs. Some had already gone on to other fields, but Mathews tried to control this exodus and to hold enough together to finish the flights. Shortly after *IX-A*, he told his staff that the Gemini Program Office, as such, would not be continued. The people would be absorbed into other MSC activities—mainly Apollo and Apollo Applications. By early August, a personnel placement committee* had begun work. It soon arranged four to six interviews for each of the 193 project office people. This allayed any

*The committee consisted of Augustine A. Verrengia (Gemini), Robert J. Bailey (Apollo), Donald T. Gregory (Flight Crew Operations), James Null (Apollo Applications), Chris C. Critzos (Flight Operations), and Elwyn H. Yeater (Engineering and Development).

immediate fears, but Mathews still warned his staff to refrain from making personal contacts for new jobs until the committee could complete its arrangements.[100] There were two more flights in the Gemini program, but it already seemed to be heading into history.

# XV

# The Final Curtain

BY the summer of 1966, other space programs—Apollo, Apollo Applications, and the Air Force's Manned Orbiting Laboratory (MOL)—were already culling Project Gemini for useful equipment and people. Engineers still working on Gemini were distracted by calls to help qualify a heatshield for the MOL, to work on airlocks for the Applications program, and to share their launch vehicle experience with Apollo. In addition, NASA Headquarters was pressing the Manned Spacecraft Center to reduce the intervals between launches again—this time from two months to six weeks. As the program neared its end, spare parts emerged as a new worry. Would there be enough hardware to finish out the missions? As Scott Simpkinson, who managed Gemini Test Operations, recalled, "It was a bit touchy, but we made it." In this hectic climate, NASA flew the last two Gemini missions.[1]

Gemini's final deadline was now flatly fixed at the end of January 1967, with Gemini XI tentatively set for 11 September and Gemini XII for 31 October 1966.[2]

Some significant goals had been set for the last two flights. For example, the Apollo Spacecraft Program Office successfully pushed for a rendezvous in the first spacecraft revolution, which would simulate lunar orbit rendezvous. There was also interest in linking an Agena to a spacecraft by a tether and then spinning the combination to produce something like artificial gravity. One short-lived proposal, a rendezvous between Gemini XII and an Apollo spacecraft, was squelched after review by both program offices. Another idea, a flyby or rendezvous of a Gemini spacecraft with an Orbiting Astronomical Laboratory, also came to nothing. And, finally, on the last mission the Air Force still hoped to fly the Astronaut Maneuvering Unit (AMU), a task that Eugene Cernan had been forced to abandon on *Gemini IX-A*[3]

On 21 March 1966, Charles Conrad and Richard Gordon were named as command pilot and pilot for Gemini XI. Neil Armstrong and William A. Anders were picked as alternates. James Lovell and Edwin Aldrin were announced as the Gemini XII crew on 17 June, with Gordon Cooper and Cernan as backups. Of the eight men, only Anders had not previously been assigned to Gemini. Crews for the ten manned flights had been drawn from three astronaut classes, with several of the pilots receiving multiple assignments.*[4]

## PLOTTING THE WAY UP

When he was training in mid-1965 as pilot for *Gemini V*, Conrad learned of a plan to fly Gemini around the Moon in a mission called LEO for Large Earth Orbit. The concept, in one form or another, had recurred sporadically (only to be scotched) ever since Gemini's first year. But LEO raised interest all the way from MSC to Congress. NASA's top leaders, James Webb and Robert Seamans, did not agree, contending that Apollo did not need a competitor. If Congress wanted to appropriate additional funds, Webb said, it would be better to spend them on the program that was designed to go to the Moon. Another idea that flourished briefly during 1965 was a possible rendezvous with a Pegasus satellite that was first considered for *Gemini VI*, then for *Gemini VIII*. When extravehicular activity (EVA) was canceled on *Gemini V, VI, and VII,* the planners realized that experience would be too limited and risks too great to have an astronaut approach a satellite in space. GPO decided in January 1966 that there would be no rendezvous with a Pegasus.

Conrad was much taken with the notion of a Gemini trip around the Moon.[5] Even after Webb dismissed the scheme, he still wanted to take Gemini as far as it would go. When he was named as command pilot, he recalled, "it didn't look like . . . [a high altitude] flight was ever going to get done on Gemini." Conrad saw a heaven-sent opportunity to resurrect the idea when he calculated that he could save some of the Agena's fuel to power a high ride.

He began a small crusade to convince NASA management that there were good reasons for going really high. Although the Weather Bureau had satellites flying at very high altitudes, their televised pictures of cloud formations had poor resolution. Moreover, the Bureau

---

*See Appendix C.

had been debating the use of a color system. Conrad argued that Gemini XI could bring back films to help them decide its worth. It was, in fact, to the experimenters that he first turned in his campaign to fly high, asking which experiments might be helped and which degraded by higher altitudes. He learned that Maurice M. Shapiro of the Naval Research Laboratory was concerned that radiation particles from the Van Allen belts might affect his nuclear emulsion experiment at the higher orbit. That almost killed Conrad's plan before it was well started. But he enlisted fellow astronaut Anders, a nuclear engineer, for a trip to Washington to argue against the threat. After Anders got friends at Goddard Space Flight Center to look into the radiation belt hazards and to devise ways of avoiding them, the high apogee excursion soon became part of Gemini XI.[6]

Another unique objective for XI, direct (first orbit) rendezvous, had been suggested before Gemini flights began. Proposed by Richard R. Carley of GPO, the idea had been put aside when interest had focused on a concentric, fourth-orbit plan. Carley's proposal revived when the Apollo office insisted on a closer simulation of lunar orbit rendezvous. With some signs of reluctance, GPO asked McDonnell to study the maneuver. The first meeting to phrase plans and ground rules for the study revealed some foot-dragging; its results included a curious stipulation: "There should be no artificial restrictions in the plan to make the mission simulate Apollo operations or to simulate lunar rendezvous conditions."[7] That position was soon reversed as Apollo interests prevailed. The first change in the flight plan to include direct rendezvous made any launch delay a reason for shifting the mission to "a modified $M = 3$ [rendezvous in the third orbit] plan," but the following version "recycled [the launch] to the next direct rendezvous launch opportunity."[8]

Although schemes for achieving artificial gravity in space preceded real manned space flight by many decades, Gemini offered the first chance to turn science fiction into fact. Half the program had passed, however, before NASA got around to planning tethered vehicle flights. GPO first asked the Engineering and Development Directorate to study the problems involved in tying the Gemini spacecraft to either the Agena or the Pegasus satellite.[9] Its backlog of Apollo work forced the directorate to decline its aid, in view of the extensive simulation required. Appeals to Flight Operations were more fruitful, however, leading to a number of tether simulations, the data from which were duly passed along to McDonnell.[10]

McDonnell's guidance and control group found that nylon or dacron tethers no longer than 50 meters and a spin rate no more than ten degrees per second produced a reasonable amount of cable tension and recommended that the pilots practice spinning on a vehicle simulator to learn how best to conserve fuel.[11]

When NASA planners listed tethered flight as a mission objective, they first thought of it as a way of evaluating the tether as an aid to stationkeeping;[12] but it might also be a means of inducing some degree of artificial gravity. The minimum spin rate depended on whether the tethered activity was intended primarily for formation flying or for achieving gravity. NASA decided to try for both, although it would settle for "an economical and feasible method of long-term, unattended station keeping," and chose a 30-meter dacron line.[13]

The Gemini Mission Review Board reviewed all these new activities in depth, especially the first-orbit rendezvous, which might be a heavy fuel user.[14] Young and Collins had expended so much fuel in the *Gemini X* rendezvous that the board was dubious about trying a first-orbit linkup, largely computed onboard, with an Agena target. But Flight Director Glynn Lunney assured the group that Mission Control could give the crew backup data on orbital insertion and on the accuracy of their first maneuver; the network would have plenty of information to help them begin the terminal phase of rendezvous. The board concluded that if the rendezvous used only half the fuel supply, about 187 kilograms, there would be ample for the rest of the mission. Some skeptics remained; William Schneider, Deputy Director for Mission Operations, bet board chairman James Elms a dollar that it could not be done that economically.[15]

The board seemed less concerned about the high apogee maneuver and the tethered vehicle exercise than about direct rendezvous. Radiation levels on *Gemini X* having been only a tenth of the preflight estimate, the board simply asked that MSC and Goddard keep track of the latest measurements. The only major question about the tether plan was the method for freeing the spacecraft from the Agena. The board was told that the plan was to fire a pyrotechnic charge, ejecting the docking bar at right angles to the spacecraft path. If that did not work, there was a break link in the tether that could be snapped by a small separation velocity.[16]

As might be expected, extravehicular activity received special attention. After the experience on *Gemini IX-A*, training methods were sought that would more closely approximate flight conditions. One likely approach simulated zero-g by putting a space-suited subject under water, where buoyancy almost balanced weight, and leaving him to cope with mass and inertia just as he would have to do in space.[17] Despite the degree of EVA success that Collins had in *Gemini X*, work on this idea went ahead. There were, as MSC Director Robert Gilruth later said, "many mixed emotions here at the Center—some of our people didn't think the neutral buoyancy work was any good." But Cernan, who checked out the method at Gilruth's request, found that moving about under water in a pressure suit closely matched his efforts in space. These findings, however, were not impressed upon Gordon in his training for Gemini XI.[18]

More was needed than a better training medium. Both equipment and body positioning aids had to be improved. Hardware changes included handholds on the target vehicle docking cone, a shorter umbilical, and better foot restraints in the spacecraft adapter. The handholds were simple to design and install. Both Collins and Young had complained about the 15-meter snake that had entangled Collins. They suggested its length be cut to 9 meters, and GPO agreed. Developing better foot restraints took a little more time. McDonnell was working on two kinds—a spring clamp like those on a ski and a bucket type. NASA chose the latter, which were nicknamed "the golden slippers."[19]

Twelve experiments were included in the Gemini XI flight plan (See Appendix D). Nine were scientific, the other three technological. Two of the science experiments—S-29, Earth-Moon libration region photography, and S-30, dim light photography/orthicon—were new to Gemini. The other seven—weather, terrain, and airglow horizon photography; radiation and zero-g effects; ion-wake measurement; nuclear emulsion; and the ultraviolet astronomical camera—and all three technological experiments—mass determination, night image intensification, and power tool evaluation—had been assigned to previous missions. The Gemini Mission Review Board concluded that they fitted properly into the Gemini XI workload. By 25 August, MSC was able to report that all experiments were ready for flight.[20]

When reduced launch intervals required faster delivery to the Cape, the challenge was met. Before the end of July, launch preparations were under way in Florida. On 11 August, NASA announced that the flight would be launched on or about 9 September, only two days after the target date set more than three months earlier.[21]

The countdown-to-launch began on schedule on 9 September 1966, but it did not finish that way. After the booster was fueled, the launch crew detected a pinhole leak in the first stage oxidizer tank, which had to be fixed. Technicians used a sodium silicate solution and an aluminum patch to plug the leak; and Mission Director Schneider reset the launch for 10 September.

Trouble for the second scheduled send-off cropped up in a different area and much later in the countdown. Conrad and Gordon had completed the required rituals and headed toward pad 19 and their spacecraft when they heard that the Atlas, only 1800 meters away, was having a problem with its autopilot. The General Dynamics test conductor called a hold in the countdown to have this suddenly wayward instrument checked. His engineers told him they were receiving faulty readings and were running checks before deciding whether to replace the part. When the delay had stretched to an hour, Schneider postponed the launch for two more days. The problem was caused by a combination of factors—a fluttering valve, unusually high winds, and a too-sensitive telemetry recorder—none of which required replacement of the autopilot. There would be no further delay.[22]

## A HIGH RIDE

On 12 September 1966, Conrad and Gordon arrived at the pad and stepped into their seats exactly on time. Guenter Wendt, Mc-Donnell pad leader, signaled his men to close the hatches, but they soon had to reopen Conrad's. He suspected that some oxygen was leaking from his side of the cabin. He was right. When the hatch had been fixed, the countdown went on. At 8:05 a.m., the Atlas roared into action. Gemini XI had its target.[23]

If ever two pilots waited anxiously for the starter's gun to crack, Conrad and Gordon did. For the first orbit catchup, the time to come out of the chute was unbelievably short. It was the shortest launch window in the Gemini program. *Gemini X*, for example, had 35 seconds in which to launch, Gemini XII would have 30 seconds. Mathews had informed McDonnell and SSD that Gemini XI's launch window was only long enough for an "on-time launch." The postlaunch mission report, however, gave two seconds as the length of the window for a first-orbit rendezvous. Rocketeers of the forties, fifties, and early sixties would have been aghast at the idea of having to launch within two ticks of the clock.[24]

Conrad chanted the count: " . . . 3, the bolts blew, and we got lift-off." This was at 9:42:26.5, just half a second into the two-second period. The Titan booster shoved *Gemini XI* toward a first-orbit rendezvous with near-perfect accuracy. At six minutes, the flight control circuit carried the glad tidings, "Gemini XI, you're GO for M equals 1." This welcome word came at booster separation, when debris could be seen out the window. Gordon had warned himself not to look, but temptation got the better of him for a brief instant.[25]

Immediately upon insertion, Conrad and Gordon performed an insertion-velocity-adjust-routine (IVAR) maneuver, to correct the flight path up or down, right or left, and add to or decrease speed as needed. During IVAR, any decrease in spacecraft speed (retrograde firing) is done with great care because of the danger of recontact with the launch vehicle. The rules, therefore, say that the pilots must have the booster in sight before they begin to cut their speed at this point. Their computer showed the crew they had made very precise insertion corrections that would help them catch a target 430 kilometers away.[26]

The first onboard calculations had succeeded; now it was time to try again. There would be no help from the ground stations, as *Gemini XI* was out of telemetry and communications range. At the appointed moment, Conrad made an out-of-plane maneuver of one meter per second. He then pitched the spacecraft nose 32 degrees up from his horizontal flight plane. Now came the test to see if their first figures had been right. They turned on the rendezvous radar—the electronic lockon signal registered immediately. Happily, the crew switched the onboard computer to the rendezvous mode and began preparing for the final part of the catchup. When they could talk to the ground again, Gordon said, "Be advised we're [within] . . . 50 [nautical] miles [93 kilometers]."[27]

Young, the Houston CapCom, then cut in over the remote line through Tananarive to give the crew some numbers for the remainder of the chase. Conrad and Gordon checked these calculations against their own and found the differences so minor they could have used either set to do the job. They decided to stick with their own solutions. Just as the spacecraft neared the high point of the orbit, Conrad fired the thrusters to produce multidirectional changes—forward, down, and to the right—to travel the remaining 39 kilometers to the target. Suddenly the Agena, whose blinking lights they had been watching in the darkness, flashed into the sunlight over the Pacific and almost blinded them. They scrambled for sunglasses, then Conrad jockeyed the spacecraft to within 15 meters of the target's docking cone. Over the coast of California, only 85 minutes after launch, rendezvous in the first orbit was achieved.[28]

A gleeful crew called out, "Mr. Kraft—would [you] believe M equals 1?" He would. Moreover, they still had 56 percent of their maneuvering fuel. This transmission made a believer out of Mission Director Schneider. He fished in his pants pocket, pulled out a one-dollar bill, and scribbled a notation for Elms: "Sep[aration] 85#, Plane Change 5#, TPI 145#, Midcourse 20#, Braking 150#, [total], 405#. I never lost a better dollar. Bill Schneider."[29]

After appropriate congratulations, Young told Conrad and Gordon to go ahead and dock. Seconds later, Conrad reported matter-of-factly, "We are docked." The *Gemini XI* crew now had an opportunity to do something else that NASA had wanted for a long time—docking and undocking practice. Each man pulled out and drove back once in daylight and once in darkness. It was easy—much easier, Conrad said, than in the translation and docking trainer on the ground. For the first time, also, a copilot was given the chance to dock with a target vehicle.[30]

Even while docking and backing away from the Agena, the crew was meeting another flight objective. Attached to the Agena target docking adapter was S-26, an experiment that studied the ion-wake structure during docking practice. Two other experiments were started at that time—S-9, nuclear emulsion, and a modified form of S-29, libration regions photography. The crew turned on the emulsion package shortly after the hookup with the target, and a telemetry check disclosed that it was working. Gordon later retrieved it from behind the command pilot's hatch. S-29, a study of dim light phenomena, could not be carried out as planned because of the three-day mission delay. The Milky Way now obscured the intended target. Instead, the crew photographed the gegenschein and two comets.

After the last docking, the crew used the main Agena engine in a test run before going to high altitude. Facing 90 degrees away from the flight path, Conrad fired the main engine, adding a velocity of 33 meters per second to pull over into a new orbital lane. This really impressed them. Gordon remarked to Young (who had flown the Agena/spacecraft combination in *Gemini X*), "I agree with you, John, riding that PPS [primary propulsion system] is the biggest thrill we've had all day."[31]

Now, after six hours of hard but frustration-free work, Conrad and Gordon powered down the spacecraft systems, ate a meal, and soon got a "good night" salutation from the network. For eight hours, they dozed and rested, awaking, as Gordon said, brighteyed and bushytailed. The only complaints the pilots had were about their dirty windows. Dirty windows had plagued all Gemini flights. Beginning with *Gemini IX-A*, all spacecraft carried covers that could be jettisoned after the launch phase, but they did not seem to help much. Earlier, Conrad had asked if Gordon could wipe his window when he went outside. Now Alan Bean, who had taken over from Young as CapCom, told the pilot to rub half the command pilot's window with a dry cloth and bring the rag back for testing.[32]

Conrad and Gordon napped and rested awhile longer, then started their next major task—preparation for EVA. Four hours before they were to open the hatch, the crew began to get their suits ready for the vacuum environment. They had practiced this so many times on the ground, Conrad said, that they soon realized they did not need all that time. Within 50 minutes, the gear was ready and running. Just a few more steps and Gordon could have gone out. So Conrad called a halt, which left them sitting there, as he later said, with all the junk on. An hour later, they hooked up Gordon's environmental support system, and he made some oxygen-flow tests. This was also a mistake, they quickly perceived. The system dumped oxygen into the cabin, which, in turn, had to vent the excess into space. They could ill afford this rate of oxygen loss, and Conrad had Gordon switch back to the spacecraft system. Gordon, uncomfortably warm, was glad to get back on the interior system. The extravehicular system's heat exchanger had been designed to operate in the vacuum of space, not in a pressurized cabin.

Briefly, the two men considered asking Flight Director Clifford Charlesworth to let Gordon go out a revolution early. But they decided to keep on schedule. As they sat and waited, they soon regretted that decision. At last it was almost time to open the hatch. Gordon began putting a sun visor on his faceplate, a real chore and one which should have been done before he put on all this extra gear. Conrad finally got the left side fastened, but he could not reach across Gordon to fasten the other side. Gordon was now getting hot and bothered and had to rest. Time had been hanging on their hands before—now it was rushing past. Gordon wrestled with the right snap for five minutes and finally got it fastened, cracking the visor in the process. He was thoroughly winded before he got out of his seat. But he opened the hatch and stood up at 24:02 hours ground elapsed time—exactly on schedule.[33]

"Here come the garbage bags," Conrad warned. Everything in the spacecraft that was not tied down began to float upward and outward—including Gordon. Outgassing of the environmental system caused this, and the crew expected it. Conrad grabbed for a strap on the leg of Gordon's suit and held him in the seat.

Gordon then deployed a handrail—this was easy. Next he picked up the S-9 nuclear emulsion package and handed it to Conrad, who shoved it down between his legs into his footwell. Gordon then tried to install a camera in a bracket to photograph his own movements, but this was more difficult. Finally, Conrad let enough of the umbilical slide through his gloved hand to let the pilot float above the camera and hit it with his fist to drive it into place.

It was now time for the spacewalker to move forward and attach a 30-meter tether, housed in the Agena target docking adapter, to the spacecraft docking bar. When Gordon pushed himself forward, he missed his goal and drifted in an arcing path above the target's adapter and around in a semicircle until he reached the adapter behind the spacecraft. But Conrad had released only 2 meters of the 9-meter umbilical, so he pulled Gordon back to the hatch to start his trek again. This time Gordon reached the target and grabbed some fixed handrails to pull himself astride the spacecraft nose.

"Ride 'em, cowboy!" Conrad shouted. Riding bareback, with his feet and legs wedged between the docked vehicles, was hard to do. In practice sessions in zero-g aircraft flights, Gordon had been able to push himself forward, straddle the reentry and recovery section, and wedge his feet and legs between the docking adapter and the spacecraft to hold himself in place, leaving his hands free to attach the tether and clamp it down. But this did not seem to work so well in the actual conditions of space. He had to fight his pressurized suit to keep from floating away, and he had neither saddle nor stirrups to help him. All he could do was hold on with one hand and try to operate the tether clamp with the other. He struggled for six minutes, finally securing the line. At least, they were ready for the tethered flight experiment that would come later in the mission. To Conrad, it was obvious that Gordon was running out of steam. What had been relatively easy in zero-g airplane flight training had become a monumental task. With

*361*

his face streaming with sweat and his eyes stinging, Gordon groped blindly about. He tried to unstow a mirror on the docking bar so Conrad could watch him when he went to the back of the spacecraft. Gordon tugged at the attachment, but it would not budge. He abandoned the frozen mirror as not worth the effort. So far, he had not had a chance to wipe Conrad's window, either.

As the pilot inched his way back to the hatch area, Conrad helped him as much as he could. They then discussed whether Gordon should go to the adapter and get the maneuvering gun stored there. His right eye was still burning, and Conrad could see just how exhausted his pilot was. The command pilot soon told Young (through the Tananarive remote station) that he had "brought Dick back in. . . . He got so hot and sweaty, he couldn't see." Gordon had no trouble getting into the spacecraft, nor did he have any difficulty closing the hatch. It had been open only 33 minutes, instead of the planned 107. One experiment (D-16, power tool evaluation) was a casualty on *Gemini XI* as it had been on *VIII*. Also scheduled for *Gemini XII*, it had been moved forward one flight because its release mechanism would interfere with that for the sensor covers on D-10 (ion-sensing attitude control); it would require additional engineering for thermal and structural impact; and it would ease the weight load (already growing too fast) on Spacecraft 12. When Gordon got so exhausted that he never reached the adapter area, the power-tool experiment that David Scott had mourned on *Gemini VIII* had to wait for Apollo. Because Conrad and Gordon were surrounded by so much loose gear, they opened the hatch an hour later and jettisoned all the umbilical extravehicular equipment.

Although there was a standup EVA period still before them, spacewalking (or swimming) on this mission was finished, and the feasiblity of working outside the spacecraft was not settled by *Gemini XI*. Cernan had told Collins and Gordon about his problems, and Collins had further emphasized his experiences to Gordon. Yet, as the flights progressed, each successive pilot continued to be amazed that the simplest tasks were so much harder than he expected. "Gene Cernan warned me about this and I took it to heart," Gordon later said. "I knew it was going to be harder, but I had no idea of the magnitude." Apparently the supporting engineers had no idea, either, since they still had not provided satisfactory restraints to help the crews.[34]

The extreme exhaustion of past EVA pilots had sometimes adversely affected the rest of the mission. But Gordon's did not. Flight planners had learned to schedule periods of lesser activity immediately after heavy workloads. Conrad and Gordon began leisurely repacking equipment and restoring order to the cabin. Communications with the ground had dwindled to brief transmissions about spacecraft systems and crew medical checks. Conrad tested a thruster that had been sluggish and found that it was working better. The crew also ate a meal and photographed the airglow horizon. Half an hour before the sleep and rest period, the *Rose Knot* Victor tracking ship flight controller sent them the numbers for their next big event—the high ride.[35]

Next day, Conrad and Gordon skipped breakfast to get the cabin ready before the hard shove in their midsections sent them upstairs. They wanted things buttoned up as though for reentry. So they suited themselves, closed their faceplates, and stowed everything they could.

As the crew made a prefiring check of the Agena, they noticed that it was not accepting their commands immediately. Orders had to be repeated before they were acknowledged. Conrad told Bean about this and learned that the Agena was responding properly. The trouble was apparently in the spacecraft displays. "It [is] a heck of a time to have a . . . glitch like that show up," Conrad complained. But the Canary Islands communicator told them everything was fine and to "GO for the burn."

At 40:30 hours into the flight, in the 26th revolution, Conrad triggered the firing signal to the target vehicle's main engine. For 26 seconds it belched a fiery stream to add 279.6 meters per second to their speed. "Whoop-de-doo!" Conrad yelled, "[that's] the biggest thrill of my life." Since they faced the Agena, the acceleration forced the crew forward onto the seat harnesses. They watched the great round ball of Earth recede. What about orbital mechanics now? they wondered. Were they going to stop? From Carnarvon, 1372 kilometers below came, "Hello, up there." Conrad answered, "I'll tell you, it's GO up here, and the world's round. . . ., you can't believe it . . . . I can see all the way from the end, around the top . . . about 150 degrees." When Bean asked him to enlarge on his impressions from his high vantage point, the command pilot continued, " . . . it really is blue. That water really stands out and everything looks blue. . . . The curvature of the earth stands out a lot. [There are] a lot of clouds... over the ocean ... [but] Africa, India, and Australia [are] clear." He went on, "Looking straight down, you can see just as clearly ... there's no loss of color and details are extremely good . . . ."

Going up, the crew had not been merely sightseers, although they had used the tourists' favorite instrument—the camera. Gordon snapped synoptic terrain and synoptic weather photographs. The weather experiment needed cloud cover, and the terrain had to have clear views of the land areas. Conrad's at-a-glance description of the eastern hemisphere thus elated the principal investigators. They eagerly awaited the more than 300 pictures clicked off.

Radiation dosage at high altitude had caused some premission concern. The Van Allen belts (two doughnut-shaped radiation zones around Earth, named for James A. Van Allen, State University of

**Gemini XI**
**12 September 1966**

*Left, astronauts Charles Conrad and Richard Gordon practice water egress; above, spectators on viewing stand cheer Gemini XI launch.*

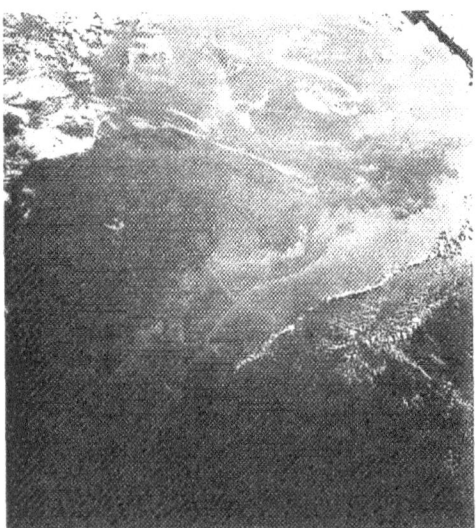

*Above left, Gemini XI photographed the Gulf Coast from Galveston to Mobile Bay at normal orbital altitude, 270 kilometers. But after the Agena had rocketed them to the record altitude of 1370 kilometers, the Earth was noticeably farther away and the curvature of the horizon was that of a huge ball, as shown right, in this photograph of the Indian Ocean near Australia.*

*Whipping around like a jump rope, the tether linking Gemini XI and Agena was unexpectedly difficult to keep under steady tension in this station-keeping exercise.*

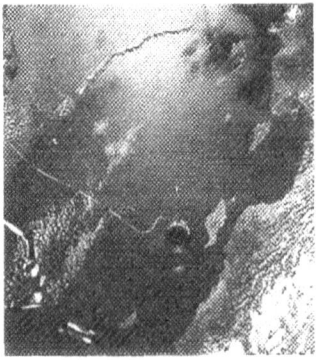

Iowa physicist) are not constant about the planet, being denser in some regions than others. High apogee orbits for *Gemini XI* were therefore planned to take place over Australia, because the level there is comparatively low. Now Conrad reported to Carnarvon, ". . . our dosimeter reads .3 rads per hour up here." Gordon amended this, saying, "Houston, radiation is revised to .2 rads per hour." To which Bean replied, "Sounds like it's safer up there than a chest x-ray." Conrad later stated that "we got less radiation in our two 850-[nautical]mile [1570-kilometer] orbits than the X crew got in their longer period of time at 450 [nautical] miles [830 kilometers]."

Over the United States in the 28th revolution, Conrad used the Agena to lower the apogee of the orbit. Firing for 23 seconds decreased speed by 280 meters per second and lowered the spacecraft orbit from 1372 to 304 kilometers. Another mission objective could be stamped "achieved."[36]

After their high-flying excursion, Conrad and Gordon were supposed to get ready for the next EVA period. Instead, Conrad told Bean, "We're trying to grab a quick bite. We haven't had anything to eat yet today." The CapCom replied, "Be our guest." After they had eaten, they still had plenty of time before the exercise was to start. In revolution 29, above Madagascar, Gordon opened the hatch and watched the sunset.

Gordon stood on the spacecraft floor, held down by a short tether like the one Collins had on *Gemini X*. It allowed him to forget about maintaining body position and left both hands free for his tasks. He mounted cameras in brackets without any difficulty. "Most enjoyable," he said of his two-hour standup period. So relaxed and well oriented was he that the monitoring physicians reported, "From a medical viewpoint, the standup EVA was relatively uneventful."

Gordon's main task during two night passes of open-hatch work was to photograph several star fields, using the S-13 ultraviolet astronomical camera. Because of his dirty window, Conrad had some difficulty in pointing the spacecraft/Agena combination in the right direction; but Gordon, with his unimpaired view into open space, coached his commander into position. Agena stabilization was somewhat erratic, but the docked vehicles were steady enough to give excellent results in about one third of the photographs.

Although neither man was really tired after the first half of the picture-snapping, Conrad considered closing the hatch and resting until the next night pass. He asked the Hawaii CapCom if there was enough oxygen. The answer was yes. But the skies were clear over the United States, and they might want to take more pictures there. In that case, said Conrad, the hatch would stay open.

Soon the crew marveled at the view of their home area—Houston. They passed quietly across Florida and out over the Atlantic with nothing to do. Suddenly, Gordon broke the silence to announce that they had just taken a catnap. "There we were. . . , he was asleep hanging out the hatch on his tether and I was asleep sitting inside the space-craft," Conrad reported. "That's a first," John Young answered, "first time sleeping in a vacuum."

"Boy, my legs are tired," Gordon said after closing the hatch. "I'm tired all over. Man, I'm beat!" Conrad answered. This time their fatigue stemmed mainly from concentration on an experiment; it bore little relation to the hard physical struggle Gordon had endured outside with the umbilical.[37]

Now the crew rested and discussed the next major mission event—the tethered vehicle exercise. There were two ways of carrying out this experiment. In the first (called gravity gradient), the docked vehicle combination assumed the position of a pole always pointing toward Earth's center. The Agena engine nozzle represented the tip, the adapter section on the spacecraft the top of the pointer. Once the pole was pointed correctly; the crew then backed the spacecraft out of the Agena docking cone slowly, until the 30-meter tether became taut. If properly positioned, a slight thrust of only three centimeters (one-tenth foot) per second would keep the line taut, and the now elongated pole would drift around Earth, with the two vehicles maintaining the same relative position and attitude.[38]

Should Conrad and Gordon fail to execute these procedures, they were then to try the spinup, or rotating, mode that had been studied by McDonnell. In this case, once the two vehicles were undocked, Conrad fired the spacecraft thrusters to induce a rotation of one degree per second to the *Gemini XI*-Agena combination. The two craft would then continue on their orbital path, with their mutual center of gravity at a specific point on the tether around which they would do a slow and continuous cartwheel. Centrifugal force would be expected to keep the line taut and the two vehicles apart, while the tether itself provided centripetal force to keep the two spacecraft in equilibrium.[39]

Over the tracking station in Hawaii, the crew separated the two vehicles cautiously to try the gravity-gradient method. There was enough initial tension in the tether to upset the Agena and to cause the Gemini spacecraft to move to the right, toward the target's docking adapter. Conrad quickly adjusted his spacecraft's motion, and the Agena righted itself without difficulty. The command pilot continued to back away from the Agena, but the tether stuck, probably in the stowage container, when about 15 meters had been released. Conrad gave a burst to his thrusters to jerk the cable free. Then, it hung up again, this time on some Velcro that had been used to hold Agena's end of the line until the spacecraft was loose. Conrad had to shift the spacecraft out of vertical alignment to peel the tether off the Velcro pad. This disturbed the Agena again, and there were still about three meters of the line to be pulled out. To do the "Non Spun Up" maneuver, as Conrad called it, the spacecraft and Agena had to be tethered and aligned vertically to Earth. The engineers expected that it would take about seven minutes for the Agena to stabilize. When the target seemed to be taking longer, they feared something was wrong with the Agena's attitude control system and told the crew to abandon the attempt and proceed to the second mode.

When Conrad tried to start the rotation, he found he had another problem. He could not get the tether taut. It seemed to rotate counterclockwise. Surprised, he reported to Young, "This tether's doing something I never thought it would do. It's like the Agena and I have a skip rope between us and it's rotating and making a big loop." He continued, "Man! Have we got a weird phenomenon going on here. This will take somebody a little time to figure out." Strangely, although the spinning line was curved, it also had tension. "I can't get it straight," Conrad muttered. For ten minutes, the crew jockeyed, using the spacecraft thrusters to straighten the arc. Finally, they got an even tether, but neither of them could ever recall exactly what they had done to stop the odd behavior of the rope.

When the tether was taut, Conrad rolled his spacecraft and blipped the thrusters to begin the slow cartwheel motion. Although this had been done gently, so to speak, Conrad felt he must have stretched the tether because it had a big loop in it when he stopped firing. The command pilot itched to do something else, but the ground engineers told him to leave it alone.

"So we really gritted our teeth" and waited, Conrad said. Sure enough, centrifugal force took over and the line smoothed out. The vehicles at either end of the rope wigwagged, but they, too, soon settled down without the pilots having to do anything. A 38-degree-per-minute rotational rate was obtained and remained steady throughout the nightside pass. The crew became so accustomed to the sight of the Agena hovering nearby that they rarely bothered to look at it. Instead, they ate their evening meal.

Conrad's satisfaction with this stationkeeping was soon disrupted. As they passed into daylight, the Hawaii CapCom told him to accelerate the spinup rate. Somewhat reluctantly, the crew agreed to try. Gordon suddenly shouted, "Oh, look at the slack! . . . It's going to jerk this thing all to heck." "That's what I was afraid of, darn it," Conrad replied. To Flight Director Charlesworth in Houston, Gordon complained, "You just ruined a good thing." When the added acceleration started, the line tightened and then relaxed. The crew felt what Conrad called "this big sling shot effect." They were being seesawed in pitch up to 60 degrees. Conrad could not accept this oscillation, so he used the spacecraft controls to steady his vehicle. To their surprise, the Agena stabilized itself again.

The rotation rate checked out at 55 degrees per minute, and the crew could now test for a minute amount of artificial gravity. When they put a camera against the instrument panel and then let it go, it moved in a straight line to the rear of the cockpit and parallel to the direction of the tether. The crew, themselves, did not sense any physiological effect of gravity. After they had been roped to the Agena for three hours, the pilots ended the exercise by jettisoning the spacecraft docking bar. All in all, it had been an interesting and puzzling experience.[40]

There had been some disappointment that the gravity-gradient mode could not be completed, but confidence rose when the spinup proved that stationkeeping could be done economically. The flight controllers had asked the crew about the remaining fuel on several occasions; they were using less fuel than had been expected. And now there was a chance for some realtime planning on the credit side of the ledger. In the past, realtime planning had been in response to such problems as degraded fuel cells, "angry alligators," or whirling spacecraft. An exercise that had been in a contingency plan, if something had gone wrong, was now fitted into the mission because almost everything had gone right.

After the two vehicles separated, Conrad had intended to decrease the spacecraft speed so *Gemini XI,* in a lower orbit, would pull ahead, leaving the Agena behind. Instead, the flight controllers told him to get ready for what was called a "coincident-orbit" (later renamed "stable-orbit") rendezvous. The spacecraft would follow the Agena by 28 kilometers and in its exact orbital path. If the plan succeeded, the crew would, in essence, be stationkeeping at very long range and with the use of very little fuel.[41]

Because of the change in plan, the separation maneuver would be different. Instead of a retrograde firing, so the Agena would trail above and behind them, Conrad and Gordon added speed and height to the spacecraft's orbit so the target passed beneath and in front of their vehicle. When the crew saw the Agena below them, moving swiftly across the South American terrain, they understood why Thomas Stafford and Cernan had trouble keeping their target in sight during the rendezvous-from-above exercise on *Gemini IX-A.*

Next they fired the thrusters to place the spacecraft in the same (coincident) orbit as the Agena and trailing it. Three-quarters of a turn around the world, Conrad decreased his forward speed and, as expected, the spacecraft dropped into the Agena's lane 30 kilometers behind the target and with no relative velocity between the vehicles.[42]

While doing their long-distance formation flying, Conrad and Gordon began to work on night image intensification (D-15), which they thoroughly enjoyed. This was a test to see if their night vision could be enhanced by equipment that scanned objects on the ground and relayed what it saw to a monitor inside the spacecraft. While Conrad aimed the spacecraft at desired targets—lights of towns and cities, cloud formations, lightning flashes, horizon and stars, airglow, coastlines, and peninsulas—Gordon watched the displays. Each pilot described what he was seeing to the spacecraft tape recorder. Conrad was handicapped by his dirty window. And the glow from the television monitor prevented him from becoming fully dark adapted. Still, the two revolutions (or about three hours) of just riding, watching, and taking pictures were very pleasant. Perhaps the most exciting sight was the lights of Calcutta, India. Outlined on the monitor was a shape almost identical to an official map of the city.

On one occasion during the experiment, the crew noticed the lights of the Agena and asked the ground how far from the target they were. The flight controller on the *Rose Knot* Victor replied that they were still 30 kilometers behind and closing very slowly. They could expect it to be about 26 kilometers away when they woke the next morning. But, when the crew broke their sleep period, in revolution 41, the target was 46 kilometers ahead. This, however, presented no problems for the re-rendezvous.[43]

The second rendezvous in *Gemini XI*, like the first, took only one orbit. At 65:27 hours of flight time, Conrad tilted the spacecraft nose 53 degrees above level flight and fired the forward thrusters. This slowed the spacecraft speed and moved it closer to Earth. Now the spacecraft was in a lower orbit than the Agena and ready for the catchup maneuver. While they waited for the final approach, the crew did the S-30 dim light photography/orthicon experiment, taking pictures of the gegenschein and zodiacal light, and completed D-15. They also turned off the switch to raise the temperature of the S-4 radiation experiment and then turned it back on. At 67:33 hours, S-4 was turned off for the last time.

An hour after the catchup maneuver began, with his ship almost level and aimed directly ahead, Conrad gave the aft thrusters a burst to raise the spacecraft orbit. Now the Agena floated just above them, its tether pointing straight up. At 66:64 hours elapsed time, Conrad began to brake his spacecraft; six minutes later, he reported that he was on station and steady with the Agena. Gordon noticed that the tether on the target had started waving slowly and surmised that this was caused by the exhaust from *Gemini XI's* thrusters. Twelve minutes later, the crew broke away from the Agena for the last time. Conrad later said, "We made the 3 foot [1 meter] per second retrograde burn and left the best friend we ever had." Gordon added, "We were sorry to see that Agena go. It was very kind to us."[44]

Conrad suggested that Flight Director Lunney might send up a tanker—the crew would be happy to refuel, remain in orbit, and do some more work. But, while this air-to-ground joking was going on, the crew was getting ready to land.[45]

There was only one significant event left before Conrad and Gordon wrapped up their mission. A secondary objective called for the crew to make an automatic reentry. The commanders of other Gemini flights had flown their spacecraft down from 120 000 meters, using the spacecraft's offset center of gravity to generate lift for changes in direction. This had enabled them to make corrections up to 550 kilometers downrange and 50 kilometers crossrange. Conrad, however, would not fly the spacecraft with his handcontroller in conjunction with computer directions; the spacecraft would follow these commands automatically.[46]

On 15 September 1966, after 70:41 hours of flight and in the 44th revolution of Earth, the retrorockets fired. Conrad and Gordon watched the computer closely. It certainly seemed to be working right. Conrad then disengaged his handcontroller and put the system on automatic. When the first crossrange errors developed, the computer commanded bank angle changes. On several occasions, the spacecraft displayed an almost human characteristic, hesitating before accepting its orders. But the system recovered quickly and performed beautifully, using a minimum of the reentry system's control fuel. The accuracy of automatic reentry was thoroughly demonstrated when the spacecraft landed within 4.6 kilometers of the U.S.S. *Guam*, the prime recovery ship, a sea-going platform for helicopters. As the spacecraft floated down to its landing, after 71:17 hours elapsed time, Young told them, "You're on TV now."[47] The *Gemini XI* flight had ended; next came the usual round of examining, debriefing, evaluating, and reporting.

## THE EVA REVIEW BOARD

When Gordon finished his postmission debriefings, he and Neil Armstrong, accompanied by MSC Deputy Director George Low and others, made a three-week, 24 000-kilometer goodwill tour of Latin America that covered 14 cities in 11 countries.[48] Meanwhile, other NASA program officials began to concentrate on getting Gemini XII ready for flight. Gordon's troubles outside the spacecraft greatly complicated premission planning, as did the lack of specific goals. Lovell complained that "essentially Gemini XII didn't have a mission. It was, I guess, by default . . . supposed to wind up the Gemini program and catch all those items that were not caught on previous flights." He added, "The only firm thing in the whole flight plan for a while was the astronaut maneuvering unit."[49]

After *Gemini IX-A*, Major General Ben Funk had begun to worry about the chances of ever flying the Air Force's AMU in the Gemini program. Gilruth assured him that it would be given every consideration because "extravehicular activity [is] a primary objective of Gemini XII." When Collins had so little trouble on the *Gemini X* EVA, hopes that the unit would get its chance to fly had revived. But when Gordon suffered exhaustion and overheating, the EVA question was again as wide open as Cernan had left it. Was there some mystery here that the Gemini engineers had not been able to unravel? Several years later, Elms said that no history of Gemini would be complete without a discussion of what he called the EVA Review Board.[50] In truth, that may well be a fitting name for the Gemini Mission Review Board before the program's final flight.

The board's first premission meeting for Gemini XII was held in Houston, where the members were being briefed on the maneuvering unit at the exact moment when Gordon was struggling with the umbilical exercise on *Gemini XI.* Although McDonnell had made all the spacecraft changes that Collins had suggested, they did not seem to be making Gordon's tasks much easier. But talking and guessing were futile, and the board soon returned to the subject on the agenda—the AMU, which, it conceded, "appeared to be a well qualified piece of space hardware . . . although complex of operation."[51]

At their next meeting, the four men* virtually became the EVA review board that Elms recalled. They "agreed that the EVA experience from previous missions was the only factor having serious potential impact on the Gemini XII Mission." Their first recommendation was to strike the AMU from Gemini XII[52] because the pilot's chance of getting into it and using it successfully seemed small, because the unit's potential value could not offset the risks involved in its use, and because the 120 minutes of EVA planned for the final mission should be devoted to a series of simple tasks that could be measured accurately in terms of workload. Mueller agreed with the board and, on 30 September, told the Air Force why the AMU was being deleted from Gemini XII:

> It is noteworthy that past EVA has revealed problems that appear less yielding to straightforward engineering solutions than other problems encountered in the Gemini Program. The EVA tasks planned for Gemini were designed to become increasingly complex and demanding on succeeding missions. And, although the experience gained on a particular mission has been carefully applied to later missions, the result has proven less than completely successful. In fact, it becomes increasingly apparent that the techniques and

---

*The membership remained the same from the beginning: Elms, Edgar Cortright, Major General Vincent Huston, and Charles Mathews.

procedures devised for EVA have evolved from analyses, theories, and experimental concepts that in certain critical instances, and for reasons currently beyond our grasp, are not entirely accurate. Consequently, I feel that we must devote the last EVA period in the Gemini Program to a basic investigation of EVA fundamentals . . . through repetitive performance of basic, easily-monitored and calibrated tasks.[53]

While the board was being briefed on the AMU at its first meeting, Aldrin was practicing with it under water in a swimming pool at McDonogh, Maryland. Later a flightready unit was installed in Spacecraft 12's adapter at Cape Kennedy. On 23 September—the day Elms sent the review board's recommendations to Mueller—it was pulled out. Aldrin, who had once worked in the Air Force experiments office in Houston, was disappointed at the loss of the AMU. He was also concerned about what was to take its place in the fast approaching mission.[54]

By July, the crew of Gemini XII was being assigned some rather precise objectives. In fact, the flight was soon extended to four days to give the crew time for experiments that depended on nighttime operations. Over the course of the program, mission planning had steadily progressed to expand manned space flight experience, but Gemini XII assumed a more conservative cast, as shown by a comparison of preliminary and final flight plans for the mission.

In July, for example, the primary objectives were rendezvous and docking, preferably in the second spacecraft orbit, and extravehicular activity with the AMU. Two of the secondary goals were repeats: re-rendezvous from above (from *Gemini IX-A*) and a tethered vehicle exercise (from *Gemini XI*). Then came the decision to delete the AMU, and Mueller told Chuck Mathews that he also opposed the re-rendezvous plan. Next, rendezvous and docking shifted from the second to the third spacecraft orbit (which had already been done). These changes, of course, affected the flight plan, delaying a final version. Mathews told MSC's senior staff as late as mid-September that the hardware would be ready for launch but that what would be done during the flight was still not firm. The final flight plan was not ready until 20 October. And it contained no surprises. Almost the only innovation was the non-spinning, gravity gradient mode of stationkeeping. But that was not really new, since Conrad and Gordon had tried it, without success, on *Gemini XI*.[55] There was to be no trail-blazing on the final mission.

If, as Lovell said, "essentially Gemini XII didn't have a mission," it did have a theme—to pierce the mystery of working in space. The strain of EVA experienced so severely by Cernan and Gordon not only clouded Gemini but raised doubts for Apollo. The lack of understanding of the difficulty emerged as a pressing concern that did much to shape Gemini's final flight. To increase the chances for success on Gemini XII, NASA now arranged to study in a careful and systematic way the basic features of EVA.[56]

Training and restraints for EVA underwent significant changes. In prior training, the crews had used zero-g aircraft flights to get the feel of weightlessness and to devise techniques for working. But experience had shown that this kind of training was useful in a very limited way, mainly for practice in getting into or out of the spacecraft. Pilots had to move fast and brace themselves before the airplane finished the Keplerian trajectory with its high-g pullout. In space, they found that everything had to be done slowly and deliberately. Nor could the kind of fatigue that Cernan and Gordon had suffered in space be assessed in zero-g flights, because the delay between successive parabolas imposed a rest period. Almost a full day had to be spent in the aircraft to accumulate 15 minutes of weightlessness.

But in mid-1966, underwater simulation had been advanced to meet these shortcomings. Moving in a viscous and buoyant fluid was very much like moving against the restraints of a pressurized suit in a weightless vacuum. Aldrin could thus get a more accurate sense of the time and physical effort required for a task on the workstands (called "busy boxes") during flight. Since the zero-g aircraft exercise did give him the feel of weightlessness, however, Aldrin continued that training also.[57]

On each of the last three missions, the pilots who went outside had complained that they needed more help in body positioning. Each spacecraft carried more restraints than the one before. The 9 restraints on *Gemini IX-A* had become 44 on Gemini XII. One helpful innovation was a waist tether that allowed the pilot to retrieve packages, turn wrenches with considerable torque, and attach the vehicle tether without undue stress. Other new features were handrails, handholds, and rings for hooking Aldrin's restraining belt to various places on the spacecraft and target vehicle. At last, an EVA pilot had all the help he would need for performing a great variety of tasks, some of considerable complexity.

After *Gemini IX-A*, MSC's Crew Systems Division puzzled over Cernan's fatigue. Collins' success in *Gemini X* suggested that the order in which he did his extravehicular tasks might have made them easier. Collins had done a standup EVA and then closed the hatch and rested before leaving the spacecraft. After Gordon had to come in early on *Gemini XI*, GPO decided that Aldrin would begin with a standup exercise and then go on to more strenuous activity.[58]

Although flight planning was the hardest part of getting ready for the final Gemini mission, hardware could have been a monumental problem—spares were becoming scarce. This danger had been foreseen and reasonable provisions made long before the scheduled launch

*373*

date, but program officials could not help being jumpy, fearing they would be unable to replace a part that had suddenly gone awry.

When the *Gemini IX* Agena had fallen into the Atlantic Ocean, Gemini XII was threatened with a major hardware shortage—an Agena and an Atlas to launch it. Replacing the Agena was no real problem. Lockheed's first production model, 5001, used for development testing at the Cape, had already been sent back to the Sunnyvale plant for refurbishment. Now it was simply a matter of tailoring it to the Gemini XII mission.[59]

Finding a new Atlas was not so easy. General Dynamics did not keep a stockpile of Atlases on the assumption that someone would come along and buy them. GPO would have to find one that had been intended for some other program. When a Lunar Orbiter flight was delayed in May, it freed an Atlas that GPO might acquire. And when Mueller approved the purchase of a replacement vehicle on 1 June, MSC was already negotiating for an Atlas at Vandenberg Air Force Base in California. But this was not the standard vehicle Gemini had been using; it was the first of a new series with some features that had never been tested in flight. Langley Research Center, in charge of the Orbiter payload, was persuaded to turn its Atlas over to Gemini in exchange for the one in California. Langley's Orbiter Atlas had only nine variances from the Gemini version, and the trade eased the minds of the MSC program engineers. By the end of September, the new Atlas waited on pad 14 at Cape Kennedy for its call to action.[60]

## THE FINALE

The final curtain snagged twice before it opened on Gemini XII. Spare parts became a problem, as had long been feared. An autopilot and a rate gyroscope in the launch vehicle had to be replaced. Then, the replacements were themselves replaced. But, on Veterans' Day—11 November—Flight Director Glynn Lunney signaled for the overture to begin.[61]

At 2:08 p.m. the substitute Atlas lifted the refurbished Agena from pad 14 and lofted it into orbit. A few minutes earlier, over on pad 19, the pressure-suited crew had shuffled up a ramp, bearing signs on their backs—"THE" and "END." This bit of humor was more than symbolism, for when launch vehicle No. 12 broke its landlock 30 seconds after 3:46 p.m., the Gemini preparations team faded into space history. Francis Carey, Martin's chief test conductor, and Colonel John Albert, Chief of the Gemini Launch Vehicle Division, 6555th Aerospace Test Wing, took justifiable pride in the 12 for 12 record, but they mourned the fact that the job had ended and the team would soon break up. That it was over could hardly have been more vividly underlined—almost at once wreckers were hacking the launch stand into scrap iron. Apollo was the future. A harbinger of this new era, *Lunar Orbiter II,* had been launched only five days earlier—6 November—on a trip to the Moon to photograph possible Apollo landing sites.[62]

Meanwhile, Lovell and Aldrin began to wonder if everybody had gone away too soon. For 25 minutes, with one brief exception, they heard nothing from the ground. The Ascension Island tracking station had the wrong acquisition time, so its communicators had not talked with the pilots. Lovell was relieved when he heard Conrad hailing him through the remote line at Tananarive with some needed data for a maneuver that was scheduled to take place within a few minutes.[63]

Things now went smoothly and, a little more than an hour after launch, Aldrin reported, "Be advised we have a solid lock-on . . . 235.52 [nautical] miles [436.18 kilometers]." From Houston, Conrad replied, "It looks like the radar meets the specs." When the spacecraft moved into a circular orbit below and behind its target, the radar showed the Agena to be 120 kilometers away. But this was the last figure the crew could trust; reception got so poor that the onboard computer refused to accept the radar's intermittent readings.

The radar failure meant that *Gemini XII* would have to rely on the backup charts it carried to complete the rendezvous. Aldrin, a member of the team that had planned and worked out chart procedures, now had a chance to see if his doctoral studies at MIT and the simulator training in St. Louis with McDonnell and MSC engineers really were practical in space.[64] The pilot, who was sometimes called "Dr. Rendezvous," had already pulled out and used the T-2 manual navigation sighting sextant to take a look at the target. When the radar went on the blink, this piece of experimental gear became operationally important.

In the automatic rendezvous mode, the radar would have fed range and range rates to the computer. Lovell would then have flown the spacecraft by the resulting numbers. This time the computer would be left in the catchup mode, and either Aldrin or Mission Control—or both—had to figure range and range rates to see if the computer was correct. For this backup method, Aldrin used the sextant to measure the angle between the local horizontal of the spacecraft and that of the Agena, ahead of and above them. He checked this information with his rendezvous chart and cranked the necessary corrections into the computer. Lovell flew the spacecraft with these numbers to rendezvous with the target, arriving there after 3 hours and 45 minutes of flight. They had used only 127 kilograms of fuel. Lovell called the *Coastal Sentry* Quebec at 4:13 hours elapsed time, saying, "We are docked." But *Gemini XII* was the fourth flight to make that announcement, and the shipboard flight controller merely replied, "Roger."[65]

For the second time, a Gemini crew was able to practice docking and undocking. They unlatched the vehicles and Lovell tried the task during the night. But the spacecraft was misaligned; the target's docking cone did not unlatch. Instead, it locked bumpers, catching on one of the three latches. Much like an automobile driver mired in the mud, Lovell fired the aft and forward thrusters, trying to rock the spacecraft free. Both vehicles were shaken, but he broke loose without damage to either. A few minutes later, Aldrin docked without difficulty.[66]

The next item on the agenda was the firing of the Agena to go to a higher altitude, but that part of the flight plan had to be changed. Eight minutes after the Agena was launched, its main engine suffered a momentary six percent decay in thrust chamber pressure and a corresponding drop in turbine speed. So, while Lovell and Aldrin chased and caught the Agena, then practiced docking, Mission Director Schneider and Flight Director Lunney had to decide whether the main engine should be fired. They soon decided that prudence was the better course—it should not.

Although the pilots missed the ride to high altitude, Lunney soon found something for them to do with their spare time. The flight plan had originally called for them to photograph a solar eclipse, if it did not conflict with the rest of the mission. This task fell by the wayside when the two-day launch delay—from 9 to 11 November—meant that the eclipse would occur during their high-altitude excursion. Canceling the main engine burn inspired two of the mission planners to thoughts of reinstating the eclipse photography. Schneider and Lunney conferred with James R. Bates, Experiments Advisory Officer for *Gemini XII*, on the effect this might have on the rest of the experiments. Since the flight plan had to be changed anyway, Bates said, why not include the eclipse?

This conference with Bates marked a significant change in mission control operations. Formerly working out of an adjacent staff support room, the experimenters' representative was now allowed by the engineers in charge to operate as a part of the flight control team in the main control room. Although there had been an experiments console in the control room by *Gemini X*, it had been only occasionally manned. Bates, on *Gemini XII*, was the first full-time experiments officer. This experience worked out so well that the custom was continued in Apollo.

Even after the eclipse became a flight-plan casualty, planners continued to plot its path. Now there was a chance to work this experiment back into the mission. The Agena's secondary propulsion system had enough power to get the spacecraft into position for an eight-second photographic pass at the proper time. Schneider and Lunney agreed that this piece of realtime planning would give an added fillip to the mission.[67]

"The eclipse got to us after all," Lovell remarked. "Yes, it looks like it," Conrad answered. Although the crew had wanted to do the experiment when it was first planned, these sudden preparations came at an inconvenient time. They were still working with the Agena and were scheduled to begin such activities as eating, sleeping, and working on other experiments.

Nevertheless, at 7:05 hours after launch, Jim Lovell fired the Agena's smaller engines to slow his speed 13 meters per second. Agena still had its doubters—Conrad had told them, "If it gets away from you . . . take it over with the [spacecraft]." But the target vehicle performed splendidly, and the crew then bedded down for the night.

The Canary Island controller greeted the crew in the morning with the news that there would be a second maneuver—5 meters forward—to line the vehicles up properly. The prospects panned out richly, and the crew reported seeing the eclipse "right on the money at 16:01:44 g.e.t." The path of the eclipse cut a swath across South America from north of Lima, Peru, nearly to the southernmost tip of Brazil. Although they thought for a moment they were slightly off track, their aim had been accurate.[68]

The sudden change in the flight plan had disturbed the crew, because of its possible interference with the first planned extravehicular exercise. After all, this objective had become the heart of their mission. Despite interruptions (especially that caused by the second maneuver), the hatch was opened on time, about 20 minutes before sunset in space. Aldrin exclaimed, in near speechless awe, "Man! Look at that!" Aldrin was amazed and impressed at seeing so much of Earth and the universe spread before his eyes.

Aldrin went about his chores slowly and deliberately, working for a short period and then resting. First, he just stood in the hatch, becoming acclimated. Then he cast loose a garbage bag. Moments later, he murmured, "Stars in the daylight? I don't think so." He soon realized that he was watching the pouch as it drifted away. He was in darkness for eight minutes before his eyes became adjusted and he could see real stars and planets. Aldrin studied his every movement— every action and reaction—so he could compare his standup experience to the umbilical period later.

He set up an ultraviolet astronomical camera. During two night passes, he photographed star fields, although Lovell had trouble turning the spacecraft in specific directions because the Agena had nearly a full load of fuel. During daylight, the pilot installed a movie camera; fixed a handrail leading to the target docking adapter cone; pulled off the ultraviolet camera, reloaded it, and put it back; retrieved a micrometeorite collection package; and took pictures. At 21:58, the crew buttoned themselves back into the spacecraft after recording their first, highly successful, 2-hour-and-20-minute exercise.[69]

The next day Lovell and Aldrin got ready for the main event of the mission—to see if a man could perform useful tasks in space at the end of an umbilical. Near the 43-hour point in the flight, Aldrin stood up in his seat and reinstalled the movie camera—just as easily as before—then removed it, stepped into space, and replaced it, using only a handrail to maintain position. The astronaut then moved, hand over hand, along the rail to the nose of the Agena docking adapter. Using his waist tether for restraint, he tied the two vehicles together for the gravity-gradient experiment without any of the problems Gordon had encountered.

The pilot floated to the hatch area and exchanged cameras with Lovell. Moving along the handrail, Aldrin went aft to the spacecraft adapter. He placed his feet in the golden slippers (overshoe-type restraints). Then he moved his body back and forth and from side to side, to see if the slippers really helped as much in holding him down as the program office had hoped. They allowed him to relax completely and to lean as much as 45 degrees to either side and 90 degrees backward.

Next he unpacked some small penlights and set to work in the busy box, torquing bolts and cutting metal. On one occasion, a bolt and washer slipped free. Aldrin maneuvered the weightless fittings into a corner, capturing one in each hand. Lovell asked him over the intercom if he was playing orbital mechanics in the adapter and the pilot replied, "Yes. I had to do a little rendezvous there." At sunrise, he returned to the open hatch. After resting for a few minutes, Aldrin again went forward to the Agena—this time to a busy box attached to the target. Lovell watched him as he pulled electrical connectors apart and put them together again. Aldrin also tried a torque wrench that had been designed for the Apollo program. For this task, he first used both waist tethers, then one, then none. On the way back to the hatch to end his second two hours of extravehicular time, Aldrin stopped to wipe the command pilot's window with a cloth. As he did, Lovell asked, "Hey, would you change the oil, too?" The "air in the tires" was "A-OK," so Aldrin climbed aboard, stood in the hatch, and watched while Lovell fired some of the thrusters. He then sat down in the spacecraft seat. The door closed easily, and Aldrin released the oxygen in his life support system to help repressurize the cabin.[70]

The third hatch opening (and the second stand-up-in-the-seat period) came on the fourth day and lasted an hour. The pilot tossed out a lot of equipment he had used during the umbilical, as well as some empty food containers. The astronauts were not really litter bugs. Discarded items from the flights, like other things in orbit around Earth, eventually reenter and burn up in the atmosphere. Aldrin then snapped several ultraviolet photographs of constellations. That finished, he went back inside and closed the hatch; the last extravehicular performance of the Gemini program was ended. But NASA engineers, mission planners, and astronauts now believed they knew much more about the fundamentals of EVA.[71]

Between the second and third hatch openings, Lovell and Aldrin went into their tethered vehicles act. Lovell backed *Gemini XII* carefully away from the Agena, forming a pole vertical to Earth. The tether deployed smoothly (with only a brief hangup) but remained slack. Lovell was exasperated at his inability to tighten it, using the spacecraft thrusters. "About this time we had a little . . . problem," he said, " . . . every time I wanted to pitch up or yaw, I would roll." Despite the control problem the crew did obtain the gravity gradient they sought. Both vehicles got upset on occasion, the spacecraft at one time wigwagging about 300 degrees. What caused these disturbances, the program office stated in its formal mission report, "is not completely understood, nor is the system behavior during and immediately following these excursions." The tether exercise lasted four hours, proving that both the controllers and the crew were confident enough to continue this form of stationkeeping through the nighttime passes.[72]

Earlier in the mission (about the time of the docking and undocking practice), the fuel cell had hinted that it might cause trouble and not last the full four days. But 30 hours passed before a power loss was actually registered. Eventually, the experts decided that there must be too much water in the tanks. Whenever the crew drank water or used it to prepare their food, the fuel cell warning light went off.

The ground controllers were not sure what had happened to the water storage system's two tanks that held the crew's drinking water and (separated by a bladder) the fuel cell product water. But, in some way or another, the astronauts had lost a place in which to store from 15 to 18 kilograms of water produced by the fuel cell. So the crew had to drink more water to make more room in the tanks and to purge the system more frequently to remove gases that accumulated in the fuel cell, if they were going to complete the mission. Drinking lots of water and watching the red warning light, they nursed the fuel cell along for more than 80 hours. The flight neared its end before the batteries had to take over the electrical load.[73]

So, even in the face of problems with the radar, the Agena main engines, and the fuel cells, *Gemini XII* had gone very well. Most of the mission objectives were accomplished, and some data were obtained from 12 of the 15 experiments assigned to the flight.[74] At times, considerable ingenuity had been required to get around the hardware difficulties.

Compared to other flights, *Gemini XII's* accomplishments tended to obscure its hardware problems, of which this final mission had more

**Gemini XII**
**11 November 1966**

*Suited up, astronauts James Lovell and Edwin Aldrin head for the elevator on pad 19 which will lift them to their waiting spacecraft. On their backs are signs, "The" and "End."*

*Learning more about the intricacies of EVA was the big assignment for Gemini XII. Above left, Aldrin is beginning his spacewalk by affixing a camera on the outside of the spacecraft. Center, he is returning from the adapter section with a micrometeoroid experiment package which will return to Earth in the spacecraft. Above right, an attempt at tethered station-keeping with Agena is more successful, thanks to what was learned from Gemini XI. With the splashdown of Gemini XII (below left), the Gemini flight program comes to an end. When the astronauts arrive back at Cape Kennedy, they are greeted with the Gemini flight "scoreboard" (below right).*

than its fair share. Some troubles that forced slight changes in the flight plan actually turned into triumphs. The failure of the radar during the terminal phase of rendezvous, for example, had underscored the fact that backup techniques, using onboard charts and computations, really worked. Radar malfunction barely caused a ripple in the routine. Other troubles nagged and frustrated the crew, and some had adverse effects on operations; but here, again, they were not able to mar the impression of success. What was remembered was Aldrin's flawless performance during the well planned extravehicular periods.

During the 59th revolution, *Gemini XII* began its controlled automatic reentry. Everything worked neatly, until the spacecraft reached its peak g loads. At that point, a pouch containing books, filters, and small pieces of equipment broke free from the Velcro on the sidewall of the cabin and landed on Lovell's lap. The pilots had unstowed the D-rings that activated the ejection seats and were holding them down between their legs. Lovell resisted the impulse to catch the pouch for fear he might "just grab ahold of the D-ring and keep pulling it." If he had, the commander, along with his pilot, would have exploded into the atmosphere, riding the ejection seats. This thought was bad news to Lovell, "because I didn't want to see myself punching out right at this high heating area." Instead, he squeezed his knees together and hoped that the pouch would not go any farther. It did not. The rest of the reentry was smooth until the moment of landing, when the spacecraft plopped down hard on the ocean.

It landed only 4.8 kilometers from the point at which it had aimed and only 5.5 kilometers from the carrier *Wasp*. A helicopter deposited the triumphant astronauts on the deck of the prime recovery vessel 28 minutes after touchdown. There, on 15 November 1966, at 2:21 p.m., e.s.t., the curtain closed on the Gemini manned space flight program.[75]

So the Gemini flag and the Gemini pennant that had flown over the Manned Spacecraft Center during each of the missions, beginning with *Gemini IV*, were lowered for the last time.[76] The manned flights had started in 1965. Gemini had succeeded in putting manned space flight on something like a routine basis, as envisioned in the Project Development Plan of 1961. This accomplishment did not go unnoticed. President Lyndon B. Johnson said:

> Ten times in this program of the last 20 months we have placed two men in orbit about the earth in the world's most advanced spacecraft. Ten times we have brought them home.

> Today's flight was the culmination of a great team effort, stretching back to 1961, and directly involving more than 25,000 people in the National Aeronautics and Space Administration, the Department of Defense, and other Government agencies; in the universities and other research centers; and in American industry.

Early in 1962, John Glenn made his historic orbital flight and America was in space. Now, nearly 5 years later, we have completed Gemini and we know that America is in space to stay.[77]

Being in space to stay rested, in part, on the shoulders of a team that was now experienced in planning, developing, managing, and operating a space flight program that had progressed far beyond the shorter flights and simpler missions of Mercury. Gemini was only the second phase of this nation's manned space flight, but its importance must not be minimized. It had dispelled most doubts about man's ability to withstand weightlessness, to operate in free space outside his spacecraft, and to seek and find another vehicle in orbital flight. Now Apollo, the third and most ambitious star, waited in the wings, and the complexities of that program dwarfed the scope of Gemini as Gemini had towered over Mercury. Only three years remained in which to accomplish the late President John F. Kennedy's "goal, before this decade is out, of landing a man on the moon and returning him safely to the earth." President Johnson warned the nation that these years might be as exasperating as the early periods of Mercury and Gemini. On 23 November he said:

> The Apollo program which follows is much more complicated. It has more elements of a yet unproven capability, and will use the larger Saturn boosters developed especially for civilian manned flight programs.
>
> The months ahead will not be easy, as we reach toward the moon. We must broaden and extend our know-how, based on the increased power of these mighty new boosters. But with Gemini as the forerunner, I am confident that we will overcome the difficulties and achieve another success.
>
> Apollo will make America truly a spacefaring Nation. The three-man Apollo is the certain forerunner of the multimanned spaceships of the not too distant future—ships that will bear the experiments and some day the experimenters of many nations—ships that will bear the hopes of all men.

About two months after the President spoke these unknowingly understated words, Apollo had to "overcome the difficulties" born of tragedy. While the NASA engineers were getting ready to report on some of the successes that had been achieved and the problems that had been solved in Gemini, a spacecraft fire on 27 January 1967 snuffed out the lives of the first Apollo crew, Virgil I. Grissom, Edward H. White II, and Roger B. Chaffee, during a test on pad 34 at Cape Kennedy.

"The months ahead will not be easy . . . . "[78]

# SUMMING UP

THE more than 1800 days that divided 7 December 1961, when Project Gemini was officially approved, from 15 November 1966, when the program's last two fliers returned from orbit, spanned a significant phase of human venture into space. Gemini provided techniques, equipment, and experience that helped bridge the difficult translation from experimental, Earth-orbiting Mercury to ambitious, lunar-landing Apollo. Gemini achieved its goals, save for land landing, quietly, systematically, and, in some degree, economically. To a large extent, at least in the general American viewpoint, the regularly flying and highly successful Gemini marked America's ascendency to first place in the space race. And its spacecraft, simpler and more efficiently designed than Apollo's (which still relied on stacked and integrated components rather than complete modules), was frequently and mistakenly cited as contributing to the Apollo concept.

For some time, the development phase of Gemini and Apollo proceeded along parallel lines, leading to a belief in some quarters that efforts devoted to Gemini were sapping Apollo's energies. Sporadically, throughout the years, a spirit of competition grew within Gemini—a feeling that its spacecraft could do more, its missions could be extended, perhaps even to lunar flight. But within Apollo doubts were increasing. Gemini had been justified partly on the basis of its contribution to Apollo experience. In 1965, Howard Tindall, whose specialty was mission planning and who had achieved local fame with his "Tindallgrams," tried to look at the question objectively and concluded that hardware and mission planning were too difficult and too concurrent for either program office to keep up with or help the other.[1]

As for the early days, Tindall's viewpoint was probably correct. Gemini had too many financial and technical problems of its own to leave much energy to worry about Apollo. Nor was the Apollo office, with its two dissimilar spacecraft, quite as cohesive an organization as it might have thought. Lunar module engineers found it equally difficult to get meaningful assistance from either Apollo command module or Gemini spacecraft people—and, no doubt, vice versa. No problem that arose on one spacecraft appeared quite like those encountered on the other two—and no one had the time to consider the problems dispassionately and apply them to their counterparts in a practical manner.[2]

Once Gemini neared its operational phase, however, things were different. Apollo managers and engineers quickly sought help in various areas. James Church wanted to learn about Gemini program control experience, especially when the Gemini people succeeded in controlling program costs. Calvin Perrine asked for information on ground test programs as the Gemini development and test experts began meeting delivery schedules more successfully. Rolf Lanzkron and Joseph Loftus were anxious to learn anything from the Gemini crews that might be applicable to Apollo flight problems. Even North American, the Apollo command module manufacturer, thought some of Gemini's checkout experience might be helpful.[3] Both North American and Grumman (the lunar module builder) had already requested manufacturing assistance from the Gemini spacecraft contractor, at one time causing William Lee, a deputy manager in the Apollo office, to caution them not to "convert McDonnell from a spacecraft manufacturer into an educational institution."[4]

Although Lee's point may have been intentionally overdrawn, Gemini manufacturing, testing, and review procedures did influence Apollo. By August 1965, many of these methods were being drawn upon to smooth the flow of hardware through the factory and on to the launch site. Of course, Gemini built upon some experience derived from Mercury—the same company manufactured the spacecraft and the same NASA group managed the project—but modular, accessible, serviceable Gemini was far more suitable for developing a systematic, if not routine, approach to getting it built, out of the factory, and onto the pad ready for launch. Gemini's vehicles, whose designers had avoided Mercury's interlocking systems, left the contractor plants much as airplanes did—all tested and nearly ready to fly. Cape Kennedy became a checkout and launch activity for Gemini, instead of the test and modification center it had been during Mercury.[5]

Besides the manufacturing and testing procedures, Gemini came to grips with several specific systems, common in one form or another to Apollo, that were new to space flight operations. Spacecraft thrusters powerful enough to alter the flight path several times and fuel cells to generate electrical energy to run the systems represented particularly impressive advances in aerospace technology. In addition, Gemini spacecraft were equipped with a computer and a radar to aid in solving the rendezvous problem. All of these systems went through troubled development and qualification periods and, in most cases, required extensive redesign. More often than not these difficulties came to the attention of NASA's top administrators. Problem-solving boards, headed by senior officials, were appointed and armed with charters to draw upon organizations and facilities in government and industry to bring about solutions. Those areas that yielded most stubbornly were aired at Gemini and Apollo executive meetings attended by NASA administrators and their staffs and company presidents and their aides—the people in charge who could bring pressures and resources to bear to fix thrusters, fuel cells, Agenas, or other recalcitrant systems.[6]

Several management bodies spawned during Gemini were not constrained to one-shot, fix-it functions but were formalized and adapted to whatever program followed. One of the more important of these dealt with manned space flight experiments. As in other cases, this activity had its origins in Project Mercury, albeit to a very limited degree. Only a few scientists gained a nodding acquaintance with NASA and industry aerospace technologists; and there had not been much interest on either side in changing that situation. Engineers had concentrated on making Mercury work, and most scientists had preferred to have their experiments ride aboard NASA's unmanned satellites. In the summer of 1963, however, science gained a permanent foothold in manned space flight operations. When the demise of the paraglider* left some unoccupied space in the vehicle, a few NASA officials saw a chance to set up an experiments program in orderly fashion. Homer E. Newell, Director of NASA's Office of Space Sciences, sent letters to more than 600 scientists, describing Gemini and inviting proposals. When the response was good, NASA established a Manned Space Flight Experiments Board in January 1964. By the fourth Gemini flight—the second manned—experiments and principal investigators had been worked into mission operations with fair success; by the last flight, procedures had been sharpened sufficiently for the board to continue in Apollo, and later in Skylab, without a break in stride.[7]

One of the quicker ways Gemini grabbed Apollo's attention, though certainly not planned that way, was its nearly catastrophic anomalies. Perhaps the most significant example was the explosion of Gemini Agena target vehicle 5002 in October 1965. The solution—to inject oxidizer into the firing chamber before the fuel—was applied to the lunar module's ascent engine simultaneously with the modifications to the Agena's primary propulsion system.[8] Visions of astronauts on

---

*Setting down on land was one goal that Gemini failed to achieve. Ironically, in 1965 there were some near-perfect tests with a limp (as opposed to an inflatable) version of the paraglider. By that time, however, the device was too far out of phase with Gemini schedules.

the lunar surface igniting their takeoff engine only to have it explode were too harrowing to be entertained.

But there were day-to-day Apollo-Gemini exchanges that did not relate to specific incidents. For example, people from the Flight Crew Support and Crew Systems Divisions worked on astronaut equipment and space suits to achieve a range of capabilities from extravehicular activity to shirtsleeve cabin operations—features of definite value to Apollo. Perhaps the group that gained the most insight into the routine operations of the two program offices was flight control. Christopher Kraft, who directed this activity, had been largely responsible for planning the old Mercury Control Center. Much improvising had been necessary to complete that project, and the facility was obviously totally inadequate to support Gemini and Apollo. NASA decided to build a control center in Houston, the new home of the Manned Spacecraft Center, and based this decision, in part, on the reasoning that flight control and spacecraft design would profit from having engineers from these two areas working together. Kraft concentrated first on Gemini requirements, partly because of manpower limitations and time constraints (Gemini would fly sooner), but mainly because of the need for Gemini experience in qualifying men, flight control equipment, and procedures to handle the far more complex missions of Apollo.[9]

Long before mission operations commenced, Kraft and his group foresaw that Gemini and Apollo flight control would require large numbers of systems, network, and trajectory specialists. Staff rooms, housing experts in these categories, were arranged around the mission operations control room. The new control center was not needed for the first two Gemini manned missions, but Kraft wanted to, and did, get it set up and operating one flight before any rendezvous maneuvers, practice or otherwise, were scheduled to take place. Kraft led his flight control team through the first rendezvous mission, as he had intended, and then withdrew to apply, in preparing for Apollo, the lessons he had learned. A major area on which he focused attention was the computer complex. Although the IBM 7094 model then in use was adequate for Gemini, it was better suited to scientific purposes. What Apollo needed, Kraft said, was a second generation model capable of supporting realtime space operations.[10] He was proved right when the flight controllers were able to change *Apollo 13*, in the middle of the mission, from a lunar landing to a circumlunar flight and thus to prevent a space tragedy.

Beginning with Gemini's sixth flight, Apollo personnel watched mission operations more closely, attending panel meetings on spacecraft systems and mission planning, observing flight control operations, and participating in mission debriefings and evaluations. On occasions when Gemini planners reacted to anomalies with a seizure of conservatism, Apollo engineers pressed to make mission activities more meaningful to the lunar program. When things really went wrong—the GATV 5002 explosion, the shutdown of Gemini launch vehicle 6 after ignition, and the stuck thruster on Spacecraft 8—systems engineering experts were assigned to determine how similar incidents might, or might not, affect Apollo.[11]

The Gemini program clung to its original flight schedule much more closely than had Mercury. Eighteen months was the lag time for the first manned Gemini mission;* the final mission, nine flights later, was still 18 months behind the schedule approved in January 1962. In contrast, the first manned orbital Mercury mission came 22 months later than scheduled, and the final mission, only three flights later, lagged more than 32 months. Mercury's period of orbital operations covered 451 days, or a flight every 112 days, to accumulate only 55 hours of crew experience. The 10 manned Gemini flights spanned 603 days, or a flight every 60 days, to accumulate 970 mission-hours and 1940 man-hours in space. Sixteen different astronauts made Gemini flights and four others trained for them. This experience was passed on to Apollo, as 15 of the 20 men subsequently flew in the lunar program. The rapid succession of Gemini missions demonstrated that it was truly a second generation spacecraft, and the length of its missions—330 hours on *Gemini VII*—allayed major medical concerns over man's ability to adapt to and function in space. More and more it became an accepted fact during Gemini that man could, should, and would fly to the Moon and back.[12]

Projects Mercury and Gemini certainly had one feature in common—both cost about double the original estimate. The best educated guess that T. Keith Glennan, the first NASA Administrator, could give for Mercury was $200 million, and its price was over $400 million at the end. Gemini started at $531 million to build what was supposed to be an improved Mercury and wound up costing $1.147 billion to cover a program that included many new developments. Unlike Mercury, Gemini had its share of financial crises; Congress and the Administration, beset with a variety of domestic and international problems, curbed the flow of money to NASA—and Gemini usually had to bear the brunt. At times the prospects must have seemed bleak to the engineers who worked on it, but the monetary cuts were never deep enough to preclude, although they often threatened, the accomplishment of Gemini's primary objectives. In what must be counted an unusual circumstance at the leading edge of technology, Gemini actually rolled back the money tide to some extent. The projected runout cost in Fiscal Year 1964 had been set at $1.354 billion, but the innovation of better test and checkout procedures that same year cut two months

---

*These scheduling figures are based on the early scheme of having only one unmanned flight in the program; the second would have carried a crew.

from the schedule and saved an estimated $200 million. Major credit for this achievement should probably go to the incentive contracts that in 1964 put Gemini procurement on a strikingly new footing.[13]

By putting manned space flight on a more routine basis, as stated in the project development plan, the rapid and successful progression of Gemini missions had a salient effect on American and international opinion—manned space flight became commonplace. During the flight period, there had been a spacecraft circling Earth about six percent of the time or, theoretically, an hour and a half for each one of the 603 days the operations covered. Not even the Wright brothers, at the dawn of powered flight, could have sustained public interest on such a regular basis. Over a thousand reporters came to Houston for *Gemini IV*, drawn by the knowledge that the new mission control center would operate for the first time and by predictions in some medical circles that the astronauts might die after being in weightless flight so long. No succeeding flight drew nearly as many until *Apollo 11*, when over a thousand reporters, cameramen, commentators, and technicians again descended on Houston, this time to write and talk about the lunar landing mission.

*Gemini VII/VI-A,* the first rendezvous mission, not only gave new life to the old saw about not being able to tell the players without a program (with four men and two spacecraft cluttering up the heavens) and proved that a 14-day flight was feasible, but it saw the Russian-American space race scorecard all but tossed aside. The fact that the Russian cosmonauts did not fly at all while American astronauts whirled about Earth at frequent intervals probably prompted the premise that the race was over. Now that the United States had gained space preeminence and international tensions seemed to be lessening despite Vietnam, the value of manned space flight was increasingly questioned when compared to the need for solving the age-old problems of hunger, housing, and education.[14]

The final event in the Gemini program took place in Houston on 1-2 February 1967, as planned, in the Manned Spacecraft Center auditorium. Some 900 people gathered from throughout the country to be greeted by Director Robert Gilruth, who asked them to divorce the recent Apollo accident from the Gemini proceedings. During the two-day conference, 21 technical papers were presented, concentrating mainly on rendezvous, extravehicular activity, and experiments.[15]

Although the summary conference was given little space by the news media, Gemini's lessons and its people, some in leadership roles, were significant factors in Apollo's recovery. Twenty-two months elapsed before America put men into space again, yet only nine months after that—in July 1969—two astronauts walked on the Moon and ten more soon followed in their steps. Gemini had contributed its share to man's quest for a better understanding and use of his environment. As it developed, the gaze was not wholly outward to the stars. Beginning with Gemini's manned missions, scientists gradually realized that photographs of Earth brought back by the astronauts could serve as a valuable tool to help identify and husband Earth's dwindling resources. Perhaps future historians will see that as Gemini's most lasting contribution.

# GEMINI EARTH PHOTOGRAPHS

## An Epilogue

THE selection of the Gemini photography for this text was not an easy task. Only a few views would be published out of the hundreds of remarkable photos worthy of such status.

As I scanned the scratched and dirty, decade old, "first master" which I have viewed a thousand times, my mind would occasionally wander back to those still vivid sights and sounds of the great epic of Gemini. I could clearly hear again the near euphoric shouts of Ed White as he stepped from *Gemini IV* into the void, coupled to the steady confident words of Jim McDivitt which told us that all was well. The deep concern in the voices of Neil Armstrong and Dave Scott describing the problems of *Gemini VIII* as they prepared to terminate their mission. The hesitant and difficult grasp for words by many crews as they attempted to describe the new phenomena and experience to an audience that would never have an opportunity to journey into space. The deep concern in voices of men trying to fix or adjust balky equipment. Gordo Cooper waking Pete Conrad as he clicked away getting more of those wonderful photos of the Himalayas. The occasional "cuss" word that would receive worldwide media coverage, ad nauseam. The last minute changes in photo equipment at the Cape. The confident joy of telling the crew a few hours before lift-off that "all photosystems are Go!" And lastly, the most unique pleasure of being the first to see that unique photography as it came out of our film processors.

I clearly remember that June 1965 day when it all began. We had just finished processing that first roll of *Gemini IV* film which showed those 16 remarkable views of Ed White's spacewalk. We stretched out the roll, and a dozen or so NASA VIPs huddled tightly around the spacewalk photos expressing elation at what they saw. Many are prominently mentioned in this history. As for myself, I stood alone at the other end of the roll quietly looking at photographs of the Earth, seeing things and places never before seen by human eyes. John Brinkmann, Photographic Technology Division Chief, called to me to come down and "see the action." My reply was, "Boss, I think the *real action* is down here."

With Mercury, space photography was born; with Gemini, it struggled toward maturity so that Apollo space photography would give you and me, indeed the whole world, an opportunity to reach out and practically touch the Moon. But Skylab would again look back toward a troubled Earth, back here, where the "real action" is.

*Richard W. Underwood*
*Thanksgiving Day 1974*

The world's second highest mountain, Mt. Godwin-Austen (K-2) in the Karakoram Range, is seen in a northerly view that looks across parts of India, Pakistan, and China. Godwin-Austen (8611 meters) is near the upper left margin. The Indus River flows in deep gorges at the left of the photograph. The deep gorge starting in a series of glaciers near Mt. Godwin-Austen and coming down the center of the photo is Syhok fork of the Indus. The mountains terminate at the upper right with the hazy Takla Makan Basin of China beyond. This overview of a largely unexplored region helps geologists better understand our restless Earth. (S65-45648; Gemini V.)

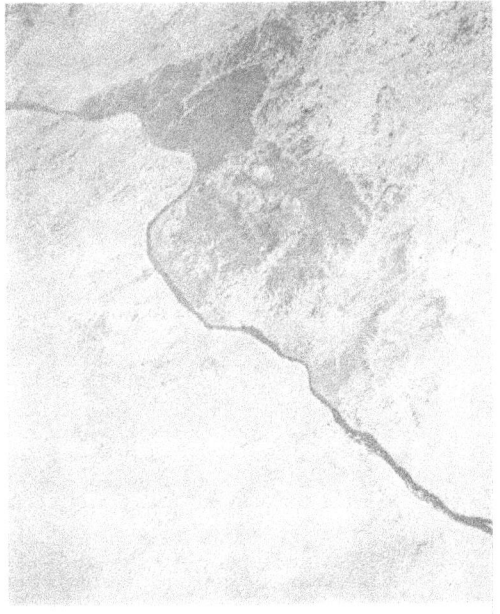

The geologic structure of the Earth controls the course of the Egyptian Nile for some 250 kilometers as the river flows along the contact between sedimentary (light) and basalt (dark) rocks. Note that one does not see a major tributary to the river in the entire 450 kilometers between the Tropic of Cancer (upper) and Wadi Halfa, Sudan, at the second cataract (lower). Today's astronaut would see man's largest creation, Lake Nasser, if he looked down from this vantage point. (S65-34780; Gemini IV.)

395

Parallel ridges of sand extend for hundreds of kilometers across the interior of Arabia in an area called Rub-al-Khali (The Empty Quarter). Well named! Seif dunes, as they are called, are rarely found. The long ridges are parallel to the prevailing winds instead of transverse, like most dunes. The photograph covers tens of thousands of square kilometers of area, but nowhere does one see any signs of life. (S65-34765; Gemini IV.)

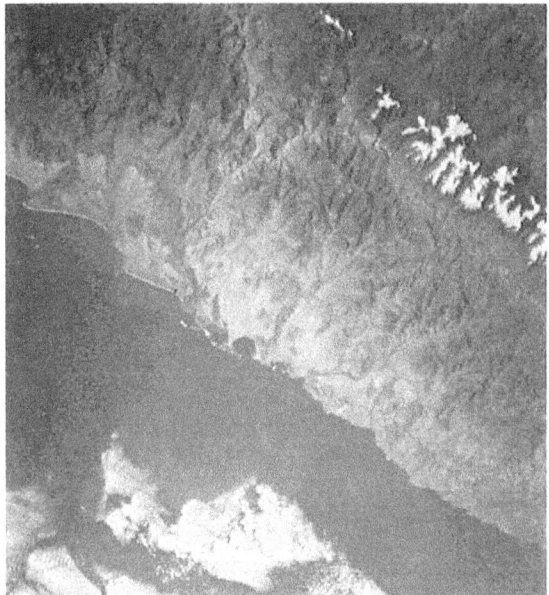

The Peru peaks of the snow-covered Cordillera Blanca (upper right) reach nearly 6800 meters in elevation, yet they are but 100 kilometers from the Pacific Ocean. Earthquakes occur with great frequency, resulting in massive loss of life and property. On the second snow-covered mountain from the right (Huascaran, Peru's highest peak) is the thin scar from a May 1962 earthquake-caused avalanche which cost 6000 deaths. (S66-38298; Gemini IX.)

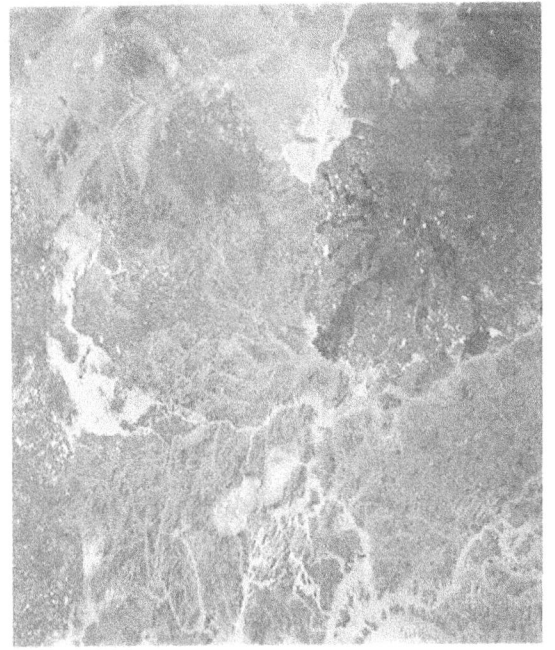

The complex geology of the poorly explored portion of Saudi Arabia known as the Hejaz, north of the Islamic holy city of Medina, is clearly seen. Geologic forms such as lava fields and flows, intrusive and extrusive basalts, extensive faulting, volcanos, playa lakes, as well as dendritic and trellis drainage, can be delineated. (S65-34665; Gemini IV.)

The sands of the Arabian Peninsula contrast with the deep blue of the Gulf of Oman and Arabian Sea at Ras al Hadd. In the lower left, seif dunes form long parallel sand ridges in the Rub-al-Khali (Empty Quarter) of Muscat and Oman. The sharp ridges of the Al Akhdar mountains are to the upper left. (S65-34661; Gemini IV.)

397

A large portion of the eastern Sahara is seen during one of the two "high apogee" revolutions of Gemini XI. The view is to the eastward looking across portions of Algeria, Niger, Chad, and most of Libya. The dark area behind the Agena S-band antenna is the Tassili-n-Ajjer Mountains. The "sand sea" Idehan Ubari (upper left) is separated from the circular "sand sea" Idehan Murzuq (right center) by the escarpments of the darker Hamadet Muhzuq. Beyond (upper right center) are the Haruj al Aswad (Black Haraj) volcanics. The dark area on the right margin is the volcanic Tibesti Mountains. Synoptic views of this type are of great value in studies of regional geology and tectonics. (S66-54525; Gemini XI.)

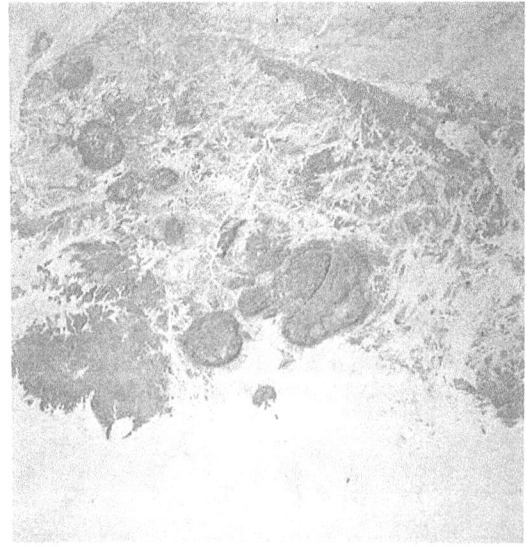

The dark domelike volcanics of Niger's Air-Au-Azbine Mountains contrast sharply with the buff-colored sands of the Tenere Sahara. A close look at the mountains show a very complex system of dikes, faults, and structural development. (S65-63158; Gemini VI.)

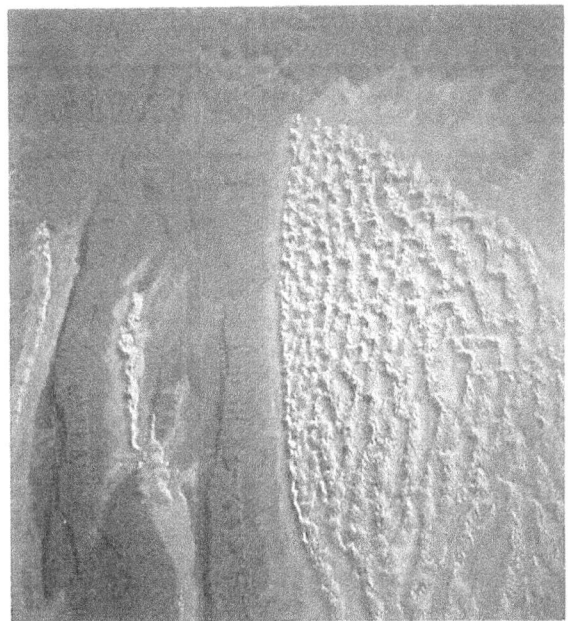

The great Sahara dominates North Africa and divides it into areas of barren mountains, plateaus, and sand deserts. An Algerian intermountain valley traps the sand and desert winds whip it up into 600-meter-high hills called the Tifernine Dunes. The location is some 150 kilometers south of Fort Flatters. Gemini photography covered most of the Sahara and was used to chart routes for exploration, locate routes for pipelines and roads, locate oases, discover mineral wealth, and better understand the climatic conditions. (S65-63829; Gemini VII.)

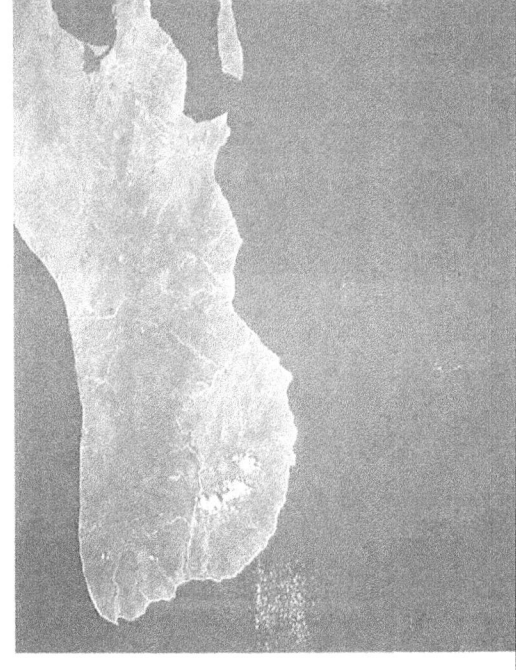

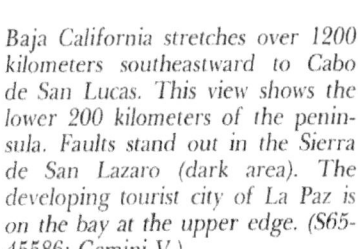

Baja California stretches over 1200 kilometers southeastward to Cabo de San Lucas. This view shows the lower 200 kilometers of the peninsula. Faults stand out in the Sierra de San Lazaro (dark area). The developing tourist city of La Paz is on the bay at the upper edge. (S65-45586; Gemini V.)

Biblical and Near East students find much of value in this photograph. Sinai is at the center, Arabia to the right, Palestine and the Levant to the upper left, and Mesopotamia to the upper right. One can trace the route of Abraham to Israel and clearly see Judea, Galilee, and Samaria. The River Jordan connects the Sea of Galilee to the Dead Sea, all in a straight line along a great rift in the Earth's crust. Mount Sinai (Gebel Musa) is seen in the Sinai. Portions of the Hejaz pilgrimage route that Moslems have used for nearly 1400 years can be seen between Damascus and Mecca. The black cloud in the upper right is the result of a fire caused by destruction on the Trans Arabian pipeline. We see the fire and smoke from over 1000 kilometers distance. (S66-54893; Gemini XI.)

The contrast between deep water (deep blue), shallow water (light blue), and small island reefs is dramatically seen in this view of the Bahamas. The deep blue circular area to the lower left is known as the Tongue of the Ocean. An underwater escarpment drops more than 1.5 kilometers to the floor of this unique area. The deep blue ellipse to the right center is Exuma Sound, equally deep. Close inspection will show that numerous islands and cays fringe much of Exuma Sound. On 12 October 1492, Christopher Columbus and 88 men first touched the New World at San Salvador Island (right center edge). (S65-45760; Gemini V.)

From 300 kilometers above the Earth, a wide-angle view of the Texas-Louisiana Gulf Coast shows almost 600 kilometers of shoreline between Aransas Bay and Vermilion Bay. The movement and distribution of waterborne sediments and pollutants are clearly visible over a wide area of the Gulf of Mexico. Regional land use can be easily delineated into such categories as forests, agricultural, grazing, wetlands, coastlines; lakes and reservoirs, and urban areas. This photo had the unique distinction of being the first space photo used in a legal case resulting in the elimination of a source of water pollution. (S66-34034; Gemini XII.)

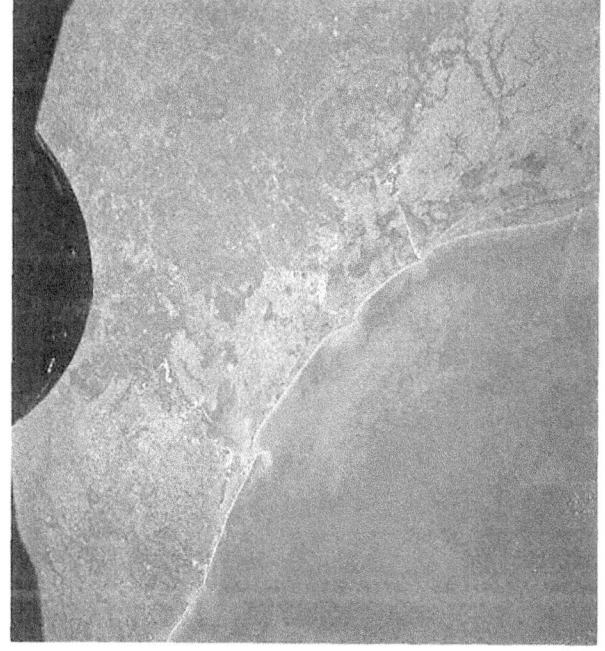

The south half of the island of Taiwan is clearly seen. The Formosa Strait is to the left. Coastal currents can be located and charted in the Pacific Ocean, Luzon Strait, South China Sea, and Formosa Strait. (S66-45868; Gemini X.)

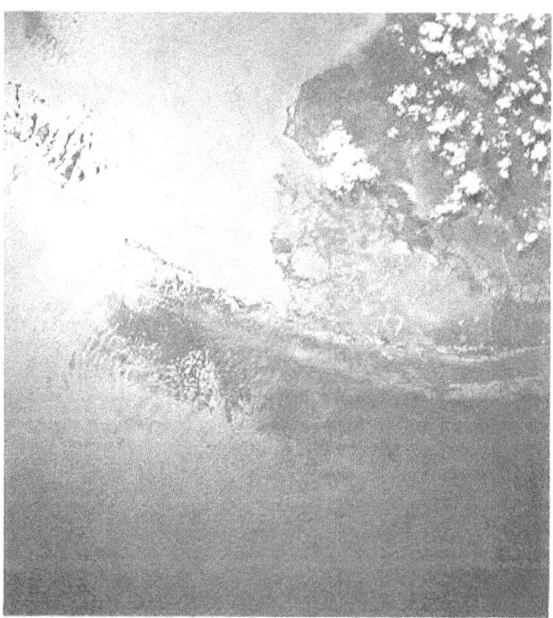

From the high vantage point of an orbiting spacecraft, one can see the bottom of the sea in many areas. The Florida Keys create a 200-kilometer-long arc from Biscayne Bay to Key West. Boat wakes are easily seen in the sun glint. The Everglades National Park is in the upper right. Florida Bay, dotted with numerous small islands, separates Cape Sable and Ponce de Leon Bay from the highway-connected Keys. The Gulf Stream flows northeastward in the lower portion of the photograph. The oceanographer has increased his knowledge of the seas by the use of space photography. (S65-34766; Gemini IV.)

*A striking photograph of a sunset seen from space. The low elevation sun lights a cloud front with gold as Gemini VII looks southward down the Andes and the Altiplano of Bolivia, Chile, and Argentina. The sun is setting in the Argentine Chaco. Sunlight barely illuminates the large salt flats of Uyuni and Atacama. Gemini photography can be beautiful as well as scientifically informative. (S65-63780; Gemini VII.)*

403

From an altitude of approximately 700 kilometers, the Indian subcontinent passes by. The Island of Ceylon is to the right, the Bay of Bengal is to the upper right, and the Arabian Sea is to the left. A narrow band of tropical rain forests stretches along the west coast of India from Cape Comorin northward, permitting one to easily differentiate rain forests from deserts and savannas. The thin band of offshore clouds running from Bombay around India into the Bay of Bengal gives new clues to India's climate. *(S66-54676; Gemini XI.)*

The navigators and explorers of 500 years ago had a great fear of sailing along the hostile Atlantic coasts of Africa. Such fears hastened the discovery and European occupation of the Americas. This view clearly shows 400 kilometers of "Skeleton Coast" of Southwest Africa. Constant northerly winds cause the development of parallel sand dunes which, in some cases, are over 200 kilometers long and over 300 meters high. Seifs, as they are called, are rarely found and parallel the wind direction; most dunes are transverse to the wind direction. A combination of wind-blown sand from the Namib Desert and the strong northward Benguela Current cause the development of sand hooks or capes. This dry and hostile area with an equally hostile name receives about 100 centimeters of rain a century. *(S65-45579; Gemini V.)*

A strange circular feature about 40 kilometers in diameter and about 300 meters deep is seen in the Sahara of Mauritania. It is called Richat. It was first believed to be caused by a meteor impact, but later geological studies and the fact that it is seen in a single space photograph indicate that it is a unique wind erosional feature caused by the abrasive forces of swirling sand. (S65-34670; Gemini IV.)

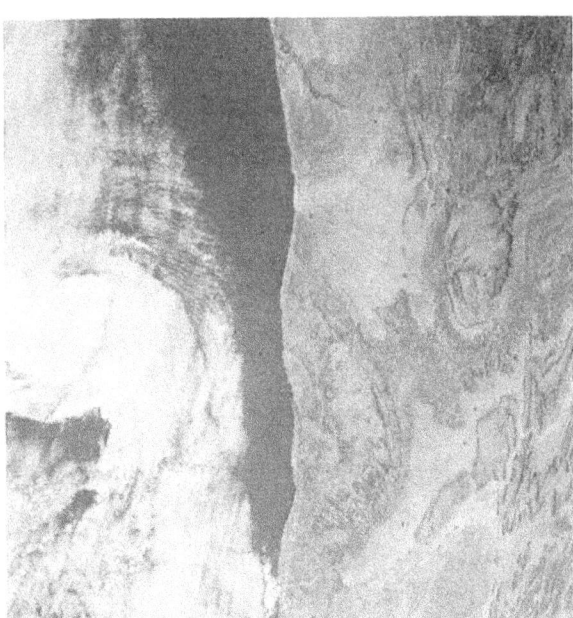

Topographic features can greatly influence the weather and the development of weather systems. A classic vortex about 50 kilometers in diameter is clearly seen off the coast of Morocco. In this case, wind shear at Ras (Cape) Rhir, a coastal promontory, caused the wind to develop its circular motion. Such phenomena can be clearly studied from the unique vantage point of space, increasing our knowledge of weather systems and their possible future control. Weather satellites have provided the information which has saved hundreds of thousands of lives in the United States and elsewhere. (S65-45665; Gemini V.)

405

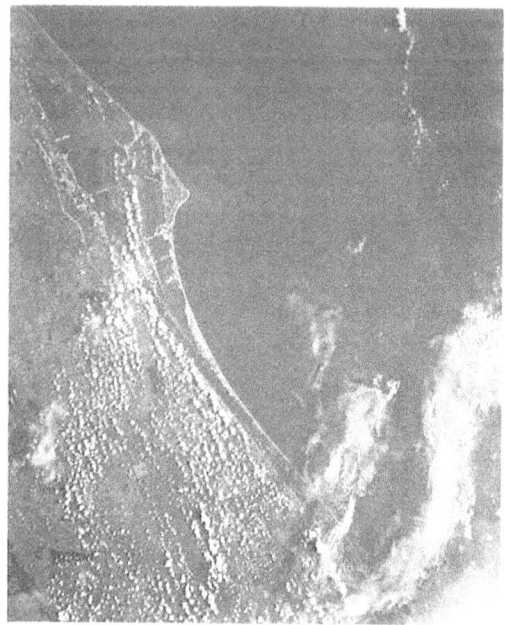

Florida's Cape Canaveral juts into the Atlantic Ocean. The various launch pads of the John F. Kennedy Space Center get larger and more complex as you proceed northward along the coast. Launch Pad 5, where the first Mercury flight was launched, is difficult to see, while the giant Launch Complex 39 built for Apollo and also used by Skylab covers several square kilometers. Cumulus clouds form in long "streets" parallel to the coast. Manmade waterways can be traced by the spoil banks which show as small white dots along the right of way. (S65-45599; Gemini V.)

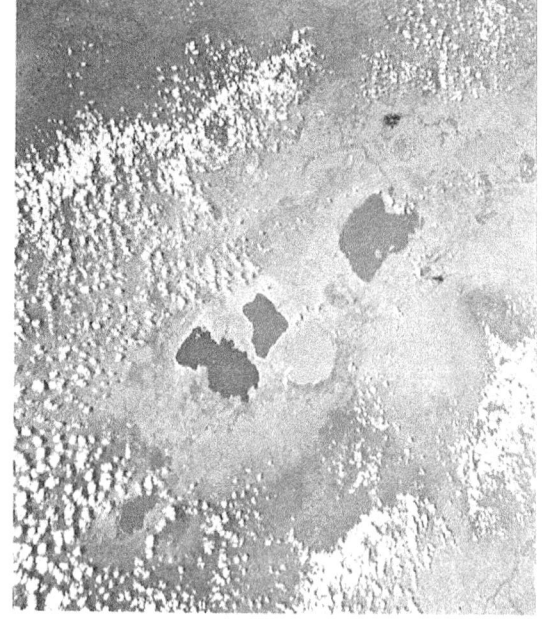

Ethiopia's Lakes Zwai, Langana, Hora Abyata, Koka, and Awusa show different colors. This is primarily due to water temperature differences resulting in each lake supporting different algae and other micromarine life. Space photography can permit a continual survey of the world's water resources and quality. (S65-63162; Gemini VI.)

406

The once mighty Colorado River separates the Mexican state of Baja California from Sonora at its mouth. It has been reduced to a sluggish, salty stream by removal of most of the water for irrigation or use by the cities of Southern California. Sediment flows can be traced in the Sea of Cortez (Gulf of California). The limits of the pre-water-use flood plain can be easily delineated. Note that the geology is completely different in Baja California (left) from that of Sonora (right). The buff sands of the Grand Sonoran Desert lend color to this extraordinary photography and contrast with the block fault mountains of Baja California. Thus, the interface of meteorology, geology, and oceanography is seen in a very remote area. (S65-34673; Gemini IV.)

The geology of south central Iran stands out clearly just east of Shiraz. Two salt lakes, Tasik and Bakhtegan, show many shorelines due to changes in level caused by spring snow melt and occasional rain. The long ridges of the Zagros Mountains run from northwest to southeast. The ancient Persian city of Persepolis is at the upper left margin. (S65-45720; Gemini V.)

407

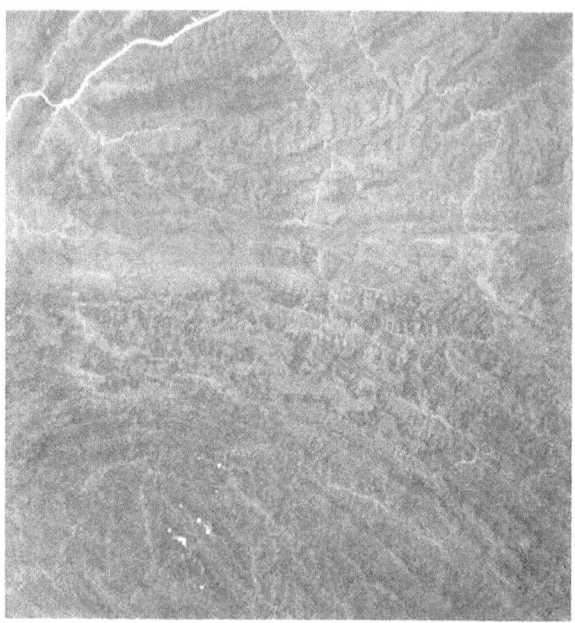

Three hundred kilometers northeast of Chunking, China, the Yangtze River flows through a deep gorge. The structural geology causes a classic display of "trellis drainage" as streams flow in parallel valleys, eroding the softer rocks, leaving the harder rock as great linear ridges. (S65-45713; Gemini V.)

This striking view of the Delta of the Nile, Suez Canal, and Sinai has become a classic space photo. Ancient Egyptian Pharaohs would tell their people that "in the autumn God will send the water down the Nile." With absolute regularity and beneath a cloudless sky, the Nile would spill over its banks and inundate this dark triangle. This was caused by rains thousands of kilometers away in Central Africa. The process would irrigate, fertilize, and clean the rich soils, generating a more than adequate food supply. Historically, nations with surplus food supplies are those that change the world. Nearly 40 000 000 people live in the 20 000-square-kilometer green triangle. (S65-34776; Gemini IV.)

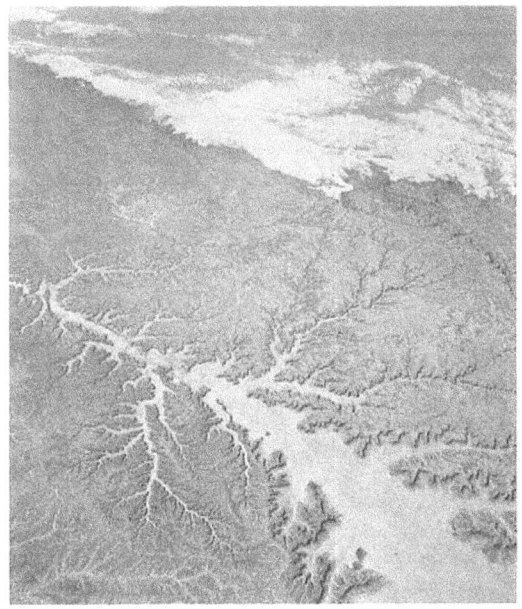

The Hadhramawt Plateau of South Yemen exhibits a complex dendritic drainage pattern and excellent examples of "stream piracy." Wadi Hadhramawt opens into the sand-filled Ramlat Sabatayn in the southwest corner of the Rub-al-Khali (The Empty Quarter), yet drainage is toward the sea. The southern coast of the Arabian Peninsula is at the upper portion of the photograph. (S65-34658; Gemini IV.)

On rare occasions water flows in the Algerian Sahara. The Oued Saoura is viewed with flowing water after a rain. The area viewed is the southwest corner of the Grand Erg Occidental about 350 kilometers SSE of Colomb Bechar. The Oued Saoura flows southeast along the right edge of the photo; just off the lower right corner it passes through a water gap, reverses direction, and flows northwest into a salt marsh called the Sebcha Melah. (S65-63830; Gemini VII)

The sharp contrast between the green irrigated Nile Delta and the arid Sahara can be clearly seen from 160 kilometers overhead. Canals, highways, railroads, towns, and villages can be located. The great city of Cairo is at the head of the delta. Just north of Cairo the Nile splits, forming the Rosetta and Damietta branches. South of Cairo, agriculture is confined to the straight, narrow, long valley to the first cataract at Aswan. Photos from space can assist in land use studies. (S65-45778; Gemini V.)

A dark arrow points westward across Texas' Edwards Plateau. Its tip is near Rankin, just east of the Pecos River, and widens to the east. It was caused by a widening rain shower several days before which brought moisture to this desert landscape and caused the plants to grow. The light area to the right is the Llano Estacado (Staked Plains) and the Permian Basin oil fields. The dark streams make up the Concho River system west of San Angelo. Photography from space can be a valuable tool in the field of agriculture. Continual monitoring from spacecraft can accurately tell the world such information as water availability, crop disease and insect infestation locations, crop identification and projected yields, grazing conditions, and much more information necessary to improve food supply for expanding populations. (S65-34704; Gemini IV.)

# SOURCE NOTES, BIBLIOGRAPHICAL NOTE, APPENDIXES, AND INDEX

# Source Notes

## Chapter I

[1] *The Houston Post*, 7 Dec. 1961.

[2] David G. McComb, *Houston, the Bayou City* (Austin, Tex., 1969), pp. 50-51, 121.

[3] Ibid., p. 257.

[4] Ibid., p. 199, quoting French journalist Pierre Voisin, as reported in *The Houston Post*, 4 May 1962.

[5] *The Houston Post*, 8 Dec. 1961; McComb, *Houston*, pp. 206-207.

[6] McComb, *Houston*, pp. 232-33.

[7] Ibid., p. 226.

[8] Ibid., p. 185.

[9] Ibid., pp. 194-96.

[10] NASA News Release 61-207, "Manned Space Flight Laboratory Location," 19 Sept. 1961; Loyd S. Swenson, Jr., James M. Grimwood, and Charles C. Alexander, *This New Ocean: A History of Project Mercury*, NASA SP-4201 (Washington, 1966), pp. 390-91; Stephen B. Oates, "NASA's Manned Spacecraft Center at Houston, Texas," *Southwestern Historical Quarterly* 67, no. 3 (January 1964), p. 355.

[11] MSC Announcement No. 2, "Designation of STG as 'Manned Spacecraft Center,'" 1 Nov. 1961.

[12] Swenson, Grimwood, and Alexander, *This New Ocean*, p. 409.

[13] Robert B. Merrifield, "Men and Spacecraft: A History of the Manned Spacecraft Center (1958-1969)," chap. IV, "Transition to Center Status (January 3—September 19, 1961)," [1972], draft MS in MSC historical archives.

[14] Ibid., chap. I, "Introduction."

[15] *The Houston Post*, 8 Dec. 1961.

[16] Ibid.; U.S. Congress, House, Committee on Science and Astronautics, *Aeronautical and Astronautical Events of 1961: Report*, 87th Cong., 2d sess., 7 June 1962, p. 71.

[17] Memo, Paul E. Purser to Robert R. Gilruth, "Log for week of December 4, 1961," 15 Dec. 1961, p. 2.

[18] Letter, Alex P. Nagy to George M. Low, 11 Dec. 1961; memo, Purser to Gilruth, "Log for week of December 11, 1961," 18 Dec. 1961, p. 2; memo, Purser to Gilruth, "Log for week of December 25, 1961," 2 Jan. 1962, p.2; memo, Robert C. Seamans, Jr., to Adm., NASA, "Membership of the NASA Project Designation Committee," 17 March 1961; memo, Harold L. Goodwin to Nagy, "Selection of the Name 'Gemini,'" 3 May 1962; letter, Seamans to Eugene M. Emme, 3 June 1969.

[19] Louis MacNeice, *Astrology* (Garden City, N. Y., 1964), pp. 85-86, 292, 294.

[20] Daniel D. McKee, interview, Los Angeles, 19 May 1967.

[21] U.S. Congress, House, Committee on Science and Astronautics, *Astronautical and Aeronautical Events of 1962: Report*, 88th Cong., 1st sess., 12 June 1963, p. 1.

[22] Testimony of James E. Webb, Seamans, and D. Brainerd Holmes in U.S. Congress, House, Subcommittee on Manned Space Flight of the Committee on Science and Astronautics, *1963 NASA Authorization: Hearings on H.R. 10100 (Superseded by H.R. 11737)*, 87th Cong., 2d sess., 1962, pp. 4-5, 102-103, 250-51, 460-62.

[23] Barton C. Hacker, "The Idea of Rendezvous: From Space Station to Orbital Opera-

tions in Space-Travel Thought, 1895-1951," *Technology and Culture* 15 (July 1974).

24 H. E. Ross, "Orbital Bases," *Journal of the British Interplanetary Society* 8 (January 1949), pp. 1-19; Kenneth W. Gatland, "Rockets in Circular Orbits," ibid., 8 (March 1949), pp. 52-59; letter, Ross to Barton C. Hacker, 9 July 1968; Michael Stoiko, *Project Gemini: Step to the Moon* (New York, 1963), pp. 34-36.

25 Memo, Hacker to David S. Akens, "Proposed Interview with Dr. [Wernher] von Braun," 8 Sept. 1968, with enclosure, "Proposed Questions for Dr. Von Braun," with von Braun's answers, n.d.

26 R. A. Smith, "Establishing Contact Between Orbiting Vehicles," presented at the Second International Congress of Astronautics, London, September 1951, published in *Journal of the British Interplanetary Society* 10 (November 1951), pp. 295-99.

27 Hacker, "The Idea of Rendezvous"; idem, "The Genesis of Project Gemini: The Idea of Rendezvous, 1929-1961," presented at the Twelfth International Congress of the History of Science, Paris, August 1968, published in *Actes* 10: *Historie des techniques* (Paris, 1971), pp. 41-46.

28 Swenson, Grimwood, and Alexander, *This New Ocean*, pp. 55-106; Robert L. Rosholt, *An Administrative History of NASA, 1958-1963*, NASA SP-4101 (Washington, 1966), pp. 19-44; Arthur L. Levine, "United States Aeronautical Research Policy, 1915-1958: A Study of the Major Policy Decisions of the National Advisory Committee for Aeronautics" (Ph. D. diss., Columbia University, 1963), pp. 124-80; Ira H. Abbott, "A Review and Commentary of a Thesis by Arthur L. Levine . . .," April 1964; Enid Curtis Bok Schoettle, "The Establishment of NASA," in Sanford A. Lakoff, ed., *Knowledge and Power: Essays on Science and Government* (New York, 1966), pp. 162-370; Mary Stone Ambrose, "The National Space Program, Phase I: Passage of the 'National Aeronautics and Space Act of 1958'" (Master's thesis, American University, 1960), part I; Elisabeth Alison Griffith, *The National Aeronautics and Space Act: A Study of the Development of Public Policy* (Washington, 1962).

29 Michael David Keller, "A History of the NACA Langley Laboratory, 1917-47," March 1968; George W. Gray, *Frontiers of Flight: The Story of NACA Research* (New York, 1948); Jerome C. Hunsaker, "Forty Years of Aeronautical Research," in *Smithsonian Report for 1955* (Washington, 1956), pp. 241-71, reprinted in *Forty-Fourth Annual Report of the National Advisory Committee for Aeronautics, 1958 (Final Report)* (Washington, 1959), pp. 3-

27; James H. Doolittle, "The Following Years, 1955-58," ibid., pp. 29-31; Levine, "United States Aeronautical Research Policy"; Rosholt, *Administrative History*, pp. 20-21; Abbott, "Review and Commentary," pp. 81, 156-57, 179, 186-87; Edwin Mansfield, *The Economics of Technological Change* (New York, 1968), pp. 45-48; Charles V. Kidd, "Basic Research—Description versus Definition," *Science*, 13 Feb. 1959, pp. 368-71, reprinted in Norman Kaplan, ed., *Science and Society* (Chicago, 1965), pp. 146-55.

30 "NACA Research in Space," in U.S. Congress, House, Select Committee on Astronautics, *Astronautics and Space Exploration: Hearings on H.R. 1181*, 85th Cong., 2d sess., 1958, pp. 404-410, public version of a classified document, same title, December 1957, describing NACA research in "Aerodynamics and Space Mechanics," "Propulsion," and "Structures and Materials."

31 Swenson, Grimwood, and Alexander, *This New Ocean*, pp. 109-16, 132.

32 D. D. Wyatt, "The Rationale of the NASA Space Program," *Advances in the Astronautical Sciences* 6 (1961): xxxiv.

33 "A National Space Vehicle Program: A Report to the President," NASA Propulsion Staff, 27 Jan. 1959. An unclassified and much condensed version of this report may be found in U.S. Congress, Senate, Subcommittee on Governmental Organization for Space Activities of the Committee on Aeronautical and Space Sciences, *Investigation of Governmental Organization for Space Activities: Hearings*, 86th Cong., 1st sess., 1959, pp. 17-24.

34 "The Ten Year Plan of the National Aeronautics and Space Administration," NASA Office of Program Planning and Evaluation, 18 Dec. 1959.

35 "A National Space Vehicle Program," pp. 3-4, 13, 23; U.S. Congress, Senate, Committee on Aeronautical and Space Sciences, *Manned Space Flight Program of the National Aeronautics and Space Administration: Projects Mercury, Gemini, and Apollo*, Staff Report, 87th Cong., 2d sess., 4 Sept. 1962, p. 158; *First Semiannual Report to the Congress, October 1, 1958-March 31, 1959*, NASA, 1959, pp. 26-27; "GAO Report on Review of Procedures Followed by the National Aeronautics and Space Administration in Awarding Contract for 1.5 Million-Pound-Thrust Single-Chamber Rocket Engine to North American Aviation, Inc." in U.S. Congress, House, Committee on Science and Astronautics, *The Production of Documents by the National Aeronautics and Space Administration for the Committee on Science and Astronautics: Hearings*, 86th Cong., 2d sess., 27, 29 Jan. 1960,

pp. 89-119.

36 *Investigation of Governmental Organization*, p. 21.

37 "Manned Space Flight Long Range Plans," NASA Office of Space Flight Development, rev. 17 Aug. 1959, p. 29.

38 Memo, John W. Crowley to Ames, Lewis, and Langley Research Centers and High Speed Flight Station, "Research Steering Committee on Manned Space Flight," 1 April 1959; memo, Crowley to Jet Propulsion Laboratory, subject as above, 8 April 1959; memo, Crowley to Dir., Space Flight Development, subject as above, 2 April 1959; Ralph W. May, Jr., "Minutes of Meeting[s] of Research Steering Committee on Manned Space Flight, May 25-26, 1959," and "June 25-26, 1959"; memo, Harry J. Goett to Abbott, "Interim Report on Operation of 'Research Steering Committee on Manned Space Flight,' " 17 July 1959; "Third Semi-Annual NASA Staff Conference: Program Formulation and Status of Activities, Monterey, California, 3-5 March, 1960," pp. 18-20; John M. Logsdon, *The Decision to Go to the Moon: Project Apollo and the National Interest* (Cambridge, Mass., 1970), pp. 56-57.

39 NASA witnesses before congressional committees during 1960 mentioned rendezvous only twice. On 1 February, Abbott referred to a ferry vehicle in describing a space laboratory concept to the House Committee on Science and Astronautics and, on 29 March, von Braun touched on orbital refueling or latching in discussing possible Saturn missions before the Senate Committee on Aeronautical and Space Sciences. U.S. Congress, House, Committee on Science and Astronautics, *Review of the Space Program: Hearings*, 86th Cong., 2d sess., 1960, pp. 304-305; U.S. Congress, Senate, NASA Authorization Subcommittee of the Committee on Aeronautical and Space Sciences, *NASA Authorization for Fiscal Year 1961: Hearings on H.R. 10809*, 86th Cong., 2d sess., 1960, p. 227.

40 The main burden of defending NASA's 1959 request for funds to study rendezvous fell to DeMarquis D. Wyatt, technical assistant to the Director of Space Flight Development in NASA Headquarters. Wyatt consistently identified the need for rendezvous techniques with the problem of logistic support for a permanent manned orbiting laboratory. Nowhere in Wyatt's testimony, in that of other NASA witnesses, nor in the prepared materials submitted to Congress by NASA was there any hint that rendezvous might serve some other purpose. U.S. Congress, Senate, NASA Authorization Subcommittee of the Committee on Aeronautical and Space Sciences, *NASA Supplemental Authorization for Fiscal Year 1959:*

*Hearings on S. 1096*, 86th Cong., 1st sess., 1959, pp. 77-80; U.S. Congress, House, Committee on Science and Astronautics and Subcommittees Nos. 1, 2, 3, and 4, *1960 NASA Authorization: Hearings on H.R. 6512*, 86th Cong., 1st sess., 1959, pp. 94-95, 97, 267-68; idem, Subcommittee on Independent Offices of the Committee on Appropriations, *National Aeronautics and Space Administration Appropriations: Hearings*, 86th Cong., 1st sess., 1959, pp. 42-45.

41 Rosholt, *Administrative History*, pp. 46-47, 107-15, 117-23; David S. Akens, *Historical Origins of the George C. Marshall Space Flight Center*, MSFC Historical Monograph No. 1 (Huntsville, Ala., December 1960), pp. 67-80, and Appendices A, B, and C; U.S. Congress, Senate, NASA Authorization Subcommittee of the Committee on Aeronautical and Space Sciences, *Transfer of von Braun Team to NASA: Hearings on H.J. Res. 567*, 86th Cong., 2d sess., 1960.

42 Willy Ley, *Rockets, Missiles, and Space Travel* (2d rev. ed., New York, 1961); Wernher von Braun and Frederick I. Ordway III, *History of Rocketry & Space Travel* (New York, 1966), pp. 86-119; Walter Dornberger, *V-2*, trans. James Cleugh and Geoffrey Halliday (New York, 1954); idem, "The German V-2," in Eugene M. Emme, ed., *The History of Rocket Technology: Essays on Research, Development, and Utility* (Detroit, 1964), pp. 29-45; Clarence Lasby, *Operation Paperclip: German Scientists in the Cold War* (New York, 1971).

43 Von Braun's early thinking is probably most clearly shown in his often reprinted "Survey of Development of Liquid Rockets in Germany and Their Future Prospects," written at the behest of his American captors immediately after the war. Nearly half the paper was devoted to the prospects for space travel. Wernher von Braun, Ernst Stuhlinger, and H. H. Koelle, "ABMA Presentation to the National Aeronautics and Space Administration," ABMA Rept. D-TN-1-59, 15 Dec. 1958, pp. 63-125; U.S. Congress, House, Committee on Science and Astronautics, *Missile Development and Space Sciences: Hearings*, 86th Cong., 1st sess., 1959, pp. 220-21.

44 "Proposal: A National Integrated Missile and Space Vehicle Development Program," Development Operations Div., ABMA Rept. D-R-37, 10 Dec. 1957, p. 6; "A National Integrated Missile and Space Vehicle Development Program," Report to the NACA Special Committee on Space Technology, 18 July 1958, p. 35.

45 H. H. Koelle et al., "Juno V Space Vehicle Development Program (Phase I): Booster

Feasibility Demonstration," ABMA Rept. DSP-TM-10-58, 13 Oct. 1958, p. 2.

46 *Manned Space Flight Program: Mercury, Gemini, and Apollo,* p. 165; *Saturn Illustrated Chronology, April 1957-June 1964,* MSFC MHR-3, updated by Evelyn Falkowski (Huntsville, Ala., 10 Aug. 1964), p. 5; "A Lunar Exploration Program Based upon Saturn-Boosted Systems," Army Ordnance Missile Command (AOMC) RCS ORDXM-C-1004, 1 Feb. 1960 (ABMA Rept. DV-TR-2-60), pp. 234-59; "Project Horizon, Phase I Report: A U.S. Army Study for the Establishment of a Lunar Military Outpost," 4 vols., AOMC, 8 June 1959, especially vol. II: "Technical Considerations and Plans," pp. 61-163.

47 Eugene M. Emme, "Historical Perspectives on Apollo," NASA Historical Note No. 75, 24 Oct. 1967, p. 17.

48 May, "Minutes, May 25-26, 1959, " p. 6, and "June 25-26, 1959," p. 10; Goett memo, 17 July 1959, p. 3.

49 James P. Gardner, Harry O. Ruppe, and Warren H. Straly, "Comments on Problems Relating to the Lunar Landing Vehicle," ABMA Rept. DSP-TN-13-58, 4 Nov. 1958, p. 2; "A Lunar Exploration Program Based upon Saturn-Boosted Systems," pp. 234-59.

50 See, however, H. H. Koelle, "On the Development of Orbital Techniques: A Classification of Orbital Carriers and Satellite Vehicles," *Proceedings of the IXth International Astronautical Congress,* Amsterdam, 25-30 Aug. 1958 (Vienna, 1959), pp. 702-46; C. L. Barker and W. H. Straly, "Rendezvous by the Chasing Technique," ABMA Rept. DSP-TM-15-59, 30 Oct. 1959; James M. Horner and Robert Silber, "Impulse Minimization for Hohmann Transfer between Inclined Circular Orbits of Different Radii," ABMA Rept. DA-TR-70-59, 2 Dec. 1959; R. F. Hoelker and Silber, "The Bi-Elliptical Transfer between Circular Coplanar Orbits," ABMA Rept. DA-TM-2-59, 6 Jan. 1959; Silber and Horner, "Two Problems of Impulse Minimization between Coplanar Orbits," ABMA Rept. DA-TM-23-59, 12 Feb. 1959.

51 May, "Minutes, May 25-26, 1959," p. 6, and "June 25-26, 1959," p. 6.

52 John M. Eggleston, interview, Houston, 7 Nov. 1966; May, "Minutes, June 25-26, 1959," p. 6, and "December 8-9, 1959," p. 6.

53 May, "Minutes, June 25-26, 1959," p. 6.

54 May, "Minutes, December 8-9, 1959," pp. 9-10; Eggleston interview; Houbolt, interview, Princeton, N.J., 5 Dec. 1966; John D. Bird, "A Short History of the Development of the Lunar-Orbit-Rendezvous Plan at the Langley Research Center," 6 Sept. 1963 (supplemented 5 Feb. 1965 and 17 Feb. 1966), p. 3;

Bird, interview, Langley Research Center, Va., 20 June 1966; John C. Houbolt, "Lunar Rendezvous," *International Science and Technology* 14 (February 1963), pp. 62-65.

55 May, "Minutes, December 8-9, 1959," p. 10.

56 Eggleston, "Inter-NASA Research and Space Development Centers Discussion on Space Rendezvous," 16-17 May 1960, pp. 1-2.

57 The papers presented are abstracted, ibid., pp. 2-6. The Langley presentation (pp. 2-4) was the most elaborate, as the Langley delegation was the largest. It included a general introduction and conclusion by Houbolt and papers by David F. Thomas, Harold D. Beck, Eggleston, Terrance M. Carney, Richard A. Hord, Bert B. Burton, and Wilford E. Sivertson, Jr.

58 Eggleston, "Inter-NASA Discussion on Rendezvous"; Eggleston and Houbolt interviews.

59 Heinz H. Koelle, "Future Projects at Marshall Space Flight Center," in "NASA-Industry Program Plans Conference, September 27-28, 1960," MSFC, pp. 39-40; Edward H. Kolcum, "NASA Re-Emphasizes Role of Contractors," *Aviation Week and Space Technology,* 3 Oct. 1960, pp. 32-33.

60 U.S. Congress, House, Committee on Science and Astronautics, *Orbital Rendezvous in Space: Hearing,* 87th Cong., 1st sess., 23 May 1961, p. 30.

61 William H. Michael, Jr., "Weight Advantages of Use of Parking Orbit for Lunar Soft Landing Mission," in Jack W. Crenshaw et al., "Studies Related to Lunar and Planetary Missions," 26 May 1960, pp. 1-2; Bird, "A Short History," pp. 1-2.

62 Houbolt, "Lunar Rendezvous," p. 63.

63 Seamans, interview, Washington, 26 May 1966; Houbolt interview.

64 "Satellite Interceptor Study System: Final Report," RCA Rept. CR-59-588-39, 31 Jan. 1960; "Saint Phase I Technical Proposal," Space Technology Laboratories, Inc., Rept. STL/TR-59-0000-09917, 21 Dec. 1959; Norman E. Sears, "Satellite-Rendezvous Guidance System," MIT Rept. R-331, May 1961, p. 295; "USAF Launches Anti-Satellite Program," *Aviation Week and Space Technology,* 14 Nov. 1960, pp. 26-27; "RCA Will Develop Saint Payload," *Aviation Week and Space Technology,* 5 Dec. 1960, p. 27; testimony of Harold Brown, DOD Director of Defense Research and Engineering, 12 June 1961, in U.S. Congress, Senate, Committee on Aeronautical and Space Sciences, *NASA Authorization for Fiscal Year 1962: Hearings on H.R. 6874,* 87th Cong., 1st sess., 1961, p. 186.

65 Thomas A Sturm, *The USAF Scientific*

Advisory Board: Its First Twenty Years, 1944-1964 (Washington, 1 Feb. 1967), p. 96; memo, J. Thomas Markley for Assoc. Dir., "Meeting between MSFC and STG on mission for Saturn Cl R and D Program and summary of MSFC trips by J. T. Markley," 8 Dec. 1960, enc. 2, "Trip Report R-60-9," 1 Dec. 1960, p. 3. For a list of Air Force projects related to rendezvous and orbital operations, see "Guidelines for a Program for Manned and Unmanned Orbital Operations," NASA Staff Paper, May 1961, p. 16; memo, Donald H. Heaton for record, "Minutes of the Executive Meeting at AFBMD on October 28, 1960," 2 Nov. 1960; Larry Booda, "Air Force Outlines Broad Space Plans," Aviation Week and Space Technology, 5 Dec. 1960, pp. 26-28.

66 Memo, Seamans to Dir., Office of Advanced Research Programs, "Programs relating to rendezvous and orbital control," 12 Sept. 1960; Seamans interview.

67 Bird, "A Short History," p. 2; Bird and Houbolt interviews; memo, Bernard Maggin to Assoc. Adm., "Staff Paper—Guidelines for a Program for Manned and Unmanned Orbital Operations,'" 23 May 1961, enclosed in staff paper.

68 Warren J. North, secretary, "Minutes of Meetings, Panel for Manned Space Flight, September 24, 30, October 1, 1960," and Appendix A, "Objectives and Basic Plan for the Manned Satellite Project."

69 Maxime A. Faget, Benjamine J. Garland, and James J. Buglia, "Preliminary Studies of Manned Satellites—Wingless Configurations: Nonlifting," in "NACA Conference on High-Speed Aerodynamics, Ames Aeronautical Laboratory, Moffett Field, Calif., 18, 19 and 20 March 1958: A Compilation of Papers Presented," pp. 19-33; Swenson, Grimwood, and Alexander, This New Ocean, pp. 86-90.

70 National Aeronautics and Space Administration Appropriations, pp. 156-58, 166-67, 186-87.

71 NASA Authorization for Fiscal Year 1961, pp. 377-78.

72 Testimony of Hugh L. Dryden in NASA Supplemental Authorization for Fiscal Year 1959, p. 50.

73 Thomas J. Wong et al., "Preliminary Studies of Manned Satellites," in "NACA Conference on High-Speed Aerodynamics," pp. 35-44; John V. Becker, "Preliminary Studies of Manned Satellites—Winged Configurations," ibid., pp. 45-47.

74 Letter, Max Rosenberg to NASA Historian, "Comments on Draft Chapter I-V and XIII-XV, Gemini History," 26 June 1970, with enclosure.

75 "Manned Space Flight Long Range Plans," pp. 15-16; memo, May to Emme, "Draft Chapters 3 and 4 of the History of Project Gemini," 29 May 1969.

76 May, "Minutes, June 25-26, 1959," p. 8.

77 Francis M. Rogallo, "Parawings for Astronautics," in Norman V. Petersen, ed., Advances in the Astronautical Sciences 16, Part 2 (1963), presented at the American Astronautical Society Symposium on Space Rendezvous, Rescue, and Recovery, Edwards AFB, Calif., 10-12 Sept. 1963, pp. 3-7; see "Man Will Conquer Space Soon," Collier's, 22 March 1952, p. 27, for the illustration.

78 Francis M. Rogallo, "Paraglider Recovery Systems," presented at the International Astronautics Society Meeting on Man's Progress in the Conquest of Space, St. Louis, Mo., 30 April, 1-2 May 1962, fig. 1—"Flexible-wing concept as presented to Langley Committee on General Aerodynamics, December 19, 1958."

79 Memo, Delwin R. Croom to Assoc. Dir., Langley, "Briefing given Space Task Group personnel concerning Rogallo's flexible lifting surface vehicle," 9 April 1959.

80 Memo, Purser to Gilruth, "Log for week of June 1, 1959," 8 June 1959, p. 2.

81 May, "Minutes, May 25-26, 1959," pp. 6, 9.

82 Goett memo, 17 July 1959, pp. 1-2, 4; May, "Minutes, May 25-26, 1959," pp. 6-8.

83 "Manned Space Flight Long Range Plans," pp. 3, 13-20; testimony of Low in U.S. Congress, Senate, NASA Authorization Subcommittee of the Committee on Aeronautical and Space Sciences, NASA Authorization for Fiscal Year 1960: Hearings on S. 1582, 86th Cong., 1st sess., 1959, pp. 333, 336; Douglas L. Worf, ibid., pp. 377, 383; Abe Silverstein in 1960 NASA Authorization, pp. 368, 379; "The Ten Year Plan of the NASA," pp. 25-26.

84 Purser, "Space Task Group Complement Analysis," 8 June 1959, pp. 1, 10; memo, Gilruth for all concerned, "Organization of Space Task Group," 26 Jan. 1959; memos, Gilruth for staff, "Organization of Space Task Group," 3 and 10 Aug. 1959; Purser, notes, "Summary of STG Organization and Mercury Management," n.d. (through 15 Jan. 1962).

85 Memo, Gilruth to Bernard Sisco, Goddard, "Langley Space Task Group FY 1961 personnel distribution," 12 June 1959, with attached chart, "Proposed FY 1961 Personnel Distribution by Division and Project."

86 Memo, H. Kurt Strass to Chief, Flight Systems Div. (FSD), "Activation of Study Program Pertaining to Advanced Manned Space Projects," 22 June 1959.

87 Memo, Strass to Chief, FSD, "First meeting of New Projects Panel, Wednesday, Aug. 12, 1959," 17 Aug. 1959.

[88] Memo, Strass to Chief, FSD, "Second meeting of New Projects Panel, Tuesday, August 18, 1959," 26 Aug. 1959, and attached chart, "Proposed Sequence of Events for Manned Lunar Mission System Analysis"; memo, Strass to Chief, FSD, "Third Meeting of New Projects Panel, Monday, September 28, 1959 (Information)," 1 Oct. 1959.

[89] "Follow On Experiments, Project Mercury Capsules," McDonnell Aircraft Corporation Engineering Report No. 6919, 1 Sept. 1959 (rev. 5 Oct. 1959); memo, Newell D. Sanders to Asst. Dir. for Propulsion, "Booster Information Requested by McDonnell Aircraft Corporation," 6 July 1959; memo, Charles H. McLellan to Assoc. Dir., Langley, "Visit of McDonnell Aircraft personnel to Aero-Physics Division," 13 Aug. 1959; Fred J. Sanders, interview, St. Louis, 14 April 1966.

[90] "Follow On Experiments, Project Mercury Capsules," p. vii.

[91] Memo, Strass to Chief, FSD, "Fourth meeting of New Projects Panel, Monday, October 5, 1959 (Action requested)," 7 Oct. 1959.

[92] "Follow On Experiments, Project Mercury Capsules," pp. 1.1-1, 2.1-1, 3.1-1, 3.2-0.

[93] Strass memo, 7 Oct. 1959; "Follow On Experiments, Project Mercury Capsules," pp. 4.1-1, 4.2-1, 5.1-1, 6.0-1, 6.1-1; Strass memo, 7 Oct. 1959.

[94] Memo, Purser to Gilruth, "Log for week of November 2, 1959," 10 Nov. 1959, p. 1.

[95] Memo, Dennis F. Hasson to Chief, FSD, "Meeting of January 7, 1960 to discuss future wind-tunnel tests for advanced Mercury projects (Information)," 11 Jan. 1960.

[96] U.S. Congress, Senate, Committee on Appropriations, *Supplemental National Aeronautics and Space Administration Appropriations, 1960: Hearings on H.J. Res. 621*, 86th Cong., 2d sess., 1960, pp. 15-16; *Review of the Space Program*, pp. 300, 674-75; U.S. Congress, House, Subcommittee on Independent Offices of the Committee on Appropriations, *Independent Offices Appropriations for 1961: Hearings*, 86th Cong., 2d sess., 1960, pp. 276-77; *NASA Authorization for Fiscal Year 1961*, pp. 143-45, 745.

[97] "Guidelines for Advanced Manned Space Vehicle Program," STG, June 1960, pp. ii, 1, 4, 6, 12, 39, 47-48. This was a compilation of papers presented to NASA Headquarters between 15 April and 3 May 1960, by STG personnel. Memo, John H. Disher to Dir., Space Flight Programs, "NASA Center Briefings on Advanced Manned Space Flight Program," 10 May 1960; Robert O. Piland, "Missions, Propulsion and Flight Time," in "Guidelines for Advanced Manned Space Vehicle Program," p. 6.

[98] Hugh L. Dryden, "NASA Mission and Long-Range Plan," in *NASA-Industry Program Plans Conference, July 28-29, 1960* (Washington, 1960), p. 8; Low, interview, Houston, 7 Feb. 1967.

[99] Charles J. Donlan, "Summary and Scheduling," in "Guidelines for Advanced Manned Space Vehicle Program," pp. 49-50.

[100] "Program Funding Requirements, Manned Space Flight Programs, March 18, 1960," STG, p. 5-1.

Fiscal Year Costs
in $ Millions

| | 1960 | 1961 | 1962 |
|---|---|---|---|
| Lifting Mercury Program | | | |
| A. System Development & Evaluation | 0.140 | 0.850 | 0.650 |
| B. Little Joe Flight Test Hardware | | .990 | 1.230 |
| C. Lifting Capsule Contracts | | 4.000 | 16.950 |
| D. Flight Operations | | .100 | .740 |
| | .140 | 5.940 | 19,570 |
| E. Supporting C & E | .250 | 1.130 | 7.100 |

[101] "Preliminary Specifications for Reentry Control Navigation System," STG, 5 April 1960; letters, Donlan to Ames and Langley, "Invitation to participate in preparing specifications and in the evaluation of proposals for a reentry guidance system for lifting Mercury," 5 April 1960.

[102] Goett, remarks on "Highlights of GSFC Program," in "Third Semi-Annual NASA Staff Conference," pp. 18-19; comments by Richard E. Horner and Dryden, ibid., pp. 20-21.

[103] Memo, Disher to Long Range Plan & Budget File, "Meeting with Dr. Glennan on 7/9/60 to discuss long range plans for Saturn utilization by OSFP," 11 July 1960, with attached chart by Low, "Flight Program, Manned Flight—Adv. Tech. Dev.," 7 July 1960; "Chronology of Budget Preparation for Fiscal Year 1962," in *NASA Authorization for Fiscal Year 1962*, p. 170.

[104] Logsdon, *The Decision to Go to the Moon*, p. 40.

[105] Swenson, Grimwood, and Alexander, *This New Ocean*, pp. 291-93.

[106] Logsdon, *The Decision to Go to the Moon*, pp. 34-38., 71-75.

## Chapter II

[1] John M. Logsdon, *The Decision to Go to*

NOTES TO PAGES 28-32

the Moon: Project Apollo and the National Interest (Cambridge, Mass., 1970), pp. 93-130; U.S. Congress, House, Committee on Science and Astronautics and Subcommittees Nos. 1, 3, and 4, 1962 NASA Authorization: Hearings on H.R. 3238 and 6029 (Superseded by H.R. 6874), 87th Cong., 1st sess., 1961, p. 380.

[2] Robert L. Rosholt, An Administrative History of NASA, 1958-1963, NASA SP-4101 (Washington, 1966), pp. 115-16, 123-27; NASA Organization Charts, 29 Dec. 1959, ibid., p. 340, and 4 April 1960, ibid., p. 343.

[3] Ibid., pp. 148-53.

[4] Minutes, Space Exploration Program Council (SEPC), 5-6 Jan. 1961, p. 2.

[5] SEPC minutes, p. 1; memo, George M. Low to Dir., Space Flight Programs, "Manned Lunar Landing Program," 17 Oct. 1960.

[6] SEPC minutes, p. 3.

[7] Ibid.; Low, "A Plan for Manned Lunar Landing," draft, 16 Jan. 1961; memo, Low to Assoc. Adm., "A Plan for Manned Lunar Landing," 24 Jan. 1961, with enclosure, "A Plan for Manned Lunar Landing," draft, 20 Jan. 1961; memo, Low to Assoc. Adm., "Transmittal of Report Prepared by Manned Lunar Working Group," 7 Feb. 1961, with enclosure, "A Plan for Manned Lunar Landing," prepared by the Lunar Landing Working Group, January 1961.

[8] "A Plan for Manned Lunar Landing," January 1961, pp. 8, 9.

[9] Paul E. Purser, "Notes on Capsule Review Board [CRB] Meeting, January 20, 1961," with enclosure, "Follow-On Mercury Missions," n.d.

[10] Ibid., p. 1; R. Cargill Hall, "The Agena-Booster Satellite," presented at American Institute of Astronautics and Aeronautics, Boston, Mass., 2 Dec. 1966, pp. 21-24; U.S. Congress, House, Committee on Science and Astronautics and Subcommittees Nos. 1, 2, 3, and 4, 1961 NASA Authorization: Hearings on H.R. 10246, 86th Cong., 2d sess., 1960, pp. 317, 345-46; U.S. Congress, House, Subcommittee on Independent Offices of the Committee on Appropriations, Independent Offices Appropriations for 1961: Hearings, 86th Cong., 2d sess., 1960, pp. 221, 236, 239, 497-98.

[11] Purser, "Notes for CRB, Jan. 20, 1961," p. 2; also pp. 15-16 above.

[12] Ibid.; memo, Purser to Robert R. Gilruth, "Log for week of January 23, 1961," 30 Jan. 1961; Purser, "Notes on Visit of Dr. Silverstein, January 26 and 27, 1961," n.d., with enclosure, "Discussion Items for Visit of Dr. Silverstein, January 26 and 27, 1961"; Purser, "Action Items from Meeting with Dr. Silverstein on January 26 and 27, 1961," p. 2.

[13] MSC Form 499, "Biographical Data [on James A. Chamberlin]," 19 June 1967.

[14] Memos, Purser to Gilruth, a series of weekly logs covering March, April, and May 1959.

[15] Memo, Gilruth to staff, "Organization of Space Task Group," 3 Aug. 1959.

[16] Purser, "Log," 30 Jan. 1961; André J. Meyer, Jr., interview, Houston, 9 Jan. 1967.

[17] "Early History of Project Gemini," no author, n.d. [Purser notes on bottom of page that this information came from McDonnell]; Chamberlin, interview, Houston, 9 June 1966.

[18] Administrator's Briefing Memorandum, Robert C. Seamans, Jr., to Adm., "Space Task Group Functions and Staffing," 30 Nov. 1960, with enclosure, memo, Abe Silverstein to Assoc. Adm., "Separation of Space Task Group from Goddard Space Flight Center," 18 Nov. 1960, with 3 enclosures; NASA General Management Instruction No. 2-2-7, "Functions and Authority—Space Task Group," 1 Jan. 1961; NASA/STG News Release, "Space Task Group Becomes Separate NASA Field Element," 3 Jan. 1961.

[19] "Manned Spacecraft Development Center: Organizational Concepts and Staffing Requirements," STG, 1 May 1961, with 3 enclosures; a second study, untitled and undated, adds a fourth enclosure, "Summary of Planning on Location of a Manned Spacecraft Development Center."

[20] Robert B. Merrifield, "Men and Spacecraft: A History of the Manned Spacecraft Center (1958-1969)," [1972], pp. 3-22 through -30; James M. Grimwood, Project Mercury: A Chronology, NASA SP-4001 (Washington, 1963), p. 147.

[21] SEPC minutes, p. 3.

[22] Agenda, NASA Inter-Center Rendezvous Discussions General Meeting—27-28 February 1961; Lindsay J. Lina and Arthur W. Vogeley, "Preliminary Study of a Piloted Rendezvous Operation from the Lunar Surface to an Orbiting Space Vehicle," 21 Feb. 1961, presented at the NASA Inter-Center Rendezvous Discussions, 27-28 Feb. 1961.

[23] U.S. Congress, Senate, Committee on Aeronautical and Space Sciences, NASA Scientific and Technical Programs: Hearings, 87th Cong., 1st sess., 1961, pp. 171-72, 439-42; 1962 NASA Authorization, pp. 805-806.

[24] U.S. Congress, House, Committee on Science and Astronautics, Orbital Rendezvous in Space: Hearings, 87th Cong., 1st sess., 23 May 1961.

[25] U.S. Congress, Senate, Committee on Aeronautical and Space Sciences, NASA Authorization for Fiscal Year 1962: Hearings on H.R. 6874, 87th Cong., 1st sess., 1961, pp. 90, 139, 171; U.S. Congress, House, Committee on

419

Science and Astronautics, *Space Orbital Rendezvous*, H.R. 909, 87th Cong., 1st sess., 15 Aug. 1961, p. 9.

[26] *Space Orbital Rendezvous*, passim.

[27] "Guidelines for a Program for Manned and Unmanned Orbital Operations," NASA Staff Paper, May 1961; memo, Bernard Maggin to Assoc. Adm., "Staff Paper—'Guidelines for a Program for Manned and Unmanned Orbital Operations,'" 23 May 1961, enclosed in staff paper.

[28] Chamberlin interview.

[29] Memo, Purser to Gilruth, "Management Meeting, March 17-20, 1961," 14 March 1961.

[30] John H. Disher, telephone interview, 16 Jan. 1969; Disher, "Notes Taken ... at March 21, 1961 Meeting at Wallops Island," pp. 5-7; letter, Chamberlin to Grimwood, 26 March 1974, with comments.

[31] "Action Items, Management Discussions, March 17-20, 1961," n.d.; Disher notes, pp. 6, 7; Chamberlin comments, 26 March 1974.

[32] Letter Contract No. 6, Glenn F. Bailey to McDonnell Aircraft Corporation, 14 April 1961; Bailey and Stephen D. Armstrong, interview, Houston, 13 Dec. 1966.

[33] Letter, John Y. Brown to Bailey, "Proposed Contract, Mercury Capsule, Firm Price and Delivery Proposal for MK II Mercury Engineering Study Program," MAC No. NASA-16-9186, 12 April 1961.

[34] "Price and Delivery Proposal for MK II Mercury Engineering Study Program," MAC No. 8185, 12 April 1961.

[35] Letter, Bailey to Brown, "Contract NAS 9-119—Engineering Study," PASO-B-1926, 14 April 1961; "Design Engineering Study for Mercury MK-II Spacecraft," Contract NAS 9-119, 24 April 1961.

[36] James T. Rose, interview, St. Louis, 13 April 1966.

[37] William J. Blatz, Winston D. Nold, and Fred J. Sanders, interviews, St. Louis, 14 April 1966; Chamberlin interview.

[38] Chamberlin interview.

[39] Ibid.; cf. Blatz and Sanders interviews.

[40] Memo, Seamans to Dirs., Offices of Space Flight, Launch Vehicles, Advanced Research, and Life Sciences Programs, "Establishment of Ad Hoc Task Group for Manned Lunar Landing Study," 2 May 1961.

[41] [William A. Fleming et al.], "A Feasible Approach for an Early Manned Lunar Landing," Report of the Ad Hoc Study Group, 16 June 1961, p. i.

[42] Memo, Fleming to Eugene M. Emme, "Comments on Gemini History, Draft Chapters I and II," 5 Aug. 1969, with enclosure, subject as above, pp. 2, 5.

[43] Fleming, interview, 6 Aug. 1968, as quoted in John M. Logsdon, "NASA's Implementation of the Lunar Landing Decision," NASA HHN-81, September 1968, p. 9.

[44] "A Feasible Approach," Part I, "Summary Report of Ad Hoc Task Group Study," p. 2.

[45] Ibid., pp. 32-47.

[46] Letter, Seamans to Emme, 8 Jan. 1969.

[47] Logsdon, "NASA's Implementation," p. 9.

[48] Letter, John C. Houbolt to Seamans, 19 May 1961; Houbolt, interview, Princeton, N.J., 5 Dec. 1966.

[49] Memo, Seamans to Dirs., Launch Vehicle and Advanced Research Programs, "Broad Study of Feasible Ways for Accomplishing Manned Lunar Landing Mission," 25 May 1961; Bruce T. Lundin et al., "A Survey of Various Vehicle Systems for the Manned Lunar Landing Mission," 10 June 1961.

[50] Letter, Lundin to Seamans, 12 June 1961.

[51] Lundin et al., "A Survey," pp. 26-27; Lundin letter, 12 June 1961.

[52] Memo, H. Kurt Strass to Dir., "Visit to NASA Headquarters, June 6, 1961, by H. Kurt Strass, Apollo Projects Office," 8 June 1961; Logsdon, "NASA's Implementation," p. 12.

[53] Memo, D. D. Wyatt for record, "Discussions with the Associate Administrator on June 15, 1961," 20 June 1961.

[54] "Orbital Operations Preliminary Project Development Plan," compiled by MSFC Committee for Orbital Operations, P. J. deFries, chairman, 15 Sept. 1961.

[55] Memo, Seamans to Dirs., Launch Vehicle, Space Flight, and Advanced Research Programs, and Acting Dir., Life Science Programs, "Establishment of Ad Hoc Task Group for Manned Lunar Landing by Rendezvous Techniques," 20 June 1961.

[56] [Donald H. Heaton et al.], "Earth Orbital Rendezvous for an Early Manned Lunar Landing," Part I, "Summary Report of Ad Hoc Task Group Study," Ad Hoc Task Group for Study of Manned Lunar Landing by Rendezvous Techniques, August 1961, pp. i, 3, 89 (emphasis in original).

[57] Ibid., pp. 8 (emphasis in original), 89.

[58] Memo, Low to Dir., Space Flight Programs, "Report of Meeting with Space Task Group on June 2, 1961," 6 June 1961.

[59] Purser, "Notes on Capsule Review Board Meeting, June 9 and June 12, 1961"; Chamberlin comments, 26 March 1974.

[60] James A. Chamberlin, "Project Gemini Design Integration," Lecture 36 in a series on engineering design and operation of manned spacecraft presented during the summer of 1963 at the Manned Spacecraft Center and to

graduate classes at Louisiana State University, the University of Houston, and Rice University. The series was later edited and published; Chamberlin's lecture became Chapter 35 in Paul E. Purser, Maxime A. Faget, and Norman F. Smith, eds., *Manned Spacecraft: Engineering Design and Operations* (New York, 1964), pp. 365-74.

[61] Ibid., p. 365.

[62] Ibid.

[63] Donald G. Wiseman, "Principles of Power Distribution and Sequencing," in Purser, Faget, and Smith, eds., *Manned Spacecraft*, pp. 195-96; William H. Allen, ed., *Dictionary of Technical Terms for Aerospace Use*, first edition, NASA SP-7 (Washington, 1965), p. 249.

[64] Chamberlin interview.

[65] Purser, "Notes on CRB, June 9 and 12, 1961"; Chamberlin, "Project Gemini Design Integration," p. 372.

[66] Philip M. Deans, "Launch-Escape Systems," in Purser, Faget, and Smith, eds., *Manned Spacecraft*, pp. 322-24.

[67] Chamberlin interview.

[68] Ibid.; Blatz interview; Chamberlin, "Project Gemini Design Integration," p. 372.

[69] Purser, "Notes on CRB, June 9 and 12, 1961."

[70] James L. Decker, "A Program Plan for a Titan Boosted Mercury Vehicle," Vol. 1, July 1961.

[71] Seamans, interview, Washington, 26 May 1966.

[72] Decker, "A Program Plan."

[73] *Gemini-Titan II Air Force Launch Vehicle Press Handbook*, published by Martin-Baltimore for issuance to news media ca. December 1964, p. II-2. This press handbook was later issued in a second edition, on manned flight, and thereafter was updated for each Gemini flight, beginning with *Gemini 3*.

[74] Chamberlin interview.

[75] Ibid.

[76] See pp. 18-19.

[77] Francis M. Rogallo et al., *Preliminary Investigation of a Paraglider*, NASA Technical Note (TN) D-443, August 1960; Rogallo and John G. Lowry, "Flexible Reentry Gliders," presented at the Society of Automotive Engineers National Aeronautics Meeting, New York, 4-8 April 1960.

[78] William W. Petynia, secretary, "Minutes of Meeting on [sic] Apollo Technical Liaison Group—Configurations and Aerodynamics, January 12, 1961," 17 Jan. 1961, pp. 12, 14; Petynia, "Minutes of Meeting of Apollo Technical Liaison Group—Configurations and Aerodynamics, April 10-12, 1961," 18 April 1961, p. 8.

[79] Memo, John W. Kiker et al. to Dir., "Interim report on paraglider—Apollo application investigation," 4 April 1961.

[80] Memo, Gilruth to Procurement Officer, "Design Study of a Paraglide Landing System for a Manned Spacecraft," 17 May 1961, with enclosure, "Statement of Work for a Design Study of a Manned Spacecraft Paraglide Landing System," 17 May 1961.

[81] See, for example, Glenn F. Bailey to North American Aviation, Inc., "Contract for Design Study of Paraglide Landing System for Manned Spacecraft," NAS 9-136, 27 May 1961; "Paraglider Development Program, Phase I—Design Study: Test Programs," STG, 30 June 1961.

[82] Letter, Bailey to North American, Attn: H. H. Cutler, "Contract NAS 9-136, Paraglide Landing System Design Study Technical Direction," PASO-B-2426, 16 June 1961; letter, Bailey to Goodyear Aircraft Corp., Attn: R. T. Madden, "Contract NAS 9-137, Paraglide Landing System Design Study Technical Direction," PASO-B-2427, 16 June 1961; letter, Bailey to The Ryan Aeronautical Co., Attn: T. Echols, "Contract NAS 9-135, Paraglide Landing System Design Study Technical Direction," PASO-B-2428, 16 June 1961; Lester A. Stewart, "Minutes of Meeting of Goodyear Aircraft Corporation (Contract NAS 9-137), Study Review Meeting, June 13, 1961," 21 June 1961; Stewart, "Minutes of Meeting of North American Aviation, Inc. (Contract NAS 9-136), Study Review Meeting, June 14, 1961," 21 June 1961; Stewart, "Minutes of Meeting of Ryan Aeronautical Company (Contract NAS 9-135), Study Review Meeting, June 15, 1961," 21 June 1961.

[83] Memo, Gilruth for Procurement Officer, "Design Study of a Paraglide Landing System for a Manned Spacecraft," 22 May 1961; letter, Brown to Bailey, "Contract NAS 9-119, MK-II Mercury Study Contract, Information Concerning," 832-16-12, 18 Aug. 1961, enclosure I, "NAS 9-119, MAC Job 832, Estimated Engineering Manhour Expenditures by Element," No. C-58496, ca. 6 Aug. 1961.

[84] Purser, "Notes on CRB, June 9 and 12, 1961."

[85] Ibid.; Chamberlin interview.

[86] Purser, "Notes on CRB, June 9 and 12, 1961."

[87] Ibid.; memo, Purser to Gilruth, "Log for week of June 5, 1961," 13 June 1961.

[88] Purser, "Notes on CRB, June 9 and 12, 1961"; "NASA Manned Space Flight, Space Task Group Financial Plan FY '62, Follow-On Mercury, June 14, 1961," pp. F-1 through F-4.

[89] Memo, Purser to Gilruth, "Log for week of June 12, 1961," 20 June 1961.

[90] Loyd S. Swenson, Jr., James M. Grimwood, and Charles C. Alexander, *This New Ocean: A History of Project Mercury*, NASA SP-4201 (Washington, 1966), p. 487.

[91] "Follow-On Experiments, Project Mercury Capsules," McDonnell Engineering Rept. 6919, 1 Sept. 1959 (rev. 5 Oct. 1959), p. 4.1-1; memo, Warren J. North to Dir., "Advanced Technology Follow-On Tests for the Mercury Capsule," 6 July 1960; memo, David L. Winterhalter to Caldwell C. Johnson, "High Performance Retrograde Rockets," 26 Jan. 1961; memo, Winterhalter to Johnson, "Higher Performance Posigrade-Retrograde Package for Mercury Follow-On Missions," 15 April 1961; memo, Winterhalter to Assoc. Dir., "Higher Performance Mercury Posigrade-Retrograde Package," 26 May 1961.

[92] Purser notes on discussion between himself, Gilruth, Faget, Chamberlin, and Charles W. Mathews on MK2 Hermes and MK1, 3 July 1961; "Spacecraft Comparisons: (1) Mark 2 Capsule—Hermes Plan, Extensive Redesign; (2) Mercury—with 18 Orbit Capability, Minimum Redesign—Part of a Program with Specialized Vehicles for Each Mission," prepared by Scheduling Section, Contracts and Scheduling Branch, ca. 1 July 1961 (see memo, Nicholas Jevas to Grimwood, 21 March 1968).

[93] Purser notes, 3 July 1961.

[94] Chamberlin comments, 26 March 1974.

[95] Ibid.; Purser notes on senior staff meeting, 7 July 1961; Jack C. Heberlig, "Notes on Senior Staff Meeting, July 7, 1961," 11 July 1961, with enclosure; "Mark II Mercury Spacecraft," McDonnell C-57342, 6 July 1961.

[96] Memo, Robert B. Voas to Meyer, "Request for feasibility study of dual seating for Redstone flights," 23 Jan. 1959.

[97] Blatz and Nold interviews.

[98] John F. Yardley, interview, St. Louis, 13 April 1966.

[99] Walter C. Williams, interview, El Segundo, Calif., 15 May 1967; Low, interview, Houston, 7 Feb. 1967.

[100] Walter F. Burke, interview, St. Louis, 15 April 1966.

[101] Discussion notes, "First Meeting of Manned Lunar Landing Steering Committee, July 6, 1961," 11 July 1961, p. 2.

[102] "Earth Orbital Rendezvous for Early Manned Lunar Landing," Part I, Fig. 4—"Master Flight Plan-Orbital Ops." pp. 17, 58.

[103] Brown letter, 18 Aug., enclosure 2, "NAS 9-119 (MAC Job 832), Estimated Tooling Manhour Expenditures by Element," C-58497, ca. 6 Aug. 1961; Low and Sanders interviews.

[104] Cf. "Mark II Mercury Spacecraft," C-57342, and "Mercury Spacecraft: Advanced Versions," McDonnell C-57978, ca. 27 July 1961.

[105] "Estimated Engineering Manhour Expenditures by Element"; memo, James I. Brownlee for record, "Negotiation of Definitive Contract NAS 9-170, 'Project Gemini Two-Man Spacecraft Development Program,'" 13 March 1963; Low and Burke interviews; Project Mercury Status Report No. 13, for period ending 31 January 1962, p. 2.

[106] Contract NAS 9-119, Amendment No. 1, 17 May 1961; Change order, Bailey to McDonnell, "Additional Engineering Manhour Requirements," 17 May 1961.

[107] Letter, Bailey to Brown, "Contract NAS 9-119, Design Engineering Study for Mercury MK-II Spacecraft," PASO-B-2791, 9 Aug. 1961; "Estimated Engineering Manhour Expenditures by Elements."

[108] Letter, Brown to Bailey, "Proposed Contract, MK-II Mercury Engineering Studies," 832-16-67, 25 Aug. 1961, with enclosure, "MK-II Mercury Engineering Studies," 8185-4, 25 Aug. 1961.

[109] Bailey to McDonnell, "Amendment Nr. 1 to Letter Contract Nr. 6," NAS 5-59, 30 Aug. 1961.

[110] NASA General Management Instruction 4-1-1, "Planning and Implementation of NASA Projects," 18 Jan. 1961. The significance of this document is discussed in Rosholt, *Administrative History*, pp. 228-29.

# Chapter III

[1] Memo, Owen E. Maynard to Chief, Flight Systems Division (FSD), "Comments on Project Mercury Mark II Progress Report Number 1," 7 July 1961; memo, [Maynard] to Chief, FSD, "Comments on Proposal for Project Mercury minimum change capsule, reconfigured capsule and two man capsule," 27 July 1961.

[2] Paul E. Purser, notes on "Titan II Manned Booster," 29 June 1961; memos, Purser to Robert R. Gilruth, "Log for week of July 3, 1961," 10 July 1961, and "Log for week of July 17, 1961," 22 July 1961; Purser, interview, Houston, 14 March 1967; Walter C. Williams, interview, El Segundo, Calif., 16 May 1967.

[3] James L. Decker, "A Program Plan for a Titan Boosted Mercury Vehicle," July 1961; Purser, notes on briefing, 3 Aug. 1961; memo, Purser to Gilruth, "Log for week of July 31, 1961," 10 Aug. 1961.

[4] Decker, "A Program Plan"; Purser notes, 3 Aug. 1961. Attached to Purser's notes are "Mercury-Titan Program, Part II: Program Cost," 2 Aug. 1961, and a proposed program

schedule, "Master Plan for Redundant System Booster," 1 Aug. 1961.

[5] Decker, "A Program Plan"; *Gemini-Titan II Air Force Launch Vehicle Press Handbook* (Martin-Baltimore, ca. December 1964), p. II-1.

[6] Jack C. Heberlig, "Notes on Senior Staff Meeting, Monday, August 7, 1961," 8 Aug. 1961, p. 3.

[7] Lloyd Mallan, *Peace Is a Three-Edged Sword* (Englewood Cliffs, N.J., 1964), pp. 190-91; letter, R. L. Tonsing to Helen T. Wells, 28 Sept. 1965.

[8] Letter, Isaac Newton to Robert Hooke, 5 Feb. 1675/6 in H. W. Turnbull and J. F. Scott, eds., *The Correspondence of Isaac Newton* 1 (Cambridge, Eng., 1959), p. 416, as cited in Alexandre Koyré, "An Unpublished Letter of Robert Hooke to Isaac Newton," *Newtonian Studies* (Cambridge, Mass., 1965), p. 227. As Koyré points out, and as many others have noted, the phrase did not originate with Newton and was, in fact, a commonplace; see Robert K. Merton, *On the Shoulders of Giants: A Shandean Postscript* (New York, 1965).

[9] Alexandre Koyré, "Newton and Descartes," *Newtonian Studies*, p. 54.

[10] Letter, D. R. Church to James A. Chamberlin, 18 Aug. 1961.

[11] R. Cargill Hall, "The Agena-Booster Satellite," presented at American Institute of Astronautics and Aeronautics, Boston, Mass., 2 Dec. 1966.

[12] James T. Rose, interview, St. Louis, 13 April 1966.

[13] Memo, Chamberlin to Dir., "Paraglider Landing System Design Studies; transmittal of final reports," 23 Aug. 1961; "Paraglider Development Program, Phase I: Final Report," North American SID 61-266, 15 Aug. 1961.

[14] Letter, J. Y. Brown to Glenn F. Bailey, "Contract NAS 9-119, MK-II Mercury Study Contract, Information Concerning," 832-16-12, 18 Aug. 1961, with enclosure, "NAS 9-119, MAC Job 832, Estimated Engineering Manhours Expenditures by Element," C-58496, ca. 6 Aug. 1961.

[15] Purser memo, 10 Aug. 1961; memo, Purser to Gilruth, "Log for week of August 7, 1961," 15 Aug. 1961; Rose, Walter D. Wolhart, and William C. Muhly, telephone interviews, 13 Oct. 1972.

[16] "Preliminary Project Development Plan for an Advanced Manned Space Program Utilizing the Mark II Two Man Spacecraft," STG, 14 Aug. 1961.

[17] Ibid., pp. 2, 20; U.S. Congress, House, Committee on Science and Astronautics, *Aeronautical and Astronautical Events of 1961: Report*, 87th Cong., 2d sess., 7 June 1962, pp.

37-38, 53; James P. Henry, *Biomedical Aspects of Space Flight* (New York, 1966), p. 82.

[18] "Preliminary Plan," 14 Aug. 1961, pp. 2-4, 8-13, 19-20.

[19] Ibid., pp. 4-5, 6-8, 11, 19, Tables 3.1, 3.2, 3.4, 5.1, Fig. 8.1.

[20] Ibid., pp. 17-18, 19, 22-23, 25-27, Fig. 8.1; Bastian Hello, interview, Baltimore, 23 May 1966.

[21] "Preliminary Plan," 14 Aug. 1961, pp. 29-33, Table A-2.2, Fig. A-5.1; cf. Figs. 5.4 with A-7.1, 8.1 with A-7.2, Table 5.1 with A-7.1.

[22] Memo, Purser to Gilruth, "Log for week of Aug. 14, 1961," n.d.; Purser notes on meeting with Gilruth, Williams, Chamberlin, Charles W. Mathews, George M. Low, and Warren J. North, 15 Aug. 1961.

[23] "Preliminary Project Development Plan for an Advanced Manned Space Program Utilizing the Mark II Two Man Spacecraft," STG, 21 Aug. 1961, p. 5.

[24] "A Lunar Landing Proposal Using Rendezvous," STG, 30 Aug. 1961.

[25] Ibid., memo, Jack Funk for Assoc. Dir., "Trip to Marshall Space Flight Center, August 1, 1961, to discuss circumlunar payload," 14 Aug. 1961.

[26] Williams interview.

[27] John C. Houbolt et al., "Manned Lunar-Landing through use of Lunar-Orbit Rendezvous" 1 (31 Oct. 1961): 5; Houbolt, interview, Princeton, N.J., 5 Dec. 1966.

[28] Memo, Robert L. O'Neal to Assoc. Dir., "Discussion with Dr. Houbolt, LRC, concerning the possible incorporation of a lunar orbital rendezvous phase as a prelude to manned lunar landing," 30 Jan. 1961.

[29] "Manned Lunar Landing Via Rendezvous," Langley Research Center, 19 April 1961; Houbolt, telephone interview, 30 Dec. 1966; letter, Chamberlin to James M. Grimwood, 26 March 1974.

[30] Letter, Gilruth to Nicholas E. Golovin, 12 Sept. 1961; Gilruth, interview, Houston, 21 March 1968.

[31] Purser, comments on Gemini draft history, 14 Jan. 1969; Chamberlin comments, 26 March 1974.

[32] Memo, Purser to Gilruth, "Log for week of September 25, 1961," 5 Oct. 1961; "Project Apollo Slides: Integrated Apollo Program," STG, 26 Sept. 1961; Rose interview.

[33] Memo, Abe Silverstein to Assoc. Adm., "Approval of Orbital Flight Development Program," 6 Oct. 1961, with Robert C. Seamans, Jr.'s signed approval, same date.

[34] Memo, Seamans to Silverstein, "Apollo Crew Selection," 13 Oct. 1961; memo, Low to Aleck C. Bond, "Possible Use of Titan II

Booster in the Manned Space Flight Program," 22 Sept. 1961.

[35] Memo, Purser to Gilruth, "Log for week of October 23, 1961," 30 Oct. 1961.

[36] "Project Development Plan for Rendezvous Development Utilizing the Mark II Two Man Spacecraft," STG, 27 Oct. 1961.

[37] Memo, Purser to Gilruth, "Log for week of October 30, 1961," 7 Nov 1961.

[38] Memo, Chamberlin to Dir., "Proposed Amendment to Letter Contract No. 6 to Contract NAS 5-59," 27 Oct. 1961, with enclosures, "Guidelines: MK-II Program Effort," n.d., and Glenn F. Bailey to McDonnell, "Amendment No. 2 to Letter Contract 6," NAS 5-59, 28 Oct. 1961.

[39] Bailey, "Amendment No. 2."

[40] "Guidelines: MK-II Program Effort."

[41] Mark II note, Jerome B. Hammack, "Meeting with Aerospace Mechanics Representatives," 19 Oct. 1961.

[42] Mark II note, "Interim Report of Proposed Rendezvous Technique, Guidance and Lunar Lander," 10 Nov. 1961, annotated to show deletion of lunar lander.

[43] Memo, Raymond D. Hill, Jr., to E. M. Flesh, "Model 133N Coordination Meeting 14-15 November 1961," PM-1467, 24 Nov. 1961.

[44] "Mercury Mark II Detail Specification," McDonnell Rept. 8356, 15 Nov. 1961, p. 8.

[45] Ibid.; Fred J. Sanders, telephone interview, 24 Feb. 1969.

[46] "Statement of Work for Phase II, Part A, System Research and Development of a Paraglider Development Program," STG, 15 Sept. 1961.

[47] Memo, Lester A. Stewart et al. to Dir., "Paraglider Development Program; Evaluation of Design Studies; Contract NAS 9-135, Ryan Aeronautical Company; Contract NAS 9-136, North American Aviation, Inc.; Contract NAS 9-137, Goodyear Aircraft Corporation," 22 Sept. 1961; TWX, Bailey to Neil C. Dopheide, 20 Nov. 1961; Bailey, "Letter Contract [NAS 9-167] between National Aeronautics and Space Administration and North American Aviation, Inc. (Contractor)," 21 Nov. 1961; TWX, Norbert F. Witte to MSC, Attn: Purser and Chamberlin, "Contract NAS 9-167, Paraglider Development Program Phase IIA: Full Scale Test Vehicle," MA14859, 13 Dec. 1961.

[48] Stewart, "Paraglider Development Program (Phase II-A): Minutes of . . . North American Aviation, Incorporated (Contract NAS 9-135 [sic]), Program Review Meeting, November 28-29, 1961," 5 Dec. 1961; Kenneth W. Christopher, telephone interview, 18 Oct. 1972.

[49] Letter, James E. Webb to Robert S. McNamara, 7 July 1961, with enclosure, memo, Seamans to Adm., "Planning of a DOD-NASA Program for Development of Large Launch Vehicles," 7 July 1961; letter, McNamara to Webb, 7 July 1961.

[50] Memo, Golovin to members of the DOD-NASA Large Launch Vehicle Planning Group (LLVPG), "Agenda—First Meeting," 20 July 1961.

[51] [Nicholas E. Golovin et al.], draft of final report of LLVPG, ca. November 1961 (see memo, Milton W. Rosen to D. Brainerd Holmes, "Recommendations for NASA Manned Space Flight Vehicle Program," 20 Nov. 1961); [Golovin et al.], "Final Report: NASA-DOD Large Launch Vehicle Planning Group," 3 vols., NASA-DOD LLVPG 105, 1 Feb. 1962. Still a third report, [Golovin et al.], "Summary Report: NASA-DOD Large Launch Vehicle Planning Group," under the same number was issued 24 Sept. 1962.

[52] TWX, Harvey Hall to Dirs., Marshall, Langley, and Jet Propulsion Laboratory (JPL), No. 128, 24 Aug. 1961; TWX, Hall to John W. Small, Jr., et al., 14 Sept. 1961; memo, Hall to LLVPG staff, "Compairson [sic] of Mission Alternatives (Rendezvous versus direct flight)," 14 Sept. 1961; "Minutes of special JPL presentation [to LLVPG] on the lunar mission, August 3, 1961"; "System Considerations for the Manned Lunar Landing Program," JPL Technical Manual TM 33-52, 3 Aug. 1961; "Man-to-the-Moon and Return Mission Utilizing Lunar-Surface Rendezvous," JPL TM 33-53, 3 Aug. 1961.

[53] Memo, Warren H. Amster to LLVPG staff, "A 'Federated' Launch Vehicle Program," 18 Sept. 1961.

[54] "Orbital Operations Preliminary Project Development Plan," compiled by MSFC Committee for Orbital Operations, P. J. deFries, chairman, 15 Sept. 1961; letter, Wernher von Braun to Harvey Hall, 15 Sept. 1961; John C. Houbolt et al., "Technical Problems of Lunar Orbit Rendezvous," unpubl. papers, draft dated September 1961; JPL contributed three papers prepared under contract: "The Lunar Surface Rendezvous Technique for Manned Lunar Landing and Return," JPL 950163, Space Technology Laboratories, Inc., 8634-0001-RC-000, 2 Oct. 1961; "Lunar Surface Assembly Techniques, A Preliminary Study of Refueling for the Lunar Surface Rendezvous," JPL 950167, Nucleonics Laboratory of Hughes Aircraft Co. FD-61-401, 2 Oct. 1961; "Analysis of a Lunar Surface Rendezvous Mission," JPL 960165, Space Craft, Inc., October 1961. (See "LLVPG Final Report," III, pp. VI-87, -88.)

[55] James F. Chalmers, LLVPG secretary, "Minutes of General Meeting, 10 October 1961," pp. 3-4.

56 Chalmers, "Minutes of a General Meeting, 23 October 1961."

57 "LLVPG Final Report," III, chap. VI.

58 Ibid., III, IX-8; Edward H. Kolcum, "Rendezvous Is Urged for Moon Flight," *Aviation Week and Space Technology*, 6 Nov. 1961, pp. 26-27.

59 For an assessment of the Golovin Committee and its results, see John M. Logsdon, "NASA's Implementation of the Lunar Landing Decision," NASA HHN-81, September 1968, pp. 32-33.

60 Memo, Rosen to Holmes, "Large Launch Vehicle Program," 6 Nov. 1961.

61 Rosen memo, 20 Nov. 1961, with enclosure, "Report of Combined Working Group on Vehicles for Manned Space Flight," n.d.

62 "Report of Combined Working Group."

63 Robert L. Rosholt, *An Administrative History of NASA, 1958-1963*, NASA SP-4101 (Washington, 1966), pp. 198-211.

64 Ibid., pp. 221-27; letter, Seamans to Holmes, 25 Oct. 1961.

65 Letter, John H. Rubel to Seamans, 7 Nov. 1961.

66 "LLVPG Final Report," III, p. IX-23.

67 Letter, Rubel to Eugene M. Emme, 12 Aug. 1969; memo, Max Rosenberg to NASA Historian, "Comments on Draft Chapter I-V and XIII-XV, Gemini History," 26 June 1970, with enclosure, "NASA Draft Gemini History, Comments on Chapters I-V; XIII-XV," n.d.

68 Memo, William A. Fleming to Assoc. Adm., "Critique on Rubel letter, dated October 20, 1961, to Seamans," 27 Oct. 1961; Chamberlin comments, 26 March 1974.

69 Letter, McNamara to Webb, 17 Nov. 1961; letter, Webb to McNamara, 28 Nov. 1961.

70 Memo, Rubel and Seamans to Dir. and Dep. Dir., LLVPG, "Request to Reconvene the Large Launch Vehicle Planning Group," 18 Nov. 1961; Chalmers, "Minutes of the Meeting of Monday November 20, 1961."

71 Memo, Rubel and Seamans to Sec. of Defense and NASA Adm., "Recommendations Relative to TITAN III and TITAN II 1/2," 5 Dec. 1961.

72 Memo, Holmes to Assoc. Adm., "Mark II Preliminary Project Development Plan," 6 Dec. 1961, with Seamans' approval, dated 7 Dec. 1961.

73 "Project Development Plan for Rendezvous Development Utilizing the Mark II Two Man Spacecraft," MSC, 8 Dec. 1961.

74 U.S. Congress, House, Committee on Science and Astronautics, *Astronautical and Aeronautical Events of 1962: Report*, 88th Cong., 1st sess., 12 June 1962, p. 1.

## Chapter IV

1 Glenn F. Bailey, interview, Houston, 13 Dec. 1966.

2 Letter, Bailey to McDonnell, "Letter Contract No. NAS 9-170," with James S. McDonnell, Jr.'s signed acceptance.

3 Letter Contract NAS 9-170, Section I, p. 1, and Exhibit "A," 11 Dec. 1961, pp. 1-2.

4 Ibid., pp. 1, 8.

5 Ibid., Section II, pp. 1, 13.

6 Memo, Robert C. Seamans, Jr., and John H. Rubel to Sec. of Defense and NASA Adm., "Recommendation relative to the division of effort between the NASA and the DOD in the development of space rendezvous and capabilities," 7 Dec. 1961.

7 Memo, Paul E. Purser to Robert R. Gilruth, "Log for week of December 11, 1961," 18 Dec. 1961; Daniel D. McKee, draft, "Instructions to the Ad Hoc Group on the Mercury Mark II," 12 Dec. 1961; "Members of Ad Hoc Working Group on Air Force Participation in the Mercury-Mark II Project," ca. 13 Dec. 1961; [Purser], draft, "NASA-DOD Operational and Management Plan for the Mercury-Mark II Program," 15 Dec. 1961.

8 Memos, Purser to Gilruth, "Log for week of December 18, 1961," 28 Dec. 1961, and "Log for week of December 25, 1961," 2 Jan. 1962; "NASA-DOD Operational and Management Plan for the Gemini Program," revised 29 Dec. 1961.

9 "NASA-DOD Plan for Gemini," pp. 5-6.

10 Ibid., p. 5; memo, William E. Lilly to dist., "Minutes of the Management Council, Manned Space Flight Program," 29 Dec 1961, with enclosure, "Minutes of the Management Council, Manned Space Flight Program, December 21, 1961"; letter, George J. Detko to Cdr. William R. Wakeland, "Management Agreements between AF/SSD and MSFC," 2 Jan. 1962, with enclosure, "National Aeronautics and Space Administration Agena B Launch Vehicle Program, Management Organization and Procedures," 14 Feb. 1961, signed by Seamans and Gen. Bernard A. Schriever; memo, Seamans to Dir., Space Sciences, Attn: Edgar M. Cortright, "Gemini-Atlas/Agena Management," 26 Jan. 1962; Robert L. Rosholt, *An Administrative History of NASA, 1958-1963*, NASA SP-4101 (Washington, 1966), pp. 274-75.

11 "NASA-DOD Plan for Gemini," pp. 6-9, 11-24; Scott H. Simpkinson, interview, Houston, 18 Jan. 1967; Harle L. Vogel, interview, Baltimore, 23 May 1966.

12 Letter, D. Brainerd Holmes to Schriever, 26 Jan. 1962, with statement of approval signed by both Holmes and Schriever.

[13] Memo, Seamans and Rubel to Sec. of Defense and NASA Adm., "NASA/DOD Operational and Management Plan for Accomplishing the GEMINI (formerly MERCURY MARK II) Program," 29 Jan. 1962.

[14] Letter, McKee to MSC, Attn: Wesley L. Hjornevik, 23 Jan. 1962, with enclosure, changed pages of "NASA-DOD Plan for Gemini"; memo, Holmes to Adm., Dep. Adm., and Assoc. Adm., "Selection of Additional Astronauts," 6 April 1962, approved 6 April by Seamans, 9 April by Hugh L. Dryden and James E. Webb, with enclosure, "Gemini and Apollo Astronaut Selection," 6 April 1962; letter, Seamans to Rubel, 10 July 1962, with enclosure, draft letter, Seamans and Rubel to Sec. of Defense and NASA Adm., "NASA/DOD Operational and Management Plan for Accomplishing the Gemini (formerly Mercury Mark II) Program," undated and signed by Seamans only, 10 July 1962; letter, Rubel to Seamans, 27 July 1962, with enclosure, Gemini agreement (retyped); memo, Seamans and Rubel to Sec. of Defense and NASA Adm., "NASA/DOD Operational and Management Plan for Accomplishing the Gemini (formerly Mercury Mark II) Program," 28 July 1962, signed as approved by Seamans and Rubel, 27 July 1962.

[15] TWX, Holt F. B. Watts, Jr., to Douglas R. Hendrickson, 26 Dec. 1961; Purser memo, 2 Jan. 1962; TWX, Walter C. Williams to Cdr., SSD, Attn: Col. Keith G. Lindell, PASO-A-6504, 27 Dec. 1961; Howard T. Harris, "Gemini Launch Vehicle Chronology, 1961-1966," AFSC Historical Publications Series 66-22-1, December 1966, p. 1.

[16] NASA-Defense Purchase Request T-2356-G, signed by Bailey, 5 Jan. 1962; Bailey to SSD, 3 Jan. 1962, with enclosure, "Statement of Work to Be Accomplished under Department of Defense Purchase Request No.  ," 3 Jan. 1961 [sic]; memo, Purser for file, "Contract NAS 9-170, Project Gemini, NASA-SSD Purchase Request No. T-2356-G, Titan II Launch Vehicles," 23 Jan. 1962.

[17] Memo, Raymond D. Hill, Jr., to E. M. Flesh, "Model 133N Coordination Meeting 14-15 November 1961," PM-1467, 24 Nov. 1961.

[18] John E. Roberts, Jr., "Agena-B requirements for Advanced Mercury rendezvous mission," Advanced Mercury note, 19 Dec. 1961.

[19] "Preliminary Report on Agena System Capabilities for Advanced Mercury Rendezvous Mission," Lockheed LMSC/A004120, 26 Jan. 1962.

[20] Memo, James T. Rose to Gemini Project Dir., "Recommendations for primary and secondary terminal phase techniques for the Gemini Rendezvous Program," 10 Jan. 1962.

[21] Letter, Gilruth to Marshall, Attn: Wernher von Braun, "Procurement of Atlas-Agena Space Vehicles," 31 Jan. 1962, with enclosures, "Exhibit 'A' to Atlas-Agena Procurement: Description of Proposed Rendezvous Techniques for Project Gemini," 30 Jan. 1962, and "Exhibit 'B': Statement of Work for Atlas-Agena Rendezvous Vehicles to be Used in Project Gemini," n.d.

[22] James A. Chamberlin, "Minutes of Meeting of Gemini Project Office and MSFC-Agena Project Office, February 28, 1962," 5 March 1962; "Gemini Agena Target Vehicle Program Progress Report, March 1965," LMSC-A605200-7, 20 April 1965, p. A-1 (hereafter cited as GATV Progress Report).

[23] Purser memo, 18 Dec. 1961.

[24] "Project Gemini Schedule Analysis," GPO, 5 Jan. 1962.

[25] MSC Announcements Nos. 8, 9, and 10, "Establishment of Gemini Project Office," "Establishment of Mercury Project Office," "Establishment of Apollo Spacecraft Project Office," 15 Jan. 1962; letter, Bailey to McDonnell, "Amendment No. 1 to Letter Contract NAS 9-170," 16 Jan. 1962.

[26] MSC Announcement No. 12, "Personnel Assignments for Mercury and Gemini Project Offices," 31 Jan. 1962; "MSC Reorganization Plans Released; Many Changes," MSC Space News Roundup, 24 Jan. 1962.

[27] "Personnel Assignments for Mercury and Gemini Project Offices"; André J. Meyer, Jr., interview, Houston, 9 Jan. 1967; memo, Meyer to Historical Office, "Comment on draft chapters of Gemini narrative history," 5 June 1969.

[28] MSC Announcement No. 21, "Relocation of Manned Spacecraft Center Headquarters," 26 Feb. 1962; Bailey letter, 16 Jan. 1962; Meyer memo, 5 June 1969.

[29] Letter, Thomas F. Dixon to Gilruth, 28 March 1962, with enclosure, Memorandum from the Administrator, "Establishment of positions and personnel under authority of Public Law 85-568, as amended," 23 March 1962, signed by Dryden; see "Telephone Directory, Manned Spacecraft Center," 15 Aug. 1962, for outline of Gemini Project Office divisions and positions, pp. xii-xiii.

[30] Brochure, "Manned Spacecraft Center, Houston, Texas: Interim Facilities," as of 1 Aug. 1962.

[31] TWX, Chamberlin to Dir., "Report of Activities for the Week Ending March 3, 1962," 5 March 1962; memo, Chamberlin to Dir., "Report of Activities for the week ending March 10, 1962," 12 March 1962.

[32] Richard L. McCreight, "Minutes of . . . McDonnell Coordination Meeting, February

19, 21, and 23, 1962," 26 Feb. 1962; Mc-
Creight, "Minutes of NASA Project Office—
McDonnell Coordination Meeting, Feb. 27, 28,
1962," 6 March 1962; Meyer interview; Willis
B. Mitchell, Jr., interview, Houston, 13 Nov.
1970; TWX, John Y. Brown to MSC, Attn:
Bailey, "Contract NAS 9-170, Gemini Project,"
16-DAH-1090, 9 Aug. 1962.

[33] Letter, Brown to MSC, Attn: Bailey,
"Contract NAS 9-170, Project Gemini, Personnel Assignments," 306-16-488, 26 Feb. 1962,
with enclosure, "Project Gemini Management
Organization Chart"; Harry W. Oldeg, "Gemini Program Management," AAS paper 66-158,
presented at the American Astronautical Society National Conference on the Management of
Aerospace Programs, University of Missouri,
Columbus, Mo., 16-18 Nov. 1966, pp. 2-3.

[34] Robert N. Lindley, "Discussing Gemini:
A 'Flight' Interview with Robert Lindley of
McDonnell," FLIGHT International, 24 March
1966, pp. 488-89.

[35] R[obert] N. Lindley, "Gemini Engineering Program, McDonnell Aircraft Corporation," presented at the Institute of Management Sciences, Dallas, 16-19 Feb. 1966, pp. 4,
6; Lindley, interview, St. Louis, 13 April 1966.

[36] Lindley, "Gemini Engineering Program,"
pp. 7-8, 10; memo, Robert F. Cortinovis,
"Procurement Review Board Meeting—MAC
Model 133N," 710-587-1, 27 Nov. 1961, with
enclosure.

[37] Letter, Brown to MSC, Attn: Bailey,
"Contract NAS 9-170, Two-Man Spacecraft,
Make or Buy Program," 301-16-349, 29 Dec.
1961, with enclosure, "Proposed 'Buy' Listing
of Major Equipment."

[38] Letter, Walter F. Burke to MSC Field
Representative, Attn: Wilbur H. Gray, "Con-
tract NAS 9-170, Procedure for Obtaining
NASA Approval of Major Subsystem Source
Selection for Project Gemini," 306-101-136, 26
Jan. 1962.

[39] Memo, Bailey to GPO, Attn: Chamberlin, "Contract NAS 9-170, Make-or-Buy Sup-
plement," 15 March 1962; memo, Cortinovis
for purchase order file, "Environmental Control System—Gemini," 710-041-2, 23 Jan.
1962, with enclosures; letter, Burke to MSC
Field Representative, Attn: Gray, "Selection of
Equipment, Contract NAS 9-170, Environmental Control System," 306-101-140, 5 Feb. 1962;
Project Gemini Quarterly Status Report No. 1,
for period ending 31 May 1962, GPO, pp. 15-
16.

[40] Letter, Gray to Burke, "Selection of
Equipment, Contract NAS 9-170, Environmental Control System," NAS/170-252, 13 Feb.
1962; Arthur H. Atkinson, "Gemini—Major

Subcontracts, McDonnell Aircraft Corporation," 3 July 1962.

[41] Quarterly Status Report No. 1, pp. 12-
13, 20; Gemini Press Reference Book, Gemini
Spacecraft Number Three, prepared by Mc-
Donnell's External Relations Division for news
media use at launch [and updated for each
mission], ca. March 1965, pp. 45-47; Quarterly
Status Report No. 3, for period ending 30
Nov. 1962, p. 15.

[42] Letter, Burke to MSC Field Office, Attn:
Atkinson, "Selection of Equipment, Contract
NAS 9-170, Reaction Control and Adapter
Propulsion Systems," 306-09-143, 8 March
1962; letter, Gray to Burke, "Selection of
Equipment, Contract NAS 9-170, Adapter
Propulsion and Reaction 'Control System,"
NAS/170-262, 21 Feb. 1962; Atkinson, "Gemi-
ni Major Subcontracts"; letter, Chamberlin to
James M. Grimwood, 26 March 1974, with
comments.

[43] Harris, "Launch Vehicle Chronology,"
p. 2; "SSVL: Gemini Launch Vehicle Directorate," 7 May 1962.

[44] Harris, "Launch Vehicle Chronology,"
pp. 5, 6; Sol Levine, "Man-Rating the Gemini
Launch Vehicle," presented at the American
Institute of Aeronautics and Astronautics 1st
Annual Meeting and Technical Display, Washington, 29 June-2 July 1964; "Aerospace Corporation Annual Report, Fiscal 1962-1963,"
draft, n.d.; letter, Col. Ralph C. Hoewing to
MSC, Attn: George F. MacDougall, Jr., "Gemini Launch Vehicle Financial Plan," 4 April
1962, with enclosures.

[45] Memo, Bailey for Project Gemini file,
"Martin-Marietta Company," 4 Jan. 1962; Col.
Robert R. Hull, interview, Los Angeles, 18
April 1966; Col. Richard C. Dineen, interview,
Huntington Beach, Calif., 15 May 1967; Bas-
tian Hello, interview, Baltimore, 23 May 1966.

[46] "Program Plan," Martin Co. ER 12255,
April 1962, pp. 3-8; J[oseph] F. Wambolt and
S[ally] F. Anderson, coordinators, "Gemini
Program Launch Systems Final Report: Gemi-
ni/Titan Launch Vehicle; Gemini/Agena Target Vehicle; Atlas SLV-3," Aerospace TOR-
1001 (2126-80)-3, January 1967, pp. H.A.-10,
-11; trip report, Richard J. Crane, "Trip Report—Modified Titan II for Gemini Program,"
13 Aug. 1962, with enclosures, esp. Appendix
A, "Martin Company Organization for Participation in Gemini Program"; Crane and John
H. Boyd, Jr., telephone interviews, 31 Oct.
1972; Hello, telephone interview, 2 Nov. 1972.

[47] Harris, "Launch Vehicle Chronology,"
pp. 6, 9; Wambolt and Anderson, "Launch
Systems Final Report," pp. II.A-I, -12, -15,
-16.-17.

[48] Letter, Detko to Gilruth, "Investigation of the Gemini 6—Agena Target Vehicle (G6-ATV) Failure," 29 Dec. 1965, with enclosure, memo, Detko to Eberhard F. M. Rees, "Gemini 6-Agena Target Vehicle (G6-ATV) Failure Investigation," n.d., with "Gemini/Agena Target Vehicle Program Historical Background" attached; Col. John G. Albert, interview, Patrick AFB, Fla., 26 May 1967; memo, Max Rosenberg to NASA Historian, "Comments on Draft Chapter I-V and XIII-XV, Gemini History," 26 June 1970, with enclosure.

[49] *World Space Directory*, I, No. 1 (Washington, 1962), p. B-180; "GATV Progress Report, September 1964," LMSC-A605200-1, 20 Oct. 1964; letter, Rockwell Hollands to Sally Gates, 26 June 1969.

[50] Chamberlin, "Minutes, February 28, 1962"; letter, Seamans to Brockway McMillan, 29 May 1962; TWX, Friedrich Duerr to SSD, Attn: Albert, 8 June 1962; "Fourth Report on MSFC Activities Covering May 26 to June 25, 1962 to Manned Spacecraft Center"; Quarterly Status Report No. 2, for period ending 31 August 1962, p. 26.

[51] "Fourth Report on MSFC Activities"; Quarterly Status Report No. 2, p. 26; memo, Homer E. Newell to Assoc. Adm., "Standard Atlas Space Launch Vehicles," 15 June 1962; letters, Seamans to Rubel, 19 June and 25 July 1962; memo of agreement, "DOD/NASA Standard ATLAS Space Booster Agreement," signed by Seamans, 5 Sept. 1962, and by Harold Brown, n.d.

[52] Wambolt and Anderson, "Launch Systems Final Report," pp. IV-1, -3.

[53] Chamberlin, interview, Houston, 9 June 1966; "Project Gemini Schedule Analysis," GPO, 4 May 1962; memo, Gerard J. Pesman to Engineering Div., "Paraglider Landing System Design Studies: review and evaluation of final report," 27 Sept. 1961.

[54] "Preliminary Project Development Plan for a Controllable Parachute-Retrorocket Landing System," STG, 21 June 1961; memo, Caldwell C. Johnson to Jack C. Heberlig, 28 Aug. 1961.

[55] Memo, William O. Armstrong to Chief, Flight Operations Div. (FOD), "Review of the development effort of the parawing landing system for the Gemini mission," 9 Feb. 1962; memo, Wayne E. Koons to Chief, FOD, "Operational problems associated with the use of ejection seats for Project Gemini," 23 Feb. 1962; memo, James M. Rutland to Chief, FOD, "Recovery system for Gemini," 21 March 1962.

[56] Memo, Christopher C. Kraft, Jr., to Mgr., GPO, "Recovery operational study and retrieval evaluation tests," 1 March 1962; memos, Kraft to Dir., "Paraglider and ejection seats for Project Gemini," 26 April and 4 May 1962, with enclosures, "The Operational Implications of Paragliders and Ejection Seats in Project Gemini," n.d.

[57] Memo, Lester A. Stewart to Procurement Officer, "Letter Contract NAS 9-167, Paraglider Development Program, Phase II-A," 22 Jan. 1962, with enclosure, "Suggested Revisions to Statement of Work for Letter Contract NAS 9-167"; Crane to North American, Change Notice No. 1, Contract NAS 9-167, 8 March 1962.

[58] Letter, D. K. Bailey to MSC, Attn: G. F. Bailey and Purser, 62MA2243, 1 March 1962, with enclosure, "Monthly Progress Letter No. 3, Paraglider Development Program Phase IIA, 20 January 1962 to 20 February 1962," p. 1; TWX, G. F. Bailey to NAA, 9 March 1962; memo, Ronald C. Bake to A. E. Hyatt, "Change Notice No. 1 to NAS 9-167," 13 March 1962; letter, Norbert F. Witte to MSC, Attn: G. F. Bailey and Purser, 62MA3530, 29 March 1962, with enclosure, "Monthly Progress Letter No. 4, Paraglider Development Program Phase IIA, 20 February 1962 to 20 March 1962," p. 1.

[59] "Monthly Progress Letter No. 4, Phase IIA," p. 1; "Final Report of Paraglider Research and Development Program, Contract NAS 9-1484," North American SID65-196, 19 Feb. 1965, p. 184.

[60] "Monthly Progress Letter No. 4, Phase IIA," pp. 1, 3; "Schedule Analysis," 4 May 1962, p. 2, Chart 4.

[61] Memo, Chamberlin to Gemini Procurement Office, Attn: Bake, "Paraglider Development Program, initiation of Phase II, Part B," GPO-00015, 19 March 1962, with enclosure, "Statement of Work for Phase II, Part B, Prototype Development in the Paraglider Development Program," 19 March 1962.

[62] Memo, Chamberlin to Gemini Procurement Office, Attn: Bake, "Paraglider Development Program, Initiation of Phase II, Part B (1)," GPO-00086, 15 May 1962; letter, Witte to MSC, Attn: Bake, "Contract NAS 9-539, Paraglider Development Program, Phase II, Part B (1), Monthly Progress Letter No. 1," 62MA9042, 8 Aug. 1962; letter, R. L. Thomas to MSC, Attn: Bake, 62MA-7227, 5 July 1962, with enclosure, "Monthly Progress Letter No. 7, Paraglider Development Program, Phase 11A, 20 May 1962 to 20 June 1962," p. 1; letter contract, NAS 9-539, for Paraglider Development Program, Phase II, Part B (1), Advanced Trainer and Prototype Wing Design, 25 June 1962; "Abstract of Meeting on Paraglider Landing System, July 19, 1962," 21 July 1962; negotiated contract, NAS 9-539, for Paraglider

Development Program, Phase II, Part B(1), 31 Oct. 1962.

[63] "Schedule Analysis," 4 May 1962, p. 2; letter, Seamans to Rubel, 10 July 1962, with enclosure, "Program Summary for Gemini," 3d, 10th, and 11th charts.

[64] "Project Gemini Schedule Analysis," GPO, 14 March 1962; "Schedule Analysis," 4 May 1962.

[65] "Schedule Analysis," 5 Jan. 1962; "Official NASA Flight Schedule," NASA Office of Management Reports, approved by Seamans and Dryden 20 March 1962.

[66] "Schedule Analysis," 4 May 1962; "Official NASA Flight Schedule," approved by Seamans and Dryden 25 July 1962.

[67] Quarterly Status Report No. 1 envisioned no serious problems.

[68] Letter, Gilruth to NASA, Attn: Ernest W. Brackett, "Transmittal of Procurement Plan for MK-II Spacecraft for Approval," 6 Dec. 1961, with enclosure, "MK-II Spacecraft Program: Procurement Plan"; Watts TWX, 26 Dec. 1961; DeMarquis D. Wyatt, interview, Washington, 13 Sept. 1966.

[69] U.S. Congress, House, Subcommittee on Manned Space Flight of the Committee on Science and Astronautics, *1963 NASA Authorization: Hearings on H.R. 10100 (Superseded by H.R. 11737)*, 87th Cong., 2d sess., 1962, Webb's remarks on 27 Feb. 1962, pp. 4, 13-14, and Seamans' on 28 Feb. 1962, pp. 102-104.

[70] Chamberlin, activity report, 28 May 1962, p. 17.

## Chapter V

[1] U.S. Congress, Committee on Science and Astronautics, *Astronautical and Aeronautical Events of 1962: Report*, 88th Cong., 1st sess., 12 June 1963, pp. 146-47, 148.

[2] "Project Development Plan for Rendezvous Development Utilizing the Mark II Two Man Spacecraft," MSC, 8 Dec. 1961, p. 21; letter, Ralph C. Hoewing to MSC, Attn: George F. MacDougall, Jr., "Gemini Launch Vehicle Financial Plan," 4 April 1962, with enclosures.

[3] NASA-Defense Purchase Request T-2356-G, signed by Leslie E. Berg, 15 May 1962, with enclosure, "Statement of Work to Be Accomplished under Department of Defense Purchase Request No. T2356G," 14 May 1962.

[4] "Project Development Plan," p. 21; memo, D. Brainerd Holmes to Adm., "Project Gemini Cost Estimates," 29 April 1963, with enclosure, "Status of Project Gemini Cost Estimates."

[5] Memo, Dave W. Lang to Charles F. Bingman, "Weekly Activity Report," 20 April 1962, with enclosure, "Weekly Activity Report, Procurement and Contracts Division, April 16-20, 1962"; "Abstract[s] of Technical Negotiation Meeting[s] on": "Simulators and Trainers, April 19, 1962," 24 April 1962; "Support Plan—MAC Report No. 8580-4, dated February 2, 1962, April 23 and 24, 1962," 2 May 1962; "Associate Contractor Coordination, Engineering Inspections and Incorporation of Government Furnished Equipment, April 24, 1962," 16 May 1962; "Gemini Facility Plans, MAC Report 8580-2, dated 15 March 1962, April 24, 1962," 4 May 1962; "Documentation Plan, MAC Report No. 8580-8, dated 29 January 1962, April 25, 1962," 4 May 1962; "Post Landing and Survival System, April 26, 1962," 27 April 1962; "Programmer/Timer (Time Reference), April 26, 1962," 1 May 1962; "Environmental Control Subsystem, April 26, 1962," 27 April 1962; "Propulsion Systems, April 26, 1962," 1 May 1962; "Environmental Criteria, April 26, 1962," 1 May 1962; "Communication System Specification, April 27, 1962," 1 May 1962; "Crew Station System Specification, April 27, 1962," 4 May 1962; "Pyrotechnics System Specification, April 27, 1962," 4 May 1962; "Guidance and Control System Specification, April 27, 1962," 9 May 1962; "Electrical System Specification, April 27, 1962," 3 May 1962; "Structural Design Criteria, April 28, 1962," 1 May 1962; "Landing System, April 28, 1962," 11 May 1962; "Gemini Spacecraft Performance Specification, revised May 1, 1962," 5 May 1962; "Program Progress Report, May 2, 1962," 8 May 1962; "Test Program, May 7, 1962," 21 May 1962; "Reliability Plan, MAC Report No. 8580-3, dated February 5, 1962, May 8, 1962," 11 May 1962; "Quality Assurance Plan, MAC Report No. 8580-7, dated January 22, 1962. May 9, 1962," 11 May 1962; "Validation Testing, May 18, 1962," 23 May 1962; André J. Meyer, Jr., interview, Houston, 9 Jan. 1967.

[6] "Gemini Spacecraft Cost and Delivery Proposal," MAC Report No. 8791, 18 April 1962, p. 18.

[7] "Project Gemini Schedule Analysis," GPO, 14 March 1962, p. 2.

[8] "Gemini Spacecraft Cost and Delivery Proposal," pp. 5-6; cf. Loyd S. Swenson, Jr., James M. Grimwood, and Charles C. Alexander, *This New Ocean: A History of Project Mercury*, NASA SP-4201 (Washington, 1966), pp. 269-70.

[9] "Abstract of Meeting on Simulators and Trainers, March 28, 1962," 3 April 1962; memo, Harold I. Johnson for all concerned, "Preliminary description of simulators and training equipment expected to be used in Project Gemini," 5 March 1962; Project Gemi-

ni Quarterly Status Report No. 1, for period ending 31 May 1962, pp. 38-39.

10 "Gemini Spacecraft Cost and Delivery Proposal," pp. 17-19.

11 Memo, Lang and Rex L. Ray to Wesley L. Hjornevik, "Appraisal of validity of McDonnell Estimates of Cost of Gemini Contract Work," 17 April 1962, with enclosures; Hoewing letter, 4 April 1962, with Paul E. Purser's annotation, undated.

12 Holmes memo, 29 April 1963.

13 Letter, R. L. Thomas to MSC, Attn: Ronald C. Bake, 62MA-7227, 5 July 1962, with enclosure, "Monthly Progress Letter No. 7, Paraglider Development Program, Phase IIA, 20 May 1962 to 20 June 1962"; letter, Thomas to MSC, Attn: Bake, 62MA-7728, 1 Aug. 1962, with enclosure, "Monthly Progress Letter No. 8, Paraglider Development Program, Phase IIA, 20 June 1962 to 20 July 1962"; letter, Norbert F. Witte to MSC, Attn: Bake, "Contract NAS 9-167, Paraglider Development Program, Phase II, Part A, Monthly Progress Letter No. 9," 62MA10200, 1 Sept. 1962; letter, George W. Jeffs to MSC, Attn: Bake, "Contract NAS9-167, Paraglider Development Program, Phase II, Part A, Monthly Progress Letter No. 10 (21 August—21 September 1962)," 62MA13775, 26 Nov. 1962; letter, H. C. Godman to NASA Office of Manned Space Flight (OMSF), "C-130 Support of NASA Gemini Program (Paraglider Development)," 18 Sept. 1962; TWX, A. A. Tischler to MSC, Attn: Bake, "Preliminary Test Evaluation Review—Full Scale Dummy Drop No. 2," MA21334, 28 Aug. 1962; Quarterly Status Report No. 2, for period ending 31 Aug. 1962, p. 13; Quarterly Status Report No. 3, for period ending 30 Nov. 1962, p. 13; letter, Jeffs to MSC, Attn: Bake, "Contract NAS 9-167, Paraglider Development Program, Phase II, Part A, Monthly Progress Letter No. 12 (21 October—20 November 1962)," 62MA15807, 31 Dec. 1962, p. 6 (with annotation, probably by Bake); memo, Lester A. Stewart to Joe W. Dodson, "Performance by Northrop Ventura in Developing Parachute Systems for Use in Project Gemini," GPO-00493, 13 Dec. 1962.

14 Letter, Paul F. Bikle to STG, Attn: Rodney G. Rose, "Synopsis of Flight Test Portion of Paraglider Development Study—Phase I," 12 Sept. 1961; memo, Stewart et al. to Dir., STG, "Paraglider Development Program; Evaluation of Design Studies; Contract NAS 9-135, Ryan Aeronautical Company; Contract NAS 9-136, North American Aviation, Inc.; Contract NAS 9-137, Goodyear Aircraft Corporation," 22 Sept. 1961.

15 NASA News Release 61-263, "Apollo Contractor Selected," 28 Nov. 1961; Rose, telephone interview, 13 June 1969. A widely known and influential RAND study first published in 1960 had pointed out the dangers of limiting competition between prospective contractors to the design phase instead of continuing it through early development; Charles J. Hitch and Roland N. McKean, *The Economics of Defense in the Nuclear Age* (New York, 1965), p. 251.

16 Witte letter, 62MA10200, 1 Sept. 1962; Jeffs letter, 62MA13775, 26 Nov. 1962, pp. 1-3; letter, Jeffs to MSC. Attn: Bake, "Contract NAS9-167, Paraglider Development Program, Phase II, Part A, Monthly Progress Letter No. 11, 20 September—20 October 1962," 62 MA13843, 26 Nov. 1962, p. 1.

17 TWX, Chamberlin to North American, Attn: Harrison A. Storms, Jr., "One-Half Scale Paraglider Program," GPO-50222, 21 Sept. 1962.

18 Witte letter, 62MA10200, 1 Sept. 1962.

19 Rose interview.

20 Jeffs letters, 62MA13843, 26 Nov. 1962, pp. 1-2, and 62MA15807, 31 Dec. 1962, p. 2.

21 "Final Report of Paraglider Research and Development Program, Contract NAS 9-1484," North American, SID65-196, 19 Feb. 1965, p. 188.

22 Quarterly Status Report No. 1, pp. 20-21; Gordon P. Cress, interview, Burbank, Calif., 5 July 1966.

23 Memo, Chamberlin to Gemini Procurement Office, Attn: James I. Brownlee, "Contract NAS 9-170, Ejection Seat Rocket Catapult—Recommendation for Authorization for Procurement," GPO-00024, 28 March 1962; Arthur H. Atkinson, "Gemini—Major Subcontracts, McDonnell Aircraft Corporation," 3 July 1962.

24 Atkinson, "Gemini Major Subcontracts"; memo, Chamberlin to Gemini Procurement Office, Attn: Berg, "Project Gemini Ejection Seat Development Test Program," GPO-00097, 21 May 1962.

25 Chamberlin, activity report, 28 May 1962, p. 1; Chamberlin memo, GPO-00097, 21 May 1962; Cress interview.

26 [Kenneth F. Hecht], "Comments on Chapter 5, 'Expansion and Crisis,'" [10 Feb. 1970], p. 1; memo, Hecht to Historical Office, "Comments on Chapter 6: The Nadir," 22 Sept. 1970; Hecht, telephone interview, 14 Nov. 1972; memo, Hecht to Mgr., GPO, "Gemini Escape System Management," 26 March 1962; "Abstract of Meeting on Ejection Seats, March 29, 1962," 3 April 1962.

27 "Abstract of Meeting on Ejection Seat Developmental Test Program, May 29, 1962," 4 June 1962; memo, Chamberlin to Dir., "Gemini Weekly Status Report (June 18,

1962)," GPO-00145, 18 June 1962; Quarterly Status Report No. 2, p. 17; Richard S. Johnston, "Life Systems Division Weekly Activities Report, 7/16/62—7/20/62," p. 3; Raymond L. Zavasky, recorder, "Minutes of Senior Staff Meeting, July 27, 1962," p. 4; memo, Richard P. Parten to Chief, Flight Operations Div., "Project Gemini Coordination Meeting on Mechanical Systems," 30 July 1962; memo, Chamberlin to Dir., "Gemini Weekly Status Report (August 6, 1962)," GPO-00257, 6 Aug. 1962; "Abstract of Meeting on Mechanical Systems, August 1-2, 1962," 7 Aug. 1962; Hecht, "Comments on Chapter 5," p. 1.

28 "Abstract of Meeting on Ejection Seats, August 3, 1962," 17 Aug. 1962; TWX, R. W. Miller to MSC, Attn: Chamberlin, "Gemini Ejection Seat Tests," 306-450-23281, 10 Aug. 1962; "Abstract of Meeting on Ejection Seats, August 6-7, 1962," 9 Aug. 1962; Weekly Activity Report for Office of the Director, Manned Space Flight, 5-11 Aug. 1962, MSC, p. 2; memo, Chamberlin to Dir., "Gemini Weekly Status Report (August 13, 1962)," GPO-00263, 13 Aug. 1962; Chamberlin, activity report, 27 Aug. 1962, p. 1.

29 "Abstract of Meeting on Ejection Seats, September 6, 1962," 11 Sept. 1962; TWXs, Miller to MSC, Attn: Chamberlin, "Gemini Ejection Seat Tests," 306-450-23965, 13 Sept. 1962, and 306-450-24240, 28 Sept. 1962; "Abstract of Meeting on Ejection Seats, September 26, 1962," 3 Oct. 1962; Quarterly Status Report No. 3, p. 18.

30 Robert Cohen, "Summary of analysis for selecting the power source for the Gemini Project," Gemini Project Note of January 23, 1962, 27 Jan. 1962.

31 Ibid., pp. 3-4; letter, Walter F. Burke to Wilbur H. Gray, "Selection of Equipment, Contract NAS 9-170, Fuel Cell System," 306-101-142, 23 Feb. 1962, with enclosures, "Chosen System Advantages, General Electric Fuel Cells" and "Substantiation of Selected Vendor Capability"; R. H. Blackmer and G. A. Phillips, "Ion-Exchange Membrane Fuel Cell for Space Vehicle Electric Power," presented at the Society of Automotive Engineers National Aerospace Engineering and Manufacturing Meeting, Los Angeles, 9-13 Oct. 1961; J. L. Schanz and E. K. Bullock, "Gemini Fuel Cell Power Source—First Spacecraft Application," ARS Paper No. 2561-62, presented at the American Rocket Society Space Power Systems Conference, Santa Monica, Calif., 25-28 Sept. 1962; "Fuel Cells for Spacecraft, Including Determination of Fuel Battery Size for Specific Application," brochure by Direct Energy Conversion Operation, General Electric, January 1964, pp. 3-4.

32 Burke letter, 306-101-142; letter, Gray to Burke, "Selection of Equipment, Contract NAS 9-170, Fuel Cell System," NAS/170-265, 21 Feb. 1962.

33 John H. Russell, interview, West Lynn, Mass., 24 April 1968.

34 Memo, George F. Esenwein to George M. Low, "Informal Visit to General Electric Direct Energy Conversion Operation on March 26, 1962 to discuss possible Apollo Fuel Cell Backup and Polymer A' Status," 2 April 1962; James F. Saunders, Jr., telephone interview, 14 Nov. 1972.

35 Letter, Chamberlin to NASA Hq., Attn: Low, "Fuel Cell for Gemini," GPO-00026, 5 April 1962, with enclosures, memo, Cohen to Mgr., Project Gemini, "Status of General Electric Co. Fuel Cell Development for Gemini," 5 April 1962, and Cohen, "Summary of Analysis."

36 Memo, Gray to Chamberlin, "Visit to Direct Energy Conversion Operation, General Electric Co., West Lynn, Mass.," NAS/170-706, 5 Sept. 1962; Quarterly Status Report No. 2, pp. 21-22.

37 R. H. Prause and R. L. Goldman, "Longitudinal Oscillation Instability Study: POGO," Martin ER-13374, December 1964, pp. 1-3; Quarterly Status Report No. 6, for period ending 31 Aug. 1963, fig. 4; Jerome B. Hammack, interview, Houston, 19 Aug. 1966.

38 Zavasky, "Minutes of Senior Staff Meeting[s], July 13, 1962," pp. 1, 3, "July 20, 1962," p. 3, and "July 27, 1962," pp. 1, 3; Prause and Goldman, "POGO Study," p. 3.

39 Zavasky, "Minutes of Senior Staff Meeting[s], July 27, 1962," p. 3, and "August 3, 1962," p. 2; Quarterly Status Report No. 6, fig. 4.

40 Quarterly Status Report No. 2, pp. 24-25; Prause and Goldman, "POGO Study," pp. 3, 20; Zavasky, "Minutes of Senior Staff Meeting, August 10, 1962," p. 4; "Joint Titan II/Gemini Development Plan on Missile Oscillation Reduction and Engine Reliability and Improvement," [Air Force Systems Command], 5 April 1963 (revised 7 May 1963), enclosure 3, "Missile Configuration/Oscillation Summary."

41 Quarterly Status Report No. 2, p. 25.

42 TWX, J. M. Gardner, Jr., to Contracting Officer, "Contract NAS 9-170, Project Gemini, Financial Reporting," 16-JMG-1000, 17 July 1962.

43 Memo, Robert L. Kline to Meyer, "Project Gemini Negotiations with MAC, Letter Contract NAS 9-170," MSC-PG-4-483, 3 Aug. 1962; "Gemini Program, MAC Estimated Cost Summary as of August 8, 1962."

[44] Memo, Kline for GPO, Attn: Mac-Dougall, "Letter Contract NAS 9-170 for Gemini Spacecraft with McDonnell Aircraft Corporation (MAC)," MSC-PG-8-843, 26 Sept. 1962.

[45] "Minutes of ... NASA-SSD Meeting on Cost of Titan II Program, March 1, 1962," 2 March 1962; "Financial Plan—Gemini," SSD, 1 March 1962, with annotations by Richard J. Crane (MSC procurement) showing revised cost estimates supplied by Maj. Roland D. Foley in a telephone call on 10 Sept. 1962.

[46] Letter, Richard C. Dineen to MSC, Attn: Chamberlin, "Budget Requirement for Gemini Launch Vehicle," 4 Oct. 1962, with enclosure, "Gemini Launch Vehicle Budget Estimate," 3 Oct. 1962.

[47] Memo, Kenneth R. Irwin to Gemini Project files, "Fund Requirements for Atlas-Agena," 19 July 1962.

[48] Memo, Holmes to Assoc. Adm., "Atlas-Agena Launch Vehicles for Gemini," 28 May 1962, with Robert C. Seamans, Jr.'s initialed approval dated 29 May 1962; letter, Crane to Marshall, Attn: Floyd M. Clark, "Procurement Request No. 100-62, dated May 1, 1962, Gemini Atlas-Agena," MSC-PG-2-116, 15 June 1962, with enclosure, "Statement of Work for Atlas-Agena Target Vehicles to Be Used in Project Gemini," 1 June 1962; letter, Daniel D. McKee to MSC, Attn: Chamberlin, "Atlas-Agena Launch Vehicles for Gemini," 25 June 1962.

[49] Holmes memo, 29 April 1963.

[50] Memo, Clyde B. Bothmer, executive secretary, to dist., 25 June 1962, with enclosure, "Minutes of the Seventh Meeting of the Management Council, Friday, June 22, 1962," p. 5.

[51] U.S. Congress, House, Report on the Activities of the Committee on Science and Astronautics, committee print, 87th Cong., 2d sess., 1962, p. 4; Astronautical and Aeronautical Events of 1962, pp. 136, 192, 200; U.S. Congress, Senate, Subcommittee of the Committee on Appropriations, Independent Offices Appropriations, 1963: Hearings on H.R. 12711, 87th Cong., 2d sess., 1962, pp. v-xxiii.

[52] TWX, William E. Lilly to MSC, M-C P 9200.023, 28 Sept. 1962.

[53] TWX, Harry W. Oldeg to Kline, "NAS 9-170, Info Regarding MAC Invoice No. 40," 306-19-804, 1 Oct. 1962; TWX, SSD to MSC, Attn: Irwin, SSVLP-2-10-1, 2 Oct. 1962; TWX, Oldeg to MSC, Attn: Kline, "Contract NAS 9-170, Request for Increase in Expenditure Limitation," 306-19-805, 4 Oct. 1962; TWX, Friedrich Duerr to Chamberlin, M-L&M-AP 10-8, 8 Oct. 1962; TWX, Storms to MSC, Attn: Bake, "Letter Contract NAS 9-539," MA26810, 19 Oct. 1962.

[54] Robert L. Rosholt, An Administrative History of NASA, 1958-1963, NASA SP-4101 (Washington, 1966), pp. 134-35.

[55] Memo, Lilly to dist., 25 Sept. 1962, with enclosure, "Minutes of the Tenth Meeting of the Management Council, Friday, September 21, 1962," pp. 6, 7, esp. enclosure 3, "Impact of FY '63 Funding Ceiling on Gemini."

[56] TWX, Holmes to MSC, M-C P 9200.028, 8 Oct. 1962.

[57] Purser, acting recorder, "Minutes of Senior Staff Meeting, October 19, 1962," p. 2.

[58] Memo, Irwin to Budget Br., Financial Mgmt. Div., Attn: Robert M. Weinert, "Comments on First Quarter Funding Review—Gemini Program," 29 Sept. 1962, with enclosure, "Gemini Status of Funds, First Quarter."

[59] TWX, William A. Parker to NASA Hq. Procurement and Supply Div., for Herbert L. Brewer, MSC-PG-4-827, 17 Sept. 1962; letter, Gray to Burke, "Instructions to Project Gemini suppliers; Contract NAS 9-170," NAS/170-770, 2 Oct. 1962; letter, Oldeg to MSC, Attn: Glenn F. Bailey, "Contract NAS 9-170, Gemini, Program Direction Subsequent to NASA/MAC Meeting in Houston on 28 September 1962," 306-16-1282, 8 Oct. 1962; letter, Gardner to MSC, Attn: Atkinson, "Contract NAS 9-170, Gemini, Reduction on Major Vendor Estimated Costs," 306-16-1296, 23 Oct. 1962, with enclosure, "Analysis of Adjustments in Major Vendor Estimated Costs."

[60] Letter, Burke to Gray, "Contract NAS 9-170 Program Direction," 306-09-93, 5 Oct. 1962; Oldeg letter, 306-16-1282, 8 Oct. 1962.

[61] Oldeg letter, 306-16-1282, 8 Oct. 1962; Gardner letter, 306-16-1296, 23 Oct. 1962.

[62] TWX, Chamberlin to Burke, "Contract NAS 9-170, Gemini Vendor Program Reviews," GPO-50242, 3 Oct. 1962; TWX, John Y. Brown to MSC, Attn: Chamberlin, 306-16-1346, 19 Oct. 1962; Purser, "Minutes of Senior Staff Meeting, October 19, 1962," p. 4.

[63] Letter, Gray to Burke, "Notification of Implementation of NASA/MAC Decisions Affecting Suppliers, Contract NAS 9-170," NAS/170-809, 15 Oct. 1962; letter, John Brown to Gray, "Contract NAS 9-170—Implementation of NASA/MAC Decisions Affecting Supplies [sic]," 306-16-1422, 2 Nov. 1962, with enclosure.

[64] TWX, John Brown to MSC, Attn: Chamberlin, "Cancellation of Static No. 1 Vehicle," 306-16-1490, 27 Nov. 1962; TWX, Chamberlin to Burke, "Gemini Reliability and Test Plan Review, December 5-7, 1962," GPO-50418, 14 Dec. 1962; TWX, Chamberlin to Burke, "Cancellation of Static No. 1 Vehicle," GPO-50436, 19 Dec. 1962.

[65] Letter, John Brown to MSC, Attn: Chamberlin, "Proposed Re-Allocation of Gemini Project Orbit Spacecraft," 306-16-1397, 6 Nov. 1962; letter, Burke to MSC, Attn: Chamberlin, "Policy with Respect to Project Orbit—Gemini," 306-09-188, 3 Jan. 1963; TWX, Chamberlin to Burke, "Spacecraft Thermal Qualification Test," GPO-50460, 3 Jan. 1963.

[66] Seventh Semiannual Report to Congress, January 1 through June 30, 1962, NASA (Washington, 1963), p. 133; Zavasky, "Minutes of Senior Staff Meeting[s], July 27, 1962," p. 5, and "August 3, 1962," pp. 1, 4.

[67] TWX, Douglas R. Lord to Chamberlin, 3 Aug. 1962.

[68] Memo, James B. Jackson, Jr., to Project Gemini files, "Telecon between Col. D. D. McKee and G. F. MacDougall on 8-3-62," 23 Aug. 1962; memo, Jackson and Galloway B. Foster, Jr., to Project Gemini files, "Additional data concerning ltr. GPO-00324 dated September 14, 1962," 19 Sept. 1962.

[69] Memo, Calvin C. Guild to Barton C. Hacker, "Gemini History," PD12/M 799-69, 17 June 1969.

[70] Memo, James E. Webb to Abraham Hyatt, no subj., 1 Nov. 1962; memo, Bothmer to dist., 2 Nov. 1962, with enclosure, "Minutes of the Eleventh Meeting of the Management Council, Tuesday, October 30, 1962," p. 5.

[71] Letter, Dineen to MSC, Attn: Chamberlin, "Coordination of Development Plan for Gemini Launch Vehicle System," 24 Oct. 1962, with enclosure, "Development Plan for Gemini Launch Vehicle System," engineering service program PS 920E (rev. of plan dated 23 March 1962), esp. Chart 7; Crane, "Titan II Gemini Launch Vehicle Status Report," 5 Jan. 1963; TWX, Chamberlin to SSD for Dineen, GPO-50302, 19 Oct. 1962.

[72] TWX, Chamberlin to SSD for Dineen, GPO-50304, 22 Oct. 1962, which revises GPO-50302.

[73] TWX, Dineen to MSC, SSVLP-23-10-6, 23 Oct. 1962.

[74] TWX, Chamberlin to SSD for Dineen, GPO-50306, 29 Oct. 1962; "Reliability Test Plan," Martin ER-12258, 15 June 1962.

[75] "Impact of FY 1963 Funding Reduction on Cost and Schedule," SSD presentation 7 Nov. 1962, to Chamberlin, MacDougall, and Hammack; Purser, recorder, "Minutes of Project Gemini Management Panel Meeting . . . , November 13, 1962," p. 3.

[76] TWX, Chamberlin to SSD for Dineen, GPO-50361, 16 Nov. 1962.

[77] Zavasky, "Minutes of Senior Staff Meeting, November 29, 1962," p. 2; "Abstract of Meeting on Launch Vehicle Reprogramming, November 27, 1962," 3 Dec. 1962. On the reliability dispute, see also Loyd S. Swenson, Jr., James M. Grimwood, and Charles C. Alexander, This New Ocean: A History of Project Mercury, NASA SP-4201 (Washington, 1966), pp. 179-80.

[78] "Abstract of Meeting on Launch Vehicle Reprogramming, November 27, 1962"; TWX, Chamberlin to SSD, Attn: Dineen, GPO-50446, 20 Dec. 1962.

[79] "Abstract of Meeting on Launch Vehicle Reprogramming, November 27, 1962"; "Review of Requirements for a Restrained Firing Program," Martin LV-114, 24 Sept. 1962; Chamberlin TWX, GPO-50446, 20 Dec. 1962.

[80] Letter, John G. Albert to Marshall, Attn: Duerr, "Gemini Propulsion," 11 Oct. 1962; letter, Duerr to Albert, "Gemini Target Vehicle Program," 17 Oct. 1962.

[81] TWX, Chamberlin to Duerr, "Atlas/Agena Program," GPO-50294, 23 Oct. 1962; memo, Floyd A. Turner to Chamberlin, "Atlas-Agena Program," 19 Oct. 1962, with enclosure.

[82] TWX, Duerr to Cdr., AF Systems Command, M-L&M-AP 10-17, 23 Oct. 1962; TWX, Duerr to Albert, M-L&M-AP 10-18, 23 Oct. 1962.

[83] Letter, Duerr to Chamberlin, "Budget Limitations for Gemini Target Vehicle," 24 Oct. 1962.

[84] "Abstract of Meeting on Reprogramming Atlas/Agena, October 25, 1962," 31 Oct. 1962; "Medium Space Vehicles Monthly Progress Report, October 1962," LMSC-447186-28, 20 Nov. 1962, p. 8.

[85] Lunar Orbit Rendezvous: News Conference on Apollo Plans at NASA Headquarters on July 11, 1962 (Washington, 1962); "Manned Lunar Landing Mode Comparison," OMSF, 24 Oct. 1962.

[86] Bothmer, "Minutes of the Eleventh Meeting of the Management Council," p. 6.

[87] "Abstract of Meeting on Reprogramming Atlas/Agena, November 2, 1962," 9 Nov. 1962; "Thirteenth Report on MSFC Activities Covering November 1 thru November 16, 1962, to Manned Spacecraft Center," 28 Dec. 1962.

[88] TWX, Duerr to Chamberlin, M-L&M-AP-11-9, 6 Nov. 1962.

[89] TWX, Duerr to Chamberlin, "Funding Requirements for Gemini," M-L&M-AS 11-59, 14 Nov. 1962.

[90] Zavasky, "Minutes of Senior Staff Meeting[s], November 9, 1962," pp. 4-5, and "November 16, 1962," pp. 3-4.

[91] Memo, Charles W. Mathews to Asst. Dir., Research and Development, "Preliminary study of possible rendezvous maneuvers which could be accomplished with a Gemini Spacecraft without the Atlas-Agena vehicle," 28

March 1962, with enclosures; "Abstract of . . . Coordination Meeting (Electrical), May 1, 1962," 2 May 1962; Peggy Dugge and Marvin R. Czarnik, "Practice Rendezvous Mission," McDonnell Guidance and Control Mechanics Design Note No. 1, 7 July 1962; memo, Carl R. Huss to Chief, Flight Operations Div., "Comments and Notes from Gemini Mission Planning and Guidance Meeting Held January 4, 1963 and January 16, 1963," 28 Jan. 1963; Quarterly Status Report No. 4, for period ending 28 February 1963, pp. 22-23.

92 Zavasky, "Minutes of Senior Staff Meeting, November 16, 1962," p. 2.

93 "Abstract of Meeting on Reprogramming Atlas/Agena, November 20, 1962," 27 Nov. 1962; "Monthly Progress Report, November 1962," LMAC-447186-29, 20 Dec. 1962, p. 3; Quarterly Status Report No. 3, p. 32; TWX, Chamberlin to Marshall, Attn: Duerr, "Atlas Agena Program," GPO-50376, 23 Nov. 1962.

94 Memo, Bothmer to dist., 1 Dec. 1962, with enclosure, "Minutes of the Twelfth Meeting of the Management Council, Tuesday, November 27, 1961," p. 2.

95 John W. Finney, "2-Man Earth Orbit Delayed until 1964," The New York Times, 28 Nov. 1962; William Hines, "Revised Gemini Space Flight Plans Could Save Both Time and Money," The Sunday Star, Washington, 2 Dec. 1962; Edward H. Kolcum, "NASA May Cut Agena from Gemini Plan," Aviation Week and Space Technology, 26 Nov. 1962; Zavasky, "Minutes of Senior Staff Meeting, November 29, 1962," p. 6; TWX, SSD to Marshall, SSVR 28-11-254, 28 Nov. 1962; TWX, Marshall to Chamberlin, M-L&M-AS 11-66, 28 Nov. 1962; TWX, Chamberlin to Low, "Gemini Atlas/Agena FY 63 Funding," GPO-50392, 29 Nov. 1962.

96 TWX, Low to MSC, "M-A S 1300.007," M-C P 9200.059, 6 Dec. 1962; Crane, "Gemini Atlas-Agena Program Status Report," 5 Jan. 1963, pp. 2-3; NASA Project Approval Document, Research and Development, 6 Dec. 1962, approved by Seamans.

97 Edward H. Kolcum, "Administration to Ask $6 Billion for NASA," Aviation Week and Space Technology, 10 Dec. 1962, p. 28.

98 TWX, Low to MSC, "M-A 1300.009," M-C P 9200.064, 21 Dec. 1962; Crane, "Gemini Atlas-Agena Program Status Report," p. 3; letter, Chamberlin to Grimwood, 25 March 1974.

99 TWX, Seamans to Marshall and Lewis, Attn: Dirs., 14 Dec. 1962.

100 Letter, Low to MSC, Attn: Robert R. Gilruth, "NASA Atlas/Agena Vehicles," M-M S 1343-540, 28 Dec. 1962; letter, Low to Maj. Gen. Osmond J. Ritland, "NASA Atlas/Agena

Vehicles," M-C S 1343-515, 28 Dec. 1962; letter, Low to Marshall, Attn: von Braun, "NASA Atlas/Agena Vehicles," M-M S 1343-541, 28 Dec. 1962; letter, Low to Kurt H. Debus, M-C S 1343-520, 9 Jan. 1963; TWX, Chamberlin to NASA Hq., Attn: Low, "NASA-Air Force Agreement on Gemini Atlas-Agena Development," GPO-50470, 9 Jan. 1963.

101 Letters, Chamberlin to Albert and Elmer P. Wheaton, GPO-00538 and -00540, 18 Jan. 1963; letter, Chamberlin to Marshall, Attn: Hans H. Hueter, "Gemini Target Vehicle Program," GPO-00531, 18 Jan. 1963; letter, Chamberlin to Lockheed, Attn: Donald E. Forney, GPO-00539, 18 Jan. 1963; memo, Crane for record, "Status Review—Coordination Conference—Atlas-Agena Program," 19 Feb. 1963; letter, Duerr to Chamberlin, 1 March 1963; "Agena Monthly Progress Report for December 1962," Marshall Light and Medium Vehicles Office, p. 1; Ninth Semiannual Report to Congress, January 1—June 30, 1963, NASA (Washington, 1964), p. 76.

102 J[oseph] F. Wambolt and S[ally] F. Anderson, coordinators, "Gemini Program Launch Systems Final Report: Gemini/Titan Launch Vehicle; Gemini/Agena Target Vehicle; Atlas SLV-3," Aerospace TOR-1001(2126-80)-3, January 1967.

103 Letter, Seamans to John H. Rubel, 25 July 1962.

104 Letter, Seamans to Harold Brown, 5 Sept. 1962, with enclosure, "Memorandum of Agreement between the Department of Defense and National Aeronautics and Space Administration: DOD/NASA Standard ATLAS Space Booster Agreement"; memo, D. L. Forsythe to Dep. Dir., Office of Space Sciences, "USAF/GDA Review for Improving Atlas Vehicles and Launch Operations for NASA Missions," 5 Sept. 1962.

105 "Abstract of Meeting on Atlas-Agena, October 18, 1962," 23 Oct. 1962; TWX, Chamberlin to Duerr and Albert, "Surplus Atlas Boosters from Project Mercury," GPO-50301, 19 Oct. 1962; "Abstract of Meeting on Atlas/Agena Reprogramming, November 20, 1962"; TWX, SSD to MSC for Kenneth S. Kleinknecht, "Refurbishing Current Mercury-Atlas Boosters for Use in the Gemini Program," SSVM-30-11-10, 1 Dec. 1962; Crane, "Gemini Atlas-Agena Program Status Report," p. 2.

106 "Abstract of Meeting on Target Vehicle Booster Conversion Study, February 13, 1963," 1 March 1963.

107 Letter, Seamans to Harold Brown, 10 Dec. 1962, with enclosure, memo of agreement, "DOD/NASA Standard ATLAS Space Booster Agreement," 10 Dec. 1962, signed by

Seamans and Brown; TWX, McKee to MSC, Attn: Chamberlain, "Gemini Arget [*sic*-Target] Booster Selection," 7 March 1963; memo, Mathews for Gemini Procurement, Attn: Stephen D. Armstrong, "Statement of Work for Atlas Standard Launch Vehicles, NASA-DOD Purchase Request T-15482-G to AFSSD," GPO-03039-A, 7 Aug. 1963, with enclosure, "Statement of Work for Atlas Standard Launch Vehicles to Be Used in Project Gemini," GP-33, 6 Aug. 1963.

108 "Gemini Spacecraft Status, December 13, 1962," prepared for presentation by Gilruth at the 13th meeting of the Management Council, 18 Dec. 1962.

109 "Project Gemini: Mission 'O' Plan," McDonnell, 12 July 1962; TWX, John Brown to MSC, Attn: Bailey, "Contract NAS 9-170, Gemini, Mission Assignment for Spacecraft No. 1," 16-DAH-1067, 24 July 1962; "Project Gemini Mission Plan: Spacecraft No. 1," McDonnell, 14 Sept. 1962; "Abstract of Meeting [s] on Mission Planning and Guidance, September 14, 1962," 26 Sept. 1962; "Electrical Systems, September 18, 1962," 26 Sept. 1962, and "Mechanical Systems, September 19, 1962," 21 Sept. 1962; "Resume of Outstanding Events in the Gemini Project for the Past Month (November 23, 1962)," prepared for Gilruth's presentation to the 12th meeting of the Management Council, 27 Nov. 1962; Quarterly Status Report No. 3, p. 39.

110 "Official Flight Schedule," NASA Office of Management Reports, approved by Seamans 20 Dec. 1962; Purser, "Minutes of Project Gemini Management Panel Meeting . . . , December 20, 1962," p. 1; "Gilruth Sees Gemini Shot Delayed 3 to 4 Months," *The Washington Post*, 17 Nov. 1962.

## Chapter VI

1 MSC Announcement No. 135, "Change in Location of Gemini Project Office," 12 Dec. 1962; MSC Announcement No. 153, "Change in Location of Chief, Mercury/Gemini Procurement and Gemini Procurement Office," 7 Feb. 1963; "Manned Spacecraft Center, Houston, Texas, Interim Facilities," MSC, as of 15 Aug. 1963, pp. 193, 201.

2 U.S. Congress, House, Committee on Science and Astronautics, *Astronautical and Aeronautical Events of 1962: Report*, 88th Cong., 1st sess., 12 June 1963, pp. 17-18; letter, Max Rosenberg to NASA Historian, "Comments on Draft Chapter I-V and XIII-XV, Gemini History," 26 June 1970, with enclosure, "NASA Draft Gemini History, Comments on Chapters I-V; XIII-XV," pp. 5-7.

3 "NASA Draft Gemini History, Comments," p. 5; letter, Brockway McMillan to Eugene M. Emme, 1 May 1970.

4 *Astronautical and Aeronautical Events of 1962*, p. 259; Phillip J. Klass, "USAF Halts Saint Work; Shifts to Gemini," *Aviation Week and Space Technology*, 10 Dec. 1962, p. 36; [Henry T. Simmons], "U.S. Space Program: Streamlined and Stretched Out," *Newsweek*, 17 Dec. 1962, pp. 72, 74.

5 Chap. IV, "The Prime Contracts."

6 Memo, Donald H. Heaton to Daniel D. McKee and John H. Disher, "SSD/Aerospace Visit to Houston," 26 Oct. 1962; "Conference on Project Gemini: Agenda; SSD-Aerospace-NASA, November 8, 1962," MSC; "Attendance List, SSD/Aerospace/NASA-Gemini Meeting, Nov. 8, 1962"; letter, Paul E. Purser to James M. Grimwood, 12 May 1970; Robert C. Seamans, Jr., interview, Washington, 26 May 1966.

7 Seamans interview; McMillan letter, 1 May 1970.

8 W. Fred Boone, "Project Gemini Talking Paper," 9 Jan. 1963; memo, Boone to Emme, "Review of the Draft Chapter 5 (Expansion and Crisis), Gemini Narrative History," 17 April 1970.

9 Memo, George L. Simpson, Jr., to Boone, no subject, 8 Jan. 1963; memo, Paul G. Dembling to Dep. Assoc. Adm. for Defense Affairs, "Gemini Program—Transfer to Air Force," 8 Jan. 1963; memo, Arnold W. Frutkin to Dep. Assoc. Adm. for Defense Affairs, "International considerations re transfer of Project Gemini to USAF," 8 Jan. 1963; memo, Edmond C. Buckley to Dep. Assoc. Adm. for Defense Affairs, "Ramifications of DOD Absorption of the GEMINI Program," 8 Jan. 1963.

10 Boone, "Project Gemini Talking Paper," p. 3.

11 McMillan letter, 1 May 1970; "NASA Draft Gemini History, Comments," p. 5.

12 Boone untitled commentary on the planning meeting of 9 January 1963, on joint DOD/NASA management of Gemini, 10 Jan. 1963.

13 Letter, Robert S. McNamara to James E. Webb, 12 Jan. 1963, with enclosure, "Agreement between the National Aeronautics and Space Administration and the Department of Defense Concerning the Gemini Program," signed by McNamara on 12 Jan.

14 Boone, draft reply (for Webb) to McNamara letter, 14 Jan. 1963; memo, William A. Fleming to Dep. Assoc. Adm. for Defense Affairs, "Comments on the Proposal by DOD for Joint Management of Gemini," 15 Jan. 1963.

15 "Agreement between the National Aeronautics and Space Administration and the Department of Defense Concerning the Gemini Program," signed by Webb and McNamara, 21 Jan. 1963; cf. memo, Seamans and John H. Rubel to Sec. of Defense and NASA Adm., "NASA/DOD Operational and Management Plan for Accomplishing the GEMINI (formerly MERCURY MARK II) Program," 29 Jan. 1962.

16 Willis H. Shapley, "Gemini Notes," [19 Jan. 1963]; Edward H. Kolcum, "Administration to Ask $6 Billion for NASA," Aviation Week and Space Technology, 10 Dec. 1962, pp. 27-28.

17 Memo, Shapley for Webb, no subject, 21 Jan. 1963, with enclosure, memo, Shapley to Dir., Bureau of the Budget, "Proposed new DOD-NASA arrangements on GEMINI program," 21 Jan. 1963; "Agreement concerning the Gemini Program," 21 Jan. 1963.

18 NASA News Release 63-12 (released simultaneously by DOD as 84-63), "NASA-DOD Gemini Agreement," 22 Jan. 1963; NASA News Release 63-11 (released simultaneously by DOD), "NASA-DOD Atlantic Missile Range Agreement," 22 Jan. 1963; "Agreement between The Department of Defense and The National Aeronautics & Space Administration Regarding Management of The Atlantic Missile Range of DoD and The Merritt Island Launch Area of NASA," signed by McNamara and Webb, 17 Jan. 1963.

19 Memo, Frutkin to Seamans, "Guidelines for Project Gemini as confirmed to Department of State," 8 Feb. 1963; letter, Seamans to McMillan, 13 March 1963.

20 Frank G. McGuire, "McNamara Spells Out AF Gemini Role," Missiles and Rockets, 1 April 1963, p. 15; Seamans, "DOD Participation in the Gemini Program," NASA Position Paper, 30 April 1963.

21 "Minutes of the First Meeting, Gemini Program Planning Board [GPPB], Friday, February 8, 1963," with enclosures.

22 NASA Negotiated Contract, NAS 9-167, "Paraglider Development Program, Phase II, Part A, System Research and Development," 9 Feb. 1962, signed by Glenn F. Bailey; letter, George W. Jeffs to Robert D. Gilruth, 28 Dec. 1962, with enclosures, briefing charts; NASA Negotiated Contract, NAS 9-539, "Paraglider Development Program, Phase II, Part B(1)," 31 Oct. 1962, signed by Bailey and (for North American) L. L. Waite.

23 Letter, Jeffs to MSC, Attn: Ronald C. Bake, "Contract NAS 9-167, Paraglider Development Program, Phase II, Part A, Monthly Progress Letter No. 12 (21 October—20 November 1962)," 62MA15807, 31 Dec. 1962, p.

2; letter, Jeffs to MSC, Attn: Bake, "Contract NAS 9-167, Paraglider Development Program, Phase II, Part A, Monthly Progress Letter No. 13 (21 November—20 December 1962)," 63MA1040, 18 Jan. 1963, p. 2; TWX, A. A. Tischler to James A. Chamberlin, "Preliminary Flight Test Report on Flight 6-3A-B/B/First HSTV Deployment," MA33082, 17 Dec. 1962; TWX, Tischler to Chamberlin, "Test Evaluation Review—Half Scale Paraglider Deployment Flight No. 1," MA33583, 21 Dec. 1962; "Flight Test No. 6, Phase II, Part A, Paraglider Development Program," North American, SID 62-1060-6, 10 Jan. 1963, pp. 9, 10.

24 Letter, Jeffs to MSC, Attn: Bake, "Contract NAS 9-167, Paraglider Development Program, Phase II, Part A, Monthly Progress Letter No. 14," 63MA3058, 28 Feb. 1963, pp. 1-2; TWX, Tischler to MSC, Attn: Chamberlin, "Preliminary Test Evaluation Review—Paraglider Deployment Flight Test No. 2," MA01569, 16 Jan. 1963; "Flight Test No. 7, Phase II, Part A, Paraglider Development Program," North American, SID 62-1060-7, 1 Feb. 1963, pp. 9-10; Project Gemini Quarterly Status Report No. 4, for period ending 28 Feb. 1963, p. 10; "Abstract of Meeting on Paraglider Landing System, January 17, 1963," 23 Jan. 1963; "MSC Status Report," prepared for Gilruth's presentation at the 14th Management Council Meeting, 29 Jan. 1963, pp. 36-37.

25 "Abstract of Meeting on Paraglider Landing System, February 6, 1963," 8 Feb. 1963; "Abstract of Meeting on Paraglider Landing System, February 13, 1963," 18 Feb. 1963; memo, Chamberlin to Robert L. Kline, "Continuation of Phase II, Part A, Paraglider Development Program, Contract NAS 9-167," GPO-00618, 20 Feb. 1963.

26 Letter, Gilruth to NASA Hq., Attn: Ernest W. Brackett, "Transmittal of a Procurement Plan and Request for Proposal for the Gemini Paraglider Program, Phase III, covering Astronaut Training, Balance of Flight Test Program and Paraglider Hardware Production," 9 Oct. 1962, with enclosures; Chamberlin, activity report, 21 Dec. 1962—24 Jan. 1963, p. 4; letter, Jeffs to MSC, Attn: Kline, "Contract NAS 9-539, Amendment No. 6, Paraglider Landing System for Project Gemini, Phase III, Monthly Progress Letter No. 1, 21 January through 20 February 1963," 63MA4873, 3 April 1963, p. 1; Bailey to North American, "Amendment Number 6 to Letter Contract NAS 9-539," GP-17, 27 Dec. 1962, with enclosure, "Exhibit 'A': Statement of Work for a Paraglider Landing System for Project Gemini"; Quarterly Status Report No. 4, p. 11; letter, Harrison A. Storms, Jr., to

MSC, Attn: Bake, "Transmittal of Cost Proposal for Gemini Paraglider Program, Phase III," 63MA142, 31 Jan. 1963, with enclosure, "Cost Proposal and Supporting Data for Gemini Paraglider Landing System, Phase III," 31 Jan. 1963; "Technical Proposal for a Paraglider Recovery System for Gemini Spacecraft," North American, SID 63-46-1, 31 Jan. 1963; "Management Proposal for a Paraglider Recovery System for Gemini Spacecraft," North American, SID 63-46-2, 31 Jan. 1963; Chamberlin, activity report, 30 Dec. 1962—5 Jan. 1963, p. 1; memo, Chamberlin to Kline, "Contract NAS 9-539," GPO-00594, 8 Feb. 1963.

27 Memo, Chamberlin to Gemini Procurement Office, "Phase III, Part 2 of the Gemini Paraglider Landing System Progress," 20 Feb. 1963.

28 "Abstract of Meeting on Paraglider Landing System, March 13, 1963," 15 March 1963; TWX, Tischler to MSC, Attn: Chamberlin, "Preliminary Flight Test Report—Half Scale Paraglider Deployment Test No. 3, Contract NAS 9-167," 14 March 1963; "Flight Test No. 8, Phase II, Part A, Paraglider Development Program," North American, SID 62-1060-8, 25 March 1963; letter, Jeffs to MSC, Attn: Kline, "Contract NAS 9-167, Paraglider Development Program, Phase II, Part A, Monthly Progress Letter No. 16, 21 February through 31 March 1963," 63MA5879, 23 April 1963, p. 3.

29 Quarterly status report No. 3, for period ending 30 Nov. 1962, p. 28; Raymond L. Zavasky, recorder, "Minutes of Senior Staff Meeting, December 14, 1962," p. 4; R. H. Prause and R. L. Goldman, "Longitudinal Oscillation Instability Study: POGO," Martin ER-13374, December 1964, p. 3.

30 Memo, Clyde B. Bothmer, executive secretary, to dist., 27 Dec. 1962, with enclosure, "Minutes of the Thirteenth Meeting of the OMSF Management Council, held on Tuesday, December 18, 1962," p. 2.

31 "Titan II Post-Flight Briefing," Martin, n.d., for Missile N-13 flight on 19 Dec. 1962; Purser, recorder, "Minutes of Project Gemini Management Panel Meeting . . . , December 20, 1962," p. 2; Zavasky, "Minutes of Senior Staff Meeting, January 4, 1963," p. 5; "Abstract of Meeting on Titan II, January 4, 1963," 9 Jan. 1963; Prause and Goldman, "POGO Study," p. 3.

32 "MSC Status Report," for 29 Jan. 1963, p. 39; "Titan II Post-Flight Briefing," Martin, n.d., for Missile N-15 flight on 10 Jan. 1963.

33 "Briefing" for Missile N-15; "Aerospace Corporation Annual Report, Fiscal 1962-1963," n.d.; letter, James A. Marsh to Richard C. Dineen, "GEMINI Launch Vehicle Engine Status and Its Effect upon the GEMINI Program," 1962.1-115, 23 Aug. 1962.

34 Zavasky, "Minutes of Senior Staff Meeting, January 11, 1963," pp. 2-3.

35 "Joint Titan II/Gemini Development Plan on Missile Oscillation Reduction and Engine Reliability and Improvement," [Air Force Systems Command], 5 April 1963 (revised 7 May 1963), p. 5; Purser, "Project Gemini Management Panel Meeting . . . , February 15, 1963," p. 4; Quarterly Status Report No. 4, pp. 27-28.

36 TWX, Chamberlin to SSD, Attn: Dineen, GPO-50671, 13 March 1963; TWX, Chamberlin to Dineen, GPO-50673, 14 March 1963.

37 "Minutes of the Third Meeting, Gemini Program Planning Board, Thursday, March 7, 1963"; memo, Seamans to D. Brainerd Holmes, "1. Titan II Booster Problem; 2. Paraglider Deployment," 21 March 1963.

38 Seamans memo, 21 March 1963; Purser, notes on meeting with McNamara and Seamans, 14 March 1963; Chamberlin, activity report, 11-17 March 1963, p. 2.

39 "Minutes of First GPPB Meeting," enclosure 1; memo, Holmes to Adm., "Project Gemini Cost Estimates," 29 April 1963, with enclosure.

40 Holmes memo, 29 April 1963, with enclosure, "Status of Project Gemini Cost Estimates [as of March 8, 1963]"; U.S. Congress, House, Committee on Science and Astronautics, *1964 NASA Authorization: Hearings on H.R. 5466 (Superseded by H.R. 7500)*, 88th Cong., 1st sess., 1963, p. 64.

41 André J. Meyer, Jr., interview, Houston, 9 Jan. 1967; memo, Holmes to Assoc. Adm., "Problems associated with Project Gemini," 25 March 1963.

42 MSC Announcement No. 168, "New Assignment of Personnel," 19 March 1963; MSC Announcement No. 125, "Appointment of Deputy Assistant Director for Engineering and Development," 30 Nov. 1962; Loyd S. Swenson, Jr., James M. Grimwood, and Charles C. Alexander, *This New Ocean: A History of Project Mercury*, NASA SP-4201 (Washington, 1966), pp. 79, 115; MSC Announcement No. 11, "Designation of Chief, Flight Operations Division," 17 Jan. 1962.

43 Meyer interview; James T. Rose, interview, St. Louis, 13 April 1966; letter, Chamberlin to Grimwood, 26 March 1974, with comments.

44 Zavasky, "Minutes of Senior Staff Meeting, March 22, 1963," p. 5; "Problems—Nov. 62 Plan," 25 March 1964, table in material compiled for "The First Gemini Executives' Meeting," 27 March 1964, Tab C.

*437*

45 Zavasky, "Minutes of Senior Staff Meeting, April 12, 1963," p. 4; Quarterly Status Report No. 5, for period ending 31 May 1963, pp. 50-51, 58; Purser, "Minutes of Project Gemini Management Panel Meeting . . . , May 2, 1963," pp. 2-3.

46 MSC Weekly Activity Report for Office of the Dir., Manned Space Flight, 2-8 June 1963, p. 2; memo, Walter C. Williams to Acting Mgr., GPO, "Third Gemini flight," 6 June 1963; "Resume of Week's Activities, [9-14 June 1963]," [GPO], p. 1; MSC Consolidated Activity Report for Office of the Dir., Manned Space Flight, 19 May—15 June 1963, p. 72; "Abstract of Meeting on Trajectories and Orbits, July 3, 1963," 9 July 1963.

47 Purser, "Management Panel Meeting, May 2, 1963," p. 3; "DOD-NASA Ad Hoc Study Group, Air Force Participation in Gemini," with errata, Final Report, 6 May 1963; memo, Seamans and McMillan for record, "Acceptance of the Joint NASA/DOD Ad Hoc Study Group Final Report on Air Force Participation in Gemini, Dated May 6, 1963," with enclosures, "Gemini Launches Master Schedule," as of 2 May 1963, and "Gemini Experiment Payload Potential."

48 Zavasky, "Minutes of Senior Staff Meeting, April 26, 1963," p. 5; Purser, "Management Panel Meeting, May 2, 1963," pp. 2-3; Zavasky, "Minutes of Senior Staff Meeting, May 3, 1963," p. 4.

49 1964 NASA Authorization, p. 584.

50 Memo, Bothmer to dist., 3 May 1963, with enclosure, "Minutes, OMSF Management Council, April 30, 1963," p. 6.

51 Memo, Holmes to Seamans, "Comments on the Status of the Gemini Development," 3 May 1963; Boone, "Statement Regarding the Revised Gemini Schedule," 10 May 1963.

52 1964 NASA Authorization, p. 1204.

53 U.S. Congress, Senate, Committee on Aeronautical and Space Sciences, NASA Authorization for Fiscal Year 1964: Hearings on S. 1245, 88th Cong., 1st sess., 1963, p. 775.

54 "Gemini Slippage Due to Variety of Causes," Aviation Week and Space Technology, 22 July 1963, p. 177.

55 TWX, R. S. Maynard to MSC, Attn: Bake, "Funding Status, Contract NAS 9-167," MA03449, 2 Feb. 1963; memo, Kline to GPO, Attn: James B. Jackson, Jr., "NAA Wire MA03449 dated February 14 [sic], 1963," APCMG-80-120, 16 Feb. 1963; Change Notice No. 6, Contract NAS 9-167, "Gemini Paraglider Program," 12 March 1963; memo, Kline to George F. MacDougall, Jr., "Contract NAS 9-167 for Phase II, Part A of the Gemini Paraglider Program," APCMG-80-305, 1 April 1963; Zavasky, "Minutes of Senior Staff Meeting,

March 29, 1963," p. 5; TWX, Maynard to MSC for Kline, MA10412, 8 April 1963.

56 Charles W. Mathews, activity report, 25-31 March 1963; Kline memo, APCMG-80-305, 1 April 1963; TWX, R. L. Stottard to Kline, "Contracts NAS 9-167 and NAS 9-539, Gemini Paraglider Program," 10 April 1963.

57 "Technical Proposal for a Paraglider Landing System," North American, SID 63-606-1, 27 May 1963; "Business Management Proposal for a Paraglider Landing System," North American, SID 63-606-2, 27 May 1963; Storms to MSC, Attn: Stephen D. Armstrong, "Contract NAS 9-1484, Paraglider Landing System Research and Development Program, Transmittal of the Final Fee Settlement Proposal," 65MA3479, 18 March 1965, with enclosure, "A Final Fee Settlement Proposal for Contract NAS 9-1484," 18 March 1965, p. V-26.

58 Purser, "Management Panel Meeting, May 2, 1963," p. 3; Zavasky, "Minutes of Senior Staff Meeting, May 3, 1963," p. 4; Quarterly Status Report No. 5, p. 51; TWX, John Y. Brown to MSC, Attn: Mathews, "Mission Definitions for Program Planning," 306-16-2199, 16 April 1963; letter, Mathews to Walter C. Burke, Attn: Brown, GPO-00887, 24 May 1963, with enclosure, launch schedule and mission definition.

59 TWX, Mathews to Jeffs, Attn: Tischler, and to Burke, Attn: Robert N. Lindley, GPO-50774, 12 April 1963; Amendment No. 8, Contract NAS 9-167, "Paraglider Development Program Phase II, Part A," 12 April 1963; memo, Mathews to Kline, "Contracts NAS 9-167 and NAS 9-539, Paraglider Landing System," GPO-00767, 24 April 1963; TWX, Jeffs to MSC for Mathews, "Re-Directed Gemini Paraglider Program," MA12552, 30 April 1963.

60 Memo, Mathews to Kline, "Initiation of a Development Program for a Paraglider Landing System," GPO-00807, 29 April 1963; Mathews, activity report, 5-11 May 1963, p. 1; letter, Kline to North American, 3 May 1963, with enclosure, letter contract NAS 9-1484 for Paraglider Landing System, accepted by Waite, 5 May 1963; Mathews, activity report, 28 April—4 May 1963, p. 2; letter, Gilruth to Dir., Flight Research Center, "Participation of Flight Research Center in Paraglider Flight Test Program," GPO-00851, 6 May 1963; memo, Bailey to Contract NAS 9-167 File, no subject, APCM-71-1507, 9 May 1963; memo, Bailey to Contract NAS 9-539 File, no subject, APCM-71-1508, 9 May 1963; Consolidated Activity Report, 28 April—18 May 1963, p. 33.

61 Kline, "Summary of Negotiations," 24 July 1963; NASA Negotiated Contract, NAS 9-

1484, "Paraglider Landing System Research and Development Program," 12 July 1963, signed by Waite (26 July) and Kline (5 Aug.); Supplemental Agreement No. 5, NAS 9-167, 12 July 1963; Supplemental Agreement No. 6, NASA 9-539, 12 July 1963.

[62] Holmes memo, 25 March 1963.

[63] Zavasky, "Minutes of Senior Staff Meeting, March 22, 1963," p. 5; TWX, Mathews to Dineen, GPO-50770, 28 March 1963.

[64] Zavasky, "Minutes of Senior Staff Meeting, March 29, 1963," p. 4; memo, Edmund E. Novotny for record, "Briefing on TITAN II/ GEMINI Problems, 29 March, 1963," 2 April 1963; memo, George M. Low to Assoc. Adm., "Titan II Fix Program," 5 July 1963; "Joint Titan II/Gemini Development Plan," p. 5.

[65] Low memo, 5 July 1963; "Joint Titan II/Gemini Development Plan."

[66] "Minutes of the Fifth Meeting, Gemini Program Planning Board, Monday, May 6, 1963"; memo, Boone to Webb and Hugh L. Dryden, "Funds for Martin Company to study Titan II difficulties," 9 May 1963; memo, Boone to Webb, subject as above, 20 May 1963; letter, Novotny to McKee, "TITAN II/ GEMINI 'Get Well' Program," 28 May 1963; memo, McKee to Dep. Dir. (Programs), OMSF, "Titan II Fixes," 28 May 1963; memo, Boone to Seamans, "Status of funds for the Titan II fixes and general improvement program," 20 Sept. 1963.

[67] Memo, Seamans and McMillan to Sec. of Defense and Adm., NASA, "Recommendations by the Gemini Program Planning Board," 29 May 1963.

[68] Memo, McNamara to Co-Chairmen of the Gemini Program Planning Board (GPPB), "Recommendation of the GEMINI Program Planning Board," 20 June 1963; memo, McNamara to Sec. of the Air Force, "Recommendation of the GEMINI Program Planning Board," 20 June 1963; memo, Webb to Co-Chairmen of the GPPB, "Recommendations by the Gemini Program Planning Board," 24 June 1963.

[69] John J. Gabrik, "Titan II Post Flight Briefing Report," n.d., for Missile N-17, flight on 24 May 1963; Quarterly Status Report No. 5, pp. 41-42; Mathews, activity report, 20-25 May 1961; memo, Sheldon Rubin to Dineen, "Results of Analysis of N-25 Configuration on Aerospace Analog Model of POGO," Aerospace 63-1944-51, 15 Oct. 1963; memo, David B. Pendley to Chief, Flight Operations Div., "Titan II Coordination Meeting of June 14, 1963," 17 June 1963; "Abstract of Meeting on Titan II, June 14, 1963," 19 June 1963.

[70] "Titan II Post Flight Briefing Report," n.d., Martin, for Missile N-20, flight on 29 May

1963; Pendley memo, 17 June 1963; "Titan II Meeting, June 14, 1963"; Mathews, activity report, 2-8 June 1963, p. 2; Prause and Goldman, "POGO Study," p. 4.

[71] "GLV Analysis of Titan II Flights," Martin, n.d., for Missile N-22, 20 June 1963; Quarterly Status Report No. 5, pp. 41-42.

[72] Purser, "Minutes of Project Gemini Management Panel Meeting . . . , June 27, 1963," pp. 4-5; Low memo, 5 July 1963, with handwritten annotation by Low; Mathews, activity report, 30 June—6 July 1963, p. 2; letter, Holmes to Osmond J. Ritland, 11 July 1963; memo, Pendley to Chief, Flight Operations Div., "Titan II Coordination Meeting of July 12, 1963," 15 July 1963; "Aerospace Annual Report, Fiscal 1962-1963."

## Chapter VII

[1] Astronautics and Aeronautics, 1963: Chronology on Science, Technology, and Policy, NASA SP-4004 (Washington, 1964), pp. 241, 244; Henry Tanner, "Record in Space Set by Bykovsky," The New York Times, 19 June 1963; Jonathan Spivak, "U.S. Scientists Believe Launching Error Aborted Soviet Plan for Space Rendezvous," The Wall Street Journal, 19 June 1963.

[2] "Minutes of the Sixth Meeting, Gemini Program Planning Board [GPPB], Friday, June 28, 1963"; letter, Robert R. Gilruth to NASA Hq., Attn: Robert C. Seamans, Jr., "Gemini Launch Vehicle Specifications and Requirements," GPO-02011-LV, 1 Aug. 1963, with enclosure, subject as above; TWX, Charles W. Mathews to SSD for Col. Richard C. Dineen, GPO-51110, 18 July 1963.

[3] L. J. Rose, "Titan II Post Flight Briefing Report," n.d., for Missile N-19, 13 May 1963; MSC Weekly Activity Report for Office of the Dir., Manned Space Flight, 28 July—3 Aug. 1963, pp. 2-3; memo, W. Fred Boone to Seamans, "August 1, 1963, Meeting on the Gemini Launch Vehicle Specifications," 2 Aug. 1963; memo, Harris F. Scherer, Jr., to GPO, Attn: Willis B. Mitchell, "Interim Report on the physiological tolerance aspects of the Gemini 'Pogo' vibration study conducted at Ames Research Center," 29 Aug. 1963; memo, Scherer to Chief, Crew Systems Div., "History of the Gemini Vibration Study," 13 Sept. 1963, with enclosure; memo, Richard S. Johnston to Dir., "Results of Gemini 'Pogo' vibration tests carried out at Ames Research Laboratory," 1 Oct. 1963; Project Gemini Quarterly Status Report No. 6, for period ending 31 Aug. 1963, p. 78; "Gemini Launch Vehicle Specifications and Requirements."

4 "Gemini Launch Vehicle Specifications and Requirements"; J[oseph] F. Wambolt and S[ally] F. Anderson, coordinators, "Gemini Program Launch Systems Final Report: Gemini/Titan Launch Vehicle; Gemini/Agena Target Vehicle; Atlas SLV-3," Aerospace TOR-1001 (2126-80)-3, January 1967, p. II.E-17.

5 "Gemini Launch Vehicle Specifications and Requirements"; Quarterly Status Report No. 5, for period ending 31 May 1963, pp. 41-42; letter, Gilruth to Maj. Gen. Ben I. Funk, 26 July 1963.

6 Gilruth letter, 26 July 1963; "Gemini Launch Vehicle Specifications and Requirements"; Walter C. Williams, interview, El Segundo, Calif., 15 May 1967; Ray C. Stiff, Jr., interview, Sacramento, Calif., 10 May 1967.

7 "Minutes of the Seventh Meeting, Gemini Program Planning Board, Monday, August 5, 1963"; letter, Seamans to Brockway McMillan, 29 Aug. 1963, with enclosure, "Gemini Launch Vehicle Specifications and Requirements," 21 Aug. 1963.

8 Letter, Funk to Gilruth, "NASA Manned Spacecraft Center Tour," 8 Aug. 1963; Donald T. Gregory, recorder, "Minutes of Senior Staff Meeting, August 9, 1963," p. 5.

9 Letter, Lt. Gen. Howell M. Estes, Jr., to Seamans, "Titan II/Gemini Program Status Summary," 18 Sept. 1963, with enclosures; letter, Brig. Gen. W. E. Leonhard to Hq. NASA (Seamans), "Titan II/Gemini Program Status Summary," 8 Oct. 1963, with enclosure; letter, Lt. Col. John J. Anderson to Seamans, 21 Oct. 1963, with enclosure, "Statement of Work, Titan II Augmented Engine Improvement Program," 3 Oct. 1963; memo, George E. Mueller to Adm., "Development of the Gemini Launch Vehicle," 6 Dec. 1965, with enclosure, "The Gemini Launch Vehicle Story," n.d.

10 "Minutes of the Eighth Meeting, Gemini Program Planning Board, Friday, September 6, 1963."

11 Raymond L. Zavasky, recorder, "Minutes of Senior Staff Meeting, July 12, 1963," p. 6; Weekly Activity Report, 28 July—3 Aug. 1963, p. 3; memo, William C. Schneider to MSC, Attn: Mathews, "Project Gemini Action Items from OMSF Status Review of June 20, 1963," 1 July 1963; TWX, Mathews to NASA Hq., Attn: Schneider, "Backup Boilerplate Spacecraft for Gemini Mission Number One," GPO-54022-A, 26 July 1963; memo, Mathews to Asst. Dirs., Engineering and Development and Administration, "Request for Engineering and Procurement Support in Preparing a Boilerplate Spacecraft as a Gemini Flight Article," GPO-04013-S, 14 Aug. 1963.

12 "Minutes of the Seventh GPPB Meeting"; TWX, Mathews to Walter F. Burke, GPO-53011-S, 6 Aug. 1963; Gregory, "Senior Staff Meeting, August 9, 1963," p. 4; letter, Col. Ralph C. Hoewing to MSC, Attn: Richard E. Lindeman, "Budget for Gemini Launch Vehicle," 11 Sept. 1963, with enclosures; memo, Lindeman to Gemini Cost Files, no subject, 9 Aug. 1963; Glenn F. Bailey to McDonnell, "Change Notice No. 2," 21 Aug. 1963; Bailey to McDonnell, "Change Notice No. 5," 11 Sept. 1963; Bailey to McDonnell, "Change Notice No. 6," 16 Sept. 1963; memo, Mathews to Gemini Procurement Office, Attn: Robert L. Kline, "Contract NAS 9-170, Change Notice No. 2," GPO-03082-A, 3 Sept. 1963.

13 "Minutes of the Eighth GPPB Meeting"; TWX, Mathews to Dineen, GPO-54150-A, 6 Sept. 1963; Zavasky, "Minutes of Senior Staff Meeting, September 13, 1963," p. 5; memo, Williams to Goddard Space Flight Center, Attn: Harry G. Gross, "Manned space projects," [25 Sept. 1963]; TWX, Williams to Maj. Gen. Leighton I. Davis, "NASA-MSC mission flight schedule," MFS 001, 8 Oct. 1963; Quarterly Status Report No. 7, for period ending 30 Nov. 1963, p. 3.

14 "Abstract of Meeting on Boilerplate Flight Article, September 24, 1963," 26 Sept. 1963; TWX, Mathews to Burke, GPO-54223-A, 30 Sept. 1963; "Abstract of Meeting on Boilerplate Flight Article Scheduling, October 8, 1963," 11 Oct. 1963; "Abstract of Meeting on Boilerplate Flight Article, November 20, 1963," 27 Nov. 1963; Weekly Activity Report, 8-14 Dec. 1963, p. 1; TWXs, Walter J. Kapryan to MSC for Mathews, AMR 01-15-74, 15 Jan., AMR 01-28-83, 28 Jan., and AMR 02-17-94, 17 Feb. 1964.

15 "The Gemini Launch Vehicle Story," p. 2; Weekly Activity Report, 18-24 Aug. 1963, p. 2; memo, Scott H. Simpkinson to Mgr. GPO, "Final MDS Piggyback Inspection on Titan #N-29, at Denver, Colo.," GPO-00749, 8 April 1963; memo, Christopher C. Kraft, Jr., to Chief, MSC Operations Support Office, "Titan II Malfunction Detection System 'Piggyback' Program," 10 April 1963, with enclosures; memo, David B. Pendley to Chief, Flight Operations Div. (FOD), "Titan II Coordination Meeting of June 14, 1963," 17 June 1963; MSC Consolidated Activity Report for Office of the Dir., Manned Space Flight, 19 May—15 June 1963, p. 27; Howard T. Harris, "Gemini Launch Vehicle Chronology, 1961—1966," AFSC Historical Publications Series 66-22-1, December 1966, p. 40; Pendley to Chief, FOD, "N-24 Malfunction Detection System (MDS) Titan II Piggyback Test," 30 Aug. 1963; Quarterly Status Report No. 6, p. 69.

16 "GLV Analysis of Titan II Launches," Martin, n.d., for Missile N-23, 23 Sept. 1963;

Quarterly Status Report No. 8, for period ending 29 Feb. 1964, p. 52.

[17] "Minutes of the Ninth Meeting, Gemini Program Planning Board, Friday, October 11, 1963."

[18] Paul E. Purser, recorder, "Minutes of Project Gemini Management Panel Meeting ..., November 13, 1963," p. 10; Dineen, interview, Huntington Beach, Calif., 15 May 1967; Funk, interview, Sunnyvale, Calif., 12 May 1967, and telephone interview, 5 Jan. 1973; memo, Sheldon Rubin to Dineen, "Results of Analysis of N-25 Configuration on Aerospace Analog Model of POGO," Aerospace 63-1944-51, 15 Oct. 1963.

[19] Estes letter, 18 Sept. 1963; letter, Leonhard to NASA Hq. (Seamans), "Titan II/Gemini Program Status Summary," 27 Sept. 1963; letter, Maj. Gen. Marcus F. Cooper to NASA Hq. (Seamans), "Titan II/Gemini Program Status Summary," 4 Oct. 1963; Leonhard letter, 8 Oct. 1963; letters, Estes to Seamans, "Titan II/Gemini Program Status Summary," 16, 24, and 30 Oct. 1963.

[20] Memo, Billy A. Neighbors to dist., "Saturn I/Agena/Gemini Vehicle Feasibility Study Initiation Meeting," R-P&VE-AV-530, 17 Oct. 1963; memo, Neighbors to dist., "Design Data, Saturn I/Agena/Gemini Vehicle Feasibility Study," R-P&VE-AV-558, 18 Nov. 1963; memo, Neighbors to dist., "Conclusions, Saturn I/Agena/Gemini Vehicle Feasibility Study," R-P&VE-AV-576, 18 Dec. 1963.

[21] Letter, George M. Low to James C. Elms, 13 April 1963; letter, Gilruth to Dir., Flight Research Center, "Participation of Flight Research Center in Paraglider Flight Test Program," GPO 00851, 6 May 1963; letter, Paul F. Bikle to MSC, Attn: Gemini-Paraglider Program Manager, "Paraglider Program status report, June 15, 1963, to July 15, 1963," 18 July 1963; Quarterly Status Report No. 7, p. 33; letter, Harrison A. Storms, Jr., to MSC, Attn: Stephen D. Armstrong, "Contract NAS 9-1484, Paraglider Landing System Research and Development Program, Transmittal of the Final Fee Settlement Proposal," 65MA3479, 18 March 1965, with enclosure, "A Final Fee Settlement Proposal for Contract NAS 9-1484," 18 March 1965, p. V-111; letter, George W. Jeffs to MSC, Attn: Kline, "Contract NAS 9-1484, Paraglider Landing System Program, Monthly Progress Report No. 5 (September 1963)," 63MA14952, 16 Oct. 1963, p. 4; letter, Jeffs to MSC, Attn: Kline, "Contract NAS 9-1484, Paraglider Landing System Program, Monthly Progress Report No. 6 (October 1963)," 63MA16325, 15 Nov. 1963, p. 3; memo, Kenneth F. Hecht to MSC Historical Office, "Comments on Chapter 6: The Nadir," 22 Sept. 1970.

[22] See chapter V, pp. 98-99; letter, Jeffs to MSC, Attn: Kline, "Contract NAS 9-1484, Paraglider Landing System Program, Monthly Progress Report No. 1 (May 1963)," 63MA8801, 15 June 1963, p. 2; Weekly Activity Report, 2-8 June 1963, p. 2; letter, Jeffs to MSC, Attn: Kline, "Contract NAS 9-1484, Paraglider Landing System Program, Monthly Progress Report No. 2 (June 1963)," 63 MA10508, 19 July 1963, pp. 2-4; Weekly Activity Report, 23-29 June 1963, pp. 1-2; Consolidated Activity Report, 16 June—20 July 1963, pp. 87-88; letter, Jeffs to MSC, Attn: Kline, "Contract NAS 9-1484, Paraglider Landing System Program, Monthly Progress Report No. 3 (July 1963)," 63MA12060, 15 Aug. 1963, p. 1; Mathews, activity report, 28 July—3 Aug. 1963, p. 1; "GPO Information for Management Council Meeting," prepared for meeting of 24 Sept. 1963; Hecht memo, 22 Sept. 1970.

[23] TWX, R. S. Maynard to MSC for Kline, MA24858, 30 Aug. 1963; letter, Jeffs to MSC, Attn: Kline, "Contract NAS 9-1484, Paraglider Landing System Program, Monthly Progress Report No. 4 (August 1963)," 63MA12926, 13 Sept. 1963, p. 1; Consolidated Activity Report, 18 Aug.—21 Sept. 1963, p. 79; Jeffs letter, 63MA16325, 15 Nov. 1963, p. 1; Consolidated Activity Report, 20 Oct.—16 Nov. 1963," pp. 20-21; Quarterly Status Report No. 7, p. 32; letter, Jeffs to MSC, Attn: Kline, "Contract NAS 9-1484, Paraglider Landing System Program, Monthly Progress Report No. 7 (November 1963)," 63MA16756, 13 Dec. 1963, pp. 1-2; Weekly Activity Report, 1-7 Dec. 1963, p. 1; letter, Jeffs to MSC, Attn: Kline, "Contract NAS 9-1484, Paraglider Landing System Program, Monthly Progress Report No. 8 (December 1963)," 64MA632, 13 Jan. 1964, p. 2; Quarterly Status Report No. 8, p. 25; Hecht memo, 22 Sept. 1970.

[24] Jeffs letter, 63MA12926, 13 Sept. 1963, p. 1; Weekly Activity Report, 28 July—3 Aug. 1963, p. 3; Jeffs letter, 63MA14952, 16 Oct. 1963, p. 1; Quarterly Status Report No. 7, p. 33; Weekly Activity Report, 27 Oct.—2 Nov. 1963, p. 1; Jeffs letter, 63MA16756, 13 Dec. 1963, p. 6; Jeffs letter, 64MA632, 13 Jan. 1964, p. 1.

[25] "Preliminary Project Development Plan for a Controllable Parachute-Retrorocket Landing System," STG, 21 June 1961; U.S. Congress, House, Committee on Science and Astronautics, *Astronautical and Aeronautical Events of 1962: Report*, 88th Cong., 1st sess., 12 June 1963, p. 256; Zavasky, "Minutes of Senior Staff Meeting, March 22, 1963," p. 2;

Consolidated Monthly Activity Report, 24 Feb.—23 March 1963, p. 39.

[26] André J. Meyer, Jr., notes on GPO staff meeting, 9 May 1963, p. 2; TWX, John Y. Brown to MSC, Attn: Mathews, "Contract NAS 9-170, Gemini, Study of Incorporation of Parasail," 16-DAH-2582, 26 May 1963; memo, Mathews to Wilbur H. Gray, "Information and Equipment Needed for Parasail Program," GPO-03044-A, 8 Aug. 1963.

[27] Memo, Maxime A. Faget to dist., "Parasail—Landing Rocket Program," 4 March 1963, with enclosure, "Parasail—Landing Rocket Program Description"; Consolidated Activity Report, 19 May—15 June 1963, p. 47; memo, Kraft to Chief, Systems Evaluation and Development Div., Attn: John W. Kiker, "Status of Pilot Visualization Program as of June 1, 1963," 17 June 1963; Consolidated Activity Report, 18 Aug.—21 Sept. 1963, p. 59; Quarterly Status Report No. 6, pp. 21-22; memo, Kiker to GPO, Attn: Mathews, "Development status of the Para-sail—landing rocket," 21 Oct. 1963, with enclosures; memo, Warren J. North and Donald K. Slayton to Dir., "Continuation of paraglider effort," 3 Sept. 1963.

[28] TWX, Brown to MSC, Attn: Mathews, "Contract NAS 9-170, Gemini, Budgetary Estimate for Production Incorporation of Parasail," 16-DAH-3393, 5 Sept. 1963; Zavasky, "Senior Staff Meeting, September 13, 1963," p. 6; Quarterly Status Report No. 6, p. 22.

[29] Memo, Low to MSC, Attn: Elms, "Paraglider development program," M-C S 1312-503, 3 Oct. 1963; letter, Gilruth to NASA Hq., Attn: Low, "Realinement of Gemini Paraglider Program," GPO-01076-M, 16 Oct. 1963; letter, Low to MSC, Attn: Gilruth, "Gemini Paraglider Program," M-C S 1312.701, 30 Oct. 1963.

[30] "A Final Fee Settlement Proposal," pp. III-1, V-36; memo, Wilburne F. Hoyler et al. to Actg. Mgr., GPO, "Paraglider Reorientation with the Gemini Program," 14 Oct. 1963; Low memo, M-C S 1312-503, 3 Oct. 1963; Quarterly Status Report No. 8, p. 58; letter, Funk to Gilruth, "Evaluation of the Paraglider," 29 Nov. 1963.

[31] Purser, "Management Panel Meeting, November 13, 1963," p. 5; "Abstract of Meetings of Gemini Launch Vehicle Panels and Coordination Committee, January 9-10, 1964," 20 Jan. 1964; memo, John A. Edwards to Dep. Dir., Gemini Program, "Gemini Water Landings," 18 Feb. 1964; Purser, "Minutes of Project Gemini Management Panel Meeting..., February 7, 1964," pp. 6, 7; letter, Holmes to Gilruth, 23 Aug. 1963; memo, Verne C. Fryklund to Dir., Office of Space Sciences, "Manned Space Flight Experiments Board," 28 Oct. 1963; memo, Willis B. Foster to Chief, Lunar and Planetary Br., "Establishment of Manned Space Flight Experiments Board," 9 Jan. 1964; NASA Management Instruction M 9000.002, "Establishment of a Manned Space Flight Experiments Board," Coordination Draft #6, 14 Jan. 1964; letter, Schneider to Mathews, 24 Jan. 1964; Hecht, telephone interview, 23 Jan. 1973.

[32] Quarterly Status Report No. 4, for period ending 28 Feb. 1963, p. 1.

[33] Zavasky, "Minutes of Senior Staff Meeting, November 16, 1962," p. 4; Quarterly Status Report No. 3, for period ending 30 Nov 1962, p. 26, Zavasky, "Minutes of Senior Staff Meeting, December 6, 1962," p. 5; "Gemini Spacecraft Status, December 13, 1962," prepared for Gilruth's presentation at the 13th Management Council Meeting, 18 Dec. 1962, pp. 1-2.

[34] Quarterly Status Report No. 4, p. 26; Consolidated Monthly Activity Report, 24 Feb.—23 March 1963, p. 56; Meyer notes, 9 May 1963; Quarterly Status Report No. 5, pp. 35-36.

[35] Quarterly Status Report No. 5, p. 36; letter, William Parker to James E. Webb, 14 July 1966, with enclosure, "United States General Accounting Office Report to the Congress of the United States: Review of the Gemini Spacecraft Fuel Cell Electrical Power Supply Systems, by the Comptroller General of the United States," draft, [July 1966], pp. 22-23.

[36] "Abstract[s] of Meeting[s] on Electrical Systems, April 30, 1963," 8 May 1963; "May 14, 1963," 21 May 1963; "May 28, 1963," 29 May 1963; "June 11, 1963," 14 June 1963; Mathews, activity report, 5-11 May 1963, p. 2; Consolidated Activity Report, 28 April—18 May 1963, p. 69; Weekly Activity Report, 16-22 June 1963, p. 2; Consolidated Activity Report, 16 June—20 July, 1963, p. 87; Weekly Activity Report, 4-10 Aug. 1963, p. 1.

[37] Quarterly Status Report No. 6, pp. 1, 63-64.

[38] TWX, Mathews to Burke, "Contract NAS 9-170 Use of Batteries on Spacecraft 3," [26 Aug. 1963]; TWX, Brown to MSC, Attn: Mathews, "Use of Batteries for Electrical Power in Spacecraft Nbr. 3," 18 Sept. 1963; G[eorge] J. Weber, "Dual Fuel Cell/Silver Zinc Battery Installation for 7 Orbit, S/C 3 Gemini Mission," McDonnell Electrical Design Note 24, 25 Sept. 1963; TWX, Mathews to Burke, GPO-54285-A, 16 Oct. 1963.

[39] TWX, Mathews to Burke, "Contract NAS 9-170, Power System Design Study," GPO-54230-A, 1 Oct. 1963.

[40] TWX, Low to MSC, Attn: Elms, "Apollo and Gemini Fuel Cells," M-C S 1000.578, 7 Oct. 1963.

[41] John W. Finney, "Power-System Snags May Cut First Gemini Flights to 2 Days," *The New York Times*, 30 Oct. 1963.

[42] Mathews, activity report, 29 Sept.—5 Oct. 1963, p. 2; Consolidated Activity Report, 22 Sept.—19 Oct. 1963, p. 95.

[43] TWX, Mathews to Burke, "Contract NAS 9-170 Visit to GE by Bell Labs Committee," GPO-54277-A, 14 Oct. 1963; N. B. Hannay, F. J. Biondi, and U. P. Thomas, "Report on Fuel-Cell Work at General Electric and Pratt & Whitney," Bell Telephone Laboratories, Inc., n.d., pp. 6-9; letter, Schneider to Burke, 8 Jan. 1964, with enclosure; G. J. Weber, J. C. Waldner, and K. A. Rogers, "Fuel Cell Interface Review," McDonnell Electrical Design Note 33, 16 Dec. 1963.

[44] Quarterly Status Report No. 7, p. 61; memo, Mathews and Wesley L. Hjornevik, "United States General Accounting Office draft report to Congress regarding fuel cells," GP-62337, 11 Aug. 1966, with enclosure, "Detailed Comments on GAO Draft Report," p. 1.

[45] TWX, Mathews to Burke, "Contract NAS 9-170, Procurement of Batteries for Spacecraft 3," GPO-54229-A, 30 Sept. 1963; TWX, Mathews to Burke, GPO-53110-S, 12 Nov. 1963; Weekly Activity Report, 17-23 Nov. 1963, p. 1; Kline to McDonnell, "Change Notice No. 16," 20 Jan. 1964; Consolidated Activity Report, 17 Nov.—21 Dec. 1963, p. 18; memo, Mathews to Chief, Gemini Spacecraft Procurement, "Contract NAS 9-170, CCP No. 16, Battery Module for Spacecraft No. 3," GPO-03307-A, 4 Dec. 1962 [sic].

[46] Consolidated Activity Report, 22 Dec. 1963-18 Jan 1964, p. 9; TWX, Mathews to Burke, GS-53158, 20 Jan. 1964; Consolidated Activity Report, 19 Jan.—15 Feb. 1964, p. 17; memo, Edward E. Winchester to Donald D. Blume, "GAO Review of the Gemini Fuel Cell Power Supply System," 15 Nov. 1965; memo, Mathews to GAO Liaison Representative, "General Accounting Office inquiry regarding fuel cell power," GS-64098, 10 Jan. 1966, with enclosure.

[47] Quarterly Status Report No. 8, p. 48; Weekly Activity Report, 19-25 Jan. 1964, p. 10; TWX, Mathews to Burke, "Contract NAS 9-170: Fuel Cell Program," GP-54541, 10 Feb. 1964; TWX, Mathews to McDonnell, Attn: Burke, GS-53211, 18 March 1964.

[48] Memo, Marlowe D. Cassetti and Richard E. Charters to Chief, FOD, "Analysis of the Effect of Adding an Escape Tower to Gemini S/C on GLV Performance," 27 March 1963; memo, Kraft to Dir., "Gemini Escape System," 10 April 1963.

[49] B. Porter Brown, "Minutes of Meeting of Gemini Egress Group of the Master Egress Committee, April 17, 1963," 17 April 1963, enclosure 1, "Gemini Ejection Seat Review Presented to Gemini Egress Group, April 17, 1963, Cape Canaveral, Florida," n.d.

[50] "Abstract of Meeting on Ejection Seats, September 26, 1962," 3 Oct. 1962; letter, Floyd L. Thompson to MSC, Attn: GPO, "Request for support of Project Gemini," 5 Dec. 1962, with enclosure, memo, Jerry L. Lowery to Assoc. Dir., Langley, "Visit of Goodyear Aircraft Corporation representatives to discuss proposed wind tunnel tests of an inflatable decelerator attached to an astronaut," 20 Nov. 1962; C. E. Heimstadt and Gordon P. Cress, interviews, Burbank, Calif., 5 July 1966; letter, Cress and Heimstadt to MSC Historical Office, 12 May 1967; letter, Cress to MSC Historical Office, "Comment Draft on Chapters 7 & 8 of Gemini Narrative History," 511/GPC/2120, 1 Dec. 1971; Hecht memo, 22 Sept. 1970.

[51] Gregory, "Minutes of Senior Staff Meeting, February 8, 1963," p. 3; Quarterly Status Report No. 4, pp. 18-19; Quarterly Status Report No. 5, p. 26; Cress interview; Cress and Heimstadt letter, 12 May 1967; Hecht memo, 22 Sept. 1970.

[52] Consolidated Monthly Activity Report, 24 Feb.—23 March 1963, p. 3; Cress interview; Quarterly Status Report No. 5, pp. 6, 26; Hecht memo, 22 Sept. 1970.

[53] Mathews, activity report, 24 April—19 May 1963, p. 1; Quarterly Status Report No. 5, p. 6; Quarterly Status Report No. 6, p. 41.

[54] [Kenneth F. Hecht], "Comments on Chapter 5, 'Expansion and Crisis,' " [10 Feb. 1970], pp. 1-2; "Abstract of Meeting on Ejection Seat Developmental Test Program, May 29, 1962," 4 June 1962; Weekly Activity Report, 27 May—2 June 1962, p. 7; Cress interview; Quarterly Status Report No. 1, for period ending 31 May 1962, pp. 2-22.

[55] Quarterly Status Report No. 3, p. 18; Zavasky, "Senior Staff Meeting, November 16, 1962," p. 3; Cress and Heimstadt letter, 12 May 1967; [Hecht], "Comments on Chapter 5," p. 2; Hecht memo, 22 Sept. 1970.

[56] Quarterly Status Report No. 5, p. 26; Weekly Activity Report, 16-22 June 1963, p. 3; Mathews, activity report, 4-10 Aug. 1963, p. 2; Quarterly Status Report No. 7, p. 42, 44; "Abstract of Meeting on Ejection Seat System, October 30, 1963," 5 Nov. 1963; W. M. Weeks, "Aerodynamic Characteristics of the Gemini Ejection Seat-Man Configuration," McDonnell Aerodynamic Information Note No. 50, 28 Oct. 1963, p. 4.

57 Weekly Activity Report, 16-22 June 1963, pp. 2-3; memo, Jack A. Kinzler to MSC Public Affairs Officer, "Comment Draft of 'Project Gemini Technology and Operations: A Chronology,' " 31 May 1967, with enclosure; Heimstadt interview; Robert Provart and John Swanson, interviews, Newbury Park, Calif., 7 July 1966.

58 Zavasky, "Minutes of Senior Staff Meeting, March 8, 1963," p. 4; William J. Blatz, interview, St. Louis, 14 April 1966; Larry E. Stewart, interview, Canoga Park, Calif., 16 May 1967.

59 "MSC Status Report," prepared for Gilruth's presentation at the 14th Management Council Meeting, 29 Jan. 1963, pp. 34-35; Quarterly Status Report No. 4, pp. 16-17; Zavasky, "Senior Staff Meeting, March 8, 1963," p. 4; memo, Meyer to MSC Historical Office, "Comments on Chapter 6 of Gemini Narrative History," 16 Nov. 1970.

60 Quarterly Status Report No. 5, pp. 23-26; Mathews, activity report 24 April—19 May 1963, p. 3; Quarterly Status Report No. 6, p. 28; Steven J. Domokos, interview, Canoga Park, Calif., 16 May 1967.

61 Quarterly Status Report No. 6, p. 29; Zavasky, "Senior Staff Meeting, July 12, 1963," p. 6; Meyer, interview, Houston, 9 Jan. 1967; Meyer memo, 16 Nov. 1970.

62 Zavasky, "Senior Staff Meeting, September 13, 1963," pp. 5-6; "Gemini Report for Management Council Meeting," prepared for meeting of 24 Sept. 1963, p. 2; Consolidated Activity Report, 22 Sept.—19 Oct 1963, p. 93.

63 "Gemini Report for Management Council Meeting," p. 2; Quarterly Status Report No. 4, pp. 11-12; Quarterly Status Report No. 5, p. 17; Zavasky, "Senior Staff Meeting, September 13, 1963," pp. 4-5; Quarterly Status Report No. 7, pp. 31-32; Bailey to McDonnell, "Change Notice No. 10," 6 Nov. 1963; Swanson interview; TWXs, Mathews to Burke, GPO-52058-LV, 9 Sept., GPO-54240-A, 7 Oct., and GPO-54437-A, 12 Dec. 1963.

64 Weekly Activity Report, 20-26 Oct. 1963, p. 2; Quarterly Status Report No. 7, pp. 17, 27-28; TWX, James R. Flanagan to McDonnell, Attn: Burke, M-C 1. 4000.532, 13 Nov. 1963; Ron Helsel, interview, Canoga Park, Calif., 16 May 1967; "Gemini Propulsion by Rocketdyne—A Chronology," 15 May 1967, p. 6; Stewart interview; memo, Robert H. Voigt to dist., "Report on Review of Business Management Activities at Rocketdyne, A Division of North American Aviation, Inc. (Report No. WR 65-12), MSC 32-0-65G," 5 May 1965, with enclosure, Raymond Einhorn, "Review of Business Management Activities at Rocketdyne . . . ," Western Region Audit Office

Report WR 65-12, April 1965, p. 67; TWX, John Brown to Armstrong, "Contract NAS 9-170, Gemini, Radiation Cooled Thrust Chamber," 306-16-6817, 13 July 1964.

65 "Official NASA Flight Schedule," NASA Office of Management Reports, approved by Seamans and Hugh L. Dryden, 20 March 1962, pp. 6, 7; memo, Seamans and McMillan for record, "Acceptance of the Joint NASA/DOD Ad Hoc Study Group Final Report on Air Force Participation in Gemini, Dated May 6, 1963," 5 Aug. 1963, enclosure 1, "Gemini Launches Master Schedule," as of 2 May 1963.

66 Consolidated Activity Report, May 1962, Tab 18, p. 11; "Gemini Agena Target Vehicle Propulsion Systems Presentation, 2 August 1962," LMSC-A057703, n.d.; "Abstract of Meeting on Atlas/Agena, August 16, 1962," 22 Aug. 1962; Quarterly Status Report No. 2, for period ending 31 Aug. 1962, p. 25; "Medium Space Vehicles Programs Monthly Progress Report, August 1962," LMSC-447186-26, 20 Sept. 1962, pp. 9-10; Quarterly Status Report No. 3, p. 31; letter, Harold T. Luskin to Maj. Charles A. Wurster, "Contract 04(695)-129, Gemini Program BAC subcontract Financial Status," LMSC/A602008, 11 Nov. 1963, enclosure A, "Model 8247 Development Program: Description of Changes and Increased Effort," pp. 1-2; Gemini Agena Target Press Handbook, published by Lockheed Missiles & Space Co. for issuance to news media, LMSC A766871, 15 Feb. 1966, pp. 1-11, -12, 4-17 to 4-22.

67 "Abstract of Meeting on Atlas/Agena, August 2, 1962," 14 Aug. 1962; Weekly Activity Report, 5-11 Aug. 1962, p. 2; TWX, Chamberlin to Friedrich Duerr, GPO 50145, 10 Aug. 1962; memo, James A. Ferrando to Chief, FOD, "Information gathered at the Atlas/Agena coordination meeting held on August 16, 1962," 17 Aug. 1962; "Monthly Progress Report, August 1962," pp. 9-10; "GATV Propulsion Systems Presentation"; Quarterly Status Report No. 2, p. 25; Luskin letter, LMSC/A602008, 11 Nov. 1963, enclosure B, "Model 8250 Development Program: Description of Changes and Increased Effort," pp. 1-2; Gemini Agena Target Press Handbook, pp. 1-13, -14, 4-17, -22 to -24.

68 Richard J. Crane, "Gemini Atlas-Agena Program Status Report," 5 Jan. 1963; memo, Crane for record, "Negotiations—Agena Contract between SSD and Lockheed at Los Angeles—Atlas/Agena Program (Jan. 21-25, 1963)," 12 Feb. 1963; "Monthly Progress Report, February 1963," LMSC-447186-32, 20 March 1963, pp. 9-10; Amendment No. 7 to Letter Contract AF 04(695)-129, 18 March 1963, with enclosure, Exhibit C, "Statement of Work,

Phase II, Target Vehicle System, Gemini Program"; Negotiated Contract AF 04(695)-129, 22 April 1963, with enclosure, "Exhibit 'B' to Contract AF 04(695)-129: Statement of Work, Target Vehicle System Development, Gemini Program: Phase I."

[69] "Monthly Progress Report, April 1963," LMSC-447186-34, 20 May 1963, pp. 2-1, -5, -6; Quarterly Status Report No. 5, p. 33; "Monthly Progress Report, May 1963," LMSC-447186-35, 20 June 1963, pp. 2-1, -2, -11; "Abstract of Meeting on Atlas/Agena, July 2, 1963," 8 July 1963; Luskin letter, LMSC/A602008, enclosure A, pp. 8-9; TWX, Mathews to SSD, Attn: Wurster, GPO-51054, 11 July 1963.

[70] Quarterly Status Report No. 6, p. 73; "Monthly Progress Report, August 1963," LMSC-447186-38, 20 Sept. 1963, p. 2-1; Luskin letter, LMSC/A602008, 11 Nov. 1963, enclosure A, p. 5; *Gemini Agena Target Press Handbook*, pp. 4-21, -22.

[71] Quarterly Status Report No. 6, p. 73; "Monthly Progress Report, August 1963," p. 2-1; Quarterly Status Report No. 7, p. 69; Consolidated Activity Report, 17 Nov.–21 Dec. 1963, p. 21; TWXs, Mathews to SSD, Attn: Wurster, GPO-52012-LV, 30 July, and GPO-52000-LV, 2 Aug. 1963.

[72] Quarterly Status Report No. 7, p. 69; Luskin letter, LMSC/A602008, 11 Nov. 1963, enclosure A, pp. 8-9; TWX, SSD to Mathews, SSVA 21-11-35, 21 Nov. 1963.

[73] Luskin letter, LMSC/A602008, 11 Nov. 1963.

[74] Ibid.

[75] Memo, V. F. Peterson to L. Orinovsky, "Letter Contract AF 04(695)-129 Reschedule ROM [rough order of magnitude] Funding Requirements," LMSC/A374952, 17 April 1963.

[76] Quarterly Status Report No. 5, p. 43, Fig. 1; letter, Herbert J. Ballard to dist., "Gemini Target Management Review—Minutes of Second Meeting," IDC 91-60/208, 24 May 1963; "Abstract of Meeting on Atlas/Agena, June 6 and 7, 1963," 12 June 1963; letter, Peterson to Mathews, "Quarterly Contractor Financial Management Report—NASA 533 Dated June 30, 1963, L/C AF 04(695)-129," LMSC/A376437, 18 July 1963, with enclosure, "NASA Contractor Financial Management Report" for quarter ending 30 June 1963; TWX, SSD to Mathews, "FY-64 Funding Requirements for Gemini Agena," SSVR 20-6-100, 21 June 1963; TWX, Mathews to SSD, Attn: Wursters [*sic*], GPO-51082, 5 July 1963; TWX, Mathews to SSD, Attn: Wurster, GPO-54018-A, 24 July 1963.

[77] TWX, Mathews to SSD, Attn: Wurster, GPO-54114-A, 27 Aug. 1963; TWX, Mathews to SSD, Attn: Wurster, GPO-54112-A, 3 Sept. 1963; letter, Wurster to MSC, Attn: Mathews, "Gemini Agena Program Cost Estimate," 10 Sept. 1963, with enclosures, (1) "Gemini Agena Program Cost Estimate—Summary," as of 6 Sept. 1963, (2) "Gemini Agena Program Cost Estimate—Detail," as of 6 Sept. 1963, and (3) memo, Lt. Col. L. D. Parsons, Jr., to Wurster, "Revised Cost Estimate for Gemini Target," 10 Sept. 1963, with enclosures; letter, Peterson to H. F. Becker, "Quarterly Financial Management Report (NASA Form 533) Dated 29 September, 1963, Contract AF 04(695)-129, Gemini," 18 Oct. 1963, with enclosure, "NASA Contractor Financial Management Report" for period ending 29 Sept. 1963.

[78] Quarterly Status Report No. 6, p. 73; "Monthly Progress Report, September 1963," LMSC-447186-39, 20 Oct. 1963, p. 2-5; Amendment No. 11 to Letter Contract AF 04(695)-129, 23 July 1963; "Monthly Progress Report, October 1963," LMSC-447186-40, 20 Nov. 1963, p. 3-1; "Abstract of Meeting on Atlas-Agena, November 6-7, 1963," 12 Nov. 1963; "Monthly Progress Report, January 1964," LMSC-447186-43, 20 Feb. 1964, p. 3-1; Purser, "Management Panel Meeting, February 7, 1964," p. 8; Consolidated Activity Report, 16 Feb.–21 March 1964, p. 21.

[79] Letter, Mathews to Luskin, GPO-03069-A, 12 Sept. 1963; memo, Mathews to dist., "Procedure for Obtaining Information and Support from Lockheed Missiles and Space Company," GPO-03066-A, 12 Sept. 1963; letters, Mathews to Wurster, GPO-02076-LV, 25 Oct. 1963, and GPO-02077-LV, 25 Oct. 1963, with enclosure, "Financial Status of Gemini-Agena Propulsion System Efforts"; letter, Mathews to Wurster, GPO-02081-LV, 7 Nov. 1963.

[80] Letter, Wurster to Ballard, "Gemini/Agena BAC Subcontract Financial Status," 30 Oct. 1963; letter, Wurster to Mathews, "Schedule Controls for Gemini Agena," 18 Nov. 1963; memo, Schneider to Mathews, "Agena Phase II Contract Negotiations," 6 Jan. 1964; memo, Armstrong to George F. MacDougall, Jr., "Financial Data on GLV Program," APCMT 87-4866, 18 Nov. 1963; memo, Armstrong to MacDougall, "Financial & Management Data from SSD concerning Booster Program," APCMT 81-4961, 9 Dec. 1963; letter, Mathews to Lt. Col. Mark E. Rivers, GV-02294, 29 July 1964.

[81] Letter, Mathews to Wurster, GPO-02012-LV, 15 Nov. 1963, with enclosure, "Statement of Work for Atlas-Agena Target Vehicles to Be Used in Project Gemini," 8

*445*

July 1963; cf. letter, Gilruth to Marshall, Attn: Wernher von Braun, "Procurement of Atlas-Agena Space Vehicles," 31 Jan. 1962, with enclosures, esp. "Exhibit 'B': Statement of Work for Atlas-Agena Rendezvous Vehicles to Be Used in Project Gemini," n.d.; William A. Summerfelt, interview, Washington, 24 Jan. 1967; Meyer interview; "Gemini Atlas Agena Target Vehicle System, Management and Responsibilities Agreement between the NASA-MSC and the USAF-AFSC-SSD," March 1965, signed by Funk on 31 March, by Col. John B. Hudson on 29 March, and by Gilruth and Mathews on 9 April 1965; Maj. Arminta Harness, interview, Los Angeles, 18 April 1966.

[82] Letter, Mathews to Wurster, GPO-02055-LV, 7 Oct. 1963; letter, Gilruth to Brig. Gen. Joseph J. Cody, Jr., GPO-01099-M, 23 Dec. 1963; Bernhard A. Hohmann and Ernst R. Letsch, interviews, El Segundo, Calif., 19 April 1966.

[83] "Abstract of Atlas/Agena Meeting, June 6 and 7, 1963"; Weekly Activity Report, 2-8 June 1963, p. 3; "Monthly Progress Report, January 1963," LMSC-447186-31, 20 Feb. 1963, p. 23; TWX, Mathews to SSD, Attn: Maj. John G. Albert, GPO-51012, 13 June 1963; TWX, Mathews to SSD, Attn: Wurster, GPO-51057, 9 July 1963; "Monthly Progress Report, November 1963," LMSC-447186-41, 20 Dec. 1963, pp. 5-9.

[84] Theodore Shabad, "Soviet Craft Put into 'Final Orbit,' " The New York Times,3 Nov. 1963; Shabad, "Soviet Satellite Run from Earth," The New York Times, 4 Nov. 1963; Astronautics and Aeronautics, 1963, p. 413.

[85] Letter, Blatz to Barton C. Hacker, 28 Sept. 1970.

[86] Hecht memo, 22 Sept. 1970.

[87] Meyer interview; memo, Meyer to Historical Office, "Comments on Chapter 5 of the Gemini Narrative History," 18 June 1970.

## Chapter VIII

[1] MSC Announcement No. 268, "Reorganization of MSC and Key Personnel Assignments," 5 Nov. 1963; MSC News Release 63-227, 5 Nov. 1963; Robert B. Merrifield, "Men and Spacecraft: A History of the Manned Spacecraft Center (1958-1969)," draft ms., [1972], pp. 5-28 to -41; NASA Release 63-237, "Williams to Head Manned Space Flight Operations for NASA," 23 Oct. 1963; "Williams Gets New NASA Job," Missiles and Rockets, 28 Oct. 1963, p. 12; MSC Announcement No. 64-2, "Resignation of J. C. Elms, Deputy Director," 17 Jan. 1964; MSC News Release 64-13, 17 Jan. 1964; NASA News Release 64-13, "NASA Names Low Deputy Director of

Manned Spacecraft Center," 19 Jan. 1964; MSC Announcement No. 64-5, "Deputy Director," 21 Jan. 1964; "Deputy Director Elms to Return to Private Industry; George M. Low, NASA Hq., Named as Replacement," MSC Space News Roundup, 22 Jan. 1964.

[2] Robert L. Rosholt, An Administrative History of NASA, 1958-1963, NASA SP-4101 (Washington, 1966), pp. 289-97; U.S. Congress, House, Committee on Science and Astronautics, 1965 NASA Authorization, Hearings on H.R. 9641 (Superseded by H.R. 10456), 88th Cong., 2d sess., 1964, pp. 69-70; NASA Release 63-225, "NASA Announces Reorganization," 9 Oct. 1963; NASA Release 63-241, "NASA Realigns Office of Manned Space Flight," 28 Oct. 1963; William C. Schneider, interview, Washington, 23 Jan. 1967; William A. Summerfelt, interview, Washington, 24 Jan. 1967; LeRoy E. Day, interview, Washington, 25 Jan. 1967; NASA Headquarters Telephone Directory, February 1964, p. 29.

[3] MSC Weekly Activity Report for Office of the Dir., Manned Space Flight, 27 Oct.-2 Nov. 1963, p. 2; memo, George E. Mueller to Adm., "Development of the Gemini Launch Vehicle," 6 Dec. 1965, with enclosure, "The Gemini Launch Vehicle Story," pp. 2-3; memo, David B. Pendley to Asst. Dir. for Flight Operations Dir. (FOD), "N-25 Titan II Piggyback Malfunction Detection System (MDS) flight," 7 Nov. 1963; memo, Pendley to Asst. Dir., FOD, "Titan II Malfunction Detection System (MDS) Piggyback Mission No. N-29," 19 Dec. 1963; TWX, Charles W. Mathews to NASA Hq. for Robert C. Seamans, Jr., GPO-52133-LV, 14 Nov. 1963; letters, Lt. Gen. Howell M. Estes, Jr., to Seamans, "Titan II/Gemini Program Status Summary," 6 and 15 Nov. 1963; memo, Mathews to Carol Sweeney, "Questions submitted by House Committee Staff Members for Dec. 11 Visit to MSC," 17 Dec. 1963; letter, Bernhard A. Hohmann to James M. Grimwood, "Comments on the Draft of 'On the Shoulders of Titans[:] a History of Project Gemini,' " 30 May 1974, with enclosures; memo, Sheldon Rubin to Col. Richard C. Dineen, "Results of Analysis of N-25 Configuration on Aerospace Analog Model of POGO," Aerospace 63-1944-51, 15 Oct. 1963, and annotated pages of draft history.

[4] Howard T. Harris, "Gemini Launch Vehicle Chronology, 1961-1966," AFSC Historical Publications Series 66-22-1, December 1966, p. 29; Weekly Activity Report, 8-14 Dec. 1963, p. 2; TWX, Mathews to NASA Hq., for Seamans, GPO-52187LV, 23 Dec. 1963; letters, Brig. Gen. W. E. Leonhard to Seamans, "Titan II/Gemini Program Status Summary," 19 Dec.

1963 and 27 Jan. 1964; Pendley memo, 19 Dec. 1963; Hohmann letter, 30 May 1974.

5 "Gemini Launch Vehicle Supplemental Specifications," NASA, Nov. 15, 1963; "Minutes of the Tenth Meeting, Gemini Program Planning Board [GPPB], Tuesday, December 3, 1963," p. 2; "Joint Titan II/Gemini Development Plan on Missile Oscillation Reduction and Engine Reliability and Improvement," [Air Force Systems Command], 5 April 1963 (revised 7 Aug. 1963); memo, Mueller to Assoc. Adm., "Memorandum of Understanding on Certain Design Requirements for the Gemini Launch Vehicle," M-C S 1370-860, 23 Nov. 1963; letter, Seamans to McMillan, M-C S 1370-821, 15 Nov. 1963.

6 "Minutes of the Tenth GPPB Meeting"; "Memorandum of Understanding on Certain Design Requirements for the Gemini Launch Vehicle," signed by Seamans and McMillan, 3 Dec. 1963.

7 "Minutes of the Eleventh Meeting, Gemini Program Planning Board, Monday, January 20, 1964"; Leonhard letter, 27 Jan. 1964.

8 Harris, "Launch Vehicle Chronology," p. 29; TWX, Mathews to SSD, Attn: Col. Richard C. Dineen, GP-54987, 24 Sept. 1964; TWX, Dineen to MSC, for Mathews, SSVLP 00009, 29 Sept. 1964; memo, Mueller to Assoc. Adm., "Status of Gemini Launch Vehicle Improvement Program," 4 March 1965; memo, Adm. W. Fred Boone to James E. Webb, "Man-rating of Titan II—Applications to Titan III," 14 Oct. 1965, with enclosure, memo, Milton W. Rosen to Boone, "Gemini Launch Vehicle Man-rating," 8 Oct. 1965; letter, Mueller to Wernher von Braun, 29 Nov. 1963; Willis B. Mitchell, Jr., telephone interview, 1 Aug. 1974.

9 Letter, Lt. Col. John J. Anderson to Seamans, 21 Oct. 1963, with enclosure, "Statement of Work, Titan II Augmented Engine Improvement Program," 3 Oct. 1963; letter, Leonhard to NASA Hq. (Seamans), "Titan II/Gemini Program Status Summary," 27 Sept. 1963; Harris, "Launch Vehicle Chronology," p. 30; Estes letter, 15 Nov. 1963; Mueller memo, 4 March 1965; Rosen memo, 8 Oct. 1965; "The Gemini Launch Vehicle Story," p. 3; Mathews TWX, GP-54987, 24 Sept. 1964; Dineen TWX, SSVLP 00009, 29 Sept. 1964.

10 Mueller memo, 6 Dec. 1965.

11 Rosen memo, 8 Oct. 1965; "Titan II Flight Summary," MG 4-8015, in "Gemini Administrator's Review, 1964" [November 1964].

12 Letter, George W. Jeffs to MSC, Attn: Robert L. Kline, "Contract NAS 9-1484, Paraglider Landing System Program, Monthly Progress Report No. 8 (December 1963)," 64MA632, 13 Jan. 1964, p. 2; letter, Harrison A. Storms, Jr., to MSC, Attn: Stephen D. Armstrong, "Contract NAS 9-1484, Paraglider Landing System Research and Development Program, Transmittal of the Final Fee Settlement Proposal," 65MA3479, 18 March 1965, with enclosure, "A Final Fee Settlement Proposal for Contract NAS 9-1484," 18 March 1965, p. V-113; "Final Report of Paraglider Research and Development Program, Contract NAS 9-1484," North American SID65-196, 19 Feb. 1965, p. 240.

13 1965 NASA Authorization, pp. 509-510; Heather David, "NASA Testifies That Rogallo Wing Is Probably Dead in Gemini Program," Missiles and Rockets, 24 Feb. 1964, p. 21.

14 Memo, John A. Edwards to Dep. Dir., Gemini Program, "Gemini Water Landings," 18 Feb. 1964; André J. Meyer, Jr., notes on GPO staff meeting, 20 Feb. 1964, p. 1; memo, Robert F. Freitag to dist., "MSF Position on Land Versus Water Landings—Apollo and Gemini," 5 March 1965.

15 Memo, Day to Dep. Dir., Gemini Program, "Future of the Paraglider," 3 March 1964, with enclosure, memo, Eldon W. Hall to Day, "Gemini Space Vehicle System Capability Using Paraglider Landing System," 27 Feb. 1964.

16 "Minutes of the Fourth Meeting, Gemini Program Planning Board, Tuesday, April 9, 1963," p. 2; Gen. Bernard A. Schriever and D. Brainerd Holmes, "Additional Guidance to Joint Ad Hoc Study Group, Air Force Participation in the Gemini Program," 12 April 1963; "DOD-NASA Ad Hoc Study Group, Air Force Participation in Gemini," Final Report, 6 May 1963, with errata; 1965 NASA Authorization, p. 508; U.S. Congress, Senate, Committee on Aeronautical and Space Sciences, NASA Authorization for Fiscal Year 1964: Hearings on S. 1245, 88th Cong., 1st sess., 1963, pp. 1027-29; Astronautics and Aeronautics, 1963: Chronology on Science, Technology, and Policy, NASA SP-4004 (Washington, 1964), pp. 473-74; John G. Norris, "Space Program Is Revised to Put Laboratory in Orbit," The Washington Post, 11 Dec. 1963; Jack Raymond, "Air Force to Loft Space Station in Place of Dyna-Soar Glider," The New York Times, 11 Dec. 1963.

17 Memo, Mueller to Assoc. Adm., "Discussion Paper for Meeting with Dr. H. Brown Regarding MOL," 13 Jan. 1964; TWX, Seamans to MSC for Robert R. Gilruth, to Western Operations Office for Mueller, to Kennedy Space Center for Kurt H. Debus, 23 Jan. 1964; letter, Harold Brown to Seamans, 28 Jan. 1964, with enclosure, memo, Brown and Seamans for record, "Gemini and Gemini-B/MOL Program," 28 Jan. 1964; memo, Mueller to

Adm., "NASA/Air Force discussions on Gemini B-MOL," 5 May 1964, with enclosure.

[18] Letter, Mueller to Alexander H. Flax, 17 March 1964; letter, Maj. Gen. Ben I. Funk to Gilruth, "Evaluation of the Paraglider," 29 Nov. 1963; letter, Gilruth to Funk, GPO-01098-M, 26 Dec. 1963; letter, Flax to Mueller, 1 May 1964; letter, Schriever to Edward C. Welsh, "Gemini Paraglider for MOL," 13 July 1964; letter, Albert C. Hall to Welsh, 13 July 1964; John W. Finney, "Johnson Orders Building of Orbiting Laboratory for Defense Experiment," *The New York Times*, 26 Aug. 1965; Philip Dodd, "1 1/2 Billion Space Lab Ok'd," *Chicago Tribune*, 26 Aug. 1965; "The President's News Conference of August 25, 1965," in *Weekly Compilation of Presidential Documents, Monday, August 30, 1965*, p. 142; Brown and Seamans memo, 28 Jan. 1964.

[19] "Paraglider Final Report," pp. 240-57; memo, Schneider to Assoc. Adm., MSF, "Fifth Deployment Test of Gemini Paraglider," 23 April 1964; TWX, Mathews to North American, Attn: Jeffs, GP-54640, 26 March 1964; Meyer, notes on GPO staff meeting, 29 April 1964; "A Final Fee Settlement Proposal," Sect. III; Meyer, notes on GPO staff meeting, 7 May 1964; memo, Mathews to Armstrong, "Contract NAS 9-170, Paraglider Recovery System, CCP No. 5," GP-03697, 12 June 1964; letter, Schneider to Mathews, 26 June 1964; Jim Maloney, "Paraglider Is Dropped from Gemini Project," *The Houston Post*, 11 Aug. 1964.

[20] "Paraglider Final Report," pp. 244-57.

[21] MSC Consolidated Activity report for Office of the Assoc. Adm., 19 July—22 Aug. 1964, p. 19; letter, Jeffs to MSC, Attn: Armstrong, "Contract NAS 9-1484, Paraglider Landing System Program, Monthly Progress Report No. 15 (July 1964)," 64MA10284, 7 Aug. 1964, p. 2; letter, Jeffs to MSC, Attn: Armstrong, "Contract NAS 9-1484, Paraglider Landing System Program, Monthly Progress Report No. 16 (August 1964)," 64MA11840, 16 Sept. 1964, p. 3; Harold Emigh, interview, Downey, Calif., 21 April 1966; "Paraglider Final Report," pp. 344, 349, 371-72; letter, Jeffs to MSC, Attn: Armstrong, "Contract NAS 9-1484, Paraglider Landing System Program, Monthly Progress Report No. 20 (December 1964)," 65MA853, 15 Jan. 1965.

[22] TWX, Armstrong to North American, Attn: R. S. Maynard, MSC-1313, 2 Dec. 1964; letter, Ralph B. Oakley to James M. Grimwood, 14 June 1972, with enclosures, "Paraglider Landing System Test Program Final Report, Contract NAS 9-5206," North American, SID 65-1638, December 1965, and Richard B. Dimon, "Dry Land Recovery! Manned Test Flight of Space Division's Paraglider System

Proves Feasibility of New Technique," *Skyline*, XXIII, No. 2 (1965), pp. 43-47; Emigh interview; letter, Mathews to Maj. E. K. Hartenberger, "Use of Paraglider Wing in North American Aviation, Inc., Sponsored Development Program for Aerial Delivery of Cargo," GPO-03270-A, 28 Nov. 1963.

[23] Memo, James E. Bone, Jr., and Walter D. Wolhart for record, "Explanations of Difference between Gemini Requirements shown in FY 63-64 Budgets and the amounts actually provided in FY 1963 and Planned for FY 1964," 30 Oct. 1963; James M. Grimwood and Barton C. Hacker, *Project Gemini Technology and Operations: A Chronology*, NASA SP-4002 (Washington, 1969), p. 283; "FY 1965 Summary Budget Back Up Data: Gemini," 2 March 1964, in "Gemini Data Book," revision of 1 Nov. 1964, p. L.4.1.

[24] Negotiated contract, "Project Gemini Two-Man Spacecraft Development Program," 27 Feb. 1963, signed by Glenn F. Bailey for MSC, by Ernest W. Brackett (29 March 1963) for NASA Hq., and by David S. Lewis for McDonnell; "Gemini Summary: NASA Administered Prior Contract Data," in "Gemini Data Book," p. J.1.5; letter, John Y. Brown to Kline, "Contract NAS 9-170, Gemini Monthly Financial Status Report; Submittal of," 306-16-4962, 17 Jan. 1964, with enclosures; "FY 1965 Summary Budget Back Up Data: Gemini Spacecraft," in "Gemini Data Book," p. L.4.3.

[25] Harris, "Launch Vehicle Chronology," pp. 28, 32, 35; "FY 1965 Summary Budget Back Up Data: Gemini—Launch Vehicles," in "Gemini Data Book," p. L.4.9; Bone and Wolhart memo, 30 Oct. 1963; TWX, Mathews to Dineen, GPO-54495-A, 13 Jan. 1964.

[26] Letter, Mathews to SSD, Attn: Maj. Charles A. Wurster, GV-02134, 5 Feb. 1964.

[27] TWX, Mathews to SSD, Attn: Wurster, 13 March 1964; memo, Richard J. Crane to Bailey, "Special Management Conference—Agena Program—Air Force—Lockheed at Sunnyvale, California, March 19, 1964," 27 March 1964, with enclosures; "FY 1965 Summary Budget Back Up Data: Gemini—Launch Vehicles."

[28] Letter, Lt. Col. Mark E. Rivers, Jr., to Mathews, "FY 1964 Fund Requirements for Gemini Agena Contract AF 04(695)-129," 4 April 1964, with enclosures.

[29] Meyer, notes on GPO staff meetings, 22 and 29 April 1964.

[30] Letter, Harold T. Luskin to Wurster, "Contract 04(695)-129, Gemini Program BAC subcontract Financial Status," LMSC/A602008, 11 Nov. 1963, enclosures A and B; "Medium Space Vehicles Programs Monthly Progress Report, April 1963," LMSC-447186-34, 20 May

1963, p. 2-6; Project Gemini Quarterly Status Report No. 7, for period ending 30 Nov. 1963, p. 69; "Abstract of Meeting on Atlas/Agena, April 7, 1964," 16 April 1964; memo, Schneider to Assoc. Adm., MSF, "Multiple Restart of Agena D Engine," 3 March 1964.

[31] "Abstract of Meeting on Atlas/Agena Coordination, May 5, 1964," 18 May 1964; Quarterly Status Report No. 9, for period ending 31 May 1964, pp. 42-43; Consolidated Activity Report, 21 June—18 July 1964, p. 16; Mathews, activity report (21 June—18 July), 21 July 1964, p. 1; TWX, Mathews to SSD, Attn: Rivers, GP-54748, 28 May 1964.

[32] Weekly Activity Reports, 5-11 April, p. 3, 19-25 April, p. 1, 6-12 Sept., p. 1, and 8-14 Nov 1964, p. 2; Quarterly Status Report No. 10, for period ending 31 Aug. 1964, p. 48; Quarterly Status Report No. 11, for period ending 30 Nov. 1964, p. 39; "GATV Progress Report, September 1964," LMSC-A605200-1, 20 Oct. 1964, pp. 2-1, -2; "GATV Progress Report, November 1964," LMSC-A605200-3, 20 Dec. 1964, pp. 2-2, 3.

[33] "Monthly Progress Report, March 1964," LMSC-447186-45, 20 April 1964, p. 3-4; Quarterly Status Report No. 9, p. 43; Weekly Activity Report, 22-28 March 1964, p. 2; Quarterly Status Report No. 10, p. 49; Weekly Activity Report, 2-8 Aug. 1964, p. 1; "Special Post Fire Functional and Leak Tests Performed on Model 8247 Rocket Engine at Santa Cruz Test Base," LMSC-A605708, 19 Oct. 1964.

[34] Paul E. Purser, recorder, "Minutes of Project Gemini Management Panel Meeting . . . , April 15, 1964," pp. 6-7, Fig. B-3-4; TWXs, Mathews to Rivers, GV-52348, 20 April, and GV-52401, 19 May 1964; letter, Mathews to Rivers, GV-02261, 28 May 1964; "Monthly Progress Report, May 1964," LMSC-447186-47, 20 June 1964, p. 3-6; letter, Mathews to Col. John B. Hudson, GP-01798, 13 Jan. 1965; memo, Charles W. McGuire to Dir., Gemini Test," Agena Target Vehicle (5001)," 5 Feb. 1965; TWX, Rivers to Mathews, 6 June 1964; TWX, Mathews to SSD, Attn: Dineen, GP-54906, 20 Aug. 1964.

[35] Weekly Activity Report, 17-23 Nov. 1963, p. 1; Gemini Press Reference Book (McDonnell, St. Louis, Mo., ca. March 1965), pp. 57-58; 1st Lt. Dennis R. Maus and A1C John R. Younger, "History of the 6511th Test Group (Parachute), 1 January—30 June 1964," AFSC Historical Publications Series 64-110-V, n.d., p. 26; Kenneth F. Hecht, telephone interview, 22 July 1971; Weekly Activity Report, 1-7 Dec. 1963, p. 1; Consolidated Activity Report, 22 Dec. 1963—18 Jan. 1964, p. 18; Weekly Activity Report, 2-8 Feb. 1964, p. 12; Consolidated Activity Report, 19 Jan.—15 Feb. 1964, p. 19; TWX,

Mathews to Walter F. Burke, GP-54571, 14 Feb. 1964; Col. Clyde S. Cherry, interview, Edwards AFB, 20 April 1966; "Just in Case of Trouble," in Goodyear Aerospace Profile, Vol. II, No. 2 (Ohio, 2d Quarter 1964), pp. 10-11.

[36] Quarterly Status Report No. 8, for period ending 29 Feb. 1964, pp. 29-30; Gordon P. Cress and C. E. Heimstadt, interviews, Burbank, Calif., 5 July 1966.

[37] Quarterly Status Report No. 8, pp. 48-49; TWX, Mathews to Burke, Attn: George J. Weber, "Contract NAS 9-170: Fuel Cell Program," GP-54541, 10 Feb. 1964; TWX, Mathews to McDonnell, Attn: Burke, GS-53211, 18 March 1964.

[38] Quarterly Status Report No. 7, for period ending 30 Nov. 1963, pp. 61-62; Weekly Activity Report, 2-8 Feb. 1964, p. 11; Quarterly Status Reports No. 8, pp. 48-49, and No. 11, pp. 15-16; "Fuel Cell Development" in "Gemini Administrator's Review, 1964."

[39] Memo, Wilbur H. Gray to Mathews, "Visit to General Electric Company, DECO, West Lynn, Mass.," GM-4224, 31 March 1964.

[40] Quarterly Status Report No. 9, p. 31.

[41] Consolidated Activity Report, 22 Dec. 1963—18 Jan. 1964, pp. 15-16; Ron Helsel, interview, Canoga Park, Calif., 16 May 1967; Schneider interview; memo, Meyer to MSC Historical Office, "Comments on Draft Chapters 7 and 8 of Gemini Narrative History," 6 Jan. 1972.

[42] Meyer, notes on GPO staff meeting, 8 Jan. 1964; memo, James R. Flanagan to Mueller, "Gemini OAMS Engines," 16 Jan. 1964; Larry E Stewart, interview, Canoga Park, Calif., 16 May 1967.

[43] Quarterly Status Report No. 8, pp. 19-20; Consolidated Activity Report, 19 Jan—15 Feb. 1964, p. 16; Meyer, telephone interview, 25 Jan. 1973; Meyer, notes on GPO staff meeting, 30 Jan. 1964, p. 1; TWX, Mathews to McDonnell, Attn: Burke, GS-53192, 25 Feb. 1964; Stephen J. Domokos, interview, Canoga Park, Calif., 16 May 1967.

[44] TWX, Mathews to McDonnell, Attn: Burke, GP-54605, 6 March 1964; memo, Flanagan to Mueller, "Gemini OAMS Engines," 20 March 1964.

[45] TWXs John Brown to MSC, Attn: Mathews, "Gemini Bidaily System Status Report[s] No. 51, RCS and OAMS," 306-16-5792, 15 April, "No. 52," 306-16-5800, 17 April, and "No. 62," 306-16-6272, 22 May 1964; Raymond L. Zavasky, recorder, "Minutes of Senior Staff Meeting, April 17, 1964," p. 4.

[46] TWX, Mathews to McDonnell, Attn: Burke, GS-53233, 16 April 1964; letter, John Brown to MSC, Attn: Mathews, "Minutes of NASA/MAC Management Meeting of 17 April

1964," 306-16-6187, 22 April 1964, with enclosure, minutes, p. 4; TWX, Brown to MSC, Attn: Mathews, "Contract NAS 9-170, Project Gemini—Incorporation of 6 Degree Ablative Material in Gemini Thrust Chamber Assemblies," 306-16-6189, 29 April 1964; TWX, Brown to MSC, Attn: Mathews, "TCA Configurations for Spacecraft 5 and Subsequent," 306-16-6195, 7 May 1964; TWX, Mathews to McDonnell, Attn: Burke, GS-53253, 2 June 1964.

[47] Purser, "Minutes of Project Gemini Management Panel Meeting ...," June 27, 1963," p. 2; "Configuration Document for Gemini Spacecraft Number 1, Preliminary," MAC No. 8611-1, 1 Oct. 1962 (revised 28 March 1963), p. 10; TWX, Mathews to Burke, GS-53188, 18 Feb. 1964; Loyd S. Swenson, Jr., James M. Grimwood, and Charles C. Alexander, *This New Ocean: A History of Project Mercury*, NASA SP-4201 (Washington, 1966), p. 228; "Abstract of Meeting on Ground Network, April 24, 1963," 7 May 1963; TWX, Mathews to Burke, GPO-50727, 21 March 1963.

[48] Letter, J. M. Gardner, Jr., to MSC, Attn: Contracting Officer, "Contract NAS 9-170, Gemini, Description of Project Orbit Testing and Spacecraft System Testing," 306-16-990, 13 July 1962, enclosure 2, "Spacecraft System Testing, Description of Work"; Quarterly Status Report No. 6, for period ending 31 Aug. 1963, p. 85; Purser, "Minutes of Project Gemini Management Panel Meeting ...," September 5, 1963," p. 2; R. W. Miller, "Monthly Progress Letter Report, 1 July 1963 thru 31 July 1963," pp. 25-26.

[49] "Spacecraft System Testing, Description," pp. 2-5; Quarterly Status Report No. 6, p. 85; Miller, "Monthly Progress Letter Report, 1 September thru 30 September 1963,"p. 24.

[50] TWX, Mathews to Burke, GPO-54222-A, 30 Sept. 1963; "Abstract of Meeting on Spacecraft No. 1 Roll-Out Inspection, October 1, 1963," 7 Oct. 1963; Mathews, activity report, 29 Sept.—5 Oct. 1963, p. 1.

[51] Purser, "Minutes of Project Gemini Management Panel Meeting ...," December 20, 1962," p. 3; "Abstract[s] of Meetings on Spacecraft Operations, August 3, 1962," 13 Aug., and "August 24, 1962," 29 Aug. 1962; Walter J. Kapryan and Wiley E. Williams, "Spacecraft Launch Preparations," in *Gemini Midprogram Conference, Including Experiment Results*, NASA SP-121 (Washington, 1966), pp. 213-14; Quarterly Status Report No. 2, for period ending 31 Aug. 1962, pp. 35-36.

[52] "Gemini Program Mission Report for Gemini-Titan 1 (GT-1)," MSC-R-G-64-1, May 1964, pp. 3-4, -5, 12-1, -11, -12, -13; Executive Order 11129, 29 Nov. 1963, and Decisions on Geographic Names in the United States, September through December 1963, Decision List No. 6303, Dept. of the Interior, 1964, as cited in Angela C. Gresser, "Historical Aspects Concerning the Redesignation of Facilities at Cape Canaveral," KSC Historical Note No. 1, April 1964, pp. 15, 18; letter, Mathews to Burke, GP-03496, 3 March 1964.

[53] "GT-1 Mission Report," pp. 3-3, 12-1, -4.

[54] "Abstract of Meeting on Titan II, March 1, 1963," 5 March 1963; Harris, "Launch Vehicle Chronology," p. 17; "GT-1 Mission Report," p. 12-6; *Gemini-Titan II Air Force Launch Vehicle Press Handbook* (Martin-Baltimore, 2d ed., Manned Space Flight, 1965), pp. B-10, D-1; Robert Goebel, Kenneth W. Graham, and David H. Baxter, "Activities Pertinent to the Technical Review of the First Gemini Launch Vehicle Propellant Tanks Inspection during the Period 11 February—6 March 1963," n.d.; Goebel and John R. Lovell, "Summary of GLV-1 Roll Out Inspection," 14 Feb. 1963; Goebel and Lovell, "Summary of GLV-001 Tank Roll Out Inspection," 5 March 1963; Haggai Cohen, interview, Baltimore, 24 May 1966; J[oseph] F. Wambolt and S[ally] F. Anderson, coordinators, "Gemini Program Launch Systems Final Report: Gemini/Titan Launch Vehicle; Gemini/Agena Target Vehicle; Atlas SLV-3," Aerospace TOR-1001(2126-80)-3, January 1967, p. II.F-1; Bastian Hello, interview, Baltimore, 23 May 1966; news release AGS-625, "Guardian Engineer Leaves People to Angels," Aerojet-General Corp., n.d.; Sol Levine, "Man-Rating the Gemini Launch Vehicle," presented at the American Institute of Aeronautics and Astronautics 1st Annual Meeting and Technical Display, Washington, 29 June—2 July 1964, pp. 24-25.

[55] "GT-1 Mission Report," p. 12-6; *Launch Vehicle Press Handbook*, p. D-1.

[56] "Gemini Launch Vehicle Familiarization Manual," Martin-Baltimore, revised February 1965, p. 1-21; *Launch Vehicle Press Handbook*, pp. 4-5, D-1; Wambolt and Anderson, "Launch Systems Final Report," p. II.F-1; Cohen interview; Willard Thackston, interview, Baltimore, 23 May 1966.

[57] Weekly Activity Reports, 2-8 June, pp. 2-3, and 16-22 June 1963, p. 3; Purser, "Management Panel Meeting, June 27, 1963," pp. 2-3; "GT-1 Mission Report," p. 12-7; Mathews, activity report, 28 July—3 Aug. 1963, pp. 1-2.

[58] "GT-1 Mission Report," p. 12-7; *Launch Vehicle Press Handbook*, p. D-7; Walter D. Smith, "Gemini Launch Vehicle Development," in *Gemini Midprogram Conference*, pp. 117-18; Wambolt and Anderson, "Launch Systems Final Report," pp. II.F-1, -2; Cohen interview.

[59] *Launch Vehicle Press Handbook*, p. D-2; TWX, Mathews to Dineen, GPO-54159-A, 9 Sept. 1963; Wambolt and Anderson, "Launch Systems Final Report," p. II.G-3; "Aerospace Corporation Annual Report, Fiscal 1963-1964," n.d.; *Weekly Activity Report*, 22-28 Sept. 1963, p. 1.

[60] Harris, "Launch Vehicle Chronology," p. 28; "GATV Progress Report, May 1965," LMSC-A605200-9, 20 June 1965, p. 2-12.

[61] "GT-1 Mission Report," p. 12-7; Harris, "Launch Vehicle Chronology," p. 28; *Launch Vehicle Press Handbook*, p. D-2; "Aerospace Annual Report, 1963-1964;" Robert R. Hull, interview, Los Angeles, 18 April 1966.

[62] "GT-1 Mission Report," p. 12-7; *Launch Vehicle Press Handbook*, pp. D-2, -3; Purser, "Minutes of Project Gemini Management Panel Meeting, November 13, 1963," p. 3.

[63] "GT-1 Mission Report," pp. 12-8, 23; Consolidated Activity Report, 20 Oct.—16 Nov. 1963, p. 84; *Launch Vehicle Press Handbook* (1964 ed.), pp. IV-10, -13, VII-10, -12, -15; Purser, "Management Panel Meeting, November 13, 1963," p. 3; Hal Taylor, ed., "Gemini Special Report: Titan II Carefully Groomed for New Role, *Missiles and Rockets*, 13 April 1964, p. 27.

[64] Letter, Mathews to Dineen, GPO-03268-A, 9 Nov. 1963; TWX, Mathews to NASA Hq. for Seamans, GPO-52121-LV, 22 Nov. 1963; Weekly Activity Report, 17-23 Nov. 1963, p. 1.

[65] Letter, Mathews to Dineen, "Gemini Launch Vehicle Coordination Committee," GPO-03277-A, 3 Dec. 1963; TWX, Mathews to Dineen, GPO-54424-A, 5 Dec. 1963; letter, Mathews to Dineen, "Gemini Launch Vehicle Coordination Committee," GP-03470, 20 Feb. 1964.

[66] TWXs, Mathews to NASA Hq. for Seamans, GPO-52179-LV, 13 Dec., and GPO-52187LV, 23 Dec. 1963; letter, Funk to Gilruth, "Gemini Launch Vehicle Meeting Regarding Operations at Cape Kennedy," 21 Jan. 1964.

[67] "GT-1 Mission Report," p. 12-8; Consolidated Activity Reports, 17 Nov.—21 Dec., pp. 21, 73, and 22 Dec. 1963—18 Jan. 1964, p. 64; Weekly Activity Reports, 5-11 Jan., p. 8, and 19-25 Jan. 1964, p. 7; *Astronautics and Aeronautics, 1964: Chronology on Science, Technology, and Policy*, NASA SP-4005 (Washington, 1965), p. 20; "Gemini-Titan II Firing a Success," *Missiles and Rockets*, 27 Jan. 1964, pp. 10-11; *Launch Vehicle Press Handbook* (1964 ed.), p. I-9; Wambolt and Anderson, "Launch Systems Final Report," p. II.F-3.

[68] "GT-1 Mission Report," p. 12-9; TWX, Kapryan to MSC for Mathews, AMR-03-105, 3 March 1964; TWX, Mathews to Hq., Attn:

Schneider, "GLV Status Reports Nos. 7 & 8, Feb. 3 & 4, 1964," GT-55036, 4 Feb. 1964.

[69] "The First Gemini Executives Meeting," 27 March 1964, Tab C, "Problems—April 63 Plan," and "Manned Space Flight Schedule: Schedule Trend, First Gemini Manned Flight"; Quarterly Status Report No. 8, pp. 80-81.

[70] Letters, Walter C. Williams to Burke, Funk, and von Braun, 12 Oct. 1962, with enclosure, "Draft Outline: Project Gemini Management Panel"; Purser, "Management Panel Meeting, November 13, 1963," pp. 3-4; Purser, "Minutes of Project Gemini Management Panel Meeting . . . , December 13, 1963," pp. 2-6.

[71] Zavasky, "Minutes of Senior Staff Meeting, November 22, 1963," p. 5; Purser, "Management Panel Meeting, December 13, 1963," pp. 5-6; Quarterly Status Report No. 8, pp. 79-80; Zavasky, "Minutes of Senior Staff Meeting, February 28, 1964," p. 6; Lewis R. Fisher, interview, Houston, 24 March 1966; John F. Yardley, interview, St. Louis, 13 April 1966; John J. Williams, interview, Cape Kennedy, Fla., 24 May 1967; MSC News Release 64-62, 30 March 1964; "MSC-Florida Operations Reorganization Announced," MSC *Space News Roundup*, 15 April 1964; G. Merritt Preston, interview, Cape Kennedy, 24 May 1967; Scott H. Simpkinson, telephone interview, 13 Aug. 1971; memo, Charles W. Mathews to Mgr., MSC-Florida Operations, "Letter of appreciation," GT-05241, 12 Nov. 1964; Yardley, "Spacecraft Check-Out," in "Gemini/Apollo Executives Meeting," 27-28 Jan. 1967, p. F-1; Mathews, "Gemini Summary," ibid., pp. G-3, 4.

[72] Purser, "Management Panel Meeting, December 13, 1963," p. 2, Fig. 2-4; TWX, Mathews to Robert C. Seamans, Jr., GPO-52159-LV, 6 Dec. 1963; "Abstract of Meeting [on] Schedule Review of Gemini Launch Vehicle, November 27, 1963," 2 Dec. 1963; "Manned Flight Schedule: Master Launch Schedule," as of 24 Dec. 1963; "First Gemini Executives Meeting," Tab C, "Features of Dec 63 Plan."

## Chapter IX

[1] MSC Announcement No. 64-64, "Reorganization and Personnel Assignments of the Gemini Program Office," 3 April 1964; MSC *Telephone Directory*, January 1964; "Major Move to Clear Lake Begins February 20," MSC *Space News Roundup*, 8 Jan. 1964.

[2] TWX, Walter J. Kapryan to Mathews, AMR-03-03-105, 3 March 1964; *Gemini-Titan II Air Force Launch Vehicle Press Handbook* (Martin-Baltimore, ca. December 1964), pp. 7-10, -13.

[3] TWX, Kapryan to Mathews, AMR 03-03-106, 3 March 1964; TWX, N. A. Mas and Maj. Carl Ausfahl to Bernhard A. Hohmann, Col. Richard C. Dineen, and Lt. Col. D. B. Ingram, AS/Cape 89-64, 6 March 1964; TWX, Mathews to NASA Hq., Attn: William C. Schneider, "GLV Status Report No. 20A," GT-55073, 27 March 1964; memo (telephone), Head, Launch Vehicle Systems Office (LVSO), to Mgr., GPO, "Launch vehicle systems daily status for March 5, 1964," 6 March 1964.

[4] Head, LVSO, memo, 6 March 1964; TWXs, Kapryan to Mathews, AMR 03-09-107, 9 March and AMR 03-11-109, 11 March 1964; memo, Leon DuGoff to Mathews, "Launch Vehicle Summary report for Period March 2 through March 6, 1964," 9 March 1964; memo, DuGoff to Mathews, "Milestone Report for March 6, 1964," 9 March 1964; memo, DuGoff to Mathews, "LVS Daily status for 3-9-64," 10 March 1964; memo, DuGoff to Mathews, "LVS Daily Status as of 1430, EST, March 11, 1964," 11 March 1964.

[5] TWXs, Kapryan to Mathews, AMR 03-13-110, 13 March, AMR 03-23-120, 23 March, and AMR 03-27-126, 27 March 1964; TWXs, Mas and Ausfahl to Hohmann, Dineen, and Ingram, AS/Cape 118-64, 13 March, and 138-64, 26 March 1964; TWX, Mathews to NASA Hq., Attn: Schneider, "GLV Weekly Summary Report No. 21," GT-55070, 24 March 1964; memo, DuGoff to Mathews, "GLV status report for March 18," n.d.; telephone call from DuGoff to GPO, 3:03, 19 March 1964; memo, DuGoff to Mgr., GPO, "Gemini launch vehicle status report," 20 March 1964; memo, DuGoff to Mgr., Test Ops., "GLV daily status for March 23, 1964," 24 March 1964; memo, DuGoff to Mgr., Test Ops., "GLV daily status for March 24, 1964," 25 March 1964; memo, DuGoff to Mgr., Test Ops., "GLV status report for March 25, 1964 through 8:00 a.m. March 26, 1964," 26 March 1964.

[6] DuGoff memo, 26 March 1964; memo, DuGoff to Mgr., Test Ops., "Launch Vehicle Systems Summary Report for Period March 23 through March 29, 1964," 30 March 1964; TWXs, Mas and Ausfahl to Hohmann, Dineen, and Ingram, AS/Cape 139-64, 27 March, and 142-64, 30 March 1964.

[7] TWX, Mathews to NASA Hq., Attn: Schneider, "GLV Status Report No. 26," GT-55089, 3 April 1964; TWX, Mas and Ausfahl to Hohmann, Dineen, and Ingram, AS/Cape 148-64, 1 April 1964; TWX, Kapryan to Mathews, AMR 03-31-134, 1 April 1964; memo, DuGoff to Mgr., Test Ops., "Combined daily status for April 1 and April 2, 1964 for GLV," 2 April 1964; "Gemini Program Mission Report for Gemini-Titan 1 (GT-1)," MSC-R-G-64-1, May 1964, pp. 12-10, -23; TWX, Mathews to NASA Hq., Attn: Schneider, "Daily GLV Status Report No. 29," GT-55096, 9 April 1964.

[8] TWX, Mathews to NASA Hq., Attn: Schneider, "Combined Daily Status for April 1 and April 2 for GLV, No. 27," GT-55092, 3 April 1964; DuGoff memo, 2 April 1964; TWX, Kapryan to Mathews, AMR 04-03-137, 3 April 1964; TWX, Mas and Ausfahl to Hohmann, Dineen, and Ingram, AS/Cape 150-64, 3 April 1964; memo, DuGoff to Mathews, "Daily status for April 2 through 1300 April 3, 1964 for GLV," 3 April 1964; Edward F. Mitros, telephone interview, 16 Oct. 1973.

[9] Letter, Mathews to Walter F. Burke, GP-03530, 24 March 1964, with enclosure, "Data Required in Support of S/C No. 1 Flight Readiness Review;" memo, Mathews to dist., "Gemini Flight Readiness Review Procedure," GT-05031, 24 March 1964, with enclosure, subject as above, 29 Feb. 1964; TWX, Mathews to Space Systems Div. (SSD), Attn: Dineen, and McDonnell, Attn: Burke, GP-54650, 30 March 1964; TWX, Mathews to SSD, Attn: Dineen, and McDonnell, Attn: Burke, GT-55081, 31 March 1964; TWX, Mas and Ausfahl to Hohmann, Dineen, and Ingram, AS/Cape 149-64, 2 April 1964; memo (telephone), Head, LVSO, to Mgr., Test Ops., "Daily Status Report from April 3 to April 5, 1964 for GLV," 6 April 1964; TWX, Kapryan to Mathews, 94-06-141, 6 April 1964; TWX, Mas and Ausfahl to Hohmann, Dineen, and Ingram, AS/Cape 153-64, 6 April 1964; W. Harry Douglas and Lester A. Stewart, telephone interviews, 23 Feb. 1973; Warren J. North, telephone interview, 28 Feb. 1973.

[10] "GT-1 Mission Report," pp. 12-13, -14; MSC News Release, GT-1, 6 April 1964.

[11] MSC News Release, GT-1, No. 2, for release at 9 a.m., 7 April 1964; "GT-1 Mission Report," pp. 12-10, -14; memo (telephone), Head, LVSO to Mgr., Test Ops., "Daily Status Report for April 6, 1964, for GLV," 7 April 1964; TWX, Mas and Ausfahl to Hohmann, Dineen, and Ingram, AS/Cape 156-64, 7 April 1964.

[12] "GT-1 Mission Report," pp. 4-1, 6-1; TWXs, Mas and Ausfahl to Hohmann, Dineen, and Ingram, AS/Cape 156-64 and 162-64, 8 April 1964; NASA News Release 64-70, "First Gemini Orbital Flight Scheduled," 30 March 1964, pp. 5-7, 8; Paul P. Haney, handwritten notes for GT-1 mission commentary, 8 April 1964; [Janet Shrum], "Weather report at 0800, April 8, [1964];" Bastian Hello, interview, Baltimore, 23 May 1966; "News Conference, Gemini/Titan-1," 8 April 1964, p. 2.

[13] "GT-1 Mission Report," pp. 4-2, -8, 5-34; *Gemini-Titan II Air Force Launch Vehicle Press Handbook* (Martin-Baltimore, 2d ed., Manned Space Flight, ca. March 1965), pp. iii, vii; [Ivan D. Ertel], *Gemini Program*, MSC Fact Sheet No. 291 (Houston, February 1965).

[14] "GT-1 Mission Report," p. 4-8; "GT-1 News Conference," pp. 2, 3; "Walter [C.] Williams Resigns to Join Private Industry," MSC *Space News Roundup*, 1 April 1964; letter, George E. Mueller to Christopher C. Kraft, Jr., "Operations Director for Gemini GT-2," 16 July 1964; "Christopher Kraft Picked to Direct Next Gemini Flight," MSC *Space News Roundup*, 5 Aug. 1964; MSC Announcement No. 64-187, "Appointment of Operations Director for GT-3," 23 Dec. 1964.

[15] Robert W. Fricke, Jr., "Mission Directive for Gemini-Titan II Mission I, GT-1 (Spacecraft No. 1)," NASA Project Gemini working paper No. 5005, 14 Nov. 1963; "GT-1 Mission Report," pp. 2-1, -2, 4-1, -3; [Ertel], *Gemini Program*; NASA News Release 64-70, p. 15; "Gemini Program Gets Off to a Successful Start," MSC *Space News Roundup*, 15 April 1964; "Manned Space Flight Network Performance Analysis for the First Gemini Mission," Goddard Space Flight Center X-552-64-206, 1 May 1964, Supplemental Report No. 12 to "GT-1 Mission Report"; William R. Corliss, "The Evolution of the Manned Space Flight Network through Gemini," 1 Dec. 1967, pp. 102-10; MSC News Release, GT-1-3, 7 April 1964.

[16] "Minutes of the Twelfth Meeting, Gemini Program Planning Board, Monday, April 20, 1964," p. 1.

[17] NASA News Release 63-246, "NASA Announces Changes in Saturn Missions," 30 Oct. 1963; memo, George M. Low to Assoc. Adm., Manned Space Flight, "Gemini Missions," 27 Nov. 1963; letter, Mathews to NASA Hq., Attn: Low, "Evaluation of Gemini Missions," GPO-01060-M, 9 Sept. 1963; memo, Mathews to Asst. Dir., Engineering and Development, "Advanced Mission Planning in Support of the Gemini Program," GPO-01063-M, 17 Sept. 1963; memo, Maxime A. Faget to Mgr., GPO, "Advance mission planning in support of the Gemini program," 30 Dec. 1963; Project Gemini Quarterly Status Report No. 8, for period ending 29 Feb. 1964, p. 74; memo, Edward Z. Gray to Dir., Gemini Program, "Gemini Lunar Mission Studies," 30 April 1964, with enclosure, memo, William B. Taylor to Dir., Adv. Manned Missions, subject as above, 21 April 1964, with enclosures; Simpkinson, interview, Houston, 18 Jan. 1967.

[18] "Gemini Circumlunar Flight 'Feasible,'" *Missiles and Rockets*, 18 May 1964, p. 17; memo, Seamans to Assoc. Dir., Manned Space Flight, "Proposed Gemini Circumlunar Mission Study," 12 June 1964; letter, Mathews to NASA Hq., Attn: Schneider, "Advanced Gemini Missions," GV-02312, 18 Sept. 1964, with enclosure, [Calvin C. Guild], "Notes on Advanced Gemini Missions, Conceptual Study July 30, 1964."

[19] Purser, "Minutes of Project Gemini Management Panel Meeting . . . , April 15, 1964," p. 2, Figs. A-2-9, A-2-12.

[20] MSC Weekly Activity Report for Office of the Dir., Manned Space Flight, 18-24 Aug. 1963, pp. 1-2; Quarterly Status Report No. 7, pp. 1-2.

[21] André J. Meyer, Jr., notes on GPO staff meeting, 11 Sept. 1963, p. 1; Zavasky, "Minutes of Senior Staff Meeting, September 13, 1963," p. 5; memo, Paul M. Sturtevant to Security Div., Public Affairs Office, and Offices Services Div., "Preliminary Review for Design Engineering Inspection (DEI) Spacecraft No. 2," GPO-01077-M, October 1964; TWX, Mathews to Dineen et al., "Postponement of the Formal Design Engineering Inspection of Spacecraft No. 2," GPO-54359-A, 13 Nov. 1963; TWX, Mathews to Dineen et al., "Rescheduling of the Formal Development Engineering Inspection of Gemini Spacecraft No. 2," GPO-54480-A, 3 Jan. 1964; memo, James C. Elms for dist., "Establishment of Development Engineering Inspection Board," GP-03426, 31 Jan. 1964; MSC Consolidated Activity Report for Office of the Assoc. Adm., Manned Space Flight, 19 Jan.—15 Feb. 1964, p. 18; memo, Mathews for dist., "Report on the Gemini Spacecraft No. 2 Development Engineering Inspection," GP-03541, 30 April 1964, with enclosure, "Gemini Spacecraft No. 2 Development Engineering Inspection Report"; memo, LeRoy E. Day to Dep. Dir., Gemini Program, "Spacecraft Schedule Status," 1 May 1964.

[22] Quarterly Status Report No. 9, for period ending 31 May 1964, pp. 1-2; Purser, "Management Panel Meeting, April 15, 1964," Fig. A-1-4.

[23] "Gemini Program Mission Report, GT-2, Gemini 2," MSC-G-R-65-1, February 1965, pp. 12-2, -3.

[24] Letter, Col. Alfred J. Gardner to dist., "Gemini Launch Vehicle Number Two Rollout Inspection," 23 July 1964, with enclosed summary, Kenneth W. Graham and David H. Baxter, "Activities Pertinent to the Technical Review of the Second Gemini Launch Vehicle Propellant Tanks during the Period 20 May—10 July 63"; "GT-2 Mission Report," pp. 12-9, -11, -12; Mathews, activity report, 7-13 July 1963, p. 2; *Launch Vehicle Press Handbook*,

2d ed., p. D-3; [Joseph] F. Wambolt and S[ally] F. Anderson, coordinators, "Gemini Program Launch Systems Final Report: Gemini/Titan Launch Vehicle; Gemini/Agena Target Vehicle; Atlas SLV-3," Aerospace TOR-100(2126-80)-3, January 1967, p. H.G-3; Quarterly Status Report No. 8, p. 3; "The First Gemini Executives Meeting," Tab D, "GLV-2 Problem Summary."

25 "GLV-2 Problem Summary"; Quarterly Status Report No. 9, p. 3; "GT-2 Mission Report," pp. 12-12 through -15; Howard T. Harris, "Gemini Launch Vehicle Chronology, 1961-1966," AFSC Historical Publications Series 66-22-1, December 1966, p. 37; Mathews, activity report, 17 May—20 June 1964, p. 1.

26 Memo, DuGoff to Mgr., Test Ops., "Gemini Launch Vehicle Activity Report No. 1," 13 July 1963; "History of Project Gemini (GT-2) at Kennedy Space Center," KSC External Affairs Office, n.d., pp. 1-16; TWX, Mathews to NASA Hq., Attn: Schneider, "Gemini Launch Vehicle (GLV) No. 2 Activity Report No. 1," GT-55168, 15 July 1964; Mathews, activity report, 21 June—18 July 1964, p. 3; memo,, DuGoff to Mgr., Test Ops., "Gemini Launch Vehicle Activity Report No. 9, period August 14 p.m.—August 16 p.m.," 17 Aug. 1964.

27 "History of GT-2 at KSC," p. 17; "GT-2 Mission Report," pp. 12-15, -16; memo, DuGoff to Asst. Mgr. for Gemini, "Thunderstorm/Lightning Incident—GLV-2," 21 Aug. 1964; Hohmann and Ernst R. Letsch, interview, El Segundo, Calif., 19 April 1966; Joseph M. Verlander, Kapryan, Dineen, and Col. Lamar D. Bowles, telephone interviews, 28 Feb. 1973.

28 DuGoff memo, 21 Aug. 1964; Wambolt and Anderson, "Launch Systems Final Report," pp. II.E-13, -14; "GT-2 Mission Report," p. 12-16.

29 "History of GT-2 at KSC," pp. 26-28; memos, DuGoff to Mgr., Test Ops., "Gemini Launch Vehicle Activity Report[s] No. 11," 26 Aug., and "No. 12," 4 Sept. 1964; "GT-2 Mission Report," p. 12-16; "First Titan 3A Vehicle Fails to Achieve Orbit," Aviation Week and Space Technology, 7 Sept. 1964, p. 32.

30 Notes, Schneider to Mueller, 19 and 24 Aug. 1964; Harris, "Launch Vehicle Chronology," p. 42; SSD Briefing to Gemini Executive Management Meeting, 4 Sept. 1964; letter, Mathews to Maj. Gen. Ben I. Funk, GA-01281, 10 Sept. 1964.

31 "GT-2 Mission Report," p. 12-16; memos, DuGoff to Asst. Mgr. for Gemini, Florida Ops., "Launch Vehicle Activity — Report #13," 10 Sept., and "Report #14," 23 Sept. 1964.

32 "GT-2 Mission Report," pp. 12-3, -4; memo, Mathews to dist., "Personnel assignments for Spacecraft 2 acceptance review," 19 Aug. 1964, with enclosure; TWX, Mathews to McDonnell, Attn: Burke, "Spacecraft 2 Acceptance Review," GT-55212-A, 25 Aug. 1964; TWX, Mathews to McDonnell, Attn: Burke, "Spacecraft 2 Acceptance Review Team and Board," GP-54939, 27 Aug. 1964; letter, Mathews to Burke, "Contract NAS 9-170; results of Spacecraft Acceptance Review, Spacecraft 2 Phase I Review," GP-03881, 12 Sept. 1964; TWX, Mathews to McDonnell, Attn: Burke, GP-54961, 11 Sept. 1964; letter, Mathews to Wilbur H. Gray, 18 Sept. 1964; letter, Mathews to Burke, "Contract NAS 9-170; results of Spacecraft Acceptance Review, Phase II for Spacecraft 2," GP-03934, 5 Oct. 1964; memo, Mathews to dist., "Results of Spacecraft Acceptance Review, Phase II for Spacecraft 2," GP-03938, 5 Oct. 1964; Mathews, activity report, 20-26 Sept. 1964, p. 2; TWX, Kapryan to Mathews, AMR 09-22-223, 22 Sept. 1964.

33 Memo, Mueller to Seamans, "Gemini Launch Schedule Approval," 15 Sept. 1964, with Seamans' concurrence the same day; Willis B. Mitchell, Jr., recorder, "Minutes of Project Gemini Management Panel Meeting . . . , September 29, 1964," p. 3, Figs. A-1-1 through -4, A-2-3; Henry Tanner, "Soviet Spaceship Is Landed Safely after 16 Circuits," The New York Times, 14 Oct. 1964.

34 DuGoff memo, 23 Sept. 1964; memos, DuGoff to Asst. Mgr. for Gemini, Florida Ops., "Launch Vehicle Activity—Report #19," 8 Oct., and "Report #21," 14 Oct. 1964; "History of GT-2 at KSC," pp. 36, 40, 41, 44, 47, 49; "GT-2 Mission Report," p. 12-16; material compiled by Alfred Rosenthal, Office of Public Affairs, Goddard; "Worldwide Tracking Network Being Tested for Manned Flights," MSC Space News Roundup, 14 Oct. 1964.

35 "GT-2 Mission Report," pp. 12-4, -5; "History of GT-2 at KSC," pp. 39-54; TWXs, Kapryan to Mathews, daily status reports: No. 9, 5 Oct., No. 10, 6 Oct., No. 11, 6 Oct., No. 13, 8 Oct., No. 14, 9 Oct., No. 15, 12 Oct., and No. 20, 19 Oct. 1964; memo, Schneider to Assoc. Adm., Manned Space Flight, "Schedule Delays at Cape Kennedy—GT-2," 4 Nov. 1964.

36 "GT-2 Mission Report," pp. 12-5, -6; "History of GT-2 at KSC," pp. 54-57, 59-60, 63-65; TWXs, Kapryan to dist., daily status reports: No. 27, 28 Oct., No. 29, 30 Oct., and No. 34, 6 Nov. 1964; TWXs, Kapryan to MSC, Attn: Mathews, daily status reports: No. 30, 2 Nov., No. 31, 3 Nov., No. 32, 4 Nov., and No. 33, 5 Nov. 1964; Verlander, interview, Cocoa Beach, Fla., 29 Aug. 1967.

[37] "GT-2 Mission Report," pp. 12-6, -7, -8, -17, -46; "History of GT-2 at KSC," pp. 66-84.

[38] "GT-2 Mission Report," pp. 12-17, 13-8, -9, -20.

[39] *Launch Vehicle Press Handbook* (1964 ed.), p. V-5; Mathews, interview, Houston, 12 Dec. 1966; Jerome B. Hammack, interview, Houston, 19 Aug. 1966; Col. John G. Albert, interview, Patrick AFB, Fla., 26 May 1967; Dineen, interview, Huntington Beach, Calif., 15 May 1967; John E. Riley, draft of press release, n.d. [9 Oct. 1964]; Hammack, telephone interview, 2 March 1973.

[40] Consolidated Activity Report, December 1964, p. 12; Kapryan, interview, Cape Kennedy, Fla., 25 May 1967; "GT-2 Mission Report," p. 12-8; "History of GT-2 at KSC," pp. 85-86; TWX, Kapryan to MSC, Attn: Mathews, daily status report No. 54, 11 Dec. 1964.

[41] Memo, Low to dist., "Transfer of MSC Florida Operations to Kennedy Space Center," 23 Dec. 1964, with enclosure, "KSC-MSC Cape Relationships," signed by MSC Director Robert R. Gilruth and KSC Director Kurt H. Debus, 21 Dec. 1964; MSC Announcement No. 64-185, "Reorganization within the Office of Manned Space Flight and Transfer of MSC Florida Operations," 24 Dec. 1964; memo, John W. Smith to Mgr., Gemini, "Actuator modification program," GV-02452, 29 Dec. 1964; "History of GT-2 at KSC," pp. 87-105; "GT-2 Mission Report," pp. 12-8, -9, -18, -21; TWX, Albert to Dineen et al., DWD 39102, 18 Jan. 1965; Albert, Hohmann, Kapryan, and Hammack interviews; Gordon T. Chambers, interview, Baltimore, 23 May 1966; Mathews, "Gemini Summary," p. G-9; memo, Kraft to dist., "GT-2, Flight Readiness and Mission Reviews," GP-01787, 8 Jan. 1965.

[42] "GT-2 Mission Report," pp. 6-2, 12-2, 13-6, -7; "History of GT-2 at KSC," p. 106; TWXs, Kapryan to MSC, Attn: Mathews, daily status reports, No. 53, 10 Dec., No. 55, 14 Dec., No. 59, 18 Dec., and No. 60, 19 Dec. 1964; note, Day to Seamans, "Gemini Spacecraft #2 Fuel Cells," 25 Jan. 1965; TWX, Mitchell to NASA Hq., Attn: James E. Webb, and MSC, Attn: Gilruth, "GT-2 Launch Summary Report," 19 Jan. 1965; Kapryan interview; H. H. Luetjen, interview, Cape Kennedy, 25 May 1967.

[43] Edward A. Armstrong and John E. Williams, "Gemini Program Flight Summary Report," MSC-G-R-66-5, July 1966, pp. 6-8; "GT-2 Mission Report," pp. 1-1, 2-1, -2, 4-2, -3; [Ertel], *Gemini Program*; TWX, Mitchell to NASA Hq., Attn: Webb, and MSC, Attn: Gilruth, "GT-2 Mission Summary Report," 20 Jan. 1965.

[44] "News Conference, Gemini-Titan-2," 19 Jan. 1965; "GT-2 Mission Report," pp. 12-31 through -34.

[45] Letter, Mathews to NASA Hq., Attn: Schneider, "Rocketdyne performance on the Gemini Program; NASA Headquarter's [sic] consideration as contractor for the Surveyor Program," GS-04072, 29 April 1964, with 18 enclosures.

[46] Memo, Mueller to Homer E. Newell, "Surveyor Vernier Engine Program," 4 May 1964; letter, Low to Gilruth, "Funds for partial support of Lunar Excursion Module descent engine project," 11 Jan. 1963, with enclosure; MSC News Release 63-92, 29 May 1963; memo, Robert H. Voigt to Asst. Mgr., Apollo Spacecraft Program Office, "Parallel Development—LM Descent Engine, Grumman Aircraft Engineering Corporation Audit Report MSC 11-67A," 8 March 1967; letter, Gilruth to Mueller, 20 Oct. 1964, with enclosures; letter, Low to Meyer, 6 Aug. 1964; letter, Gilruth to Burke, 12 Aug. 1964; letter, Gilruth to Samuel K. Hoffman, 12 Aug. 1964, with enclosure, "Suggested Actions for Rocketdyne's Consideration"; letter, Low to Mueller, 28 Sept. 1964, with enclosure, "Recommendations."

[47] Low letter, 28 Sept. 1964; memo, Voigt to dist., "Report on Review of Business Management Activities at Rocketdyne, A Division of North American Aviation, Inc. (Report No. WR 65-12) MSC 32-0-65G," 5 May 1965, with enclosure, Raymond Einhorn, "Review of Business Activities at Rocketdyne, a Division of North American Aviation, Inc.," Western Region Audit Office Report No. WR 65-12, April 1965, passim., but esp. pp. 4-5, 13-14, 59-60, 72-74.

[48] Memo, Day to E. Z. Gray, "Backup 25-pound thruster development," 2 June 1964; Gilruth letter, 20 Oct. 1964; "Gemini Propulsion by Rocketdyne—A Chronology," 15 May 1967, p. 5; Weekly Activity Report, 21-27 March 1965, p. 1.

[49] Weekly Activity Reports, 31 May—6 June, p. 1, and 28 June—4 July 1964, p. 1.

[50] Memo, Clarence C. Gay, Jr., to Day, "Status Review of Pyrotechnic Program," 5 Oct. 1964; Quarterly Status Report No. 10, for period ending 31 Aug. 1964, pp. 19, 24-26; Weekly Activity Report, 1-7 Nov. 1964, p. 2; "Abstract of Meeting on Mode I Abort Test Program, June 16, 1964," 18 June 1964; TWX, Mathews to McDonnell, Attn: Burke, GP-54793, 19 June 1964; memo, Schneider to Assoc. Adm., Manned Space Flight, "Miscellaneous Points You Have Raised," 10 Aug. 1964; memo, Stephen D. Armstrong to dist., "NASA Defense Purchase Request Number T-28860-G," BG6-976, 14 Sept. 1964, with enclosure;

Gordon P. Cress, interview, Burbank, Calif., 5 July 1966; Mathews, "Gemini Summary," pp. G-5, -6; Weekly Activity Report, 27 Sept.—3 Oct. 1964, p. 2; Quarterly Status Report No. 11, for period ending 30 Nov. 1964, p. 19; letter, George W. Jeffs to Dir., MSC, Attn: Armstrong, "Contract NAS 9-1484, Paraglider Landing System Program, Monthly Progress Report No. 19 (November 1964)," 64MA15681, 11 Dec. 1964, p. 2.

51 Cress interview; Quarterly Status Report No. 11, p. 18; Consolidated Activity Report, December 1964, p. 25; letter, Cress to MSC Historical Office, "Comment Draft on Chapters 7 & 8 of Gemini Narrative History," 511/GPC/2120, 1 Dec. 1971.

52 Weekly Activity Report, 10-16 Jan. 1965, p. 2; Quarterly Status Report No. 12, for period ending 28 Feb. 1965, p. 10; Col. Clyde S. Cherry, interview, Edwards AFB, Calif., 20 April 1966.

53 Quarterly Status Report No. 12, p. 9; Weekly Activity Report, 28 Feb.—6 March 1965, p. 2; Cress letter, 1 Dec. 1971.

54 Quarterly Status Report No. 12, pp. 9-11; Hilary A. Ray, telephone interview, 6 March 1973; Quarterly Status Report No. 13, for period ending 31 May 1965, pp. 8-9; Cress letter, 1 Dec. 1971.

55 Weekly Activity Report, 19-25 July 1964, p. 1; Meyer, notes on NASA/MAC management meeting, 17 July 1964, pp. 4-5; letter, Schneider to Mathews, 10 July 1964; Meyer, notes on GPO staff meeting, 14 July 1964, pp. 1, 3; "Gemini V Mission Directive," NASA Program Gemini working paper No. 5028, 21 July 1965, pp. 2-1, 3-4.

56 TWXs, Mathews to McDonnell, Attn: Burke, GV-52498, 5 Aug., and GS-53297, 7 Aug. 1964; Meyer, notes on NASA/MAC management meeting, 14 Aug. 1964, pp. 4-6; memo, Mathews to dist., "Responsibility assignments," 20 Aug. 1964, with enclosure, "NASA/MAC Management Meeting 14 August 1964," p. 4; letter, John Y. Brown to MSC, Attn: Mathews, "Minutes of NASA/MAC Management Meeting of 18 September 1964," 306-16-7784, 28 Sept. 1964, with enclosure, p. 2; Mitchell, "Management Panel Meeting, September 29, 1964," p. 5; TWX, Schneider to MSC, Attn: Mathews, "Fuel Cell in Spacecraft Five (5) and Six (6)," 1 Oct. 1964.

57 Meyer notes, 14 Aug. 1964, pp. 3-6; Quarterly Status Report No. 10, pp. 40-41.

58 "NASA/MAC Management Meeting 14 August 1964," p. 3; Meyer, notes on GPO staff meeting, 6 Oct. 1964, p. 2; Robert A. Dittman, "Gemini Program Office Staff Meeting Minutes, October 6, 1964," p. 3; Quarterly Status Reports: No. 10, pp. 40-41, and No. 11, pp. 11, 15-16.

59 Letter, Schneider to Mathews, 22 Sept. 1964, with enclosure, draft memo, Mathews to McDonnell, "Request for Information Concerning the General Electric Effort on Subcontract PO Y 20153R," 22 Sept. 1964.

60 "NASA/MAC Management Meeting 18 September 1964," p. 2; Mitchell, "Management Panel Meeting, September 29, 1964," p. 5; TWX, Mathews to NASA Hq., Attn: Schneider, GP-51508, 7 Oct. 1964; TWX, Schneider to MSC, Attn: Mathews, "Spacecraft #6 Configuration," 13 Oct. 1964; memo, Eldon W. Hall to Dep. Dir., Gemini, "Power supply for GTA-6," 12 Oct. 1964; memo, Day to Schneider, "Use of Fuel Cell Spacecraft #6," 19 Oct. 1964; TWX, Schneider to MSC, Attn: Mathews, 6 Nov. 1964.

61 Letter, Brown to MSC, Attn: Mathews, "Minutes of NASA/MAC Management Meeting 18 December 1964," 306-16-8320, 23 Dec. 1964, with enclosure, p. 4; memo, Mathews and Wesley L. Hjornevik to Dep. Dir., "United States General Accounting Office draft report to Congress regarding Gemini fuel cells," GP-62337, 11 Aug. 1966, with enclosure, "Detailed Comments on GAO Draft Report;" Mathews, "Gemini Summary," p. G-5.

62 "GATV Progress Report, October 1964," LMSC-A605200-2, 20 Nov. 1964, pp. 2-3, -4; "GATV Progress Report, February 1965," LMSC-A605200-6, 20 March 1965, p. 4-1; "GATV Progress Report, March 1965," LMSC-A605200-7, 20 April 1965, p. 4-1; Wambolt and Anderson, "Launch Systems Final Report," p. III.E-1; Mathews, "Gemini Summary," p. G-6.

63 "GATV Progress Report, November 1964," LMSC-A605200-3, 20 Dec. 1964, pp. 2-3, -5, 7-8; Quarterly Status Report No. 11, pp. 4, 37; "GATV Progress Report, December 1964," LMSC-A605200-4, 20 Jan. 1965, pp. 2-1, -3, -5; Consolidated Activity Reports: 18 Oct.—30 Nov., p. 16, and December 1964, p. 14; "GATV Progress Report, January 1965," LMSC-A605200-5, 20 Feb. 1965, p. 2-1; Harold W. Nolan, interview, Sunnyvale, Calif., 1 July 1966; TWX, Mathews to McDonnell, Attn: Burke, "Contract NAS 9-170 Gemini TDA-1 Delivery," GP-54878, 3 Aug. 1964.

64 Quarterly Status Report No. 12, pp. 32-34; "GATV Progress Report, January 1965," pp. 2-1, -2; "GATV Progress Report, February 1965," pp. 2-1, -4, -5, -8.

65 "GATV Progress Report, March 1965," pp. 2-1 through -6; "GATV Progress Report, April 1965," LMSC-A605200-8, 20 May 1965, p. 2-3; "Abstract of Meeting of [sic] Atlas/

Agena Coordination, April 13, 1965," 5 May 1965.

66 "GATV Progress Report, April 1965," p. 2-1; "GATV Progress Report, May 1965," LMSC-A605200-9, 20 June 1965, pp. 2-1, -2, -3, -4, -6; Wambolt and Anderson, "Launch Systems Final Report," p. III.G-3.

67 Weekly Activity Reports: 29 Nov.—5 Dec., p. 3, and 6-12 Dec. 1964, p. 4; "Abstract of Meeting on Atlas/Agena Coordination, January 12, 1965," 20 Jan. 1965; "Abstract of Meeting on Atlas Agena Coordination, February 16, 1965," 1 March 1965; Quarterly Status Report No. 12, p. 52.

68 Day, interview, Washington, 25 Jan. 1967; Lt. Col. W. A. Cobb, "Acceleration of Launch Plan to Two Months Centers," reprogramming presentation, 18 Sept. 1964; letter, John A. Edwards to Kraft, 3 Dec. 1964; memo, Charles W. McGuire to Dir., Gemini Test, "GLV Two-month Launch Interval," 28 Dec. 1964, with enclosure, memo, Col. Robert R. Hull for record, "Test Philosophy Conference," 16 Dec. 1964; memo, Robert F. Freitag to William D. Putnam, "Two-Month Launch Interval Study," 26 July 1967, with enclosure, subject as above, 14 Jan. 1965; Lt. Col. Alexander C. Kuras and Col. John G. Albert, "Gemini-Titan Technical Summary," 24 Jan. 1967, p. 138; memo, Day to MSC, Attn: Historical Office, "Comments re . . . Gemini History," 11 March 1969; Funk, interview, Sunnyvale, Calif., 12 May 1967; note, Schneider to Mueller, "Second Launch Pad for Gemini," 17 Sept. 1964; note, Simpkinson for James M. Grimwood, 14 May 1971; letter, Day to MSC, Attn: Grimwood, "Comments on the Final Manuscript of the Gemini History," 23 June 1971; Mathews, "Gemini Summary," pp. G-3, -4.

69 Purser, "Minutes of Project Gemini Management Panel Meeting . . . , February 4, 1965," p. 3, Fig. A-3-1; M. Scott Carpenter, recorder, "Minutes of Senior Staff Meeting, February 12, 1965," p. 2.

## Chapter X

1 "News Conference, GT-3 Crew Selection," 13 April 1964, pp. 1, 2, 5.

2 U.S. Congress, House, Committee on Science and Astronautics, Astronautical and Aeronautical Events of 1962: Report, 88th Cong., 1st sess., 12 June 1963, p. 122; William Hines, "Life-Sized Grissom Rides Again," The Evening Star, Washington, 30 March 1965; U.S. Congress, Senate, Committee on Aeronautical and Space Sciences, United States Astronauts: Staff Report, Senate Doc. No. 42, 88th Cong., 1st sess., 15 Nov. 1963, pp. 7, 11, 31, 35; Astronautics and Aeronautics, 1964:

Chronology on Science, Technology, and Policy, NASA SP-4005 (Washington, 1965), pp. 15-16; NASA Release No. 65-81, "Project: Gemini-Titan 3," press kit, 11 March 1965, pp. 43-49; NASA News Release 62-200-A, "Nine New Pilots Selected for Space Flight Training," 17 Sept. 1962.

3 "GT-3 Crew Selection," p. 5.

4 "Gemini Program Mission Report, GT-3, Gemini 3," MSC-G-R-65-2, April 1965, pp. 7-7, -45; "GT-3 Crew Selection," p. 6; memo, Harold I. Johnson to dist., "Preliminary description of simulators and training equipment expected to be used in Project Gemini," 5 March 1962, with enclosures; TWX, Charles W. Mathews to McDonnell, Attn: Walter F. Burke, GP-54718, 11 May 1964; MSC News Release No. 63-13, 26 Jan. 1963; "Abstract of Meeting on Crew/Hatch Clearance, July 30-31, 1963," 8 Aug. 1963; André J. Meyer, Jr., notes on GPO staff meeting, 6 Aug. 1963, p. 3; Raymond L. Zavasky, recorder, "Minutes of Senior Staff Meeting, August 16, 1963," p. 4; TWXs, Mathews to Burke, GPO-54046-A, 12 Aug., and GPO-54094-A, 20 Aug. 1963; MSC Consolidated Activity Report for Office of the Dir., Manned Space Flight, 21 July—17 Aug. 1963, p. 23; Project Gemini Quarterly Status Report No. 7, for period ending 30 Nov. 1963, p. 5; memo, Peter J. Vorzimmer, "Memorandum on the Gemini Project History," [June 1967]; TWX, Mathews to Burke, GP-7049, 8 April 1965.

5 Mathews, activity reports: 31 May—2 June, p. 1, 4-10 Oct., p. 1, and 18 Oct.—27 Nov. 1964, pp. 3-4; MSC Consolidated Activity Report for Office of the Assoc. Adm., Manned Space Flight, 21 June—18 July 1964, p. 17.

6 TWX, Mathews to McDonnell, Attn: Burke, GP-54686, 28 April 1964; "GT-3 Mission Report," pp. 7-7, -45; MSC Weekly Activity Report for Office of the Assoc. Adm., 15-21 Nov. 1964, p. 1.

7 Consolidated Activity Report, 17 May—20 June 1964, p. 30; Zavasky, "Minutes of Senior Staff Meeting, June 19, 1964," p. 3; Quarterly Status Report No. 10, for period ending 31 August 1964, p. 56.

8 "GT-3 Mission Report," pp. 7-9, -44; Consolidated Activity Report 23 Aug.—19 Sept. 1964, p. 50.

9 TWX, Wilbur H. Gray to GPO, "Gemini Weekly Activity Report No. 80," 713-488-0454, 25 Aug. 1964; Paul T. Chaput, "Crew Egress Procedures Developed during the Qualification Test Program for the Gemini Spacecraft At-Sea Operations," NASA Program Gemini working paper No. 5015, 26 Aug. 1964; "GT-3 Mission Report," pp. 7-8, -44; Consolidated Activity Report, 20 Sept.—17 Oct. 1964, p. 32.

[10] "GT-3 Mission Report," pp. 7-8, -44; "GT-3 Prime Crew Train at Sea," MSC *Space News Roundup*, 3 March 1965; "GT-3 Crews Perform Parachute and Egress Training," MSC *Space News Roundup*, 17 Feb. 1965.

[11] Quarterly Status Report No. 6, for period ending 31 Aug. 1963, pp. 77-78; "GT-3 Mission Report," pp. 7-8, -44; "News Conference, GT-4 Crew Selection," 29 July 1964; Consolidated Activity Reports, 18 Oct.—30 Nov., p. 28, and December 1964, p. 25.

[12] Mathews, activity report, 18 Oct—27 Nov 1964, pp. 2-3; TWX, Mathews to SSD, Attn: Richard C. Dineen, GV-52670, 13 Jan. 1965; "GT-3 Mission Report," pp. 7-7, -8, -45.

[13] Consolidated Activity Report, January 1965, p. 16; "GT-3 Mission Report," pp. 7-8, -44.

[14] "GT-3 Mission Report," pp. 7-7, -43, -44, 12-21.

[15] "Astronauts Ready Now," *The Kansas City Times*, 5 Feb. 1965; Ellis Rall, "Gemini Shot May Come in March," *Evening World-Herald*, Lincoln, Neb., 5 Feb. 1965.

[16] Gray, "[Weekly] Activity Report [No. 43], Gemini Program Office, December 10, 1963," p. 3; Quarterly Status Report No. 9, for period ending 31 May 1964, p. 47; Zavasky, "Minutes of Senior Staff Meeting, April 10, 1964," p. 7; TWX, Mathews to McDonnell, Attn: Burke, "Development Engineering Inspection of Gemini Spacecraft No. 3," GP-54685, 23 April 1964; TWX, Mathews to DOD Rep., Project Gemini Support Ops., "Development Engineering Inspection of Gemini Spacecraft No. 3," GP-54729, 14 May 1964; TWX, Mathews to Langley, Ames, and Flight Research Centers and Goddard Space Flight Center, "Development Engineering Inspection of Gemini S/C No. 3," GP-54734, 14 May 1964; memo, Mathews to dist., "Development Engineering Inspection of Gemini Spacecraft No. 3," GP-03651, 15 May 1964; memo, Mathews to Chief, AFSC Field Office, "Development Engineering Inspection of Gemini Spacecraft No. 3," GP-03655, 15 May 1964; letter, Mathews to NASA Hq., Attn: George E. Mueller, "Development Engineering Inspection of Gemini Spacecraft No. 3," GP-03652, 18 May 1964; TWX, Mathews to McDonnell, Attn: Burke, "Contract NAS 9-170; Development Engineering Inspection of Gemini S/C No. 3," GP-54761, 2 June 1964; "Gemini DEI: Development Engineering Inspection, Spacecraft No. 3, 9-10, June 1964," McDonnell, n.d.; memo, Mathews to dist., "Report on the Gemini Spacecraft No. 3 Development Engineering Inspection," GP-03724, 23 June 1964, with enclosure; memo, James C. Elms to dist., "Establishment of Development Engineering

Inspection Board," GP-03426, 31 Jan. 1964; memo, George M. Low to dist., "Development Engineering Inspection Board for Gemini Spacecraft No. 3," GP-03690, 8 June 1964.

[17] Mathews, activity report, 26 July—1 Aug. 1964, p. 1; letter, Mathews to Burke, GV-02254, 12 May 1964; memo, Mathews to dist., "Organization of Gemini Configuration Control Board," 6 July 1963; letter, Mathews to Burke, GV-02351, 1 Sept. 1964.

[18] Memo, James I. Brownlee for record, "Contract NAS 9-170, Gemini Incentive Arrangement," 14 Jan. 1964, with enclosures; memo, A. E. Wyatt to Glenn F. Bailey, "Cost Incentive Plan for Contract NAS 9-170, Gemini Program," 12 Feb. 1964, with enclosure; memo, Mueller to Assoc. Adm., "Gemini Contract Status," 2 March 1964; memo, Mathews to dist., "Manned Spacecraft Center Contract NAS 9-170, establishment of NASA Task Force Team to study and to develop a plan for conversion of Gemini Contract NAS 9-170 from cost-plus-fixed-fee to cost-plus-incentive-fee," GP-03504, 11 March 1964; Weekly Activity Report, 5-11 April 1964, pp. 4-5; Quarterly Status Report No. 9, pp. 63-64.

[19] Memo, Dave W. Lang to NASA Hq., "Conversion of Gemini Spacecraft Contract NAS 9-170 from CPFF to CPIF—Request for Headquarters' Approval of RFP," with enclosure, draft letter, Stephen D. Armstrong to Harry W. Oldeg, "Contract NAS 9-170, Request for Incentive Proposal (RFP)," 4 May 1964; memo, Richard C. Henry to dist., "Review of RFP for MAC Contract CPIF Conversion," 11 May 1964; letter, Bailey to Oldeg, "Contract NAS 9-170, Request for Incentive Proposal (RFP)," 19 May 1964, with enclosures; TWX, Armstrong to William A. Summerfelt, 1 May 1964; memo, Bailey to Wesley L. Hjornevik, "Contract NAS 9-170—MAC-Gemini Spacecraft Incentive Status Report," 15 May 1964; Kenneth S. Kleinknecht, "Minutes of Incentive Task Group Meeting," 18 May 1964, pp. 1-2; memo, Kleinknecht to dist., "Contract NAS 9-170—Gemini Incentive Plan—Issuance of RFP," 25 May 1964, with enclosure; memo, Kleinknecht to Mathews, "Contract NAS 9-170—Gemini Incentive Implementation Task—Status Report," 28 May 1964; Kleinknecht, "Minutes of Incentive Task Group Meeting," 27 May 1964, with enclosures; minutes of Incentive Task Group meeting, 7 July 1964.

[20] Memo, Richard J. Crane to Bailey, "Procurement—GPO Attendance at Negotiations—Gemini Launch Vehicle (Titan II)—Air Force—Martin-Marietta at SSD, Los Angeles Beginning March 16, 1964," 24 March 1964; George F. MacDougall, Jr., and John B. All-

dredge, "Contract NAS 9-170—Incentive Implementation Plans Visit by George F. MacDougall and Brooks Alldredge to SSD on March 19, 20, 21, 1964," n.d.; memo, Anthony L. Liccardi to Dep. Dir., Gemini, "Information on Martin CPIF Contract," 14 May 1964, with enclosure, memo, E. L. Christianson to dist., "Report of Trip to the Martin Company, Baltimore, Maryland on 13 May 1964," 14 May 1964; Howard T. Harris, "Gemini Launch Vehicle Chronology, 1961-1966," AFSC Historical Publications Series 66-22-1, December 1966, pp. 39-40, E-2, -3; memo, Maj. Robert A. Krahn for record, "Negotiation of Contract AF 04(695)-129, Phase II," 10 Aug. 1964; memo, Crane to Chief, Gemini Spacecraft Procurement, "Negotiations—Agena Contract—SSD—Lockheed—NASA at Los Angeles—Beginning July 14, 1964," 22 July 1964.

[21] Memo, Kleinknecht to Gemini Incentive Task Group, "Contract NAS 9-170—Incentive Negotiation Plans," 16 Nov. 1964; TWX, J. Pemble Field, Jr., to MSC, Attn: Kleinknecht, 16 Nov. 1964; memo, William C. Schneider to Assoc. Adm., Manned Space Flight, "MAC Gemini Contract Conversion," 23 Nov. 1964; letter, Schneider to Mathews, 24 Nov. 1964; Consolidated Activity Reports, December 1964, p. 35, and January 1965, p. 28; memo, Mathews to GPO personnel, "Gemini Spacecraft Contract, NAS 9-170, January 28, 1965," GP-01894, 16 Feb. 1965.

[22] Memo, Mathews to dist., "Module Test Review of Spacecraft 3 at McDonnell Aircraft Corporation, St. Louis, Missouri on September 21 and 22, 1964," GP-03891, 16 Sept. 1964; TWX, Mathews to McDonnell, Attn: Burke, "Contract NAS 9-170, Module Test Review of Spacecraft 3," GT-55233, 16 Sept. 1964; Mathews, activity report, 20-26 Sept. 1964, p. 1; memo, Mathews to dist., "Results of Module Test Review of Spacecraft 3," GP-01530, 16 Oct. 1964; letter, Mathews to Burke, "Contract NAS 9-170; results of the Module Test Review of Spacecraft 3 on September 22, 1964," GP-01522, 27 Oct. 1964.

[23] Letter, Mathews to Burke, "Contract NAS 9-170, Gemini Spacecraft Acceptance Review," GT-05137, 25 Aug. 1964; memo, John A. Edwards to Assoc. Adm., Manned Space Flight, "Gemini Mission Readiness Acceptance Procedures," 6 Nov. 1964; memo, Mathews to dist., "Spacecraft Acceptance Review Phase I of Spacecraft 3," GP-01681, 27 Nov. 1964; letter, Mathews to Burke, GP-01673, 27 Nov. 1964, with enclosures; letter, Mathews to Burke, "Contract NAS 9-170; results of the Spacecraft Acceptance Review, Phase I of Spacecraft 3, on December 3-4, 1964," GP-01721, 18 Dec. 1964; letter, Mathews to Burke, GP-01729, 18 Dec. 1964; memo, Scott H. Simpkinson to Mgr., GPO, "Test Objectives of 'Simulated Flight'—SEDR H-431-3," GT-05153, 15 Sept. 1964; memo, Mathews to dist., "Results of Spacecraft Acceptance Review, Phase II, for Spacecraft 3," GP-01785, 13 Jan. 1965; letter, Mathews to Burke, "Contract NAS 9-170, results of the Spacecraft Acceptance Review, Phase II, of Spacecraft 3 on December 22, 1964," GP-01786, 14 Jan. 1965; memo, Mathews to NASA Resident Mgr., MAC-St. Louis, "Acceptance of Spacecraft 3," GP-01734, 24 Dec. 1964.

[24] Consolidated Activity Report, December 1964, p. 13; TWXs, Walter J. Kapryan to MSC, Attn: Mathews, daily status reports, Nos. 1-4, 5-8 Jan., Nos. 5-9, 10-15 Jan., Nos. 10-14, 18-22 Jan., Nos. 15-19, 25-29 Jan., and Nos. 20-24, 1-5 Feb. 1965; "GT-3 Mission Report," p. 12-23.

[25] *Gemini-Titan II Air Force Launch Vehicle Press Handbook* (Martin-Baltimore, 2d ed., Manned Space Flight, ca. March 1965), pp. D-6, -7; Harris, "Launch Vehicle Chronology," p. 43; TWX, Mathews to SSD, Attn: Dineen, GV-52559, 2 Oct. 1964.

[26] Memo, Leon DuGoff to Asst. Dir., Launch Ops., launch vehicle activity report No. 1, 20 Jan. 1965; memos, DuGoff to Dep. Dir., Launch Ops., "Arrival 1st Stage GLV-3," 23 and 25 Jan. 1965; TWXs, Lt. Col. John G. Albert to Dineen et al., DWD 39110, 26 Jan., and DWD 39124, 2 Feb. 1965; memos, DuGoff to Dep. Dir., Launch Ops., launch vehicle activity reports, No. 11, 17 Feb., and No. 27, 19 March 1965: "Project Gemini (GT-3) Chronology of Technical Progress at Kennedy Space Center," KSC Planning and Technical Support Office, 7 May 1965, pp. 7, 12, 13-29.

[27] "GT-3 Mission Directive," NASA Program Gemini working paper No. 5017A, 15 Feb. 1965, p. 2-1.

[28] Letter, Mathews to NASA Hq., Attn: Schneider, "Gemini Mission Assignments," GV-02183, 13 March 1964; Mathews, activity report, 28 April—4 May 1964, p. 1; memo, Walter C. Williams to Actg. Mgr., GPO, "Third Gemini Flight," 6 June 1963; "Abstract of ... Coordination Meeting (Electrical), May 1, 1962," 2 May 1962; "Abstract of Meeting on Trajectories and Orbits, July 3, 1963," 9 July 1963; letter, Low to Elms, 19 July 1963.

[29] Memo, Christopher C. Kraft, Jr., to dist., "Proposed Mission Plan for GT-3," 25 Oct. 1963, with enclosure, "Proposed Mission Plan for the GT-3 Gemini Flight," 18 Oct. 1963; Meyer, notes on GPO staff meeting, 2 Jan. 1964; memo, Low to MSC, Attn: Mathews, "Configuration of Gemini Spacecrafts #2, 3, and 4," 4 Jan. 1964.

30 Harold R. Williams, "18 Obits [sic] Urged for Gemini Trip with Two Astronauts," *The Times-Picayune*, New Orleans, La., 13 Aug. 1964; "U. S. Astronauts Seeking Longer Gemini Flight," *St. Louis Post-Dispatch*, 27 Sept. 1964; "Gemini Astronauts Want First Flight of 18 Orbits," *The Hartford* (Connecticut) *Courant*, 27 Sept. 1964; Frank Macomber, "Grissom, Young to Orbit Thrice," *The Indianapolis Star*, 17 Jan. 1965; "Astronauts Ready Now."

31 "GT-3 Mission Directive," p. 2-1; memo, Mathews to Chief, Propulsion and Power Div., "GT-3 Flight Plan dated February 3, 1965," GP-01993, 9 March 1965; memo, Mathews to Asst. Dir., Flight Ops., and Asst. Dir., Flight Crew Ops., "OAMS Insertion Maneuver for GT-3," GV-02526, 17 March 1965; TWX, Mathews to KSC, Attn: Kraft, GV-12014, 18 March 1965.

32 Eldon W. Hall and Vearl N. Huff, interview, Washington, 24 Jan. 1967; memo, Edwards to Hall, "Letter from the Republican Conference to Mr. Webb," 15 May 1964; Howard W. Tindall, Jr., interview, Houston, 16 Dec. 1966; Tommy W. Holloway, "GT-3 Flight Plan," Preliminary B, 20 Sept. 1964; memo, Hall to Schneider, "Interim Status Report on Decay Safe Orbits," 11 Dec. 1964; letter, Edwards to Kraft, 5 Jan. 1965; memo, Tindall to Chief, Mission Planning and Analysis Div., "Complete revision of the GT-3 flight plan," 7 Jan. 1965; memo, Hall to Dep. Dir., Gemini, "Fail-Safe Orbits," 11 Jan. 1965; memo, Robert O. Aller to file, "Fail-safe orbit for GT-3," 15 Jan. 1965; TWX, Mathews to McDonnell, Attn: Burke, GV-52676, 15 Jan. 1965; TWX, Mathews to McDonnell, Attn: Burke, "Special Trajectory and Orbits Meeting," GP-51709, 21 Jan. 1965; "Abstract of Meeting on Trajectories and Orbits, January 27, 1965," 5 Feb. 1965; memo, Aller for record, "Mission Planning for GT-3," 29 Jan. 1965; Frank J. Suler and Bobbie D. Weber, "A Proposed Mission Plan for the First Manned Gemini Flight (GT-3) Utilizing a Retrograde Maneuver Prior to Retrofire," MSC Internal Note No. 65-FM-11, 9 Feb. 1965; [Holloway], "GT-3 Flight Plan," Final, 4 March 1965; Martin Caidin, *Marooned* (New York, 1964).

33 "Abstract of Trajectories and Orbits Meeting, January 27, 1965"; Suler and Weber, "A Proposed Mission Plan," p. 1; "GT-3 Mission Directive," p. 3-1; "GT-3 Mission Report," p. 4-1.

34 Letter, Homer E. Newell to "Dear Colleague," 20 Aug. 1963, with enclosures; memo, Willis B. Foster to Dir., Program Review and Resources Management, "Submission for 1964 President's Annual Report," 30 Oct. 1964, with enclosures; memo, Foster to Chief, Lunar and Planetary Br., "Establishment of Manned Space Flight Experiments Board," 5 Jan. 1963 [sic-1964]; memo, Verne C. Fryklund, Jr., to Dir., Space Sciences, "Manned Space Flight Experiments Board," 28 Oct. 1963; letter, Schneider to Mathews, 24 Jan. 1964, with enclosures, letter, D. Brainerd Holmes to Robert R. Gilruth, 23 Aug. 1963, and Assoc. Adm., Manned Space Flight, to dist., "Establishment of a Manned Space Flight Experiments Board," NASA Management Instruction M 9000.002, 14 Jan. 1964; memo, Mueller to dist., "Manned Space Flight Experiments Board," 17 March 1964.

35 MSC Management Instruction No. 37-1-1, "In-Flight Experimental Programs," 18 July 1963; MSC Management Instruction No. 2-3-1, "Manned Spacecraft Center In-Flight Scientific Experiments Coordination Panel," 15 Oct. 1962; letter, Mathews to Schneider, GP-61010, 15 March 1965; memo, Newell to Assoc. Adm., Manned Space Flight, "Proposed scientific experiments for the Gemini 3 flight," 14 Nov. 1963; memo, Jocelyn R. Gill to Schneider, "Gemini Scientific Proposals," 14 Nov. 1963; Warren Gillespie, Jr., acting secretary, "Minutes of In-Flight Scientific Experiments Coordination Panel, January 16, 1964"; letter, Mathews to NASA Hq., Attn: Low, "Scientific investigations during the GT-3 missions [sic]," GPO-01101-M, 10 Jan. 1964; letter, Mathews to NASA Hq., Attn: Low, "Scientific Experiments for Early Gemini Missions," GP-03439, 5 Feb. 1964; memo, Foster to Assoc. Adm., "Recommended scientific investigations during the GT-3 Mission," 11 Feb. 1964; memo, Mathews to Gemini Procurement, Attn: Larry G. Damewood, "Integration of two NASA experiments on GT-3," GP-03511, 12 March 1964.

36 Gillespie, "In-Flight Experiments Panel Meeting"; Gordon C. Hrabal, "Experiments for GT-3 Mission," NASA Program Gemini working paper No. 5014, 22 Sept. 1964, pp. 5-1 through -11; "GT-3 Mission Report," p. 8-3.

37 Hrabal, "Experiments for GT-3," pp. 4-1 through -10; "GT-3 Mission Report," pp. 8-1, -2.

38 Hrabal, "Experiments for GT-3," pp. 6-1 through -17; Loyd S. Swenson, Jr., James M. Grimwood, and Charles C. Alexander, *This New Ocean: A History of Project Mercury*, NASA SP-4201 (Washington, 1966), pp. 431-32, 453-56, 597.

39 Memo, Clifford H. Nelson to Eugene C. Draley, "Proposed Reentry Experiment on the Mercury Capsule for Studying Radio Frequency Transmission Blackout," 12 June 1962; memo, Nelson to Draley, subject as above, 18 June 1962; memo, Nelson to Assoc. Dir., "Pro-

posed Mercury Capsule 'Blackout' Elimination Experiment," 27 June 1962, with enclosure, memo, William F. Cuddihy to Kleinknecht, "Proposed Reentry Experiment on the Mercury Capsule for Studying Radio Frequency Transmission Blackout," 27 June 1962; memo, Kleinknecht to Langley, Attn: Axel T. Mattson, "Proposed reentry experiment on the Mercury spacecraft for studying radio frequency transmission blackout," 27 July 1962; memo, Cuddihy to Mattson, "Blackout Info," 8 Oct. 1962, with enclosure, letter, Cuddihy to Lewis R. Fisher, n.d., with enclosure, "Reentry Communication Methods," 2 Oct. 1962; memo, [Cuddihy] to Norman G. Foster, "Information requested for Reentry Communication Experiment," 29 Jan. 1963; memo, Fisher to Kleinknecht, "Summary of February 25, 1963 Meeting of In-Flight Scientific Experiments Coordination Panel," 26 Feb. 1963, with enclosure; memo, John H. Kimzey to Chief, Systems Evaluation and Development Div., "In-Flight Scientific Experiments for MA-10," 27 Feb. 1963; memo, Mattson to MSC, Attn: William O. Armstrong, "Proposal for Reentry Communications Experiment to Be Flown on Gemini Spacecraft," 1 Oct. 1963, with enclosure; memo, Lyle C. Schroeder to Clinton E. Brown, "Status of Proposed Gemini Reentry Communication Experiment," 21 Jan. 1964; memo, Mathews to Actg. Chief, Gemini Procurement, "Reentry Communications Experiment for GT-3," GP-03518, 16 March 1964; memo, Schroeder to Brown, "Status, Reentry Communications Experiment on Gemini," 17 March 1964; memo, Boyd C. Myers II to Dir., MSF Program Control, "Request for Additional Information on Costs for Incorporating Reentry Communications Experiment on Gemini," 1 May 1964; TWX, Mathews to NASA Hq., Attn: Schneider, "NASA Gemini Experiment POISE 3, Reentry Communications," GP-54726, 13 May 1964; memo, Cuddihy to Assoc. Dir., "Reentry Communications Experiment for GT-3 Flight," 13 May 1964; Mathews, activity report, 17 May—20 June 1964, p. 1; memo, William E. Lilly to Dep. Assoc. Adm., Adv. Research and Technology, "Reentry Communications Experiment (T-1) on Gemini," 15 May 1964; memo, Myers to Dir., MSF Program Control, "Reentry Communications Experiment on Gemini," 25 May 1964; William Armstrong, "Notes on Reentry Communications Experiment," 10 June 1964; TWX, W. G. Robinson to MSC, Attn: Stephen Armstrong, "Contract NAS 9-170, Experiment Order 63-05, NASA Re-Entry Communications," MAC 306-16-6880, 26 June 1964; memo, Myers to Dir., MSF Program Control, "Reentry Communications Experiment on

Gemini," 29 June 1964; memo, Mathews to Chief, Gemini Spacecraft Br., "Statement of Work, GP-44, dated April 3, 1964, for the integration of the Reentry Communications Experiment into Spacecraft 3," GP-03760, 10 July 1964; memo, Schroeder to MRB files, "Status, Gemini Reentry Communication Experiment," 17 July 1964; memo, Mathews to Langley, Attn: Floyd L. Thompson, "Langley Research Center Reentry Communications Experiment, T-1, Gemini Mission GT-3," GP-61140, 23 April 1965.

40 Hrabal, "Experiments for GT-3," pp. 6-1 through -13; memo, Schroeder to Assoc. Dir., "Flight Crew Support Requested during the Gemini Reentry Communications Experiment," 9 April 1964; Cuddihy memo, 27 June 1962; Schroeder memo, 21 Jan. 1964.

41 TWX, Mathews to SSD, Attn: Dineen, "GT-3 Design Certification Review," GV-52697, 28 Jan. 1965; letter, Mathews to Burke, GP-01874, 4 Feb. 1965; memo, Mathews to Astronaut Office, Attn: Virgil I. Grissom, "Design Certification Review," GP-01878, 4 Feb. 1965; memo, Mathews to dist., "Design Certification Review," GP-01873, 4 Feb. 1965, with enclosure; letter, Mathews to Dineen, GP-01879a, 8 Feb. 1965; memo, Mathews to Mgr., Ops. Planning and Development, "Design Certification Review," GP-01881a, 8 Feb. 1965; letter, Schneider to Donald K. Slayton, 17 Feb. 1965; memo, Mathews to Chief, Spacecraft Section, "Contract NAS 9-170, Gemini, Design Certification Review, CCP No. 70," GP-61043, 17 March 1965; Walter D. Smith, "Martin/Titan," in "Gemini/Apollo Executives Meeting," 27-28 Jan. 1967, p. D-8; memo, Mathews to dist., "Spacecraft 3 Flight Readiness Review," GP-01972, 2 March 1965; TWX, Mathews to McDonnell, Attn: Burke, GP-51798, 2 March 1965; letter, Mathews to Burke, GP-61033, 15 March 1965; memo, Mathews to Spacecraft 3 Flight Readiness Review Board, "Action Items from Flight Readiness Review," GP-61057, 19 March 1965; TWX, Mathews to SSD, Attn: Dineen, GP-51805, 8 March 1965; TWX, Mathews to SSD, Attn: Dineen, GP-7023, [ca. 17 March 1965]; TWX, Mathews to McDonnell, Attn: Burke, "Contract NAS 9-170, GT-3, Mission Review," GP-70013, 10 March 1965; TWX, Mathews to KSC, Attn: G. Merritt Preston et al., GP-71017, 16 March 1965; TWX, Albert to Dineen et al., DWD 39171, 19 March 1965; DuGoff report No. 27; Roy B. Carthen, notes taken from DuGoff, 29 March 1965.

42 Memo, Mathews and Preston to Dep. Dir., Mission Requirements and Flight Ops., "Static Fire of the Gemini RCS and OAMS," 14 Oct. 1963; memo, Schneider to Assoc. Adm., Manned Space Flight, "RCS-OAMS Stat-

ic Firing at Cape Kennedy," 27 May 1964; memo, Mathews to Chief, Propulsion and Power Div., "Thrust chamber assembly hot firing check-out procedures during prelaunch activities," GP-61376, 8 June 1965; TWX, Mathews to McDonnell, Attn: Burke, "Static Fire Tests as Part of Gemini Spacecraft Check-Out at the Eastern Test Range," GP-7140, 13 May 1965; "GT-3 at KSC," p. 31; "GT-3 Mission Report," pp. 6-1, -2, 12-1.

43 "GT-3 Mission Report," pp. 6-2, 7-10, -29; GT-3 mission commentary transcript, 23 March 1965, pp. 1-5; "GT-3 Flight Crew Technical Debriefing," NASA Program Gemini working paper No. 5025, 3 June 1965, pp. 2-1 through -3; DuGoff, "GT-3 Launch Summary Report Notes," 23 March 1965; memo, E. E. Christensen to Schneider, "Leak around Transducer Fitting Resulting in GT-3 Hold," 25 March 1965; memo, Schneider to Christensen, subject as above, 25 March 1965; "Air to Ground Transmission," GT-3 News Center, 23 March 1965, p. 1; "Air-to-Ground Voice Transcription," Supplemental Report No. 5 to "GT-3 Mission Report," 23 June 1965, p. 2.

44 Swenson, Grimwood, and Alexander, *This New Ocean*, pp. 341, 368, 420, 446, 470, 492; Dora Jane Hamblin, "Spacecraft Anonymous," *Life*, 11 Oct. 1968, pp. 112-13; TWX, Henry W. Suydam to Will Lang, 26 March 1965; Grissom and John W. Young, "Molly Brown was OK from the first time we met her," *Life*, 2 April 1965, p. 41.

45 "GT-3 Debriefing," pp. 3-2 through -6; "GT-3 Flight Crew Self-Debriefing," NASA Program Gemini working paper No. 5026, 3 June 1965, pp. 2-1, 3-2; "GT-3 Mission Report," pp. 7-11, -20, -21; "Air-to-Ground Voice," p. 3; Grissom and Young, "Molly Brown was OK," p. 41.

46 "GT-3 Self-Debriefing," p. 2-1; "GT-3 Debriefing," p. 3-8; "GT-3 Mission Report," p. 7-12; "Air-to-Ground Voice," p. 5.

47 "Air-to-Ground Voice," p. 6; Grissom and Young, "Molly Brown was OK," p. 42; "GT-3 Mission Report," p. 7-22; "GT-3 Self-Debriefing," pp. 2-2, 3-5.

48 "GT-3 Self-Debriefing," pp. 2-2, 3-4; "GT-3 Mission Report," p. 7-25; "GT-3 Debriefing," pp. 4-39, -40; Grissom and Young, "Molly Brown was OK," p. 42; R[obert] O. Piland and P[aul] R. Penrod, "Experiments Program Summary," in *Gemini Midprogram Conference, Including Experiment Results*, NASA SP-121 (Washington, 1966), p. 309; Michael A. Bender, P. Carolyn Gooch, and Sohei Kondo, "Experiment S-4, Zero g and Radiation on Blood during Gemini III," in "Manned Space Flight Experiments Symposium: Gemini Missions III and IV," presented in Washington, 18-19 Oct. 1965, pp. 217-36; memo, Mathews to Chief, Quality Assurance Br., "Gemini Experiments S-2 and S-3," GP-61111, 3 April 1965.

49 "GT-3 Self-Debriefing," pp. 2-3, -4; Grissom and Young, "Molly Brown was OK," p. 42; "GT-3 Mission Report," pp. 7-15, -16; "Air-to-Ground Voice," pp. 19, 25, 42-43; [Ivan D. Ertel], *Gemini 3 Flight*, MSC Fact Sheet No. 291-A (Houston, April 1965), unpaged; memo, Carl T. Rowan, U. S. Information Agency, to the President, "Daily Reaction Report," 24 March 1965.

50 "GT-3 Debriefing," pp. 5-1 through -5; "GT-3 Mission Report," pp. 7-16, -17, -25; "Air-to-Ground Voice," pp. 45-46.

51 "GT-3 Mission Report," p. 7-26; memo, Schroeder to MSC Historical Office, "Comments on Draft Chapters of Gemini Narrative History . . . ," 9 Jan. 1972, with enclosure, "Results of Gemini 3 Reentry Communications Experiment," n.d.; Lyle P. Schroeder and Francis P. Russo, "Flight Investigations and Analysis of Alleviation of Communications Blackout by Water Injection during Gemini 3 Reentry," NASA TM X-1521, 18 Aug. 1967; Piland and Penrod, "Experiments Program Summary," p. 312.

52 "GT-3 Mission Report," pp. 7-17 through -19, 7-25 through -27; Grissom and Young, "Molly Brown was OK," p. 42; "GT-3 Debriefing," pp. 6-1 through -5, 7-4, -5; memo, Duncan R. Collins to Mgr., GPO, "Gemini landing system, spacecraft repositioning," GS-64011, 7 April 1965; TWX, Mathews to McDonnell, Attn: Burke, GS-10006, 14 April 1965.

53 Grissom and Young, "Molly Brown was OK," pp. 41-42; "GT-3 Debriefing," pp. 7-5 through -7, 7-9 through -11; "GT-3 Mission Report," pp. 7-19, -20, -27; "GT-3 Self-Debriefing," pp. 2-7, -8; "GT-3 Post-Launch Press Conference," 23 March 1965; "GT-3 Press Conference with Astronauts," 25 March 1965, p. 23; "DOD Support of Project Gemini, GT-3," n.d.

54 Memo, Mathews to Chief, Spacecraft Ops. Br., "GT-3 postflight scientific debriefing of the crew," GP-61018, 10 March 1965; memo, Brig. Gen. David M. Jones to Assoc. Adm., "GT-3 Post Mission Cape-Related Activities," 17 March 1965; memo, Mary-Frances Thompson to Gill, "Debriefing of Astronauts, Gemini-GT-3, Manned Spacecraft Center, Houston, Texas, April 12, 1965," 14 April 1965; letter, Gilruth to NASA Hq., Attn: Mueller, "Training for and conduct of manned space flight experiments," 19 April 1965; memo, Slayton to Mgr., GPO, "Crew pre-flight scientific briefings," 23 April 1965; letter, Gill

to Rita Rapp, 27 April 1965; letter, Willis Foster to MSC, Attn: Mathews, "Astronaut Scientific Debriefing," 11 May 1965; letter, Gilruth to Mueller, 21 May 1965; "GT-3 Self-Debriefing," p. 2-4; "GT-3 Astronaut Conference," pp. 14-15; Earl Ubell and David Hoffman, "Two Astronauts Team Up as Comics," *The Washington Post,* 26 March 1965; memo, Robert C. Seamans, Jr., to Mueller, "Gemini Operations," 15 April 1963; William Hines, " '$30 Million Sandwich': House Doesn't Relish Astronauts' Snack," *The Evening Star,* Washington, 15 April 1965; letter, Julian Scheer to James A. McDivitt, 24 May 1965, with enclosures; U.S. Congress, House, Subcommittee of the Committee on Appropriations, *Independent Offices Appropriations for 1966,* Part 2, *Hearings,* 89th Cong., 1st sess., 1965, pp. 912-13; memo, Mueller to Assoc. Adm., "Gemini III Operations," 4 May 1965; "Cassandra," "That Corned-Beef Sandwich in Space," *San Francisco Chronicle,* 5 May 1965; "Snacks in Space Are Out," *The Houston Post,* 7 May 1965; letter, Dick C. Nooe to Webb, n.d.; letter, Scheer to Nooe, 26 May 1965; Walter M. Schirra, Jr., interview, Houston, 4 May 1967; Lola H. Morrow, interview, Cocoa Beach, Fla., 24 May 1967.

[55] Memo, Mathews to GPO staff, "Gemini III Success," GA-60165, 29 March 1965.

## Chapter XI

[1] NASA News Release No. 62-172, "NASA Mission Control Center to Be at Houston, Texas," 20 July 1962; "News Conference, Manned Spacecraft Center Mission Control Announcement," 9 April 1965.

[2] "McDivitt, White Named Prime Crew for Second Manned Gemini Flight," MSC *Space News Roundup,* 5 Aug. 1964; "News Conference, GT-4 Crew Selection," 29 July 1964.

[3] Andre' J. Meyer, Jr., notes on NASA/MAC management meeting, 17 July 1964, p. 3; "NASA/MAC Management Meeting 16 October 1964," p. 6.

[4] James M. Grimwood and Barton C. Hacker, *Project Gemini Technology and Operations: A Chronology,* NASA SP-4002 (Washington, 1969), p. 277.

[5] MSC Weekly Activity Report for Office of Assoc. Adm., Manned Space Flight, 6-12 Dec. 1964, p. 3; James A. McDivitt, interview, Houston, 7 April 1967.

[6] Reginald M. Machell, interview, Houston, 18 April 1967; TWX, Charles W. Mathews to McDonnell, Attn: Walter F. Burke, "Contract NAS 9-170, Astronaut Harness Fit and Contouring," 14 May 1965.

[7] Union of Soviet Socialist Republics, USSR Aero Sports Federation, "Records File

on the First in the World Space Flight in the 'Voskhod-2' Spacecraft Including the Man's Emerging into Outer Space, March 18-19, 1965. The Crew of the Spacecraft Consists of the USSR Citizens: 1. Pilot-Cosmonaut Pavel Ivanovich Belyayev, Spacecraft Commander; 2. Pilot-Cosmonaut Alexei Arkhipovich Leonov, Second Pilot," Moscow, 1965; " 'Space Walk' Is Still Year Away for U.S.," *The Evening Star,* Washington, 18 March 1965; "Long Stride into Space," *The New York Herald Tribune,* 19 March 1965; "Over and Out," *The Sun,* Baltimore, 19 March 1965; Earl Ubell, "Gemini," *New York Herald Tribune,* 21 March 1965; Wayne Thomis, "The Space Race: How Russia and U.S. Stand Now," *Chicago Tribune,* 28 March 1965; Marvin Miles, "Open-Hatch Space Test Undecided," *Los Angeles Times,* 18 April 1965; William Hines, "New Gemini Crew on TV, Tell of 4-Day-Flight Plans," *The Evening Star,* Washington, 30 April 1965; Frank Macomber, "Stroll in Space Definite for U.S. Gemini Astronaut," *The San Diego Union,* 25 April 1965; Jim Maloney, "Shepard on Job 4 Years after Flight," *The Houston Post,* 10 May 1965; Evert Clark, "4-Day Gemini Trip Is Set for June 3: Opening of Hatch on 2-Man Flight Still Undecided," *The New York Times,* 22 May 1965; "American Space Walk," *The Evening Star,* Washington, 28 May 1965.

[8] "Abstract of Meeting on Crew Support Systems, May 10-11, 1962," 14 May 1962; TWX, Mathews to Burke, "Contract NAS 9-170; Provision of Extra Vehicular Operations Capability in Spacecraft Number 2 and Up," GPO-50633, 28 Feb. 1963.

[9] "GT-4 Crew Selection"; *Astronautics and Aeronautics, 1964: Chronology on Science, Technology, and Policy,* NASA SP-4005 (Washington, 1965), p. 265; "Gemini Astronauts May Stick Heads out Craft's Window," *The Washington Post,* 30 July 1964; McDivitt interview.

[10] "Program Plan for Gemini Extravehicular Operation," [GPO], 31 Jan. 1964; Warren Gillespie, Jr., acting secretary, "Minutes of In-Flight Scientific Experiments Coordination Panel, January 16, 1964," p. 5; Machell interview.

[11] MSC Consolidated Activity Report for Office of the Assoc. Adm., Manned Space Flight, 19 Jan.—15 Feb. 1964, p. 39; letter, Mathews to Burke, "Contract NAS 9-170, spacecraft provisions for Gemini extravehicular operation," GP-03573, 15 April 1964; letter, Mathews to Burke, GS-64025, 14 May 1965; "Abstract of Meeting on Extravehicular Activity, July 14, 1964," 27 July 1964; Meyer, notes on GPO staff meeting, 1 April 1964.

[12] Robert A. Dittman, "Gemini Program

Office Staff Meeting Minutes, October 6, 1964," p. 3.

[13] John W. Young, interview, Houston, 28 April 1967.

[14] Kenneth S. Kleinknecht, interview, Houston, 5 Dec. 1966; George M. Low, interview, Houston, 7 Feb. 1967; TWX, Mathews to McDonnell, Attn: Burke, GP-51575, 12 Nov. 1964; Young interview.

[15] Memo, Duncan R. Collins to Mgr., GPO, "Review of effectivity of Spacecraft 6 for extravehicular operation," GS-04151, 14 Jan. 1965; Project Gemini Quarterly Status Report No. 12, for period ending 28 Feb. 1965, p. 43.

[16] TWX, Mathews to McDonnell, Attn: Burke, GP-51747, 10 Feb. 1965; memo, Robert L. Frost for record, "Gemini Spacecraft 4 altitude chamber test plan briefing," GS-64006, 25 March 1965; Weekly Activity Report, 21-27 March 1965, p. 1.

[17] Low interview; Larry E. Bell, interview, Houston, 10 Sept. 1968; letter, Paul E. Purser to John W. Macy, 18 Aug. 1965, with enclosure, David L. Schwartz, "EVA—The Story of a Team Effort by Civil Service Employees," n.d.

[18] Schwartz, "A Team Effort."

[19] Ibid.; Eugene F. Kranz, interview, Houston, 28 April 1967.

[20] Low interview; Robert R. Gilruth, interview, Houston, 21 March 1968; William C. Schneider, interview, Washington, 23 Jan. 1967.

[21] Bell interview; Quarterly Activity Report for Office of the Assoc. Adm., Manned Space Flight, for period ending 30 April 1965, p. 38; Quarterly Activity Report, for period ending 31 July 1965, p. 31; Quarterly Status Report No. 13, for period ending 31 May 1965, pp. 9, 10.

[22] Robert C. Seamans, Jr., to William D. Putnam, comments on narrative history of Gemini, comment edition, 3 April 1969; memo, Seamans to Adm., "Extra Vehicular Activity for Gemini IV," 24 May 1965, with Webb's signed approval, 25 May 1965.

[23] Bell interview; Paul P. Haney, interview, Houston, 16 Sept. 1968; NASA Release No. 65-158, "Project: Gemini 4," press kit, for release 21 May 1965, p. 5; William Hines, "NASA Opens 'Open Secret' on Plans for Gemini 4 Flight," The Evening Star, Washington, 21 May 1965; Evert Clark, "U.S. 'Space Walk' Planned in June," The New York Times, 25 May 1965; "Press Conference on Extravehicular Activity," 25 May 1965.

[24] Low interview; Meyer, notes on GPO staff meeting, 2 Jan. 1964; letter, Low to MSC, Attn: Mathews, "Configuration of Gemini Spacecrafts #2, 3, and 4," 4 Jan. 1964.

[25] "[Gemini III] Air to Ground Transmission," GT-3 News Center, 23 March 1965, p. 14.

[26] Low interview; Howard W. Tindall, Jr., interview, Houston, 16 Dec. 1966; Wyendell B. Evans, telephone interview, 1 Oct. 1968; TWXs, Mathews to SSD, Attn: Richard C. Dineen, GV-12035, 9 April, and GV-12084, 25 May 1965.

[27] Robert L. Sharp, interview, St. Louis, 14 April 1966; "Gemini Program Mission Report, Gemini IV," MSC-G-R-65-3, July 1965, p. 7-14.

[28] Charles A. Berry, interview, Houston, 18 March 1968; Quarterly Status Reports: No. 5, for period ending 31 May 1965, pp. 50-51; No. 7, for period ending 30 Nov. 1963, p. 79; No. 10, for period ending 31 Aug. 1964, pp. 3, 59; Consolidated Activity Report 19 Jan.—15 Feb. 1964, p. 17; memo, Mathews to Gemini Procurement, Attn: James I. Brownlee, "Contract NAS 9-170; Gemini, installation of batteries in Spacecraft 4, CCP No. 20," GP-03532, 20 March 1964; Eldon W. Hall and Vearl N. Huff, interview, Washington, 24 Jan. 1967; Raymond D. Hill, Jr., interview, Titusville, Fla., 23 May 1967; Kleinknecht interview.

[29] Berry interview.

[30] James Waggoner, interview, Los Angeles, 8 July 1966; Lawrence F. Dietlein, "Experiment M-3, Inflight Exerciser on Gemini IV," in "Manned Space Flight Experiments Symposium: Gemini Missions III and IV," presented in Washington, 18-19 Oct. 1965, pp. 41-48; Dietlein, interview, Houston, 22 March 1968.

[31] Transcribed segments of Leonov's space walk during Voskhod II and at postflight conferences in Perm, Baykonur, and Moscow, however, do not mention disorientation. Astronautics and Aeronautics, 1965: Chronology on Science, Technology, and Policy, NASA SP-4006 (Washington, 1966), pp. 225-26; "Continued Reportage on Flight of Voskhod 2," 22 March 1965 (14 pages), and "Reportage on Moscow Welcome to Cosmonauts," 23 March 1965 (13 pages), U.S.S.R. National Affairs; Berry interview.

[32] Memo, M. Scott Carpenter to dist., "Cosmonaut Training," 24 Nov. 1964; Margaret M. Jackson and M/Sgt. C. W. Sears, "The Effect of Weightlessness upon the Normal Nystagmic Reaction," Aerospace Medical Association, 36th Annual Scientific Meeting, New York, 26-29 April 1965, preprint of scientific program, pp. 138-39.

[33] Letter, John H. Boyd, Jr., to Jack E. Riley, 10 May 1965, with enclosure, "Launch Countdown"; "Preliminary GT-4 Flight Crew Debriefing Transcript," Part I, 16 June 1965, pp. 6-7; "Gemini Status Report," Gemini 4

News Center, 2 June 1965, 5:00 p.m.; [Ivan D. Ertel], *Gemini 4 Flight*, MSC Fact Sheet No. 291-B (Houston, 1965); "Gemini IV Mission Report," p. 7-43.

34 [Ertel], *Gemini 4 Flight*; Boyd letter, 10 May 1965; "Preliminary Debriefing," Part I, pp. 1, 7.

35 Gemini 4 mission commentary transcript, 3 June 1965, tape 2, pp. 5-8; J[oseph] F. Wambolt and S[ally] F. Anderson, "Gemini Launch Systems Final Report: Gemini/Titan Launch Vehicle; Gemini/Agena Target Vehicle; Atlas SLV-3," Aerospace TOR-1001(2126-80)-3, January 1967, p. II.G-5; Lt. Col. Alexander C. Kuras and Col. John G. Albert, "Gemini-Titan Technical Summary," 24 Jan. 1967, pp. 140-41; Albert, interview, Patrick AFB, Fla., 26 May 1967; Joseph M. Verlander, interview, Cocoa Beach, Fla., 29 Aug. 1967; "Number of Holds Manned Space Flight Launches," compiled by MSC Historical Office, ca. 1964.

36 "Europeans See Shot on Early Bird," *The Washington Post*, 4 June 1965.

37 "NASA Manned Spacecraft Center Master Plan & Architectural Concept," Brown and Root, Inc., [April 1962]; Howard I. Gibbons, telephone interview, 31 Oct. 1968; Quarterly Activity Report, 30 April 1965, p. 20; Jim Maloney, "MSC Will Lease Building for News," *The Houston Post*, 11 May 1965; Jim Schefter, "MSC to Spend $223,000 on News Center on Outside," *Houston Chronicle*, 11 May 1965; letter, W. H. Sheley, Jr., to Gilruth, 16 May 1966.

38 Stanley P. Weiss, recorder, "Minutes of Senior Staff Meeting, May 21, 1965," p. 2; "Houston Gemini News Center Accreditation List," 11 June 1965.

39 "Preliminary Debriefing," Part I, pp. 17-18, 20-21, 23-25, 31; "Gemini IV Mission Report," p. 4-1.

40 "Preliminary Debriefing," Part I, pp. 38, 50-57.

41 Ibid., pp. 54-55, 58-69, 72; Gemini 4 mission commentary, tape 7, p. 1; Meyer, comments on draft chapter of Gemini narrative history, 28 Feb. 1969.

42 "Preliminary Debriefing," Part I, pp. 85-95, 96-98; Gemini 4 mission commentary, tape 9, pp. 3-6.

43 "Preliminary Debriefing," Part I, pp. 100-103.

44 Ibid., pp. 108, 109-16.

45 Ibid., pp. 133-45; Gemini 4 mission commentary, tape 11, EVA-1 through -14.

46 "Preliminary Debriefing," Part I, pp. 145-54; "Composite Air-to-Ground and Onboard Voice Tape Transcription of the GT-4 Mission," NASA Program Gemini working paper No. 5035, 31 Aug. 1965, pp. 56-62; Frederick T. Burns et al., "Gemini Extravehicular Activities," in Reginald M. Macheil, ed., *Summary of Gemini Extravehicular Activity*, NASA SP-149 (Langley, Va., 1967), p. 3-3.

47 "Preliminary Debriefing," Part I, p. 186; "GT-4 Air-to-Ground," pp. 83, 90, 92; McDivitt interview; "GT-4 Flight Crew Report," 11 June 1965, tape 5, p. 3.

48 MSC Gemini 4 Release No. 3, 28 May 1965; memo, Ertel to Public Affairs Officer, "Release on GT-4 Flight Controller and Recovery Personnel," 14 May 1965, with enclosure; Kranz interview.

49 "Building 30 inhouse facilities for support personnel for MCG Test 617," 1 June 1965; memo, Mathews to members of Mission Evaluation Team, "Gemini IV Mission Evaluation Team," GT-65141, 28 May 1965, with enclosures; Scott H. Simpkinson, interview, Houston, 18 Jan. 1967; memo, Mathews to Special Asst., "Technical monitor and principal investigator staff for support for experiments in the Mission Control Center," GP-61328, 25 May 1965, with enclosure; memo, Mathews to Chief, Flight Support Div., "Gemini experiments," GP-61329, 2 June 1965; memo, Purser to Mgr., GPO, "Experimenter staff support for GT-4," 2 June 1965; memo, Mathews to Asst. Dir., Flight Crew Ops., Attn: Helmut A. Kuehnel, "Gemini IV experiments inflight consultation plan," GP-61355, 3 June 1965; MSC Announcement No. 65-81, "Designation of Manager, Experiments, in E&D, and Establishment of the Experiments Program Office," 21 June 1965.

50 "Gemini IV Mission Report," pp. 8-2, -4, -7, -8, -9; TWX, Mathews to NASA Hq., Attn: Schneider, "Gemini Experiment D-9, Simple Navigation," GP-7126, 12 May 1965; Capt. E. M. Vallerie, "Experiment D-9, Simple Navigation on Gemini IV," in "Gemini III and IV Experiments Symposium," pp. 105-109; Lts. M. F. Schneider, J. F. Janni, and G. E. Radke, "Experiment D-8, Radiation in Spacecraft Gemini IV," ibid., pp. 171-216; R[obert] O. Piland and P[aul] R. Penrod, "Experiments Program Summary," in *Gemini Midprogram Conference, Including Experiment Results*, NASA SP-121 (Washington, 1966), pp. 310-11; "Preliminary Debriefing," Part II, pp. 219-28.

51 "Preliminary Debriefing," Part I, p. 222, Part II, p. 219; Paul D. Lowman, Jr., "Experiment S-5, Synoptic Terrain Photography during Gemini IV," in "Gemini III and IV Experiments Symposium," pp. 19-32; Gordon C. Hrabal, "Experiments for GT-4 Mission," NASA Program Gemini working paper No. 5023, 14 May 1965, p. 13-3; see also color photographs taken by *Gemini IV* crew in

*Earth Photographs from Gemini III, IV, and V*, NASA SP-129 (Washington, 1967), pp. 13-108.

[52] "Preliminary Debriefing," Part II, pp. 230-32; Lawrence F. Dietlein, "Experiment M-3" and "Experiment M-4, Inflight Phonocardiogram," in "Gemini III and IV Experiments Symposium," pp. 40-48, 49-59; Pauline B. Mack et al., "Experiment M-6, Bone Demineralization on Gemini IV," ibid., pp. 61-80; Dietlein and Rita M. Rapp, "Experiment M-3, Inflight Exercise—Work Tolerance," in *Gemini Midprogram Conference*, p. 394; Mack et al., "Experiment M-6, Bone Demineralization," ibid., pp. 413-14; "Gemini IV Mission Report," p. 7-46.

[53] TWX, Mathews to SSD, Attn: Major Charles A. Wurster, GV-52310 (ca. May 1964); TWX, Mathews to McDonnell, Attn: Burke, "Contract NAS 9-170; Experiment Order 63-04, Electrostatic Charge Experiment," GA-51182, 27 Aug. 1964; memo, Mathews to Asst. Dir., E and D, "GT-4 Experiment MSC-1, Electrostatic Charge Measurement," GA-01360, 13 Jan. 1965; memo, Clifford M. Jackson to Mgr., Spacecraft, "Status of electrostatic charge measurement experiment, MSC-1," GS-04161, 16 Feb. 1965; "Preliminary Debriefing," Part II, pp. 217-18, 229; "Gemini IV Mission Report," pp. 8-12 to -14, -16, -17; Patrick Lafferty, "Experiment MSC-1, Electrostatic Charge on Gemini III and Gemini IV," in "Gemini III and IV Experiments Symposium," pp. 137-39; James Marbach, "Experiment MSC-2, Proton-Electron Measurement on Gemini IV," ibid., pp. 149-51; William D. Womack, "Experiment MSC-3, Tri-Axis Magnetometer on Gemini IV," ibid., pp. 161-63; Max Petersen, "Experiment MSC-10, Two-Color Earth Limb Photography," ibid., pp. 121-31.

[54] "Gemini IV Mission Report," p. 4-3; "Preliminary Debriefing," Part I, pp. 243-45, Part II, pp. 90-92; "GT-4 Air-to-Ground," pp. 305-10, 313-14, 319-22, 332, 335.

[55] "GT-4 Air-to-Ground," pp. 384, 386; "Gemini IV Mission Report," pp. 4-3, -4; "Preliminary Debriefing," Part I, pp. 260-61, 269-74, 280-81.

[56] "Preliminary Debriefing," Part I, pp. 284-93; "Gemini IV Mission Report," p. 7-27.

[57] "Preliminary Debriefing," Part I, pp. 297-306, 310-13; "GT-4 Air-to-Ground," pp. 390-92, 394.

[58] "GT-4 Air-to-Ground," p. 395; Gemini 4 mission commentary, tape 183, pp. 4-7, tape 184, pp. 1-3, tape 186, p. 2; "Recovery Mission Chronology," [Naval History Office, TF-140, December 1966].

[59] Bennett W. James et al., interview, Houston, 29 June 1967; "Gemini IV Mission Report," pp. 7-47 through -49; Charles A. Berry et al., "Man's Response to Long-Duration Flight in the Gemini Spacecraft," in *Gemini Midprogram Conference*, pp. 241, 247, 248, 250; Berry interview.

[60] See "Current News," 30 July 1965, a compilation of news stories and front pages of newspapers all over the world, prepared by NASA's Office of Public Information, especially *The Washington Daily News*, 4 June 1965; *The Boston Globe*, 8 June 1965; *The Miami Herald*, 8 June 1965; *New York Herald Tribune*, 4 June 1965; *The New York Times*, 15 June 1965; *The Washington Post*, 18 June 1965; *The Yomiuri Shimbun*, Tokyo, 8 and 9 June 1965; *The Daily Telegraph*, London, 8 June 1965; White House News Release, "Text of the Remarks by the President at the Manned Spacecraft Center, Houston, Texas, June 11, 1965;" "President Visits Center, Honors Gemini IV Crew," MSC *Space News Roundup*, 25 June 1965; TWX, Alex P. Nagy to Purser, AF-1-866, 30 July 1965; George Sherman, "French Crowds Cheer America's Space Twins," *The Sunday Star*, Washington, 20 June 1965; "Space Twins Steal Some Red Thunder," *Washington Daily News*, 19 June 1965.

[61] Memo, Seamans to Assoc. Adm., Advanced Research and Technology, "Orbital Mechanics," 9 June 1965; letter, Floyd L. Thompson to NASA, Attn: Alfred J. Eggers, Jr., "Rendezvous and station-keeping studies," 6 Oct. 1965, with enclosure, memo, W. Hewitt Phillips to Assoc. Dir., "Orbital mechanics associated with Gemini flights," 30 Sept. 1965. Enclosed with Phillips' memo are papers presented to MSC on 11 Aug. 1965: Gary P. Beasley, "Digital Computer Analysis in Support of GT-5 Mission"; Kenneth R. Garren, "Use of Visual Cues for Determining Range (with Optical Aids) and Direction of Motion of a Flashing Light"; Jack E. Pennington, "Range Estimation Studies Using Only Apparent Object Size;" Alfred J. Meintel, Edward R. Long, and Pennington, "GT-5 Piloted Simulation"; Purser, comments on draft chapter, 13 Feb. 1969.

[62] Memo, Simpkinson to Mgr., GPO, "Failure analysis of the spacecraft 4 IGS malfunction and corrective action for spacecraft 5," GT-62576, 6 Aug. 1965; letter, Mathews to NASA Hq., Attn: Mueller, "Analysis and corrective action for the computer power-down failure and the possible eye irritation problem experienced during the Gemini IV mission," GT-62580, 9 Aug. 1965; Conrad D. Babb and Charles E. Dunn, interview, Owego, New York, 25 April 1968; John F. Yardley, "Spacecraft Check-Out," in "Gemini/Apollo Executives Meeting," 27-28 Jan. 1967, p. F-36.

63 See chap. X, pp. 228-29, McDivitt interview; TWX, Mathews to McDonnell, Attn. Burke, "Contract NAS 9-170, Gemini V Mission Plan," GV-12129, 13 July 1965; memo, Everett E. Christensen to dist., "GT-5 Operations Readiness Review," 13 July 1965.

64 "GT-4 Flight Crew Report," Tape 5, p. 1; "GT-4 Air-to-Ground," pp. 318, 340; "NASA/MAC Management Meeting 18 September 1964," p. 2; "NASA/MAC Management Meeting 18 December 1964," p. 3; memo, Collins to Mgr., GPO, "Status of space suits for GT-5, GT-6, and GT-7," GS-64059, 11 Aug. 1965; Quarterly Status Report No. 14, for period ending 31 Aug. 1965, p. 9; memo, Mathews to Chief, Crew Systems Div., "Delivery of Gemini space suits for Spacecraft 5 altitude chamber tests," GS-64003, 18 March 1965; TWX, Schneider to Mathews, "Deletion of EVA," MG-595, 12 July 1965.

65 LeRoy E. Day, interview, Washington, 25 Jan. 1967; Meyer, notes on GPO staff meeting, 6 Jan. 1965; "News Conference, GT-5 Crew Selection," 8 Feb. 1965.

66 Riley D. McCafferty, interview, Cape Kennedy, Fla., 25 May 1967; Meyer, notes on GPO staff meeting, 29 June 1965; Meyer comments, 28 Feb. 1969.

67 TWX, Mathews to SSD, Attn: Lt. Col. Mark E. Rivers, GT-55372, 3 Feb. 1965; memo, Mathews to dist., "Simultaneous launch demonstration between the Gemini Atlas Agena Target Vehicle and the Gemini Launch Vehicle/Spacecraft on F-10 days for the GT-5 mission," GT-05346, 19 Feb. 1965; TWX, Mathews to SSD, Attn: Rivers, GT-55388, 23 Feb. 1965; "Abstract of Meeting on GT-5 Simultaneous Launch Demonstration, March 2, 1965," 8 March 1965; "Abstract of Meeting on Rendezvous and GT-5 Simultaneous Launch Demonstration, May 18, 1965," 25 May 1965; TWX, Mathews to SSD, Attn: Dineen, GP-7182, 26 May 1965; TWX, Mathews to KSC, Attn: Paul C. Donnelly, "Launch Preparation Schedule—Spacecraft 5," GP-7222, 9 June 1965; "Abstract of Meeting on Atlas/Agena Coordination, June 16, 1965," 29 June 1965; Quarterly Status Report No. 14, p. 18; "GAATV Launch Demonstration with GT-5 GLV," Lockheed Missiles & Space Co., LMSC-273407, 3 Sept. 1965, pp. iii, 7-2; TWX, Mathews to Dineen, GP-54516, 23 Jan. 1964; TWX, Mathews to SSD, Attn: Dineen, GV-12146, 23 July 1965; Albert interview; "Abstract of Meeting on Trajectories and Orbits, December 4, 1964," 23 Dec. 1964; Bobby K. Culpepper, "Partial Proposed Mission Plan for the GT-5 Gemini Flight, REP Plan IV," MSC Internal Note No. 64-FM-87, 1 Dec. 1964; memo, Mathews to Mgr., Florida Ops., Attn: J. T. Garo-

falo, Jr., "Cherry Picker Modification for Gemini Egress," GV-02263, 25 May 1964; Schneider letter, 10 July 1964; letter, Mathews to NASA Hq., Attn: Schneider, "Gemini Mission Assignments," GV-02183, 13 March 1964.

68 Weiss, "Minutes of Senior Staff Meeting, June 18, 1965," p. 2; Meyer notes, 29 June 1965, p. 1; Neil A. Armstrong, interview, Houston, 6 April 1967; Weiss, "Minutes of Senior Staff Meeting, July 2, 1965," p. 2; "Manned Space Flight Schedules," Vol. I, "Level 1 Schedules and Resources Summary," April, pp. 1-3, -6, June, p. 1-6, and July 1965, p. 1-6.

69 TWX, Mathews to McDonnell, Attn: Burke, "Contract NAS 9-170, Gemini, Flight Readiness Review for Spacecraft 5," GP-7274, 13 July 1965; TWX, Mathews to McDonnell, Attn: Burke, "Contract NAS 9-170, Gemini, Spacecraft 5 Flight Readiness Review Data Requirements," GP-7283, 20 July 1965; TWX, Mathews to McDonnell, Attn: Burke, "Contract NAS 9-170, Gemini, Revision of Agenda for Flight Readiness Review for Spacecraft 5," GP-7295, 27 July 1965; TWX, Mathews to McDonnell, Attn: Burke, "Contract NAS 9-170, Spacecraft 5 Flight Readiness Review Action Items," GP-7304, 4 Aug. 1965; "Gemini Program Mission Report, Gemini V," MSC-G-R-65-4, October 1965, p. 12-19; TWX, Mathews to McDonnell, Attn: Burke, "Contract NAS-170, Gemini V Mission Review," GA-6017, 6 Aug. 1965; TWX, Mathews to SSD, Attn: Dineen and Lt. Col. Fountain M. Hutchison, GP-7309, 11 Aug. 1965; Gemini News Center Release No. 9, 17 Aug. 1965; [Gemini News Center Release], "Air Force OKs Gemini Booster for Cooper-Conrad Flight," 18 Aug. 1965.

70 Ernest A. Amman, interview, Cape Kennedy, Fla., 24 May 1967; "Gemini V Preliminary Scrub Press Conference," 19 Aug. 1965; "3:00 P.M. Cape Press Briefing," 19 Aug. 1965; Gemini News Center Release No. 17, 20 Aug. 1965.

71 Memo, Webb to Donald K. Slayton, 14 Aug. 1965.

72 "Gemini V Technical Debriefing," Part I, 1 Sept. 1965, pp. 1, 6, 10-12, "Gemini V Mission Report," pp. 1-1, 4-18, -19, 5-117; Wambolt and Anderson, "Launch Systems Final Report," p. II.G-5; Gemini 5 mission commentary transcript, 21 Aug. 1965, tape 10, p. 1; memo, John J. Turner to Mgr., GPO, "Oxidizer charging equipment and procedures," GV-66231, 8 Oct. 1965.

73 "Gemini V Air-to-Ground Transcription," 5 Oct. 1965, pp. 22, 25, 27; "Gemini V Debriefing," Part I, p. 48.

74 "Gemini V Air-to-Ground," pp. 24, 26-

467

33; "Gemini V Debriefing," Part I, pp. 56-61, 63, 65; "Gemini V Mission Report," pp. 4-2, 7-7; John D. Hodge and Jones W. Roach, "Flight Control Operations," in *Gemini Midprogram Conference*, p. 184; briefing, first shift change, 4:00 p.m., 21 Aug. 1965, pp. 1-2, 4-5, 8.

[75] "Gemini V Air-to-Ground," p. 54; Kranz interview; Hodge and Roach, "Flight Control Operations," p. 184; "Gemini V Debriefing," Part I, pp. 65-66; "Gemini V Mission Report," pp. 5-68, -69; first shift briefing, pp. 1, 5-7.

[76] "Gemini V Air-to-Ground," pp. 64-65, 67; Kranz interview; "Gemini V Mission Report," pp. 4-2, -3; Gemini 5 mission commentary, tape 61, p. 1, tape 63, p. 1, tape 86, pp. 1, 2; shift change conference No. 2, 11:30 p.m., 21 Aug. 1965, pp. 1, 3-4.

[77] "Gemini V Debriefing," Part I, pp. 65, 72; "Gemini V Mission Report," pp. 4-2, 7-9; Gemini 5 mission commentary, tape 81, p. 1.

[78] "Gemini V Debriefing," Part I, pp. 80-82; "Gemini V Mission Report," pp. 5-19, 7-47; "Gemini V Air-to-Ground," p. 155.

[79] "Gemini V Air-to-Ground," p. 187; "Gemini V Debriefing," Part I, pp. 74-75; "Gemini V Mission Report," pp. 7-9, -48.

[80] "Gemini V Mission Report," pp. 4-3, 6-4, 7-7, -8; "Gemini V Debriefing," Part I, p. 102; Tindall interview; change of shift press briefing No. 4, 2:40 p.m., 22 Aug. 1965; change of shift press briefing, 2:45 p.m., 23 Aug. 1965, pp. 2-4.

[81] Memo, John A. Edwards to dist., "Gemini 5 24-hour report for 1500 August 24 to 1500 August 25," 26 Aug. 1965, with enclosure; "Gemini V Mission Report," p. 4-5; "Gemini V Debriefing," Part I, p. 117; Charles Conrad, Jr., interview, Houston, 31 March 1967.

[82] "Gemini V Mission Report," pp. 7-8, 8-1, -4 through -10, -53, -54, -55; Col. D. McKee, "Experiments D-1, D-2, and D-6, Basic Object, Nearby Object, and Surface Photography," in "Manned Space Flight Experiments Interim Report, Gemini V Mission," presented in Washington, 6 Jan. 1966, pp. 169-81; Burden Brentnall, "Experiment D4/D7, Celestial Radiometry and Space-Object Radiometry," in *Gemini Midprogram Conference*, pp. 356-77; Seibert Q. Duntley et al., "Experiments S-8/D-13, Visual Acuity and Astronaut Visibility," ibid., pp. 329-46.

[83] "Gemini V Mission Report," pp. 8-11 through -15; Lawrence F. Dietlein and William V. Judy, "Experiment M-1, Cardiovascular Conditioning," in *Gemini Midprogram Conference*, pp. 381-92; Earl Miller, "Experiment M-9, Human Otolith Function," ibid., pp. 431-36; Mack et al., "Experiment M-6, Bone Demineralization," ibid., pp. 413-14.

[84] E. P. Ney and W. F. Huch, "Experiment S-1, Zodiacal Light Photography," in "Gemini V Experiments Interim Report," pp. 1-8; Paul D. Lowman, Jr., "Experiment S-5, Synoptic Terrain Photography," ibid., pp. 9-17; Kenneth M. Nagler and Stanley D. Soules, "Experiment S-6, Synoptic Weather Photography," ibid., pp. 19-30; F. Saiedy, D. Q. Wark, and W. A. Morgan, "Experiment S-7, Cloud-Top Spectrometer," ibid., pp. 31-44; *Earth Photographs from Gemini III, IV, and V*, pp. 111-255, esp. p. 201.

[85] Amman interview; "Gemini V Mission Directive," NASA Program Gemini working paper No. 5028, 21 July 1965, pp. 3-6, -7; Edward F. Mitros, telephone comment on draft chapter, 29 July 1969; Gemini 5 mission commentary, tape 413, p. 1.

[86] Gemini 5 mission commentary, tape 426, pp. 4-6; "Gemini V Air-to-Ground," pp. 642, 646; "Gemini V Debriefing," Part I, p. 170.

[87] "Gemini V Mission Report," p. 4-6; "Gemini V Debriefing," Part I, pp. 173-78.

[88] "Gemini V Mission Report," p. 4-6; "Gemini V Debriefing," Part I, pp. 184-86; Tindall interview.

[89] "Gemini V Mission Report," pp. 6-14, -15, 7-10, -11; "Gemini V Debriefing," Part I, pp. 197-99, 203-204, 206, 213-14; Gemini 5 mission commentary, tape 454, p. 1, tape 455, p. 1, tape 458, p. 1, tape 459, p. 1, tape 461, p. 1; Tindall interview.

[90] James interview.

[91] Berry interview.

[92] Memo, Hugh L. Dryden, Cabinet Report to the President, "Significance of Gemini V Accomplishments," 11 Sept. 1965.

[93] "Washington Ceremonies Honor Gemini V Crew, Dr. Berry; Thirteen-Day Goodwill Tour of Six Foreign Nations Begins," MSC *Space News Roundup*, 17 Sept. 1965; Neocosmos Tzallas, "Russian Cosmonaut Greets Cooper and Conrad in Athens," *The Washington Post*, 18 Sept. 1965; photo, *The Evening Bulletin*, Philadelphia, 18 Sept. 1965, captioned, "Soviet, American Spacemen meet prior to a dinner in their honor last night . . . ."

## Chapter XII

[1] Walter M. Schirra, Jr., interview, Houston, 4 May 1967; Warren Burkett, "Schirra Expects to Get Flight of His Own after Gemini Test," *The Houston Chronicle*, 2 Feb. 1965; "Schirra Aiming for 6th Flight," *The Sun*, Baltimore, 2 Feb. 1965; Jim Maloney, "Goal for Age 46: The Moon," *The Houston Post*, 2 Feb. 1965; "News Conference, GT-6 Crew Selection," 5 April 1965, p. 2.

2 NASA News Release No. 65-218, "NASA Names Borman, Lovell Gemini 7 Crew, "1 July 1965; MSC News Release No. 65-66, 1 July 1965; Ivan D. Ertel, *Gemini XII Flight and Gemini Program Summary*, MSC Fact Sheet No. 291-I (Houston, December 1966); MSC News Release No. 66-20, "Gemini and Apollo Crews Selected," 21 March 1965; MSC News Release 180-63, 18 Oct. 1963.

3 Project Gemini Quarterly Status Report No. 7, for period ending 30 Nov. 1963, p. 86; Quarterly Status Report No. 8, for period ending 29 Feb. 1964, pp. 75, 77; "Manned Space Flight Schedules," Vol. I, "Level 1 Schedules and Resources Summaries," April 1964, p. 1-3; André J. Meyer, Jr., interview, Houston, 9 Jan. 1967; Eldon W. Hall, interview, Washington, 24 Jan. 1967; Raymond L. Zavasky, recorder, "Minutes of Senior Staff Meeting, July 10, 1964," pp. 3-4.

4 TWX, Charles W. Mathews to SSD, Attn: Lt. Col. Mark E. Rivers, GV-12088, 27 May 1965; Quarterly Status Report No. 13, for period ending 31 May 1965, p. 20; "GATV Progress Report, May 1965," LMSC-A605200-9, 20 June 1965, pp. 2-1, -3, -4; memo, James E. Powers, Jr., to Mgr., GPO, "Gemini Agena Target Vehicle modified command and communication hardware," GP-61396, 8 June 1965; memo, William C. Schneider to Assoc. Adm., Manned Space Flight, "Use of Agena 5002 for GTA-6," 9 June 1965; "GATV Progress Report, July 1965," LMSC-A605200-11, 20 Aug. 1965, pp. 2-7, 4-11, -12; Schirra interview; Tommy W. Holloway, "GTA-6 Flight Plan," Preliminary, 2 Aug 1965; "Abstract of Meeting of [sic] Atlas/Agena Coordination, August 11, 1965," 20 Aug. 1965; "Gemini VI Astronaut Prelaunch Press Conference," 11 Sept. 1965; TWX, George E. Mueller to MSC, Attn: Mathews, "Use of Agena Propulsion during Emini [sic] Mission 6," MGS-969, 22 Sept. 1965; Holloway, "Gemini VI Flight Plan," Final, 1 Oct. 1965.

5 Edwin E. Aldrin, Jr., "Line of Sight Guidance Techniques for Men in Orbital Rendezvous" (Ph.D. dissertation, Massachusetts Institute of Technology, 1964); Schirra interview; Dean F. Grimm, interview, Houston, 13 April 1967; Aldrin, interview, Houston, 4 April 1967.

6 Grimm interview (additional information from telephone interview, 12 Feb. 1969); Schirra interview; Marvin R. Czarnik, interview, St. Louis, 15 April 1966; "Preflight Training Plan for Fourth Manned Gemini Flight Crew (GTA-6)," NASA Program Gemini working paper No. 5031, 23 Aug. 1965.

7 "Manned Space Flight Schedules," Vol. I, June 1965, p. 1-7; J[oseph] F Wambolt and S[ally] F. Anderson, coordinators, "Gemini Program Launch Systems Final Report: Gemini/Titan Launch Vehicle; Gemini/Agena Target Vehicle; Atlas SLV-3," Aerospace TOR-1001(2126-80)-3, January 1967, p. II.G-5; MSC Weekly Activity Report for Office of Assoc. Adm., Manned Space Flight, 1-7 Aug. 1965, p. 1; "Gemini Program Mission Report, Gemini VI-A [sic]," MSC-G-R-65-5, October 1965, p. 12-3; James M. Grimwood and Barton C. Hacker, *Project Gemini Technology and Operations: A Chronology*, NASA SP-4002 (Washington, 1969), pp. 279, 281; "Medium Space Vehicles Programs Monthly Progress Report, August 1963," LMSC-447186-38, 20 Sept. 1963, p. 2-2; "Aestract[s] of Meeting[s] on Atlas/Agena, March 20, 1963," 27 March 1963; "April 3, 1963," 9 April 1963; "June 6 and 7 1963," 12 June 1963; "July 2, 1963," 8 July 1963; "August 7, 1963," 20 Aug. 1963; "September4, 1963," 10 Sept. 1963; "October 2, 1963," 8 Oct. 1963; memo, G. Merritt Preston to GPO, Attn: William R. Wakeland and Walter J. Kapryan, "Gemini/Agena Combined RF Compatibility and Functional Compatibility Test, 'Plan X,'" 28 Oct. 1963; TWX, Mathews to SSD, Attn: Rivers, GT-55349, 8 Jan. 1965; TWXs, Mathews to Dep. Dir., Launch Ops., KSC, Attn: Preston, SSD, Attn: Rivers, and McDonnell, Attn: Walter F. Burke, "Definition of Responsibilities for Cape GATV Testing," GT-55395, 12 March, and "Plan-X Test Objectives," GT-55394, 17 March 1965; letter, Mathews to Col. Alfred J. Gardner, GP-61612, 20 Aug. 1965; memo, Scott H. Simpkinson to Mgr., GPO, "Plan 'X' testing of Spacecraft 6 and GATV 5002," GT-65374, 13 Sept. 1965.

8 "Gemini VI Mission Directive," NASA Program Gemini working paper No. 5037, 20 Sept. 1965, p. 3-2; Schirra interview; NASA Release No. 65-327, "Project: Gemini 6," press kit, n.d. (for release 20 Oct. 1965), p. 1; Meyer, notes on GPO staff meeting, 25 Aug. 1965, p. 1; MSC News Release No. 65-85, 25 Sept. 1965; memo, Mathews to Thomas P. Stafford, "Experiments scheduled for Spacecraft 6," GP-61135, 9 April 1965; memo, Mathews to Chief, Flight Crew Support Div., Attn: Chief, Spacecraft Ops. Br., "Gemini Experiment MSC-4, Optical Communication," GP-61324, 26 May 1965; letter, Robert R. Gilruth to Ames Research Center, Attn: Smith J. DeFrance, "Request for training of GTA-6 flight crews," EG27-65-626, 18 Aug. 1965; letter, Mueller to Gilruth, 13 Oct. 1965.

9 John Miller, interview, San Diego, Calif., 18 March 1967; "Lockheed Agena in U. S. Space Programs," Lockheed Fact Sheet (ca. October 1965); "Spacecraft Prelaunch Test Procedure: Outline for the Rendezvous Mis-

sion Countdown," McDonnell SEDR RMC-6, 23 Oct. 1965; NASA News Release No. 65-331, "NASA Appoints Schneider Deputy Director, Mission Operations," 15 Oct. 1965; memo, LeRoy E. Day to William D. Putnam, "Comments on Gemini History ...," 28 July 1969, with enclosure.

[10] Gemini 6 mission commentary transcript, 25 Oct. 1965, tape 26, p. 1; "Outline for the Rendezvous Mission Countdown"; "Gemini VI-A [sic] Mission Report," p. 6-2.

[11] Miller interview; "Gemini VI-A [sic] Mission Report," pp. 4-2, 6-2; Gemini 6 mission commentary, tape 28-1.

[12] Jack L. Shoenhair, interview, Sunnyvale, Calif., 11 May 1967; Schneider, interview, Washington, 23 Jan. 1967; "Gemini VI-A [sic] Mission Report," p. 6-3.

[13] Gemini 6 mission commentary, tapes 30-36; "Gemini VI-A [sic] Mission Report," p. 6-3; letter, George M. Low to Gilruth, 20 June 1963; memo, Mathews to Asst. Dir., Flight Ops., "Radar Skin-tracking Requirement for Gemini Missions," GV-02264, 3 June 1964; letter, Christopher C. Kraft, Jr., to Goddard Space Flight Center, Attn: Niles R. Heller, "Radar skin tracking support for project Gemini," 27 Aug. 1964; letters, Gilruth to NASA Hq., Attn: Mueller, "Orbital tracking of expended stages of Gemini launch vehicle," 18 Sept., and GA-01286, 2 Oct. 1964; letter, Mueller to Gilruth, 19 Nov. 1964; memos, Mathews to Asst. Dir., Flight Ops., "Radar skin-tracking requirements for Gemini Program," GV-02421, 2 Dec. 1964, GV-02495, 16 Feb., and GV-02514, 4 Mar. 1965; Quarterly Status Report No. 14, for period ending 31 Aug. 1965, p. 24.

[14] Jerome B. Hammack, interview, Houston, 18 April 1966; "Gemini 6 Scrub Briefing," 25 Oct. 1965, tape 1, p. 1, tape 1-A, p. 1; Earl Ubell, "'No Joy, No Joy': Agena Failure Is Probed in Great Secrecy," New York Herald Tribune, 26 Oct. 1965.

[15] Frank Borman, interview, Houston, 18 April 1967; John F. Yardley, interview, St. Louis, 13 April 1966; letter, Paul P. Van Riper to Eugene M. Emme, 31 July 1969; memo, James E. Webb to the President, "Space Rescue," 2 June 1965; memo, Carl B. Peterson to Dep. Dir., "Martin rescue study," 4 Aug. 1965.

[16] Ted A. Guillory, "Gemini VII Flight Plan," Preliminary, 4 Oct. 1965, Section II, "Flight Plan."

[17] Memo, Schneider to Mueller, "Second Launch Pad for Gemini," 17 Sept. 1964; memo, Hall to Dep. Dir., Gemini Program, "Simultaneous Launch of Two Gemini Spacecraft," 19 Feb. 1965.

[18] Low, interview, Houston, 7 Feb. 1967; Mathews, interview, Houston, 2 Dec. 1966; Meyer, notes on GPO staff meeting, 7 April 1965, p. 2.

[19] Joseph M. Verlander, interview, Cocoa Beach, Fla., 29 Aug. 1967; Col. John G. Albert, interview, Patrick AFB, Fla., 26 May 1967; Raymond D. Hill, Jr., interview, Titusville, Fla., 23 May 1967; Yardley interview; Walter D. Smith, interview, Baltimore, 23 May 1966; Preston, interview, Cape Kennedy, Fla., 24 May 1967; H. H. Luetjen, interview, Cape Kennedy, 25 May 1967; J. Carroll Curlander, interview, Baltimore, 24 May 1966; "Dual Countdown for Gemini," April, and "Rapid Fire Gemini," 28 July 1965, revised 20 Aug. 1965, Martin Co. studies (Van Riper letter, 31 July 1969).

[20] "Gemini Program Mission Report, Gemini VI-A," MSC-G-R-66-2, January 1966, p. 3-28; "Gemini Program Mission Report, Gemini VII," MSC-G-R-66-1, January 1966, p. 3-40; Yardley, Preston, and Albert interviews; "Transcription of Spacecraft Test Conductor's Log during period October 25 to October 27, 1965 (following GT-6 mission scrub resulting from Agena failure)," n.d., p. 141.

[21] There is little or no documentation of the events covering the VII/VI decision. Most of the material came from desk calendars, note pads, and interviews with the chief actors. The MSC History Office is deeply indebted to Paul P. Van Riper, Professor of Public Administration, Cornell University, who did a study on major NASA decisions and was kind enough to allow us the use of his research and to offer his personal assistance in piecing together what actually happened and when. Memo, Nina Scrivener (Webb's secretary) to Van Riper, 25 March 1967, with enclosure, "Summary of Telephone Conversations re Gemini 7/6," n.d.; Van Riper, notes on interview with Mary S. Turner (Robert C. Seamans, Jr.'s secretary), 17 March 1967, using her 1965 desk calendar pad; Van Riper, notes on interview with Col. Lawrence W. Vogel, 14 March 1967, based on notes taken at meetings in Webb's office; Van Riper letter, 31 July 1969.

[22] Yardley and Low interviews; Gilruth, interview, Houston, 21 March 1968; Meyer, notes on GPO staff meeting, 27 Oct. 1965, p. 1.

[23] Memo, Mueller to Seamans, [GLV performance analysis], 27 Oct. 1965; Kapryan, interview, Cape Kennedy, Fla., 25 May 1967; Preston interview; Van Riper, notes on interview with Preston, 5 Jan. 1967; Meyer notes, 27 Oct. 1965, p. 1; W. A. Krzywicki (McDonnell), "S/C #6 Pad Schedule," 30 Nov. 1965 (annotated, "official release date [30 Nov.] not day plan made" and "plan developed 10/26 or 27"); William Hines, "Space Aides

Map Plans after Gemini 6 Setback," *The Evening Star*, Washington, 26 Oct. 1965; John Troan, "Space Shot in November?" *The Washington Daily News*, 26 Oct. 1965.

[24] Low interview; John D. Hodge, interview, Houston, 12 March 1968; Gerald M. Truszynski, interview, Washington, 13 Sept. 1966; memo, Mathews to Asst. Dirs., Flight Ops. and Flight Crew Ops., "Real and delayed time telemetry data recording, Gemini VI-A and Gemini VII," GT-65197, 3 Dec. 1965; TWXs, Mathews to dist., "Gemini Mission Designations," GT-11165, 20 Nov., and GT-11168, 26 Nov. 1965; memo, Mueller to Adm., "Gemini VI-A Mission," 18 Nov. 1965, with enclosure, "Mission Operation Report: Gemini VI-A Flight," M-913-65-08, 18 Nov. 1965; routing slip, Kraft to [MSC Historical Office], 2 July [1969], with enclosure.

[25] Gilruth, Low, and Yardley interviews.

[26] Low interview; letter, Day to MSC, Attn: Grimwood, "Comments on the Final Manuscript of the Gemini History," 23 June 1971; Putnam, notes on interview with Seamans, 20 July 1967; Van Riper, notes on Turner interview; letter, Seamans to Emme, 30 July 1969; Van Riper letter, 31 July 1969.

[27] Van Riper, notes on interview with Willis B. Shapley, n.d. (probably March 1967); Vogel, notes taken at meetings in Webb's office, 27 Oct. 1965; Putnam notes, 20 July 1967; Van Riper, notes on Turner interview; "Summary of Telephone Conversations re Gemini 7/6"; Van Riper, notes on Vogel interview; Seamans letter, 30 July 1969; Van Riper letter, 31 July 1969; Gilruth and Low interviews; letter, Scrivener to Van Riper, 8 Aug. [1969]; letter, Alice McGilvra to Van Riper, 5 Aug. 1969.

[28] Putnam notes, 20 July 1967; Van Riper, notes on interview with Julian Scheer, 8 Dec. 1966; "Summary of Telephone Conversations re Gemini 7/6."

[29] TWX, Webb to Joseph Laitin, The White House (draft press release and memorandum for the President), 27 Oct. 1965, 7:10 p.m., e.s.t.

[30] News Conference #176-A at the White House (Austin, Tex.), 28 Oct. 1965; Low interview; Meyer notes, 27 Oct. 1965, p. 1.

[31] Letter, Bernhard A. Hohmann to MSC, Attn: Grimwood, 12 Aug. 1969, with enclosure, annotated pages of draft chapter; "Gemini Launch Vehicle Operations, GT-7 & GT-6B," n.d.; Yardley interview.

[32] Hill and Luetjen interviews; Guenter F. Wendt, interview, Cape Kennedy, Fla., 25 May 1967; Krzywicki, "S/C #6 Pad Schedule"; "Transcription of Spacecraft Test Conductor's Log," p. 144; TWX, Mathews to McDonnell,

Attn: Burke, and KSC, Attn: Preston, "Contract NAS 9-170, Gemini, Disposition of Spacecraft 6 and All Equipment Assigned Thereto," GP-7386, 28 Oct. 1965.

[33] Riley D. McCafferty, interview, Cape Kennedy, Fla., 25 May 1967; TWX, Mathews to McDonnell, Attn: Burke, "Contract NAS 9-170, GT-7 Crew Training," GV-12185, 26 Sept. 1965; "Gemini VI-A Mission Report," pp. 7-8, -9; "Gemini VII Mission Report," pp. 7-9, -10; Truszynski and Grimm interviews; Howard W. Tindall, Jr., interview, Houston, 16 Dec. 1966; "Gemini 7/6 Mission Control Procedures Briefing," 24 Nov. 1965.

[34] Schirra, Borman, and Low interviews; Larry E. Bell, interview, Houston, 10 Sept. 1968; Vogel, notes on meetings in Webb's office, 29 Oct. and 2 Nov. 1965; TWX, Mathews to McDonnell, Attn: Burke, "Extravehicular Life Support System Installation in Spacecraft 6," GS-53397, 29 Dec. 1964.

[35] James A. Lovell, Jr., interview, Houston, 15 April 1967; memo, D. Owen Coons to Dep. Dir., "Medical Operations Directive GT-6," 27 Sept. 1965; Guillory, "Gemini VII Preliminary Flight Plan"; Guillory, "Gemini VII Flight Plan," Final, 15 Nov. 1965; Gemini 7/6 News Center Release No. 10, "Gemini 7/6 Flight Controllers," 2 Dec. 1965.

[36] Borman and Lovell interviews; "Gemini VII Flight Crew Press Conference," 1 Nov. 1965; TWX, Mathews to McDonnell, Attn: Burke, "Spacecraft 7 Stowage Review," GP-51766, 25 Feb. 1965; TWX, Mathews to McDonnell, Attn: Burke, "Contract NAS 9-170, Gemini Spacecraft 7 Crew Station Stowage Inspection," GP-7235, 17 June 1965; memo, Mathews to dist., "Gemini Spacecraft 7 Crew Station Stowage Review, Phase I, June 29-30, 1965," GS-64044, 8 July 1965; memo, Mathews to dist., "Gemini Spacecraft 7 Crew Station Stowage Review, Phase II, July 15, 1965," GS-64054, 28 July 1965.

[37] "Gemini Suit Requirements: NASA visit—September 26, 1962," unsigned report [probably Forrest R. Poole]; "National Aeronautics & Space Administration Conference, Manned Spacecraft Center, Houston, Texas, January 30-31, 1963," 21 Feb. 1963; Borman and Lovell interviews; L. Gordon Cooper, Jr., interview, Houston, 11 April 1967; Charles Conrad, Jr., interview, Houston, 31 March 1967; Poole, interview, Houston, 1 May 1968; Meyer, notes on NASA/MAC management meeting, 16 July 1965, p. 1; "NASA/MAC Management Meeting 11 July 1965, Preliminary," p. 1; "Evaluaton of Modified Flight Suit ('Shirt Sleeve') configuration for the Gemini VII Mission," McDonnell report No. B948, 9 Aug. 1965, pp. 1-3, 14-16, 19, 31, 50, 52, 53;

memo, James V. Correale to Historical Office, "Comments to draft chapter of Gemini narrative history . . . ," EC11BE-69-098, 3 Oct. 1969; memo, Mathews to dist., "Suit configuration for Gemini VII," GS-64055, 27 July 1965; John B. Lee, recorder, "Minutes of Senior Staff Meeting, August 6, 1965," p. 1; "NASA/MAC Management Meeting 12 August 1965," p. 2; Meyer, notes on NASA/MAC management meeting, 12 Aug. 1965, p. 3; Richard S. Johnston, James V. Correale, and Mathew I. Radnofsky, *Space Suit Development Status,* NASA TN D-3291 (Langley, Va., February 1966), pp. 2-16; TWX, Vogel to MSC, Attn: Gilruth, AO-628N, 15 Sept. 1965; letter, Gilruth to NASA Hq., Attn: Seamans, "Gemini space suit development," 22 Sept. 1965, with enclosures; letter, Mueller to Gilruth, 23 Oct. 1965; "Light Weight Suit Briefing," 4 Nov. 1965; TWX, Mathews to NASA Hq., Attn: Day, GP-7405, 18 Nov. 1965; letter, Mathews to NASA Hq., Attn: Mueller, "Light-weight suit evaluation," GT-65490, 19 Nov. 1965, with enclosure, "Design Certification Report on the Lightweight Space Suit, G-5C, for Gemini VII Mission," 19 Nov. 1965.

38 Letter, Mueller to Gilruth, 23 Sept. 1965; "Gemini VII Medical Experiments Briefing," 1 Nov. 1965; memo, Robert O. Piland to Asst. Dir., Flight Ops., "Gemini VII Experiment Priorities," EX1365-0239, 18 Nov. 1965, with enclosures, "Experiments Priority—Gemini VII" and "GT-7 Experiment Activities Priorities"; NASA Release No. 65-362, "Project: Gemini 7/6," press kit, n.d. (for release 29 Nov. 1965), pp. 7-20; TWX, Mueller to MSC, Attn: Mathews, "Conduct of Experiments and Operational Test on Gemini VII and Gemini VIA Missions," MGO-30, 1 Dec. 1965.

39 Borman interview; Russell A. Schweickart, interview, Houston, 1 May 1967; memo, Schneider to dist., "Transmittal of Report on GEMINI Experiments," 12 May 1964, with enclosure, "Description of Gemini Experiments: Flights GT-3 through GT-7," OMSF, 13 April 1964, pp. 5, 20-23; Peter Kellaway, "Experiment M-8, Inflight Sleep Analysis," in *Gemini Midprogram Conference, Including Experiment Results,* NASA SP-121 (Washington, 1966), pp. 423-29.

40 Borman interview; G. D. Whedon et al., "Experiment M-7, Calcium and Nitrogen Balance," in *Gemini Midprogram Conference,* pp. 417-21.

41 Lawrence F. Dietlein and E[lliott] S. Harris, "Experiment M-5, Bioassays of Body Fluids," in *Gemini Midprogram Conference,* pp. 403-406; "Abstract of Meeting on [sic] Gemini VI Experiments Board, September 15, 1965," 30 Sept. 1965; *Gemini VI* press kit, pp.

17-18; memo, Mathews to Asst. Dir., Flight Ops., "Gemini VI experiments and associated equipment," EX1365-0223, 17 Nov. 1965.

42 "Gemini 7 Communications Experiment (Laser)," 8 July 1965; letter, John M. Walker to Langley, Attn: George B. Graves, Jr., "Optical communication experiments using MSC-4 experiment on GT-7," 1 July 1965; letter, Floyd L. Thompson to NASA, Attn: Walker, "Manned Spacecraft Center optical communications experiment on GT-7," 2 Sept. 1965; DeFrance to MSC, Attn: Douglas S. Lilly, "Requirements for support on Ames participation in MSC-4 experiment," 12 Oct. 1965; DeFrance to MSC, "Request for astronaut training for participation in MSC-4 laser communications experiment on GT-7," 12 Oct. 1965; letter, Thompson to Ralph Hicks, "Range support for forthcoming MSC Gemini-Titan 7 Laser Experiment," 20 Oct. 1965, with enclosure; letter, Russell G. Robinson to MSC, Attn: Edward O. Zeitler, "Gemini MSC-4 Experiment," 29 Dec. 1965; memo, Robert L. Jones to Mgr., Experiments Program Office (EPO), "Launch azimuth for Gemini VII," 8 Sept. 1965; memo, Mathews to Mgr., EPO, "Launch Azimuth for Gemini VII," GV-66192, 14 Sept. 1965; "Gemini VII Flight Crew Press Conference," tape 2, p. 5; *Gemini VII* press kit, pp. 12-14, 16-18; TWX, Mathews to NASA Hq., Attn: Day, GV-12277, 26 Nov. 1965; R[obert] O. Piland and P[aul] R. Penrod, "Experiments Program Summary," in *Gemini Midprogram Conference,* pp. 305-12; Wilbur A. Ballentine, "DOD/NASA Gemini Experiments Summary," ibid., pp. 307-17; Charles E. Manry, telephone interview, 22 May 1973.

43 Memo, Mathews to dist., "Gemini VII Mission Planning," GV-66188, 21 Sept. 1965; memo, Mathews to dist., "Mission Planning," GV-66198, 25 Sept. 1965; memo, Mathews to dist., "Mission Planning for Gemini VI through XII," GV-66208, 1 Oct. 1965; "Gemini VII Flight Crew Press Conference," tape 1, pp. 1-3; TWX, Day to MSC, Attn: Piland, MGS-421, 9 Nov. 1965; memo, Piland to Asst. Dirs., Flight Ops., Flight Crew Ops., and Mgr., GPO, "Gemini VII Experiments, Final Flight Plan," EX4/M36-65, 29 Nov. 1965; memo, Piland to Dep. Dir., no subject, 30 Nov. 1965; memo, Mathews to Asst. Dir., Flight Crew Ops., "Gemini VI and VII Flight Plans," GV-66282, 30 Nov. 1965; TWX, Mathews to NASA, Attn: Day, "Gemini VII Experiments," EX1365-98, 30 Nov. 1965; Richard T. Hamm, "Description of the Gemini VII Station Keeping Hybrid Simulation," McDonnell, Gemini Guidance and Control Design Note No. 388, 13 July 1966; Meyer notes, 27 Oct. 1965, p. 3; memo, Simpkinson to MSC Historical Office,

Attn: Grimwood, "Review of Gemini narrative history . . . ," 19 Aug. 1969.

⁴⁴ Loyd S. Swenson, Jr., James M. Grimwood, and Charles C. Alexander, *This New Ocean: A History of Project Mercury,* NASA SP-4201 (Washington, 1966), pp. 402-407; "Launch Vehicle No. 7 Flight Evaluation," Martin Engineering Report No. 13227-7 and Supplemental Report No. 2 to "Gemini VII Mission Report," January 1966, pp. vii-viii; "Gemini VII Technical Debriefing," 23 Dec. 1965, pp. 1-3.

⁴⁵ "Gemini VII Voice Communications (Air-to-Ground, Ground-to-Air, and On-Board Transcription)," McDonnell Control No. 115308, Vol. 1, n.d., pp. 1-3; C. E. Agajanian, "Launch Vehicle Flight Evaluation Report, NASA Mission Gemini/Titan, GT-7,"Aerospace TOR-6696(6126-42)-10 and Supplemental Report No. 1 to "Gemini VII Mission Report," February 1966, p. 4-1; "Gemini VII Technical Debriefing," p. 4; TWX, Kenneth S. Kleinknecht to NASA Hq., Attn: Webb, and MSC, Attn: Gilruth, "Launch Summary Report, Gemini Mission VII," GT-11118, 4 Dec. 1965.

⁴⁶ "Gemini VII Voice," I, pp. 5-19; "Gemini VII Debriefing," pp. 12-19; "Gemini VII Mission Report," pp. 7-2, -3; Borman and Lovell interviews.

⁴⁷ "Gemini VII Voice," I, pp. 24, 29, 33-36, 42, 44, 48, 58, 59, 62, 66, 68, 73, 74; TWX, Kleinknecht to NASA Hq., Attn: Webb, and MSC, Attn: Gilruth, "Daily Report Number 1—Gemini Mission VII," GT-11119, n.d. [5 Dec. 1965], pp. 6, 8-10, 12.

⁴⁸ "Gemini VII Voice," I, pp. 125, 126, 133; "Gemini VII Debriefing," p. 27; TWX, Kleinknecht to NASA Hq., Attn: Webb, and MSC, Attn: Gilruth, "Daily Report No. 2," GT-11120, 6 Dec. 1965, pp. 6-7.

⁴⁹ Borman and Lovell interviews; letter, Gilruth to NASA Hq., Attn: Mueller, "Use of G-5C suits on Gemini VII," 29 Nov. 1965; letter, Mueller to Gilruth, 3 Dec. 1965; letter, Mathews to NASA Hq., Attn: Mueller, "Removal of space suits during Gemini VII," GS-64097, 4 Dec. 1965, with enclosures, Gilruth letter, 29 Nov. 1965, and "Suit Procedures," n.d.; memo, Low for record, "Gemini 7 suit configuration," 7 Dec. 1965.

⁵⁰ Borman interview; "Gemini VII Voice," I, pp. 134, 140, 173, 174, 179, 183, 216-18, 235-36, 296, 299, 303, 319, 323-24, 329-30, Vol. II, 341, 343, 441, 444-47, 453; TWX, Mathews to McDonnell, Attn: Burke, "Contract NAS 9-170, Gemini, Personalized Equipment for Spacecraft 7 Crew Members," GS-10038, 8 July 1965; TWX, Kleinknecht to NASA Hq., Attn: Webb, and MSC, Attn: Gilruth, "Daily Report No. 5," GT-11123, 9 Dec. 1965, p. 9.

⁵¹ TWX, Kleinknecht to NASA Hq., Attn: Webb, and MSC, Attn: Gilruth, "Daily Report No. 7," GT-11125, 11 Dec. 1965, pp. 8-9; memo, Day to Mueller, "Gemini VII Suit Configuration," 9 Dec. 1965; TWX, Mueller to MSC, Attn: Schneider, "Suit Operation for Gemini VII," M-468, 10 Dec. 1965; memo, Mueller to Gilruth, "G-5C Operational Test Procedure," 12 Dec. 1965; "Gemini VII Voice," II, p. 580; "Gemini VII Debriefing," p. 30.

⁵² "Gemini VII Debriefing," p. 175; "Gemini VII Mission Report," pp. 7-50, -51.

⁵³ Kleinknecht, "Daily Report[s] No. 1," pp. 4, 6, 8, 9, "No. 3," GT-11121, 7 Dec., pp. 4-5, 6-7, "No. 5," pp. 4-5, 7, "No. 9," GT-11128, 14 Dec. 1965, p. 5; "Gemini VII Voice," II, pp. 359, 365; TWX, Mathews to McDonnell, Attn: Burke, GV-52652, 28 Dec. 1964.

⁵⁴ Albert and Verlander interviews; "Gemini VII Voice," I, pp. 6, 136; Kleinknecht, "Daily Report No. 1," p. 13; Alexander C. Kuras and John G. Albert, "Gemini-Titan Technical Summary," 24 Jan. 1967, pp. 144-45; TWX, Mathews to McDonnell, Attn: Burke, "Contract NAS 9-170, Gemini, Need for Radar and Transponder for Tests at the Cape," GP-7392, 5 Nov. 1965; memo, Mathews to Asst. Dir., Flight Crew Ops., "Rendezvous radar transponder test for the Gemini VII mission," GV-66279, 27 Nov. 1965.

⁵⁵ "Gemini VII Voice," I, pp. 178, 183; Kleinknecht, "Daily Report[s] No. 2," pp. 11, 12, "No. 4," GT-11122, 8 Dec. 1965, "No. 5," pp. 13-14; "Gemini VI-A Mission Report," p. 12-21.

⁵⁶ TWX, Kleinknecht to NASA Hq., Attn: Webb, and MSC, Attn: Gilruth, "Daily Report No. 8," GT-11127, 13 Dec. 1965, pp. 15-16; Wambolt and Anderson, "Launch Systems Final Report," p. II.E-19; Kuras and Albert, "Gemini/Titan Technical Summary," p. 145; John J. Williams, interview, Cape Kennedy, Fla., 24 May 1967; memo, Mueller to Adm., "Gemini VI-A Mission, Post Launch Report No. 1," 23 Dec. 1965, with enclosure, subject as above, M-913-65-08, same date.

⁵⁷ Kenneth F. Hecht, interview, Houston, 23 Feb. 1967; Howard Simons, "Reset for Wednesday: Plug Trouble Delays Gemini 6 Liftoff," *The Washington Post,* 13 Dec. 1965; Verlander interview; Eugene F. Kranz, interview, Houston, 28 April 1967; Simpkinson, interview, Houston, 18 Jan. 1967; "Gemini VI-A Post Launch Report No. 1," p. 1a; [Gemini 7/6] mission commentary transcript, 12 Dec. 1965, tape 339, p. 2.

⁵⁸ "Gemini VI Debriefing," p. 251; Wendt interview.

[59] Seamans letter, 30 July 1969.

[60] "Gemini VI-A Post Launch Report No. 1," p. 1a; Albert interview; Kleinknecht, "Daily Report No. 8," p. 16.

[61] Preston interview; "Gemini 6 Scrub Press Conference," 12 Dec. 1965.

[62] Hohmann, interview, El Segundo, Calif., 19 April 1966; Albert interview; Maj. Gen. Ben I. Funk, interview, Sunnyvale, Calif., 12 March 1967; E. Douglas Ward and Louis D. Wilson, interview, Sacramento, Calif., 30 June 1966; Lt. Col. Fountain M. Hutchison and Capt. Howard T. Harris, interview, Los Angeles, 19 April 1966; Haggai Cohen, interview, Baltimore, 24 May 1966; Bastian Hello, interview, Baltimore, 23 May 1966; Kleinknecht, "Daily Report No. 8," pp. 16-17; Wambolt and Anderson, "Launch Systems Final Report," p. II.E-19; Kuras and Albert, "Gemini-Titan Technical Summary," p. 145; Hohmann letter, 12 Aug. 1969; "Gemini VI-A Post Launch Report No. 1," p. 1a; letter, Hutchison to MSC Historical Office, 11 Aug. 1969; Hammack, telephone interview, 2 March 1973.

[63] Kleinknecht, "Daily Report No. 9," p. 12; Hutchison and Harris interview.

[64] "Gemini VII Voice," II, pp. 560, 585, 647, III, pp. 738, 750, 751; Kuras and Albert, "Gemini-Titan Technical Summary," pp. 145, 146; "Launch Operations Techniques, Manned Space Flight," Gemini Launch Vehicle Div., 6555th Aerospace Test Wing, 22 Dec. 1966, p. 168.

[65] Schirra interview; Thomas P. Stafford, interview, Houston, 3 April 1967; Warren J. North, interview, Houston, 10 Jan. 1967.

[66] Tindall and Czarnik interviews; Dean F. Grimm, Thomas P. Stafford, and Walter M. Schirra, Jr., "Report on Gemini VI Rendezvous," 28 Feb. 1966, p. 1; Henry L. Richter, Jr., Instruments and Spacecraft, October 1957—March 1965, NASA SP-3028 (Washington, 1966), pp. 313-15.

[67] "Gemini VI-A Mission Report," p. 1-1, -2; Evert Clark, "At Last, Gemini 6 Day Is Perfect As Even Sun Comes Out in Time," The New York Times, 16 Dec. 1965; "Gemini VI Debriefing," pp. 13, 18; "Gemini VII Voice," III, pp. 751, 752, 755; "Gemini VII Debriefing," p. 145; "Gemini 7/6 Flight Controllers," [p. 15]; "Gemini VI-A Post Launch Report No. 1," p. 1b.

[68] Grimm, Stafford, and Schirra, "Gemini VI Rendezvous," pp. 1-2; "Gemini VI-A Mission Report," p. 1-2.

[69] [Ivan D. Ertel], Gemini VII/Gemini VI: Long Duration/Rendezvous, MSC Fact Sheet No. 291-D (Houston, Jan. 1966), p. 9; "Gemini VI-A Mission Report, pp. 4-15, -16, -18; "Gemini VI-A Post Launch Report No. 1," p. 1b; Astronautics and Aeronautics, 1965:

Chronology on Science, Technology, and Policy, NASA SP-4006 (Washington, 1966), p. 551.

[70] "Gemini VI-A Mission Report," pp. 4-16, -18, 7-17; TWX, Kleinknecht to NASA Hq., Attn: Webb, and MSC, Attn: Gilruth, "Special Rendezvous Report—Gemini Mission VII/VI," 15 Dec. 1965; Grimm, Stafford, and Schirra, "Gemini VI Rendezvous," pp. 1-2; Astronautics and Aeronautics, 1965, p. 551.

[71] "Gemini VI-A Mission Report," pp. 1-2, 4-12, -16, -19; "Gemini VI-A Post Launch Report No. 1," p. 1c; Kleinknecht, "Special Rendezvous Report."

[72] "Gemini VI-A Mission Report," pp. 4-12, -19, 7-2, -20; Grimm, Stafford, and Schirra, "Gemini VI Rendezvous," p. 2; memo, Tindall to dist., "Rendezvous odds and ends," 65-FM1-212, 30 Dec. 1965; "Gemini VI Debriefing," pp. 27, 37-38.

[73] Thomas P. Stafford, Walter M. Schirra, and Dean F. Grimm, "Rendezvous of Gemini VII and Gemini VI-A," in Gemini Midprogram Conference, p. 291; "Gemini VI-A Mission Report," p. 7-21; "Gemini VII Voice," III, p. 766; Grimm, Stafford, and Schirra, "Gemini VI Rendezvous," p. 14.

[74] Kleinknecht, "Special Rendezvous Report"; Grimm, Stafford, and Schirra, "Gemini VI Rendezvous," p. 11; Stafford, Schirra, and Grimm, "Rendezvous of Gemini VII and Gemini VI-A," p. 291; "Gemini VI-A Mission Report," pp. 7-23, -24; Hodge interview; Tindall memo, 30 Dec. 1965; "Gemini VII Voice," III, p. 769; "Jubilation," caption of photo in MSC Space News Roundup, 23 Dec. 1965.

[75] U.S. Congress, House, Committee on Science and Astronautics, Astronautical and Aeronautical Events of 1962: Report, 88th Cong., 1st sess., 12 June 1963, pp. 146-47, 148; Robert Korengold, "2 Reds Go on Orbiting As Observers Report Signs of Rendezvous," The Washington Post, 14 Aug. 1962; David Miller, "Split-Second Precision Put 2 Vostoks Close Together," New York Herald Tribune, 14 Aug. 1962; Korengold, "Both Reds Pass Million Miles Travel in Orbit," The Washington Post, 14 Aug. 1962; TWX, Rhett Turnipseed to NASA, Houston, "Text of an Interview by an Izvestia Correspondent with the Soviet Cosmonaut Pavel Romanovich Popovich [21 Dec. 1965]," 29 Dec. 1965; "Gemini 7/6 Astronaut Post Flight Press Conference," 30 Dec. 1965, tape 8, p. 2; James M. Grimwood and Ivan D. Ertel, "Project Gemini," Southwestern Historical Quarterly, 81, no. 3 (January 1968), p. 407; [Ertel], Gemini VII/Gemini VI, p. 16.

[76] "Gemini VII Debriefing," pp. 137-38; "Gemini VI Debriefing," pp. 59-60; "Gemini VII Voice," III, pp. 767-68, 771.

[77] "Gemini VI-A Mission Report," pp. 7-2, -25, -26, -27; "Gemini VII Voice," III, pp. 763, 774; "Gemini VI Debriefing," pp. 69, 70, 71, 80, 84, 100; memo, Duncan R. Collins to Chief, Mission Planning Office, "Recommended activities for Gemini VI/Gemini VII mission," GS-64090, 3 Nov. 1965; Tindall memo, 30 Dec. 1965.

[78] "Gemini VI Debriefing," pp. 113, 114, 117-18; "Gemini Mission Rules, Gemini VIA," 4 Nov. 1965, p. 2-1.

[79] "Gemini VII Voice," III, p. 825; letters, Michael Kapp to Schirra, 8 Dec. 1965, and Grimwood, 29 March 1967; Sarah W. Lopez and Riley D. McCafferty, telephone interviews, 9 June 1969; Frances Slaughter, telephone interview, 10 June 1969; Stafford, telephone interview, 2 Oct. 1969; Schirra interview.

[80] "Gemini VI Voice Communications (Air-to-Ground, Ground-to-Air and On-Board Transcription)," McDonnell Control No. C-115269, n.d., p. 124.

[81] "Gemini VI-A Mission Report," pp. 7-29, -30; "Gemini VI Debriefing," pp. 155, 169, 171-72; notes, James L. Gibson, "Crew Activities during Reentry Phase of Gemini VI-A," 17 March 1966.

[82] TWX, Office of Sec. Defense to DOD Mgr. for MSF Support Ops., Andrews AFB, Md., 19 Oct. 1965; letter, John S. Foster, Jr., to Seamans, 19 Oct. 1965; Grimwood and Hacker, *Gemini Chronology*, p. 265; memo, Philip H. Bolger to Actg. Dep. Dir., Gemini, "World Records to Be Certified during Gemini VII/VI-A," 2 Dec. 1965; Carl R. Huss, telephone interview, 7 June 1973.

[83] Borman and Lovell interviews.

[84] Lovell interview.

[85] "Gemini VII Voice," III, pp. 827, 830-31, 839, 863, 865; "Gemini VII Mission Report," pp. 6-8, -9; TWXs, Kleinknecht to NASA Hq., Attn: Webb, and MSC, Attn: Gilruth, "Daily Report[s] No. 10," GT-11200, 16 Dec. 1965, pp. 4, 7, and "No. 11," GT-11201, 17 Dec. 1965, pp. 4, 6; Meyer, notes on GPO staff meeting, 25 Jan. 1966, pp. 2-3; TWX, Mathews to McDonnell, Attn: Burke, "Contract NAS 9-170, Gemini," GP-7468, 7 Feb. 1966; memo, Clarence C. Gay, Jr., to Day, "Use of 90° Billet on S/C #7," 15 Feb. 1966.

[86] Borman and Lovell interviews; "Gemini VII Voice," III, pp. 862-64, 868, 869, 873-78; "Gemini VII Debriefing," pp. 95-99; "Gemini VII Mission Report," p. 6-7.

[87] "Gemini VII Voice," III, pp. 950, 959, 960, 968, 971, 973, 978; "Gemini VII Mission Report," pp. 7-7, -57.

[88] Lovell interview.

[89] Ibid.; "Gemini VII Debriefing," pp. 35, 36; "Gemini VII Mission Report," pp. 7-8, -58;

"Gemini VII Voice," III, pp. 979, 994.

[90] "Gemini VII Voice," III, pp. 980, 986, 991; "Gemini VII Mission Report," pp. 7-57, -58; "Gemini VII Debriefing," pp. 43,44.

[91] "Gemini VII Debriefing," pp. 45-52; "Gemini VII Voice," III, pp. 816, 996, 1007; "Gemini VII Mission Report," pp. 7-58, -59, -74.

[92] "Gemini VII Voice," III, pp. 952-53, 957, 964-65, 1008, 1011; "Gemini VII Debriefing," pp. 53, 56, 61; "Gemini VII Mission Report," pp. 4-20, 7-9.

[93] Charles A. Berry, interview, Houston, 18 March 1968; Schweickart interview.

[94] "Gemini VII Mission Report," pp. 7-75 through -77, 8-21; Charles A. Berry et al., "Man's Response to Long-Duration Flight in the Gemini Spacecraft," in *Gemini Midprogram Conference*, pp. 253-61.

[95] "Project Development Plan for Rendezvous Development Utilizing the Mark II Two Man Spacecraft," MSC, 8 Dec. 1961.

[96] "Post Recovery Press Conference," 18 Dec. 1965, tape 41A, pp. 1-2.

[97] Letter, Mueller to Gilruth, 29 Nov. 1965; memo, Mathews to dist., "Preparation for Gemini Mid-Program Conference," GA-60237, 2 Dec. 1965; letters, Mathews to Col. Richard C. Dineen, GP-61889, and Burke, GP-61885, 13 Dec. 1965; letter, Edward Z. Gray to Maxime A. Faget, "Report on Gemini Experiments at the Gemini Mid-Program Conference," 21 Dec. 1965; memo, Gilruth to Slayton, "Far Eastern tour of Astronauts Borman and Schirra," 9 Feb. 1966.

## Chapter XIII

[1] Letter, LeRoy E. Day to Charles W. Mathews, 11 Jan. 1966, with enclosure, [George E. Mueller], "Statement Regarding the Remainder of the Gemini Program," 27 Dec. 1965; André J. Meyer, Jr., notes on GPO staff meeting, 25 Jan. 1966, p. 2.

[2] Meyer, notes on GPO staff meeting, 31 Aug. 1965, p. 4; TWXs, Mathews to SSD, Attn: Col. Alfred J. Gardner, GV-12216, 28 Sept., and GV-12227, 24 Oct. 1965.

[3] Horace E. Whitacre, "A Development History Summary of the Agena Target Vehicle Primary Propulsion System," n.d., p. 9; F[rederick] A. Boorady and D. A. Douglass, "Agena Gemini Rocket Engine Hypergolic Ignition—Hard Start Problem Solved during Project Surefire," n.d., p. 9.

[4] William C. Schneider, interview, Washington, 23 Jan. 1967; memo, Gardner to MSC Historical Office, "Comments on draft of Gemini narrative history," 14 Oct. 1969; Whitacre, interview, Houston, 3 June 1969.

[5] Gardner memo, 14 Oct. 1969; A. J. Steele, "Summary of Alternate Failure Hypotheses for GATV 5002," LMSC-A778486, 5 Jan. 1966; Harold W. Nolan, interview, Sunnyvale, Calif., 1 July 1966; Jerome B. Hammack, interview, Houston, 19 Aug. 1966; "GATV Progress Report, November 1965," LMSC-A605200-15, 20 Dec. 1965, pp. 2-1, -2; Project Gemini Quarterly Status Report No. 15, for period ending 30 November 1965, p. 21.

[6] Whitacre, "A Development History," pp. 1, 3; "Symposium on Hypergolic Rocket Ignition at Altitude," LMSC-A776842, 1 Dec. 1965, p. 3-6; Boorady and Douglass, "Agena Hypergolic Ignition," pp. 3, 4; Whitacre and Nolan interviews; Frederick A. Boorady and Jerome Salzman, "Modification of the Agena Rocket Engine for Gemini Target Vehicle Multiple Restart Capability," September 1965, pp. 2, 4, 15-17; Richard M. Spath, interview, Sunnyvale, Calif., 1 July 1966.

[7] Hammack interview; Quarterly Status Report No. 15, p. 21.

[8] NASA Management Instruction No. 4-1-7, "Mission Failure Investigation Policy and Procedures," 24 March 1964; memo, Robert C. Seamans, Jr., to Assoc. Adm., Manned Space Flight, "Gemini VI Mission Failure Investigation," 27 Oct. 1965; NASA News Release No. 65-342, "NASA Names Panel to Review Agena Failure," 27 Oct. 1965; letter, Mueller to Robert R. Gilruth, 29 Oct. 1965, with enclosure, Mueller, "Gemini Agena Target Vehicle (GATV) Review Board," 29 Oct. 1965; memo, Eberhard F. M. Rees to dist., "Gemini/Agena Target Vehicle Program investigation," 8 Nov. 1965.

[9] Nolan and Hammack interviews; "Symposium on Hypergolic Rocket Ignition," p. ii.

[10] "Symposium on Hypergolic Rocket Ignition," p. A-1.

[11] Ibid., pp. 3-1, -2, -6, -10; Boorady and Douglass, "Agena Hypergolic Ignition," p. 5; Whitacre, "A Development History," pp. 9, 10; Wulfgang C. Noeggerath, "Symposium on Hypergolic Rocket Engine Ignition at Altitude, Lockheed Missiles & Space Company, 12 to 13 November 1965," n.d., p. 1; memo, W. R. Abbott to Jack L. Shoenhair, "Comments on draft of chapter . . . : 'Trials of Agena and Gemini VIII,' " 8 Oct. 1969; letter, Charles E. Feiler to Noeggerath, 18 Nov. 1965; draft memo, [Whitacre] for record, "Scientific meeting at LMSC to discuss possible cause of 8247 hard start," n.d.; memo, [Whitacre] for record, "Specific recommendations for activities relating to reconfiguration of the 8247 engine and verification testing," 19 Nov. 1965.

[12] "GATV Progress Report, November 1965," pp. 2-3, -4; [Whitacre] memo, 19 Nov. 1965; [Whitacre] draft memo, n.d.; "Symposium on Hypergolic Rocket Ignition," pp. 5-4, -5; Gardner memo, 14 Oct. 1969; letter, Seymour C. Himmel to MSC, Attn: Mathews, "Investigation of GATV failure by LeRC Agena Project," 7 Dec. 1965, with enclosure, memo, Channing C. Conger and Robert E. Alexovich for record, "Analysis of GATV Flight Data," 30 Nov. 1965; letter, George J. Detko to MSC, Attn: Gilruth, "Investigation of the Gemini 6-Agena Target Vehicle (G6-ATV) failure by MSFC," 23 Dec. 1965, with enclosure, "G6-ATV Failure Investigation by MSFC," 23 Dec. 1965; letter, Robert H. Gray to Dir., MSC, "Analysis of GATV-1 Flight Data," 30 Dec. 1965, with enclosure, "GATV-1 Flight Analysis," n.d.; TWX, Mathews to SSD, Attn: Gardner, GV-12344, 27 Jan. 1966; letter, Himmel to MSC, Attn: Mathews, "GATV Failure Analysis," 14 Feb. 1966; letter, Detko to MSC, Attn: Gilruth, "Investigation of the Gemini 6-Agena Target Vehicle (G6-ATV) failure by MSFC," R-AS-VG-86-66, 9 March 1966; "A Chronology of the Arnold Engineering Development Center," AFSC Historical Publications Series 62-101 [probably 1968], p. 81.

[13] Day, interview, Washington, 25 Jan. 1967; memo, Day to Mueller, "Gemini Augmented Target Docking Adapter," 9 Dec. 1965; R[obert] N. Lindley, "Gemini Engineering Program, McDonnell Aircraft Corporation," presented at the Institute of Management Sciences, Dallas, Tex., 16-19 Feb. 1966, p. 18.

[14] Richard W. Keehn, interview, San Diego, Calif., 18 May 1967.

[15] Day interview; Day memo, 9 Dec. 1965; TWX, Day to MSC, Attn: Mathews, 10 Dec. 1965; memo, Mueller to Assoc. Adm., "Revised Gemini Project Approval Document, Change 4," 9 Dec. 1965, with enclosure, "Project Approval Document 91-1, Research and Development," n.d.; memo, Meyer to Chief, Gemini Spacecraft Procurement Sec., "Contract NAS 9-170, Gemini, Statement of Work for the Augmented Target Docking Adapter," GP-61893, 15 Dec. 1965, with enclosure; memo, Meyer to Chief, Gemini Spacecraft Procurement Br., "Statement of work for NASA Defense Purchase Request T-53291-G," GP-61894, 15 Dec. 1965, with enclosure; memo, Mathews to Asst. Dir., Flight Ops., "Augmented Target Docking Adapter for Gemini Rendezvous Missions," GV-66297, 17 Dec. 1965; Hammack interview.

[16] Hammack interview; memo, Mathews to Asst. Dir., Flight Crew Ops., "Flight Plan for the Gemini VIII-A Mission," GV-66321, 19 Jan. 1966.

[17] Whitacre, "A Development History," p.

12; [Whitacre], notes on "Action Item from MSF Program Review held November 23, 1965 on Agena High Altitude Testing of Bell Engine," n.d.; Gardner memo, 14 Oct. 1969; letter, Gen. Bernard A. Schriever to Maj. Gen. Ben I. Funk and Brig. Gen. Lee V. Gossick, 22 Nov. 1965.

[18] Memo, William A. Lee to Mgr., ASPO, "An AEDC facility conflict with Gemini," PA/M11-65-261, 16 Nov. 1965, with enclosure, "Derivation of Minimum Time Estimates," n.d.; Walter C. Williams and Bernhard A. Hohmann, interview, El Segundo, Calif., 15 May 1967; TWX, Schriever to NASA, for Mueller, SCG 42245, 15 Nov. 1965; Whitacre, "A Development History," p. 12; "GATV Progress Report, December 1965," LMSC-A605200-16, 20 Jan. 1965, pp. 1-1, 3-4; TWX, L. Eugene Root et al. to Mueller and Gilruth, "Gemini Agena Vehicle 5003 Milestone," LMSC/A781 408/64-60/537, 15 Jan. 1965.

[19] Gardner memo, 14 Oct 1969; TWX, Gardner to AFSC et al., "Gemini Agena Target Vehicle Flight Safety Review Board (GATV FSRB)," SSVT 36402, 22 Dec. 1965; "Gemini Agena Target Vehicle Program Status, 5 January 1966," LMSC-A777567, 5 Jan. 1966.

[20] Meyer, notes on GPO staff meeting, 11 Jan. 1966, p. 1; TWX, Mathews to SSD, Attn: Col. Richard C. Dineen and Gardner, GV-12347, 28 Jan. 1966; memo, Duncan R. Collins to Mgr., GPO, "Augmented target docking adapter," GS-64106, 4 Jan. 1966; TWX, Mathews to McDonnell, Attn: Walter F. Burke, "Contract NAS 9-170, Gemini, Monthly NASA/MAC Management Meeting," GP-7449, 11 Jan. 1966; TWX, Mathews to McDonnell, Attn: Burke, "Contract NAS 9-170, Gemini, Design and Acceptance Reviews of the Augmented Target Docking Adapter," GP-7450, 11 Jan. 1966; TWX, Mathews to McDonnell, Attn: Burke, "Contract NAS 9-170, Augmented Target Docking Adapter A-900-20 Meeting," GV-12319, 12 Jan. 1966; memo, Mathews to Chief, Gemini Spacecraft Procurement Sec., "Implementation of Augmented Target Docking Adapter," GP-61954, 24 Jan. 1966; TWX, Mathews to McDonnell, Attn: Burke, "Augmented Target Docking Adapter, Contract NAS 9-170," GV-12324, 19 Jan. 1966; letter, Mathews to Burke, "Contract NAS 9-170, Gemini, results of Augmented Target Docking Adapter Design Review," GP-61947, 20 Jan. 1966.

[21] Memo, Mathews to Asst. Dir., Flight Ops., Attn: Chief, Landing and Recovery Div., "Recovery of Gemini Rendezvous and Recovery Section," GV-66187, 15 Sept. 1965; "Gemini Program Mission Report, Gemini VI-A," MSC-G-R-66-2, January 1966, p. 6-12; memo,

James E. Bost to GPO files, "Contract NAS 9-170, results of negotiations of Contract Change Proposal 100, Augmented Target Docking Adapter . . . ," GP-61936, 13 Jan. 1966; TWX, Mathews to SSD, Attn: Col. B. J. McCarroll, GV-12312, 20 Jan. 1966; TWX, Mathews to McDonnell, Attn: Burke, "Contract NAS 9-170, Gemini, NASA/McDonnell Management Meeting on Mission Planning," GP-7459, 27 Jan. 1965; TWX, Mathews to SSD, Attn: McCarroll, GV-12346, 27 Jan. 1966; TWX, Mathews to McDonnell, Attn: Burke, "Contract NAS 9-170, Gemini, Augmented Target Docking Adapter Acceptance Review, Phase II," GP-7461, 28 Jan. 1966; letter, Mathews to Burke, "Contract NAS 9-170, Gemini, results of Augmented Target Docking Adapter Acceptance Review," GP-61988, 10 Feb. 1966; "Gemini Program Mission Report, Gemini IX-A," MSC-G-R-66-6, n.d., p. 12-8.

[22] Note, Schneider to Mueller, "Miscellaneous Items," 21 Jan. 1966; Whitacre, "A Development History," p. 14; Boorady and Douglass, "Agena Hypergolic Ignition," p. 13.

[23] Letter, Hohmann to SSD, Attn: Col. John B. Hudson, "Static Test Firings of GATV," 12 Jan. 1966, with enclosures, (1) letter, E. B. Doll to STG, Attn: Gilruth, "Requirements for Flight Readiness Firings on Mercury/Atlas Missiles," 23 March 1960, (2) TWX, Lt. Gen. Howell M. Estes to MSC, Attn: Gilruth, "Atlas Sustainer Engine Turbopump Problems," SSG-13-8-9, 13 Aug. 1962, and (3) "GLV Static Firing Charts;" letter, Lawrence A. Smith to SSD, Attn: Gardner, "Contract AF 04(695)-545, Gemini Program Vehicle 5003 Static Firing Recommendations," 11 Jan. 1966.

[24] Letter, Funk to Mueller, "Static-Fire Test," 14 Jan. 1966, with Hohmann and Smith letters; letter, G. Merritt Preston to NASA Hq., Attn: Mueller, "Agena Static Firing," 11 Jan. 1966; memo, Charles W. McGuire to Actg. Dep. Dir., Gemini, "Balance Sheet on Static Firing of GATV-5003," 13 Jan. 1966; memo, Day to Assoc. Adm., Manned Space Flight, "GATV 5003 Static Fire Evaluation," 14 Jan. 1966; memo, Clarence C. Gay, Jr., to dist., "Agena Design Certification," 3 Feb. 1966; letter, Wernher von Braun to Mueller, 17 Feb. 1966; letter, Kurt H. Debus to Mueller, 23 Feb. 1966; letter, Gilruth to NASA Hq., Attn: Mueller, "Agena Design Certification," GV-66369, 12 March 1966, with Smith letter.

[25] Whitacre, "A Development History," p. 14; Meyer, notes on NASA/MAC management meeting, 11 Feb. 1966, p. 2; Whitacre, telephone interview, 5 March 1971; Day interview; TWX, Mathews to McDonnell, Attn: Burke, "Contract NAS 9-170, Augmented Target Docking Adapter," GV-12359, 17 Feb. 1966.

26 Memo, Mathews to dist., "Gemini Design Certification Review, March 6 and 7, 1966," GP-62028, 1 March 1966; TWX, Gardner to NASA et al., "Gemini Agena Target Vehicle 5002 [sic] Flight Safety Review Board and Design Certification Review Board Meeting," SSVT 39094, 1 March 1966; Day interview; letter, Mueller to Gilruth, 2 March 1966.

27 Boorady and Douglass, "Agena Hypergolic Ignition," p. 13; Whitacre, "A Development History," p. 15; TWX, Mathews to SSD, Attn: Gardner, GP-7494, 10 March 1966; memo, Mathews to NASA Hq., Attn: Mueller, "Action items resulting from the Gemini VIII Design Certification Board," GA-60,161, 13 March 1966, with 13 enclosures; memo, Mathews to NASA Hq., "Gemini Agena Target Vehicle action items resulting from the Gemini VIII Design Certification Review," GP-62071, 15 March 1966, with enclosures, (1) memo, Gardner to Mueller and Funk, 15 March 1966, (2) memo, Douglass and Boorady, "Primary Propulsion Subsystem," 14 March 1966, and (3) Reservations, Gerald M. Forslund, 14 March 1966.

28 "Gemini IX-A Mission Report," p. 12-8; TWX, Mathews to McDonnell, Attn: Burke, "Contract NAS 9-170, Gemini Augmented Target Docking Adapter Acceptance Review," GP-7488, 15 March 1966.

29 Reginald M. Machell, interview, Houston, 18 April 1967; Harold I. Johnson, interview, Houston, 10 Feb. 1967; Johnson, David C. Schultz, and William C. Huber, "Maneuvering Equipment," in Reginald M. Machell, ed., Summary of Gemini Extravehicular Activity, NASA SP-149 (Langley, Va., 1967), pp. 6-28, -29; letter, Col. Daniel D. McKee to George M. Low, "NASA/DOD Mission Planning," 13 June 1965.

30 Astronautics and Aeronautics, 1965: Chronology on Science, Technology, and Policy, NASA SP-4006 (Washington, 1966), p. 444; "Mission Operations Report, Gemini Flight Number Eight," M-913-66-09, 3 March 1966, pp. 31-33; David R. Scott, interview, Houston, 5 April 1967.

31 "Abstract of Meeting on Extravehicular Life Support System and Extravehicular Support Package Testing, September 2, 1965," 14 Sept. 1965, with enclosure, "Extravehicular Life Support System and Extravehicular Support Package Spacecraft 8 Test Plan," n.d.; Frederick T. Burns et al., "Gemini Extravehicular Activities," in Machell, ed., Summary of Extravehicular Activity, p. 3-6; Larry E. Bell et al., "Life Support Systems for Extravehicular Activity," ibid., pp. 4-29, -58, -59; TWX, Mathews to McDonnell, Attn: Burke, "Contract NAS 9-170, Gemini, Fit Check of Extravehicu-

lar Support Pack in Spacecraft 8 Adapter," GS-10089, 30 Dec. 1965.

32 "Meeting on Extravehicular . . . Testing, September 2, 1965."

33 "Gemini 8 Astronaut Crew Press Conference," 26 Feb. 1966, pp. 2-6; Machell interview; Neil A. Armstrong, interview, Houston, 6 April 1967; Johnson, Schultz, and Huber, "Maneuvering Equipment," p. 6-7.

34 Harold I. Johnson et al., "Extravehicular Maneuvering about Space Vehicles," in Gemini Summary Conference, NASA SP-138 (Washington, 1967), pp. 92-94; "Rendezvous and Extravehicular Systems," Gemini Design Certification Report, February 1966, p. 2.1-2; MSC Quarterly Activity Report for Office of Assoc. Adm., Manned Space Flight, for period ending January 1966, p. 44; memo, Mathews to Asst. Dir., Flight Crew Ops., Attn: Chief, Flight Crew Support Div., "Qualification of the hand-held maneuvering unit," GT-65457, 5 Nov. 1965; Johnson interview.

35 Machell interview; Bell et al., "Life Support Systems for EVA," p. 4-47; Quarterly Activity Report for period ending January 1966, p. 44.

36 Machell interview; "Abstract of Meeting on Extravehicular Life Support Systems/Extravehicular Support Package, January 5 and 6, 1966," 7 Jan. 1966; "Abstract of Meeting on Extravehicular Life Support System and Extravehicular Support Package, January 28, 1966," n.d.; Bell et al., "Life Support Systems for EVA," pp. 4-41, -43, -44; "Gemini Program Mission Report, Gemini VIII," MSC-G-R-66-4, 29 April 1966, p. 12-2.

37 Memos, Mathews to dist., "Mission Planning," GV-66170, 2 Sept., and GV-66198, 25 Sept. 1965; memo, Mathews to dist., "Mission Planning for Agena," GV-66245, 21 Oct. 1965; memo, Whitacre to Mgr., GPO, "Astronaut sleeping during the Gemini-Agena docked-mode operation," GV-66359, 12 March 1966; TWX, Mathews to SSD, Attn: Gardner, "Crew Sleeping on Docked Agena," GV-12375, n.d. [probably 12 March 1966].

38 Mathews memos, GV-66170, GV-66198, GV-66245; memo, Mathews to dist., "Mission Planning for Gemini VI through XII," GV-66208, 1 Oct. 1965; memo, Mathews to Mgr., EXPO, "Mission duration," GP-62009, 21 Feb. 1966.

39 Memo, Mathews to dist., "Mission Planning Gemini VIII," GV-66252, 4 Nov. 1965.

40 Tommy W. Holloway, "Gemini VIII Flight Plan," Final, 24 Feb. 1966.

41 Memo, George C. Franklin to James M. Grimwood, "Power Tool Experiment (D-16), Gemini VIII," 11 July 1969; Franklin, telephone interview, 11 July 1969; "Abstract of

Meeting on Experiments for Gemini VIII, October 8, 1965," 15 Nov. 1965; "Abstract of Meeting on Gemini Experiments Status Review, November 9 and 10, 1965," 23 Nov. 1965; TWX, Mathews to NASA, Attn: Day, "Gemini Experiment D-3," EX4/T3-65, 17 Dec. 1965; memo, Mathews to Chief, Gemini Spacecraft Procurement Sec., "Engineering Review of McDonnell Aircraft Corporation Report A492-16A, Cost and Delivery Proposal for Experiment D-16 (Minimum Reaction Power Tool), dated April 9, 1965," GP-61342, 2 June 1965; "D-16 Experiment Briefing," 10 March 1966; TWX, Mathews to NASA, Attn: Day, "Gemini VIII Experiment S-9, Nuclear Emulsion," EX42/T14-66, 16 Feb. 1966; memo, Schneider to Willis B. Foster, "Use of Agena for Experiment Package," 21 June 1965; TWX, Mathews to SSD, Attn: Gardner, "Stability of Non-Powered Agena," GV-12186, 6 Sept. 1965; memo, Mathews to Mgr., EXPO, Attn: Norman G. Foster, "Incorporation of Experiment S-10 on Gemini Agena Target Vehicles," GV-66182, 17 Sept. 1965; Jocelyn R. Gill and Willis B. Foster, "Science Experiments Summary," in *Gemini Summary Conference*, pp. 301-302, 303; Norman G. Foster and Olav Smistad, "Gemini Experiments Program Summary," ibid., p. 225; NASA Release No. 66-52, "Project: Gemini 8," press kit, 8 March 1966, pp. 29-30; "Gemini 8 Experiments Briefing at Cape Kennedy," 14 March 1966.

[42] Note, Schneider to Mueller, 9 Feb. 1966, with enclosure, memo, Robert O. Aller to John A. Edwards, "RKV Support of AS 201 on Gemini VIII," 4 Feb. 1966; MSC News Release No. 66-13, 11 Feb. 1966; MSC News Release No. 66-16, 17 Feb. 1966; "Postlaunch Report for Mission AS-201 (Apollo Spacecraft 009)," MSC-A-R-66-4, 6 May 1966.

[43] MSC Announcement No. 64-120, "Designation of Flight Directors," 31 Aug. 1964; memo, Low to Christopher C. Kraft, Jr., "MSC Apollo Operations Plan," 1 Feb. 1966; memo, Kraft to dist., "MSC Apollo Operations Plan," 3 Feb. 1966, with enclosures, (1) Low letter, 1 Feb. 1966, and (2) "Manned Spacecraft Center Apollo Operations Plan," February 1966; William Hines, "Launch Now Slated Wednesday Morning," *The Evening Star*, Washington, 14 March 1966; letter, Low to James C. Elms, 25 March 1966; note, Schneider to Elms, "The Role of Mr. Kraft during Gemini VIII," 31 March 1966; William M. Bland, Jr., telephone interview, 1 Aug. 1969.

[44] TWX, Mathews to SSD, Attn: Dineen and Gardner, GP-7479, 2 March 1966; TWX, Mathews to McDonnell, Attn: Burke, "Contract NAS 9-170, Gemini, Results of Spacecraft 8 Flight Readiness Review," GP-7482, 7 March

1966; memo, Mathews to dist., "Spacecraft 8 Extravehicular Equipment Flight Readiness Review," GP-62075, 9 March 1966; letter, Mathews to Burke, "Contract NAS 9-170, Gemini, results of Spacecraft 8 Extravehicular Equipment Flight Readiness Review," GP-62069, 9 March 1966; memo, Mathews to dist., "Gemini Spacecraft 8 Design Certification Review," GP-62077, 9 March 1966; Mathews memo, GA-60,161; transcript of telephone conversation, Mueller and Seamans, 14 March 1966, 9:45 a.m.; Howard Simons, "Capsule Leaks Delay 3-Day Gemini Flight," *The Washington Post*, 15 March 1966; Albert Sehlstedt, Jr., "Gemini 8 Delayed at Least a Day," *The Sun*, Baltimore, 15 March 1966.

[45] See pp. 483-86 above; "Gemini VIII Mission Report," p. 1-1.

[46] Gemini 8 mission commentary transcript, 16 March 1966, tape 20, pp. 2-3, tape 21, p. 1; "Gemini VIII Mission Report," pp. 6-3, -4; "Atlas SLV-3, Space Launch Vehicle Flight Evaluation Report, SLV-3 5302," General Dynamics GDC/BKF66-012 and Supplemental Report No. 7 to "Gemini VIII Mission Report," 17 June 1966, p. 10; "Gemini Agena Target Vehicle 5003, Systems Test Evaluation (45-Day Report)," LMSC-A817204 and Supplemental Report No. 6 to "Gemini VIII Mission Report," 5 May 1966, p. 2-12.

[47] "Gemini VIII Technical Debriefing," 21 March 1966, pp. 1-2; Scott interview; Gemini 8 mission commentary, tape 23, p. 1.

[48] Gemini 8 mission commentary, tape 30, p. 1, tape 31, p. 1, tape 34, p. 2; "Gemini VIII Debriefing," pp. 5, 6; "Gemini VIII Mission Report," pp. 2-1, 6-4, -5; "Launch Vehicle No. 8 Flight Evaluation," Martin Co. Engineering Report No. 13227-8 and Supplemental Report No. 2 to "Gemini VIII Mission Report," April 1966, pp. vii, II-1, -2.

[49] "Gemini VIII Mission Report," p. 4-2; "Gemini VIII Debriefing," p. 13; Gemini 8 mission commentary, tape 37, p. 4; "Gemini VIII Voice Communications (Air-to-Ground, Ground-to-Air and On-Board Transcription)," McDonnell Control No. C-115471, n.d., pp. 8-10.

[50] "Gemini VIII Debriefing," pp. 18-20; "Gemini VIII Mission Report," pp. 4-2, 7-2; "Gemini VIII Voice," pp. 15-18.

[51] "Gemini VIII Debriefing," pp. 21, 22; "Gemini VIII Voice," p. 25.

[52] "Gemini VIII Debriefing," pp. 22-23; "Gemini VIII Mission Report," pp. 4-2, 7-2.

[53] "Gemini VIII Debriefing," pp. 23-25; "Gemini VIII Mission Report," p. 7-25.

[54] "Gemini VIII Debriefing," pp. 27-29; "Gemini VIII Mission Report," pp. 4-2, 7-3; Gemini 8 mission commentary, tape 44, pp. 4-

6; "Gemini VIII Voice," pp. 34-35.

55 "Gemini VIII Debriefing," pp. 29-30; "Gemini VIII Voice," pp. 39-40; Ben Vester et al., interview, Baltimore, 25 May 1966.

56 "Gemini VIII Debriefing," p. 30; "Gemini VIII Mission Report," pp. 4-2, -3, 7-3, -4; "Gemini VIII Voice," p. 40.

57 "Gemini VIII Debriefing," pp. 34, 35, 36, 37-38; "Gemini VIII Mission Report," pp. 4-3, 7-3, -4, -8, -19; "Gemini VIII Voice," pp. 43, 44, 45, 47.

58 "Gemini VIII Debriefing," pp. 36-40, 41, 42, 43-47; "Gemini VIII Mission Report," pp. 4-3, 7-4, -19; "Gemini VIII Voice," pp. 55-60.

59 "Gemini VIII Mission Report," pp. 4-3, 7-4, -19; "Gemini VIII Debriefing," pp. 47-49; "Gemini VIII Voice," pp. 60, 61, 64, 65, 67, 68, 70; "Air-Ground Playback Briefing," 17 March 1966, tape 6A, p. 1.

60 "Gemini VIII Voice," pp. 70, 71; Gemini 8 mission commentary, tape 58, p. 1.

61 "Gemini VIII Mission Report," p. 6-1; "Gemini VIII Voice," p. 74; Scott and Armstrong interviews; Gemini 8 mission commentary, tape 60, p. 2.

62 Scott and Armstrong interviews; "Gemini VIII Debriefing," pp. 54, 55-56.

63 Scott and Armstrong interviews; "Gemini VIII Debriefing," pp. 55-59; "Gemini VIII Mission Report," pp. 4-4, 7-6, -7, -20, -21.

64 "Gemini VIII Debriefing," pp. 59-60; Scott and Armstrong interviews.

65 "Air-Ground Playback Briefing," tape 6A, p. 1; Scott interview; James R. Fucci, telephone interview, 12 Aug. 1969; "Gemini VIII Voice," p. 75.

66 [Ivan D. Ertel], Gemini VIII: Rendezvous and Docking Mission, MSC Fact Sheet No. 291-E (Houston, April 1966); Charles A. Berry, interview, Houston, 18 March 1968; Armstrong and Scott interviews; "Gemini VIII Debriefing," pp. 60, 61; "Gemini VIII Voice," pp. 76, 80, 81.

67 John D. Hodge, interview, Houston, 12 March 1968; Mathews, interview, Houston, 2 Dec. 1966; Armstrong and Scott interviews; "Gemini VIII Voice," pp. 77, 78; "Air-Ground Playback Briefing," tape 6A, p. 1.

68 Hodge and Schneider interviews; Eugene F. Kranz, interview, Houston, 28 April 1967; Clifford E. Charlesworth, interview, Houston, 13 Dec. 1966; "Gemini VIII Voice," p. 81; see pp. 228-29, above.

69 Nolan interview; Preston, interview, Cape Kennedy, Fla., 24 May 1967.

70 Letter, Seamans to Eugene M. Emme, 3 Oct. 1969; NASA Release No. 65-388, "Seamans Takes Oath as Webb's Deputy at Space Agency," 21 Dec. 1965.

71 Raymond D. Hill, Jr., interview, Titusville, Fla., 24 May 1967; TWX, Mathews to McDonnell, Attn: Burke and Lindley, "Contract NAS 9-170, Gemini, Support of Gemini VIII Mission," GS-10100, 8 March 1966.

72 [Ertel], Gemini VIII; "Gemini 8 Post Recovery Press Briefing," 16 March 1966, tape 4A, pp. 4-5; Gemini 8 mission commentary, tape 63, p. 1; "Gemini 8 Pilot's [sic] Report," 26 March 1966, tape A, p. 1.

73 "Recovery Requirements, Gemini VIII," 31 Jan 1966; Kranz interview.

74 "Gemini VIII Voice," pp. 93, 98, 102, 103; Armstrong interview; Encyclopedia Americana, International ed., s.v. "Disasters."

75 "Gemini VIII Voice," pp. 109, 111; "Gemini VIII Mission Report," p. 4-11.

76 "Gemini VIII Debriefing," pp. 90-91; "Gemini VIII Voice," pp. 96, 109, 111, 112.

77 "Post Recovery Press Briefing," tape 4A, p. 5, tape 4B, p. 1; "Gemini VIII Mission Report," pp. 6-32, -33, 7-23, -36; "Gemini VIII Debriefing," pp. 94, 95, 96; "Gemini VIII Voice," p. 93; [Ertel], Gemini VIII; Lt. Jerry Poppink, "Pararescueman!" The Airman Magazine, August 1966, pp. 13, 14, 15.

78 "Gemini VIII Mission Report," pp. 6-33, -34, -36, 7-11, -36, -37, -38; Poppink, "Pararescueman!" p. 14; "Gemini VIII Debriefing," pp. 96-97; [Ertel], Gemini VIII; "Air-Ground Playback Briefing," tape 6A, p. 2; Don J. Green interview, Houston, 29 June 1967; Toni Zahn, telephone interview, 13 Aug. 1969.

79 "Gemini VIII Agena Target Vehicle Flight Plan," February 1966, p. 1; memo, Whitacre to Mgr., GPO, "Post-rendezvous Gemini Agena target vehicle maneuvers," GV-66350, 8 March 1966; [Ertel], Gemini VIII; "Agena Press Conference," 24 March 1966, pp. 9-10; "Gemini Agena Target Vehicle 5003," LMSC-A817204, p. vii.

80 Letter, Mathews to Gardner, GV-66301, 5 Jan. 1966, with enclosures, (1) memo, Kraft to GPO, Attn: Whitacre, "Request for information for the generation of Propellant Remaining Computer Program," 21 Dec. 1965, and (2) memo, Kraft to Mgr., GPO, "SPS thruster alignment and Agena vehicle errors resulting from c.g. offsets," 21 Dec. 1965; "Agena Press Conference," pp. 10, 11; "Gemini VIII Mission Report," p. 1-3.

81 "Agena Press Conference," pp. 13, 14.

82 "Gemini VIII Mission Report," p. 5-187; "Gemini Agena Target Vehicle 5003," LMSC-A817204, p. A-1; TWX, Mathews to SSD, Attn: Gardner, "Commands to Agena Vehicle 5003 during the Gemini VIII Mission," GV-12397, 4 April 1966; TWX, Network Ops. to Walter H. Wood, "GT-8 Agena," 23 March 1966.

83 "Gemini VIII Voice," p. 63; "Gemini

VIII Mission Report," p. 5-88; MSC Gemini News Center Release, "Short in Circuitry Blamed for Gemini 8 Mission Termination," 19 March 1966; Robert L. Sharp, interview, St. Louis, 14 April 1966; memo, Mathews to Chief, Flight Safety Office, "Control system modifications," GP-62154, 22 April 1966; "Gemini IX-A Mission Report," p. 3-8.

[84] Gilruth, interview, Houston, 21 March 1968; Meyer, interview, Houston, 9 Jan. 1967; Mathews interview.

## Chapter XIV

[1] Eugene A. Cernan, interview, Houston, 1 May 1967.

[2] Gemini News Center Release No. 10, "Gemini 7/6 Flight Controllers," 2 Dec. 1965; Cernan interview.

[3] Memo, Robert R. Gilruth to NASA Hq., Attn: NASA Safety Dir., "Aircraft Accident," 25 May 1966, with enclosures; TWX, Gilruth to NASA Hq., Attn: George E. Mueller, 1 March 1966; "Space: Rendezvous in St. Louis," Time, 11 March 1966, p. 27; memo, Thomas P. Stafford to MSC Historical Office, "Comment draft chapter of Gemini narrative history," 22 May 1970, with annotated pages attached; Edward F. Mitros, telephone interview, 16 March 1970.

[4] Gilruth memo, 25 May 1966; Theodore P. Wagner, "Jet Crash Kills 2 Gemini-9 Astronauts," The Washington Post, 1 March 1966; Stafford memo, 22 May 1970.

[5] John H. Bickers, interview, St. Louis, 13 April 1966; memo, Bickers to Michael Witunski, "Activities of John Bickers on Morning of 28 February 1966," No. 716, 1 March 1966; Gilruth memo, 25 May 1966; Wagner, "Jet Crash Kills 2"; "Plane Hits Building, Killing Gemini 9 Crew Bassett, See," The Houston Post, 1 March 1966.

[6] Bickers memo, 1 March 1966; Jack Amerine, "Shepard Leads NASA Probe of Jet Crash," Houston Chronicle, 1 March 1966; Jim Maloney, "Flyers Died on Way to Work," The Houston Post, 1 March 1966.

[7] UPI telephoto, "Flags at half-staff at McDonnell Aircraft corporation plant in St. Louis . . . ," Chicago Tribune, 2 March 1966; "Astronauts to Be Buried in Arlington," The Sun, Baltimore, 2 March 1966; "Memorials Today for 2 Astronauts," The Houston Post, 2 March 1966; memo, Julian Scheer to Lt. Gen. Frank A. Bogart et al., "Attendance at Funerals of Astronauts," 11 April 1966; letter, George M. Low to NASA Hq., Attn: Bogart, "Attendance at funerals of astronauts," 28 April 1966.

[8] Loyd S. Swenson, Jr., James M. Grimwood, and Charles C. Alexander, This New Ocean: A History of Project Mercury, NASA SP-4201 (Washington, 1966), p. 443; MSC News Release No. 66-27, 18 April 1966; MSC News Release 66-20, "Gemini and Apollo Crews Selected," 21 March 1966.

[9] G. Merritt Preston, interview, Cape Kennedy, Fla., 24 May 1967.

[10] Memo, Charles W. Mathews to Chief, Gemini Spacecraft Procurement Sec., "Spacecraft inspection requirement," GP-62115, 4 April 1966; memo, Mathews to Chief, Flight Safety Office, "Control system modifications," GP-62154, 22 April 1966; "Gemini Program Mission Report, Gemini IX-A," MSC-G-R-66-6, n.d., p. 3-8.

[11] Memo, Col. Daniel D. McKee to MSC, Attn: Paul E. Purser, "DOD/NASA Gemini Experiments," 26 Aug. 1963, with enclosures; Robert B. Voas, exec. sec., "Minutes of In-Flight Scientific Experiments Coordination Panel, September 23, 1963," n.d., p. 7; Stafford memo, 22 May 1970; Col. Wilbur A. Ballentine, interview, Houston, 16 Jan. 1967; "Rendezvous and Extravehicular Systems," Gemini Design Certification Report, February 1966.

[12] McKee, interview, Los Angeles, 19 May 1967; Ballentine interview; unsigned draft memo to Mgr., GPO, "Tether and Stabilization Requirements for Extravehicular Mission," n.d.

[13] McKee and Ballentine interviews; Warren J. North, interview, Houston, 10 Jan. 1967; "Technical Development Plan for DOD/NASA Gemini Experiments, 631A," SSD, 23 Sept. 1963, pp. 5-17, -18.

[14] [Reginald M. Machell], "EVA Possibilities for Gemini 8," annotated, "10, 11, 12," 9 July 1965; TWX, William C. Schneider to MSC, Attn: Mathews, MG-317, 9 Sept. 1965.

[15] Letter, Gilruth to Robert F. Freitag, 2 Feb. 1966, with enclosure, draft, [McKee], "NASA/DOD Position on Untethered Extravehicular Activity," n.d.

[16] Letter, Mueller to Gilruth, 17 March 1966, with enclosure, MSF Position Paper, "Tethered Extravehicular Activity in the Gemini Program," n.d.; letter, Mueller to Brig. Gen. Harry L. Evans, 17 March 1966, with enclosure as above.

[17] Letter, Evans to Mueller, "Tethered vs Untethered Extravehicular Activity," 25 March 1966; memo, LeRoy E. Day to Assoc. Adm., Manned Space Flight, "General Evans' letter on untethered EVA," 14 April 1966; letters, Mueller to Gilruth and Evans, 27 April 1966, with enclosures; memo, Mathews to Asst. Dir., Flight Crew Ops., "Gemini IX Extravehicular Flight Plan," GS-64127, 14 April 1966; memo,

Alfred P. Alibrando to Scheer, 23 March 1966.

[18] McKee and Ballentine interviews.

[19] Vearl N. Huff, interview, Washington, 24 Jan. 1967.

[20] Memo, Mathews to dist., "Mission Planning," GV-66198, 25 Sept. 1965; memo, Mathews to dist., "Mission Planning for Gemini IV through XII," GV-66208, 1 Oct. 1965; Huff interview; memo, Mathews to dist., "Mission Planning for Gemini IX, X, XI, XII," GV-66289, 2 Dec. 1965.

[21] "Abstract of Meeting on Trajectories and Orbits, January 20, 1966," 31 Jan. 1966; "Gemini Rendezvous Summary," MSC Internal Note No. 67-FM-128 (TRW Systems Group No. 05952-H281-R0-00), 1 Nov. 1967, pp. 2-2, B-2 through -5, C-2; W. Bernard Evans and Marvin R. Czarnik, "Summary of Rendezvous Operations," in Gemini Summary Conference, NASA SP-138 (Washington, 1967), pp. 10-11; P. W. Malik and G.A. Souris, Project Gemini: A Technical Summary, NASA CR-1106 (Langley, Va., 1968), p. 274; memo, Carl R. Huss to MSC Historical Office, Attn: Grimwood, "Comments on draft chapter of Gemini narrative history . . . ," 70-FM-H-29, 3 June 1970; memo, Robert E. Prahl to dist., "Gemini IX Insertion Velocity Adjust Routine (IVAR) Study," 66-FM32-56, 26 April 1966; memo, Ben F. McCreary to Chief, Mission Planning and Analysis Div., "Gemini IX booster recontact study for overspeed insertions requiring IVAR corrections," 66-FM34-25, 20 April 1966; memo, Mathews to dist., "Gemini IX M=3 differential altitude," GV-66434, 23 May 1966; memo, Mathews to Asst. Dirs., Flight Ops. and Flight Crew Ops., "Gemini IX Mission Activities Priorities," GV-66415, 3 May 1966; letter, Mathews to NASA Hq., Attn: Schneider, "Gemini IX Mission Activities Priorities," GV-66416, 3 May 1966.

[22] Note, Schneider to Mueller, 4 Feb. 1966, annotated, "OK if no impact on F.O. [flight operations]. G."

[23] A. L. Brady, "Configuration Control Board No. 4 Minutes, February 3, 1965"; memo, Jackson B. Craven to Chief, Apollo Flight Systems Br., "CSM Rendezvous Radar and LEM Weight Study," 9 Feb. 1965; memo, R. Wayne Young to Chief, Guidance and Navigation (G&N) Contract Engineering Br., "G&N Configuration Control Panel Meeting No. 5," 12 Feb. 1965; memo, Owen E. Maynard to Chief, Instrumentation and Electronic Systems Div. (IESD), "Requirement for VHF ranging capability between CSM and LEM," PS6/65M182, 15 Feb. 1965; memo, Maynard to Chief, IESD, "Maximum acceptable ambiguity for CSM-LEM VHF ranging system," PS6/65M201, 25 Feb. 1965; Cline W. Frasier,

"LEM Rendezvous Radar vs. Optical Tracker Study," 16 March 1965; André J. Meyer, Jr., notes on GPO staff meeting, 11 Jan. 1966, p. 1; memo, Robert C. Duncan to Chief, IESD, "Request for support: Evaluation board for LORS [lunar optical rendezvous system]—RR [rendezvous radar] 'Olympics,' " EG4-66-80, 25 Jan. 1966, with enclosure, memo, Wayne Young to Grumman, Attn: Robert S. Mullaney, "Contract NAS 9-1100, Rendezvous Radar Testing," EG4-3-66-70, 25 Jan. 1966; memo, Donald K. Slayton to Chief, Guidance and Control Div., "LORS—RR 'Olympics,' " 1 Feb. 1966; MSC News Release No. 66-38, 2 June 1966; Apollo Spacecraft Program Quarterly Status Report No. 16, for period ending 30 June 1966, p. 53; Quarterly Activity Report for Office of the Assoc. Adm., Manned Space Flight, for period ending 31 July 1966 p. 55.

[24] Stafford, interview, Houston, 3 April 1967; NASA News Release No. 66-97, "Project: Gemini 9," press kit, 4 May 1966, pp. 13-14; "Gemini IX-A Mission Report," p. 2-2; "Gemini Program/Mission Directive," NASA Program Gemini working paper No. 5039, 19 Nov. 1965, Appendix A, "Gemini Missions," p. A-9-2.

[25] "Gemini IX-A Mission Report," pp. 7-10, -11.

[26] "Gemini Program/Mission Directive," Appendix A, Sec. A-9; Ted A. Guillory, Charles L. Stough, and Lt. Charles F. Davis, Jr., "Gemini IX Flight Plan," Final, 18 April 1966; Kenneth A. Young, telephone interview, 19 March 1970; memo, Mathews to dist., "Mission Planning for Agena," GV-66245, 21 Oct. 1965; Meyer notes, 11 Jan. 1966; "Trajectories and Orbits Meeting, January 20, 1965"; memo, Mathews to Asst. Dirs., Flight Ops. and Flight Crew Ops., "Gemini IX Mission Plan and Flight Plan," GV-66380, 24 March 1966.

[27] Evans and Czarnik, "Summary of Rendezvous Operations," pp. 12-14; Malik and Souris, Gemini Technical Summary, pp. 288-90; "Gemini IX-A Mission Report," p. 4-17; TWX, Mathews to NASA Hq., Attn: Day, GV-12393, 24 March 1966; TWX, Mathews to McDonnell, Attn: Walter F. Burke, "Contract NAS 9-170, Spacecraft Consumables Loading for Gemini IX," GV-12410, 22 April 1966; TWX, Mathews to McDonnell, Attn: Burke, "Contract NAS 9-170, Spacecraft Consumables Loading for Gemini X," GV-12453, 21 June 1966.

[28] Memo, Mathews to Asst. Dirs., Flight Ops. and Flight Crew Ops., "Gemini IX Mission activities priorities," GV-66390, 11 April 1965; Mathews memo, GV-66415, 3 May 1966.

[29] McKee memo, 26 Aug. 1963; "Techni-

cal Development Plan for DOD/NASA Gemini Experiments," p. 4-25; "DOD/NASA Gemini Experiments Study," Interim Report No. SSD-TDR-63-406 (McDonnell Report No. A358), 24 Jan. 1964, pp. 1.6.1 through 1.6.77; Frederick T. Burns et al., "Gemini Extravehicular Activities," in Reginald M. Machell, ed., *Summary of Gemini Extravehicular Activity*, NASA SP-149 (Langley, Va., 1967), p. 3-8; North interview; Harold I. Johnson, interview, Houston, 10 Feb. 1967.

[30] Ronald C. Croston and James B. Griffin, "Manned Flight Simulation of the Air Force Modular Maneuvering Unit," printed in *AIAA Fourth Manned Space Flight Meeting*, St. Louis, 11-13 Oct. 1965 (New York, 1965), pp. 118-26; Edwin E. Aldrin, Jr., interview, Houston, 4 April 1967; Griffin, telephone interview, 10 Oct. 1969; Herbert E. Smith, telephone interview, 10 Oct. 1969; Harold I. Johnson, David C. Schultz, and William C. Huber, "Maneuvering Equipment," in Machell, ed., *Summary of Gemini Extravehicular Activity*, p. 6-36.

[31] Memo, Mathews to dist., "Gemini extravehicular equipment integration," GS-64108, 17 Jan. 1966; Larry E. Bell et al., "Life Support Systems for Extravehicular Activity," in Machell, ed., *Summary of Gemini Extravehicular Activity*, p. 4-8; Elton M. Tucker, telephone interview, 24 March 1970; James W. McBarron II, telephone interview, 25 July 1973.

[32] David C. Schultz and Hilary A. Ray, Jr., "Body Positioning and Restraints," in Machell, ed., *Summary of Gemini Extravehicular Activity*, pp. 5-1, -8; Johnson, Schultz, and Huber, "Maneuvering Equipment," p. 6-33; "Additional EVA Provisions," McDonnell Engineering Change Proposal No. 597, 14 Oct. 1965; Cernan interview.

[33] Memo, Christopher C. Kraft, Jr., to dist., "Assignment of Flight Directors," 23 March 1966; "Houston Gemini News Center Accreditation List," *Gemini IV*, 11 June 1965; "Gemini News Center Accreditation List," *Gemini IX-A*, 1 June 1966; Anne Thompson, "On-the-Ground Teams: Blacks, Greens, Whites at Controls for Gemini-9," *Houston Chronicle*, 13 May 1966.

[34] "Atlas SLV-3, Space Launch Vehicle Flight Evaluation Report, SLV-5303," General Dynamics GDC/BKF-66-029 and Supplemental Report No. 6 to "Gemini IX-A Mission Report," 27 June 1966, p. 1-1; Flora Lewis, "Gemini 9 Postponed 3 Weeks," *The Washington Post*, 18 May 1966; Project Gemini Quarterly Status Report No. 17, for period ending 31 May 1966, p. 16.

[35] Richard W. Keehn, interview, San Diego, 18 May 1967; Harold W. Nolan and Rich-

ard M. Spath, interview, Sunnyvale, Calif., 1 July 1966; Griffin, interview, Canoga Park, Calif., 16 May 1967; "Gemini 9 Mission, Public Information Operations," NASA/MSC working paper, 5 April 1966, p. 1; Evert Clark, "Failure of Agena Bars Gemini Trip," *The New York Times*, 18 May 1966; Meyer, notes on GPO staff meeting, 18 May 1966; "Gemini Agena Target Vehicle 5004 Systems Test Evaluation (45-Day Report)," LMSC-A819881 (Supplemental Report No. 7 to "Gemini IX-A Mission Report") 30 June 1966, p. x; letter, V. F. Peterson to AFSSD, Attn: Michael Aftanas, Jr., "LMSC Product Assurance and Reliability Program Plan for Project Sure Fire, Contract AF 04(695)-545, Gemini, S.A. 25," 29 April 1966, with enclosure; TWX, Mathews to SSD, Attn: Col. Alfred J. Gardner, GV-12395, 1 April 1966; "Project Sure Fire, GATV Engine Modification and Test Program: Final Report," SSD-545-66-10 (LMSC-A818110), 1 July 1966, 2 vols.; "GATV Design Certification Report for Gemini VIII Mission," LMSC-A794903, 26 Feb. 1966; letters, Mathews to Brig. Gen. Lee V. Gossick and W. H. Gisel, GV-66384 and GV-66383, 19 April 1966; *The Lockheed Star*, Vol. XIII, No. 7 (1 April 1966); "Gemini Agena Target Vehicle 5004, NASA Mission Gemini IX Flight Safety Review at ETR," Aerospace TOR-669(6183)-12, 16 May 1966; Swenson, Grimwood, and Alexander, *This New Ocean*, pp. 299-300, 335-37, 383, for example.

[36] Meyer, notes on GPO staff meeting, 25 Jan. 1966, p. 3; TWXs, Mathews to SSD, Attn: Col. Billy J. McCarroll, GP-7506, 31 March, and GP-7544, 6 May 1966; TWX, Mgr., Manned Space Flight Support Ops. [Maj. Gen. Leighton I. Davis], Patrick AFB, Fla., to AIG 7106 [Address Indicator Grouping, which indicates distribution No. 7016, in this case about 50 addressees], DDMA 63161, 17 May 1966; Kathryn A. Lansdowne, telephone interview, 7 Aug. 1973; Meyer notes, 18 May 1966, p. 1; letter, Schneider to Preston, 11 April 1966.

[37] TWX, Mathews to SSD, Attn: Col. John B. Hudson, GV-12434, 18 May 1966; TWX, Alibrando to MSC for Paul P. Haney and to KSC for John W. King and Gordon L. Harris, M-N-382, 19 May 1966.

[38] Meyer notes, 18 May 1966, p. 1; Meyer, notes on NASA/MAC management meeting, 19 May 1966, p. 1.

[39] Meyer, notes on GPO staff meeting, 26 April 1966, p. 2; memo, Mathews to dist., "Gemini IX Agena parking orbit," GV-66431, 13 May 1966; memo, Mathews to Dep. Dir., "Notes regarding reentry from orbit for proposed Gemini IX-A backup mission," GV-66447, 26 May 1966; TWX, Mathews to SSD,

Attn: Col. Robert R. Hull, and McDonnell, Attn: Burke, "Gemini IX-A and Gemini IX-B (Backup Mission) Targeting Conditions," GV-12447, 27 May 1966.

40 Memo, Robert C. Seamans, Jr., for record, "Selection of Gemini IX Mission Objectives," 31 May 1966.

41 Keehn interview.

42 Ibid.; "SLV-3, Flight Evaluation Report," pp. 2-1, - 3; Dugald O. Black, telephone interview, 30 March 1970.

43 Nolan and Keehn interviews; Jerome B. Hammack, interview, Houston, 19 Aug. 1966; "Gemini Agena Target Vehicle 5004," pp. vii, viii, 3-12, 4-1.

44 "Gemini IX-A Mission Report," pp. 5-149, -150, -151; Gemini 9A News Center Releases Nos. 17, 18, and 19, 1 June 1966; "Gemini 9A Postponement Press Briefing," 1 June 1966; "Gemini 9A Pre-Mission Update Conference," 2 June 1966; Malik and Souris, Gemini Technical Summary, p. 343; [Ivan D. Ertel], Gemini IX-A: Rendezvous Mission, MSC Fact Sheet No. 291-F (Houston, August 1966); Lt. Col. Alexander C. Kuras and Col. John G. Albert, "Gemini-Titan Technical Summary," 24 Jan. 1967, pp. 147-48; Gemini-Titan II Air Force Launch Vehicle Press Handbook (Martin-Baltimore, Manned Space Flight, 2d ed. rev., 1966), p. D-16; TWX, Edmond C. Buckley to Goddard, Attn: John F. Clark, to MSC, Attn: Gilruth, and Patrick AFB, Attn: Maj. Gen. Vincent G. Huston, "Guidance System Update Failure on GT-9 Mission," TD-7122, 3 June 1966; quote confirmed by telephone with Stafford, 26 July 1973.

45 [Ertel], Gemini IX-A; "Gemini IX-A Voice Communications (Air-to-Ground, Ground-to-Air, and On-Board Transcription," McDonnell Control No. C-115803, n.d., pp. 8-10; "Gemini IX A Technical Debriefing," 11 June 1966, pp. 9-11; "Gemini IX-A Mission Report," pp. 4-14, -15; TWX, Mathews to SSD, Attn: Hull, and McDonnell, Attn: Burke, "Gemini IX ATDA Ephemeris Data and GLV Insertion Parameters," GV-12437, 23 May 1966.

46 "Gemini IX-A Voice," pp. 13, 15, 16; "Gemini IX-A Mission Report," p. 4-15; "Gemini IX A Debriefing," p. 15.

47 "Gemini IX A Debriefing," pp. 16, 17; "Gemini IX-A Mission Report," pp. 4-15, -16; "Gemini IX-A Voice," pp. 20, 21, 25, 26, 32.

48 "Gemini IX-A Voice," pp. 28, 29, 30, 32, 33; George Towner, interview, Baltimore, 25 May 1966; "Gemini IX-A Mission Report," p. 7-2.

49 "Gemini IX-A Mission Report," pp. 7-2, -3; "Gemini IX-A Voice," pp. 37-51; "Gemini IX A Debriefing," pp. 30, 31, 34-37.

50 "Gemini IX-A Mission Report," p. 4-16; "Gemini IX-A Voice," pp. 55, 56, 57.

51 "Gemini IX-A Voice," pp. 58-64; Stafford and Cernan interviews.

52 Schneider, interview, Washington, 23 Jan. 1967; Cernan and Stafford interviews; "Gemini IX-A Voice," pp. 65, 66, 112.

53 "Gemini IX-A Mission Report," pp. 5-161 through -164; TWX, Mathews to SSD, Attn: Hudson, GV-12356, 10 Feb. 1966; TWX, Mathews to McDonnell, Attn: Burke, "Contract NAS 9-170, Augmented Target Docking Adapter," GV-12359, 17 Feb. 1966; memo, Purser to Gilruth and Low, 7 June 1966; TWX, Mathews to McDonnell, Attn: Burke, "Contract NAS 9-170, Gemini, Definition of Augmented Target Docking Adapter/Shroud Problems Relative to Gemini IX-A Mission," GP-7579, 9 June 1966; letter, Burke to MSC, Attn: Mathews, "Contract NAS 9-170, Project Gemini, Definition of ATDA Shroud Separation Problem during GT-IX-A," 306-09-244, 20 June 1966; letter, Edward C. Welsh to Mueller, 21 June 1966; letter, Mueller to Welsh, 7 July 1966; memo, Mathews to Chief, Gemini Spacecraft Procurement Sec., "Investigative testing of Augmented Target Docking Adapter shroud release mechanism," GP-62268, 13 July 1966; Scott H. Simpkinson, interview, Houston, 18 Jan. 1967; H. H. Luetjen, interview, Cape Kennedy, Fla., 25 May 1967; memo, Simpkinson to Grimwood, "Orbital Operations Perfected," 14 May 1970, with annotated pages of comment draft attached; Simpkinson, telephone interview, 29 March 1971.

54 Memo, Bobby K. Culpepper to dist., "Trajectory information for the Gemini IX/ATDA mission," 66-FM6-43, 24 May 1966; memo, Mathews to NASA Hq., Attn: Schneider, "Gemini IX-A Mission Activities Priorities," GV-66437, 26 May 1966; "Gemini IX-A Mission Report," pp. 4-16, -17, 6-12, 7-3, -4, -25, -26; "Gemini IX A Debriefing," pp. 43-56; "Gemini IX-A Voice," pp. 67, 71, 72, 73, 76, 84.

55 "Gemini IX-A Mission Report," p. 4-16; Paul C. Kramer, Edwin E. Aldrin, and William E. Hayes, "Onboard Operations for Rendezvous," in Gemini Summary Conference, pp. 37-38.

56 "Gemini IX-A Voice," pp. 85, 86, 89, 92, 94-96; "Gemini IX A Debriefing," pp. 57, 61, 62, 63, 64.

57 [Ertel], Gemini IX-A.

58 "Gemini IX-A Mission Report," pp. 4-18, -19, -32, -33, 6-13, -15, 7-4, -5; "Gemini IX-A Voice," pp. 108, 110, 114, 119, 120, 124, 127-28.

59 "Gemini IX-A Mission Report," pp. 7-5,

-6, -27, -28; "Gemini IX-A Voice," pp. 120, 121, 122, 125, 127; "Gemini IX A Debriefing," p. 80; Stafford interview.

[60] Gemini 9-A mission commentary transcript, 4 June 1966, tape 88, pp. 2-3, tape 91, p. 1; TWX, Kenneth S. Kleinknecht to NASA Hq., Attn: James E. Webb, and MSC, Attn: Gilruth, "Daily Report Number 1—Gemini IX-A Mission," GT-11212, 4 June 1966, pp. 5, 12; "Gemini IX-A Voice," pp. 128, 129, 130, 131, 136.

[61] Mathews memo, GS-64127, 14 April 1966; [Ertel], Gemini IX-A; "Gemini IX A Debriefing," pp. 105-108, 114, 115, 117; "Gemini IX-A Voice," pp. 181, 182, 187, 189-94; Burns et al., "Gemini Extravehicular Activities," p. 3-8.

[62] Burns et al., "Gemini Extravehicular Activities," p. 3-8; Schultz and Ray, "Body Positioning and Restraints," pp. 5-1, -6; Cernan interview; "Gemini IX-A Debriefing," pp. 112, 126, 127-28, 129, 130, 132, 133, 137; "Gemini IX-A Voice," pp. 196, 201, 203, 206, 207, 209, 212, 213, 215; "Gemini IX-A Mission Report," pp. 7-8, -9, -10.

[63] "Gemini IX-A Voice," pp. 217-20, 222-27, 230-44; Burns et al., "Gemini Extravehicular Activities," pp. 3-8, -9; Bell et al., "Life Support Systems," pp. 4-52, -53; Schultz and Ray, "Body Positioning and Restraints," pp. 5-1, -2; Johnson, Schultz, and Huber, "Maneuvering Equipment," pp. 6-39, -40; "Gemini IX A Debriefing," pp. 145-96, 199; "Gemini IX-A Mission Report," pp. 7-8, -9, -10; Cernan and Stafford interviews; "Abstract[s] of Meeting[s] on Extravehicular Activity on Spacecraft 10, 11, and 12, June 21, 1966," 27 June 1966, and "June 30, 1966," 6 July 1966; "Gemini 9 Pilot's [sic] Report," 17 June 1966, tape F, p. 7.

[64] "Abstract of Meeting on Experiments for Gemini VIII and IX, January 7, 1966," 11 Jan. 1966; "Abstract of Meeting on Experiments for Gemini IX and X, March 3 and 4, 1966," 5 April 1966; "Abstract of Meeting on Gemini IX Experiments, Functional Verification Review Board, April 20, 1966," 27 April 1966; Mathews memo, GV-66437, 26 May 1966; "Gemini 9 Experiments Briefing," 16 May 1966; Gemini 9A News Center Release No. 2, "Gemini 9A Experiments," 26 May 1966; "Gemini IX-A Mission Report," pp. 7-6, -7, -8, 8-1 through -85; "Gemini IX-A Voice," pp. 156-71.

[65] "Gemini IX-A Mission Report," pp. 7-6, 8-59, -60, -61; "Gemini IX A Debriefing," pp. 290-91.

[66] "Gemini IX-A Mission Report," pp. 7-8, 8-49 through -52; "Gemini IX A Debriefing," pp. 90, 91; Malik and Souris, Gemini Technical Summary, pp. 303-304; Jocelyn R. Gill and

Willis B. Foster, "Science Experiments Summary," in Gemini Summary Conference, p. 316.

[67] "Gemini IX-A Mission Report," pp. 7-6, -7, 8-79, -80.

[68] "Gemini IX-A Mission Report," pp. 7-6, -7, 8-63, -64, -73, -74; "Gemini IX A Debriefing," pp. 93-98, 100-103; Gill and Foster, "Science Experiments Summary," p. 299; Malik and Souris, Gemini Technical Summary, pp. 313-14, 318-19.

[69] Joe D. St. Clair, Edward A. Armstrong, and John E. Williams, "Gemini Program Flight Summary Report: Gemini Missions I through X," MSC-G-R-66-5 (Revision A), September 1966, p. 44; Warren R. Young, ed., To the Moon, Sect. II, The story in pictures and text (New York, 1969), p. 109; Don J. Green, interview, Houston, 29 June 1967.

[70] Bell, interview, Houston, 10 Sept. 1968.

[71] Memo, Seamans for Mueller, "Gemini IX Review," 9 June 1966; Meyer, notes on GPO staff meeting, 14 June 1966, p. 1.

[72] Memo, Mueller to Dep. Adm., "Review of Gemini Missions X, XI, and XII," 23 June 1966, annotated: "Action approved. Robert C. Seamans, Jr. June 24, 1966"; letter, Mueller to Gilruth, 23 June 1966, with enclosure, "Gemini Mission Review Board," 22 June 1966; letter, Seamans to Harold Brown, 21 June 1966; letter, Brown to Seamans, 29 June 1966; letter, Gilruth to NASA Hq., Attn: Mueller, "Gemini Mission Review Board," GP-62275, 30 June 1966.

[73] James C. Elms, telephone interview, 29 Oct. 1969; [Elms], "Committee Action on Gemini X Mission," 5 July 1966.

[74] Elms interview; Elms, "Interim Report—Gemini Mission Review Board," 15 July 1966, p. 3.

[75] "Gemini Program/Mission Directive," Appendix A, Sec. 10; Meyer notes, 18 May 1966, p. 1.

[76] Jack Amerine, "Young, Collins Named Crew of Gemini-10," Houston Chronicle, 25 Jan. 1966; Jim Maloney, "Young and Collins Gemini 10 Crew," The Houston Post, 25 Jan. 1966; TWX, Mathews to McDonnell, Attn: Burke, "Contract NAS 9-170, Gemini, Astronaut Fit Check," GP-7513, 5 April 1966; "Gemini and Apollo Crews Selected;" John W. Young, interview, Houston, 8 May 1967.

[77] John Young interview; Meyer, notes on GPO staff meeting, 1 Feb. 1966, p. 2; "Gemini Program/Mission Directive," Change No. 2, 15 Feb. 1966, p. D-10; memo, Mathews to Asst. Dir., Flight Crew Ops., Attn: Chief, Flight Crew Support Div., "Fifty-foot umbilical for Spacecraft 10," GS-64116-A, 25 Feb. 1966; memo, James V. Correale to GPO, "Gemini extra-vehicular life support system," 17 Sept.

1963; "General Requirements for an Engineering Study and Preliminary Design of a One-Man Propulsion Device for the Gemini Program," Exhibit A, "Statement of Work," 19 Sept. 1963; memo, Richard S. Johnston to Mgr., GPO, "Gemini Extravehicular Life Support System Development," 25 March 1964; TWX, Mathews to McDonnell, Attn: Burke, "Contract NAS 9-170, Gemini Configuration Control Board Meeting Number 94, 3-12-66," GV-12376, 14 March 1966; TWX, Mathews to McDonnell, Attn: Burke, "Contract NAS 9-170, Gemini—Action Items of NASA/McDonnell Management Meeting, 3-23-66," GP-7510, 1 April 1966; Burns et al., "Gemini Extravehicular Activities," pp. 3-12, -13; Bell et al., "Life Support Systems for Extravehicular Activity," pp. 4-88, -89; Johnson, Schultz, and Huber, "Maneuvering Equipment," p. 6-4.

[78] Mathews memo, GV-66208, 1 Oct. 1965; "Abstract of Meeting on Trajectories and Orbits, February 16, 1966," 3 March 1966; TWX, Mathews to McDonnell, Attn: Burke, "Contract NAS 9-170, Gemini, Meeting on Gemini X Onboard Rendezvous Procedures," GS-10104, 22 March 1966; TWX, Mathews to McDonnell, Attn: Burke, "Contract NAS 9-170, Gemini, Onboard Procedures for Gemini X Primary Rendezvous," GS-10107, 7 April 1966; memo, Mathews to Asst. Dir., Flight Ops., "Gemini X Mission Plan," GV-66389, 7 April 1966; "Abstract of Meeting on Trajectories and Orbits, April 28, 1966," 31 May 1966; TWX, Mathews to McDonnell, Attn: Burke, "Contract NAS 9-170, Gemini Configuration Control Board Meeting Number 102, May 9, 1966," 10 May 1966; Meyer notes, 14 June 1966.

[79] "Trajectories and Orbits Meeting, February 16, 1966"; "Trajectories and Orbits Meeting, April 28, 1966"; Howard W. Tindall, Jr., telephone interview, 5 Nov. 1969; TWX, Networks Ops. to W. H. Wood, "GT-8 Agena," 25 March 1966; Mathews memo, GV-66447, 26 May 1966; memo, Mathews to Asst. Dirs., Flight Ops. and Flight Crew Ops., "Gemini X Mission Changes and Priorities," GV-66454, 10 June 1966; Elvin B. Pippert, Jr., J. V. Rivers, and Tommy W. Holloway, "Gemini X Flight Plan," Final, 22 June 1966; TWX, Mathews to SSD, Attn: Gardner and Hull, "Revisions to the Gemini Program Mission Directive for Gemini X," GV-12456, 27 June 1966; memo, Mathews to NASA Hq., Attn: Schneider, "Gemini X Mission Changes and Priorities," GV-66458, 27 June 1966; "Abstract of Meeting on Trajectories and Orbits, June 29 and July 8, 1966," 14 July 1966; Larry D. Davis, telephone interview, 12 Nov. 1969.

[80] Elms, "Interim Report," pp. 5-6; "Ab-

stract of Meeting on Gemini IX Agena Real Time Mission Evaluation Support, April 27, 1966," 6 May 1966; letter, Mathews to NASA Hq., Attn: Schneider, "Contingency Missions for Gemini X, XI, and XII," GV-66461, 23 June 1966; TWX, Mathews to SSD, Attn: Hull, and McDonnell, Attn: Burke, "Pertinent Gemini X Information," GV-12455, 24 June 1966; letter, Schneider to Kraft, 7 July 1966; TWX, Mathews to SSD, Attn: Hull, and McDonnell, Attn: Burke, "Gemini X Alternate Mission," GV-12461, 8 July 1966; Gemini 10 News Center Release No. 10, "Alternate Gemini 10 Plans," 16 July 1966.

[81] NASA News Release No. 66-155, "Gemini 10 Launch Set for July 18," 17 June 1966; TWX, Mathews to SSD, Attn: Hull and Lt. Col. Fountain M. Hutchison, "Flight Safety Review for Gemini Launch Vehicle 10," GP-7613, 12 July 1966; TWX, Mathews to SSD, Attn: Gardner, "Gemini X Atlas-Agena Target Vehicle System Flight Safety Review," GP-7614, 12 July 1966; Gemini 10 News Center Release No. 12, "Status Report," 17 July 1966; [Ivan D. Ertel], *Gemini X: Multiple Rendezvous, EVA Mission*, MSC Fact Sheet No. 291-G (Houston, September 1966); Gemini 10 mission commentary transcript, 18-21 July 1966, tape 3, p. 1, tape 4, p. 1, tape 5, p. 1; "Gemini Program Mission Report, Gemini X," MSC-G-R-66-7, August 1966, pp. 1-1, 4-1; TWX, Kleinknecht to NASA Hq., Attn: Webb, and MSC, Attn: Gilruth, "Launch Summary Report, Gemini X Mission," GT-11215, 19 July 1966; "Gemini X Technical Debriefing," 26 July 1966, pp. 1-12; Michael Collins, interview, Houston, 17 March 1967; Frank Thistle, *Rocketdyne: The First 25 Years* . . . . (Van Nuys, Calif., 1970).

[82] "Gemini X Debriefing," pp. 12-19; "Gemini X Mission Report," pp. 5-14, -15, 7-2, -23, -24; "Gemini X Voice Communications (Air-to-Ground, Air-to-Air and On-Board Transcription)," McDonnell Control No. C-115883, n.d., pp. 10-20, passim; William H. Allen, ed., *Dictionary of Technical Terms for Aerospace Use*, 1st ed., NASA SP-7 (Washington, 1965), p. 9.

[83] "Gemini X Voice," pp. 57, 61-67, 69, 71, 72, 73, 76; "Gemini X Debriefing," pp. 28, 29, 32-41; John Young and Collins interviews; memo, Mathews to Asst. Dir., Flight Ops., "Ground support required for the Gemini X onboard M=4 rendezvous," GS-64121, 19 April 1966; Elms, "Second Interim Report, Gemini Mission Review Board, August 18, 1966," n.d., p. 1; memo, Mathews to Chief, Gemini Spacecraft Procurement Sec., "Contract NAS 9-6408, additional postflight analysis on Gemini X mission," GP-62332, 3 Aug.

1966; Meyer, notes on GPO staff meeting, 18 Aug. 1966, p. 2.

[84] "Gemini X Voice," p. 78; "Gemini X Mission Report," p. 1-2.

[85] "Gemini X Mission Report," pp. 1-2, 4-8, -25, 5-151; "Gemini X Voice," pp. 85, 89, 90, 91, 92; "Gemini X Debriefing," pp. 47, 50; Gemini 10 News Center Release No. 3, "Gemini 10 Flight Controllers," 13 July 1966.

[86] James M. Grimwood and Barton C. Hacker, *Project Gemini Technology and Operations: A Chronology,* NASA SP-4002 (Washington, 1969), pp. 266-67; "Gemini X Voice," pp. 93-99; "Gemini X Debriefing," p. 51; "Gemini X Mission Report," pp. 4-8, -21, -26.

[87] "Gemini X Technical Debriefing," pp. 52-53; "Gemini X Voice," pp. 99-107; "Gemini X Mission Report," pp. 4-8, -25.

[88] "Gemini X Voice," pp. 112, 113; "Gemini X Debriefing," pp. 59, 60; "Gemini X Mission Report," pp. 4-8, -21, -26.

[89] "Interim Report, Manned Space Flight Experiments, Gemini X Mission, July 18-21, 1966," MSC TA-R-67-1, March 1967, pp. 1, 5; William Dan Womack, "Experiment M405 (MSC-3), Triaxis Magnetometer," ibid., pp. 9-19; James Marbach, "Experiment M408 (MSC-6), Beta Spectrometer," ibid., pp. 21-36; Reed S. Lindsey, "Experiment M409 (MSC-7), Bremsstrahlung Spectrometer," ibid., pp. 37-48; Frank B. Newman, "Lesson Plan and Handout, Special Technical Data, Gemini X Experiments," SCD, No. M35, 17 June 1966, pp. 37-39, 53-60; Malik and Souris, *Gemini Technical Summary,* pp. 310, 311-12; "Abstract of Meeting on Experiments for Gemini X and XI, May 11 and 12, 1966," 1 June 1966; "Abstract of Meeting on Gemini X Experiments Functional Verification Review Board, June 20, 1966," 24 June 1966; memo, Mathews to Asst. Dir., Flight Crew Ops., "Gemini VI experiments and associated equipment," EX1365-0223, 17 Nov. 1965; memo, Mathews to Mgr., EXPO, "M-5 Experiment on Gemini VIII," GP-61958, 24 Jan. 1966; letter, Day to Piland, 23 June 1966; TWX, Gilruth to NASA Hq., Attn: Mueller, EX42/T65-66, 12 July 1966; TWX, Charles A. Berry to NASA Hq., Attn: Actg. Dir., Space Medicine, 29 June 1966; Gemini 10 News Center Release No. 5, "Changes in the experiment schedule," 14 July 1966; letter, Berry to NASA Hq., Attn: Brig. Gen. Jack Bollerud, "Medical reasons for deletion of Experiment M-5 from the remaining Gemini and first manned Apollo flights," 25 July 1966; TWX, Kleinknecht to NASA Hq., Attn: Webb, and MSC, Attn: Gilruth, "Daily Report Number 1—Gemini X Mission," GT-11216, 19 July 1966, p. 13; "Gemini X Mission Report," pp. 8-5, -7.

[90] "Gemini X Debriefing," pp. 60-68; "Gemini X Voice," pp. 118-25, 131, 132-49; "Gemini X Mission Report," pp. 7-5, -6, -31, -32, -47, -48, 8-62, -63, -64, -65; Karl G. Henize and Lloyd R. Wakerling [sic], "Experiment S013 (S-13), Ultraviolet Astronomical Camera," in "Gemini X Experiments Interim Report," pp. 97-104; John R. Brinkmann and Robert L. Jones, "Experiment M410 (MSC-8), Color Patch Photography," ibid., pp. 49-53; TWX, Mathews to McDonnell, Attn: Burke, "Contract NAS 9-170, Gemini, Extravehicular Activity Meeting," GP-7586, 16 June 1966; memo, Mathews to Dirs., Flight Crew Ops. and Flight Ops., "Gemini X extravehicular flight plan," GS-64160, 15 July 1966.

[91] "Gemini X Voice," pp. 151-59, 165, 167, 168, 171, 174, 175, 182; "Gemini X Debriefing," pp. 68, 69, 71, 72; Gemini 10 mission commentary, tape 109, pp. 2-3; TWX, Kleinknecht to NASA Hq., Attn: Webb, and MSC, Attn: Gilruth, "Daily Report Number 2—Gemini X Mission," GT-11217, 20 July 1966, pp. 7, 10, 12; Mathews memo, GS-64160, 15 July 1966.

[92] "Gemini X Voice," pp. 183-90, 192, 194, 196-97; "Gemini X Mission Report," pp. 4-8, -21, -26, 6-17, -18, 7-27; "Gemini X Debriefing," pp. 75, 76; Kleinknecht TWX, GT-11217, 20 July 1966, pp. 13-14; "Abstract of Meeting on Gemini Experiment S-26, Ion Wake Measurement, Gemini X and XI, September 13, 1965," 23 Sept. 1965; TWX, Electro-Optical Systems, Inc., to MSC, Attn: James W. Campbell, No. 15, 1 Aug. 1966; David B. Medved, "Experiment S026 (S-26), Ion-Wake Measurement," in "Gemini X Experiments Interim Report," pp. 105-14; Paul D. Lowman, Jr., and Herbert A. Tiedemann, "Experiment S005 (S-5), Synoptic Terrain Photography," ibid., pp. 63-72; Kenneth M. Nagler, "Experiment S006 (S-6), Synoptic Weather Photography," ibid., pp. 73-79.

[93] "Gemini X Mission Report," pp. 1-3, 7-27; "Gemini X Voice," pp. 202, 208-11, 218, 219; "Gemini X Debriefing," pp. 79-81; memo, Mathews to dist., "Mission Planning," GV-66170, 2 Sept. 1965; TWX, Mathews to SSD, Attn: Gardner, "Stability of Non-Powered Agena," GV-12186, 6 Sept. 1965; "Abstract of Meeting on Atlas/Agena Coordination, April 21, 1966," 10 May 1966; Gemini 10 News Center Release No. 13, "NORAD role in Gemini 10," 17 July 1966.

[94] "Gemini X Mission Report," p. 4-9; "Gemini X Voice," pp. 219-34, 236-38; Gemini 10 mission commentary, tape 165, p. 1, tape 167, p. 1, tape 169, p. 3.

[95] "Gemini X Mission Report," pp. 7-32, -33, 8-59, -60; "Gemini X Debriefing," pp. 95-

107; Kleinknecht TWX, GT-11216, 19 July 1966, p. 13; "Gemini X Voice," pp. 170, 238-50; Mathews TWX, GV-12376, 14 March 1966; memo, Mathews to Chief, Engineering Div., Attn: Head, Test Systems Sec., "Government-furnished aeronautical equipment 50-foot umbilical support to McDonnell Aircraft Corporation for Gemini X," GP-62088, 22 March 1966; "Abstract of Meeting on Experiment S-10 (Agena Micrometeorite Collector), August 9, 1965," 18 Aug. 1965; TWX, Gilruth to NASA Hq., Attn: Mueller, "T-017 Meteoroid Impact," 8 July 1966; Curtis L. Hemenway, "Experiment S010 (S-10), Agena Micrometeorite Collection," in "Gemini X Experiments Interim Report," pp. 81-95; Burns et al., "Gemini Extravehicular Activities," pp. 3-12, -13; Collins, John Young, and Johnson interviews.

[96] "Gemini X Voice," pp. 251-54, 255-57; "Gemini X Debriefing," pp. 108-13, 115-16; Gemini 10 mission commentary, tape 112, p. 1, tape 113, p. 1, tape 174, p. 1; TWX, Mathews to McDonnell, Attn: Burke, "Contract NAS 9-170, Gemini Hatch Closing Loads," GS-10108, 7 April 1966; TWX, Mathews to McDonnell, Attn: Burke, "Contract NAS 9-170, Gemini, Zero G Flight Article," GS-10119, 1 June 1966; Grimwood and Hacker, *Gemini Chronology*, p. 270.

[97] "Gemini X Debriefing," pp. 116-17, 120, 121, 122; "Gemini X Mission Report," pp. 4-2, -13, -33; "Gemini X Voice," pp. 273, 275, 280-83; TWX, Kleinknecht to NASA Hq., Attn: Webb, and MSC, Attn: Gilruth, "Gemini X Mission Summary Report," GT-11218, 21 July 1966, pp. 6, 9, 10.

[98] Kleinknecht TWX, GT-11218, 21 July 1966, pp. 1-3, 11-12; "Gemini X Mission Report," pp. 1-3, 5-28, -29, 6-14, -15, 7-37, -38, -39; Grimwood and Hacker, *Gemini Chronology*, p. 265; [Ertel], *Gemini X*; "Gemini X Debriefing," pp. 128-42; Gemini 10 mission commentary, tape 236, pp. 1-2.

[99] Memo, Mathews to Dir., Flight Ops., "Post-reentry Agena testing," GV-66479, 13 July 1966; "Gemini X Mission Report," pp. 5-140, -142, -143; TWX, Kleinknecht to NASA Hq., Attn: Webb, and MSC, Attn: Gilruth, "Gemini X Interim Report," GT-11219, 30 July 1966, pp. 19, 27, 33, 39.

[100] Letter, Mueller to Gilruth, 1 Aug. 1966, with enclosure, "Gemini Objectives and Accomplishments"; Meyer, notes on GPO staff meetings, 8 and 14 June 1966; memo, Mathews to GPO personnel, "GPO Phasedown," GA-60428, 15 Aug. 1966; memo, Mathews to Albert B. Triche, "Phasedown planning," GA-60422, 12 Aug. 1966; memo, Mathews to Harle L. Vogel, "Phasedown planning," GA-

60423, 12 Aug. 1966; Augustine A. Verrengia, telephone interview, 10 Aug. 1973.

## Chapter XV

[1] Gemini Program Office Reading Files, 1966, passim. (esp. letter, Charles W. Mathews to Walter F. Burke, "Contract NAS 9-4412, Procedures for acceptance reviews of heat. shield qualification spacecraft," GP-62118, 4 April 1966; letter, Mathews to Col. Russel M. Herrington, Jr., "Heat shield qualification spacecraft acceptance procedure," GP-62120, 6 April 1966; letter, Mathews to Burke, "Contract NAS 9-4412, Heat Shield Qualification Spacecraft Acceptance Review, Phase II, Board," GP-62260, 5 July 1966; memo, Scott H. Simpkinson to Dep. Mgr., GPO, "Saturn SIV-B Airlock Experiment proposal," GT-65558, 26 Jan. 1966; memo, Simpkinson to Dep. Mgr., GPO, "Airlock contract requirement negotiations," GT-65778, 7 Sept. 1966); letter, Gen. Samuel C. Phillips to Marshall, Attn: Saturn V Program Mgr., "Application of Gemini Experience to Apollo," 7 April 1966; TWX, Phillips to Marshall et al., "Meeting on Application of Gemini Launch Vehicle Experience to Apollo," MAT-2-460, 22 July 1966; letter, Phillips to MSC, Attn: Mgr., GPO, "Gemini/Apollo Launch Vehicle Meeting," 15 Sept. 1966; John B. Lee, recorder, "Minutes of Senior Staff Meeting, February 4, 1966," p. 4; André J. Meyer, Jr., notes on GPO staff meetings, 1 Feb., p. 1, 18 May 1966, p. 2; Simpkinson, interview, Houston, 18 Jan. 1967; Col. John G. Albert, interview, Patrick AFB, Fla., 26 May 1967.

[2] LeRoy E. Day, interview, Washington, 25 Jan. 1967; Lee, "Minutes of Senior Staff Meeting, April 22, 1966," p. 3; Meyer, notes on GPO staff meeting, 26 April 1966, p. 1; TWX, Mathews to McDonnell, Attn: Burke, "Contract NAS 9-170, Gemini, Delivery Schedules for Spacecraft 11 and 12," GP-7530, 25 April 1966.

[3] Memo, Mathews to dist., "Establishment of a Committee for Apollo/Gemini Mission Planning Coordination," GV-02466, 18 Jan. 1965; memo, Mathews to dist., "Mission Planning Gemini XI and XII (Onboard Direct Rendezvous)," GV-66300, 27 Dec. 1965, with enclosure, "Ground Rules for McDonnell Aircraft Company [sic] Onboard Direct Rendezvous Study," Ref. ECP 659, n.d.; memo, Mathews to dist., "Mission Planning Gemini VIII through XII (Tethered Vehicle Studies)," GV-66296, 16 Dec. 1965; Meyer, notes on GPO staff meeting, 25 Jan. 1966, p. 1; "Abstract of Meetings on Trajectories and Orbits, June 29 and July 8, 1966," 14 July 1966; Lee, "Minutes of Senior

Staff Meeting, April 15, 1966," p. 3; memo, Mathews to dist., "Gemini XII," GV-66284, 2 Dec. 1965; Meyer, notes on GPO staff meeting, 11 Jan. 1966, p. 1; memo, John A. Edwards to Assoc. Adm., Manned Space Flight, "OAO-A1/ Gemini XII," 2 Sept. 1966; letter, Maj. Gen. Ben I. Funk to Robert R. Gilruth, 12 July 1966.

[4] MSC News Release No. 66-20, "Gemini and Apollo Crews Selected," 21 March 1966; MSC News Release No. 66-46, 17 June 1966.

[5] Charles Conrad, Jr., interview, Houston. 31 March 1967; "Preliminary Project Development Plan for an Advanced Manned Space Program, Utilizing the Mark II Two Man Spacecraft," STG, 14 Aug. 1961, App. A; "Gemini Large Earth Orbit," McDonnell Control No. C-100858, Report B743, 19 June 1965; Meyer, interview, Houston, 9 Jan. 1967; Meyer, notes on GPO staff meetings, 29 June, p. 1, and 27 July 1965, p. 1; letter, Rep. Olin E. Teague (D.-Tex.) to James E. Webb, 18 Aug. 1965; letter, Webb to Teague, 10 Sept. 1965; letter, George E. Mueller to Abe Silverstein, 1 Sept. 1965; memo, Edgar L. Harkleroad to Chief, Gemini Mission Planning, "Gemini-Pegasus Rendezvous Summary to date," 25 May 1965; Meyer notes, 25 Jan. 1966, p. 1.

[6] Conrad interview; "Summary Minutes, Seventh Inflight Experimenters Meeting, 67-1," held 25-26 Aug. 1966, p. 5; "Trajectories and Orbits Meetings, June 29 and July 8, 1966"; "Gemini Program/Mission Directive," NASA Program Gemini working paper No. 5039, 19 Nov. 1965, Appendix A, "Gemini Missions," Change 4, 12 July 1966, pp. A-11-1 through -5; letter, Mueller to Gilruth, 1 Aug. 1966, with enclosure; memo, John M. Eggleston to dist., "Information on Radiation Hazard to Gemini and Apollo from the Solar Flare Particle Event of September 2, 1966," 9 Sept. 1966.

[7] Memo, Christopher C. Kraft, Jr., to dist., "Second meeting of Mission Planning Coordination Group," 22 Oct. 1963; Raymond L. Zavasky, recorder, "Minutes of Senior Staff Meeting, June 12, 1964," p. 3; TWX, Mathews to SSD, Attn: Col. Richard C. Dineen, "Direct Ascent Rendezvous Guidance for Gemini," GP-51690, 12 Feb. 1965; memo, Mathews to dist., "Mission Planning for Gemini IX, X, XI, XII," GV-66289, 2 Dec. 1965; Mathews memo, GV-66300, 27 Dec. 1965; TWX, Mathews to McDonnell, Attn: Burke, "Contract NAS 9-170, Gemini, Development of Gemini Computer Math Flows," GS-10090, 5 Jan. 1966; TWX, Mathews to McDonnell, Attn: Burke, "Contract NAS 9-170, Gemini, Initial Conditioning for First Apogee Rendezvous Analyses," GS-10095,

2 Feb. 1966; TWX, Mathews to SSD, Attn: Dineen, "Mission Planning Information," GV-12386, 16 March 1966; "Abstract of Meeting on Trajectories and Orbits, April 19, 1966," 6 May 1966; Wyendell B. Evans, telephone interview, 20 Aug. 1973.

[8] Mathews memos, GV-66289, 2 Dec. 1965, and GV-66300, 27 Dec. 1965; Meyer notes, 25 Jan. 1966, p. 1; "Gemini Program/ Mission Directive," p. A-11-1, Change 1, 1 Jan. 1966, and Change 2, 15 Feb. 1966.

[9] Eugene M. Emme, A History of Space Flight (New York, 1965), p. 86; Willy Ley, "Station in Space," in a symposium entitled "Man Will Conquer Space Soon," Collier's, 22 March 1952, p. 30; Wernher von Braun, "Crossing the Last Frontier," ibid., pp. 29, 72; memo, Mathews to Asst. Dir., E and D, Attn: William E. Stoney, "Tethered vehicle studies," GV-66209, 1 Oct. 1965.

[10] "Tethered Vehicle Study," [GPO], 26 Oct. 1965; memo, Robert C. Duncan to Mgr., GPO, through Asst. Dir., E and D, "Tethered vehicle studies," EG-65-921, 26 Oct. 1965; memo, Eldon W. Hall to Actg. Dep. Dir., Gemini Program, "Tether Studies," 4 Nov. 1965; David D. Lang, telephone interview, 22 June 1970; memo, Lang to Grimwood, "Comment Draft Chapter . . . of Gemini Narrative History," 22 June 1970.

[11] Phillip McLaughlin, "Spin Up Studies for the Gemini-Agena System in a Tethered Configuration," McDonnell Gemini Design Note No. 356, 23 March 1966.

[12] "Gemini Program/Mission Directive," p. A-11-2, Changes 2 and 4.

[13] McLaughlin, "Spin Up Studies," p. 4; "Gemini Program Mission Report, Gemini XI," MSC-G-R-66-8, October 1966, pp. 1-3, 5-86; NASA Release No. 66-226, "Project: Gemini 11," press kit, 24 Aug. 1966, p. 11.

[14] James C. Elms, telephone interview, 29 Oct. 1969.

[15] Elms, "Second Interim Report—Gemini Mission Review Board, August 18, 1966;" Elms interview.

[16] Elms, "Second Interim Report"; TWX, Mathews to McDonnell, Attn: Burke, "Contract NAS 9-170, Gemini Configuration Control Board Meeting Number 96, March 28, 1966," GV-12389, 31 March 1966; TWX, Mathews to McDonnell, Attn: Burke, "Contract NAS 9-170, Gemini/Agena Tether Break Link," GV-12482, 28 July 1966.

[17] "Abstract of Meeting on Extravehicular Activity on Spacecraft 10, 11, and 12, June 30, 1966," 6 July 1966; letter, Mathews to Langley Research Center, Attn: Dir., "Simulation support for Gemini extravehicular activities," GV-66466, 30 June 1966, with enclosure, "State-

ment of Work"; letter, G. Samuel Mattingly to MSC Historical Office, 5 Oct. 1970.

[18] Memo, Donald K. Slayton to Mgr., GPO, "Gemini Extravehicular Operation," CF40-4M-51, 23 March 1964, with enclosure, Appendices, esp. Appendix B, "Astronaut Training Program," p. 6; "Abstract of Meeting on Extravehicular Activity on Spacecraft 11 and 12, August 2, 1966," 4 Aug. 1966; Gilruth, interview, Houston, 21 March 1968; Reginald M. Machell, interview, Houston, 18 April 1967; Richard F. Gordon, Jr., interview, Houston, 20 March 1967.

[19] "Extravehicular Activity Meeting, June 30, 1966"; memo, Mathews to Mgrs., EXPO and Spacecraft, "Action items resulting from the Design Certification Review of July 11, 1966," GP-62286, 18 July 1966; TWX, Mathews to McDonnell, Attn: Burke, "Contract NAS 9-170, Gemini, Delivery of 30-Foot Umbilicals," GP-7620, 28 July 1966; "Meeting on Extravehicular Activity, August 2, 1966;" "Abstract of Meeting on Extravehicular Activity on Spacecraft 11 and 12, August 16, 1966," 22 Aug. 1966; TWX, Mathews to McDonnell, Attn: Burke, "Contract NAS 9-170, Gemini Configuration Control Board Meeting Number 117, August 15, 1966," GV-12497, 17 Aug. 1966; TWX, Mathews to McDonnell, Attn: Burke, "Contract NAS 9-170, Gemini, Thermal Performance Testing on 30-Foot Umbilicals," GS-10126, 19 Aug. 1966; letter, Mathews to NASA Hq., Attn: Mueller, "Open item, Design Certification Review Board, Gemini XI; extravehicular feet restraints," GA-60465, 27 Aug. 1966; memo, Clarence C. Gay, Jr., to dist., "July 11, 1966 Design Certification Review Action Item," 7 Sept. 1966; Frederick T. Burns et al., "Gemini Extravehicular Activities," in Reginald M. Machell, ed., Summary of Gemini Extravehicular Activity, NASA SP-149 (Langley, Va., 1967), p. 3-19; Larry E. Bell et al., "Life Support Systems for Extravehicular Activity," ibid., pp. 4-88, -89; David C. Schultz and Hilary A. Ray, Jr., "Body Positioning and Restraints," ibid., pp. 5-2, -3.

[20] Memo, Mathews to NASA Hq., Attn: Schneider, "Gemini XI Mission Priorities," GV-66488, 2 Aug. 1966; memo, Mueller to Adm., "Gemini Mission XI," 6 Sept. 1966, with enclosure, "Mission Operations Report, Gemini XI Mission," M-913-66-13, pp. 2-11; Gemini 11 press kit, p. 16; "Seventh Inflight Experimenters Meeting," p. 7; "Interim Report, Manned Space Flight Experiments, Gemini XI Mission, September 12-15, 1966," MSC-TA-R-67-2, May 1967; Elms, "Second Interim Report."

[21] TWX, Mathews to SSD, Attn: Col. John B. Hudson, GP-7536, 29 April 1966; letter, Mathews to Albert, GP-62168, 3 May 1966;

TWX, Mathews to SSD, Attn: Capt. Davis B. Conkling, "Major Hardware Shipment for Gemini XI Mission," GA-6022, 20 June 1966; NASA News Release No. 66-215, "Gemini Eleven to Fly Sep. 9 on 3-Day Plan," 11 Aug. 1966.

[22] [Ivan D. Ertel], Gemini XI Mission: High Altitude, Tethered Flight, MSC Fact Sheet No. 291-H (Houston, October 1966); "Gemini XI Mission Report," pp. 5-106, -107; Alexander C. Kuras and John G. Albert, "Gemini-Titan Technical Summary," 24 Jan. 1967, p. 149; Gemini 11 News Center Release No. 9, "Status Report," 9 Sept. 1966; Gemini 11 mission commentary transcript, 9 Sept. 1966, tape 1, p. 1, 10 Sept. 1966, tape 3, p. 1, tape 5, p. 1, tape 7, p. 1, tape 17, p. 1, tape 18, p. 1, tape 28, p. 1; "Gemini 11 Scrub Press Conference," 10 Sept. 1966; Gatha F. Cottee, telephone interview, 23 Aug. 1973; "Gemini XI Mission Report," p. 6-2; Gemini 11 News Center Release No. 13, "Status Report," 11 Sept. 1966.

[23] "Gemini XI Technical Debriefing," 19 Sept. 1966, pp. 1-2; Gemini 11 mission commentary, 12 Sept. 1966, tape 3, p. 1, tape 23, p. 1; [Ertel], Gemini XI; Guenter F. Wendt, interview, Titusville, Fla., 23 May 1967; G. Merritt Preston, interview, Cape Kennedy, 24 May 1967.

[24] "Gemini XI Mission Report," p. 4-1; [Ertel], Gemini XI; Gemini 11 mission commentary, tape 37, p. 1; TWX, Mathews to SSD, Attn: Col. Robert R. Hull, and McDonnell, Attn: Burke, "Gemini X and X-A planning information," GV-12473, 12 July 1966; TWX, Mathews to SSD, Attn: Hull, and McDonnell, Attn: Burke, "Gemini XI Launch Windows," GV-12487, 3 Aug. 1966; TWX, Mathews to SSD, Attn: Hull, and McDonnell, Attn: Burke, "Gemini XII Launch Windows," GV-12517, 6 Oct. 1966.

[25] "Gemini XI Debriefing," pp. 3, 6, 8, 9; [Ertel], Gemini XI; "Gemini XI Voice Communications (Air-to-Ground, Ground-to-Air and On-Board Transcription)," McDonnell Control No. C-11598, n.d., pp. 2, 4, 5, 6.

[26] "Gemini XI Mission Report," pp. 4-13, 5-8, -9; Paul C. Kramer, Edwin E. Aldrin, and William E. Hayes, "Onboard Operations for Rendezvous," in Gemini Summary Conference, NASA SP-138 (Washington, 1967), pp. 27-28.

[27] "Gemini XI Mission Report," pp. 4-13, -24, 5-12, 7-18; "Gemini XI Debriefing," p. 12; "Gemini XI Voice," p. 13.

[28] "Gemini XI Mission Report," pp. 4-13, -14, -24, 5-12, -13, 7-18, -19, -20; "Gemini XI Debriefing," pp. 14, 20, 21; "Gemini XI Voice," pp. 14, 17, 18-23.

[29] "Gemini XI Voice," p. 24; Elms inter-

view; TWX, Mathews to McDonnell, Attn: Burke, "Contract NAS 9-170, Spacecraft Consumable Loadings for Gemini XI," GV-12494, 12 Aug. 1966.

30 "Gemini XI Debriefing," pp. 207-209; "Gemini XI Voice," pp. 29, 36, 37; "Gemini XI Mission Report," pp. 4-1, 7-20.

31 "Gemini XI Debriefing," pp. 24-25, 26, 27-28, 184, 185; "Gemini XI Voice," pp. 29, 31-34, 40, 42, 45, 46; "Abstract of Meeting on Gemini Experiment S-26, Ion Wake Measurement, Gemini X and XI, September 13, 1965," 23 Sept. 1965; David B. Medved and Ballard E. Troy, Jr., "Experiment S026, Ion-Wake Measurement," in "Gemini XI Experiments Report," pp. 119-20; F. W. O'Dell et al., "Experiment S009, Nuclear Emulsion," ibid., p. 85.

32 "Gemini XI Voice," pp. 49, 50, 51, 54, 59, 60, 69-70; Meyer, notes on GPO staff meeting, 8 June 1966, p. 1; Gemini 11 mission commentary, 13 Sept. 1966, tape 87, p. 1.

33 "Gemini XI Voice," pp. 73, 75, 76, 77, 78, 80, 81, 83; "Gemini XI Debriefing," pp. 32-35; "Gemini XI Mission Report," pp. 1-2, 4-1, 5-83, -84, 7-1, -22; Burns et al., "Gemini Extravehicular Activities," p. 3-20; Bell et al., "Life Support Systems," p. 4-11; Bell, interview, Houston, 10 Sept. 1968; Gordon interview.

34 "Gemini XI Voice," pp. 83-89, 91, 92-94, 97; "Gemini XI Debriefing," pp. 35-41; "Gemini XI Mission Report," pp. 1-2, 4-1, 5-84, -86, -88, -90, 7-22, -23, -24, -25; TWX, Mathews to McDonnell, Attn: Burke, "Contract NAS 9-170, Gemini Configuration Control Board Meeting Number 119, August 31, 1966," GV-12502, 1 Sept. 1966; Burns et al., "Gemini Extravehicular Activities," p. 3-19; Bell et al., "Life Support Systems," pp. 4-54, -55; David C. Schultz et al., "Extravehicular Training and Simulation," in Machell, ed., Summary of Gemini Extravehicular Activity, p. 7-22; G. Fred Kelly and D. Owen Coons, "Medical Aspects of Extravehicular Activity," ibid., p. 9-2; Reginald M. Machell, Larry E. Bell, and David C. Schultz, "Results and Conclusions," ibid., p. 10-3; TWX, Mathews to NASA Hq., Attn: Schneider and John A. Nicholas, GP-7197, 8 June 1965.

35 "Gemini XI Voice," pp. 95, 99-104, 107, 108, 112, 116; Gemini 11 News Center Release No. 2, "Gemini 11 Flight Controllers and Recovery Personnel," 6 Sept. 1966; M. J. Kooman, R. T. Seal, Jr., and John Lintott, "Experiment S011, Airglow Horizon Photography," in "Gemini XI Experiments Report," p. 96.

36 "Gemini XI Voice," pp. 119-26, 128-31, 140, 141; "Gemini XI Debriefing," pp. 44-46, 47, 48-49; "Gemini XI Mission Report," pp. 1-2, 4-2, -30, -32, 5-14, -28, -121, -123, -126,

-131, 6-10, -11, -17, -18, 7-3, -4, -20, -21, -46, -47, 8-49; Paul D. Lowman, "Experiment S005, Synoptic Terrain Photography," in "Gemini XI Experiments Report," p. 68; Kenneth M. Nagler and Stanley D. Soules, "Experiment S006, Synoptic Weather Photography," ibid., p. 76; M. A. Bender et al., "Experiment S004, Radiation and Zero-G on Human Blood and Neurospora," ibid., pp. 45-50; "Gemini II Pilot's [sic] Report," 26 Sept. 1966, tape 10D, p. 1; Earth Photographs from Gemini VI through XII, NASA SP-171 (Washington, 1968); U.S. Congress, House, Committee on Science and Astronautics, Astronautical and Aeronautical Events of 1962: Report, 88th Cong., 1st sess., 12 June 1963; Loyd S. Swenson, Jr., James M. Grimwood, and Charles C. Alexander, This New Ocean: A History of Project Mercury, NASA SP-4201 (Washington, 1966), p. 52.

37 "Gemini XI Voice," pp. 146-49, 151-56, 158, 159-78; "Gemini XI Debriefing," pp. 49-53, 179-82; "Gemini 11 Pilot's Report," tape 10C, p. 6; "Gemini XI Mission Report," pp. 1-3, 4-2, 6-11, -12, 7-25, -26, -44, 8-50, -51; Gemini 11 mission commentary, 14 Sept. 1966, tape 183, pp. 2-4; Burns et al., "Gemini Extravehicular Activities," p. 3-19; Schultz and Ray, "Body Positioning and Restraints," pp. 5-12, -27; Lowman, "Synoptic Terrain Photography," pp. 67-68; Karl G. Henize and Lloyd R. Wackerling, "Experiment S013, Ultraviolet Astronomical Camera," in "Gemini XI Experiments Report," pp. 105, 106.

38 TWX, Mathews to McDonnell, Attn: Burke, "Contract NAS 9-170, Gemini XI Flight Plan Procedures for Tether Evaluation," GV-12496, 16 Aug. 1966; memo, Mathews to Dir., Medical Research and Ops., Attn: Chief, Medical Ops. Office, "Gemini XI Tethered Maneuver Acceleration," GS-64164, 9 Sept. 1966; David D. Lang and Roger K. Nolting, "Operations with Tethered Space Vehicles," in Gemini Summary Conference, pp. 60-63; Gemini 11 press kit, pp. 13-14; John H. Boynton, telephone interview, 13 May 1970; Lang, telephone interview, 31 July 1974.

39 Lang and Nolting, "Operations with Tethered Space Vehicles," pp. 55-58; Gemini 11 press kit, p. 14; McLaughlin, "Spin Up Studies;" Boynton interview.

40 "Gemini XI Voice," pp. 188-91, 192-216, 219-21, 228; "Gemini XI Debriefing," pp. 54-58; "Gemini XI Mission Report," pp. 1-3, 7-26 through -28.

41 Gemini 11 press kit, p. 14; "Gemini XI Mission Report," p. 6-12, 7-28; "Change of Shift Briefing," 14 Sept. 1966, tape 7A, pp. 1, 2; Kramer, Aldrin, and Hayes, "Onboard Operations for Rendezvous," p. 39.

42 "Gemini XI Mission Report," pp. 1-3, 4-

15, -26, 5-29, 6-12, -13, 7-28, -29; "Gemini XI Debriefing," pp. 60, 61; "Gemini XI Voice," pp. 241, 243-44, 245, 250-51; Carl R. Huss, Kenneth A. Young, and James D. Alexander, telephone interviews, 15 May 1970.

43 "Gemini XI Mission Report," pp. 4-15, 8-13 through -16; "Gemini XI Voice," pp. 263, 266-85, 288-303; "Gemini XI Debriefing," pp. 64-68; Thomas J. Shopple, George F. Eck, and Albert R. Prince, "Experiment D015, Night Image Intensification," in "Gemini XI Experiments Report," pp. 17, 21-23, 28, 30, 31-34.

44 Kramer, Aldrin, and Hayes, "Onboard Operations for Rendezvous," p. 39; "Gemini XI Mission Report," pp. 1-3, 4-15, -16, -27, 6-14, -15, 7-29, -30, 8-4 through -6, -13, -25, -65; "Gemini XI Voice," pp. 308, 321-24; "Gemini XI Debriefing," pp. 69-74.

45 "Gemini XI Voice," pp. 315, 321.

46 "Gemini XI Mission Report," p. 2-1; Gemini 11 press kit, p. 15; David M. Box et al., "Controlled Reentry," in Gemini Summary Conference, pp. 159-65; P. W. Malik and G. A. Souris, Project Gemini: A Technical Summary, NASA CR-1106 (Langley, Va., 1968), pp. 295-96; G. W. Knori, "The Gemini XI Re-entry Monitoring and Control Procedure," McDonnell Design Note No. 394, 18 Aug. 1966.

47 "Gemini XI Mission Report," pp. 1-3, 4-16, 5-14 through -17, 6-15, -26, -27, 7-30, -31; "Gemini XI Voice," p. 345.

48 George M. Low's written report of the trip, in journal form, provides some interesting insight and commentary about behind-the-scene and in-the-scene activities involved in a goodwill tour. His comments on the leaders of the countries (in which they met the president or acting president, cabinet members, legislators, governors, and mayors) indicate that these officials were surprisingly familiar with the details of Gemini. Low said the press of South America also seemed to be well informed about the program. Low, "Latin American Tour with Astronauts Armstrong and Gordon, October 7-11, 1966," 16 Nov. 1966.

49 "Gemini XII Technical Debriefing," 22 Nov. 1966, p. 428; James A. Lovell, Jr., interview, Houston, 15 April 1967.

50 Funk letter, 12 July 1966; letter, Gilruth to Funk, GS-64161, 26 July 1966; Elms, conversation with Grimwood, 14 Oct. 1968; Elms interview.

51 Elms, "Third Interim Report, Gemini Mission Review Board, October 25, 1966"; memo, Samuel H. Hubbard to Chairman [Elms], "Gemini Mission Review Board Meeting of September 13, 1966," 18 Oct. 1966.

52 Elms, "Third Interim Report;" memo, Elms to Assoc. Adm., OMSF, "Review of Gemini XII Mission Elements," 23 Sept. 1966.

53 Elms memo, 23 Sept. 1966; letter, Mueller to Ferguson, 30 Sept. 1966.

54 TWX, Mathews to Langley Research Center, Attn: Floyd L. Thompson, Otto F. Trout, Jr., and Robert R. Moore, Jr., GV-12504, 1 Sept. 1966; TWX, Edwards to MSC, Attn: Mathews, "Experiment Deletion from Gemini XII," MGS-522, 27 Sept. 1966; datafax transmission, William J. O'Donnell to MSC, Attn: Howard I. Gibbons, MSFC, Attn: Joe M. Jones, and KSC, Attn: John W. King, "Draft release #3," 27 Sept. 1966; letter, Mathews to McDonogh School, Attn: Robert L. Lamborn, GV-66555, 9 Nov. 1966; "Gemini Program Mission Report, Gemini XII," MSC-G-R-67-1, January 1967, p. 12-5; Mattingly letter, 5 Oct. 1970.

55 "Gemini Program/Mission Directive," p. A-12-1, Change 4; memo, Gill to dist., "Summary Minutes of 6th Inflight Experiments Meeting, 66-3, held at NASA Headquarters, Washington, D.C., April 28, 1966, April 29, 1966," 16 June 1966, p. 4; Lee, "Minutes of Senior Staff Meeting, August 12, 1966," p. 3; TWX, Mathews to McDonnell, Attn: Burke, "Contract NAS 9-170, Gemini Configuration Control Board Meeting Number 116, August 8, 1966," GV-12492, 11 Aug. 1966; Youngblood, "Minutes of Senior Staff Meeting, September 30, 1966," p. 3; memo, Mathews to NASA Hq., Attn: Schneider, "Gemini XII Mission Priorities," GV-66519, 7 Oct. 1966; Youngblood, "Minutes of Senior Staff Meeting, September 16, 1966," p. 4; E[lvin] B. Pippert, Jr., and T[ommy] W. Holloway, "Gemini XII Flight Plan," Final, 20 Oct. 1966; Meyer, notes on GPO staff meeting, 20 Oct. 1966, p. 1; NASA Release No. 66-272, "Project: Gemini 12," press kit, 28 Oct. 1966, pp. 1, 11.

56 TWX, Mueller to MSC, Attn: Gilruth, "Gemini XII Flight Plan," M/473, 26 Sept. 1966; TWX, Gilruth to NASA Hq., Attn: Mueller, "Gemini Extravehicular Reporting," GP-7658, 4 Oct. 1966; "Gemini XII Debriefing," p. 428; Elms memo, 23 Sept. 1966.

57 [Warren J. North], "EVA Position Paper," n.d., pp. 2, 3; Schultz et al., "Extravehicular Training and Simulation," pp. 7-23, -35; Reginald M. Machell et al., "Summary of Gemini Extravehicular Activity," in Gemini Summary Conference, pp. 139-46; Gemini II press kit, p. 22.

58 David C. Schultz et al., "Body Positioning and Restraints during Extravehicular Activity," in Gemini Summary Conference, pp. 79, 83-84, 85, 86-87; TWX, Mathews to McDonnell, Attn: Burke, "Contract NAS 9-170, Gemini, Action Items Resulting from Spacecraft 12 Flight Readiness Review," GP-7679, 31 Oct. 1966; Bell interview.

59 TWXs, Mathews to SSD, Attn: Col. Alfred J. Gardner, GV-12384, 14 March, and GP-7509, 1 April 1966; TWX, Mathews to SSD, Attn: Conkling, "Major Hardware Shipment for Gemini XII Mission," GA-6023, 11 Aug. 1966; "Abstract of Meeting on Atlas/Agena Coordination, August 24, 1966," 14 Sept. 1966; James M. Grimwood and Barton C. Hacker, *Project Gemini Technology and Operations: A Chronology,* NASA SP-4002 (Washington, 1969), p. 279.

60 Meyer notes, May 18, 1966, p. 1; TWX, Mathews to NASA Hq., Attn: Day, "Procurement of an Atlas Launch Vehicle for Gemini XII Mission," GP-7582, 14 June 1966; TWX, Mathews to SSD, Attn: Gardner, GV-12368, 8 March 1966; TWX, Mathews to SSD, Attn: Gardner, "Disadvantages of Flying Atlas 5307 in Its Planned Configuration," GP-7610, 8 July 1966; TWX, Mathews to SSD, Attn: Gardner, "Atlas Vehicle for Gemini XII," GV-12468, 19 July 1966; TWX, Mathews to Lewis Research Center, Attn: Edward F. Baehr and Henry W. Plohr, "Atlas Vehicle for Gemini XII," GV-12469, 18 July 1966; Lee, "Minutes of Senior Staff Meeting, July 8, 1966," p. 4.

61 NASA News Release No. 66-272, "Gemini Finale 98-Hour Flight Begins Nov. 9," 28 Oct. 1966; Gemini 12 News Center Releases Nos. 5, 6, and 7, "Status Report[s]," 8 Nov. 1966, and Nos. 9 and 11, same title, 9 Nov. 1966; TWX, Mathews to SSD, Attn: Hull and Gardner et al., "Revised Gemini XII Launch Times," GV-12529, 9 Nov. 1966; Kuras and Albert, "Gemini-Titan Technical Summary," p. 117; "Launch Operations Techniques, Manned Space Flight," Gemini Launch Vehicle Div., 6555th Aerospace Test Wing, 22 Dec. 1966, p. 172; Howard T. Harris, "Gemini Launch Vehicle Chronology, 1961-1966," AFSC Historical Publications Series 66-22-1, December 1966, pp. 83-84; "Launch Vehicle No. 12 Flight Evaluation," Martin Engineering Report No. 13227-12 and Supplemental Report No. 2 to "Gemini XII Mission Report," December 1966, p. XVI-1.

62 "Gemini XII Mission Report," p. 1-1; photograph, "The End," *The Sun,* Baltimore, 12 Nov. 1966; Jim Strothman, "Gemini Gang Is Breaking Up," *The Miami Herald,* 30 Oct. 1966; "Historic Gemini Pad Bowing to Wreckers," *The Sunday Star,* Washington, 13 Nov. 1966; Bruce K. Byers, "Lunar Orbiter: A Preliminary History," NASA HHN-71, August 1969, pp. 168-72.

63 "Gemini XII Voice Communications (Air-to-Ground, Ground-to-Air and On-Board Transcription)," McDonnell Control No. C-116106, n.d., pp. 12-15; "Gemini XII Debriefing," pp. 17-18; "Gemini XII Mission Report," p. 5-4.

64 "Gemini XII Mission Report," pp. 1-2, 4-1, -15, 5-20, 6-6 through -8, 7-1; "Gemini XII Voice," pp. 26, 30; "Gemini XII Debriefing," pp. 23, 28-31; Edwin E. Aldrin, Jr., "Line of Sight Guidance Techniques for Men in Orbital Rendezvous" (Ph.D. dissertation, Massachusetts Institute of Technology, 1964); Dean F. Grimm, interview, Houston, 13 April 1968.

65 "Gemini XII Debriefing," pp. 21, 27, 31, 32, 34, 37, 49; "Gemini XII Mission Report," pp. 1-2, 4-17, 6-8, 7-25, -26; "Gemini XII Voice," pp. 33-44; "Interim Report, Manned Space Flight Experiments, Gemini XII Mission, November 11-15, 1966," MSC-TA-R-67-3, August 1967, p. 149; Gemini 12 mission commentary transcript, 11 Nov. 1966, tape 56, p. 3.

66 "Gemini XII Mission Report," pp. 4-1, 5-1, -15, 7-2, -26; "Gemini XII Debriefing," pp. 53, 54-55, 57, 61; "Gemini XII Voice," pp. 67, 75.

67 Pippert and Holloway, "Gemini XII Flight Plan," Section II, p. 78; "Gemini XII Mission Report," pp. 4-1, -2, 6-2, -9, 7-1, -27; memo, Willis B. Foster to Gemini Mission Dir., "Astronaut Observations from Gemini XII of the November 12th Total Solar Eclipse," 11 Oct. 1966, with enclosure; letter, Gardner to Grimwood, 24 June 1970; Gemini 12 press kit, p. 11; "Gemini XII Prelaunch Press Briefing," 10 Nov. 1966, tape 3A, p. 6; Albert Sehlstedt, Jr., "Sun Eclipse Photo Sought," *The Sun,* Baltimore, 3 Nov. 1966; "Gemini 12 Flight Postponed; Photos of Eclipse Canceled," *The Evening Bulletin,* Philadelphia, 9 Nov. 1966; "Change of Shift Press Briefing," 11 Nov. 1966, tape 5B, p. 5; "Gemini XII Voice," pp. 65, 73, 87; James R. Bates, telephone interviews, 10 April 1970, 27 Aug. 1973; Young, telephone interview, 14 April 1970; William O. Armstrong, interview, Washington, 24 Jan. 1967; Gemini 10 News Release No. 3, "Gemini 10 Flight Controllers," 13 July 1966.

68 "Gemini XII Voice," pp. 73-74, 77, 78, 81, 82, 86; "Gemini XII Debriefing," pp. 69, 73, 74, 76; "Gemini XII Mission Report," pp. 4-17, -31, 6-9, -10, 7-4, -27; Young interview, 14 April 1970; "Path—The shaded area in which next Saturday's eclipse of the sun will be total for a two-minute period," *The Washington Post,* 6 Nov. 1966.

69 "Gemini XII Mission Report," pp. 4-2, 7-2, -3, -6, -27, -28, -33; "Gemini XII Voice," pp. 97-148; "Gemini XII Debriefing," pp. 79, 88-90, 92, 94, 97-99, 102-104, 106, 111, 113-14; Burns et al., "Gemini Extravehicular Activities," p. 3-25; Machell et al., "Summary of Extravehicular Activity," p. 141.

70 "Gemini XII Mission Report," pp. 4-2,

7-3, -4, -28 through -30; "Gemini XII Voice," pp. 222-87; "Gemini XII Debriefing," pp. 130-234; Machell et al., "Summary of Extravehicular Activity," p. 142.

[71] "Gemini XII Mission Report," pp. 4-2, 5-88, 7-30; "Gemini XII Debriefing," pp. 284-93; Machell et al., "Summary of Extravehicular Activity," pp. 139-43.

[72] "Gemini XII Voice," pp. 311, 312-21; "Gemini XII Mission Report," pp. 4-3, 5-17 through -19, 6-12, 7-30, -31; Ivan D. Ertel, *Gemini XII Flight and Gemini Program Summary*, MSC Fact Sheet No. 291-I (Houston, December 1966); TWX, Mathews to McDonnell, Attn: Burke, "Gemini XII Flight Data for Tether Exercise," GT-11307, 27 Dec. 1966.

[73] "Gemini XII Mission Report," pp. 5-63 through -66; "Gemini XII Debriefing," pp. 433-34; "Gemini 12 in Home Stretch," *The Washington Daily News*, 14 Nov. 1966; Malik and Souris, *Gemini Technical Summary*, pp. 51, 53, 58-59; Percy S. Miglicco, telephone interview, 3 June 1970; Miglicco comments on draft chapter, n.d.

[74] "Gemini XII Mission Report," p. 8-3.

[75] "Gemini XII Mission Report," pp. 1-4, 2-1, 4-5, 5-99, 6-15, -16, -30, 7-7, -8, -32; "Gemini XII Debriefing," pp. 301, 303-305, 310-11, 315, 322; "Gemini XII Voice," pp. 494-95.

[76] "Last Flight for Gemini Flags," MSC *Space News Roundup*, 9 Dec. 1966.

[77] "Project Development Plan for Rendezvous Development Utilizing the Mark II Two Man Spacecraft," MSC, 8 Dec. 1961; *Weekly Compilation of Presidential Documents, Monday, November 21, 1966*, p. 1702.

[78] Letter, Vice President Hubert H. Humphrey to Webb, 8 Dec. 1966; memo, Robert C. Seamans, Jr., to Assoc. and Asst. Adms., Field Center Dirs., "Gemini Program Record of Accomplishments, attached," 17 Jan. 1967, with enclosures; White House News Release, "John F. Kennedy, President of the United States, Special Message to Congress, 25 May 1961," p. 9; *Weekly Compilation, November 28, 1966*, p. 1719; Al Rossiter, Jr., "3 Apollo Astronauts Meet Death Sealed in Blazing Space Capsule," *Houston Chronicle*, 28 Jan. 1967.

## Summing Up

[1] Memo, Howard W. Tindall, Jr., to Chief, Mission Planning and Analysis Div., "Can Gemini contribute to Apollo?" 8 Jan. 1965.

[2] William F. Rector III, interview, Redondo Beach, Calif., 27 Jan. 1970.

[3] Courtney G. Brooks, James M. Grimwood, and Loyd S. Swenson, Jr., "Chariots for Apollo: A History of Lunar Spacecraft," draft, 1976, p. 349, note.

[4] Memo, William A. Lee to ASPO Div. Chiefs, Asst. Div. Chiefs, and Br. Chiefs, "Data Exchange Meetings with McDonnell Aircraft Corporation," PA/M9-65-252, 27 Sept. 1965.

[5] NASA OMSF Apollo Program Directive No. 6, "Sequence and Flow of Hardware Development and Key Inspection, Review, and Certification Checkpoints," 12 Aug. 1965.

[6] TWX, George M. Low to MSC, Attn: James C. Elms, "Apollo and Gemini Fuel Cells," M-C S 1000.578, 7 Oct. 1963; memo, André J. Meyer, Jr., to MSC Historical Office, "Comments on Draft Chapters 7 and 8 of Gemini Narrative History," 6 Jan. 1972; William C. Schneider, interview, Washington, 23 Jan. 1967.

[7] Grimwood, "Planning the Experiments," unpublished draft chapter of Gemini history, 31 July 1968; memo, George E. Mueller to Mac C. Adams, no subj., 20 Oct. 1965; memo, Adams to Assoc. Adm., Manned Space Flight, "Possible application of limp paraglider to Apollo," n.d.; Robert R. Gilruth, conversation with Grimwood, 18 Aug. 1972.

[8] Memo, William A. Lee to Low, "LEM Ascent Engine 'Fuel Lead,' " 11 April 1966.

[9] Christopher C. Kraft, Jr., interview, Houston, 5 Oct. 1967.

[10] Ibid.

[11] Letter, Samuel C. Phillips to MSFC, Attn: Saturn V Program Mgr., "Application of Gemini Experience to Apollo," 7 April 1966, with enclosures; letter, Phillips to MSC, Attn: Mgr., GPO, "Gemini/Apollo Launch Vehicle Meeting," 15 Sept. 1966.

[12] Appendixes B and C, this volume; Loyd S. Swenson, Jr., James M. Grimwood, and Charles C. Alexander, *This New Ocean: A History of Project Mercury*, NASA SP-4201 (Washington, 1966), Appendix D; memo, Hugh L. Dryden, Cabinet Report to the President, "Significance of Gemini V Accomplishments," 11 Sept. 1965; "Project Gemini Schedule Analysis," GPO, 5 Jan. 1962.

[13] Appendix E, this volume; Swenson, Grimwood, and Alexander, *This New Ocean*, Appendix F; letter, Paul E. Purser to Grimwood, 12 June 1974.

[14] Appendix B, this volume; MSC Public Affairs Office accreditation lists, *Gemini IV* through *Apollo 11*; "World Press Reaction to Gemini VI and VII," United States Information Agency Research and Reference Series R-189-65, 22 Dec. 1965.

[15] *Gemini Summary Conference*, SP-138 (Washington, 1967); Jim Maloney, "Gemini Conference Delegates Silenced about Apollo Fire," *The Houston Post*, 2 Feb. 1967.

# Bibliographical Note

THIS history of Project Gemini rests ultimately on the paperwork generated by the project itself. Virtually all the documents cited in the notes are available in the History Archives, Johnson Space Center, Houston, Texas. Gathered over the past 12 years, these archives now comprise over 200 linear meters of filed and shelved documents, bearing not only on Gemini but on all American manned space flight programs—Mercury, Apollo, Skylab, Apollo-Soyuz, and Shuttle—as well as the institutional history of the Center and some special topics such as space-suit development.

Most of the material on which this book is based does not lend itself easily to listing in a formal bibliography. The published and unpublished documents listed below thus represent, in a sense, only the tip of the iceberg—those items that may be conveniently cited; as the notes clearly show, they in no way approach a description of the sources. In fact, they tend to be the most peripheral. In the ongoing work of Project Gemini, whose fallout provided most of the evidence for our attempt to tell the story, there was little time or opportunity for the writing, much less the publication, of the more formal books or articles that lend themselves to citation. The nature of Project Gemini, as sketched in the introduction and displayed in the book, has also meant that the years since its completion have added little to the story.

In this note, we shall make some effort to describe the nature of the sources that we have used. The backbone of the Gemini history was chiefly provided by a class of material that might be labeled serial documents. Probably the most important of these were the regular, recurring progress, status, and activity reports submitted by contractors to NASA and by lower NASA elements to higher. They vary greatly in content, format, quality, and usefulness, but they often provide the major, sometimes the only, basis for reconstructing the sequence and significance of particular events, especially during Gemini's developmental period. Among the more helpful of these reports were:

Lockheed Missiles & Space Co. "Gemini Agena Target Vehicle Program Progress Report." LMSC-A605200-1 to -16. Sunnyvale, Calif., 20 Oct. 1964—20 Dec. 1965.
———. "Medium Space Vehicles Programs Monthly Progress Report." LMSC-447186-26 to -47. Sunnyvale, Calif., 20 Sept. 1962—20 May 1964.
North American Aviation, Inc. "Contract NAS 9-167, Paraglider Development Program, Phase II, Part A, Monthly Progress Letter No. 1," 20 Nov. 1961, through ". . . No. 16," 31 March 1963.
———. "Contract NAS 9-539, Paraglider Development Program, Advanced Trainer and Prototype Wing Design, Phase II, Part B(1), Monthly Progress Letter No. 1," 20 June 1962, through ". . . No. 9," 31 March 1963.
———. "Contract NAS 9-1484, Paraglider Landing System Program, Monthly Progress Report No. 1," for May 1963 through ". . . No. 21," for Jan. 1965.
U.S., Air Force, Space Systems Division (SSD). "Titan II/Gemini Program Status Summary." Weekly letters, SSD to NASA Assoc., Adm., 18 Sept. 1963 to 27 Jan. 1964.
U.S., NASA, Manned Spacecraft Center. "Weekly Activity Report for the Office of Director [later Associate Administrator], Manned Space Flight." Houston, 27 May 1962—7 Aug. 1965.
———. "Consolidated [monthly] Activity Report for the Office of the Director [later Associate Administrator], Manned Space Flight." Houston, May 1962—Jan. 1965.
———. "Quarterly Activity Report for the Office of the Associate Administrator, Manned Space Flight." Houston, 30 April 1965—30 July 1966.
———. Gemini Project [later Program] Office. "GPO Weekly Activity Report for the Director, MSC." Houston, 5 March 1962—10 April 1965.
———. "Project Gemini Quarterly Status Report No. 1, for period ending May 31, 1962," through ". . . No. 17, for period ending May 31, 1966."
U.S., NASA, Marshall Space Flight Center. "First Report on MSFC Activities Covering February 26 thru March 25, 1962, to Manned Spacecraft Center," through "Thirteenth . . . November 1 thru November 16, 1962, . . . "

A second group of serial documents comprised the official minutes and sometimes the informal notes of meetings of the boards and panels that supervised or dealt with various aspects of the Gemini program. The most directly involved were the coordination panels, which were largely responsible for the day-to-day decision-making in Gemini development (as discussed in Chapter IV). The abstracts of these coordination panel meetings record the decisions taken and, sometimes, the reasons for them; ordinarily, though, the reasons for the decisions must be sought elsewhere, since the abstracts tend to be brief to the point of being cryptic. The six panels first set up early in 1962—spacecraft mechanical systems, electrical systems, operations; Atlas-Agena, Gemini launch vehicle, and paragliders—were later joined by others to deal with particular areas as they became important. Among these were launch guidance and control, rendezvous and reentry guidance and control, trajectories and orbits, launch integration, range safety, network integration, and experiments.

Besides the coordination panels, Gemini was served by a number of other boards and panels, the minutes of which were often helpful, particularly in pinning down the precise nature of problems as understood at particular times, since the experts were faced with explaining their piece of the program to what were essentially knowledgeable outsiders. The most useful records were those of:

Gemini Mission Review Board. July-Oct. 1966.
Gemini Project/Program Office Staff Meetings. 1963-1966.
Manned Spacecraft Center Senior Staff Meetings. July 1961-Sept. 1966.
Manned Space Flight Experiments Board. 1964-1966.
NASA-DOD Gemini Program Planning Board. Feb. 1963-April 1964.
NASA/MAC Management Meetings. 1964-1966.
NASA Management Council. Dec. 1961—Sept. 1963.
Project Gemini Management Panel. Nov. 1962—Feb. 1965.

Another class of relatively formal documents that were indispensable in writing the Gemini history were the contracts between NASA and the organizations that did most of the actual work of development and operation. This is not the place for a treatise on contracting, but a few points are worth noting. The initial agreement usually took the form of a letter contract, a means of getting work started before or while negotiations took place. Eventually the letter contract gave way to the negotiated or final contract. The basic document normally included a "statement of work," particularly useful for the historian in furnishing a clear and direct statement of what the contractor agreed to provide. Major contracts were regularly changed, supplemented, amended, etc., each producing a notice added to the basic contract. A complete list of major Gemini contractors, subcontractors, and vendors may be consulted in James M. Grimwood and Barton C. Hacker, *Project Gemini Technology and Operations: A Chronology,* NASA SP-4002 (Washington, 1969), pp. 284-89.

Useful for following the changes in organization and administration of the Manned Spacecraft Center and of the Gemini Program Office were three sets of internal publications. MSC Announcements, numbered serially for each year, were the means of notifying Center employees of changes as they occurred. The semi-official MSC *Space News Roundup* was a bi-weekly newspaper focused on local news, which also contained stories about the Center's programs as well. The periodically revised MSC telephone directories were particularly helpful in determining the exact positions of people working on Project Gemini at particular times.

As the notes to the text should make clear, our major reliance was on the working documents directly related to the conduct of Project Gemini—the memorandums, letters, teletype communications, and other messages that described, explained, ordered, informed, coordinated, and otherwise kept the several parts of Gemini in touch with each other and with the outside world. One point that should probably be made is that the person who signs a message is often not its author. This is almost invariably true for interagency communications, less commonly true internally. This is the major reason we have usually preferred to identify organizations—e.g., Gemini Project Office, Space Systems Division—as the actors in our history. This trait is not unique to NASA, of course, but it clearly influences the kind of history that may be written of a NASA program.

Of considerable value as background material were a variety of documents related to NASA's efforts to maintain its public image. The MSC fact sheets, printed at the Center and distributed throughout the world, may have been the most widely read source of public information on Gemini. Among them were a series on the Gemini missions by Ivan D. Ertel, beginning with MSC Fact Sheet No. 291, *Gemini Program*, and followed by Fact Sheet Nos. 291-A through 291-I, April 1965 to December 1966, dealing with Gemini missions from the third through the twelfth. NASA also prepared and distributed to reporters a press kit for each mission. These kits were substantial compilations running to dozens of pages, intended to provide a comprehensive background for news stories about the missions. Other press materials were also helpful: the press handbooks prepared by some Gemini contractors (Martin, McDonnell, and Lockheed, in particular, which are cited here); transcripts of NASA-conducted press conferences during missions and on some other occasions (e.g., the introduction of newly selected astronauts); and such regular mission-related briefings as the one at each change of shift. These materials were often helpful in filling out the more technical record provided by the Gemini mission evaluation team in the Mission Report (this and the following are cited in full in the bibliography), as supplemented by the technical debriefings of the crew, by special detailed studies on particular aspects of a mission (e.g., launch vehicle performance), and by the transcript of all communications between ground and flight crews during the course of a mission.

NASA distributes internally a daily compilation of current news, photoduplicated articles on space-related topics from a broad spectrum of newspapers. The JSC History Office has a file of this compilation beginning in 1958. Another useful source of reaction to NASA activities is the trade press. Numerous journals are devoted to the doings of the aerospace industry; the two we found most consistently useful were *Aviation Week and Space Technology* and *Missiles and Rockets.*

Interviews were a major source for this history. The chance to put questions to the people who actually did what we were writing about went a long way to compensate for the difficulties of studying contemporary history. Cooperation was general, whether in small matters or large. Two types of interviews appear in the following list. Most were lengthy conversations that were tape-recorded and subsequently transcribed; the typescripts of these interviews are on file in the JSC History Office. We also conducted much briefer interviews by telephone; these were usually addressed to relatively specific matters of fact or information and were not recorded, although notes may have been taken. Interviews in this latter category are marked by an asterisk in the following list.

# People Interviewed

1. Albert, John G.
2. Aldrin, Edwin E., Jr.
3. Alexander, James D.*
4. Alphin, James H.*
5. Amman, Ernest A.
6. Andrich, Stephen M.
7. Armstrong, Neil A.
8. Armstrong, Stephen D.
9. Armstrong, William O.
10. Babb, Conrad D.
11. Bachman, Dale
12. Bailey, Glenn F.
13. Bake, Ronald C.*
14. Ballentine, Wilbur A.
15. Barton, John
16. Bates, James R.*
17. Bell, Larry E.
18. Berry, Charles A.
19. Bickers, John H.
20. Bird, John D.
21. Black, Dugald O.*
22. Black, Stanley
23. Blackert, Robert S.
24. Bland, William M., Jr.*
25. Blatz, William J.
26. Borman, Frank
27. Bost, James E.*
28. Bowles, Lamar D.*
29. Boyd, John H., Jr.*
30. Boynton, John H.*
31. Bratton, R. Dean*
32. Buhler, Cary
33. Burke, Walter F.
34. Byerly, Kirk L.
35. Byrnes, Martin A., Jr.
36. Cernan, Eugene A.

37. Chamberlin, James A.
38. Chambers, Gordon T.
39. Charlesworth, Clifford E.
40. Cherry, Clyde S.
41. Christopher, Kenneth W.
42. Church, John
43. Clements, Henry E.*
44. Cohen, Haggai
45. Cohen, Robert
46. Collins, Michael
47. Conrad, Charles, Jr.
48. Cooper, L. Gordon, Jr.
49. Correale, James V.*
50. Cottee, Gatha F.*
51. Crane, Richard J.*
52. Cress, Gordon P.
53. Curlander, J. Carroll
54. Czarnik, Marvin R.
55. Davis, Larry D.*
56. Day, LeRoy E.
57. Deans, Philip M.*
58. Decker, James L.*
59. Dietlein, Lawrence F.
60. Dineen, Richard C.
61. Disher, John H.*
62. Domokos, Steven J.
63. Dotts, Homer W.*
64. Douglas, W. Harry*
65. Duggan, Orton L.*
66. Dunkelman, Lawrence
67. Dunn, Charles E.
68. Eggleston, John M.
69. Ellmer, Paul
70. Elms, James C.*
71. Emigh, Harold
72. Engstrom, Bert

73. Evans, Tom
74. Evans, W. B.*
75. Everline, Robert T.*
76. Farguson, Dale
77. Fisher, Lewis R.
78. Fore, Wallace
79. Foster, Norman G.
80. Franklin, George C.*
81. Friedman, Stanley
82. Fucci, James R.*
83. Funk, Ben I.
84. Furman, Francis O.
85. Gellerman, Joseph B.
86. Gerathewohl, Siegfried J.
87. Gibbons, Howard I.*
88. Gill, Jocelyn R.
89. Gilruth, Robert R.
90. Gordon, Richard F., Jr.
91. Gray, Wilbur H.
92. Green, Don J.
93. Griffin, James J.
94. Grimm, Dean F.
95. Hahn, Jack R.
96. Hall, Eldon W.
97. Hammack, Jerome B.
98. Haney, Paul P.
99. Harness, Arminta
100. Harris, Howard T.
101. Hauger, Lloyd
102. Hecht, Kenneth F.
103. Heimstadt, C. E.
104. Hello, Bastian
105. Helsel, Ron
106. Henry, James P.*
107. Hill, Raymond D., Jr.
108. Hobokan, Andrew
109. Hodge, John D.
110. Hohmann, Bernhard A.
111. Hollands, Rockwell
112. Houbolt, John C.
113. Huff, Vearl N.
114. Hull, Robert R.
115. Huss, Carl R.*
116. Hutchison, Fountain M.
117. Hutchison, Homer W.

118. Jackson, James B., Jr.
119. Jackson, Lee
120. James, Bennett W.
121. James, George
122. Jeffs, George W.
123. Jevas, Nicholas*
124. Jimerson, Leroy S.
125. Joachim, James W.
126. Johnson, Harold I.
127. Kapp, Michael
128. Kapryan, Walter J.
129. Keehn, Richard W.
130. King, John W.*
131. Kleinknecht, Kenneth S.
132. Koons, Wayne E.*
133. Kranz, Eugene F.
134. Kuehnel, Helmut A.*
135. Lang, Dave W.
136. Lang, David D.*
137. Lansdowne, Kathryn A.*
138. Ledlie, James
139. Lenz, James
140. Letsch, Ernst R.
141. Lindley, Robert N.
142. Lineberry, Edgar C.
143. Lopez, Sarah W.*
144. Lovell, James A., Jr.
145. Low, George M.
146. Luetjen, H. H.
147. Lunney, Glynn S.*
148. Lutz, Charles C.
149. McBarron, James W., II*
150. McCabe, Robert
151. McCafferty, Riley D.
152. McCreavey, William
153. McDivitt, James A.
154. McFadden, Eugene R.
155. McKee, Donald D.
156. McMann, Harold J.
157. MacDougall, George F., Jr.*
158. Machell, Reginald M.
159. Manry, Charles E.*
160. Marbach, James*
161. Mathews, Charles W.
162. May, Bill

163. Mayer, John P.*
164. Meyer, André J., Jr.
165. Miglicco, Percy S.
166. Miller, John
167. Mitchell, Willis B., Jr.
168. Mitros, Edward F.*
169. Morgan, Frank G., Jr.
170. Morrow, Lola H.
171. Mueller, George E.
172. Muhly, William C.*
173. Nagler, Kenneth M.*
174. Nolan, Harold W.
175. Nold, Winston D.
176. North, Warren J.
177. Oldeg, Harry W.
178. Petersen, Jean L.*
179. Poole, Forrest R.
180. Preston, G. Merritt
181. Provart, Robert
182. Purchase, Alan
183. Purser, Paul E.
184. Raines, Ray
185. Rapp, Rita M.*
186. Ray, Hilary A.*
187. Ringer, Jerome
188. Rose, James T.
189. Rose, Rodney G.*
190. Russell, John H.
191. Samonski, Joan P.*
192. Sanders, Fred J.
193. Sanderson, Alan N.*
194. Satterfield, James M.*
195. Saunders, James F., Jr.*
196. Schirra, Walter M., Jr.
197. Schlicker, Albert
198. Schmitt, Joe W.*
199. Schneider, William C.
200. Schroeder, Lyle C.*
201. Schweickart, Russell L.
202. Scott, David R.
203. Seamans, Robert C., Jr.
204. Sharp, Robert L.
205. Sheckells, George
206. Shoaf, Harry C.*
207. Shoenhair, Jack L.
208. Shuck, Lowell
209. Simpkinson, Scott H.
210. Sims, John R.
211. Slaughter, Frances*
212. Smistad, Olav*
213. Smith, Herbert E.*
214. Smith, Walter D.
215. Spath, Richard M.
216. Stafford, Thomas P.
217. Stewart, Larry E.
218. Stewart, Lester A.*
219. Stiff, Ray C., Jr.
220. Stullken, Donald E.*
221. Summerfelt, William A.
222. Swanson, John
223. Sweeney, John L.
224. Tenebaum, Dan M.
225. Thackston, Willard
226. Tindall, Howard W., Jr.
227. Tomlinson, Charles C.
228. Towner, George
229. Trombka, Jacob L.
230. Truszynski, Gerald M.
231. Tucker, Elton M.*
232. Van Bockel, John J.
233. Van Riper, Paul P.
234. Verlander, Joseph M.
235. Verrengia, Augustine A.*
236. Verrier, Don
237. Vester, Ben
238. Vogel, Harle L.
239. Waggoner, James
240. Wambolt, Joseph F.
241. Ward, E. Douglas
242. Weber, George J.*
243. Wendt, Guenter F.
244. Westkaemper, Robert M.
245. Whitacre, Horace E.
246. Williams, John J.
247. Williams, Walter C.
248. Williams, Wiley E.
249. Wilson, Louis D.
250. Wolhart, Walter D.
251. Workman, Robert O.
252. Wyatt, DeMarquis D.

253. Yardley, John F.
254. York, Irving
255. Young, John W.
256. Young, Kenneth A.*
257. Young, Richard S.*
258. Young, Robert B.
259. Younger, George B.
260. Zahn, Toni*
261. Zeitler, Edward O.*


Still another group of sources deserves special mention—the comments we received on draft chapters of this history and on draft versions of *Project Gemini Technology and Operations*. These comments varied considerably in scope, format, and value, but a number were substantial and documented critiques on the text. The relevant comments are cited in the notes.

The bulk of the remaining sources are listed in the following bibliography. Any classification must inevitably be arbitrary, at least in part. We have divided the primary sources into four classes: (1) Studies, Proposals, Long-Range Plans, and other documents mostly related to Gemini's formative stages; (2) Gemini Plans, Procedures, Working Papers, Design Notes, and other materials related directly to the operation of the program; (3) Gemini Reports, Reviews, Evaluations and other assessments of the conduct of the project; and (4) Printed Primary Sources. Secondary sources have merely been separated into two classes: (5) Unpublished Secondary Sources and (6) Published Secondary Sources.


## I. PRIMARY SOURCES: STUDIES, PROPOSALS, LONG-RANGE PLANS

Aldrin, Edwin E., Jr. "Line of Sight Guidance Techniques for Men in Orbital Rendezvous." Ph.D. dissertation, Massachusetts Institute of Technology, 1964.

Barker, C. L., and Straly, W. H. "Rendezvous by the Chasing Technique." ABMA Report DSP-TM-15-59. Huntsville, Ala., 30 Oct. 1959.

Blackmer, R. H., and Phillips, G. A. "Ion-Exchange Membrane Fuel Cell for Space Vehicle Electric Power." Paper presented at Society of Automotive Engineers, National Aerospace Engineering and Manufacturing Meeting, Los Angeles, 9-13 Oct. 1961.

Brown and Root, Inc. "NASA Manned Spacecraft Center Master Plan & Architectural Concept." Houston, July 1961.

California Institute of Technolgy, Jet Propulsion Laboratory. "Man-to-the-Moon and Return Mission Utilizing Lunar-Surface Rendezvous." JPL TM 33-53. Pasadena, Calif., 3 Aug. 1961.

————. "System Considerations for the Manned Lunar Landing Program." JPL TM 33-52. Pasadena, Calif., 3 Aug. 1961.

Decker, James L. "A Program Plan for a Titan Boosted Mercury Vehicle." Martin Co., Baltimore, July 1961.

Eggleston, John M. "Inter-NASA Research and Space Development Centers Discussion on Space Rendezvous, May 16-17, 1960." Washington, n.d.

Frasier, Cline W. "LEM Rendezvous Radar vs. Optical Tracker Study." Grumman Aircraft Engineering Corp., Bethpage, N.Y., 16 March 1965.

Gardner, James P.; Ruppe, Harry O.; and Straly, Warren H. "Comments on Problems Relating to the Lunar Landing Vehicle." ABMA Report DSP-TN-13-58. Huntsville, Ala., 4 Nov. 1958.

General Electric Co., Direct Energy Conversion Operation. "Fuel Cells for Spacecraft, Including Determination of Fuel Battery Size for Specific Application." West Lynn, Mass., January 1964.

Grumman Aircraft Engineering Corp. "Apollo Mission Planning Task Force," 3 vols. LED-540-7. Bethpage, N.Y., 4 May 1964.

Hoelker, R. F., and Silber, Robert. "The Bi-Elliptical Transfer between Circular Co-Planar Orbits." ABMA Report DA-TM-2-59. Huntsville, Ala., 6 Jan. 1959.

Horner, James M., and Silber, Robert. "Impulse Minimization for Hohmann Transfer between Inclined Circular Orbits of Different Radii." ABMA Report DA-TR-70-59. Huntsville, Ala., 2 Dec. 1959.

[Houbolt, John C., et al.] "Manned Lunar-Landing through the Use of Lunar-Orbit Rendezvous." 2 vols. Langley, Va., ca. August 1961.

————. "Technical Problems of Lunar Orbit Rendezvous." Langley, Va., September 1961.

Hughes Aircraft Co., Nucleonics Laboratory. "Lunar Surface Assembly Techniques: A Preliminary Study of Refueling for the Lunar Surface Rendezvous." Hughes Report FD-61-401 (JPL 950167), 2 Oct. 1961.

Koelle, Heinz H. "Future Projects at Marshall Space Flight Center." In U.S., NASA, MSFC, "NASA-Industry Program Plans Conference, September 27-28, 1960." Huntsville, Ala., n.d.

————; Williams, F. L.; Huber, W. G.; and Callaway, R.C. "Juno V Space Vehicle Development Program (Phase I): Booster Feasibility Demonstration." ABMA Report DSP-TM-10-58. Huntsville, Ala., 13 Oct. 1958.

Lina, Lindsay J., and Vogeley, Arthur W. "Preliminary Study of a Piloted Rendezvous Operation from the Lunar Surface to an Orbiting Space Vehicle." Paper presented at NASA Inter-Center Rendezvous Discussions, Washington, 27-28 Feb. 1961.

Lundin, Bruce T.; Downhower, Walter J.; Eggers, A. J., Jr.; Johnson, Lt. Col. George W. S.; Loftin, Laurence K., Jr.; Ruppe, Harry O.; Escher, William J. D.; and May, Ralph W., Jr. "A Survey of Various Vehicle Systems for the Manned Lunar Landing Mission." NASA, Washington, 10 June 1961.

McDonnell Aircraft Corp. "Follow On Experiments, Project Mercury Capsules." MAC Engineering Report 6919, St. Louis, 1 Sept. 1959; rev. 5 Oct. 1959.

————. "Gemini Large Earth Orbit." MAC Report B743; Control No. C-100858. St. Louis, 19 June 1965.

————. "Mark II Mercury Spacecraft." MAC Control No. C-57342. St. Louis, 6 July 1961.

————. "Mercury Spacecraft: Advanced Versions." MAC Control No. C-57978. St. Louis, ca. 27 July 1961.

————. "Price and Delivery Proposal for MK II Mercury Engineering Study Program." MAC Report 8185. St. Louis, 12 April 1961.

Michael, William H., Jr. "Weight Advantages of Use of Parking Orbit for Lunar Soft Landing Mission," in Jack W. Crenshaw; John P. Gapcynski; Wilbur L. Mayo; and Michael, "Studies Related to Lunar and Planetary Missions." Langley, Va., 26 May 1960.

Radio Corp. of America. "Satellite Interceptor Study System: Final Report." RCA Report CR-59-588-39. Burlington, Mass., 31 Jan. 1960.

Rogallo, Francis M. "Paraglider Recovery Systems." Paper presented at International Astronautics Society, Meeting on Man's Progress in the Conquest of Space, St. Louis, 30 April—2 May 1962.

————, and Lowry, John G. "Flexible Reentry Gliders." Paper presented at Society of Automotive Engineers, National Aeronautics Meeting, New York, 4-8 April 1960.

Sears, Norman E. "Satellite-Rendezvous Guidance System." MIT Report R-331. Cambridge, Mass., May 1961.

Silber, Robert, and Horner, James M. "Two Problems of Impulse Minimization between Coplanar Orbits," ABMA Report DA-TM-23-59. Huntsville, Ala., 12 Feb. 1959.

Space Craft, Inc. "Analysis of a Lunar Surface Rendezvous Mission. JPL 960165. Pasadena, Calif., October 1961.

Space Technology Laboratories, Inc. "The Lunar Surface Rendezvous Technique for Manned Lunar Landing and Return." STL Report 8634-0001-RC-000 (JPL 950163). Redondo Beach, Calif., 2 Oct. 1961.

————. "Saint Phase 1 Technical Proposal." STL/TR-59-0000-09917. Redondo Beach, Calif., 21 Dec. 1959.

U.S., Army Ballistic Missile Agency. "Proposal: A National Integrated Missile and Space Vehicle Development Program." ABMA Report D-R-37. Huntsville, Ala., 10 Dec. 1957.

U.S., Army Ordnance Missile Command. "A Lunar Exploration Program Based upon Saturn-Boosted Systems." AOMC RCS ORDXM-C-1004 (ABMA Report DV-TR-2-60). Huntsville, Ala., 1 Feb. 1960.

————. "Project Horizon, Phase I Report: A U.S. Army Study for the Establishment of a Lunar Military Outpost." 4 vols. Huntsville, Ala., 8 June 1959.

U.S., National Advisory Committee on Aeronautics. "NACA Conference on High-Speed Aerodynamics, Ames Aeronautical Laboratory, Moffett Field, California, March 18, 19 and 20, 1958: A Compilation of Papers Presented." Moffett Field, n.d.

U.S., NASA. "Guidelines for a Program for Manned and Unmanned Orbital Operations." Washington, May 1961.

————. "Lunar Orbit Rendezvous: News Conference on Apollo Plans at NASA Headquarters on July 11, 1962." Washington, 1962.

————. "Third Semi-Annual NASA Staff Conference: Program Formulation and Status of Activities, Monterey, California, 3-5 March 1960." Washington, 1960.

————. Ad Hoc Task Group on a Feasible Approach for an Early Manned Lunar Landing. "A Feasible Approach, . . . ." Part I: "Summary Report of Ad Hoc Task Group Study." Washington, 16 June 1961.

————. Ad Hoc Task Group for Study of Manned Lunar Landing by Rendezvous Techniques. "Earth Orbital Rendezvous for an Early Manned Lunar Landing," Part I: "Summary Report of Ad Hoc Task Group Study." Washington, August 1961.

————. Combined Working Group on Vehicles for Manned Space Flight. "Report." Washington, 20 Nov. 1961.

————. Lunar Landing Working Group. "A Plan for Manned Lunar Landing." Washington, January 1961.

————. Office of Program Planning and Evaluation. "The Ten Year Plan of the National Aeronautics and Space Administration." Washington, 18 Dec. 1959.

————. Office of Space Flight Development. "Manned Space Flight Long Range Plans." Washington, 17 Aug. 1959.

————. Propulsion Staff. "A National Space Vehicle Program: A Report to the President." Washington, 27 Jan. 1959.

U.S., NASA, Manned Spacecraft Center. "Project Development Plan for Rendezvous Development Utilizing the Mark II Two Man Spacecraft." Langley, Va., 8 Dec. 1961.

U.S., NASA, Marshall Space Flight Center, Committee for Orbital Operations. "Orbital Operations Preliminary Project Development Plan." Huntsville, Ala., 15 Sept. 1961.

U.S., NASA, Space Task Group. "Guidelines for Advanced Manned Space Vehicle Program." Langley, Va., June 1961.

————. "Manned Spacecraft Development Center; Organizational Concepts and Staffing Requirements." Langley, Va., 1 May 1961.

————. "Preliminary Project Development Plan for a Controllable Parachute-Retrorocket Landing System." Langley, Va., 21 June 1961.

————. "Preliminary Project Development Plan for an Advanced Manned Space Program Utilizing the Mark II Two Man Spacecraft." Langley, Va., 14 Aug. 1961; rev. 21 Aug. 1961.

————. "Project Development Plan for Rendezvous Development Utilizing the Mark II Two Man Spacecraft." Langley, Va., 27 Oct. 1961.

U.S., NASA and Department of Defense, Large Launch Vehicle Planning Group. "Final Report." 3 vols. NASA-DOD LLVGP 105. Washington, 1 Feb. 1962.

————. "Summary Report." NASA-DOD LLVPG 105. Washington, 24 Sept. 1962.

von Braun, Wernher; Stuhlinger, Ernst; and Koelle, H. H. "ABMA Presentation to the National Aeronautics and Space Administration." ABMA Report D-TN-1-59. Huntsville, Ala., 15 Dec. 1958.

## 2. PRIMARY SOURCES: GEMINI PLANS, PROCEDURES, WORKING PAPERS, DESIGN NOTES

Beasley, Gary P. "Digital Computer Analysis in Support of GT-5 Mission." Paper presented at

MSC Meeting on Orbital Mechanics Associated with Gemini Flights, Houston, 11 Aug. 1965.
Chaput, Paul T. "Crew Egress Procedures Developed during the Qualification Test Program for the Gemini Spacecraft At-Sea Operations." Gemini Working Paper 5015. Houston, 26 Aug. 1964.
Cohen, Robert. "Summary of Analysis for Selecting the Power Source for the Gemini Project." Gemini Project Note. Houston, 23 Jan. 1962.
Culpepper, Bobby K. "Partial Proposed Mission Plan for the GT-5 Gemini Flight, REP Plan IV." MSC Internal Note 64-FM-87. Houston, 1 Dec. 1964.
Dugge, Peggy, and Czarnik, Marvin R. "Practice Rendezvous Mission." MAC Guidance and Control Mechanics Design Note 1. St. Louis, 7 July 1962.
Fricke, Robert W., Jr. "Mission Directive for Gemini-Titan II Mission 1 (Spacecraft No. 1)." Gemini Working Paper 5005. Houston, 14 Nov. 1963.
Garren, Kenneth R. "Use of Visual Cues for Determining Range (with Optical Aids) and Direction of Motion of a Flashing Light." Paper presented at MSC Meeting on Orbital Mechanics Associated with Gemini Flights, Houston, 11 Aug. 1965.
Guillory, Ted A.; Stough, Charles L.; and Davis, Charles F., Jr. "Gemini IX Flight Plan, Final." Houston, 18 April 1966.
Hamm, Richard T. "Description of the Gemini VII Station Keeping Hybrid Simulation." MAC Gemini Guidance and Control Design Note 388. St. Louis, 13 July 1966.
Holloway, Tommy W. "GT-3 Flight Plan, Preliminary B." Houston, 20 Sept. 1964.
———. "GT-3 Flight Plan, Final." Houston, 4 March 1965.
———. "GTA-6 Flight Plan, Preliminary." Houston, 2 Aug. 1966.
———. "Gemini VI Flight Plan, Final." Houston, 1 Oct. 1965.
———. "Gemini VIII Flight Plan, Final." Houston, 24 Feb. 1966.
Hrabel, Gordon C. "Experiments for GT-3 Mission." Gemini Working Paper 5014. Houston, 22 Sept. 1964.
———. "Experiments for GT-4 Mission." Gemini Working Paper 5023. Houston, 14 May 1965.
Knori, G. W. "The Gemini XI Re-entry Monitoring and Control Procedure." MAC Design Note 394. St. Louis, 18 Aug. 1966.
Lockheed Missiles & Space Co. "Gemini Agena Target Vehicle Program Status, 5 January 1966." LMSC-A777567. Sunnyvale, Calif., 5 Jan. 1966.
———. "Preliminary Report on Agena System Capabilities for Advanced Mercury Rendezvous Missions." LMSC-A004120. Sunnyvale, Calif., 26 Jan. 1962.
———. "Symposium on Hypergolic Rocket Ignition at Altitude." LMSC-A776842. Sunnyvale, Calif., 1 Dec. 1965.
McDonnell Aircraft Corp. "Configuration Document for Gemini Spacecraft Number 1, Preliminary." MAC Report 8611-1, St. Louis, 1 Oct. 1962; rev. 28 March 1963.
———. "DOD/NASA Gemini Experiments Study." MAC Report A358 (Interim Report SSD-TOR-63-406). St. Louis, 24 Jan. 1964.
———. "Gemini Spacecraft Cost and Delivery Proposal." MAC Report 8791. St. Louis, 18 April 1962.
———. "Mercury Mark II Detail Specification." MAC Report 8356. St. Louis, 15 Nov. 1961.
———. "Project Gemini: Mission 'O' Plan." St. Louis, 12 July 1962.
———. "Project Gemini Mission Plan: Spacecraft No. 1." St. Louis, 14 Sept. 1962.
———. "Spacecraft Prelaunch Test Procedure: Outline for the Rendezvous Mission Countdown." MAC SEDR RMC-6. St. Louis, 23 Oct. 1965.
McLaughlin, Phillip. "Spin Up Studies for the Gemini-Agena System in a Tethered Configuration." MAC Gemini Design Note 356. St. Louis, 23 March 1966.
Martin Co. "Program Plan." ER 12255. Baltimore, April 1962.
———. "Reliability Test Plan." ER 12258. Baltimore, 15 June 1962.
———. "Review of Requirements for a Restrained Firing Program." LV-114. Baltimore, 24 Sept. 1962.
Meintel, Alfred J.; Long, Edward R.; and Pennington, Jack E. "GT-5 Piloted Simulation." Paper presented at MSC Meeting on Orbital Mechanics Associated with Gemini Flights, Houston, 11 Aug. 1965.
Mueller, Donald D. "Zero Gravity Indoctrination for the Gemini/Apollo Astronauts." Aviation Medical Research Laboratory Memo P-31. Johnsville, Pa., March 1963.
Newman, Frank B. "Lesson Plan and Handout, Special Technical Data, Gemini X Experiments." MAC SCD M35. St. Louis, 17 June 1966.
Noeggerath, Wulfgang C. "Symposium on Hypergolic Rocket Ignition at Altitude, Lockheed Mis-

siles & Space Company, 12 to 13 November 1965." Sunnyvale, Calif., n.d.

North American Aviation, Inc. "Business Management Proposal for a Paraglider Landing System." SID 63-606-2. Downey, Calif., 27 May 1963.

———. "Management Proposal for a Paraglider Recovery System for Gemini Spacecraft." SID 63-46-2. Downey, Calif., 31 Jan. 1963.

———. "Technical Proposal for a Paraglider Landing System." SID 63-606-1. Downey, Calif., 27 May 1963.

———. "Technical Proposal for a Paraglider Recovery System for Gemini Spacecraft." SID 63-46-1. Downey, Calif., 31 Jan. 1963.

Pennington, Jack E. "Range Estimation Studies Using Only Apparent Object Size." Paper presented at MSC Meeting on Orbital Mechanics Associated with Gemini Flights, Houston, 11 Aug. 1965.

Pippert, Elvin B., Jr., and Holloway, Tommy W. "Gemini XII Flight Plan, Final." Houston, 20 Oct. 1966.

———; Rivers, J. V.; and Holloway, Tommy W. "Gemini X Flight Plan, Final." Houston, 22 June 1966.

Roberts, John E., Jr. "Agena-B Requirements for Advanced Mercury Rendezvous Mission." Advanced Mercury Note. Langley, Va., 19 Dec. 1961.

Suler, Frank J., and Weber, Bobbie D. "A Proposed Mission Plan for the First Manned Gemini Flight (GT-3) Utilizing a Retrograde Maneuver Prior to Return." MSC Internal Note 65-FM-11. Houston, 9 Feb. 1965.

U.S., Air Force Systems Command. "Joint Titan II/Gemini Development Plan on Missile Oscillation Reduction and Engine Reliability and Improvement." Andrews AFB, Md., 5 April 1963; rev. 7 May and 7 Aug. 1963.

———, Space Systems Division, "Technical Development Plan for DOD/NASA Gemini Experiments, 631A." Los Angeles, 23 Sept. 1963.

U.S., NASA, Manned Spacecraft Center. "Gemini 9 Mission, Public Information Operations." MSC Working Paper. Houston, 5 April 1966.

———, Flight Crew Operations Div. (FCOD). "Composite Air-to-Ground and Onboard Voice Transcription of the GT-4 Mission." Gemini Working Paper 5035. Houston, 31 Aug. 1965.

———, FCOD. "Gemini Extravehicular Activity Program: Mission Planning, Crew Procedures and Training." Houston, 30 Jan. 1964.

———, FCOD. "GT-3 Flight Crew Self-Debriefing." Gemini Working Paper 5026. Houston, 3 June 1965.

———, FCOD. "GT-3 Flight Crew Technical Debriefing." Gemini Working Paper 5025. Houston, 3 June 1965.

———, FCOD. "Preflight Training Plan for Fourth Manned Gemini Flight Crew (GTA-6)." Gemini Working Paper 5031. Houston, 23 Aug. 1965.

———, Gemini Project (Program) Office. "GT-3 Mission Directive." Gemini Working Paper 5017A. Houston, 15 Feb. 1965.

———, GPO. "Gemini V Mission Directive." Gemini Working Paper 5028. Houston, 21 July 1965.

———, GPO. "Gemini VI Mission Directive." Gemini Working Paper 5037. Houston, 20 Sept. 1965.

———, GPO. "Gemini Program/Mission Directive." Gemini Working Paper 5039. Houston, 19 Nov. 1965.

———, GPO. "Program Plan for Gemini Extravehicular Operation." Houston, 31 Jan. 1964.

Weber, George J. "Dual Fuel Cell/Silver Zinc Battery Installation for 7 Orbit S/C 3 Gemini Mission." MAC Electrical Design Note 24. St. Louis, 25 Sept. 1963.

———; Waldner, J. C.; and Rogers, K. A. "Fuel Cell Interface Review." MAC Electrical Design Note 33. St. Louis, 16 Dec. 1963.

Weeks, W. M. "Aerodynamic Characteristics of the Gemini Ejection Seat-Man Configuration." MAC Aerodynamic Information Note 50. St. Louis, 28 Oct. 1963.

## 3. PRIMARY SOURCES: GEMINI REPORTS, REVIEWS, EVALUATIONS

Aerospace Corp. "Gemini Agena Target Vehicle 5004, NASA Mission Gemini IX Flight Safety Review at ETR." TOR-669(6183)-12. Los Angeles, 16 May 1966.

Agajanian, C. E. "Launch Vehicle Flight Evaluation Report, NASA Mission Gemini/Titan, GT-7." Aerospace Report TOR-669 (6126-42)-10. Los Angeles, February 1966.

General Dynamics Corp. "Atlas SLV-3, Space Launch Vehicle Flight Evaluation Report, SLV-3 5302." GDC/BKF66-012. San Diego, Calif., 17 June 1966.

———. "Atlas SLV-3, Space Launch Vehicle Flight Evaluation Report, SLV-5303." GDC/BKF66-029. San Diego, Calif., 27 June 1966.

Grimm, Dean F.; Stafford, Thomas P.; and Schirra, Walter M., Jr. "Report on Gemini VI Rendezvous." MSC, Houston, 28 Feb. 1966.

Hannay, N. B.; Biondi, F. J.; and Thomas, U. P. "Report on Fuel-Cell Work at General Electric and Pratt & Whitney." Bell Telephone Laboratories, Inc., Murray Hill, N.J., 1963.

Lockheed Missiles & Space Co. "GAATV Launch Demonstration with GT-5 GLV." LMSC-273407. Sunnyvale, Calif., 3 Sept. 1965.

———. "GATV Design Certification Report for Gemini VIII Mission." LMSC-A794903. Sunnyvale, Calif., 26 Feb. 1966.

———. "Gemini Agena Target Vehicle 5003, Systems Test Evaluation (45-Day Report)." LMSC-A817204. Sunnyvale, Calif., 5 May 1966.

———. "Gemini Agena Target Vehicle 5004 Systems Test Evaluation (45-Day Report)." LMSC-A819881. Sunnyvale, Calif., 30 June 1966.

———. "Project Sure Fire, GATV Engine Modification and Test Program: Final Report." 2 vols. LMSC-A818110 (SSD-546-66-10). Sunnyvale, Calif., 1 July 1966.

———. "Special Post Fire Functional and Leak Tests Performed on Model 8247 Rocket Engine at Santa Cruz Test Base." LMSC-A605708. Sunnyvale, Calif., 19 Oct. 1964.

Low, George M. "Latin American Tour with Astronauts Armstrong and Gordon, October 7-11, 1966." Houston, 16 Nov. 1966.

McDonnell Aircraft Corp. "Evaluation of Modified Flight Suit ('Shirt Sleeve') Configuration for the Gemini VII Mission." MAC Report B948. St. Louis, 9 Aug. 1965.

———. "Gemini VI Voice Communications (Air-to-Ground, Ground-to-Air and On-Board Transcription)." MAC Control No. C-115269. St. Louis, n.d.

———. "Gemini VII Voice Communications. . . ." MAC Control No. 115308. St. Louis, n.d.

———. "Gemini VIII Voice Communications . . . ." MAC Control No. C-115471. St. Louis, n.d.

———. "Gemini IX-A Voice Communications. . . ." MAC Control No. C-115803. St. Louis, n.d.

———. "Gemini X Voice Communications. . . ." MAC Control No. C-115883. St. Louis, n.d.

———. "Gemini XI Voice Communications. . . ." MAC Control No. C-115958. St. Louis, n.d.

———. "Gemini XII Voice Communications. . . ." MAC Control No. C-116106. St. Louis, n.d.

Martin Co. "Launch Vehicle No. 7 Flight Evaluation." ER 13227-7. Baltimore, January 1966.

———. "Launch Vehicle No. 8 Flight Evaluation." ER 13227-8. Baltimore, April 1966.

———. "Launch Vehicle No. 12 Flight Evaluation." ER 13227-12. Baltimore, December 1966.

North American Aviation, Inc. "Flight Test No. 6, Phase II, Part A, Paraglider Development Program." SID 62-1060-6. Downey, Calif., 10 Jan. 1963.

———. "Flight Test No. 7, Phase II, Part A, Paraglider Development Program." SID 62-1060-7. Downey, Calif., 1 Feb. 1963.

———. "Flight Test No. 8, Phase II, Part A, Paraglider Development Program." SID 62-1060-8. Downey, Calif., 25 March 1963.

———. "Paraglider Development Program, Phase I: Final Report." SID 61-266. Downey, Calif., 15 Aug. 1961.

Prause, R. H., and Goldman, R. L. "Longitudinal Oscillation Instability Study: POGO." Martin ER 13374. Baltimore, December 1964.

Schanz, J. L., and Bullock, E. K. "Gemini Fuel Cell Power Source—First Spacecraft Application." ARS Paper 2561-62. Presented at American Rocket Society, Power Systems Conference, Santa Monica, Calif., 25-28 Sept. 1962.

Schroeder, Lyle C., and Russo, Francis P. "Flight Investigations and Analysis of Alleviation of Communications Blackout by Water Injection during Gemini 3 Reentry." NASA TM X-1521. Langley, Va., 18 Aug. 1967.

Steele, A. J. "Summary of Alternate Failure Hypotheses for GATV 5002." Lockheed LMSC-A778486. Sunnyvale, Calif., 5 Jan. 1966.

U.S., Information Agency. "World Press Reaction to Gemini VI and VII." USIA Research and Reference Series R-189-65. Washington, 22 Dec. 1966.

U.S., NASA, Goddard Space Flight Center. "Manned Space Flight Network Performance Analysis for the First Gemini Mission." GSFC X-552-64-206. Greenbelt, Md., 1 May 1964.

U.S., NASA, Manned Spacecraft Center. "Interim Report, Manned Space Flight Experiments, Gemini X Mission, July 18-21, 1966." MSC TA-R-67-1. Houston, March 1967.

———. "Interim Report, Manned Space Flight Experiments, Gemini XI Mission, September 12-

15, 1966." MSC-TA-R-2. Houston, May 1967.

———. "Interim Report, Manned Space Flight Experiments, Gemini XII Mission, November 11-15, 1966." MSC-TA-R-67-3. Houston, August 1967.

———. "Manned Space Flight Experiments Symposium, Gemini Missions III and IV." Washington, 18-19 Oct. 1965.

———. "Manned Space Flight Experiments Symposium Interim Report: Gemini V Mission." Washington, 6 Jan. 1966.

———, GPO. "Gemini Program Mission Report for Gemini-Titan 1 (GT-1)." MSC-R-G-64-1. Houston, May 1964.

———, GPO. "Gemini Program Mission Report, GT-2, Gemini 2." MSC-G-R-65-1. Houston, February 1965.

———, GPO. "Gemini Program Mission Report, GT-3, Gemini 3." MSC-G-R-65-2. Houston, April 1965.

———, GPO. "Gemini Program Mission Report, Gemini IV." MSC-G-R-65-3. Houston, July 1965.

———, GPO. "Gemini Program Mission Report, Gemini V." MSC-G-R-65-4. Houston, October 1965.

———, GPO. "Gemini Program Mission Report, Gemini VI-A." MSC-G-R-66-2. Houston, January 1966.

———, GPO. "Gemini Program Mission Report, Gemini VII." MSC-G-R-66-1. Houston, January 1966.

———, GPO. "Gemini Program Mission Report, Gemini VIII." MSC-G-R-66-4. Houston, 29 April 1966.

———, GPO. "Gemini Program Mission Report, Gemini IX-A." MSC-G-R-66-6. Houston, n.d.

———, GPO. "Gemini Program Mission Report, Gemini X." MSC-G-R-66-7. Houston, August 1966.

———, GPO. "Gemini Program Mission Report, Gemini XI." MSC-G-R-66-8. Houston, October 1966.

———, GPO. "Gemini Program Mission Report, Gemini XII." MSC-G-R-67-1. Houston, January 1967.

———, GPO. "Rendezvous and Extravehicular Systems." Gemini Design Certification Report. Houston, February 1966.

U.S., NASA, Office of Manned Space Flight. "Mission Operations Report, Gemini Flight Number Eight." M-913-66-09. Washington, 3 March 1966.

———. "Mission Operations Report, Gemini XI Mission." OMSF M-913-66-13. Washington, September 1966.

## 4. PRINTED PRIMARY SOURCES

Croston, Ronald C., and Griffin, James B. "Manned Flight Simulations of the Air Force Modular Maneuvering Unit." AIAA Fourth Manned Space Flight Meeting, St. Louis, October 11-13, 1965. New York, 1965, pp. 118-26.

Dryden, Hugh L. "NASA Mission and Long-Range Plan." NASA-Industry Program Plans Conference, July 28-29, 1960. Washington, 1960, pp. 6-9.

Gatland, Kenneth W. "Rockets in Circular Orbits." Journal of the British Interplanetary Society 8 (March 1949): 52-59.

Koelle, Heinz H. "On the Development of Orbital Techniques: A Classification of Orbital Carriers and Satellite Vehicles." In IXth International Astronautical Congress, Amsterdam, August 1958, Proceedings, pp. 702-56. Vienna, 1959.

Rogallo, Francis M. "Parawings for Astronautics." Advances in the Astronautical Sciences 16, Part 2 (1963): 3-7.

———; Lowry, John G.; Croom, Delwin R.; Taylor, Robert T. Preliminary Investigation of a Paraglider. NASA Technical Note D-443. Langley, Va., August 1960.

Ross, H. E. "Orbital Bases," Journal of the British Interplanetary Society 8 (January 1949): 1-19.

Smith, R. A. "Establishing Contact between Orbiting Vehicles." Journal of the British Interplanetary Society 10 (November 1951): 295-99.

U.S., Congress, House, Committee on Appropriations, Subcommittee on Independent Offices. National Aeronautics and Space Administration Appropriations, Hearings. 86th Cong., 1st sess., 1959.

———. Independent Offices Appropriations for 1961, Hearings. 86th Cong., 2d Sess., 1960.

———. *Independent Offices Appropriations for 1966*, Part 2, *Hearings*. 89th Cong., 1st sess., 1965.

———. Committee on Science and Astronautics. *Missile Development and Space Sciences, Hearings*. 86th Cong., 1st sess., 1959.

———. *1960 NASA Authorization, Hearings on H. R. 6512*. 86th Cong., 1st sess., 1959.

———. *1961 NASA Authorization, Hearings on H.R. 10246*. 86th Cong., 2d sess., 1960.

———. *The Production of Documents by the National Aeronautics and Space Administration for the Committee on Science and Astronautics*. 86th Cong., 2d sess., 1960.

———. *Review of the Space Program, Hearings*. 86th Cong., 2d sess., 1960.

———. *1962 NASA Authorization, Hearings on H.R. 3238 and 6029 (Superseded by H.R. 6874)*. 87th Cong., 1st sess., 1961.

———. *Orbital Rendezvous in Space, Hearings*. 87th Cong., 1st sess., 23 May 1961.

———. *Space Orbital Rendezvous*. 87th Cong., 1st sess., 16 Aug. 1961, H. Rept. 909.

———. *1963 NASA Authorization, Hearings on H.R. 10100 (Superseded by H.R. 11737)*. 87th Cong., 2d sess., 1962.

———. *Report on the Activities of the Committee on Science and Astronautics*. 87th Cong., 2d sess., 1962.

———. *1964 NASA Authorization, Hearings on H.R. 5466 (Superseded by H.R. 7500)*. 88th Cong., 1st sess., 1963.

———. *1965 NASA Authorization, Hearings on H.R. 9641 (Superseded by H.R. 10456)*. 88th Cong., 2d sess., 1964.

———, Select Committee on Astronautics. *Astronautics and Space Exploration, Hearings on H.R. 1181*. 85th Cong., 2d sess., 1958.

U.S., Congress, Senate, Committee on Aeronautical and Space Sciences, Subcommittee on Governmental Organization for Space Activities. *Investigation of Governmental Organization for Space Activities, Hearings*. 86th Cong., 1st sess., 1959.

———, NASA Authorization Subcommittee. *NASA Supplemental Authorization for Fiscal Year 1959, Hearings on S. 1096*. 86th Cong., 1st sess., 1959.

———. *NASA Authorization for Fiscal Year 1960, Hearings on S. 1582*. 86th Cong., 1st sess., 1959.

———. *NASA Authorization for Fiscal Year 1961, Hearings on H.R. 10809*. 86th Cong., 2d sess., 1960.

———. *Transfer of von Braun Team to NASA, Hearings on H.R. Res. 567*. 86th Cong., 2d sess., 1960.

———. *NASA Scientific and Technical Programs, Hearings*. 87th Cong., 1st sess., 1961.

———. *NASA Authorization for Fiscal Year 1962, Hearings on H.R. 6874*. 87th Cong., 1st sess., 1961.

———. *NASA Authorization for Fiscal Year 1964, Hearings on S. 1245*. 88th Cong., 1st sess., 1963.

———. Committee on Appropriations. *Supplemental National Aeronautics and Space Administration Appropriations, 1960, Hearings on H. J. Res. 621*. 86th Cong., 2d sess., 1960.

———. *Independent Offices Appropriations, 1963, Hearings on H.R. 12711*. 87th Cong., 2d sess., 1962.

Wyatt, D. D. "The Rationale of the NASA Space Program." *Advances in the Astronautical Sciences* 6 (1961): xxix-xxx.

## 5. UNPUBLISHED SECONDARY SOURCES

Abbott, Ira H.A., "A Review and Commentary of a Thesis by Arthur L. Levine. . . ." Washington, April 1964.

Akens, David S. "Historical Origins of the George C. Marshall Space Flight Center." MSFC Historical Monograph No. 1, Huntsville, Ala., December 1960.

Ambrose, Mary Stone. "The National Space Program, Phase I: Passage of the 'National Aeronautics and Space Act of 1958.' "Master's thesis, American University, Washington, 1960.

Armstrong, Edward A., and Williams, John E. "Gemini Program Flight Summary Report." MSC-G-R-66-5. Houston, July 1966.

Bird, John D. "A Short History of the Development of the Lunar-Orbit-Rendezvous Plan at the Langley Research Center." Langley, Va., 6 Sept. 1963; supplemented 5 Feb. 1965, and 17 Feb. 1966.

Boorady, Frederick A., and Douglass, D. A. "Agena Gemini Rocket Engine Hypergolic Ignition—Hard Start Problem Solved during Project Surefire," Lockheed, Sunnyvale, Calif., n.d.

Boorady, Frederick A., and Salzman, Jerome. "Modification of the Agena Rocket Engine for Gemini Target Vehicle Multiple Restart Capability." Lockheed, Sunnyvale, Calif., September 1965.

Byers, Bruce K. "Lunar Orbiter: A Preliminary History." NASA Historical Note No. HHN-71. Washington, August 1969.

Corliss, William R. "The Evolution of the Manned Space Flight Network through Gemini." GSFC, Greenbelt, Md., 1 Dec. 1967.

Einhorn, Raymond. "Review of Business Management Activities at Rocketdyne. . . ." Western Region Audit Office Report WR 65-12. San Francisco, April 1965.

Emme, Eugene M. "Historical Perspectives on Apollo." NASA Historical Note No. 75. Washington, 24 Oct. 1967.

Gresser, Angela C. "Historical Aspects concerning the Redesignation of Facilities at Cape Canaveral." KSC Historical Note No. 1. Cape Kennedy, Fla., April 1962.

Hall, R. Cargill. "The Agena-Booster Satellite." Paper presented at AIAA Meeting, Houston, 2 Dec. 1966.

Harris, Howard T. "Gemini Launch Vehicle Chronology, 1961-1966." AFSC Historical Publications Series 66-22-1. Los Angeles, December 1966.

Harwood, William B., and Benson, Taylor G. "Means of Motivating the Individual Worker Involved in the Spacecraft Programs." Paper presented at NASA-Industry Information Procedures Conference, Houston, 22-23 April 1964.

Keller, Michael David. "A History of the NACA Langley Laboratory, 1917-1947." Ph.D. dissertation, University of Arizona, 1968.

Kuras, Alexander C., and Albert, John G. "Gemini-Titan Technical Summary." Cape Kennedy, Fla., 24 Jan. 1967.

Levine, Arthur L. "United States Aeronautical Research Policy, 1915-1958: A Study of the Major Policy Decisions of the National Advisory Committee for Aeronautics." Ph.D. dissertation, Columbia University, 1963.

Levine, Sol. "Man-Rating the Gemini Launch Vehicle." Paper presented at AIAA 1st Annual Meeting and Technical Display, Washington, 29 June—2 July 1964.

Lindley, Robert N. "Gemini Engineering Program, McDonnell Aircraft Corporation." Paper presented at Meeting of Institute of Management Sciences, Dallas, 16-19 Feb. 1966.

Lockheed Missiles & Space Co. "Gemini Agena Target Press Handbook." LMSC A766871. Sunnyvale, Calif., 15 Feb. 1966.

———. "Gemini Agena Target Vehicle Familiarization Handbook." LMSC-A602521. Sunnyvale, Calif., 1 April 1964.

Logsdon, John M. "NASA's Implementation of the Lunar Landing Decision." NASA Historical Note No. HHN-81. Washington, September 1968.

McDonnell Aircraft Corp. "Gemini Press Reference Book: Gemini Spacecraft Number Three." St. Louis, ca. March 1965; with rev. ed. for later spacecraft.

———. "NASA Project Gemini Familiarization Manual," Vol. I: "Long Range and Modified Configurations," and Vol. II: "Rendezvous and Docking Configurations." MAC SEDR 300. St. Louis, 15 March 1964; and later rev.

Malik, P. W., and Souris, G. A. "Gemini Final Summary Report." MAC Report F169. St. Louis, 20 Feb. 1967.

Martin Co. "Gemini Launch Vehicle Familiarization Manual." Baltimore, February 1965.

———. "Gemini-Titan II Air Force Launch Vehicle Press Handbook." Baltimore, ca. December 1964; and later rev.

Mathews, Charles W. "Gemini Summary." Paper presented at Gemini/Apollo Executives Meeting, Washington, 27-28 Jan. 1967.

Maus, Dennis R., and Younger, John R. "History of the 6511th Test Group (Parachute), 1 January - 30 June 1964." AFSC Historical Publications Series 64-110-V. El Centro, Calif., n.d.

Merrifield, Robert B. "Men and Spacecraft: A History of the Manned Spacecraft Center (1958-1969)." Houston, 1972.

North American Aviation, Inc. "A Final Fee Settlement Proposal for Contract NAS 9-1484." Downey, Calif., 18 March 1965.

———. "Final Report of Paraglider Research and Development Program, Contract NAS 9-1484." SID65-196. Downey, Calif., 19 Feb. 1965.

———. "Paraglider Landing System Test Program Final Report, Contract NAS 9-5206." SID 65-

1638. Downey, Calif., December 1965.

Oldeg, Harry W. "Gemini Program Management." AAS paper 66-158. Presented at American Astronautical Society, National Conference on the Management of Aerospace Programs, University of Missouri, Columbus, 16-18 Nov. 1966.

Rocketdyne Div., North American Aviation, Inc. "Gemini Propulsion by Rocketdyne—A Chronology." Canoga Park, Calif., 15 May 1967.

St. Clair, Joe D.; Armstrong, Edward A.; and Williams, John E. "Gemini Program Flight Summary Report: Gemini Missions I through X." MSC-G-R-66-5. Houston, September 1966.

Schwartz, David L. "EVA—The Story of a Team Effort by Civil Service Employees." Houston, ca. August 1965.

Thistle, Frank. "Rocketdyne: The First 25 Years . . ." Van Nuys, Calif., 1970.

TRW. "Gemini Rendezvous Summary." TRW Systems Group No. 05952-H281-RO-00 (MSC Internal Note No. 67-FM-128). Redondo Beach, Calif., 1 Nov. 1967.

U.S., Air Force, Arnold Engineering Development Center. "A Chronology of the Arnold Engineering Development Center." AFSC Historical Publications Series 62-101. Tullahoma, Tenn., ca. 1968.

———, 6555th Aerospace Test Wing, Gemini Launch Vehicle Div. "Launch Operations Technique, Manned Space Flight." Cape Kennedy, Fla., 22 Dec. 1966.

U.S., Department of Defense, Manager for Manned Space Flight Support Operations. "Summary Report: Department of Defense Support of Project Gemini, January 1963-November 1966." Patrick AFB, Fla., 6 March 1967.

U.S., General Accounting Office. "Report to the Congress of the United States: Review of the Gemini Spacecraft Fuel Cell Electrical Power Supply Systems." Washington, July 1966.

U.S., NASA, Goddard Space Flight Center. "The Manned Space Flight Tracking Network." Greenbelt, Md., 1965.

U.S., NASA, Kennedy Space Center. "History of Project Gemini (GT-2) at Kennedy Space Center." Cape Kennedy, Fla., n.d.

———. "Project Gemini (GT-3) Chronology of Technical Progress at Kennedy Space Center." Cape Kennedy, Fla., 7 May 1965.

U.S., NASA, Manned Spacecraft Center. "Manned Spacecraft Center, Houston, Interim Facilities." Houston, 1 Aug. 1962; rev. 15 Aug. 1963.

———, Historical Office. "Number of Holds, Manned Space Flight Launches." Houston, ca. 1964.

U.S., NASA, Marshall Space Flight Center. "Saturn Illustrated Chronology, April 1957—June 1964." Updated by Evelyn Falkowski. MSFC MHR-3. Huntsville, Ala., 10 Aug. 1964.

U.S., Navy, Naval History Office. "Recovery Mission Chronology [TF-140]." Washington, December 1966.

Wambolt, J[oseph] F., and Anderson, S[ally] F., coordinators. "Gemini Program Launch Systems Final Report: Gemini/Titan Launch Vehicle; Gemini/Agena Target Vehicle; Atlas SLV-3." Aerospace TOR-1001(2126-80)-3. El Segundo, Calif., January 1967.

Whitacre, Horace E. "A Development History Summary of the Agena Target Vehicle Primary Propulsion System." Houston, n.d.

Yardley, John F. "Spacecraft Check-Out." Paper presented at Gemini/Apollo Executives Meeting, Washington, 27-28 Jan. 1967.

Zeitler, Edward O., and Rogers, Thomas G. "The Gemini Program: Biomedical Sciences Experiments Summary." NASA TM X-58074. Houston, September 1971.

———. "The Gemini Program: Physical Sciences Experiments Summary." NASA TM X-58075. Houston, September 1971.

## 6. Published Secondary Sources

Allen, William H., ed. Dictionary of Technical Terms for Aerospace Use. NASA SP-7. Washington, 1965.

Caidin, Martin. Marooned. New York, 1964.

Doolittle, James H. "The Following Years, 1955-58." In Forty-Fourth Annual Report of the National Advisory Committee for Aeronautics, 1958 (Final Report), pp. 29-31. Washington, 1969.

Dornberger, Walter. "The German V-2." In Eugene M. Emme, ed., The History of Rocket Technology: Essays on Research, Development, and Utility, pp. 29-45. Detroit, 1964.

———. V-2. Translated by James Cleugh and Geoffrey Halliday. New York, 1954.

Emme, Eugene M. Aeronautics and Astronautics: An American Chronology of Science and Technology in the Exploration of Space, 1915-1960. Washington, 1961.

————. A History of Space Flight. New York, 1965.

Gray, George W. Frontiers of Flight: The Story of NACA Research. New York, 1948.

Griffith, Elisabeth Allison. The National Aeronautics and Space Act: A Study of the Development of Public Policy. Washington, 1962.

Grimwood, James M. Project Mercury: A Chronology. NASA SP-4001. Washington, 1963.

————, and Ertel, Ivan D. "Project Gemini." Southwestern Historical Quarterly 81 (January 1968): 393-418.

————; and Hacker, Barton C.; with Vorzimmer, Peter J. Project Gemini Technology and Operations: A Chronology. NASA SP-4002. Washington, 1969.

Hacker, Barton C. "The Genesis of Project Gemini: The Idea of Rendezvous, 1929-1961." In XIIth International Congress of the History of Science, Paris, 1968, Actes, Vol. X: Histoire des techniques, pp. 41-46. Paris, 1971.

————. "The Idea of Rendezvous: From Space Station to Orbital Operations in Space-Travel Thought, 1895-1951." Technology and Culture 15 (July 1974): 373-88.

Henry, James P. Biomedical Aspects of Space Flight. New York, 1966.

Hitch, Charles J., and McKean, Roland N. The Economics of Defense in the Nuclear Age. New York, 1965.

Houbolt, John C. "Lunar Rendezvous." International Science and Technology 14 (February 1963): 62-65.

Hunsaker, Jerome C. "Forty Years of Aeronautical Research." Forty-Fourth Annual Report of the National Advisory Committee for Aeronautics, 1958 (Final Report). Washington, 1959, pp. 3-27.

Johnston, Richard S.; Correale, James V.; and Radnofsky, Matthew I. Space Suit Development Status. NASA TN D-3291. Langley, Va., February 1966.

Kidd, Charles V. "Basic Research—Description versus Definition." In Norman Kaplan, ed., Science and Society, pp. 146-55. Chicago, 1965.

Koyre, Alexandre. Newtonian Studies. Cambridge, Mass., 1965.

Lasby, Clarence. Operation Paperclip: German Scientists in the Cold War. New York, 1971.

Ley, Willy. Rockets, Missiles, and Space Travel. 2d rev. ed. New York, 1961.

Lindley, Robert N. "Discussing Gemini: A 'Flight' Interview with Robert Lindley of McDonnell." FLIGHT International, 24 March 1966, pp. 488-89.

Logsdon, John M. The Decision to Go to the Moon: Project Apollo and the National Interest. Cambridge, Mass., 1970.

McComb, David G. Houston, the Bayou City. Austin, Tex., 1969.

MacNeice, Louis. Astrology. Garden City, N.Y., 1964.

Machell, Reginald M., ed. Summary of Gemini Extravehicular Activity. NASA SP-149. Langley, Va., 1967.

Malik, P. W., and Souris, G. A. Project Gemini: A Technical Summary. NASA CR-1106. Langley, Va., 1968.

Mallan, Lloyd. Peace Is a Three-Edged Sword. Englewood Cliffs, N.J., 1964.

Mansfield, Edwin. The Economics of Technological Change. New York, 1968.

Merton, Robert K. On the Shoulders of Giants: A Shandean Postscript. New York, 1965.

Morse, Mary Louise, and Bays, Jean Kernahan. The Apollo Spacecraft: A Chronology, Vol II: November 8, 1962—September 30, 1964. NASA SP-4009, Washington, 1973.

Oates, Stephen B. "NASA's Manned Spacecraft Center at Houston, Texas." Southwestern Historical Quarterly 67 (January 1964): 350-75.

Purser, Paul E.; Faget, Maxime A.; and Smith, Norman F., eds. Manned Spacecraft: Engineering Design and Operations. New York, 1964.

Richter, Henry L., Jr., ed. Instruments and Spacecraft, October 1957—March 1965. NASA SP-3028. Washington, 1966.

Rosholt, Robert L. An Administrative History of NASA, 1958-1963. NASA SP-4101. Washington, 1966.

Schoettle, Enid Curtis Bok. "The Establishment of NASA." In Sanford A. Lakoff, ed., Knowledge and Power: Essays on Science and Government, pp. 162-370. New York, 1966.

Sturm, Thomas A. The USAF Scientific Advisory Board: Its First Twenty Years, 1944-1964. Washington, 1 Feb. 1967.

Swenson, Loyd S., Jr; Grimwood, James M.; and Alexander, Charles C. This New Ocean: A History of Project Mercury. NASA SP-4201. Washington, 1966.

U.S., Congress, House, Committee on Science and Astronautics. Aeronautical and Astronautical

*Events of 1961.* 87th Cong., 2d sess., 7 June 1962.

———. *Astronautical and Aeronautical Events of 1962.* 88th Cong., 1st sess., 12 June 1963.

U.S., Congress, Senate, Committee on Aeronautical and Space Sciences. *Manned Space Flight Programs of the National Aeronautics and Space Administration: Projects Mercury, Gemini, and Apollo: Staff Report.* 87th Cong., 2d sess., 4 Sept. 1962.

———. *United States Astronauts: Staff Report.* 88th Cong., 1st sess., 1963, S. Doc. 42.

U.S., NASA. *Astronautics and Aeronautics, 1963: Chronology on Science, Technology, and Policy.* NASA SP-4004. Washington, 1964.

———. *Astronautics and Aeronautics, 1964:* . . . . NASA SP-4005. Washington, 1965.

———. *Astronautics and Aeronautics, 1965:* . . . . NASA SP-4006, Washington, 1966.

———. *Earth Photographs from Gemini III, IV, and V.* NASA SP-129. Washington, 1967.

———. *Earth Photographs from Gemini VI through XII.* NASA SP-171, Washington, 1968.

U.S., NASA, Manned Spacecraft Center. *Gemini Midprogram Conference, Including Experiment Results.* NASA SP-121. Washington, 1966.

———. *Gemini Summary Conference.* NASA SP-138. Washington, 1967.

Young, Warren R., ed. *To the Moon,* Sect. II: *The Story in Pictures and Text.* New York, 1969.

# Appendix A

# Glossary of Abbreviations and Acronyms

| | |
|---|---|
| AAS | American Astronautical Society |
| ABMA | Army Ballistic Missile Agency |
| AEDC | Arnold Engineering Development Center (Air Force test organization and facilities) |
| AEIP | Augmented Engine Improvement Program |
| AF | Air Force |
| AFB | Air Force Base |
| AFSC | Air Force Systems Command (formerly ARDC) |
| AGE | Aerospace ground equipment |
| AIAA | American Institute of Astronautics and Aeronautics |
| AMR | Atlantic Missile Range |
| AMRL | Aerospace Medical Research Laboratories |
| AMU | Astronaut maneuvering unit |
| A-OK | Everything in good working order |
| AOMC | Army Ordnance Missile Command |
| ARDC | Air Research and Development Command |
| ARPA | Advanced Research Projects Agency |
| AS | Apollo-Saturn (used until mid-1967 as Apollo mission designations; e.g., AS-201) |
| ASPO | Apollo Spacecraft Program Office (NASA-MSC, Houston) |
| ATC | Air Training Command |
| ATDA | Augmented target docking adapter |
| ATV | Agena target vehicle |
| BAC | Bell Aerosystems Company |
| Ballute | Balloon parachute |
| BSD | Ballistic Systems Division, Air Force |
| C & C | Command and communications (Agena system) |
| CapCom | Capsule communicator |
| CCP | Contract change proposal |
| c.g. | Center of gravity |
| CR | Contractor report |
| CRB | Capsule Review Board |
| CSAT | Combined systems acceptance test |
| CSM | Command and service modules (Apollo) |
| CSQ | *Coastal Sentry* Quebec tracking ship |

| | |
|---|---|
| D- | Department of Defense experiments prefix |
| DCR | Design certification review |
| DEI | Design (development) engineering inspection |
| DOD | Department of Defense |
| E and D | Engineering and Development |
| ECS | Environmental control system |
| EEG | Electroencephalogram |
| ELSS | Extravehicular life support system |
| EP | Educational Publication (NASA) |
| EPO | Experiments Program Office (also EXPO) |
| ER | Engineering Report |
| ESP | Extravehicular support package |
| ETR | Eastern Test Range (Air Force) |
| EVA | Extravehicular activity |
| EXPO | Experiments Project Office (also EPO) |
| FACI | First article configuration inspection |
| FO | Flight Operations |
| FOD | Flight Operations Division; Flight Operations Directorate |
| FOP | Financial operating plan |
| FSD | Flight Systems Division |
| FSRB | Flight Safety Review Board |
| g | Gravity |
| G and N, G&N | Guidance and navigation |
| GAO | General Accounting Office |
| GATV | Gemini Agena target vehicle |
| GDC | General Dynamics Convair (usually GD/C) |
| GE | General Electric |
| GEMSIP | Gemini Stability Improvement Program |
| g.e.t. | Ground elapsed time |
| G4C | Gemini pressure suit, 4th model, David Clark Company (Grissom and Young wore G3C suits, which were not capable of supporting extra-vehicular activity; Borman and Lovell wore G5C soft suits) |
| GIE | Ground interface equipment |
| GLV | Gemini launch vehicle |
| GOSS | Ground operational support system |

| | | | |
|---|---|---|---|
| GPO | Gemini Project/Program Office | MCC | Mission Control Center |
| GPPB | Gemini Program Planning Board | MDS | Malfunction detection system |
| GSFC | Goddard Space Flight Center (NASA) | MFS | Mission flight schedule |
| GT | Gemini-Titan | MHR | Marshall Historical Report |
| GTA | Gemini-Titan-Agena | MIT | Massachusetts Institute of Technology |
| HHMU | Handheld maneuvering unit | MK, Mk | Mark |
| HHN | Headquarters Historical Note (NASA) | MOCR | Mission Operations Control Room |
| HSTV | Half-scale test vehicle | MODS | Manned orbital development system |
| IBM | International Business Machines | MOL | Manned Orbiting Laboratory |
| ICBM | Intercontinental ballistic missile | MORAD | Manned Orbital Rendezvous and Docking |
| IESD | Instrumentation and Electronic Systems Division, MSC | MR | Mercury-Redstone |
| IPST | Israel Program for Scientific Information | MSC | Manned Spacecraft Center; also prefix for technological experiments |
| IVAR | Insertion velocity adjust routine (spacecraft maneuver) | MSF | Manned space flight |
| JPL | Jet Propulsion Laboratory | MSFC | George C. Marshall Space Flight Center (NASA) |
| K | Kelvin (measurement of temperature—metric system) | MSFEB | Manned Space Flight Experiments Board |
| KSC | John F. Kennedy Space Center | NACA | National Advisory Committee for Aeronautics |
| laser | Light Amplification by Stimulated Emission of Radiation | NASA | National Aeronautics and Space Administration |
| LEM | Lunar excursion module (later changed to LM—lunar module—Apollo) | n.d. | No date |
| | | NORAD | North American Air Defense |
| LEO | Large Earth Orbit | NTIS | National Technical Information Service, Department of Commerce |
| LeRC | Lewis Research Center (NASA) | | |
| LLVPG | Large Launch Vehicle Planning Group | OAMS | Orbit attitude and maneuvering system |
| LMSC | Lockheed Missiles & Space Company | OMSF | Office of Manned Space Flight, NASA |
| LORS | Lunar optical rendezvous system | POISE | Panel on In-Flight Scientific Experiments |
| LRC | Langley Research Center (NASA) | PSAC | President's Science Advisory Committee |
| L/V | Launch vehicle | rad | Unit of absorbed dose of radiation equal to an energy of 100 ergs per gram of irradiated material |
| LVS | Launch vehicle systems | | |
| LVSO | Launch Vehicle Systems Office | | |
| M- | Medical experiments prefix | | |
| M = 1, = 2, etc. | Revolution in which rendezvous is to take place | REP | Rendezvous evaluation pod |
| | | RFP | Request for proposal |
| MA | Mercury-Atlas | RKV | *Rose Knot* Victor tracking ship |
| MAC | McDonnell Aircraft Corporation | ROM | Rough order of magnitude |
| | | RR | Rendezvous radar |
| MALLIR | Manned Lunar Landing Involving Rendezvous | S- | Science experiments prefix |
| | | Saint | Satellite interceptor |
| | | SAR | Spacecraft acceptance review |
| | | S/c | Spacecraft |

| | |
|---|---|
| SCD | Specification control drawing |
| SEDR | Spacecraft Engineering Design (Development) Report |
| sep | Separation |
| SEPC | Space Exploration Program Council |
| SID | Space & Information Systems Division, North American Aviation, Inc. |
| SLV | Standard (Atlas) launch vehicle |
| SNORT | Supersonic Naval Ordnance Research Track |
| Sope | Simulated off-the-pad ejection |
| SP | Special Publication (NASA) |
| SPS | Secondary propulsion system (Agena) |
| SSD | Space Systems Division, Air Force |
| STG | Space Task Group |
| STL | Space Technology Laboratories |
| T- | Aircraft designation (T-38—trainer-38); also technical experiments prefix |
| TCA | Thrust chamber assembly |
| TDA | Target docking adapter |
| TETS | Thursday Evening Tanking Society |
| TM | Technical memorandum |
| TN | Technical note |
| TPI | Terminal phase initiation |
| TRW | Thompson Ramo Wooldridge Inc. |
| TT | Technical translation |
| TWX | Teletype message |
| UV | Ultraviolet |
| VAT | Vehicle acceptance team |
| VHF | Very high frequency |
| WETS | Wednesday Evening Tanking Society |
| WRE | Weapons Research Establishment, Department of Supply, Commonwealth of Australia |
| WSMR | White Sands Missile Range |

Appendix B
Flight Data Summary

Flight Data Summary

| Mission | Description | Primary Objectives | Result | Secondary Objectives | Result |
|---|---|---|---|---|---|
| *Gemini 1*<br>8 April 1964 | Unmanned, not recovered (mission terminated after 3 orbits), reentered 12 April, during 64th revolution; launch time 11:00:01.69 a.m., e.s.t.; apogee, 320.3 km (173 n.m.), perigee, 160.3 km (86.6 n.m.). | Demonstrate GLV performance; flight-qualify subsystems | A | Evaluate operational procedures for GLV trajectory and cutoff conditions | A |
| | | Determine exit heating on GLV and spacecraft | A | Verify orbital insertion by tracking C-band transponder in spacecraft | A |
| | | Demonstrate structural integrity of GLV and spacecraft | A | Demonstrate performance of launch and tracking networks | A |
| | | Demonstrate GLV and ground guidance systems performance in achieving proper orbital insertion | A | Provide training for flight controllers and prelaunch and launch crews and facilities | A |
| | | Monitor, evaluate GLV switchover circuits | A | | |
| *Gemini 2*<br>19 Jan. 1965 | Unmanned, suborbital; launch, 9:03:59.861 a.m., e.s.t.; altitude, 171.1 km (92.4 n.m.); range, 3422.4 km (1848 n.m.); duration, 18 mins, 16 secs.; landing, 16°36'N. 49°46'W; miss distance (from planned landing point), 62.9 km (34 n.m.). | Demonstrate reentry heat protection during maximum heating reentry | A | Obtain test results on fuel cell and reactant supply, cryogenics, and communications systems | PA[1] |
| | | Demonstrate structural integrity of spacecraft | A | Demonstrate and further flight-quality GLV and spacecraft from countdown through insertion | A |
| | | Demonstrate satisfactory performance of major subsystems | A | | |
| | | Demonstrate checkout and launch procedures | A | | |

A—Achieved; PA—Partially achieved; NA—Not achieved

[1] Fuel cell deactivated before liftoff

523

*Flight Data Summary—Continued*

| Mission | Description | Primary Objectives | Result | Secondary Objectives | Result |
|---|---|---|---|---|---|
| | | Evaluate backup guidance steering signals through launch | A | Train flight controllers and qualify ground communications and tracking systems | A |
| *Gemini 3* 23 March 1965 | Grissom and Young; 3 orbits; launch, 9:24:00.064 a.m., e.s.t.; highest apogee, 224 km (121 n.m.), lowest perigee, 158.5 km (85.6 n.m.); duration, 4 hours, 52 mins., 31 secs.; landing, 22°26' N, 70°51' W; miss distance, 111.1 km (60 n.m.) | Demonstrate manned orbital flight; evaluate two-man design | A | Evaluate flight crew equipment, biomedical instrumentation, and personal hygiene system | PA[3] |
| | | Demonstrate and evaluate tracking network | A | Perform 3 experiments | PA[4] |
| | | Demonstrate OAMS capability in orbital maneuvers and in retrofire backup | A | Evaluate low-level longitudinal oscillations (Pogo) of the GLV | A |
| | | Demonstrate controlled reentry and landing | PA[2] | General photographic coverage in orbit | PA[5] |
| | | Evaluate major spacecraft subsystems | A | | |
| | | Demonstrate systems checkout, prelaunch, and launch procedures | A | | |
| | | Demonstrate and evaluate recovery procedures and systems | A | | |
| *Gemini 4* 3-7 June 1965 | McDivitt and White; 62 revolutions; launch, 10:15:59.562 a.m., | Evaluate effects of prolonged space flight | A | Demonstrate and evaluate EVA and control by use of HHMU and tether | A |
| | | Demonstrate and evaluate | PA[6] | | |

*Flight Data Summary—Continued*

| Mission | Description | Primary Objectives | Result | Secondary Objectives | Result |
|---|---|---|---|---|---|
| | e.s.t.; highest apogee, 296.1 km (159.9 n.m.), lowest perigee, 159.4 km (86.1 n.m.); duration, 97 hours, 56 mins, 12 secs.; EVA time, 36 mins.; landing, 27°44'N, 74°11'W; miss distance, 81.4 km (44 n.m.) | performance of spacecraft and systems in 4-day flight / Evaluate procedures for crew rest and work cycles, eating schedules, and realtime flight planning | A | Stationkeep and rendezvous with second stage of GLV | PA[7] |
| | | | | Evaluate spacecraft systems | A |
| | | | | Make in- and out-of-plane maneuvers | A |
| | | | | Further test OAMS retro backup capability | A |
| | | | | Perform 11 experiments | A |
| Gemini V 21-29 Aug. 1965 | Cooper and Conrad; 120 revolutions; launch, 8:59:59.518 a.m.. e.s.t.; highest apogee, 349.8 km (188.9 n.m.), lowest perigee, 161.8 km (87.4 n.m.); duration, 190 hours, 55 mins, 14 secs.; landing, 29°44'N, 69°45'W; miss distance, 170.3 km (92 n.m.) | Evaluate rendezvous G&N system with REP | NA[8] | Demonstrate controlled reentry guidance | NA[9] |
| | | Demonstrate 8-day capability of spacecraft and crew | A | Evaluate fuel cell | A |
| | | Evaluate effects of weightlessness for 8-day flight | A | Demonstrate all phases of guidance and control system operation needed for rendezvous | A |
| | | | | Evaluate capability of both crewmen to maneuver spacecraft to rendezvous | NA[10] |
| | | | | Checkout rendezvous radar | A |
| | | | | Execute 17 experiments | PA[11] |

[2] Angle of attack during reentry lower than expected
[3] Personal hygiene system only partially tested
[4] Operating mechanism failed on S-2, Synergistic Effect of Zero Gravity on Sea Urchin Eggs
[5] Improper lens setting on 16mm camera
[6] Computer-controlled reentry not flown because of inadvertent alteration of computer memory
[7] Separation and rendezvous not attempted because of fuel consumption
[8] REP rendezvous not attempted because of decision to power down fuel cells
[9] 89-mile overshoot caused by incorrect navigation coordinates transmitted to spacecraft computer by ground
[10] Depended on rendezvous with REP
[11] D-2, Nearby Object Photography, not conducted when REP rendezvous was canceled

## Flight Data Summary—Continued

| Mission | Description | Primary Objectives | Result | Secondary Objectives | Result |
|---|---|---|---|---|---|
| *Gemini VI-A*[12] 15-16 Dec. 1965 | Schirra and Stafford; 16 revolutions; launch, 8:37:26.471 a.m., e.s.t.; highest apogee, 311.3 km (168.1 n.m.), lowest perigee, 160.9 km (86.9 n.m.); duration, 25 hours, 51 mins., 24 secs.; landing, 23°35′N, 67°50′W; miss distance, 12.9 km (7 n.m.) | Rendezvous with *Gemini VII* | A | Perform closed-loop rendezvous in fourth orbit | A |
| | | | | Stationkeep with *Gemini VII* | A |
| | | | | Evaluate reentry guidance capability | A |
| | | | | Conduct visibility tests for rendezvous, using *Gemini VII* as target | A |
| | | | | Perform 3 experiments | PA[13] |
| *Gemini VII* 4-18 Dec. 1965 | Borman and Lovell; 206 revolutions; launch, 2:30:03.702 p.m., e.s.t.; highest apogee, 327.9 km (177.1 n.m.), lowest perigee, 161.4 km (87.2 n.m.); duration, 330 hours, 35 mins., 1 sec.; landing, 25°25.1′N, 70°6.7′W; miss distance, 11.8 km (6.4 n.m.) | Conduct 14-day mission and evaluate effects on crew | A | Provide target for *Gemini VI-A* | A |
| | | | | Stationkeep with VI-A and with second stage of GLV | A |
| | | | | Conduct 20 experiments | A |
| | | | | Evaluate lightweight pressure suit | A |
| | | | | Evaluate spacecraft reentry capability | A |
| | | | | Conduct systems tests | A |
| *Gemini VIII* 16-17 March 1966 | Armstrong and Scott; 7 revolutions; launch, 11:41:02.389 a.m., e.s.t.; highest apogee, 298.7 km (161.3 n.m.); lowest perigee, 159.8 km (86.3 n.m.); | Rendezvous and dock with GATV | A | Rendezvous and dock in 4th revolution | A |
| | | Conduct EVA | NA[14] | Perform docked-vehicle maneuvers | NA[14] |
| | | | | Evaluate systems and conduct 10 experiments | PA[14] |

*Flight Data Summary—Continued*

| Mission | Description | Primary Objectives | Result | Secondary Objectives | Result |
|---|---|---|---|---|---|
| | duration, 10 hours, 41 mins., 26 secs.; landing, 25°13.8'N, 136°0'E; mission terminated early for electrical short in control system | | | Conduct docking practice and re-rendezvous | NA[14] |
| | | | | Evaluate auxiliary tape unit | A |
| | | | | Demonstrate controlled reentry | A |
| | | | | ParkGATV in 407.4-km (220-n.m.) circular orbit | A |
| *Gemini IX-A* [15] 3-6 June 1966 | Stafford and Cernan; 45 revolutions; launch, 8:39:33.335 a.m., e.s.t.; highest apogee, 311.5 km (168.2 n.m.), lowest perigee, 158.7 km (85.7 n.m.); duration, 72 hours, 20 mins, 50 secs.; EVA time, 2 hours, 7 mins.; landing, 27°52'N, 75°0.4'W; miss distance, 704 m (0.38 n.m.) | Rendezvous and dock. Conduct EVA | PA[16] A | Rendezvous with ATDA in 3rd revolution | A |
| | | | | Conduct systems evaluation and equi-period rendezvous | A |
| | | | | Execute 7 experiments | PA[17] |
| | | | | Practice docking | NA[16] |
| | | | | Rendezvous from above | A |
| | | | | Demonstrate controlled reentry | A |
| *Gemini X* 18-21 July 1966 | Young and Collins; 43 revolutions; launch, 5:20:26.648 p.m., e.s.t.; highest apogee, 753.3 km (412.2 n.m.), lowest | Rendezvous and dock with GATV | A | Rendezvous and dock in 4th revolution | A |
| | | | | Rendezvous with GATV 8, using Agena propulsion systems | A |

[12]GATV propulsion failure on 25 Oct. 1965; mission rescheduled

[13]Stationkeeping with *Gemini VII* interfered with conduct of D-8, Radiation in Spacecraft

[14]Mission terminated early

[15]Atlas failure on 17 May 1966, mission rescheduled

*Flight Data Summary—Continued*

| Mission | Description | Primary Objectives | Result | Secondary Objectives | Result |
|---|---|---|---|---|---|
|  | perigee, 159.8 km (86.3 n.m.); duration, 70 hours, 46 mins., 39 secs.; EVA time, 1 hour, 29 mins.; landing, 26°44.7'N, 71°57'W; miss distance, 6.2 km (3.4 n.m.) |  |  | Conduct EVA | A |
|  |  |  |  | Practice docking | NA[18] |
|  |  |  |  | Perform 14 experiments | PA[18] |
|  |  |  |  | Systems evaluation: bending-mode tests; docked maneuvers; static discharge monitoring; post-docked Agena maneuvers; reentry guidance; park GATV in 352.4 km (190.3 n.m.) orbit | A |
| *Gemini XI* 12-15 Sept. 1966 | Conrad and Gordon; 44 revolutions; launch, 9:42:26.546 a.m., e.s.t.; highest apogee, 1368.9 km (739.2 n.m.), lowest perigee, 160.3 km (86.6 n.m.); duration, 71 hours, 17 mins., 8 secs.; EVA time, 2 hours, 43 mins.; landing, 24°15.4'N, 70°0.0'W; miss distance, 4.9 km (2.65 n.m.) | Rendezvous and dock with GATV in 1st revolution | A | Practice docking | A |
|  |  |  |  | Perform EVA | A |
|  |  |  |  | Conduct 11 experiments | PA[19] |
|  |  |  |  | Maneuver while docked (high apogee excursion) | A |
|  |  |  |  | Conduct tethered vehicle test | A |
|  |  |  |  | Demonstrate automatic reentry | A |
|  |  |  |  | Park GATV 10 in 352.4 km orbit | A |

Flight Data Summary—Continued

| Mission | Description | Primary Objectives | Result | Secondary Objectives | Result |
|---------|-------------|--------------------|--------|---------------------|--------|
| *Gemini XII* 11-15 Nov. 1966 | Lovell and Aldrin; 59 revolutions; launch, 3:46:33.419 p.m., e.s.t.; highest apogee, 301.3 km (162.7 n.m.), lowest perigee, 160.7 km (86.8 n.m.); duration, 94 hours, 34 mins., 31 secs.; EVA time, 5 hours, 30 mins.; landing, 24°35′ N, 69°57′W; miss distance, 4.8 km (2.6 n.m.) | Rendezvous and dock Evaluate EVA | A A | Conduct tethered vehicle operation<br>Perform 14 experiments<br>Rendezvous and dock in 3rd revolution<br>Demonstrate automatic reentry<br>Perform docked maneuvers<br>Practice docking<br>Conduct systems tests<br>Park GATV 12 in 555.6 km (300 n.m.) orbit | A<br>A<br>A<br><br>A<br><br>A[20]<br>NA[20]<br>A<br>A<br>NA[21] |

[16] Docking impossible when shroud failed to jettison

[17] S-10, Agena Micrometeorite Collection, not attempted because EVA did take place near GATV

[18] Fuel consumption already too high

[19] D-16, Power Tool Evaluation, canceled when EVA was terminated early

[20] Docked maneuvers canceled because of a propulsion anomaly during GATV insertion

[21] GATV attitude control gas depleted by earlier maneuvers

This Appendix is based on Edward A. Armstrong and John E. Williams, "Gemini Program Flight Summary Report," MSC-G-R-66-5, July 1966, and Changes 1, November 1966, and 2, January 1967.

# Appendix C

# Astronaut Flight Assignments

# Astronaut Flight Assignments

| | Mercury | Gemini | Apollo | Skylab |
|---|---|---|---|---|
| *First Group (7) Selected April 1959* | | | | |
| M. Scott Carpenter | MA-6* MA-7 | | | |
| L. Gordon Cooper, Jr. | MA-8* MA-9 | V XII* | 10* | |
| John H. Glenn, Jr. | MR-3* MR-4* MA-6 | | | |
| Virgil I. Grissom | MR-3* MR-4 | III VI-A* | 1[1] | |
| Walter M. Schirra, Jr. | MA-7* MA-8 | III* VI-A | 7 | |
| Alan B. Shepard, Jr. | MR-3 MA-9* | | 14 | |
| Donald K. Slayton[2] | MA-7[3] | | | |
| *Second Group (9) Selected September 1962* | | | | |
| Neil A. Armstrong | | V* VIII XI* | 8* 11 | |
| Frank Borman | | IV* VII | 8 | |
| Charles Conrad, Jr. | | V VIII* XI | 9* 12 | 2 |
| James A. Lovell, Jr. | | IV* VII IX-A*[4] X*[4] XII | 8*[5] 11* 13 | |
| James A. McDivitt | | IV | 1*[1] 9 | |
| Elliot M. See, Jr. | | V* IX[6] | | |
| Thomas P. Stafford[2] | | III* VI-A IX-A*[6] | 7* 10 | |
| Edward H. White II | | IV VII* | 1[1] | |
| John W. Young | | III VI-A* X | 7* 10 13* 16 17*[7] | |
| *Third Group (14) Selected October 1963* | | | | |
| Edwin E. Aldrin, Jr. | | IX-A*[4] X*[4] XII | 8* 11 | |
| William A. Anders | | XI* | 8 11* | |
| Charles A. Bassett II | | IX[6] | | |
| Alan L. Bean[2] | | X*[8] | 9* 12 | 3 |
| Eugene A. Cernan | | IX-A*[6] XII* | 7* 10 14* 17 | |
| Roger B. Chaffee | | | 1[1] | |
| Michael Collins | | VII* X | 8[5] 11 | |
| R. Walter Cunningham | | | 7 | |
| Donn F. Eisele | | | 7 10* | |
| Richard F. Gordon, Jr. | | VIII* XI | 9* 12 15* | |
| Russell L. Schweickart | | | 1*[1] 9 | 2* |
| David R. Scott | | VIII | 1*[1] 9 12* 15 17*[7] | |
| Clifton C. Williams, Jr. | | X*[8] | | |
| *Fourth Group (6) Selected June 1965 (Scientist Astronauts)* | | | | |
| Owen K. Garriott | | | | 3 |
| Edward G. Gibson | | | | 4 |
| Joseph P. Kerwin | | | | 2 |
| Harrison H. Schmitt | | | 15* 17 | |

*Backup crew

## Astronaut Flight Assignments—Continued

| | Mercury | Gemini | Apollo | Skylab |
|---|---|---|---|---|
| *Fifth Group (19) Selected April 1966* | | | | |
| Vance D. Brand[2] | | | 15* | 3* 4* |
| Gerald P. Carr | | | | 4 |
| Charles M. Duke, Jr. | | | 13* 16 17*[7] | |
| Joe H. Engle | | | 14* | |
| Ronald E. Evans[2] | | | 14* 17 | |
| Fred W. Haise, Jr. | | | 11* 13 16* | |
| James B. Irwin | | | 12* 15 17*[7] | |
| Don L. Lind | | | | 3* 4* |
| Jack R. Lousma[2] | | | | 3 |
| Thomas K. Mattingly II | | | 13[9] 16 | |
| Bruce McCandless II | | | | 2* |
| Edgar D. Mitchell | | | 10* 14 16* | |
| William R. Pogue | | | | 4 |
| Stuart A. Roosa | | | 14 16* 17*[7] | |
| John L. Swigert | | | 13*[9] | |
| Paul J. Weitz | | | | 2 |
| Alfred M. Worden | | | 12* 15 17*[7] | |
| *Sixth Group (11) Selected August 1967 (Scientist Astronauts)* | | | | |
| William B. Lenoir | | | | 3* 4* |
| F. Story Musgrave | | | | 2* |

[1]Grissom, White, and Chaffee killed in fire on pad; flight canceled

[2]Stafford, Slayton, and Brand announced as crew of Apollo Soyuz Test Program, 30 Jan. 1973; Bean, Evans, and Lousma selected as backup crew

[3]Slayton removed from flight for heart condition, replaced by Carpenter

[4]Originally backup to *Gemini X*; moved to same position on *IX* when See and Bassett were killed

[5]Replaced Collins who had undergone surgery

[6]See and Bassett killed in aircraft accident, replaced by Stafford and Cernan

[7]When Irwin resigned on 23 May 1972, he, Worden, and Scott were replaced by Young, Roosa, and Duke

[8]Replaced original *Gemini X* backup crew

[9]Swigert replaced Mattingly, who had been exposed to measles, just before the flight

*Backup crew

# Appendix D

D-1. *Experiments by Flight*

| Experiment | Objective | Equipment | Result |
|---|---|---|---|
| *Gemini III* | | | |
| T-1, Reentry Communications | To see if fluid injected into ionized plasma during reentry would reduce blackout to the point where communications were possible | Water expulsion system on the inside surface of spacecraft right landing gear door; self contained except for activating switch in cabin; weighing about 39 kg (85 lb) | Increased C-band and UHF telemetry signals |
| S-2, Sea Urchin Egg Growth | To explore gravitational field effect on cells exposed to low gravity conditions | Metal cylinder containing 8 separate samples of sea urchin eggs, sperm, and a fixative solution; cylinder—8.2 × 17.1 cm (3.25 × 6.75 in), 721 gm (25.4 oz); handle on one end activated either fertilization or fixative | Not completed; handle broke near end of mission |
| S-4, Zero G and Radiation on Blood | To examine biological effects of radiation by measuring changes in human blood samples exposed to known quantity and quality of radiation | Radiation source, Phosphorus 32, housed in hermetically sealed aluminum box, 9.3 × 3.3 × 9.6 cm (3.7 × 1.3 × 3.8 in), weighing 0.45 kg (1 lb), and located inside cabin on right hatch; identical package operated | No apparent effect |

*Experiments by Flight*—Cont.

| Experiment | Objective | Equipment | Result |
|---|---|---|---|
| | | in laboratory at Cape Kennedy during flight agreed with flight findings | |
| *Gemini IV* | | | |
| M-3, Inflight Exerciser | To evaluate the general day-to-day physical condition of crew | A pair of rubber bungee cords attached to a nylon foot strap at one end and a nylon handle at the other | Little difference from preflight reactions to exercise |
| M-4, Inflight Phonocardiogram | To measure heart muscle deterioration against a simultaneous electrocardiogram | Heart sounds picked up by a microphone attached to each astronaut's chest and recorded on biomedical recorder | No significant changes from ground tests |
| M-6, Bone Demineralization | To investigate effects of prolonged weightlessness and immobilization associated with confinement for a period of days | X-rays taken before and after flight—especially the heel bone and the end bone of the fifth finger of the right hand of each crew member | Distinct losses in bone mass compared to bed-rested patients for the same time period |
| MSC-1, Electrostatic Charge | To detect and measure any accumulated electrostatic charge on the surface of the spacecraft | Electric field sensor, mounted in spacecraft retrograde section, controlled from cabin by a | Readings were higher than expected, but this was caused by sensitivity of instrument to other |

Experiments by Flight—Cont.

| Experiment | Objective | Equipment | Result |
|---|---|---|---|
| | | switch and weighing 0.81 kg (1.8 lb) | influences; sensor was modified for later flights |
| MSC-2, Proton-Electron Measurement | To measure radiation environment immediately outside spacecraft, correlate radiation measurements inside spacecraft, and predict radiation levels on future missions | Proton-electron measuring device mounted in equipment adapter section, with sensor face toward rear of spacecraft, operated by the pilot with a switch and weighing 5.6 kg (12.5 lb) | Operated completely successfully; all data telemetered to ground |
| MSC-3, Tri-Axis Magnetometer | To monitor direction and amplitude of Earth's magnetic field with respect to spacecraft | Tri-axis flux-gate magnetometer, consisting of an electronics unit and sensors, located in equipment adapter section, with sensors facing aft; sensors mounted on boom that could be extended beyond end of adapter; operated by the pilot with two switches (one to extend boom and other to activate both MSC-2 and 3) located in the cabin and weighing 1.5 kg (3.5 lb) | Successful; all data telemetered to ground |

*Experiments by Flight—Cont.*

| Experiment | Objective | Equipment | Result |
| --- | --- | --- | --- |
| MSC-10, Two-Color Earth Limb Photography | To determine if the Earth limb can be used in future guidance and navigation sightings | 70mm Hasselblad camera with black and white film and a special filter mosaic to allow each picture to be taken partly through a red and partly through a blue filter; the experimental film magazine weighed about 0.45 kg (1 lb) | 30 good pictures |
| S-5, Synoptic Terrain Photography | To get high-quality pictures of large land areas that have been previously well mapped by aerial photography for comparison and to serve as a standard for interpretation of pictures of unknown areas of Earth, the Moon, and other planets; to obtain high-quality photographs of relatively poorly mapped areas of Earth, to answer such questions as continental drift, structure of Earth's mantle, and overall structure of the continents | 70 mm modified Hasselblad camera, model 500C, with 55 frames per roll of film | 100 usable terrain study photographs |

Experiments by Flight—Cont.

| Experiment | Objective | Equipment | Result |
|---|---|---|---|
| S-6, Synoptic Weather Photography | To augment information from meteorological satellites; satellites usually take photos from altitude of 643.7 km (400 n.m.) or more; Gemini photos can be taken from altitudes of about 161 km (100 n.m.) | Camera used in MSC-10 and S-5, with color film | About 200 pictures, half of which were useful for weather studies |
| D-8, Radiation in Spacecraft | To measure radiation level and distribution inside spacecraft | 7 sensors inside spacecraft, 5 on wall of pressure vessel, 2 inside cockpit; 1 shielded to simulate amount of radiation crew received beneath skin; shield removed during pass through South Atlantic Anomaly | Radiation doses within acceptable levels |
| D-9, Simple Navigation | To gather information on phenomena that could be used for autonomous space navigation | Handheld sextant containing natural density, blue haze, and green emission filters; weighing 3.6 kg (8 lb) | Information good but statistical data lacking to evaluate |
| *Gemini V* | | | |
| M-1, Cardiovascular Conditioning | To determine effectiveness of pneumatic cuffs in preventing heart and blood | Pneumatic cycling system and a pair of venous cuffs worn on pilot's legs, alternatively | Cuffs (scheduled to work for full 8 days) stopped operating when oxygen in |

*Experiments by Flight*—Cont.

| Experiment | Objective | Equipment | Result |
|---|---|---|---|
| | distribution system deterioration induced by prolonged weightlessness | deflating and inflating to 80 mm of mercury | storage tank dropped below operational levels; limited results showed pilot's overall condition, postflight, better than commander's, with significantly less blood pooling in legs |
| M-3, Inflight Exerciser | See *Gemini IV* | See *Gemini IV* | See *Gemini IV* |
| M-4, Inflight Phonocardiogram | See *Gemini IV* | See *Gemini IV* | See *Gemini IV* |
| M-6, Bone Demineralization | See *Gemini IV* | See *Gemini IV* | Command pilot showed greater changes than bed-rested patients for same period; pilot showed equivalent changes to same patients |
| M-9, Human Otolith | To evaluate capability of astronaut to orient himself during flight; to measure changes in otolith (gravity gradient sensors in inner ear) functions | Special goggles, one eye piece containing light source in the form of movable white line; crewman positioned line with a calibrated screw to what he judged to be right pitch axis of space-craft | In general, coordinate space sense existed even in weightlessness if contact cues were adequate |

*Experiments by Flight*—Cont.

| Experiment | Objective | Equipment | Result |
|---|---|---|---|
| Cardiovascular Effects of Space Flight | To evaluate effects of prolonged weightlessness on the cardiovascular system (no number as it became operational procedure rather than experiment on future flights) | Comparison of preflight and postflight blood pressures, blood volumes, pulse rates, and electrocardiograms | On all flights, data revealed little change from preflight to postflight |
| MSC-1, Electrostatic Charge | See *Gemini IV* | See *Gemini IV* | Insufficient time after *Gemini IV* flight to modify instrument; shield placed on sensor had little effect and readings were high; measurement became operational procedure in rendezvous flights |
| S-1, Zodiacal Light Photography | To photograph the zodiacal light (in the west after twilight and in the east before sunrise), to try to determine its origin, minimum angle from Sun at which it could be studied without twilight interference, and whether the gegenschein could be | 35mm Widelux camera with high speed color film | 14 usable frames |

*Experiments by Flight*—Cont.

| Experiment | Objective | Equipment | Result |
|---|---|---|---|
| | detected and measured above the airglow layer | | |
| S-5, Synoptic Terrain Photography | See *Gemini IV* | See *Gemini IV* | 170 usable pictures, a large proportion of excellent quality |
| S-6, Synoptic Weather Photography | See *Gemini IV* | See *Gemini IV* | 250 excellent pictures |
| S-7, Cloud-Top Spectrometer | To measure altitude of clouds | 35 mm camera fitted with defraction grating and containing infrared film | Results good enough to warrant design of second generation weather satellite instrument |
| S8/D13, Visual Acuity and Astronaut Visibility | To test crew visual performance during flight and ability to detect and recognize objects on Earth's surface | Inflight vision tester—small, self-contained, binocular optical device with transilluminated array of 36 high contrast and low contrast rectangles, half oriented vertically and half horizontally; rectangle size, contrast, and orientation were random; presentation was sequential; and sequences were nonrepetitive; visual acuity | Crew showed no degradation of visibility during 8-day flight; land observations were partially obscured by weather conditions and fuel cell troubles; when weather was good, thruster problems prevented crew from orienting spacecraft properly, although smoke markers were sighted in each pass; during revolution 92, Texas site |

Experiments by Flight—Cont.

| Experiment | Objective | Equipment | Result |
|---|---|---|---|
| | | equipment consisted of inflight photometer to monitor spacecraft window, test patterns at 2 ground observation sites, and instrumentation for measuring atmosphere, lighting, and patterns | was glimpsed and photographed, and crew reported seeing this test area in revolution 107 |
| D-1, Basic Object Photography | To determine man's ability to acquire, track, and photograph objects in space | 35mm Zeiss contarex camera, mounted on pilot's window | Presented no problems |
| D-2, Nearby Object Photography | To obtain high resolut on pictures of orbiting object, while maneuvering, stationkeeping, and observing in a manual control mode | Same as D-1 | When rendezvous evaluation pod (REP) was abandoned, experiment could not be carried out |
| D-4/D-7, Celestial Radiometry and Space-Object Radiometry | To provide information on spectral analysis of regions of interest, supplied by star fields, principal planets, Earth and Moon, and other objects, such as satellites and REP | Radiometric measuring devices using common mirror optics that can measure radiant intensity from the ultraviolet through infrared as a function of wave lenghts— | 3 hrs and 10 min of data gathered—21 measurements of 30 objects; demonstrated advantages of using man to obtain basic data, thus permitting identification |

545

*Experiments by Flight*—Cont.

| Experiment | Objective | Equipment | Result |
|---|---|---|---|
| | | radiometer, interferometer, and cryogenic interferometers | and selection of target, choice of equipment mode, ability to track effectively, and augmenting, validating, and coordinating of data through on-the-spot voice comments |
| D-6, Surface Photography | To study problems associated with acquiring, tracking, and photographing terrestrial objects | Same as D-1 | Equipment performed successfully, but weather hampered much of the experiment, with some of the planned areas covered by clouds |
| *Gemini VI-A* | | | |
| S-5, Synoptic Terrain Photography | See *Gemini IV* | See *Gemini IV* | 28 fair to excellent pictures |
| S-6, Synoptic Weather Photography | See *Gemini IV* | See *Gemini IV* | 100 high quality pictures |
| D-8, Radiation in Spacecraft | See *Gemini IV* | See *Gemini IV*, with addition of removable brass shield on the tissue equivalent | On one run, the survey was performed by pilot, but command pilot was |

Experiments by *Flight*—Cont.

| Experiment | Objective | Equipment | Result |
|---|---|---|---|
| | | ionization chamber of command pilot's hatch | stationkeeping and failed to remove shield from sensor; on second run, both crewmen were busy stationkeeping; although additional data were attained, primary objectives were not achieved |
| *Gemini VII* | | | |
| M-1, Cardiovascular Conditioning | See *Gemini V* | See *Gemini V* | Operated for 311 hr, turned off 3 hr before reentry; significantly less blood pooling in pilot's postflight tests than in command pilot's |
| M-3, Inflight Exerciser | See *Gemini IV* | See *Gemini IV* | See *Gemini IV* |
| M-4, Inflight Phonocardiogram | See *Gemini IV* | See *Gemini IV* | Confirmed findings of *Gemini IV* and *V* |
| M-5, Bioassays of Body Fluids | To study astronaut reactions to stress | Intake and output of body fluids measured and analyzed preflight, inflight, and postflight | No gross changes noted |

Experiments by Flight—Cont.

| Experiment | Objective | Equipment | Result |
|---|---|---|---|
| M-6, Bone Demineralization | See *Gemini IV* | See *Gemini IV* | Significantly smaller loss in bone masses than in *Gemini IV* and *V*, probably because crew ate and exercised more and slept better and longer |
| M-7, Calcium Balance Study | To evaluate effects of 14-day flight on bones and muscles of crew | Intake and output of both fluid and solid matter (including perspiration) were measured and analyzed preflight, inflight, and postflight | Inflight urine collection was unsatisfactory because of leakage, 1 bag broken, and 4 not labeled; however, command pilot showed marked increase in calcium excretion starting on 8th day of flight |
| M-8, Inflight Sleep Analysis | To assess crew state of alertness, levels of consciousness, and depth of sleep during flight | Electroencephalograph recorded on biomedical recorder by 2 pair of scalp electrodes (command pilot only) | Results showed poor sleep on first night (expected since first night in strange surroundings usually disrupts sound sleep); until 54 hr, 20 min after liftoff, when sensors were dislodged, commander's sleep appeared normal after first night |
| M-9, Human Otolith Function | See *Gemini V* | See *Gemini V* | See *Gemini V* |

*Experiments by Flight*—Cont.

| Experiment | Objective | Equipment | Result |
|---|---|---|---|
| MSC-2, Proton-Electron Measurement | See *Gemini IV* | See *Gemini IV* | Erratic response in equipment indicated failure in proton mode—data inconclusive |
| MSC-3, Tri-Axis | See *Gemini IV* | See *Gemini IV* | Z-axis detector failed before launch; X- and Y-axis performed as expected during flight |
| MSC-4, Optical Communications | To evaluate optical communication system (laser), to check crew ability as pointing element, and to probe atmosphere, using an optical coherent radiator from outside atmosphere | Flight transmitter and ground-based receiver-transmitter system | Unfavorable cloud conditions and operating difficulties with the ground-based equipment yielded little data, but laser beacon visible from orbital altitudes |
| MSC-12, Landmark Contrast Measurement | To measure visual contrast of land-sea boundaries and other types of terrain for onboard Apollo guidance and navigation | Star occultation photometer—single-unit, dual-mode, handheld, externally powered instrument, 127 × 127 × 76.2 cm (5 × 5 × 3 in), weighing 1.1 kg (2½ lb), for measuring contrast of Sun-illuminated ground target and to determine extent to | No information because instrument malfunctioned |

549

*Experiments by Flight*—Cont.

| Experiment | Objective | Equipment | Result |
|---|---|---|---|
| | | which sight line to selected star penetrates planetary atmosphere | |
| S-5, Synoptic Terrain Photography | See *Gemini IV* | See *Gemini IV* | 250 useful pictures; cloud cover over many areas and dirty spacecraft windows accounted for poor quality of some |
| S-6, Synoptic Weather Photography | See *Gemini IV* | See *Gemini IV* | 240 exposures, some of which were not usable because of coating on windows |
| S8/D13, *Visual Acuity and Astronaut Visibility* | See *Gemini V* | See *Gemini V* | Patterns seen on revolutions 17 and 31; no apparent change in crew visual performance |
| D-4/D-7, Celestial Radiometry and Space-Object Radiometry | See *Gemini V* | See *Gemini V* (with minor variations) | 37 separate measurements taken; 3 hr, 6 min, 19 sec of data gathered, all satisfactory |
| D-5, Star-Occultation Navigation | To investigate feasibility and operational value of star occulting measurements in | See MSC-12 | No useful information because of instrument malfunction |

Experiments by Flight—Cont.

| Experiment | Objective | Equipment | Result |
|---|---|---|---|
|  | development of a simple, accurate, and self-contained navigational capability |  |  |
| D-9, Simple Navigation | See *Gemini IV* | See *Gemini IV* | 37 star-to-horizon, 5 planet-to-Moon (or star-to-Moon) limb, 6 star-to-star, and 8 zero measurements to stars were made; crew performance and equipment excellent |
| *Gemini VIII* |  |  |  |
| M-5, Bioassays of Body Fluids | See *Gemini VII* | See *Gemini VII* | Preflight and postflight samples obtained; one sample from command pilot before early end of flight |
| S-1, Zodiacal Light Photography | See *Gemini V* | See *Gemini V* | No results—mission terminated early |
| S-3, Frog Egg Growth | To study effects of subgravity on development in a | Two units, one mounted on each hatch sill structure, | First and second chambers activated correctly (though |

*Experiments by Flight—Cont.*

| Experiment | Objective | Equipment | Result |
|---|---|---|---|
| | biological system that is gravity oriented | each having four two-celled chambers, one for frog eggs and one for fixative; weight of each is 1.8 kg (4 lb); at 40 min after liftoff, pilot would turn handle, letting fixative (formalin) into righthand chambers 1 and 2, killing eggs and preserving them for microscopic study; fixative would be released into righthand chambers 3 and 4 at 2 hr, 10 min; two of the chambers on the lefthand side would be fixed just before reentry and the last two would be left alive for comparison | second was 15 min late); mission ended before time to activate the others |
| S-7, Cloud-Top Spectrometer | *See Gemini V* | *See Gemini V* | Mission ended early |
| S-9, Nuclear Emission | To study cosmic radiation at orbital altitudes | Nuclear emulsion package 21.5 × 15.2 × 7.6 cm (8.5 × 6 × 3 in), weighing 5.9 kg (13 lb) stowed in | Not recovered because of early end of mission |

*Experiments by Flight*—Cont.

| Experiment | Objective | Equipment | Result |
|---|---|---|---|
| | | spacecraft retrograde adapter; a spring loaded fairing would jettison at insertion, exposing package; EVA astronaut would retrieve | Mission ended early; no EVA |
| S-10, Agena Micrometeorite Collection | To collect plates that had been exposed to micrometeorite impact and debris and return them to Earth for study | Package, 13.9 × 15.8 × 2.5 cm (5.5 × 6.25 ×1 in), weighing 1.8 kg (4 lb), mounted on TDA of Agena; hinged to fold open and expose 8 plates of highly polished surfaces, such as metal, plastic, glass, etc., to be opened by pilot during EVA and left for retrieval on later mission | |
| D-3, Mass Determination | To determine technique and accuracy of direct contact method of measuring the mass of an orbiting object | No special equipment needed; after docking with the orbiting object, Gemini would push the docked combination with a known thrust; from the change in velocity of the orbiting object, its mass could be computed | Mission ended early; no docked maneuvers |

*Experiments by Flight*—Cont.

| Experiment | Objective | Equipment | Result |
|---|---|---|---|
| D-14, UHF/VHF Polarization | To obtain information on communication systems through the ionosphere | UHF/VHF transmitter with 2.4 m (8-ft) extendable antenna mounted on top centerline of retrograde adapter section | Mission ended early |
| D-15, Night Image Intensification | To develop system for night surveillance of Earth features | Image-orthicon camera, portable viewing monitor, recording monitor, 16mm camera, TV camera control unit, and equipment control unit; one crewman would look directly at scene, the other through the TV monitor; crew comments would later be compared with the scene as recorded on film | Mission ended early |
| D-16, Power-Tool Evaluation | To determine man's ability to perform specified work tasks under zero gravity and in pressurized suit | Minimum reaction, battery-powered tool, 27.1 cm (10.7 in) long, weighing 3.4 kg (7.6 lb), hand wrench, work plate with 7 nondetachable bolts (4 on face and 3 on reverse side), and knee tether; mounted | Mission ended early; no EVA |

Experiments by Flight—Cont.

| Experiment | Objective | Equipment | Result |
|---|---|---|---|
| *Gemini IX-A* | | in retro adapter to be operated by pilot during EVA | |
| M-5, Bioassay of Body Fluids | See *Gemini VII* | See *Gemini VII* | See *Gemini VII* |
| S-1, Zodiacal Light Photography | See *Gemini V* | See *Gemini V*; this time, however, experiment was planned for EVA | EVA ended early because of faceplate fogging; instead, 17 pictures were taken from inside spacecraft |
| S-10, Agena Micrometeorite Collection | See *Gemini VIII* | See *Gemini VIII* | EVA postponed to 3d day; not performed in vicinity of ATDA; experiment not retrieved |
| S-11, Airglow Horizon Photography | To photograph Earth's airglow in the atomic oxygen and sodium light spectra to study character and dynamics of upper atmosphere | 70mm Maurer camera, extended exposure timer, illuminated camera sight, and 2-point variable pitch bracket for mounting camera in pilot's window | 44 pictures, 3 of dayglow |
| S-12, Spacecraft Micrometeorite Collection | To determine micrometeorite activity in near-Earth environment; to expose microbiological specimens | Aluminum collection box, $27.9 \times 13.9 \times 3.1$ cm (11 $\times 5.5 \times 1.25$ in), weighing 2.9 kg (6 lb 8 oz), with two | Successfully recovered after exposure of over 16 hrs; penetration holes, some fractions of the biological |

*Experiments by Flight—Cont.*

| Experiment | Objective | Equipment | Result |
|---|---|---|---|
| | to space to determine survivability in vacuum, extreme temperatures, and radiation; and to search for any organisms capable of living on micrometeorites in space | collection compartments and an internal electric motor and thermally insulated batteries; one compartment to be sterilized for analysis to see if any non-terrestrial organisms are present; the other will contain bacteria, molds, and spores, to see if they survive space flight | organisms survived; no evidence of non-terrestrial organisms |
| D-12, Astronaut Maneuvering Unit | To provide EVA mobility and control in attitude and translation and to provide oxygen supply and communications | Rectangular aluminum back-pack weighing 75.2 kg (166 lb) fully loaded; 81.2 × 55.8 × 48.2 cm (32 × 22 × 19 in), with form-fitting cradle where pilot sits during flight; 4 forward- and 4 aft-firing thrusters, and 2 up- and down-firing; stores 10.8 kg (24 lb) of hydrogen peroxide; thrusters controlled by two sidearm supports; lefthand assembly gives translation control in 4 directions, a | When pilot got overheated and his faceplate fogged over repeatedly, EVA was called off before AMU exercise could be carried out |

Experiments by Flight—Cont.

| Experiment | Objective | Equipment | Result |
|---|---|---|---|
| | | switch for selecting manual or automatic stabilization, and volume control of communications; righthand arm contains controls for positioning pilot in pitch, roll, and yaw; also stores 3.4 kg (7.5 lb) oxygen and a battery-powered UHF transceiver to provide communications with spacecraft | |
| D-14, UHF/VHF Polarization | See *Gemini VIII* | See *Gemini VIII* | Performed 6 times; 3 more scheduled but antenna broken off by pilot during EVA; since limited number of measurements acquired, only partially successful |
| *Gemini X* | | | |
| MSC-3, Tri-Axis Magnetometer | See *Gemini IV* | See *Gemini IV* | Data not conclusive |
| MSC-5, Lunar UV Spectral Reflectance | To determine the UV spectral reflectance of the lunar surface between 2000 and 3200 angstrom | 70mm Maurer camera with UV lens | Canceled before flight when launch date slipped |

557

*Experiments by Flight—Cont.*

| Experiment | Objective | Equipment | Result |
|---|---|---|---|
| MSC-6, Beta Spectrometer | To predict as accurately as possible, for Apollo, radiation doses crews will be subjected to so degree of hazard can be assessed and preventive measures taken | Similar in function to proton-electron spectrometer used for MSC-4, but different in design; consists of 2 containers, one housing detector and analyzer system, the other, data processing system; total weight, 7.2 kg (16 lb) and located in retrograde section of spacecraft adapter; protected during launch by half-hinged door that is automatically jettisoned during separation from booster | Unexpectedly high fuel usage during first 2 days of mission eliminated controlled attitude passes; on 3d day, spacecraft was flown in tumbling mode through South Atlantic Anomaly, resulting in one good transversal of magnetic field; location of data points within anomaly were good and provided a good picture of the electron distributional direction |
| MSC-7, Bremsstrahlung Spectrometer | To measure the bremsstrahlung (braking radiation) flux-energy spectra inside spacecraft while passing through South Atlantic Anomaly | X-ray detection system mounted on the inner wall of pressurized cabin behind command pilot's seat about shoulder height and weighing less than 3.4 kg (7.5 lb) | Measurement of radiation is possible with this spectrometer |
| MSC-8, Color Patch Photography | To determine whether existing color film can take true-color pictures in space | Color patch/slate, 20.3 × 20.3 × .015 cm (8 × 8 × 1/16 in), supporting 4 color | Because of trouble with spacecraft ECS, EVA terminated after only 4 of |

Experiments by Flight—Cont.

| Experiment | Objective | Equipment | Result |
|---|---|---|---|
| | | targets (red, blue, yellow, and gray) in a matte finish ceramic; a 0.91-m (3-ft) extension rod to hold the patch 0.91 m in front of 70 mm Maurer Camera | the planned 9 pictures had been taken; color patch and rod were discarded; but enough data obtained to determine, by comparison of film and backup color patch, that commercial color films were suitable for photography in space |
| MSC-12, Landmark Contrast Measurement | See *Gemini VII* | See *Gemini VII* | Not performed, because of fuel-usage and time limitations |
| S-1, Zodiacal Light Photography | See *Gemini V* | See *Gemini V* | 20 pictures, difficult to use quantitively—film only half as sensitive as that used on *Gemini IX-A*; observations of same star field in various exposures shows that dirty windows cause variance in light transmission by a factor of at least 6; Earth horizon not seen in any of the pictures |

*Experiments by Flight*—Cont.

| Experiment | Objective | Equipment | Result |
|---|---|---|---|
| S-5, Synoptic Terrain Photography | See *Gemini IV* | See *Gemini IV* | Approximately 75 pictures, most of good quality, though some were affected by dirty spacecraft windows and others by cloud cover over areas photographed |
| S-6, Synoptic Weather Photography | See *Gemini IV* | See *Gemini IV* | Over 200 high quality pictures |
| S-10, Agena Micrometeorite Collection | See *Gemini VIII* | See *Gemini VIII* | Pilot recovered package from *Agena 8*; only four outer panels exposed, as package had been in closed position; micrometeorite-flux values agreed generally with known values from other experiments; microorganisms on exposed areas were dead, but those inside had good survival rates; pilot planned to leave similar package on *Agena 8* for future retrieval but did not want to risk getting his umbilical tangled on the |

Experiments by Flight—Cont.

| Experiment | Objective | Equipment | Result |
|---|---|---|---|
| S-12, Spacecraft Micrometeorite Collection | See *Gemini IX-A* | See *Gemini IX-A* | target vehicle<br><br>Pilot retrieved package during egress for EVA; it apparently floated up out of spacecraft later and was lost |
| S-13, Ultraviolet Astronomical Camera | To obtain data on UV radiation of hot stars and to develop and evaluate basic techniques for photography of celestial objects from manned spacecraft | 70mm Maurer camera, with UV lens; since spacecraft did not have UV windows, pictures would be taken through opened hatch | 22 frames exposed on southern Milky Way; 4 problems: 12 frames marred by vertical streak, probably caused by static electricity from camera operation in vacuum; poor image quality in center of field and good quality away from center, possibly resulted from film being too close to lens (bowing toward lens in vacuum); cable release broken during assembly of camera; and bracket screw backed out, preventing proper insertion into mount; considered successful, however, as it provided useful scientific |

Experiments by Flight—Cont.

| Experiment | Objective | Equipment | Result |
|---|---|---|---|
| | | | data and showed need for better equipment on future flights |
| S-26, Ion-Wake Measurement | To investigate ion and electron wake structure | Inboard and outboard ion detectors, electron detector, and data programmer on GATV adapter; inboard collected data when vehicle was parallel to flight path; outboard, when GATV was yawed at right angles to path; programmer sent realtime telemetry to ground during undocking—crucial since GATV delayed-time tape recorder inadvertently cut off during undocking | Limited results because of fuel usage; electron and ion temperatures higher than expected; registered shock effects during docking and undocking |
| D-5, Star-Occultation Navigation | See *Gemini VII* | See *Gemini VII* | Difficulty with attitude control while docked; only 5 stars tracked to total occultation (6 needed); undocked configuration, 7 stars tracked but problems encountered with entering visual occultation data into |

*Experiments by Flight—Cont.*

| Experiment | Objective | Equipment | Result |
|---|---|---|---|
| | | | computer; technique is accurate and flexible, useful for automatic, semiautomatic, or aided manual-navigation applications |
| D-10, Ion-Sensing Attitude Control | To investigate feasibility of attitude control system using environmental positive ions and an electrostatic detection system to measure spacecraft pitch and yaw | Two sensors, mounted on booms 0.91-m (3-ft) long, each 27.9 × 16.5 × 15.2 cm (11 × 6.5 × 6 in) and weighing 3.1 kg (7 lb), with 7 computed data points and operating at an angle of $\pm 15°$ | Comparison of system with inertial guidance system showed agreement in measurement of both pitch and yaw angles; response of system to variations in position was rapid, on the order of milliseconds |
| *Gemini XI* | | | |
| S-4, Zero G and Radiation Effects on Human Blood and *Neurospora* | See *Gemini III* | See *Gemini III*, with addition of bread mold *Neurospora* and thermoelectric cooler | Neither orbital space flight nor any stresses connected with it produced significant, unpredicted genetic damage, insofar as chromosomal aberration production is valid measure of this type of effect; no synergistic effect exists between radiation and factors associated with |

563

*Experiments by Flight—Cont.*

| Experiment | Objective | Equipment | Result |
| --- | --- | --- | --- |
| | | | space flight |
| S-5, Synoptic Terrain Photography | See *Gemini IV* | See *Gemini IV*; this time the crew would use the 70 mm general purpose Maurer camera as well | 145 pictures of excellent quality; all planned areas photographed plus some additional |
| S-6, Synoptic Weather Photography | See *Gemini IV* | See *Gemini IV*; both cameras used in this experiment as in S-5 | 180 good quality pictures |
| S-9, Nuclear Emulsion | See *Gemini VIII* | See *Gemini VIII* | Retrieved by EVA pilot; measurements and extrapolated results obtained higher in the atmosphere on very high altitude balloon flights are consistent with data from S-9 |
| S-11, Airglow Horizon Photography | See *Gemini IX-A* | See *Gemini IX-A* | 25 useful pictures; films show variations in altitude and intensity of airglow |
| S-13, Ultraviolet Astronomical Camera | See *Gemini X* | See *Gemini X*, except that carbon dioxide cartridge added to eliminate streaking | 39 frames exposed—5 excellent, 6 good, 8 fair, 13 poor, 2 bad, and 5 useless |

Experiments by *Flight*—Cont.

| Experiment | Objective | Equipment | Result |
|---|---|---|---|
| S-26, Ion-Wake Measurement | See *Gemini X* | See *Gemini X* | Radar, onboard voice tape recorder (for recording start and stop times), and auxiliary receptacle (to provide time markers) not operating; thruster firings in adapter-south configuration decrease ion flux to outboard sensor and increase it to inboard ion sensor and enhance electron concentration to outboard electron sensor; strip-chart data shows that definitive wake-cone angles can be determined; in many cases, electron distribution follows ion depletion effects, indicating wake is plasma rather than ion |
| S-29, Earth-Moon Libration Region Photography | To investigate L4 and L5 libration points of Earth-Moon system to determine possible existence of clouds of particulate matter orbiting Earth in these | 70mm Maurer camera | Because of 3-day mission delay, could not be carried out as planned; instead crew took pictures of gegenschein and 2 comets. |

*Experiments by Flight*—Cont.

| Experiment | Objective | Equipment | Result |
|---|---|---|---|
| S-30, Dim Light Photography/Orthicon | To obtain pictures of faint and diffuse astronomical phenomena, such as airglow layer in profile, brightest Milky Way, zodiacal light at 60° elongation, gegenschein, and libration points of Earth-Moon system regions | D-15 low light TV system plus spacecraft optical sight | 400 frames recorded; about 30 percent of film for D-15 and S-30 not exposed; camera recording cathode ray shorted out and failed during final sequence |
| D-3, Mass Determination | See *Gemini VIII* | See *Gemini VIII* | Successfully completed and method feasible, but additional statistical samples needed before system is adopted for use in future missions |
| D-15, Night Image Intensification | See *Gemini VIII* | See *Gemini VIII* | Of 42 sequences recorded, 13 were of medium to heavy cloud formations and 14 over open ocean areas; conclusions: cities easily identifiable by lights; cloud formations prominent, even at night, |

Experiments by *Flight*—Cont.

| Experiment | Objective | Equipment | Result |
|---|---|---|---|
| | | | as were lightning flashes, horizon and stars, and airglow; coastlines gave good to poor contrast; peninsulas were most significant geographic features seen; pilot stated that scenes viewed on the monitor were superior to film sequences of same features |
| D-16, Power Tool Evaluation | See *Gemini VIII* | See *Gemini VIII* | Not attempted because EVA ended early |
| *Gemini XII* | | | |
| MSC-3, Tri-Axis Magnetometer | See *Gemini IV* | See *Gemini IV* | Successfully accomplished; magnitude of geomagnetic fields, measured during 10th revolution, compared well with theoretically calculated magnitude, using McIlwain computer codes |
| MSC-6, Beta Spectrometer | See *Gemini X* | See *Gemini X* | Omnidirectional flux apparently consistent with |

Experiments by Flight—Cont.

| Experiment | Objective | Equipment | Result |
|---|---|---|---|
| | | | earlier measurements; representative electron spectra established apparent decay of artificially injected electrons (from Starfish high altitude nuclear test of July 1962) to such low levels that natural trapped electrons were becoming detectable |
| MSC-7, Bremsstrahlung Spectrometer | See *Gemini* X | See *Gemini* X | Crew turned equipment on and off 4 times for total of 32 hr; data indicated that electrons did penetrate spacecraft wall; bremsstrahlung-count-rate energy distribution was within reasonable estimation of such distributions |
| S-3, Frog Egg Growth | See *Gemini* VIII | See *Gemini* VIII, except that there was only 1 unit this time, mounted on pilot's hatch, instead of 2 | All phases of experiment performed, with good results; apparently gravitational field not necessary for eggs to divide normally, nor for |

Experiments by Flight—Cont.

| Experiment | Objective | Equipment | Result |
|---|---|---|---|
| | | | later stages of development |
| S-5, Synoptic Terrain Photography | See Gemini IV | See Gemini XI | 130 usable pictures, most with Hasselblad |
| S-6, Synoptic Weather Photography | See Gemini IV | See Gemini XI | 200 pictures show cloud patterns and are of excellent quality |
| S-10, Agena Micrometeorite Collection | See Gemini VIII | See Gemini VIII | EVA pilot removed protective fairing and exposed both interior and exterior collection surfaces; package left on GATV 12 for possible retrieval during later orbital flight |
| S-11, Airglow Horizon Photography | See Gemini IX-A | See Gemini IX-A | 23 good pictures of sunlight and night airglow |
| S-12, Spacecraft Micrometeorite Collection | See Gemini IX-A | See Gemini IX-A | Recovered after 6 hr, 20 min exposure; fewer penetration holes than on IX-A; no living organisms from space on sterile collection surfaces; confirmed sounding rocket findings that solar ultraviolet radiation and soft X-rays responsible for death of |

Experiments by Flight—Cont.

| Experiment | Objective | Equipment | Result |
|---|---|---|---|
| | | | microorganisms exposed to space |
| S-13, Ultraviolet Astronomical Camera | See *Gemini X* | See *Gemini XI* | 30 frames exposed—3 excellent, 7 good, 9 fair, 8 poor, 1 bad, and 2 lightstruck; troubles with focus, static marks, and light streaks persisted, but center images improved, indicating that increased tension of film-retaining spring eliminated warping (or bowing) of film |
| S-29, Earth-Moon Libration Region Photography | See *Gemini XI* | See *Gemini XI* | Of 11 pictures of L4, only 3 were properly exposed; mechanical failure of shutter mechanism in red-lens assembly caused over-exposure; unknown amount of double exposures caused by failure of film-advance at end of first roll; no conclusive results possible |
| S-51, Sodium Cloud Photography | To measure daytime wind velocity of Earth's high | 70mm Maurer camera | Crew did not see frings, but took 26 pictures of area |

Experiments by Flight—Cont.

| Experiment | Objective | Equipment | Result |
|---|---|---|---|
| | atmosphere as a function of altitude between 55.5 and 148.1 km (30 and 80 n.m.) by use of rocket-made vertical sodium clouds | | during firings; all were overexposed because camera shutter locked in open position |
| D-10, Ion-Sensing Attitude Control | See Gemini X | See Gemini X | Offered proof that it is possible to measure pitch and yaw to within fraction of a degree; could reduce time required for such maneuvers as docking, photography, and reentry (crew reduced time to align inertial platform from 40 min to 5 min by using pitch and yaw sensors as reference); could, with addition of horizon sensor, give complete description of spacecraft position and attitude; and could, with addition of servosystem, be used as complete automatic attitude control system applicable at altitudes of the lowest satellites up to at least 10 Earth radii. |

*Experiments by Flight*—Cont.

| Experiment | Objective | Equipment | Result |
|---|---|---|---|
| T-2, Manual Midcourse Space Navigation | To evaluate astronaut ability to make navigational measurements through handheld sextant | Line-of-sight optical sextant, 17.7 × 18.4 × 15.3 cm (7×7-¼ × 6-1/16 in), weighing 2.8 kg (6.25 lb) | Based on learning-curve data during initial period of familiarization and training, baseline data for comparison with flight results, and data obtained during flight; standard deviation of inflight measurements was ±9 arc sec, indicating that handheld sextant may be useful for navigational measurements during midcourse phase of lunar or interplanetary flight; pilot performance was the same in space as on the ground |

## D-2. Experiments by Number

| | Experiment | Principal Investigator | Affiliation | Missions |
|---|---|---|---|---|
| M-1 | Cardiovascular Conditioning | Lawrence F. Dietlein William V. Judy | NASA/MSC NASA/MSC | V, VII |
| M-3 | Inflight Exercise and Work Tolerance | Lawrence F. Dietlein Rita M. Rapp | NASA/MSC NASA/MSC | IV, V, VII |
| M-4 | Inflight Phonocardiogram | Lawrence F. Dietlein C. Vallbona | NASA/MSC Texas Institute for Rehabilitation and Research | IV, V, VII |
| M-5 | Biochemical Analysis of Body Fluids | Harry S. Lipscomb Elliot Harris Lawrence F. Dietlein | Baylor College of Medicine NASA/MSC NASA/MSC | VII, VIII, IX-A |
| M-6 | Bone Demineralization | Pauline Beery Mack | Texas Woman's University | IV, V, VII |
| M-7 | Calcium and Nitrogen Balance | G. Donald Whedon Leo Lutwak William F. Neuman | National Institutes of Health Cornell University University of Rochester | VII |
| M-8 | Inflight Sleep Analysis | Peter Kellaway | Baylor College of Medicine | VII |
| M-9 | Human Otolith Function | Robert L. Maulsby Ashton Graybiel | Baylor College of Medicine Naval Aerospace Medical Institute | V, VII |
| | | Earl F. Miller II | Naval Aerospace Medical Institute | |
| MSC-1 | Electrostatic Charge | Patrick E. Lafferty | NASA/MSC | IV, V |
| MSC-2 | Proton-Electron Spectrometer | James R. Marbach | NASA/MSC | IV, VII |
| MSC-3 | Tri-Axis Flux-Gate Magnetometer | William D. Womack | NASA/MSC | IV, VII, X, XII |
| MSC-4 | Optical Communications | Douglas S. Lilly | NASA/MSC | VII |
| MSC-5 | Lunar UV Spectral Reflectance | Roy C. Stokes | NASA/MSC | X |

*Experiments by Number—Continued*

| | Experiment | Principal Investigator | Affiliation | Missions |
|---|---|---|---|---|
| MSC-6 | Beta Spectrometer | James R. Marbach | NASA/MSC | X, XII |
| MSC-7 | Bremsstrahlung Spectrometer | Reed S. Lindsay, Jr. | NASA/MSC | X, XII |
| MSC-8 | Color-Patch Photography | John R. Brinkmann Robert L. Jones | NASA/MSC NASA/MSC | X |
| MSC-10 | Two-Color Earth Limb Photography | Max Petersen | Massachusetts Institute of Technology | IV |
| MSC-12 | Landmark Contrast Measurements | Charles E. Manry | NASA/MSC | VII, X |
| T-1 | Reentry Communications | Lyle C. Schroeder Theo E. Sims William F. Cuddihy | NASA/Langley NASA/Langley NASA/Langley | III |
| T-2 | Manual Navigation Sightings | Donald W. Smith Brent Y. Creer | NASA/Ames NASA/Ames | XII |
| D-1 | Basic Object Photography | AF Avionics Laboratory (H. T. Kozuma, monitor) | Wright-Patterson AFB | V |
| D-2 | Nearby Object Photography | AF Avionics Laboratory (H. T. Kozuma, monitor) | Wright-Patterson AFB | V |
| D-3 | Mass Determination | AFSC Field Office (Rudolph J. Hamborsky, monitor) | NASA/MSC-DOD | VIII, XI |

Experiments by Number—Continued

| | Experiment | Principal Investigator | Affiliation | Missions |
|---|---|---|---|---|
| D-4 | Celestial Radiometry | AF Cambridge Laboratory (Burden Brentnall, monitor) | USAF-Hanscom Field | V, VII |
| D-5 | Star-Occultation Navigation | Robert M. Silva Terry R. Jorris | AF Avionics Laboratory AF Avionics Laboratory | VII, X |
| D-6 | Surface Photography | AF Avionics Laboratory (H. T. Kozuma, monitor) | Wright-Patterson AFB | V |
| D-7 | Space Object Radiometry | AF Cambridge Laboratory | USAF-Hanscom Field | V, VII |
| D-8 | Radiation in Spacecraft | M. F. Schneider | AF Weapons Lab., Kirkland AFB | IV, VI-A |
| | | J. F. Janni | AF Weapons Lab., Kirkland AFB | |
| | | G. E. Radke | AF Weapons Lab., Kirkland AFB | |
| D-9 | Simple Navigation | Robert M. Silva Terry R. Jorris | AF Avionics Laboratory AF Avionics Laboratory | IV, VII |
| D-10 | Ion-Sensing Attitude Control | Rita C. Sagalyn | AF Cambridge Laboratory | X, XII |
| D-12 | Astronaut Maneuvering Unit | AFSC Field Office (Edward G. Givens, project officer) | NASA/MSC-DOD | IX-A |
| D-13 | Astronaut Visibility | Seibert Q. Duntley | University of California | V, VII |
| D-14 | Ultrahigh and Very High Frequency Polarization | Robert E. Ellis | US Naval Research Laboratory | VIII, IX-A |

*Experiments by Number—Continued*

| | Experiment | Principal Investigator | Affiliation | Missions |
|---|---|---|---|---|
| D-15 | Night Image Intensification | Thomas J. Shopple | Naval Air Development Center | VIII, XI |
| | | George F. Eck | Naval Air Development Center | |
| | | Albert R. Prince | Naval Air Development Center | VIII, XI |
| D-16 | Power-Tool Evaluation | AF Avionics Laboratory (Victor L. Etredge, monitor) | Wright-Patterson AFB | |
| S-1 | Zodiacal Light Photography | Edward P. Ney | University of Minnesota | V, VIII, IX-A, X |
| S-2 | Sea Urchin Egg Growth | Richard S. Young | NASA/Ames | III |
| S-3 | Frog Egg Growth | Richard S. Young | NASA/Ames | VIII, XII |
| S-4 | Radiation and Zero-G Effects on Blood and Neurospora | Michael A. Bender | Atomic Energy Commission | III, XI |
| S-5 | Synoptic Terrain Photography | Paul D. Lowman, Jr. | NASA/Goddard | IV, V, VI-A, VII, X, XI, XII |
| S-6 | Synoptic Weather Photography | Kenneth M. Nagler Stanley D. Soules | US Weather Bureau Environmental Science Services Administration | IV, V, VI-A, VII, X, XI, XII |
| S-7 | Cloud-Top Spectrometer | Fuad Saiedy | University of Maryland | V, VIII |

*Experiments by Number—Continued*

| | Experiment | Principal Investigator | Affiliation | Missions |
|---|---|---|---|---|
| S-8 | Visual Acuity | Seibert Q. Duntley | University of California | V, VII |
| S-9 | Nuclear Emulsion | Maurice M. Shapiro<br>Carl E. Fichtel | Naval Research Laboratory<br>NASA/Goddard | VIII, XI |
| S-10 | Agena Micrometeorite | Curtis L. Hemenway | Dudley Observatory | VII, IX-A, X, XII |
| S-11 | Airglow Horizon<br>Photography | Martin J. Kooman | Naval Research Laboratory | IX-A, XI, XII |
| S-12 | Micrometeorite Collection | Curtis L. Hemenway | Dudley Observatory | IX-A, X, XII |
| S-13 | Ultraviolet Astronomical<br>Camera | Karl G. Henize | Dearborn Observatory | X, XI, XII |
| S-26 | Ion-Wake Measurement | David B. Medved | Electro-Optical Systems, Inc. | X, XI |
| S-29 | Earth-Moon Libration<br>Regions Photography | Elliott C. Morris | US Geological Center | XII |
| S-30 | Dim Sky Photography/<br>Orthicon | Curtis L. Hemenway | Dudley Observatory | XI |
| S-51 | Daytime Sodium Cloud<br>Photography | Jacques-Emile Blamont | Centre National de la<br>Recherche Scientifique | XII |

Information for both D-1 and D-2 from Edward O. Zeitler and Thomas G. Rogers, compilers, "The Gemini Program: Biomedical Sciences Experiments Summary," NASA TM X-58074, September 1971; Zeitler and Rogers, compilers, "The Gemini Program: Physical Sciences Experiments Summary," NASA TM X-58075, September 1971; "A Review of Medical Results of Gemini 7 and Related Flights," OMSF, held at KSC, 23 Aug. 1966; press kits; interim experiments reports; midprogram report; final summary report; telephone conversations with Zeitler, R. Dean Bratton, Olav Smistad, and Lawerence F. Dietlein, all of JSC, and with Richard S. Young of Ames.

# Appendix E

# Costs

*E-1. Projected Program Cost Trend*

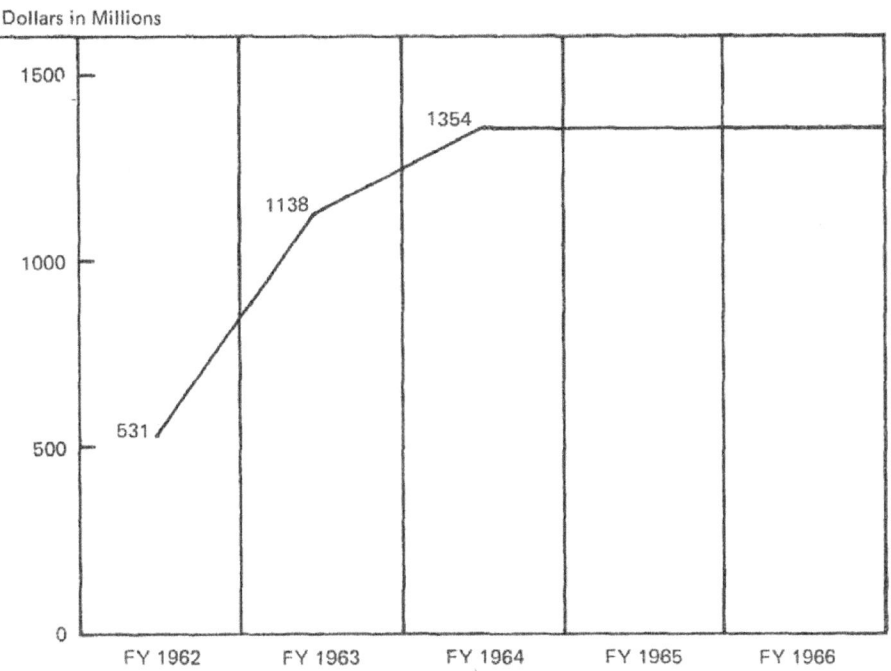

(MILLIONS OF DOLLARS) (as of approximately Nov. 1963)

E-2. *Manned Space Flight Schedule*
*Gemini Program (GPO Responsibility)*
*Total Accrued Cost — All Years*
(millions of dollars)

| | FY62 | FY63 | FY64 | FY65 | FY66 | FY67 | TOTAL |
|---|---|---|---|---|---|---|---|
| Spacecraft | 20.0 | 178.6 | 223.5 | 177.0 | 77.5 | 19.5 | 696.1 |
| Paraglider | -0- | 9.1 | 15.8 | 2.3 | -0- | -0- | 27.4 |
| Atlas | -0- | -0- | 4.7 | 10.1 | 11.4 | 4.9 | 31.1 |
| Agena | .9 | 14.5 | 26.1 | 21.7 | 31.1 | 5.8 | 100.1 |
| GLV | 12.8 | 71.2 | 77.5 | 60.9 | 39.6 | 21.3 | 283.3 |
| Support | -0- | .6 | .9 | 2.3 | 4.6 | .9 | 9.3 |
| Total | 33.7 | 274.0 | 348.5 | 274.5 | 164.2 | 52.4 | 1,147.3 |
| Cum | 33.7 | 307.7 | 656.2 | 930.7 | 1,094.9 | 1,147.3 | |

Information supplied by Stephen D. Armstrong, 3 Oct. 1973

Status as of 12-31-66

# Appendix F

# Worldwide Tracking Network

NETWORK FUNCTIONS:

Communications between network stations and control center
Tracking and control of two vehicles simultaneously
Voice and telemetry communications with spacecraft
Dual command data to two orbiting vehicles simultaneously
Reliability of all onsite systems for extended periods of time

NETWORK EQUIPMENT SYSTEMS:

Acquisition Aid
Radar Tracking
Telemetry
Remote Site Data Processors
Command
Communications
Consoles

REMOTE STATION CONSOLES:

Maintenance and Operations
Gemini and Agena Systems Monitors
Command Communicator
Aeromedical Monitor

MISSION CONTROL CENTER (MCC) FUNCTIONS:

Direct overall mission
Issue guidance parameters and monitor guidance computations
  and propulsion capability
Evaluate performance and capabilities of space vehicle equipment
  systems
Evaluate capabilities and status of spacecraft crew and life support
  systems
Direct and supervise activities of ground support systems
Direct recovery activities
Conduct simulation and training exercises
Schedule and regulate transmission of recorded data from sites
Support postmission analyses

MISSION CONTROL CENTER EQUIPMENT SYSTEMS:

Real-Time Computer Complex
Communications
Display
  Computer Interface Subsystem

Timing Subsystem
Television Subsystem
Group Display Subsystem
Console Subsystem
Command
Gemini Launch Data
Simulation Checkout and Training

MISSION OPERATIONS CONTROL ROOM (MOCR):

Fig. F-1 shows the location of the key personnel in the MOCR.

1. Mission Director—has overall mission responsibility and control of test operations, including scrubbing and rescheduling missions and making real time decisions on alternates when problems arise.

2. Department of Defense representatives—have overall control of DOD support forces, including deployment of recovery forces, operation of recovery communications network, and search, location, and retrieval of crew.

3. Public Affairs Officer—provides mission status information to the public.

4. Flight Director—handles detailed control of mission from liftoff through splashdown, takes over for Mission Director in his absence.

5. Assistant Flight Director—assists Flight Director and handles Flight Director's duties during his absence.

6. Network Controller—has detailed operational control of Ground Operational Support System (GOSS) network.

7. Operations and Procedures Officer—handles detailed implementation of MCC/GOSS mission control procedures.

8. Vehicle Systems Engineers—monitor and evaluate performance of all electrical, mechanical, and life support equipment aboard spacecraft and, during rendezvous missions, Agena.

9. Flight Surgeon—directs all operational medical activities and monitors crew status.

10. Spacecraft Communicator—handles voice communications with crew, exchanging information on the progress of the mission.

11. Flight Dynamics Officer—monitors and evaluates flight parameters required to achieve successful orbital flight; gives Go or No/Go recommendations to Flight Director.

12. Retrofire Officer—monitors impact prediction displays and determines retrofire times.

13. Guidance Officer—detects Stage I and Stage II booster slow-rate deviations and other programmed events, verifies proper performance of Gemini Inertial Guidance System, and recommends action to Flight Director.

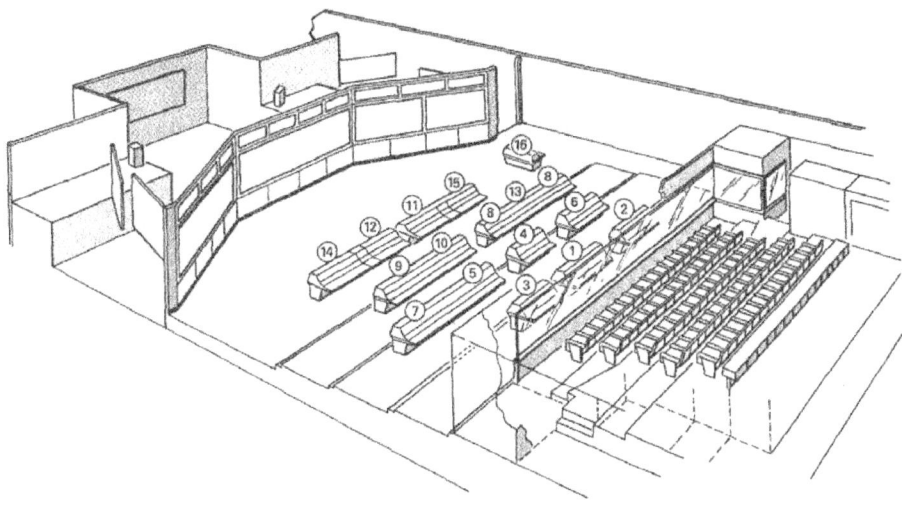

*Figure F-1. Mission Operations Control Room*

14. Booster Systems Engineer—monitors propellant tank pressurization systems and advises flight crew and Flight Director of systems abnormalities.

Experiments Officer—after booster/spacecraft separation replaces booster systems engineer and monitors and updates experiments until reentry.

15. Assistant Flight Dynamics Officer—monitors and evaluates Gemini launch vehicle systems and reports any abnormalities to Flight Director.

16. Maintenance and Operations Supervisor—monitors performance of MCC equipment and its ability to support mission in progress, sees that any problems are cleared up expeditiously.

Fig. F-2 shows the location of the Staff Support Rooms, where technical specialists analyze data and long-term performance-trends, compare trends with baseline data, and relay information and recommendations to MOCR personnel:

Flight Dynamics—monitor and evalute all aspects of powered flight that concern crew safety and orbital insertion, evaluate and recommend modification of trajectories to meet mission objectives, and investigate and study potential maneuver requirements and actual or potential contingency situations.

Vehicle Systems—monitor detailed status of trends of flight sys-

587

GEMINI

# MCC-H FIRST FLOOR PLAN

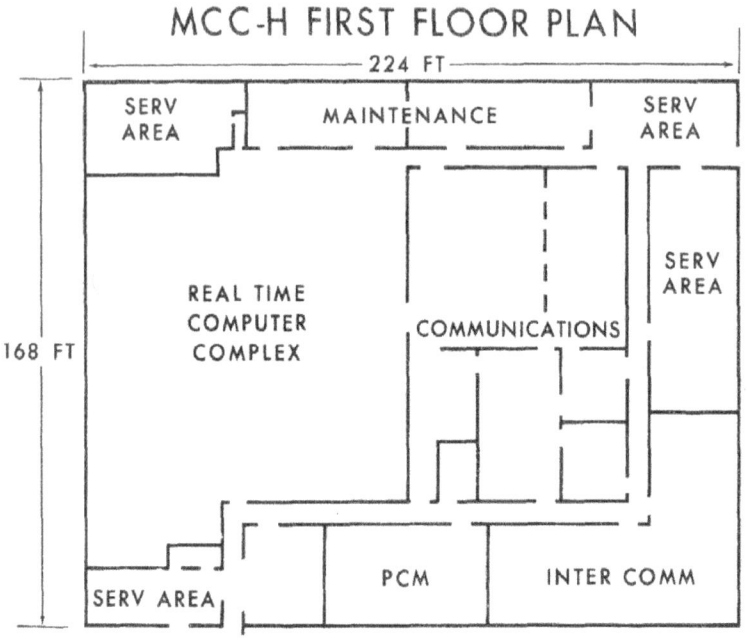

GEMINI

# MCC-H SECOND FLOOR PLAN

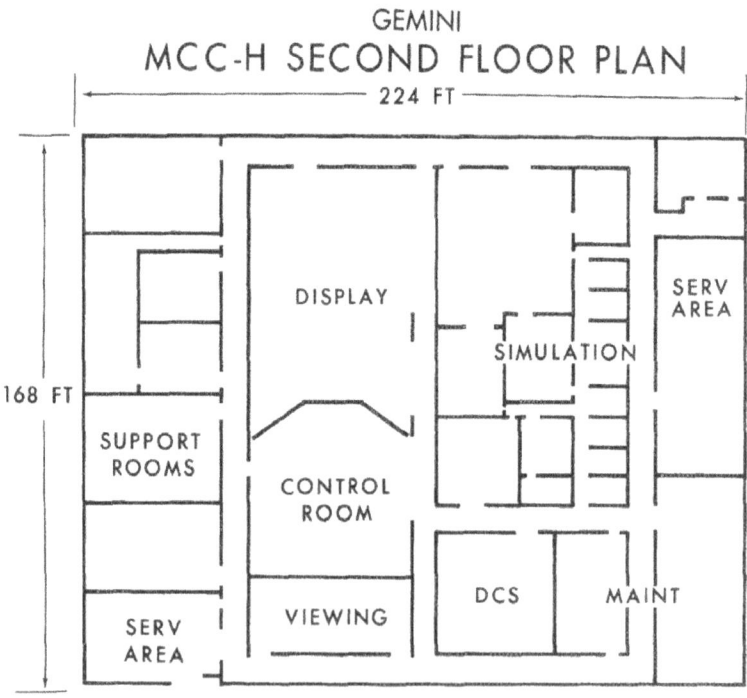

*Figure F-2. Location of Staff Support Rooms*

## GEMINI
# MCC-H THIRD FLOOR PLAN

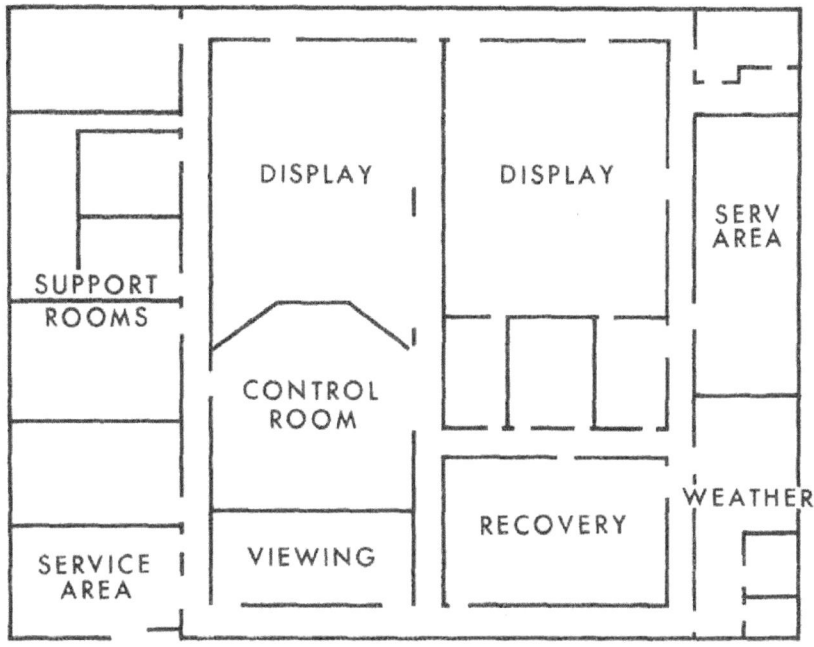

tems and components of spacecraft and attempt to avoid, correct, or circumvent failure of equipment onboard spacecraft.

Life Systems—monitor and evaluate physiological and environmental data telemetered from spacecraft.

Flight Crew—coordinate non-medical flight crew activities that involve effective control of spacecraft, as well as any scientific experiments attempted during flight.

Network—schedule, monitor, and direct network activities and readiness checks, verify remote site prepass equipment checks, and direct all network handover operations.

Operations and Procedures—provide detailed technical and administrative support, including mission plans and procedures, mission control communications plans and procedures, and generate documentation change notices to networks and MCC flight controllers.

NETWORK RESPONSIBILITY:

Manned Spacecraft Center—overall management of Gemini; direction and mission control of network immediately preceding and

589

during mission simulation or actual mission.

Goddard Space Flight Center—planning, implementation, and technical operation of manned space flight tracking and data acquisition (technical operation is defined as operation, maintenance, modification, and augmentation of tracking and data acquisition facilities to function as an instrumentation network in response to mission requirements).

Weapons Research Establishment (WRE)—Department of Supply, Commonwealth of Australia, maintenance and operation of network stations in Australia.

Department of Defense—maintenance and operational control of DOD assets and facilities required to support Gemini, including stations at Eastern Test Range (ETR), Western Test Range, White Sands Missile Range (WSMR), and Air Proving Ground Center.

## GEMINI TRACKING STATIONS

Antigua—call signal, ANT; DOD range station in the British West Indies; secondary*

Ascension—call signal, ASC; DOD range station on a British island in the South Atlantic; secondary

*Figure F-3. Gemini Tracking Stations*

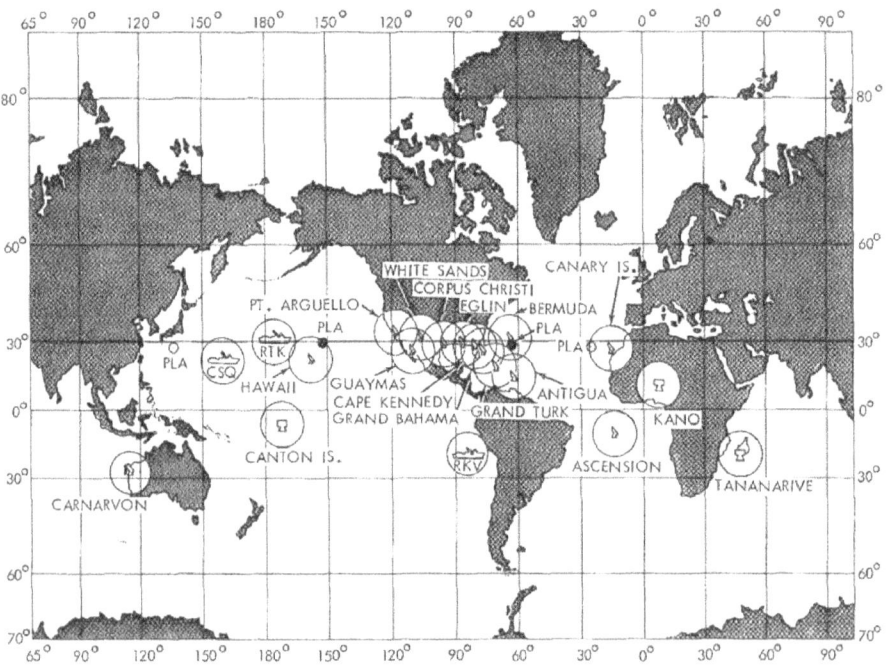

Bermuda—call signal, BDA; confirmed orbits and recommended go/no go decision; British owned; primary*

Canton Island—call signal, CTN; small coral atoll about halfway between Australia and Hawaii; co-dominion status under U.S. and British commissioners; secondary

Cape Kennedy Launch Control Center—call signal, CNV; controlled all launches; primary

Carnarvon—call signal, CRO; in northwestern Australia; operated by WRE personnel; primary

*Coastal Sentry* Quebec—call signal, CSQ; originally a C1-M-AV1 class freighter, considerably modified as tracking ship; primary

Corpus Christi—call signal, TEX; located at Rodd Field, Tex.; primary

Eglin—call signal, EGL; 76 km (47 mi) northwest of Panama City, Fla.; on the Air Force Eglin Gulf Test Range; secondary

Goddard—call signal, GSFC; NASA center located at Greenbelt, Md.; secondary

Grand Bahama—call signal, GBI; one of the Bahama Islands, British owned; almost due east from West Palm Beach, Fla.; secondary

Grand Canary—call signal, CYI; 193 km (120 mi) off the coast of Africa and 45 km (28 mi) north of the equator; Spanish owned; essential for tracking if an abort was commanded by Bermuda; primary

Grand Turk—call signal, GTK; one of the Turks and Caicos Islands in the British West Indies; radar coverage during final phase of reentry; secondary

Guaymas—call signal, GYM; in Mexico on the Gulf of California; primary

Kano—call signal, KNO; in northern Nigeria, about 845 km (525 mi) from the major seaport, Lagos, on the Gulf of Guinea; secondary

Kauai—call signal, HAW; the farthest north of the major islands that make up the state of Hawaii; primary

Mission Control Center—call signal, HOU; at MSC, in Houston, Tex., complete mission control responsibility; primary

Perth—call signal, MUC**; in western Australia; operated by WRE personnel; secondary

Point Arguello—call signal, CAL; about 64 km (40 mi) north of Santa Barbara; part of Navy-operated Pacific Missile Range; primary

Pretoria—call signal, PRE; north of Johannesburg, South Africa; secondary

*Range Tracker*—call signal, RTK; usually located in Pacific, west of Midway; operated by AF Western Test Range; secondary

*Rose Knot* Victor—call signal, RKV; operated by AF Eastern Test Range; primary

Tananarive—call signal, TAN; in the Malagasy Republic; secondary

Wallops Island—call signal, WLP; off the coast of Virginia; secondary

White Sands—call signal, WHS; located north of El Paso, Tex., on Army's White Sands Missile Range in New Mexico; secondary

Woomera—call signal, WOM; in south Australia at a rocket test facility; operated by WRE personnel; secondary

*Primary stations are those that can give direct commands to spacecraft systems; secondary stations are those used mostly for radar and telemetry information.

**Early in the program, station equipment was late in arriving, so the Mercury station at Muchea was used; when Perth was activated, the same call signal was retained.

Information from William R. Corliss, "The Evolution of the Manned Space Flight Network through Gemini," comment draft, 1 Dec. 1967; "The Manned Space Flight Tracking Network," GSFC, 1965; James M. Satterfield, JSC, 10-11 Sept. 1973; [Ivan D. Ertel], "MCC: Mission Control Center," MSC brochure, ca. 1966.

# Appendix G

# Department of Defense Support

G-1. Financial Support
($ in Thousands)

| Support Category | Other Agencies | | | Army | | | Air Force | | | Navy | | | Total | | |
|---|---|---|---|---|---|---|---|---|---|---|---|---|---|---|---|
| | NASA | DOD | Total | NASA | DOD | Total | NASA | DOD | Total | NASA | DOD | Total | NASA | DOD | Total |
| Airlift | | | | | | | 641.9 | 1 736.7 | 2 378.6 | | 37.8 | 37.8 | 641.9 | 1 774.5 | 2 416.4 |
| Research & Technology | | | | 6.2 | 4.0 | 10.2 | 1 789.1 | 3 624.2 | 5 413.3 | 5 483.7 | 267.0 | 5 750.7 | 7 279.0 | 3 895.2 | 11 174.2 |
| Launch Veh Acquisition | | | | | | | 194 372.9 | 5 741.2 | 200 114.1 | | | | 194 372.9 | 5 741.2 | 200 114.1 |
| Mapping & Charting | 13.9 | | 13.9 | | | | 10.1 | 11.2 | 21.3 | | | | 24.0 | 11.2 | 35.2 |
| Range Operations | | | | 1 236.3 | 158.6 | 1 394.9 | 13 395.5 | 22 004.4 | 35 399.9 | 5 062.0 | 631.0 | 5 693.0 | 19 693.8 | 22 794.0 | 42 487.8 |
| Recovery Operations | 5.8 | 13.9 | 19.7 | 3.2 | 51.1 | 54.3 | 6 787.0 | 2 492.9 | 9 279.9 | 5 112.1 | 30 077.9 | 35 190.0 | 11 908.1 | 32 635.8 | 44 543.9 |
| Launch Operations | | | | | | | 217 006.0 | 18 229.0 | 235 235.0 | | | | 217 006.0 | 18 229.0 | 235 235.0 |
| General Support | | | | | | | 1 561.2 | 615.5 | 2 176.7 | 2 653.0 | 76.0 | 2 729.0 | 4 214.2 | 691.5 | 4 905.7 |
| Total | 19.7 | 13.9 | 33.6 | 1 245.7 | 213.7 | 1 459.4 | 435 563.7 | 54 455.1 | 490 018.8 | 18 310.8 | 31 089.7 | 49 400.5 | 455 139.9 | 85 772.4 | 540 912.3 |

Extracted from "Summary Report: DOD Support of Project Gemini, Jan 1963—Nov 1966," submitted to Secretary of Defense by Lt. Gen. Leighton I. Davis, USAF, 6 March 1967

NASA—Costs Reimbursed to the DOD by NASA
DOD—Costs Absorbed by the DOD

## G-2. DOD Resources by Flight

| Mission | Personnel | Aircraft | Ships |
|---------|-----------|----------|-------|
| Gemini 1 | 5 176 | 11 | 3 |
| Gemini 2 | 6 562 | 67 | 16 |
| Gemini 3 | 10 185 | 126 | 27 |
| Gemini IV | 10 249 | 134 | 26 |
| Gemini V | 10 265 | 114 | 19 |
| Gemini VII/VI-A | 10 125 | 125 | 16 |
| Gemini VIII | 9 655 | 96 | 16 |
| Gemini IX-A | 11 301 | 92 | 15 |
| Gemini X | 9 067 | 78 | 13 |
| Gemini XI | 9 054 | 73 | 13 |
| Gemini XII | 9 775 | 65 | 12 |

Excerpted from "Summary Report of DOD Support"

# Index

# THE AUTHORS

BARTON C. Hacker has been Assistant Professor of History and Mechanical Engineering at Iowa State University since 1970. Born in Chicago, Illinois (1935), he received his B.A. in Liberal Arts from the University of Chicago (1955). After serving in the U.S. Army, he returned to the University of Chicago, which awarded him his B.A., with honors, in History (1960), his M.A. (1962), and his Ph.D. (1968). He taught history of science at the University of Chicago (1965-1966) and was a Research Associate in the Department of History at the University of Houston (1966-1969). He has published articles in *Technology and Culture, Military Affairs,* and *Journal of Humanistic Psychology.* He is co-author of *Project Gemini Technology and Operations: A Chronology* (1969). Married, with one son, he won the Robert H. Goddard Historical Essay Award of the National Space Club in 1972.

James M. Grimwood has been NASA Johnson Space Center Historian since 1962. He was born in Lincoln, Alabama (1922), receiving his A.B. degree in History from Howard College, Birmingham, Alabama (1948), and his M.A. from the University of Alabama (1950). He taught history in secondary schools (1950-1952) and at San Antonio College in Texas (1958-1960). Grimwood was an Air Force historian in South Carolina and Texas (1953-1960). Prior to joining JSC, he was historian with the Army Missile Command, Huntsville, Alabama, preparing histories of Army missile systems. He is joint author of *This New Ocean: A History of Project Mercury* (1966), co-author of *Project Gemini Technology and Operations: A Chronology* (1969), and author of *Project Mercury: A Chronology* (1963). He is married and has two children.

# NASA HISTORICAL PUBLICATIONS

HISTORIES

- Frank W. Anderson, Jr., *Orders of Magnitude: A History*, NASA SP-4403, 1976, GPO.*
- William R. Corliss, *NASA Sounding Rockets, 1958—1968: A Historical Summary*, NASA SP-4401, 1971, GPO.
- Constance McL. Green and Milton Lomask, *Vanguard—A History*, NASA SP-4202, 1970; also Washington: Smithsonian Institution Press, 1971.
- Edwin P. Hartman, *Adventures in Research: A History of the Ames Research Center, 1940-1965*, NASA SP-4302, 1970, NTIS.**
- Mae Mills Link, *Space Medicine in Project Mercury*, NASA SP-4003, 1965, NTIS.
- Alfred Rosenthal, *Venture into Space: Early Years of Goddard Space Flight Center*, NASA SP-4301, 1968, NTIS.
- Robert L. Rosholt, *An Administrative History of NASA, 1958-1963*, NASA SP-4101, 1966, NTIS.
- Loyd S. Swenson, Jr., James M. Grimwood, and Charles C. Alexander, *This New Ocean: A History of Project Mercury*, NASA SP-4201, 1966, NTIS.

REFERENCE WORKS

- *The Apollo Spacecraft: A Chronology*, NASA SP-4009: Volume I, 1969, NTIS. Volume II, 1973, GPO. Volume III, 1976, GPO. Volume IV, forthcoming.

---

*GPO: Available from Superintendent of Documents, Government Printing Office, Washington, D.C. 20402.

**NTIS: Available from National Technical Information Service, Springfield, Virginia 22161.

- *Astronautics and Aeronautics: A Chronology of Science, Technology, and Policy,* annual volumes 1961-73, with an earlier summary volume, *Aeronautics and Astronautics, 1915-1960.* Early volumes available from NTIS; recent volumes from GPO. *Astronautics and Aeronautics, 1974,* in press.
- Katherine M. Dickson (Library of Congress), *History of Aeronautics and Astronautics: A Preliminary Bibliography,* NASA HHR-29, NTIS.
- *Project Gemini Technology and Operations: A Chronology,* NASA SP-4002, 1969, NTIS.
- *Project Mercury: A Chronology,* NASA SP-4001, 1963, NTIS.
- *Project Ranger: A Chronology,* JPL/HR-2, 1971, NTIS.
- *Skylab: Preliminary Chronology,* NASA HHN-130, May 1973, NTIS.
- Jane Van Nimmen and Leonard C. Bruno, *NASA Historical Data Book, 1958-1968,* Vol. I, *NASA Resources,* NASA SP-4012, 1976, NTIS.
- Helen T. Wells, Susan H. Whiteley, and Carrie E. Karegeannes, *Origins of NASA Names,* NASA SP-4402, in press.

☆ U.S. GOVERNMENT PRINTING OFFICE: 1978 O—204-020

www.ingramcontent.com/pod-product-compliance
Lightning Source LLC
Chambersburg PA
CBHW081425170526
45166CB00008B/2108